Anatomical Basis of Cranial Neurosurgery

Wolfgang Seeger • Josef Zentner

Anatomical Basis of Cranial Neurosurgery

Springer

Wolfgang Seeger
Linden
Germany

Josef Zentner
Department of Neurosurgery
University Medical Center Freiburg
Freiburg
Germany

ISBN 978-3-030-09699-1 ISBN 978-3-319-63597-2 (eBook)
https://doi.org/10.1007/978-3-319-63597-2

Softcover re-print of the Hardcover 1st edition 2018

Printed on acid-free paper

This Springer imprint is published by the registered company Springer International Publishing AG part of Springer Nature
The registered company address is: Gewerbestrasse 11, 6330 Cham, Switzerland

Dedicated to
Gazi M. Yaşargil

Preface

The main intention for writing this book is to help training neurosurgeons and their colleagues of related disciplines to combine the modern conventional anatomy with the special anatomical aspects of brain surgery. Imaging methods, especially neuronavigation, make better and more spatial presentations in vivo possible than by using conventional anatomical dissections. Nevertheless, many anatomical details, which are essential for neurosurgery, are not sufficiently presentable by imaging methods and need supporting by modern anatomical presentations. This may be perceptible by all five parts of this book:

- Angiographic presentations prefer exact imagings of vessels, omitting adjacent structures. For training neurosurgeons, topographical landmarks and common variants of vessels should be added, combined with presentations of surrounding cerebral structures. After this, vessels and cranial nerves should be presented together with cisternal walls. At an early stage of the anatomo-surgical training, young neurosurgeons would more easily understand the topography, especially during its assisting at operations, performed by experienced neurosurgeons.
- Further aspects are topographical errors at the interpretation of imaging findings. Using topographical landmarks, some errors could be avoided.
- A third aspect are anatomical details which are important during surgery, but not exact to define by imagings. These details can be presented by illustrations of anatomical dissections.

It seems to be useful at the beginning of anatomo-surgical descriptions to present an intracranial region, which is often approached by multiple surgical routes. These preconditions are given by presenting the frontal region. After this will follow further parts about the temporal, parietal, occipital, and suboccipital region.

Special aspects of the frontal part are as follows:

- The motor cortex and the pyramidal tract
- The frontal limbic cortex
- Cornu anterius of the lateral ventricle and third ventricle and surrounding structures

Special aspects of the temporal part are as follows:

- Limbic and neocortical structures according to amygdalohippocampectomy and other methods of the temporal epilepsy surgery
- Temporobasal approaches for cisternal approaches (and amygdalohippocampectomy)

Special aspects of the parietal part are as follows:

- The Sylvian region of the parietal lobe
- The midline region according to the region of sinus sagittalis superior and to surgical approaches transpassing interfornical to the third ventricle

Special aspects of the occipital part are as follows:

- Confluens sinuum and adjacent structures
- Galenic area
- The optic system

Special aspects of the infratentorial part are as follows:

- Extradural structures for avoiding bleeding, air embolism, and problems of wound healing after surgery
- The central cerebellar region
- The intrasurgical landmarks at the surface of the middle and caudal brainstem
- The microanatomy of the middle and caudal brainstem with presentations of the relationships between nuclei, fibers, and intrarhombencephalic arteries

This book is a summary of life-long surgical experiences of Wolfgang Seeger and of his experiences by giving anatomical training lectures on surgery at numerous neurosurgical departments. It was supported by anatomical brain dissections and by the drawings of Wolfgang Seeger. The drawings in this book are mainly developed in the last 10 years and in the publications of Wolfgang Seeger since 1978. All drawings are now presented schematized in order to facilitate understanding by younger neurosurgeons.

After that, this book has been improved considerably by Josef Zentner. He and his co-workers demonstrated and discussed surgical approaches in the operating room, especially before and after surgery. In this book, some surgical approaches were presented by both authors, especially surgery at the limbic areas and frontal, temporal, and parietal craniotomies.

Both authors are thankful for many motivations by many colleagues and their co-workers, especially of the Neurosurgical Clinic, Freiburg, and further H.R. Eggert and his successor W. Deinsberger, J.M. Gilsbach, A.G. Harders, R. Oeckler, and J.P. Warnke. Karen Seeger always has been a helpful corrector of imperfections in the English language of this book.

Freiburg, Germany Wolfgang Seeger
Freiburg, Germany Josef Zentner
October 2016

Contents

Part V Infratentorial Region

Part I

Frontal Region

1.1 Introduction

This introduction to Part I contains a short summary of each chapter and of the completion to Chapters 1–3.

1.2 Chapter 1 Diagnostic Base

It presents an introduction for diagnostic principles:

- For planning surgical approaches are presented extra- and intracranial *landmarks.* It may help to define craniocerebral structures and of the motor cortex.
- *Vessels* are simplified presented for understanding its variability later on.
- *MRTs and neuronavigation* are combined with landmarks. Fine structures are added which are essential for surgery but often not sufficient defined at imaging.

1.3 Chapter 2 Anatomical Base of Surgery

It contents extracranial layers, dural structures, intracerebral structures, and cisterns according to surgery:

- *Galea aponeurotica* and its adjacent extracranial layers are presented according to the preservation of its nerves and vessels at surgical approaches, especially N. VII and supraorbital vessels and nerves.
- The numerous variants of bridging veins are essential anatomical components between *dura and cortex*. Interruptions of veins and its variants could be dangerous even for cerebral areas, which may be located far distant to the surgical approach. Another aspect is the avoiding damage of N. olfactorius by considering its subdural and subarachnoidal course and endangering bulbus olfactorius by overlying bones during surgery.
- Defining of the *motor cortex* using landmarks: The topography of the motor cortex is better to define, if landmarks and imaging are presented first. After this, exact neurophysiological methods can be performed easier.
- *Preservation of allocortical (limbic) structures* surrounding corpus callosum. Lesions could be followed by psychological deficits.
- An essential surgical problem is the consequent preservation of cisternal *arachnoid membranes* and of fine *perforating feeding vessels* of basal ganglia, pyramidal tract, and hypothalamus. This principle must be taken in consideration at surgical approaches to all basal cisterns, not only at the frontobasal region.

1.4 Chapter 3 Special Surgical Aspects

- Operations at the frontal lobe are often the first large surgical operations which are performed by training neurosurgeons. Structures in the depth must be well known. Here is presented the example of a surgical approach between insula and cornu anterius, close to basal ganglia and other essential adjacent structures in the depth.
- Midline approaches to cisterna corporis callosi and the third ventricle are presented. Favorable and less favorable aspects of midline approaches and transcortical endoscopic approaches to the third ventricle are discussed according to the preservation of allocortex (stria longitudinalis cinguli and fornix).
- Callosotomy is a palliative treatment modality for severe epilepsies which prevents propagation of epileptic activity from one to the other hemisphere.
- The principles of frontotemporal (pterional) basal approaches are presented. Difficult to understand are the sylvian arachnoid walls, Liliequist's membrane, the topography of the carotid cistern, and adjacent structures. For surgery, the high variability of arteries has to be considered, especially Heubner's artery and A. corporis callosi mediana, according to its perforating branches. Heubner's artery may be mixed with other arteries. For a better understanding of these aspects, the structures are presented by schematic drawings on the base of Yaşargil's findings at surgery and of Rhoton's photographs of anatomical dissections.

1.5 Chapter 4 Completion for Chapters 1 to 3: Topographical Areas of Rhinoliquorrhea

1 Diagnostic Base

1.1 Landmarks (Figs. 1.1 and 1.2)

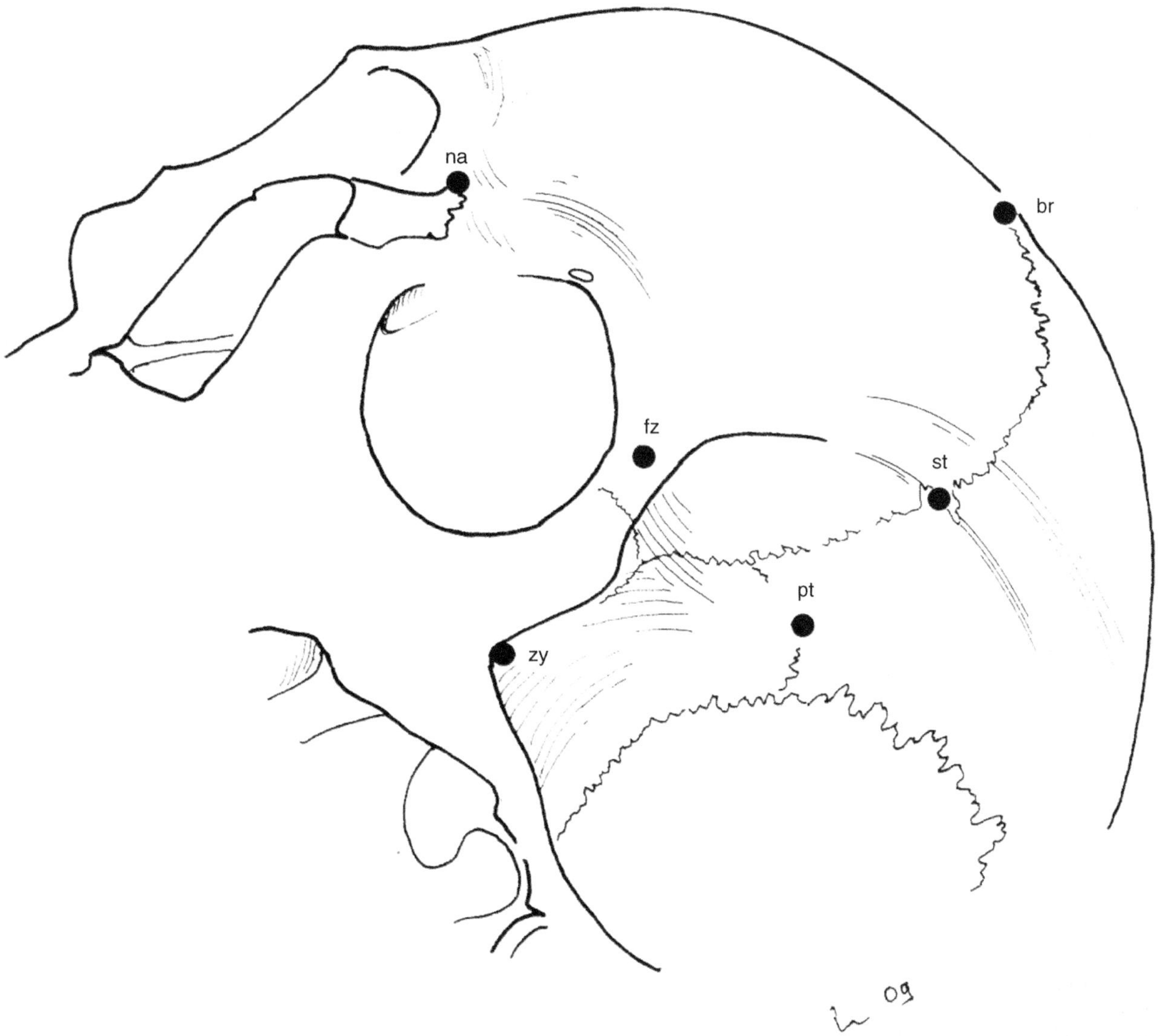

Fig. 1.1 Lateral and dorsal bony landmarks for planning frontal and frontotemporal surgical approaches. Defining by palpation: *fz* frontozygomatic point, *na* nasion, *st* stephanion (palpation of the margin of M. temporalis during innervation and palpation of sutura frontalis, which is usually thickened at this point), *zy* angle of os zygomaticum. Defining by imaging: *br* bregma, *pt* pterion

W. Seeger, J. Zentner, *Anatomical Basis of Cranial Neurosurgery*, https://doi.org/10.1007/978-3-319-63597-2_1

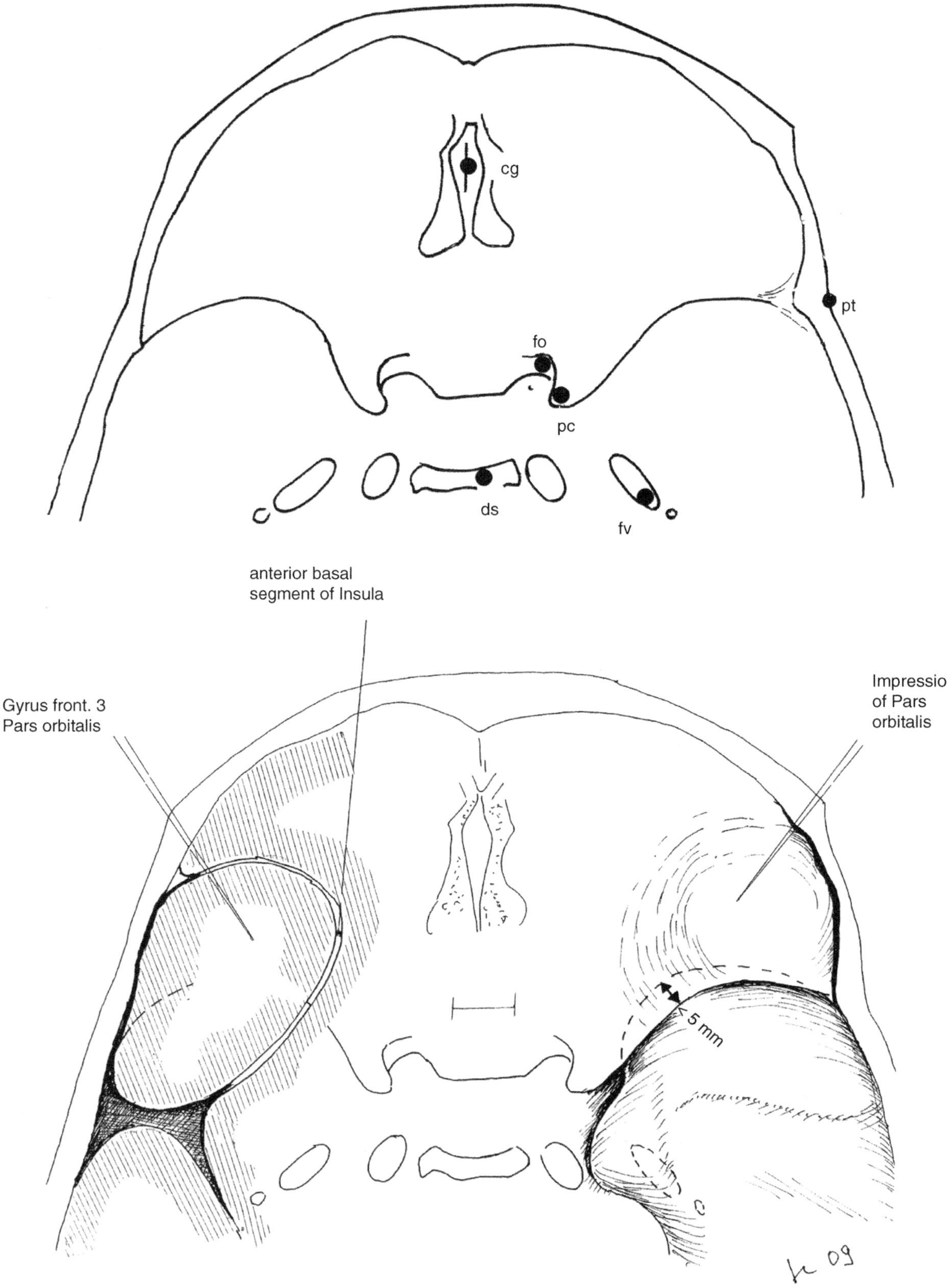

Fig. 1.2 Fossa cranii anterior and frontal cerebral base are not congruent. Gyrus frontalis 3 (pars orbitalis) is overlying ala minor and the temporal pole. *cg* crista galli, *ds* dorsum sellae, *fo* foramen opticum, *fv* foramen ovale, *pc* processus clinoideus anterior, *pt* pterion

1.2 Radiology (Figs. 1.3, 1.4, 1.5, and 1.6)

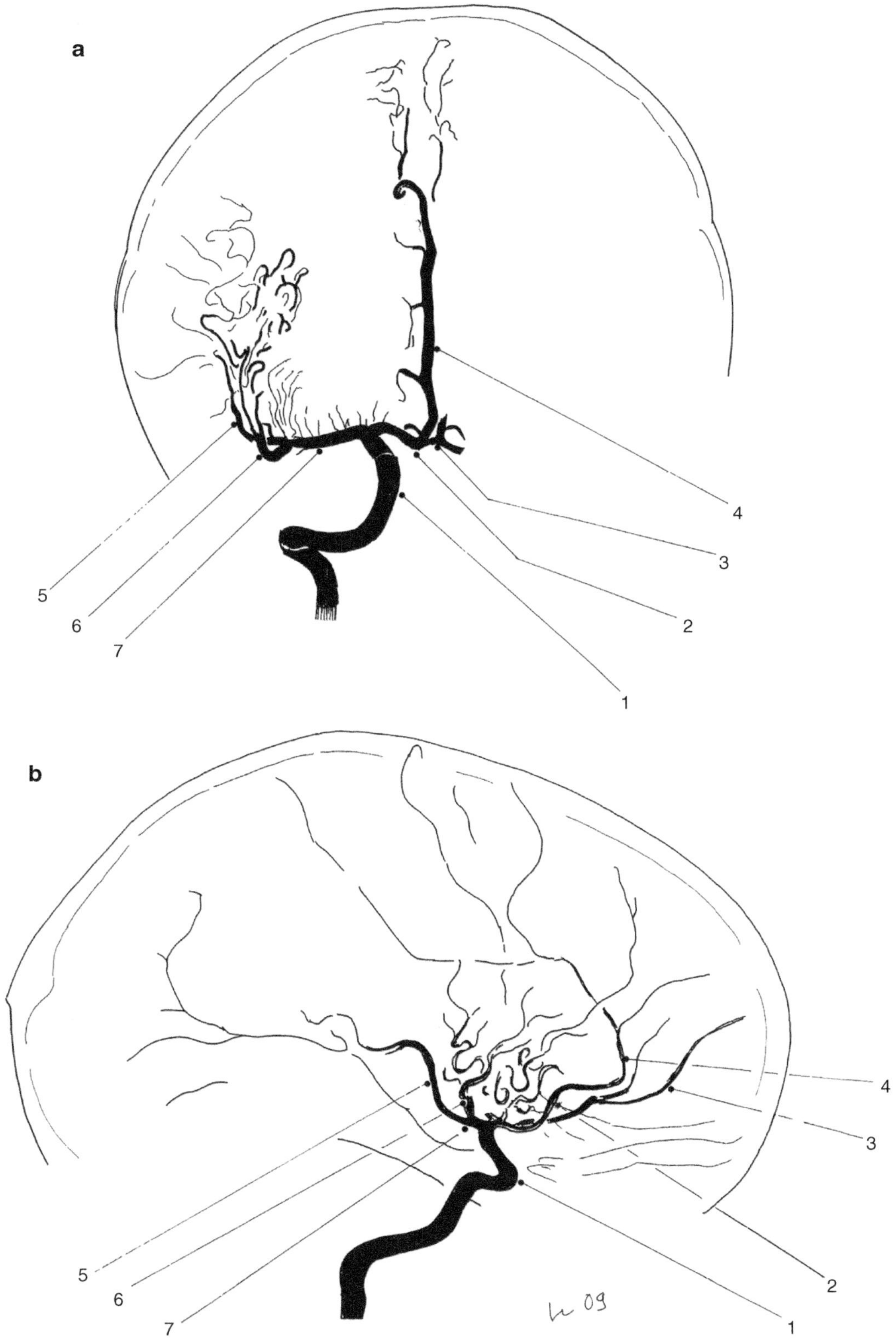

Fig. 1.3 Arteriography. (**a**) *1* A. carotis interna, *2* A1 (proximal segment of A. cerebri anterior), *3* A. communicans anterior, *4* A2 (distal segment of A. cerebri anterior), *5* A. cerebri media, distal branch, *6* A. operculofrontalis, *7* A. cerebri media, proximal trunk; (**b**) *1* siphon of A. carotis interna; *2* A2; *3* A. callosomarginalis; *4* A. pericallosa; *5* A. cerebri media, distal trunk; *6* A. operculofrontalis; *7* A. cerebri media, proximal trunk

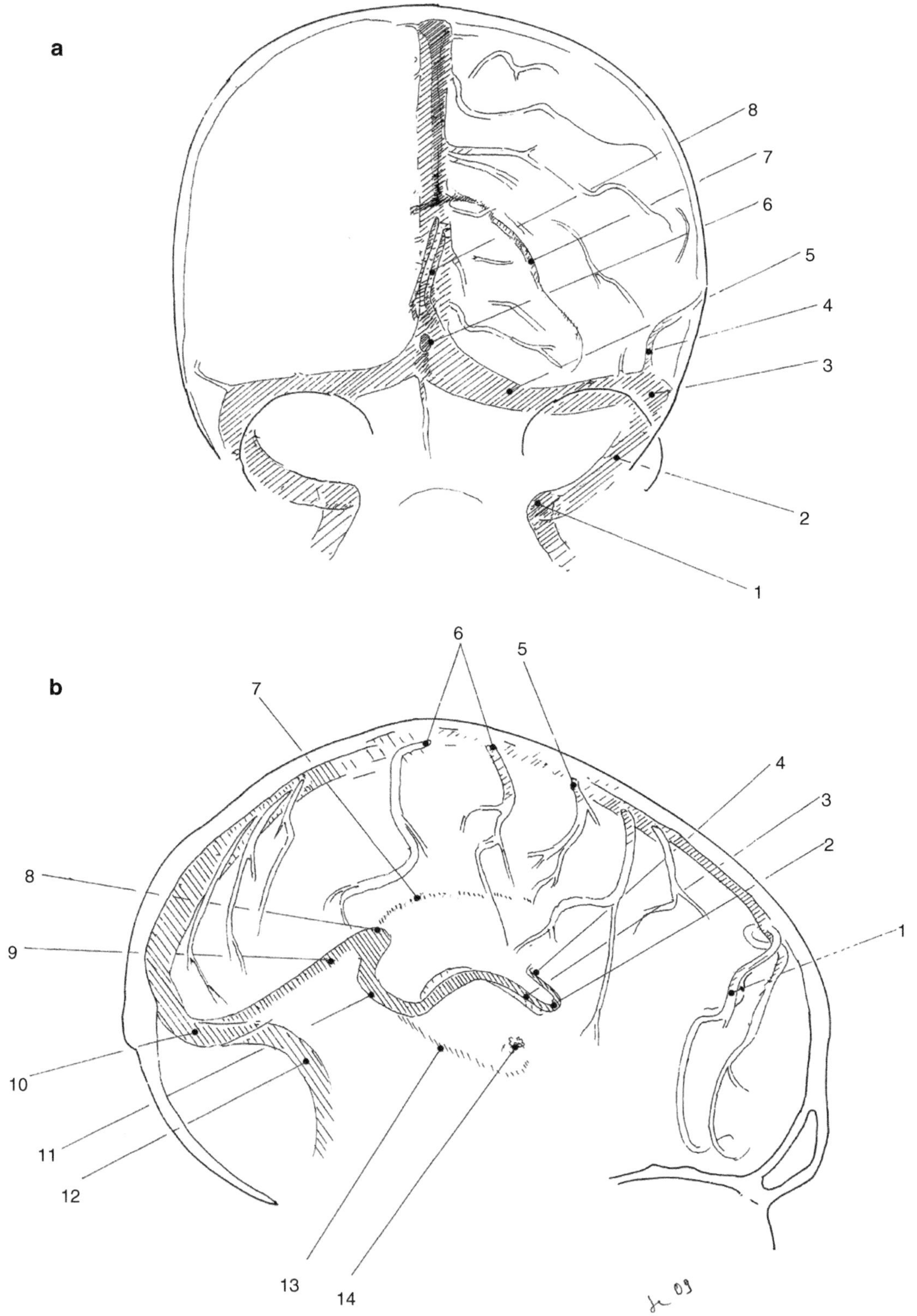

Fig. 1.4 Phlebography. Interruption of larger veins could be dangerous, not only interruptions of well-known veins, e.g., the rolandic vein, vein of Labbé, and inner cerebral veins. (**a**) *1* Bulbus superior of V. jugularis interna, *2* sinus sigmoideus, *3* knee of sinus transversus, *4* vein of Labbé, *5* sinus transversus, *6* confluens sinuum, *7* V. basalis of Rosenthal, *8* V. cerebri interna; (**b**) *1* frontal bundle of bridging veins (Browder and Kaplan 1976) (variant, dangerous for surgery), *2* angulus venosus, *3* V. cerebri interna, *4* V. thalamostriata, *5* vein(s) of the precentral region, *6* central and postcentral veins, *7* sinus sagittalis inferior, *8* galenic point, *9* sinus rectus, *10* confluens sinuum, *11* galenic vein, *12* sinus sigmoideus, *13* V. basalis (Rosenthal), *14* temporal choroid point

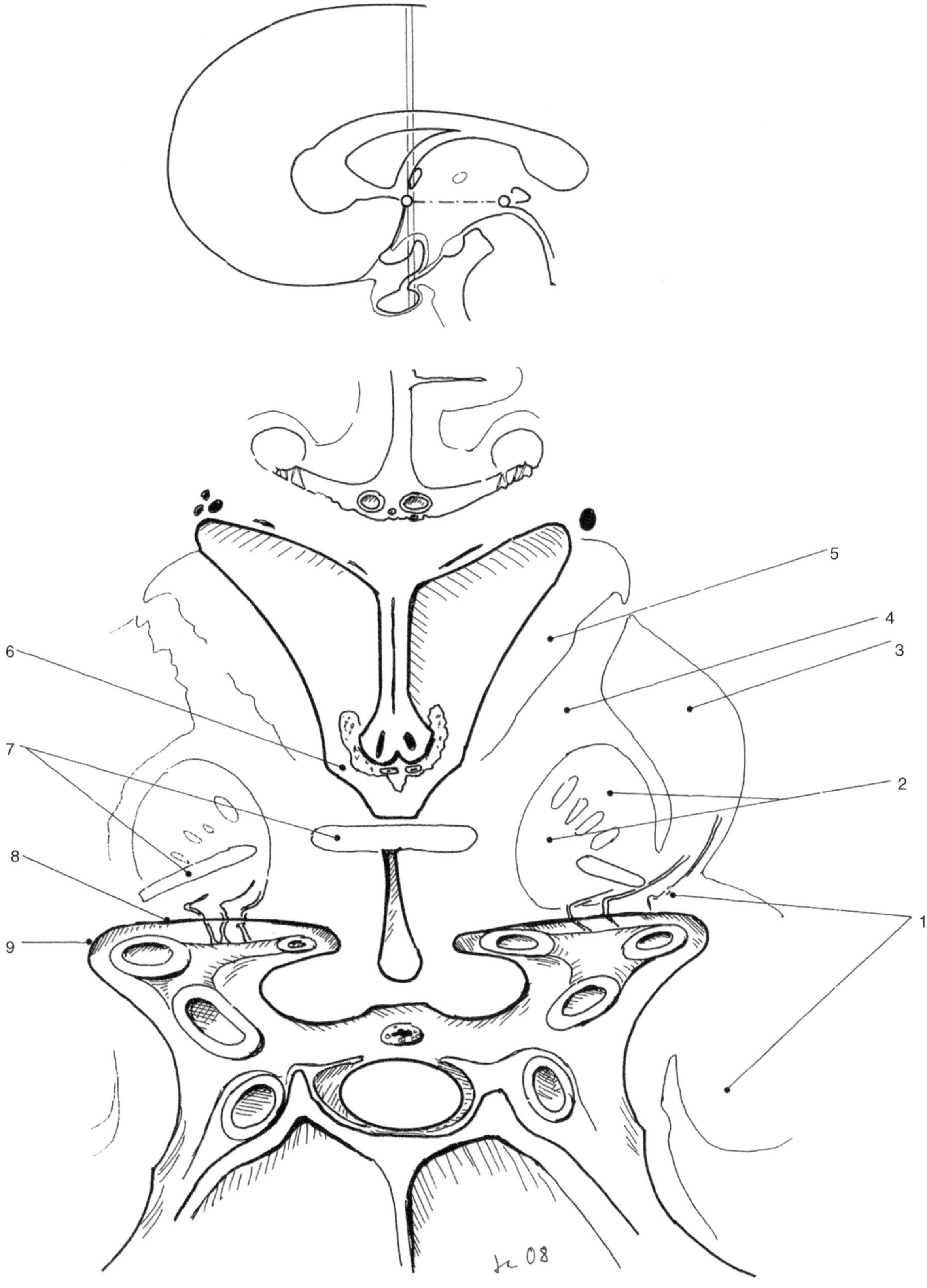

Fig. 1.5 MRT, coronal level of Commissura ant. Details added, which are not always sufficient defined by imaging methods. *1* Amygdala, *2* globus pallidus, *3* putamen, *4* knee of capsula interna, *5* nucleus caudatus, *6* foramen interventriculare (Monroi), *7* commissura anterior, *8* substantia perforata anterior, *9* limen insulae

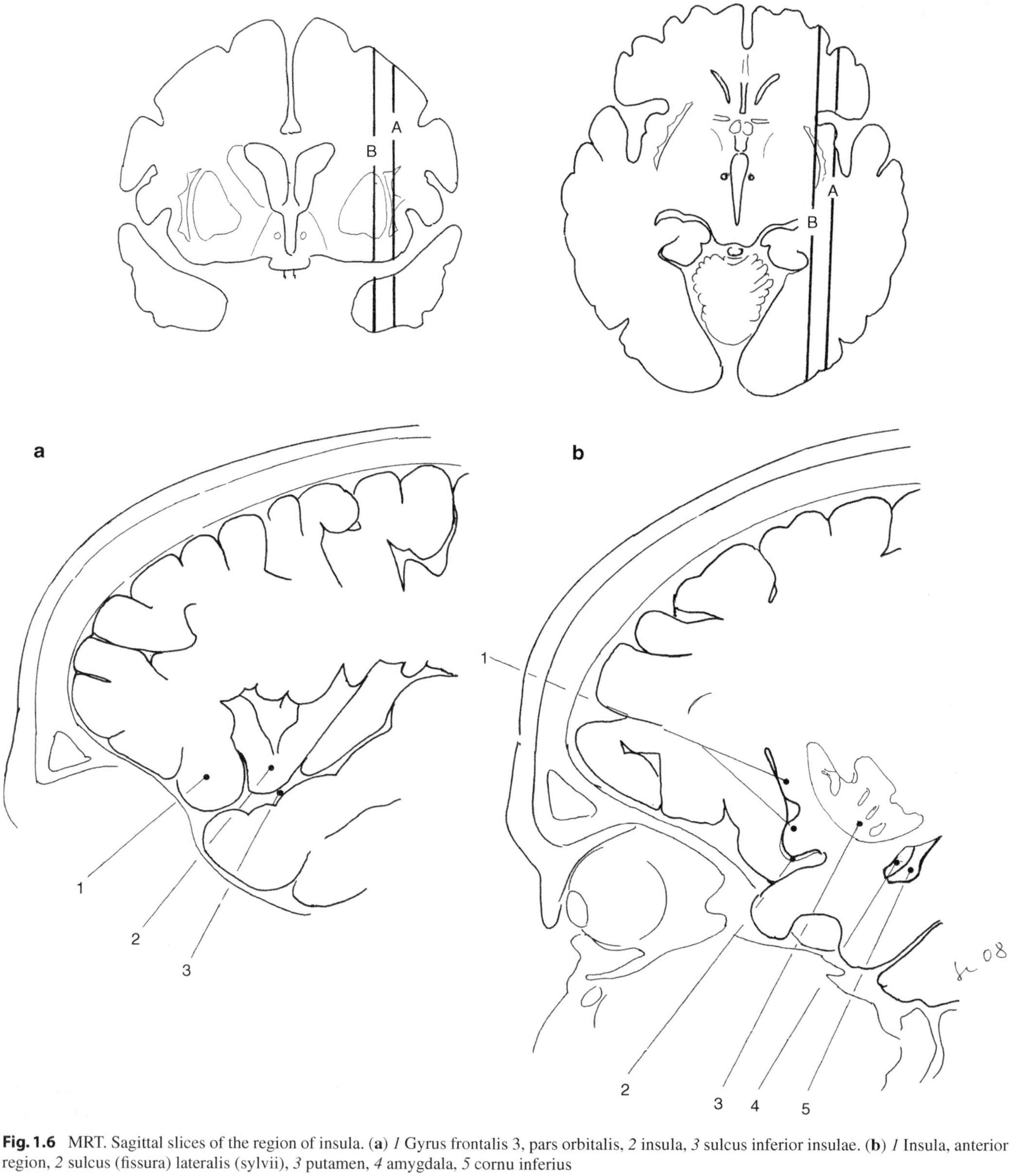

Fig. 1.6 MRT. Sagittal slices of the region of insula. (**a**) *1* Gyrus frontalis 3, pars orbitalis, *2* insula, *3* sulcus inferior insulae. (**b**) *1* Insula, anterior region, *2* sulcus (fissura) lateralis (sylvii), *3* putamen, *4* amygdala, *5* cornu inferius

1.3 Further Imaging (Figs. 1.7, 1.8, 1.9, 1.10, 1.11, 1.12, 1.13, 1.14, and 1.15)

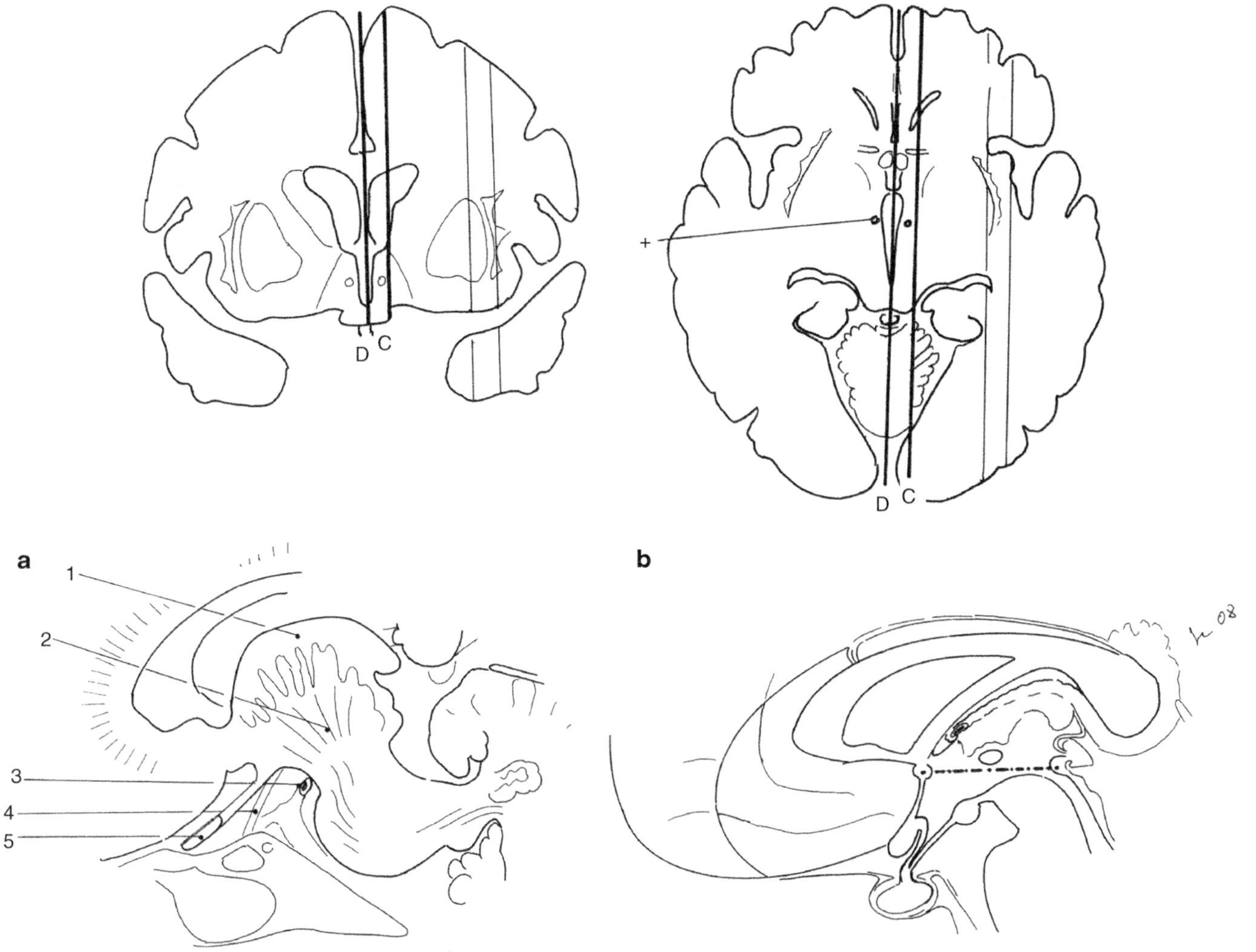

Fig. 1.7 Continuation of Fig. 1.6. Region of cornu anterius and midline structures. (**a**) *1* Thalamus, *2* capsula interna, *3* A. cerebri posterior, *4* N. oculomotorius, *5* N. opticus; (**b**) *dotted*: intercommissural line

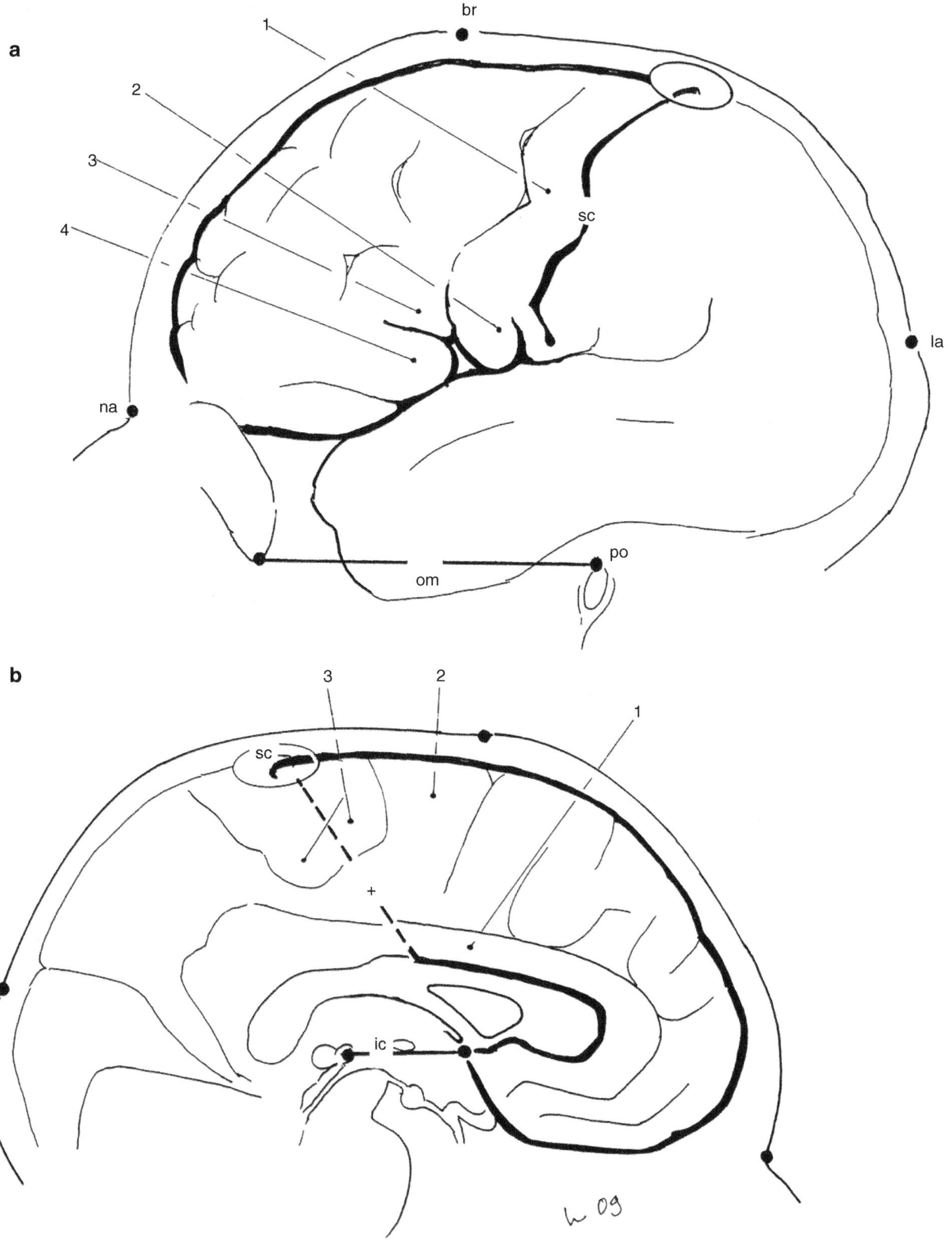

Fig. 1.8 Lobus frontalis is far extended to the parietal area of the skull. This can be presented by neuronavigation and marked by landmarks. (**a**) *br* bregma, *la* lambda, *na* nasion, *om* orbitomeatal line, *po* porion, *sc* sulcus centralis, lateral limit between lobus frontalis and parietalis. *1* Gyrus praecentralis; *2* gyrus frontalis 3, pars opercularis; *3* as 2, pars triangularis; *4* as 2, pars orbitalis. (**b**) *ic* intercommissural line, *sc* sulcus centralis. *1* Gyrus cinguli, *2* gyrus frontalis 1, *3* lobulus paracentralis. + The medial limit between lobus frontalis and parietalis is exact to define at the dorsal point of sulcus centralis – sc

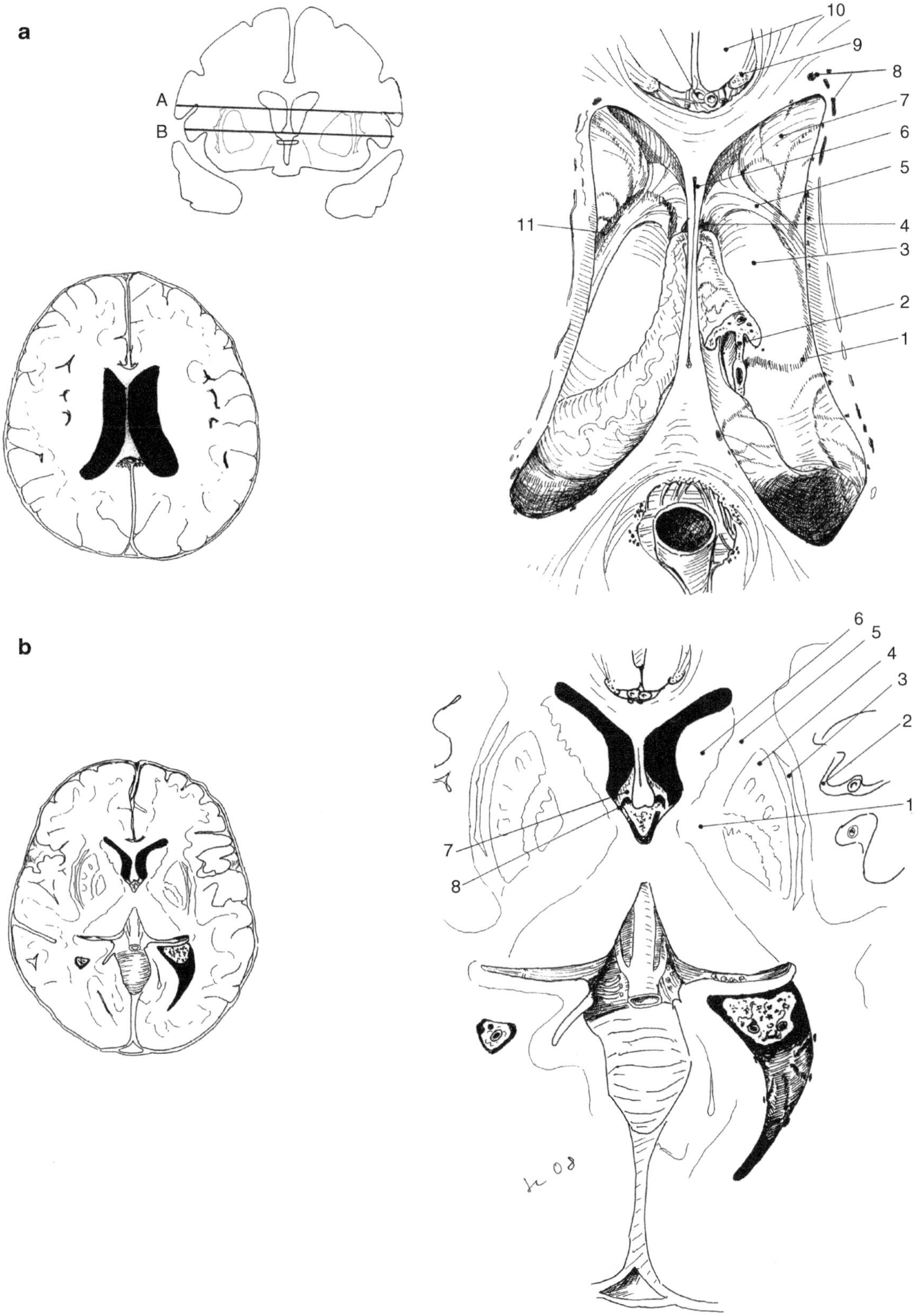

Fig. 1.9 Ventriculus lateralis neuronavigation modified and details added. (**a**) *1* V. ventriculi lateralis directa, common variant (Huang and Wolf 1974b), *2* taenia chorioidea ventriculi III, *3* thalamus, *4* For. interventriculare (Monroi), *5* stria terminalis, *6* cavum septi pellucidi, *7* caput nuclei caudati, *8* Vv. longitudinales, *9* stria longitudinalis cinguli, *10* gyrus cinguli, *11* V. thalamostriata; (**b**) *1* knee of capsula interna, *2* sulcus lat. (Sylvii), *3* claustrum, *4* putamen**,** *5* crus anterius of capsula interna, *6* caput nuclei caudati, *7* columna fornicis, *8* For. interventriculare (Monroi)

a

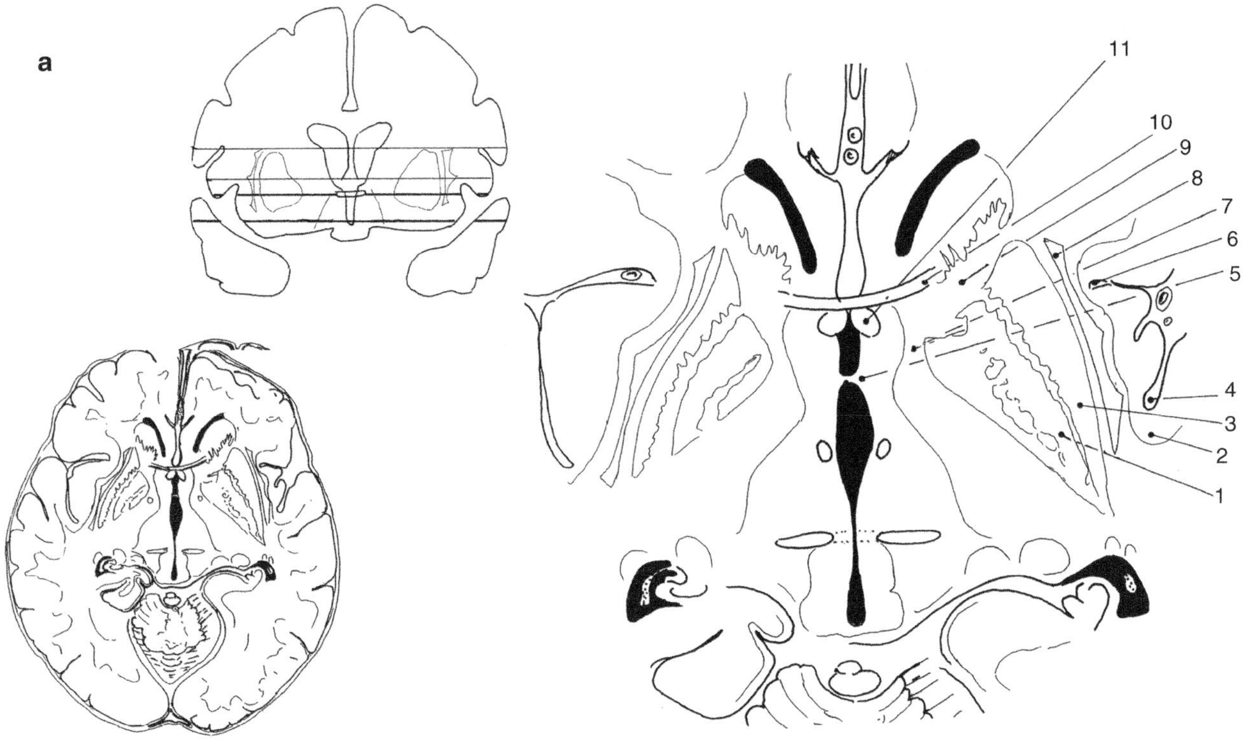

b

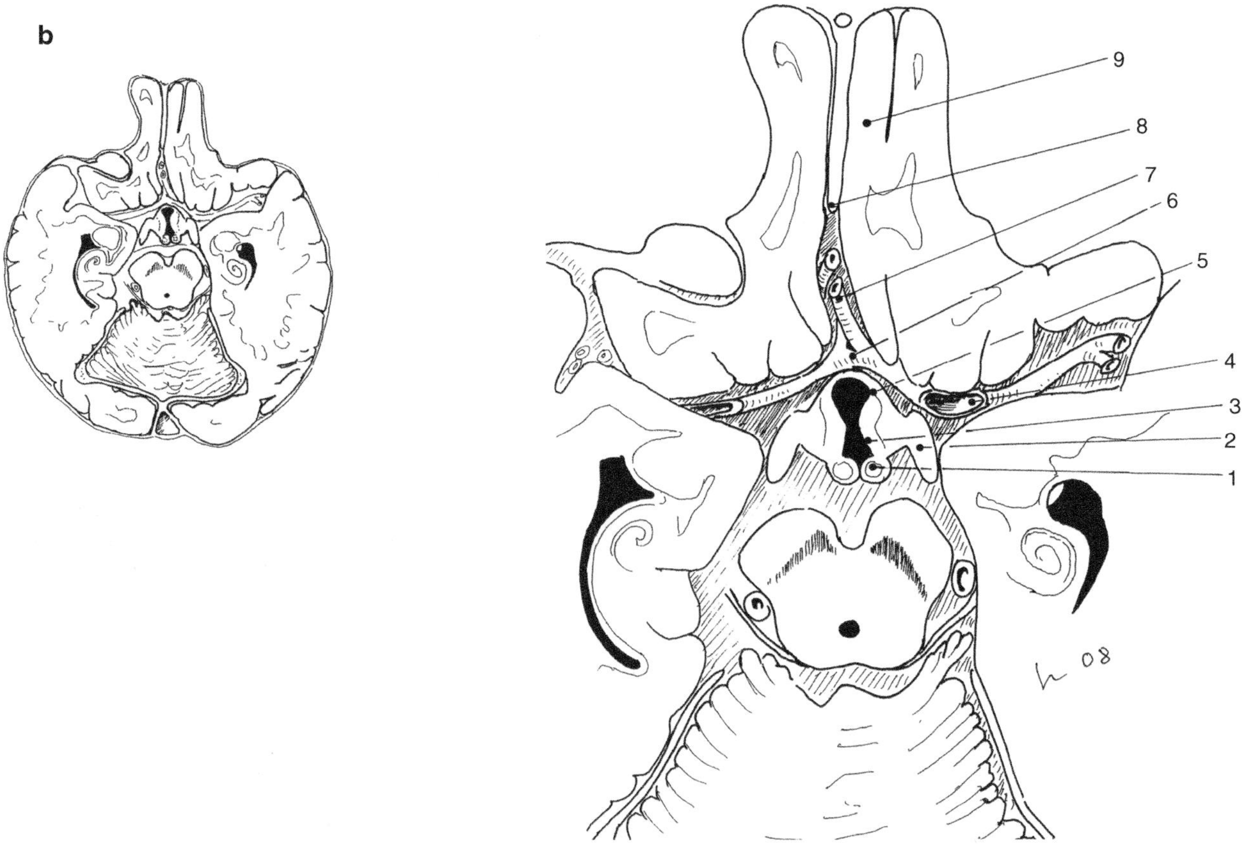

Fig. 1.10 Continuation of Fig. 1.9. (**a**) *1* Globus pallidus, *2* cortex of insula, *3* putamen, *4* sulcus inferior insulae, *5* adhesio interthalamica, *6* sulcus superior insulae, *7* knee of capsula interna, *8* claustrum, *9* crus anterius of capsula interna, *10* commissura anterior, *11* columna fornicis. (**b**). *1* Corpus mamillare, *2* tractus opticus, *3* recessus infundibuli, *4* carotid bifurcation, *5* recessus opticus, *6* A. communicans anterior, *7* A2, *8* margin of falx, *9* gyrus rectus

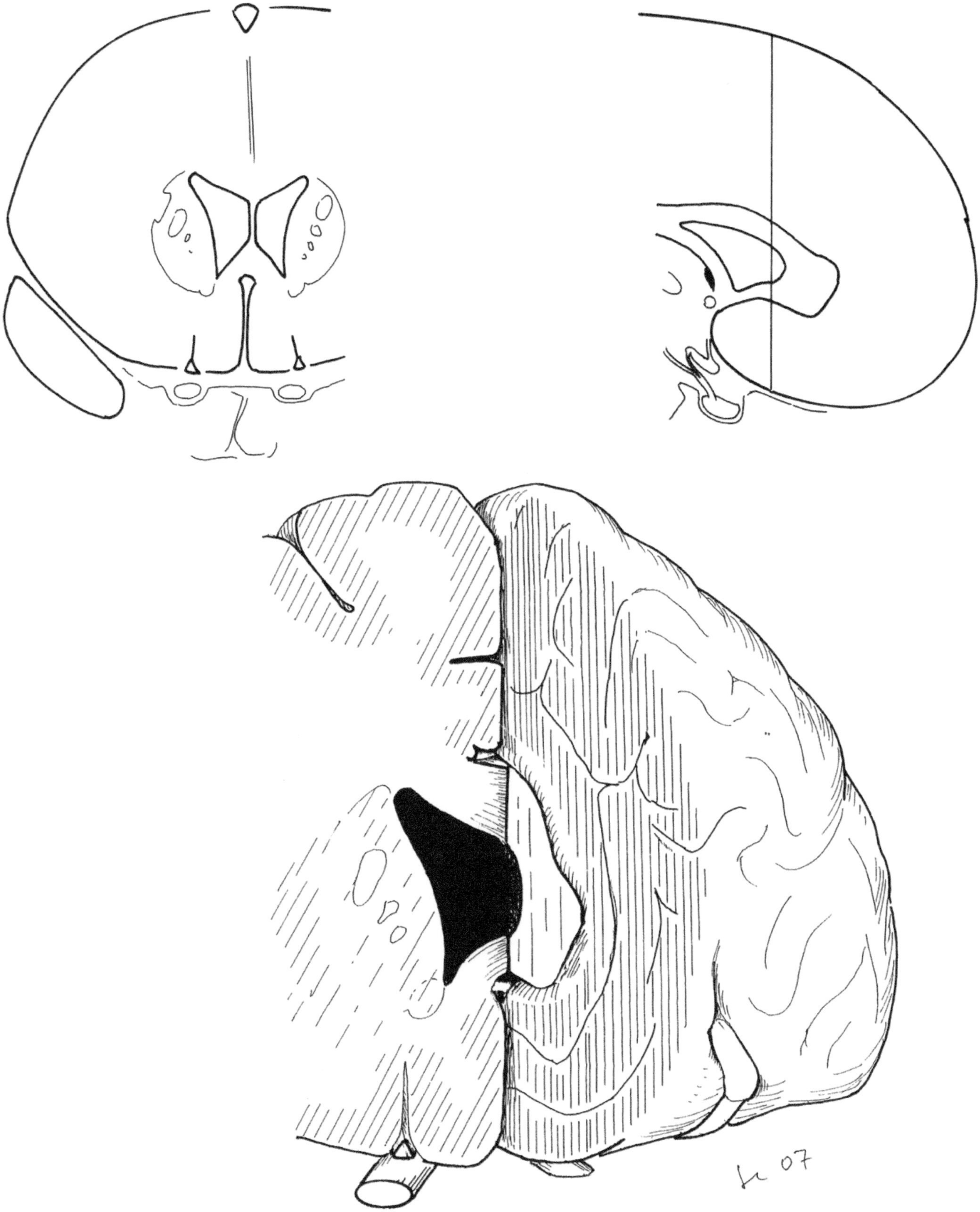

Fig. 1.11 *Anatomical dissection* similar to neuronavigation

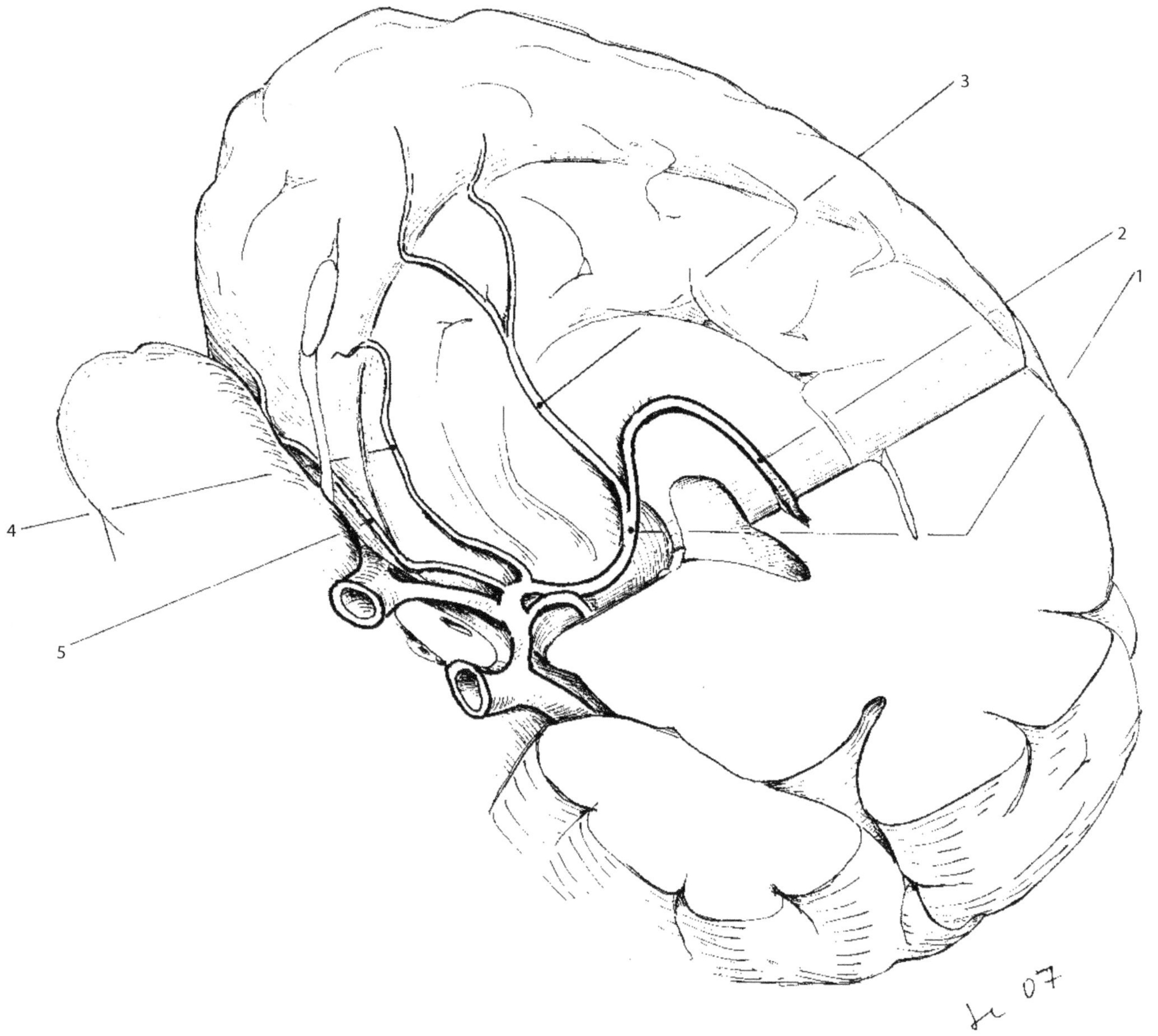

Fig. 1.12 Addendum for Fig. 1.11. *1* A2, *2* A. pericallosa, *3* A. callosomarginalis, *4* A. frontopolaris, *5* medial frontobasal artery

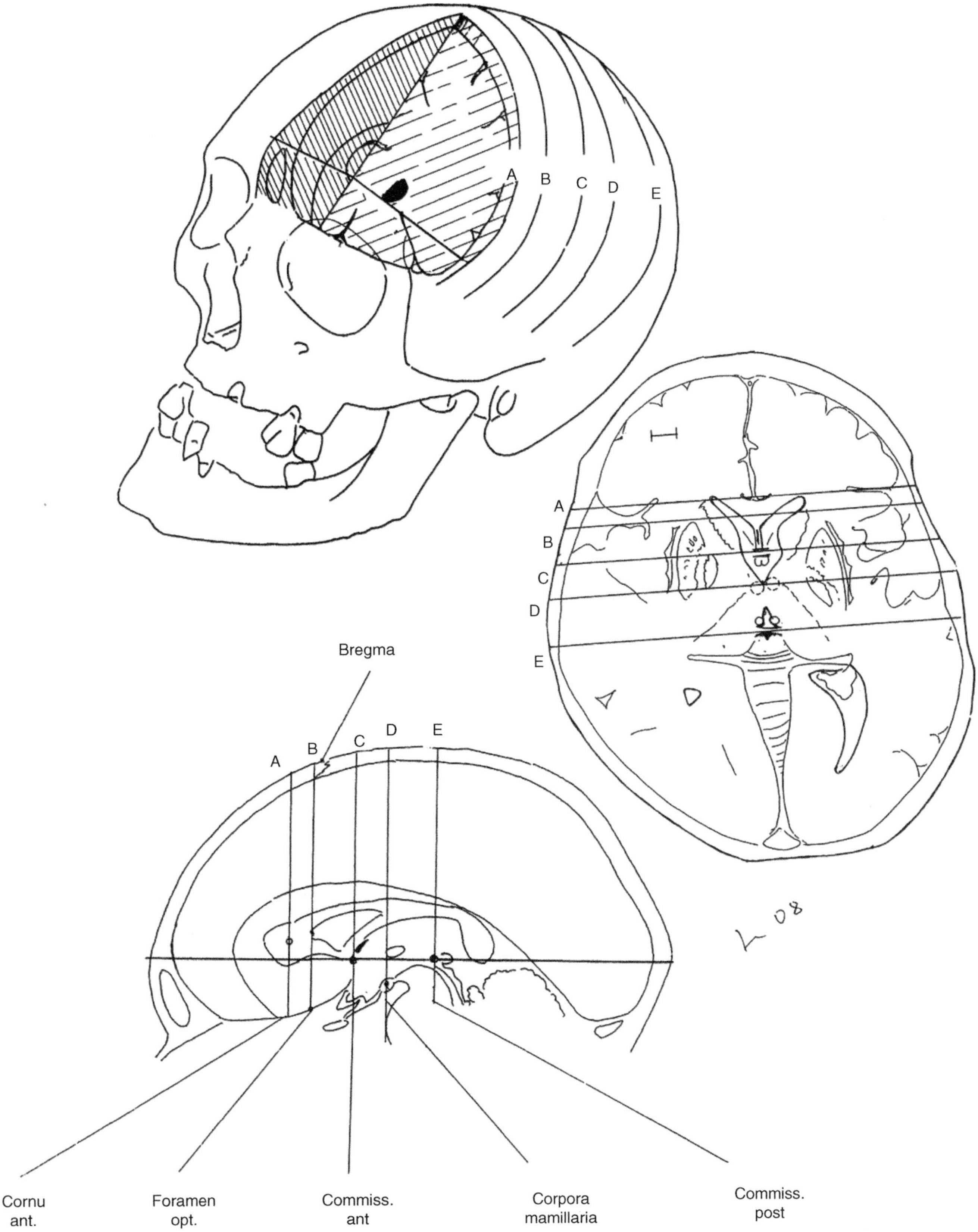

Fig. 1.13 Drawings of the skull and brain according to neuronavigation. Topograms for Figs. 1.14 and 1.15

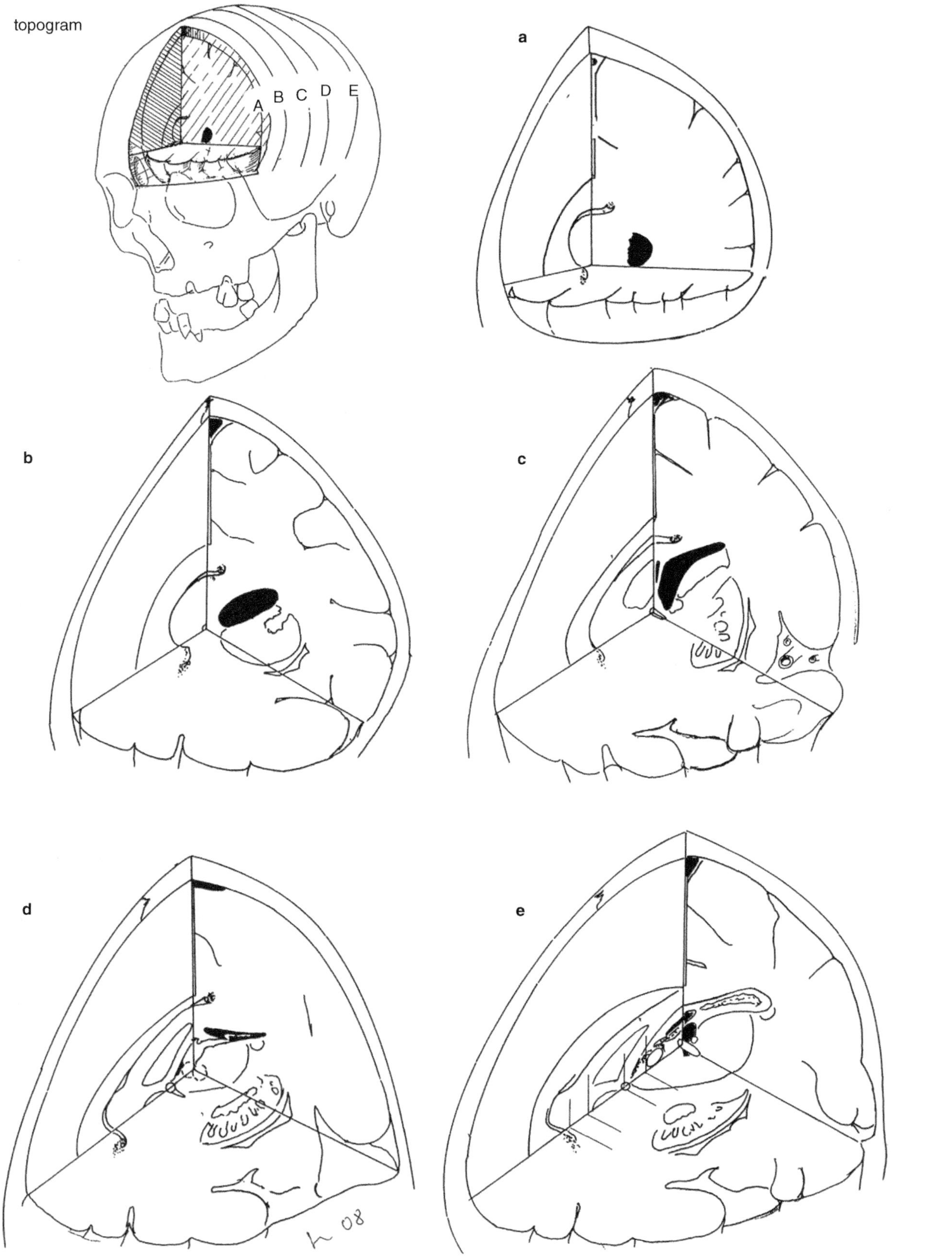

Fig. 1.14 Continuation of Fig. 1.13. Coronal, axial, and sagittal planes combined. (**a**) Tip of cornu anterius, (**b**) level between bregma and foramina optica, (**c**) commissura anterior, (**d**) corpora mammillaria, (**e**) commissura posterior

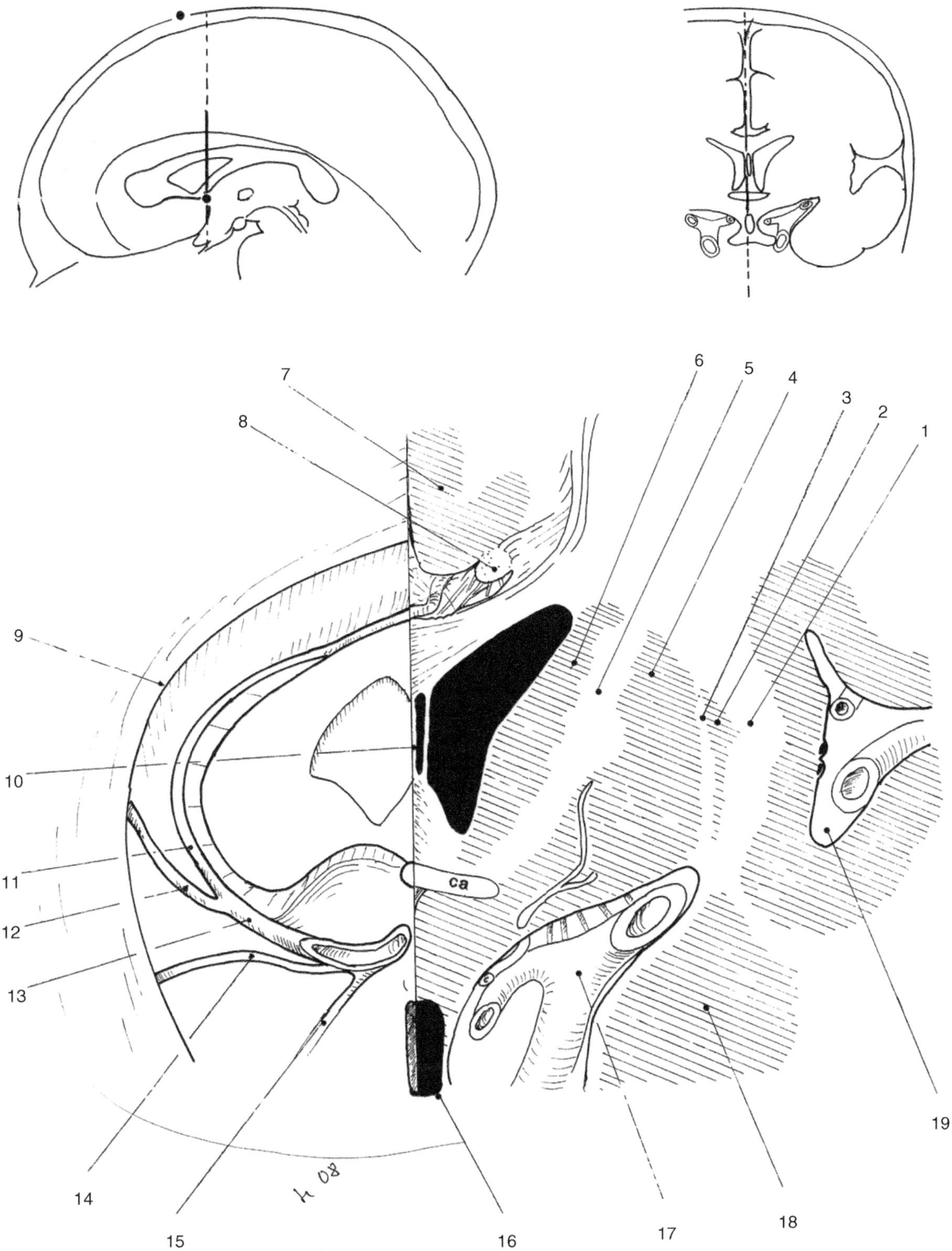

Fig. 1.15 As Fig. 1.14c. Sectional enlargement. *1* capsula extrema, *2* claustrum, *3* capsula externa, *4* putamen, *5* knee of capsula interna, *6* nucleus caudatus, *7* gyrus cinguli, *8* stria longitudinalis cinguli, *9* edge of falx, *10* septum et cavum septi pellucidi, *11* A. pericallosa, *12* A. callosomarginalis, *13* A2, *14* A. frontopolaris, *15* medial frontobasal artery, *16* third ventricle, *17* carotid bifurcation, *18* amygdala/uncus, *19* sulcus lateralis (Sylvii)

Anatomical Base of Surgery

2

2.1 Extradural Topography and Dural Veins (Figs. 2.1, 2.2, 2.3, 2.4, 2.5, 2.6, and 2.7)

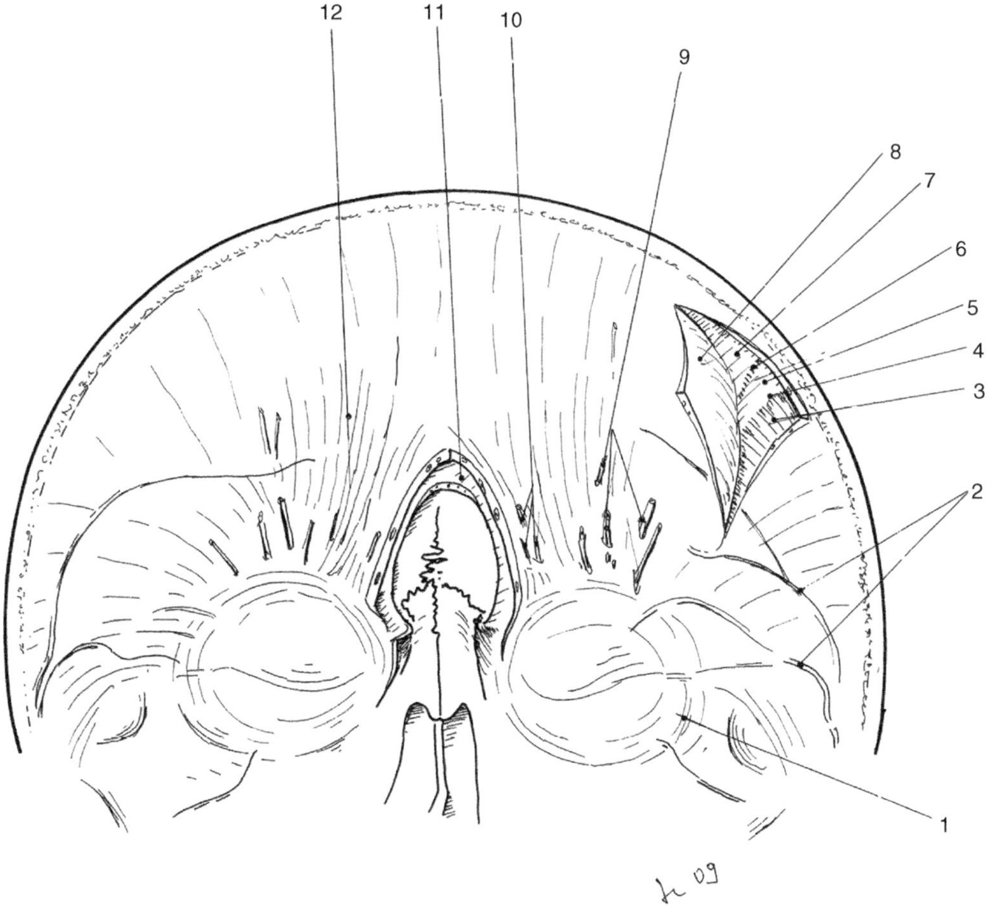

Fig. 2.1 *Galea aponeurotica* and adjacent structures. *1* M. orbicularis oculi; *2* N. facialis, frontal ramification; *3* M. temporalis; *4* as *3*, insertion at linea temporalis inferior; *5* fascia temporalis; *6* as *5*, its insertion at linea temporalis superior; *7* periosteum; *8* galea flap; *9* supraorbital vessels and nerves; *10* frontal vessels and nerves; *11* periosteum; *12* M. epicranius, frontal portion

W. Seeger, J. Zentner, *Anatomical Basis of Cranial Neurosurgery*, https://doi.org/10.1007/978-3-319-63597-2_2

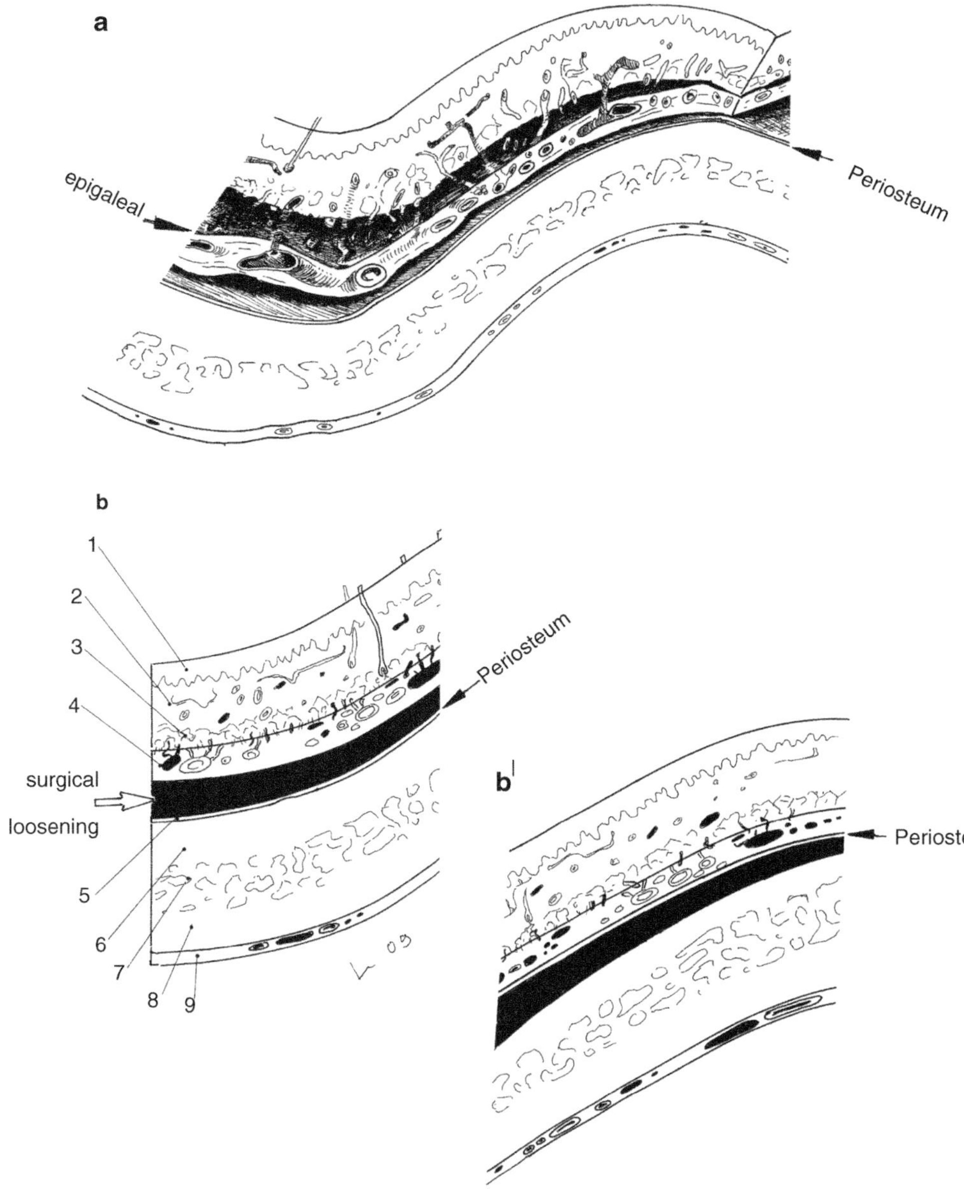

Fig. 2.2 Continuation of Fig. 2.16. *Cutis, galea, periosteum, and tabula externa and interna* according to surgical loosenings for presentations of arteries (e.g., for surgical anastomosis between A. temporalis superficialis and a branch of A. cerebri media) (**a**) and according to usual frontal approaches (**b**) or (**b'**). *1* Epidermis, *2* cutis, *3* subcutis, *4* galea, *5* periosteum, *6* tabula externa, *7* diploe, *8* tabula interna, *9* dura

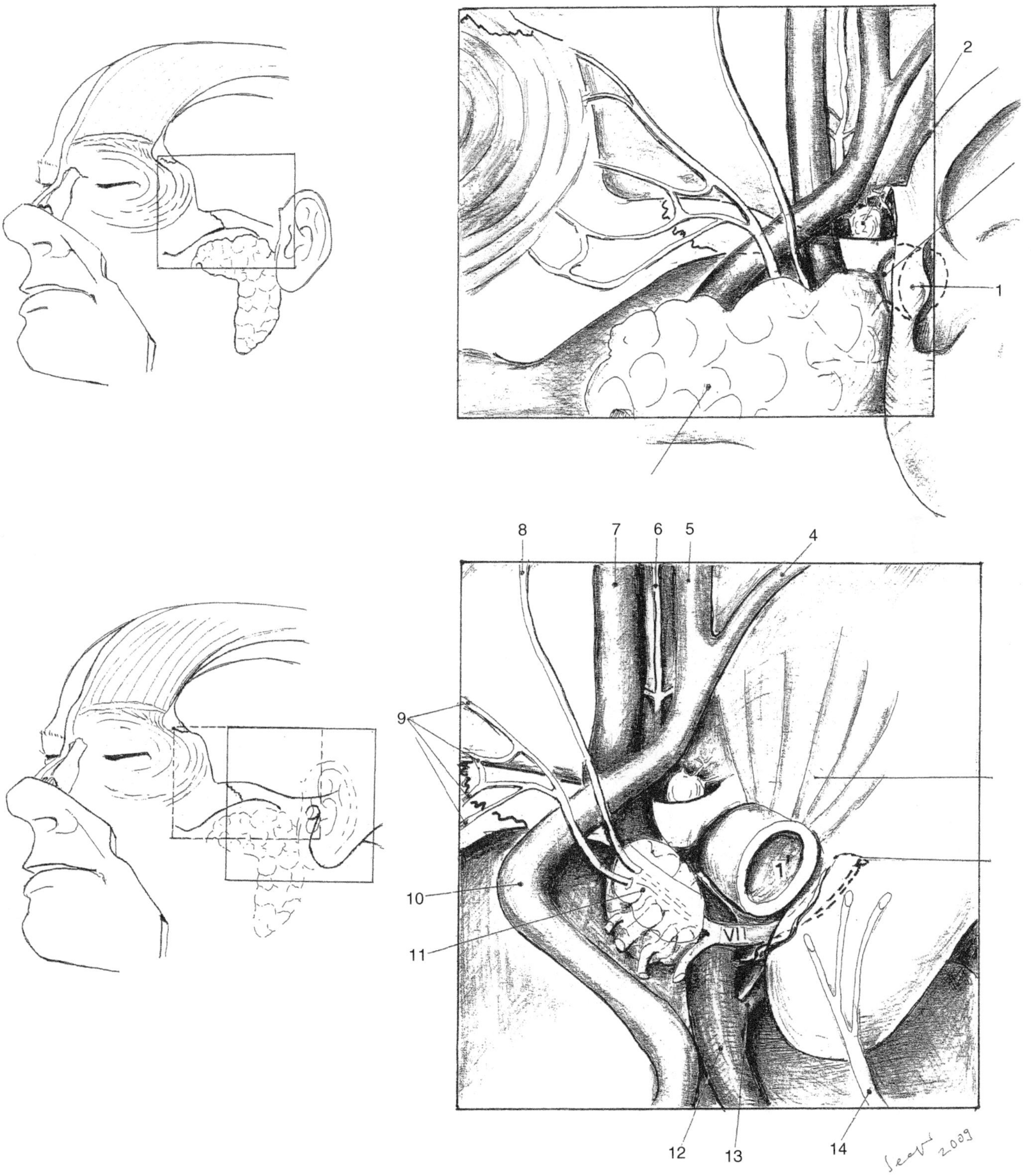

Fig. 2.3 Ramifications of N. facialis and adjacent structures which could be endangered at frontotemporal skin incision. *1* Meatus acusticus externus, projection dotted; *1'* as *1*, pars cartilaginea; *2* lymphonodus; *3* glandula parotis; *4* V. temporalis superficialis, ramus parietalis; *5* as *4*, ramus frontalis; *6* N. auriculotemporalis; *7* A. temporalis superficialis; *8* N. facialis, R. (Rr.) frontalis; *9* as *8*, Rr. zygomatici; *10* V. retromandibularis; *11* glandula parotis, residual; *12* A. carotis externa; *13* A. occipitalis; *14* N. auricularis magnus; *15* For. stylomastoideum, projection; *16* M. temporoparietalis, portion of M. epicranius

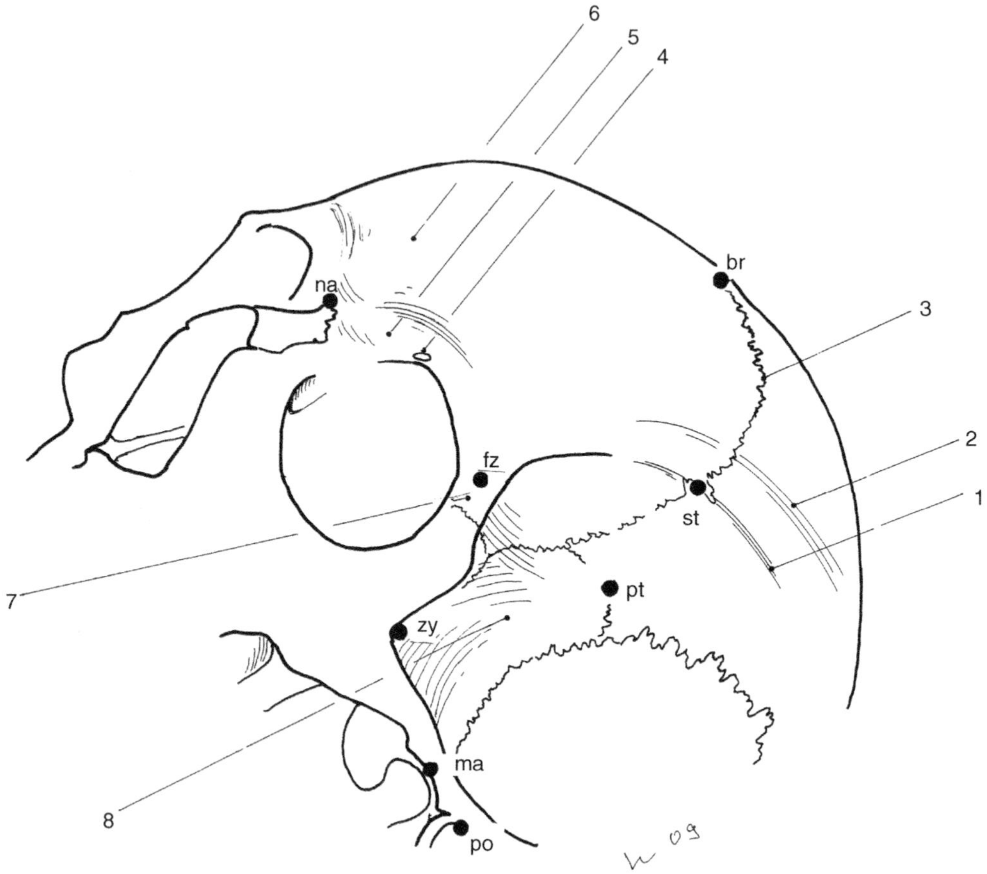

Fig. 2.4 Relief of the skull. Landmarks added for planning frontal and frontotemporobasal (pterional) surgical approaches. *br* bregma, *fz* frontozygomatic point, *ma* fossa mandibularis, *na* nasion, *pt* pterion, *po* porion, *st* stephanion, *zy* angle of os zygomaticum. *1* Linea temporalis inferior, *2* linea temporalis superior, *3* sutura coronalis, *4* For. supraorbitale, *5* arcus superciliaris, *6* glabella, *7* processus zygomaticus, *8* ala major

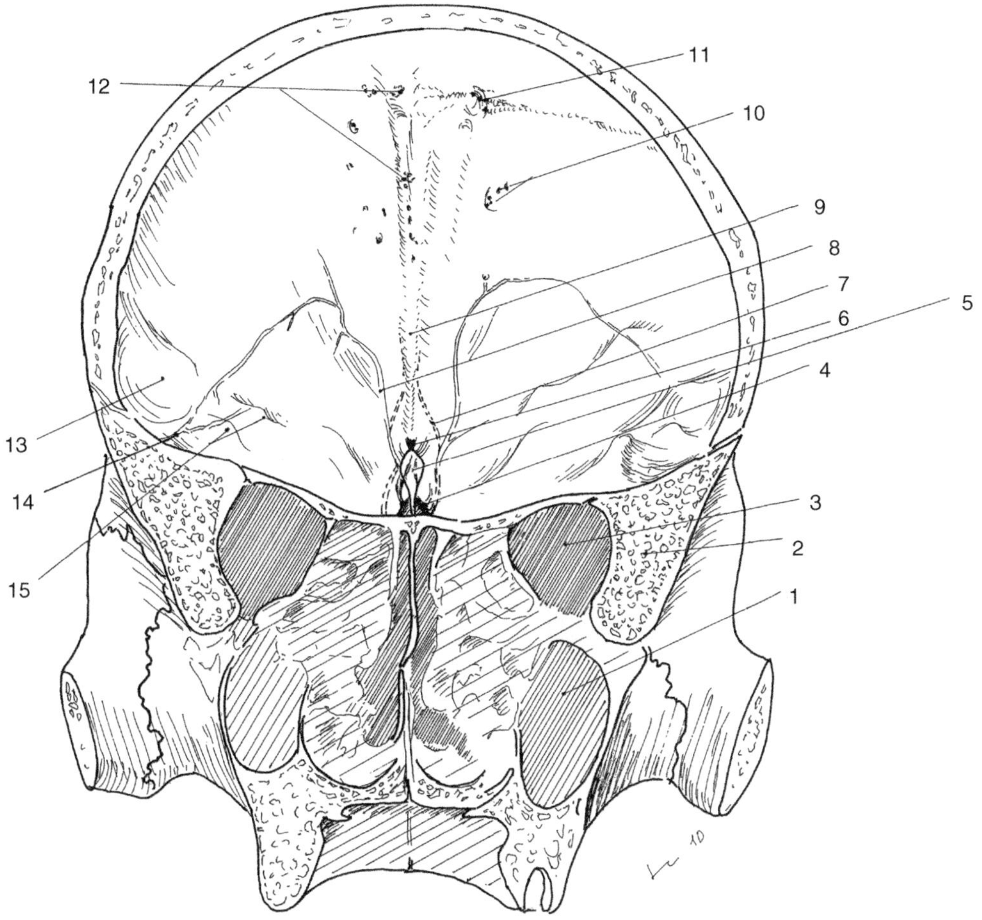

Fig. 2.5 *Bony sulci of meningeal arteries widened* at hypervascularized lesions (e.g., frontobasal meningiomas). Feeding arteries are A. carotis externa, A. meningea media, A. carotis interna, A. ophthalmica, and A. ethmoidalis anterior et posterior. These should be considered at interventional embolizations. *1* Sinus maxillaris; *2* ala major; *3* orbita; *4* lamina cribrosa; *5* crista galli; *6* For. caecum; *7* sulci of arteries (arteries originating from falx); *8* sulci of further dural vessels; *9* sulcus sagittalis; *10* foveolae arachnoidales of Pacchioni (containing granulationes arachnoidales); *11* as *10* and sulci of dural veins (sinus sphenoparietalis); *12* as *10*, close to sinus sagittalis superior; *13* part of impressiones gyrorum; *14* sulcus of R. anterior of A. meningea media; *15* Juga cerebralia

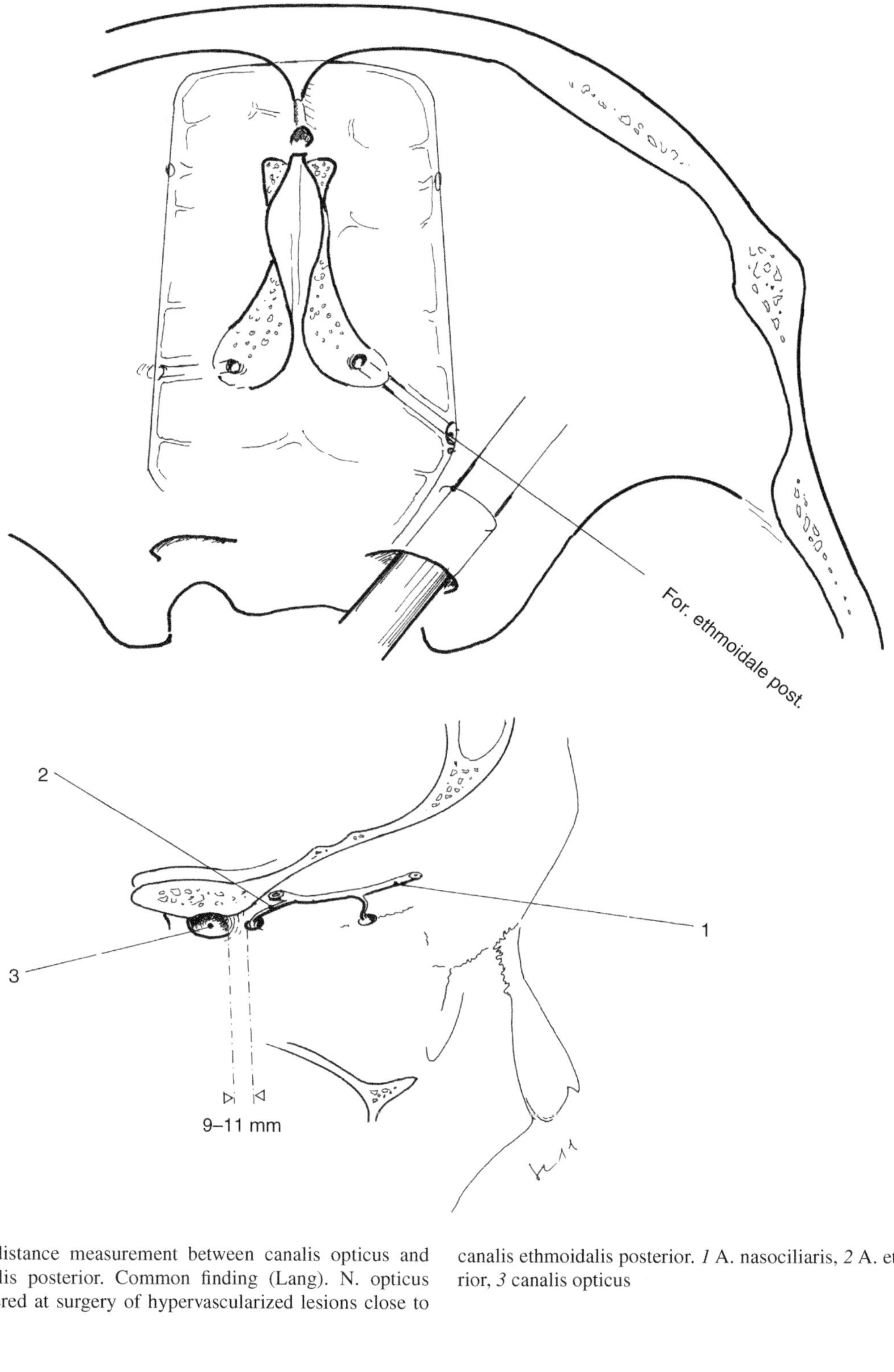

Fig. 2.6 Short distance measurement between canalis opticus and canalis ethmoidalis posterior. Common finding (Lang). N. opticus could be endangered at surgery of hypervascularized lesions close to canalis ethmoidalis posterior. *1* A. nasociliaris, *2* A. ethmoidalis posterior, *3* canalis opticus

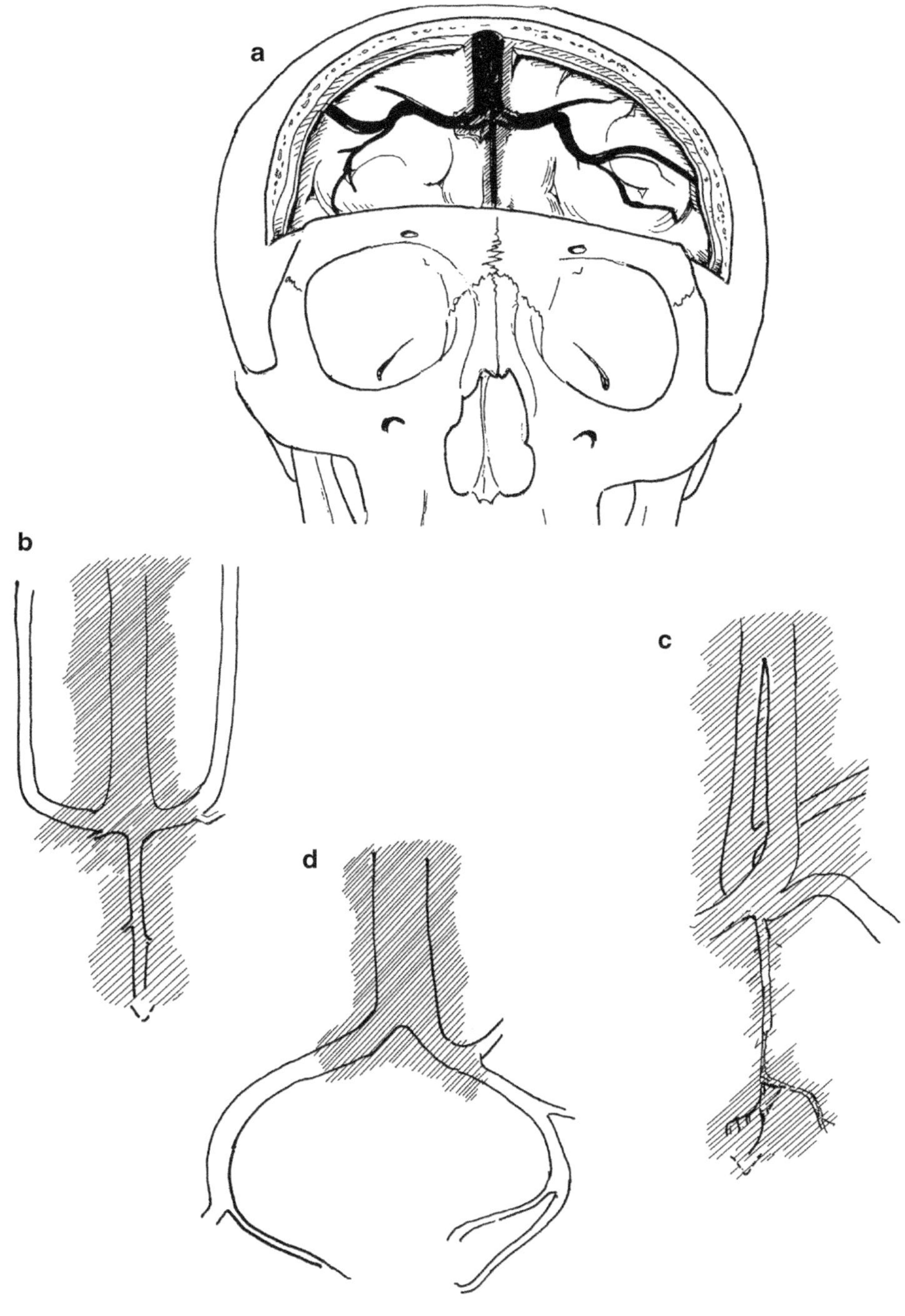

Fig. 2.7 Variants of frontal bridging veins and sinus sagittalis superior (drawing according to casts of Browder and Kaplan 1976, schematized). **a** Usual finding. **b** This variant could be a draining vein of occipital to frontal regions. After its interruption at surgery, multiple located cortical encephalomalacia is possible (experiences of the author). **c** Veins crossing to the contralateral side. After interruption, contralateral cortical encephalomalacia is possible. **d** At this bilateral type of frontal bridging veins, usually the rostral segment of sinus sagittalis superior is absent. Hatched: fine dural venous network. It is usually to be found adjacent to all dural sinuses

2.2 Intradural Topography

2.2.1 Survey (Figs. 2.8, 2.9, 2.10, 2.11, 2.12, 2.13, 2.14, and 2.15)

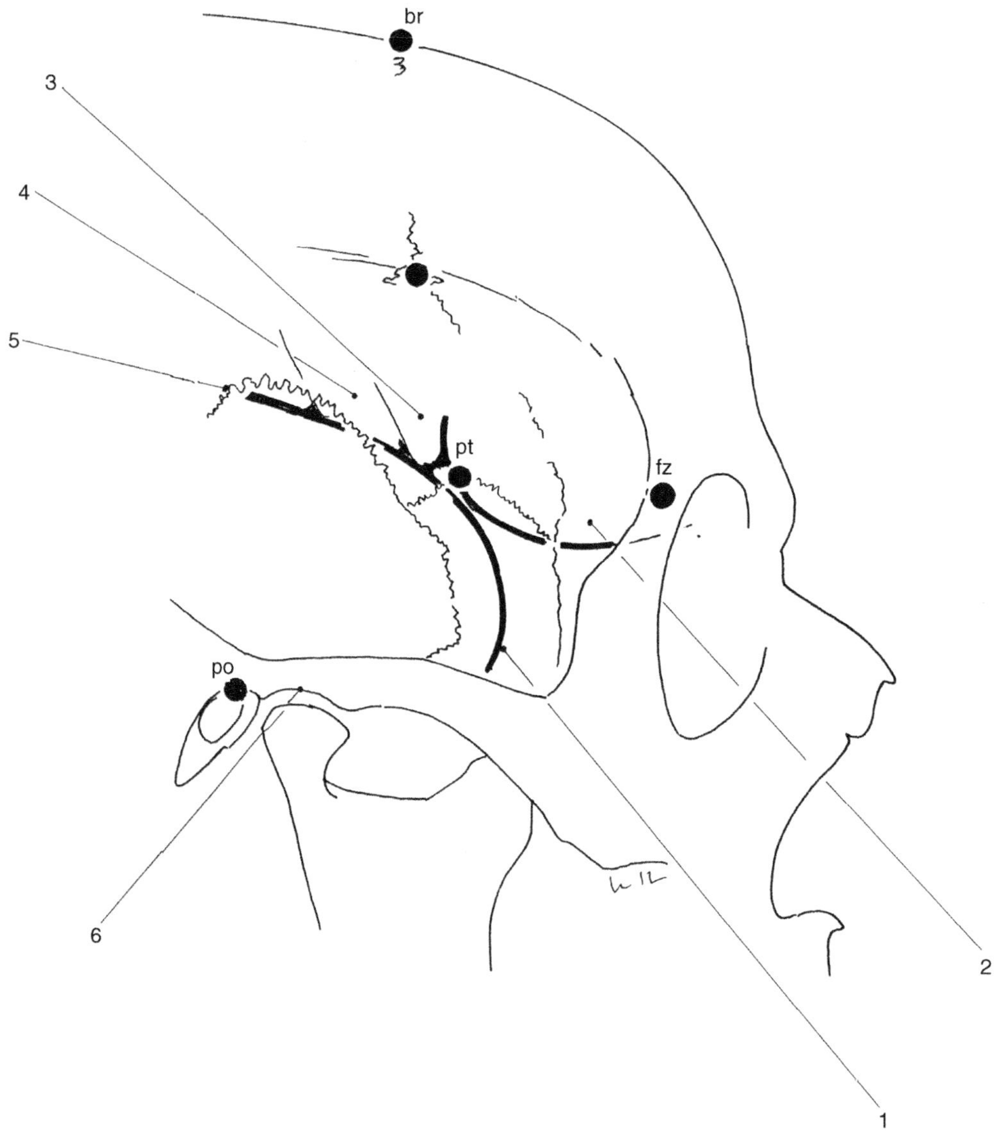

Fig. 2.8 Skull. Landmarks added. Projection of the sylvian region drawn in *br* bregma, *fz* frontozygomatic point, *po* porion, *pt* pterion, and *st* stephanion. *1* Temporal pole; *2* third frontal gyrus, pars orbitalis; *3* as *2*, pars triangularis; *4* as *2*, pars opercularis; *5* sutura squamosa; *6* fovea mandibularis

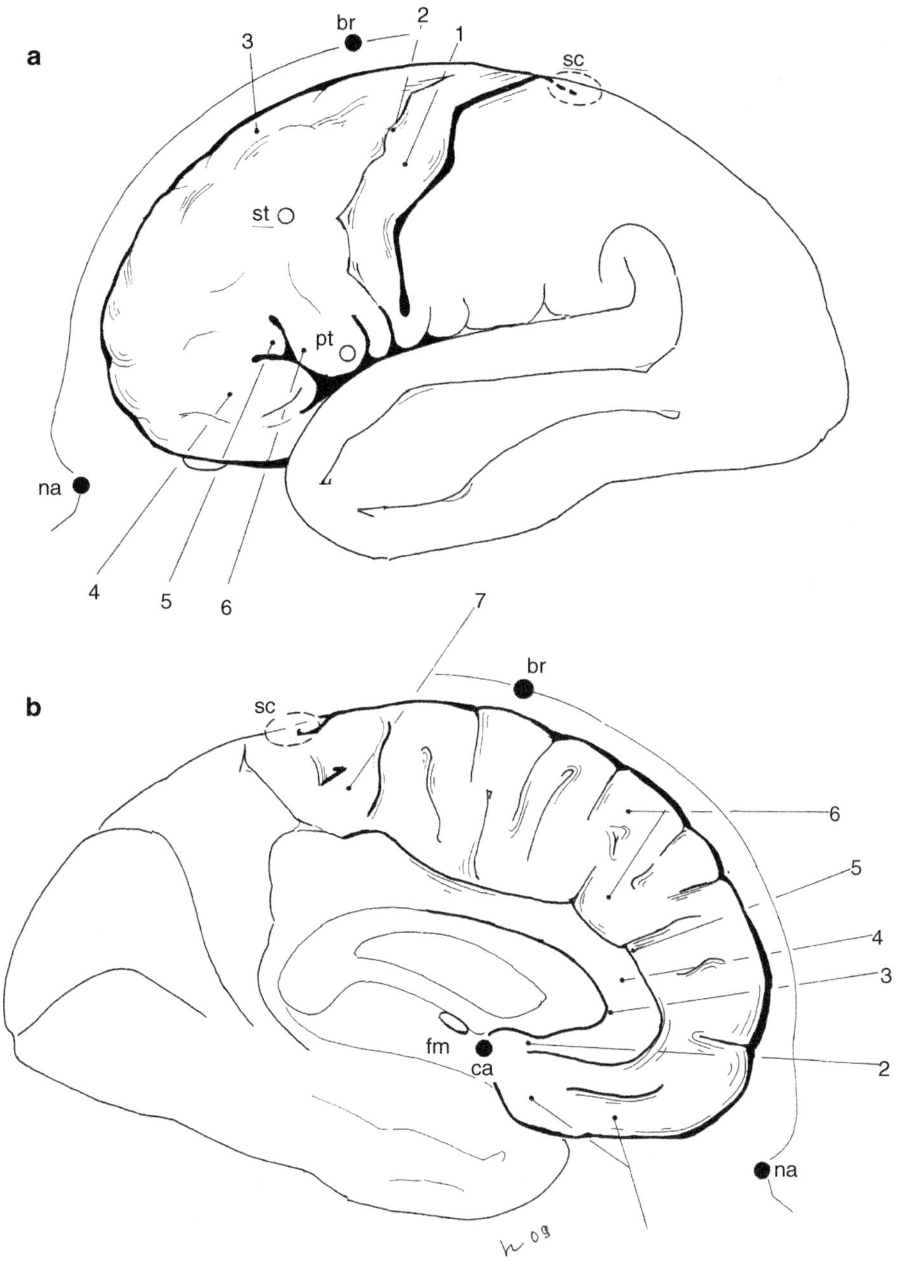

Fig. 2.9 Typical lateral **a** and medial **b** cortical relief, schematized. Frontal lobe renders prominent. Landmarks added. Typical reliefs present the so-called opercular part of the third frontal gyrus: pars orbitalis, pars triangularis, pars opercularis, and the base of gyrus precentralis (the region between the posterior segment of pars opercularis and sulcus centralis is usually motor cortex. It will be exactly defined using evoked potentials or magnetoencephalogram (MEG; e.g., see Kristewa et al. 1991)); see A. **a** *br* bregma; *na* nasion; *pt* pterion; *sc* sulcus centralis, upper point; *st* stephanion. *1* Gyrus precentralis; *2* sulcus precentralis; *3* first frontal gyrus; *4* third frontal gyrus, pars orbitalis; *5* as *4*, pars triangularis; *6* as *4*, pars opercularis. **b** *br* bregma, *ca* commissura anterior, *fm* foramen interventriculare, *sc* sulcus centralis. *1* Gyrus rectus, *2* area subcallosa, *3* sulcus corporis callosi, *4* gyrus cinguli, *5* sulcus cinguli, *6* first frontal gyrus, *7* lobulus paracentralis (anterior region: part of the motor area)

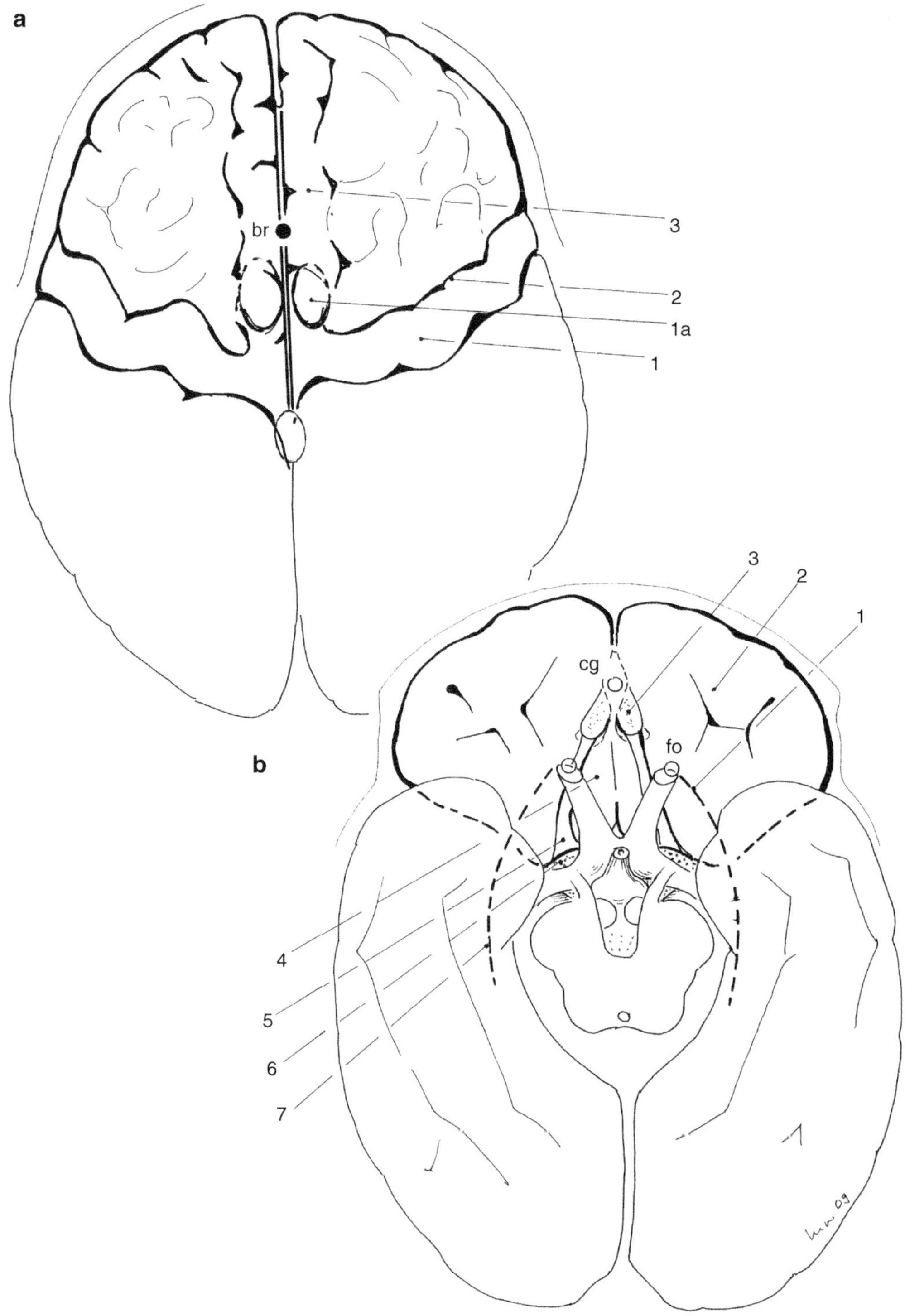

Fig. 2.10 Dorsal **c** and basal cortical relief **d**. Projections of basal dural duplications and landmarks added. The supplementary motor area (SMA) of one side participates in all bilateral motor functions. Surgical resections would not entail motor deficits. This is an important aspect at resections of the frontal lobe, especially at eliminations of epilepsy foci of SMA. **c** *br* bregma, *sc* sulcus centralis. *1* Gyrus precentralis, *1a* supplementary motor area (SMA) (Rauber-Kopsch (1987) Thieme, p. 436. Its posterior and medial limits (close to gyrus precentralis) are well to define at surgery, but not rostral and lateral), *2* sulcus precentralis, *3* first frontal gyrus. **d** *cg* crista galli, projection; *fo* foramen opticum projection. *1* Plica petroclinoidea anterior, projection; *2* gyri orbitales; *3* bulbus olfactorius; *4* gyrus rectus; *5* trigonum olfactorium; *6* substantia perforata anterior; *7* tentorial fold, projection

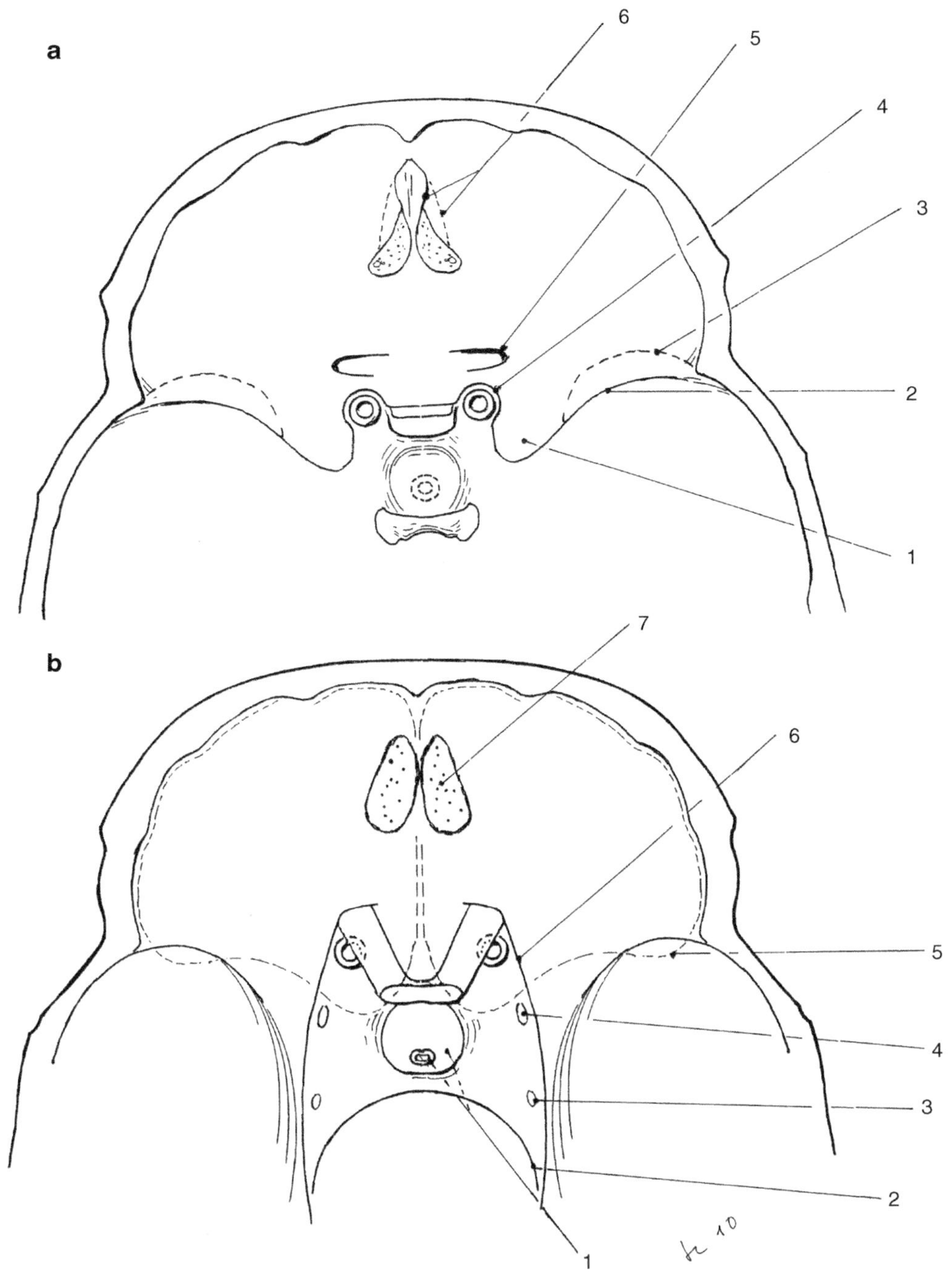

Fig. 2.11 Cranial base **a** and dural penetration points of cranial nerves **b**. **a** *1* Processus clinoideus anterior; *2* ala minor; *3* fossa cranii media overlaid by ala minor; *4* siphon of A. carotis interna added; note bony incisure; *5* foramen opticum; *6* overlaying of lamina cribrosa (and bulbus olfactorius) by the medial part of os ethmoidale; this has to be taken in consideration at surgical elevation of lobus frontalis (see Figs. 2.42 and 3.35). **b** *1* Infundibulum and diaphragma sellae; *2* plica petroclinoidea posterior; *3* dural gap for N. trochlearis; *4* dural gap for N. oculomotorius; *5* frontal lobe, projection; *6* plica petroclinoidea anterior; *7* lamina cribrosa, adjacent bony structures omitted

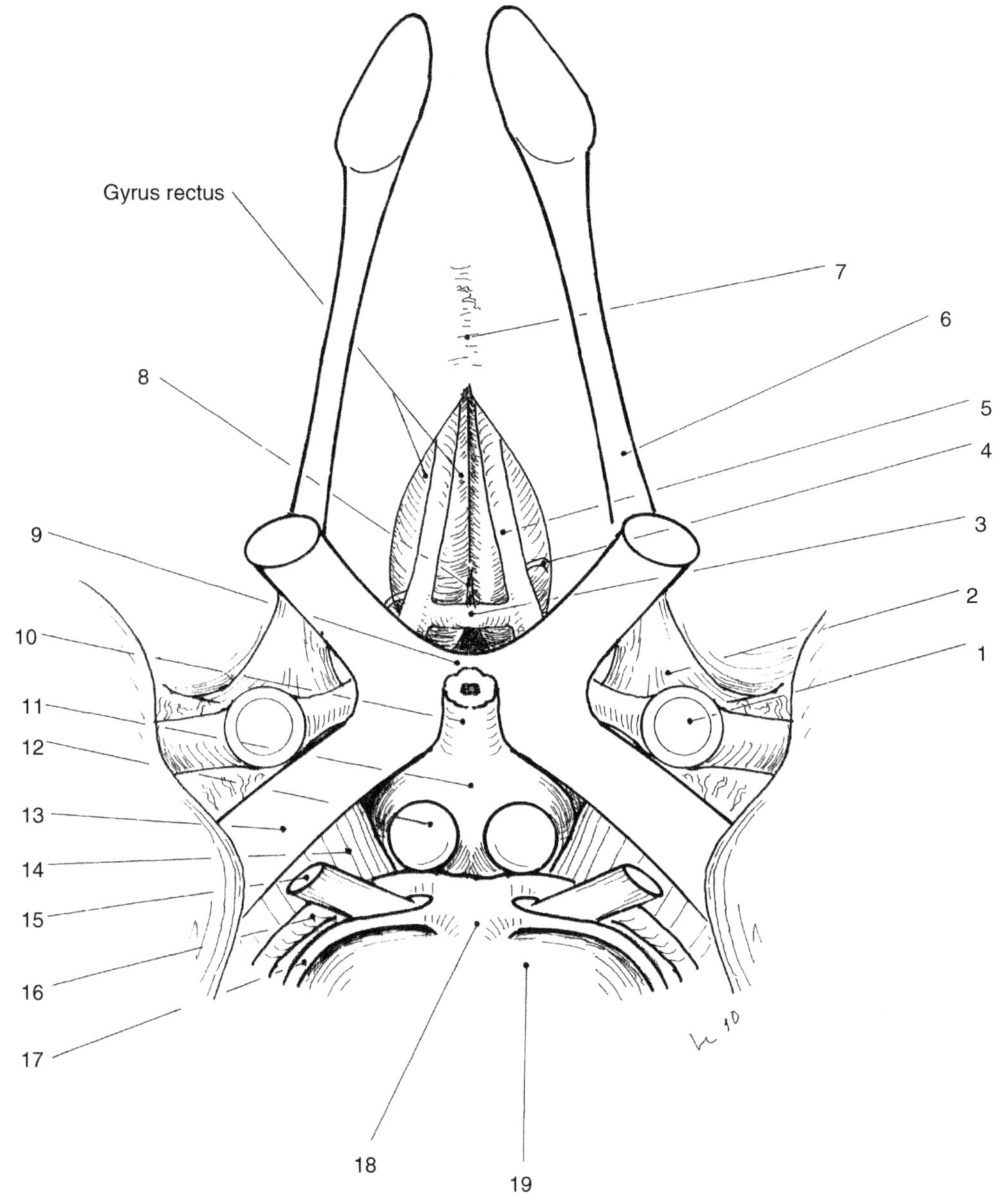

Fig. 2.12 Contents of suprasellar cisterns according to surgery, e.g., of aneurysms of A. cerebri anterior and A. communicans anterior. Gyri recti and Aa. cerebri anterior shifted aside after splitting the basal arachnoidea. Sulcus medianus sometimes cannot be defined during surgery, because it is masked by a thickened arachnoidea. Gyrus rectus could be endangered uni- or bilaterally, with following psychological deficits (Hütter 1999, 2000). *1* Carotid bifurcation; *2* trigonum olfactorium; *3* A. communicans anterior; *4* A. recurrens (Heubner) must be preserved at surgery; *5* A2; *6* tractus olfactorius; *7* interhemispheric basal arachnoidea often not transparent; *8* A. corporis callosi mediana, cave perforating branches feeding the hypothalamus; *9* chiasma; *10* infundibulum; *11* tuber cinereum; *12* corpus mammillare; *13* tractus opticus; *14* crus cerebri; *15* N. oculomotorius; *16* A. cerebri posterior; *17* A. cerebelli superior; *18* A. basilaris; *19* pons

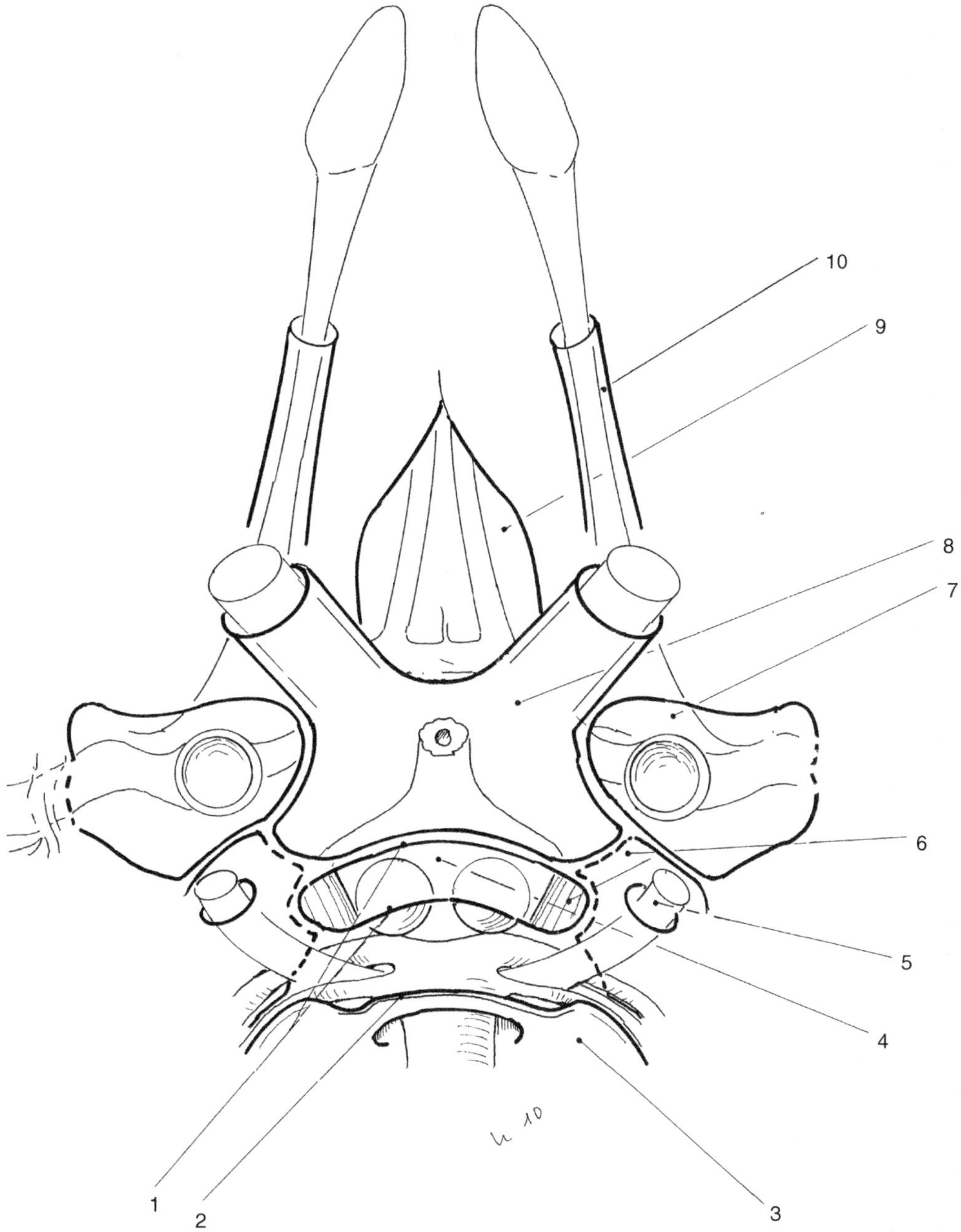

Fig. 2.13 Continuation of Fig. 2.12. *Arachnoid membranes* (walls) *of cisterns*. *1* Liliequist's membrane, diencephalic portion (Rhoton 2000); *2* as *1*, mesencephalic portion; *3* cisterna pontis medialis; *4* cisterna interpeduncularis; 5 N. oculomotorius; *6* cisterna cruralis; *7* carotid cistern; *8* cisterna chiasmatis; *9* cisterna laminae terminalis; *10* cisterna olfactoria

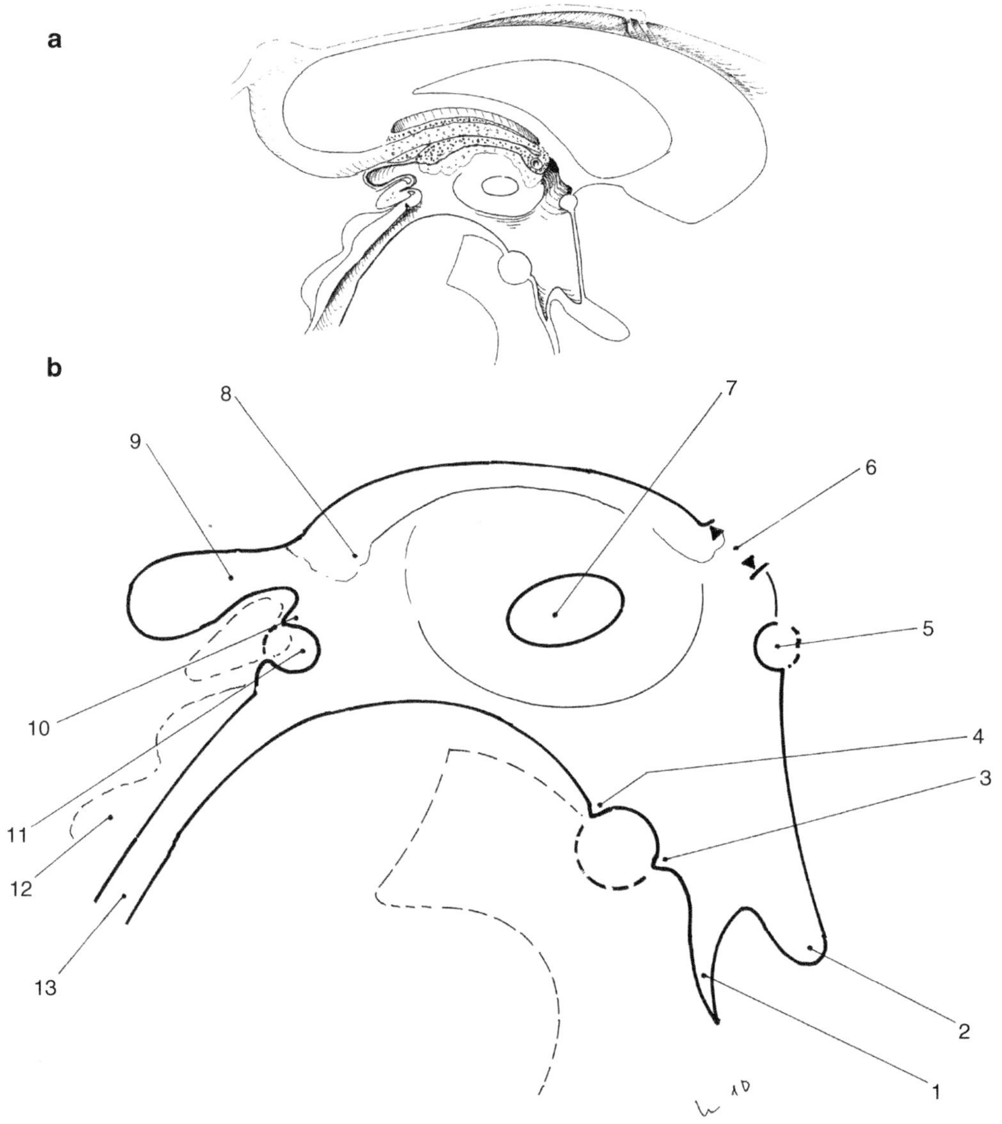

Fig. 2.14 Third ventricle. Surrounding structures **a** and lumen **b**. *1* Recessus infundibuli; 2 recessus opticus; *3* recessus praemamillaris; *4* recessus postmamillaris; *5* bulging of commissura anterior; *6* foramen interventriculare (Monroi); *7* adhesio interthalamica; *8* plexus choroideus; *9* recessus suprapinealis; *10* recessus pinealis; *11* bulging of commissura posterior; *12* tectum, projection; *13* aquaeductus (sylvii)

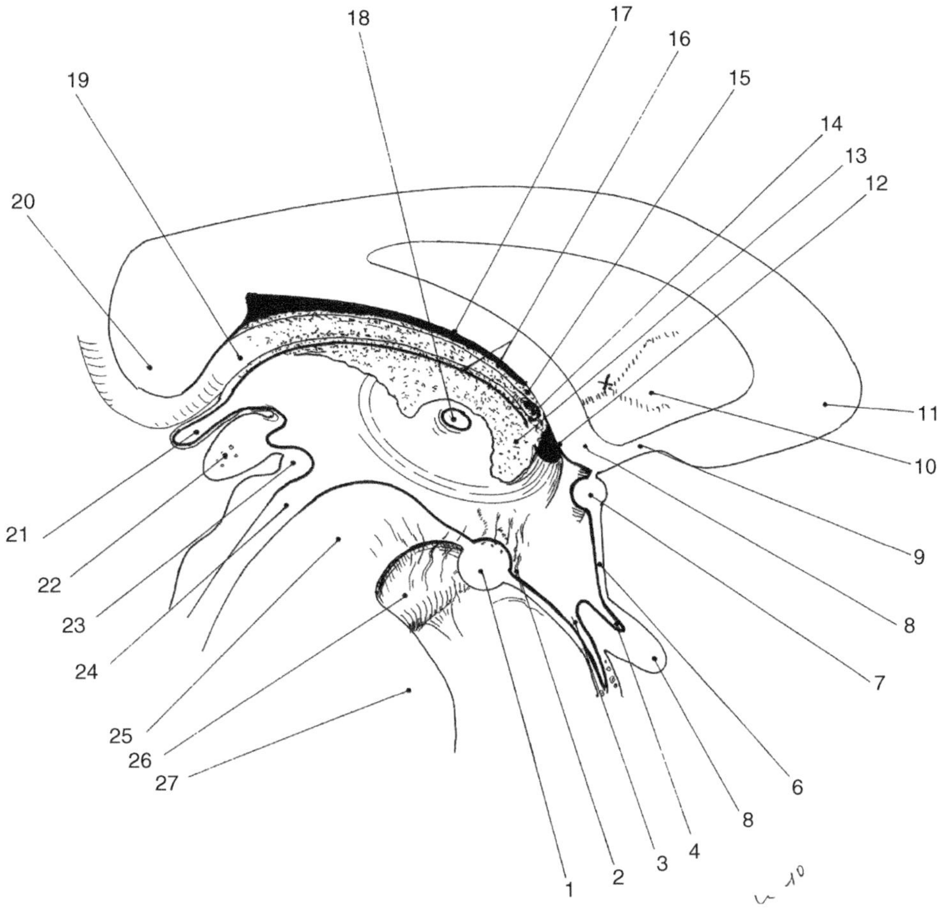

Fig. 2.15 Third ventricle. Details of surrounding structures. *1* Corpus mammillare, *2* recessus praemamillaris, *3* recessus infundibuli, *4* recessus opticus, *5* chiasma, *6* lamina terminalis, *7* commissura anterior, *8* columna fornicis, *9* rostrum corporis callosi, *10* septum pellucidum, *11* genu corporis callosi, *12* For. interventriculare (Monroi), *13* plexus choroideus, *14* angulus venosus, *15* fibrous wall between foramen interventriculare and fissura transversa, *16* velum interpositum, *17* fissura transversa, *18* adhesio interthalamica, *19* V. magna (Galeni), *20* splenium corporis callosi, *21* recessus suprapinealis, *22* corpus pineale, *23* commissura posterior, *24* aquaeductus (sylvii), *25* tegmentum mesencephali, *26* crus cerebri, *27* pons, x V. septi pellucidi

2.2.2 Superficial Structures and Structures in the Depth (Figs. 2.16, 2.17, 2.18, 2.19, 2.20, 2.21, and 2.22)

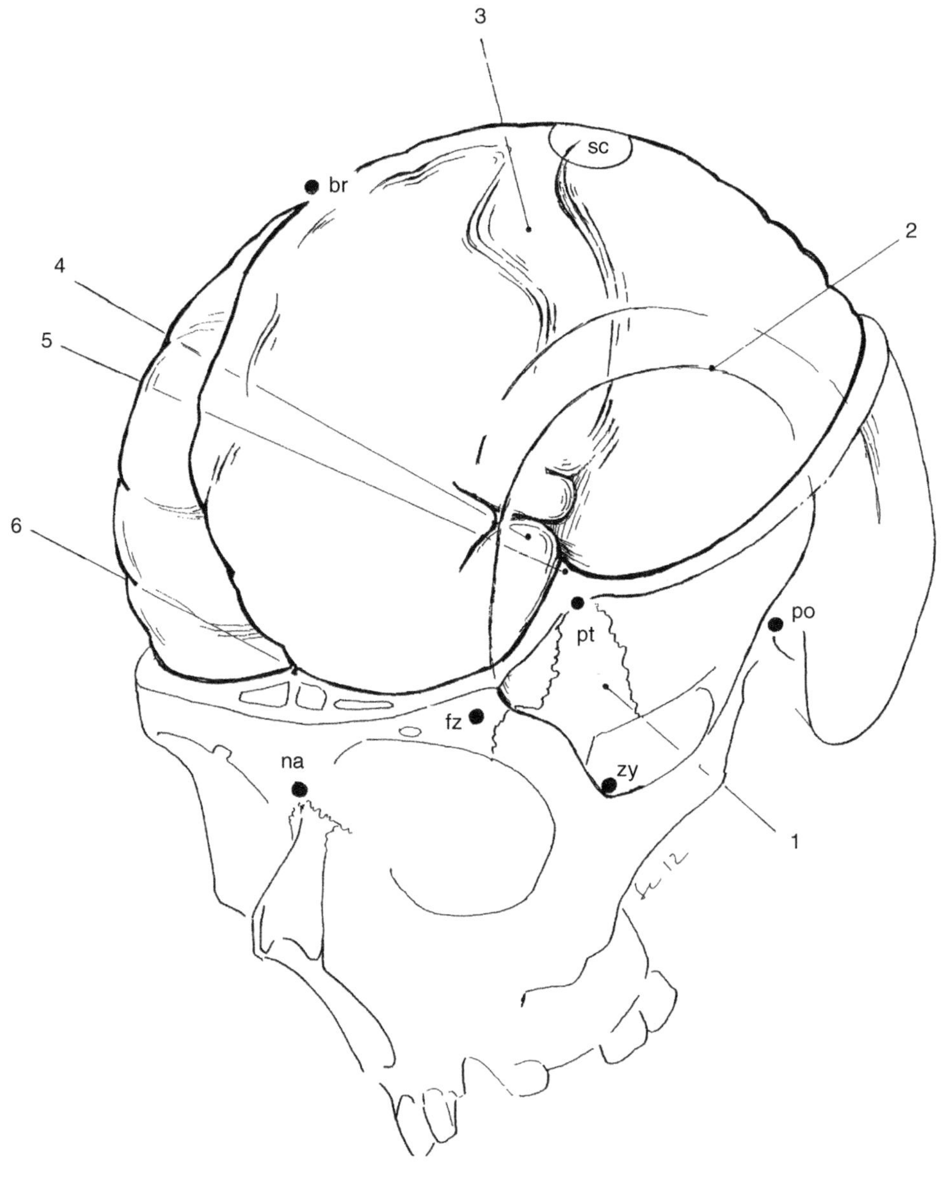

Fig. 2.16 Superficial structures of the skull and brain. *br* bregma, *fz* frontozygomatic point, *na* nasion, *po* porion, *pt* pterion, *sc* sulcus centralis, *zy* angle of os zygomaticum. *1* Ala major; *2* linea temporalis inferior, projection; *3* gyrus precentralis; *4* pars orbitalis of gyrus frontalis 3; *5* ala major et minor; *6* crista frontalis

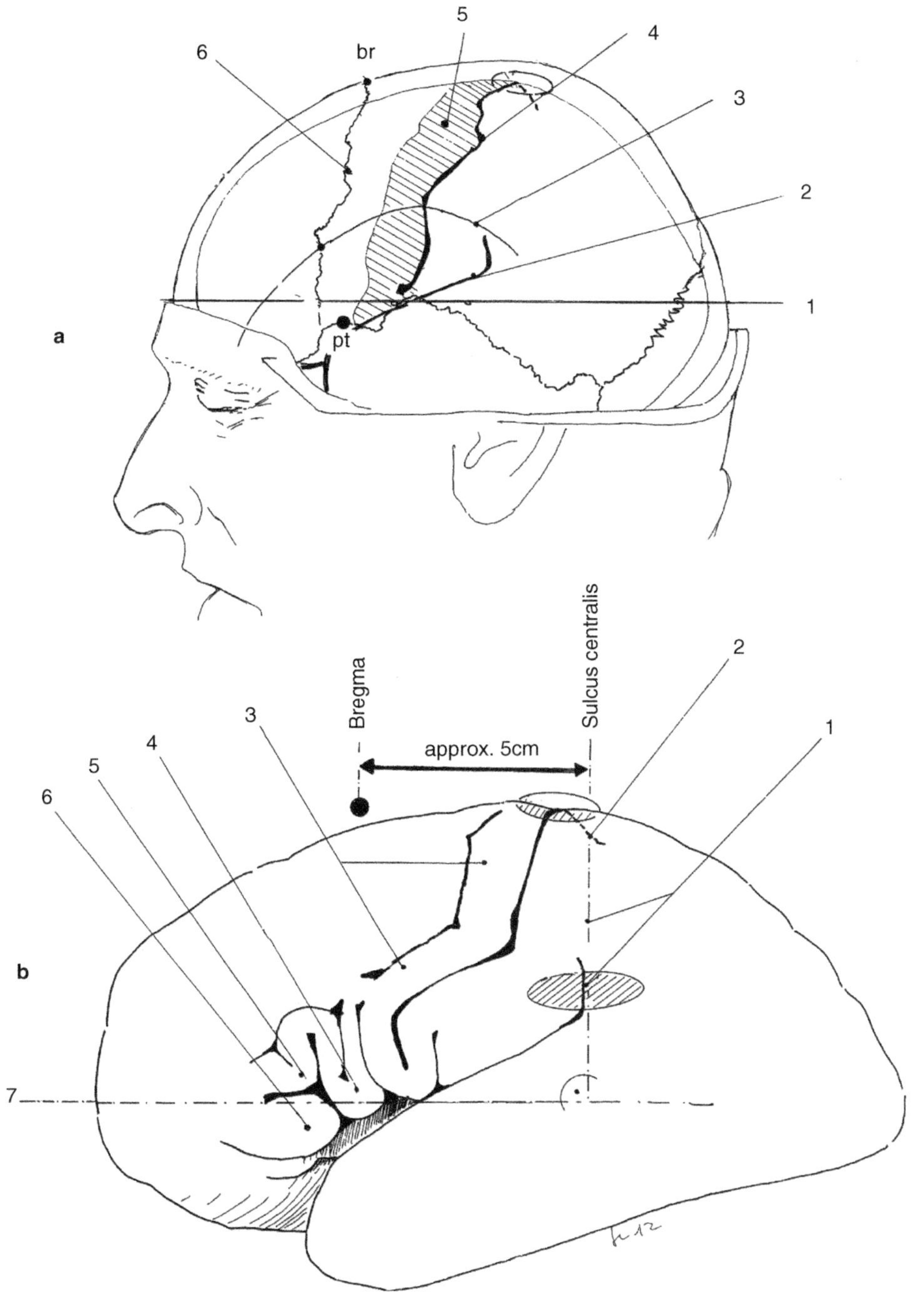

Fig. 2.17 Gyrus precentralis and landmarks. **a** Gyrus precentralis and the skull, **b** gyrus precentralis and surrounding cortex. **a** *br* bregma, *pt* pterion, *st* stephanion. *1* Intercommissural level; *2* pars ascendens of sulcus lateralis (sylvii); *3* linea temporalis inferior, projection; *4* sulcus centralis; *5* gyrus precentralis; *6* sutura coronalis, projection. **b** *1* Angularis region; *2* sulcus centralis; *3* gyrus precentralis; *4* third frontal gyrus, pars opercularis; *5* as *4*, pars triangularis; *6* as *4*, pars orbitalis; *7* intercommissural level

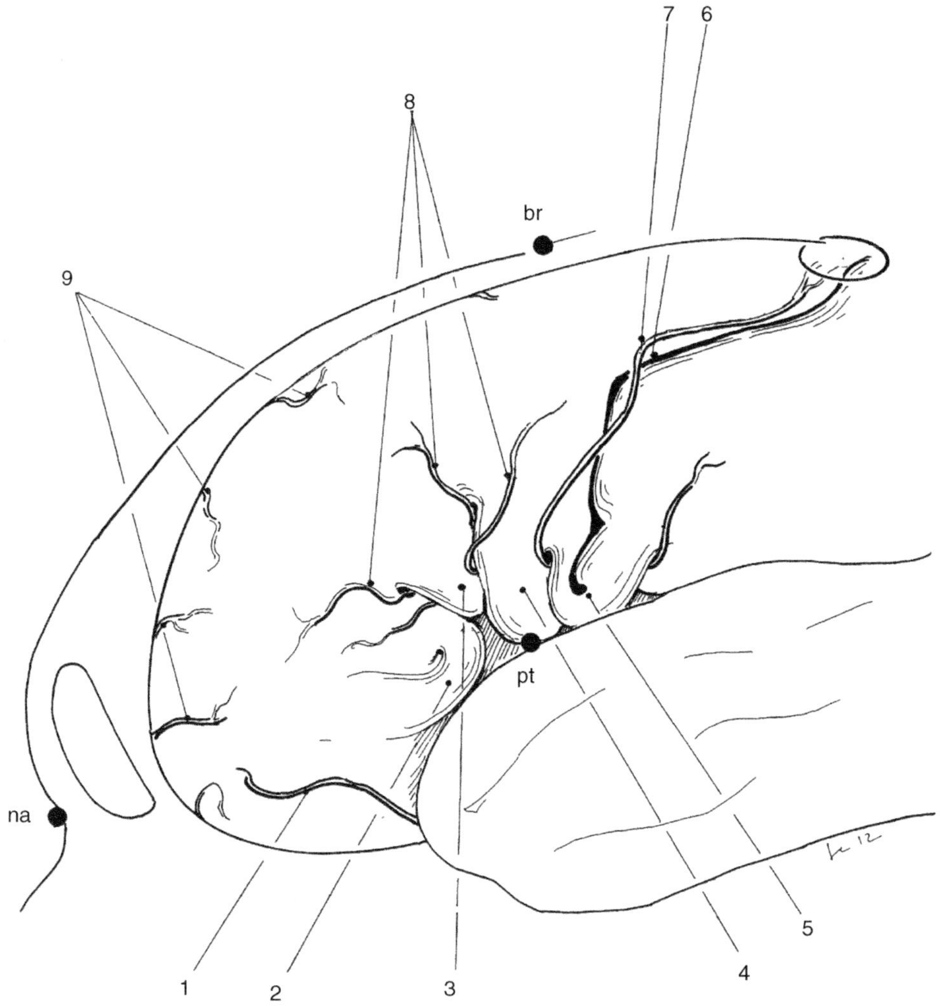

Fig. 2.18 Arteries and surrounding structures. Landmarks added for assisting neuronavigation, if necessary. *br* bregma, *na* nasion, *pt* pterion. *1* Frontobasal lateral artery; *2* pars orbitalis of gyrus frontalis 3; *3* pars triangularis; *4* pars opercularis; *5* sulcus centralis; *6* as 5, typical bending; *7* A. sulci centralis; *8* branches of A. operculofrontalis; *9* ramification of A. callosomarginalis and of A. pericallosa

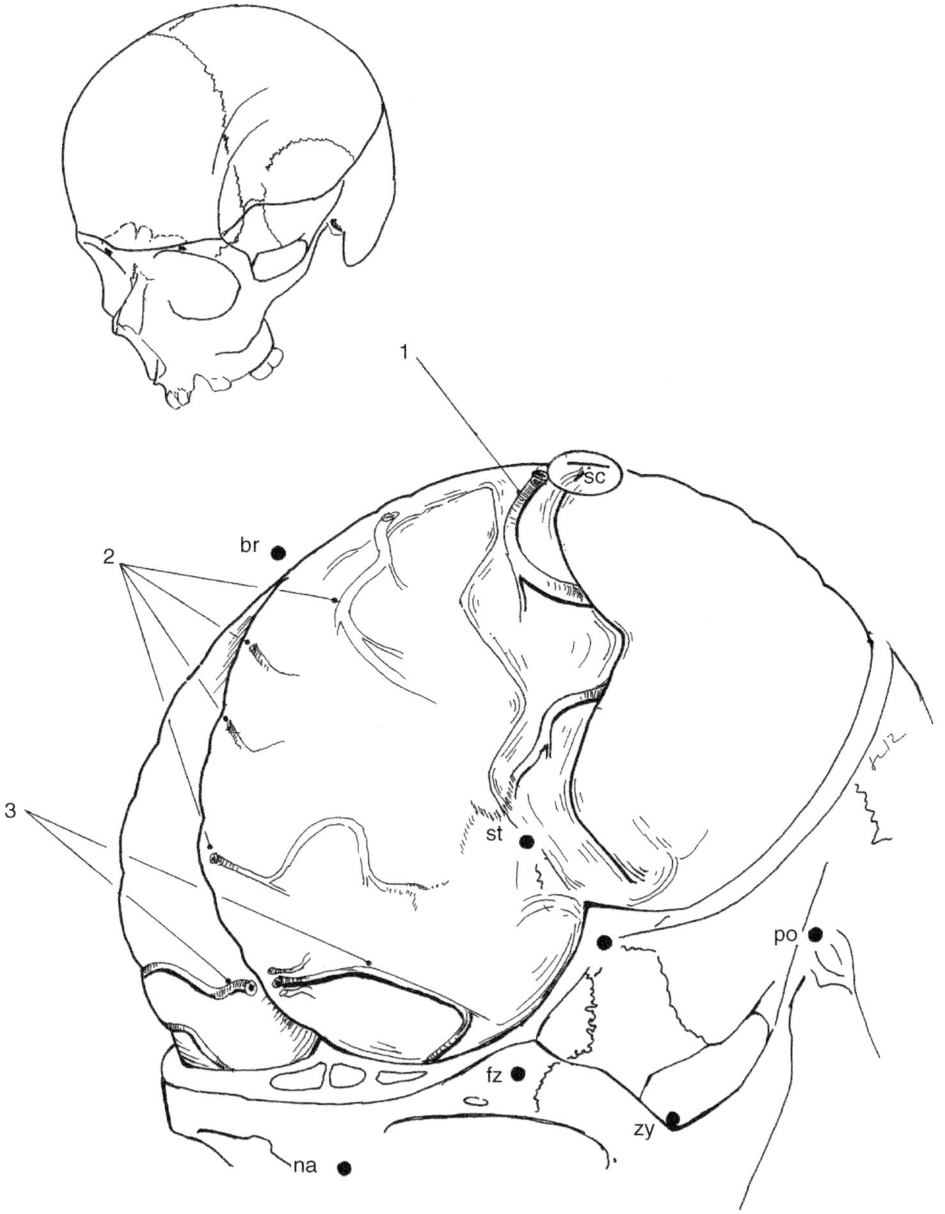

Fig. 2.19 Continuation of Fig. 2.18. *Veins*. *br* bregma, *fz* frontozygomatic point, *na* nasion, *po* porion, *pt* pterion, *sc* sulcus centralis, *st* stephanion, *zy* angle of os zygomaticum. *1* V. rolandica, *2* further bridging veins, *3* bundles of frontal bridging veins

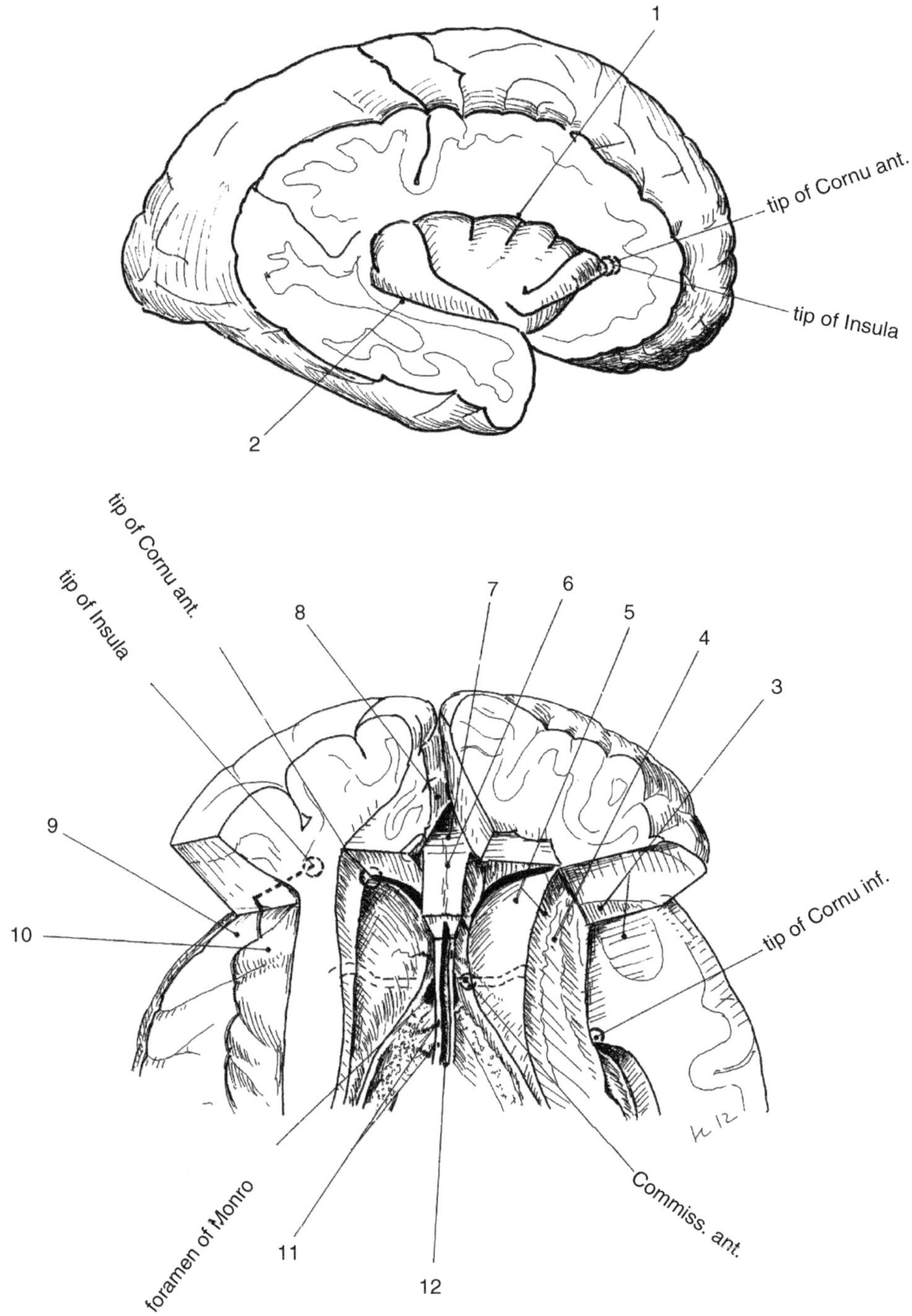

Fig. 2.20 Structures in the depth. *1* Sulcus superior insulae, *2* sulcus inferior insulae, *3* putamen, *4* corona radiata, *5* caput nuclei caudati, *6* indusium griseum, *7* genu corporis callosi, *8* gyrus cinguli, *9* Heschl's gyri, *10* insula, *11* medial septal veins, *12* cavum septi pellucidi

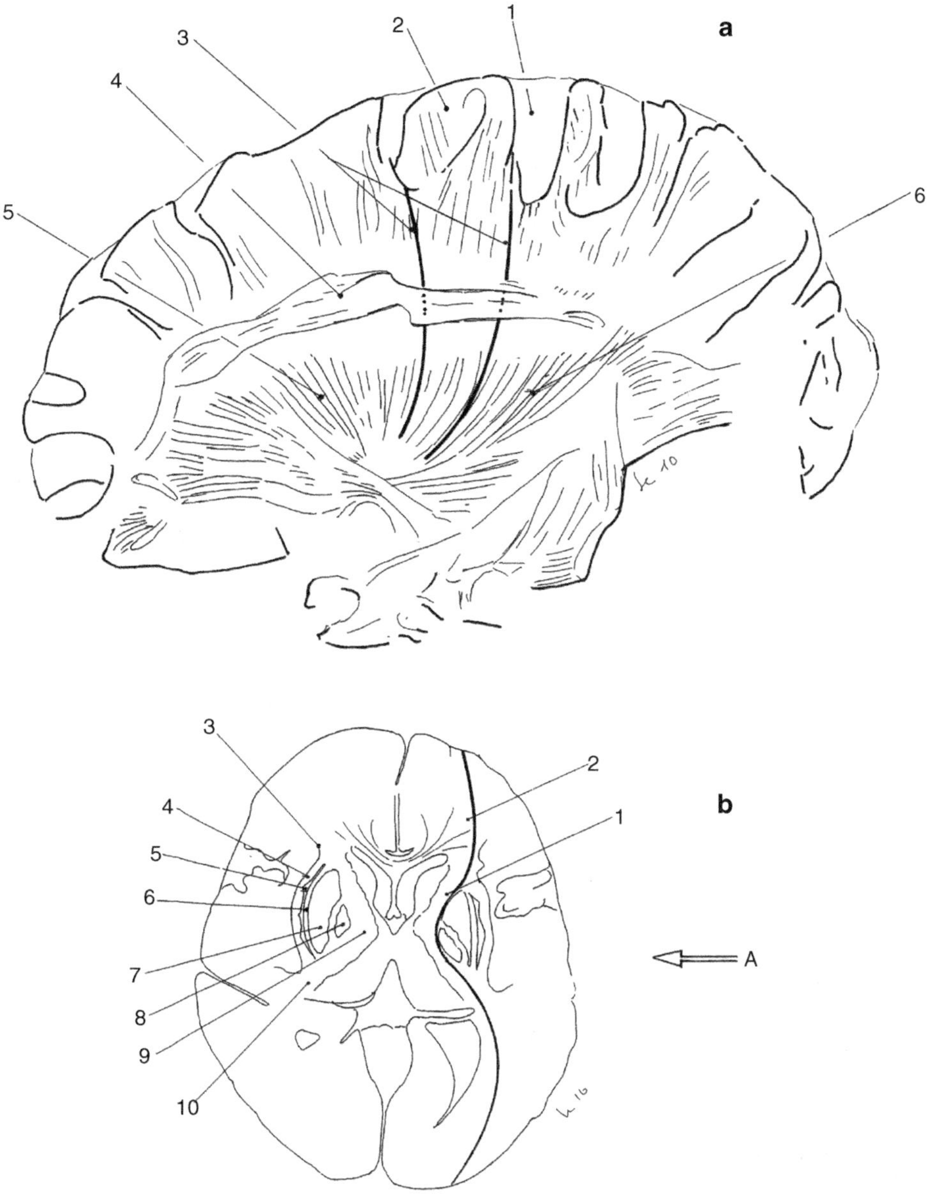

Fig. 2.21 Tractus pyramidalis (segment of corona radiata). **a** Historical illustration of a fiber brain dissection (according to Ludwig and Klingler 1956). Tractus pyramidalis here renders prominent. **b** Topogram added. Limit of resection renders prominent. **a** *1* Sulcus centralis, *2* gyrus precentralis, *3* tractus corticospinalis of corona radiata, *4* fasciculus longitudinalis superior, *5* tractus frontopontinus of corona radiata, *6* tractus parietotemporopontinus of corona radiata. **b** *1* Capsula interna, crus anterior; *2* fiber bundles of corpus callosum, forceps anterior; *3* insula, anterior tip; *4* insula and capsula extrema; *5* claustrum; *6* capsula externa; *7* putamen; *8* globus pallidus; *9* capsula interna, knee; *10* capsula interna, crus posterior

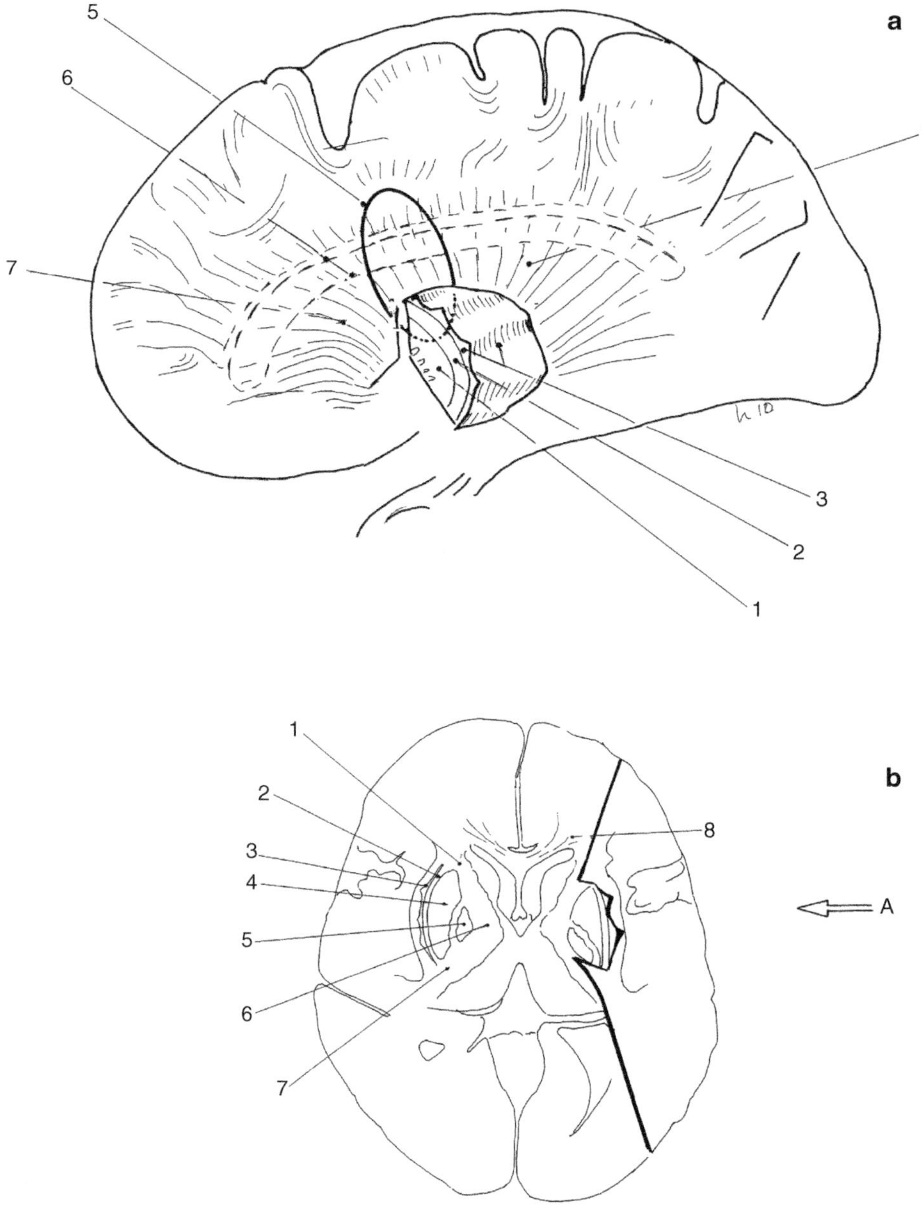

Fig. 2.22 Continuation of Fig. 2.21 (Ludwig and Klingler 1956). **a** Pyramidal tract close to capsula interna. **b** Topogram added. Resection renders prominent. **a** *1* Putamen; *2* capsula externa; *3* claustrum, capsula extrema and cortex of insula omitted; *4* corona radiata; *5* transitional region between corona radiata and capsula interna; *6* corpus callosum fibers penetrating corona radiata; *7* corona radiata, frontal segment. **b** *1* Capsula interna, crus anterior; *2* capsula externa; *3* claustrum; *4* putamen; *5* globus pallidus; *6* capsula interna, knee (motor fibers); *7* capsula interna, crus posterior (sensoric fibers)

2.2.3 Midline Structures (Figs. 2.23, 2.24, 2.25, 2.26, 2.27, 2.28, 2.29, 2.30, 2.31, and 2.32)

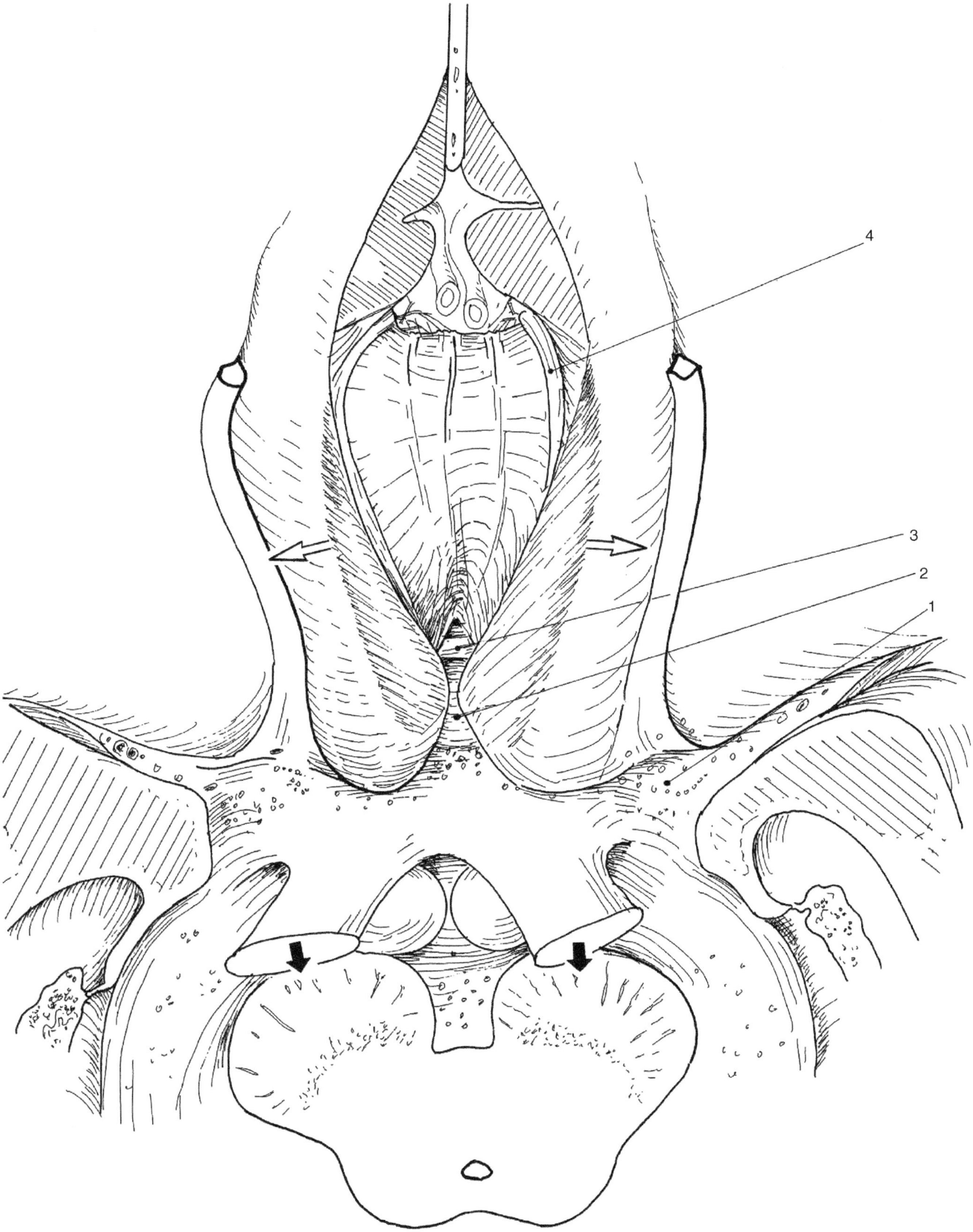

Fig. 2.23 Suprachiasmal interhemispheric structures. *1* Roof of the carotid cistern (substantia perforata anterior), *2* lamina terminalis, *3* commissura anterior, *4* stria longitudinalis cinguli. *Light arrows*: sulcus longitudinalis dilated. *Black arrows*: chiasma retracted

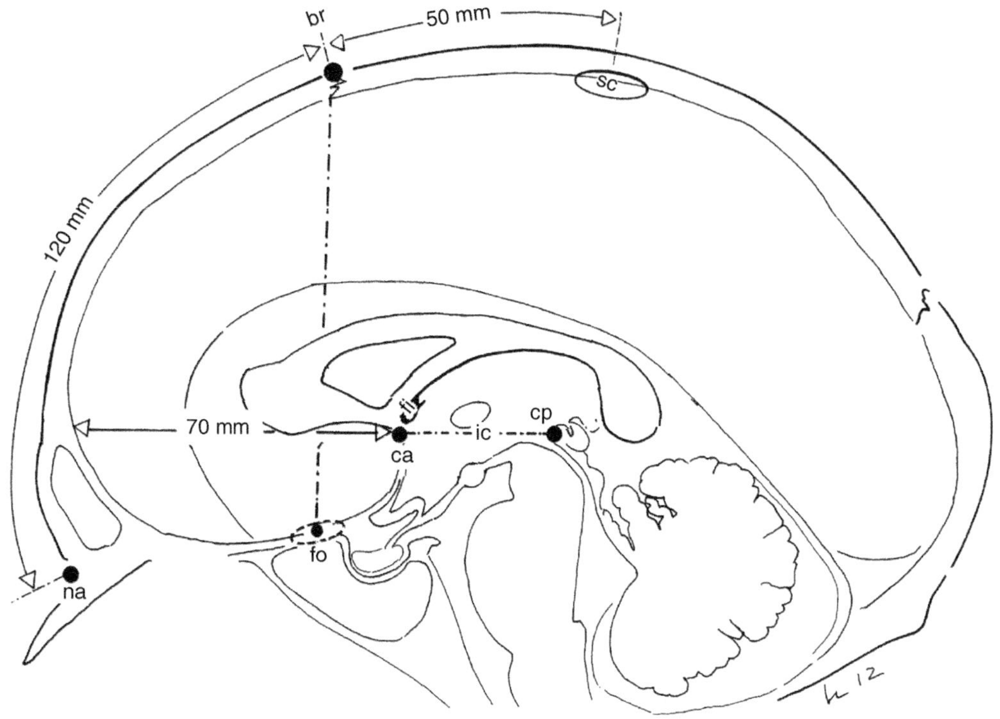

Fig. 2.24 *Medial landmarks* could be helpful, if neuronavigation is not available. Topographical relationships of the intercommissural line to the coronal plane between foramina optica (projection) and bregma and to the nasion-bregma distance measurement (distance measurement between nasion at the bony surface 12 cm and at the skin surface 13 cm). Upper point of sulcus centralis added. Cave. *br* bregma, *ca* commissura anterior, *cp* commissura posterior, *fi* foramen interventriculare (Monroi), *fo* foramina optica (projection), *ic* intercommissural line, *na* nasion, *sc* sulcus centralis

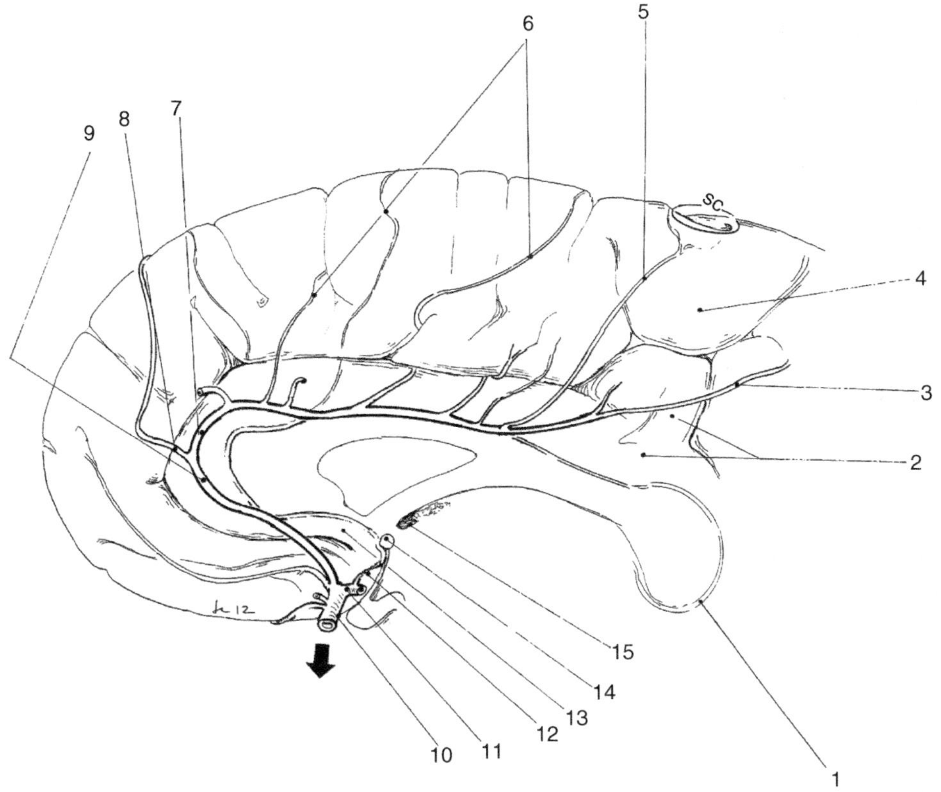

Fig. 2.25 Medial arteries and adjacent structures. Anatomical dissection (A1 shifted and extended for presenting subcallosal structures (*arrow*)). *1* A. corporis callosi mediana, right-sided branch; *2* gyrus cinguli; *3* A. praecunealis; *4* lobulus paracentralis; *5* A. paracentralis; *6* Aa. frontales internae; *7* A. pericallosa; *8* A. callosomarginalis; *9* A2; *10* A1; *11* A. communicans anterior; *12* as 1, common trunk; *13* subcallosal areas; *14* commissura anterior; *15* foramen interventriculare (Monroi)

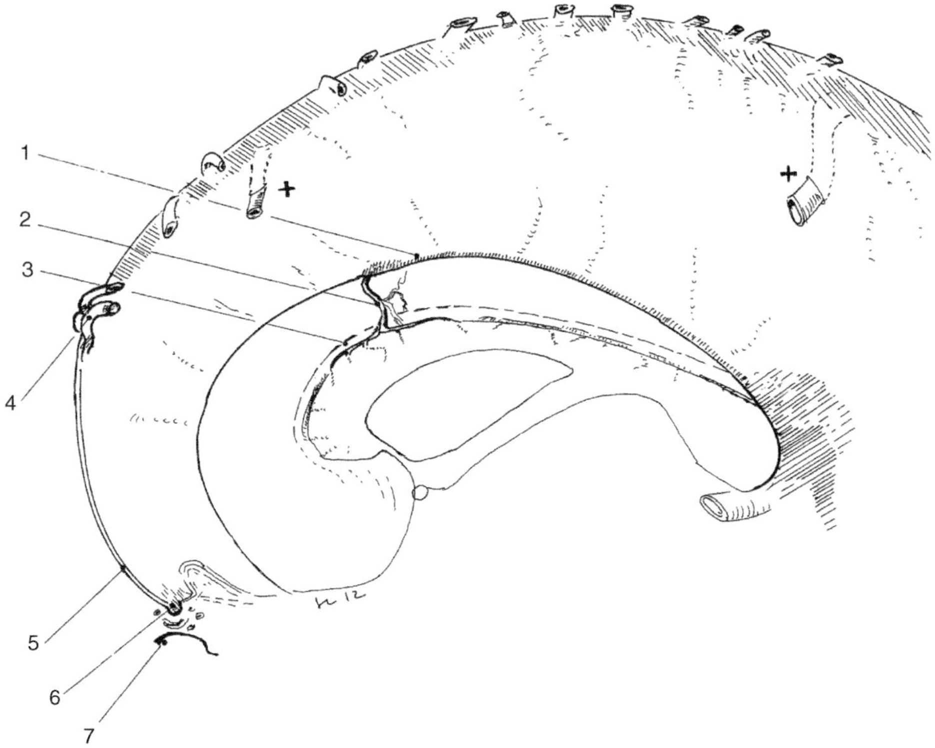

Fig. 2.26 Medial bridging veins at the sinuses. Note distance measurements between veins for planning surgical midline approaches. *1* Sinus sagittalis inferior; *2* bridging vein(s), variable; *3* stria longitudinalis cinguli, projection; *4* bridging veins at the beginning of the main segment of sinus sagittalis superior; *5* hypo- or aplastic segment of sinus sagittalis superior; *6* beginning of sinus sagittalis superior at For. caecum; *7* cavum nasi; + bridging veins, rare variants (Experience of the author)

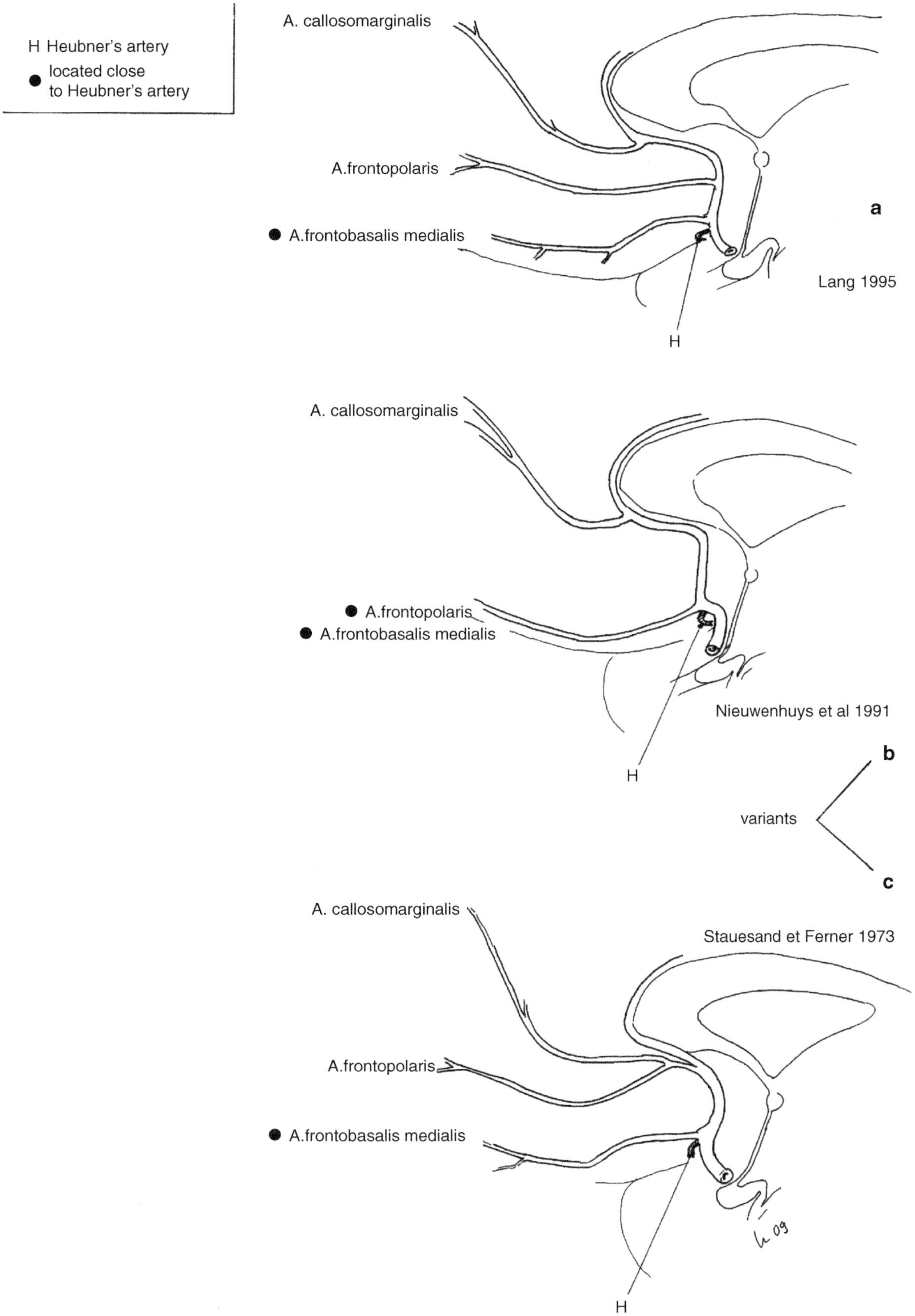

Fig. 2.27 Variable frontomedial arteries could be mixed with Heubner's artery. This has to be considered at surgery of aneurysms of A. communicans anterior (Yaşargil)

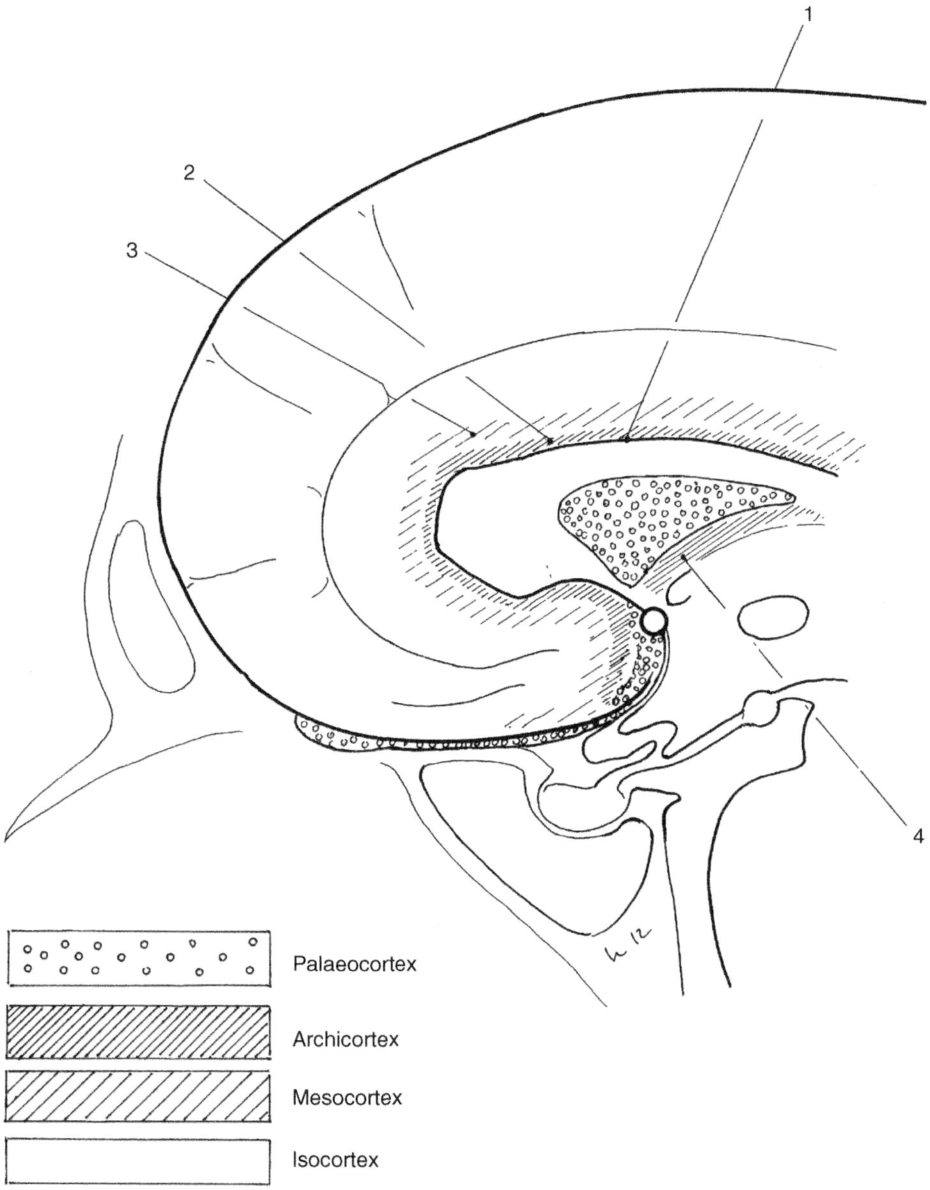

Fig. 2.28 Frontomedial allocortex (according to Stephan 1975, schematized). *1* Indusium griseum at the dorsal surface of corpus callosum; *2* stria longitudinalis cinguli, enclosed by gyrus cinguli at the lateral rim of cisterna corporis callosi; *3* gyrus cinguli; *4* columna fornicis. Uni- or bilateral lesions of stria longitudinalis cinguli, of fornix, of subcallosal gyri, or of the posterior segment of gyrus rectus (e.g., cave at surgical eliminations of aneurysms of A. communicans anterior) could be followed by psychological deficits

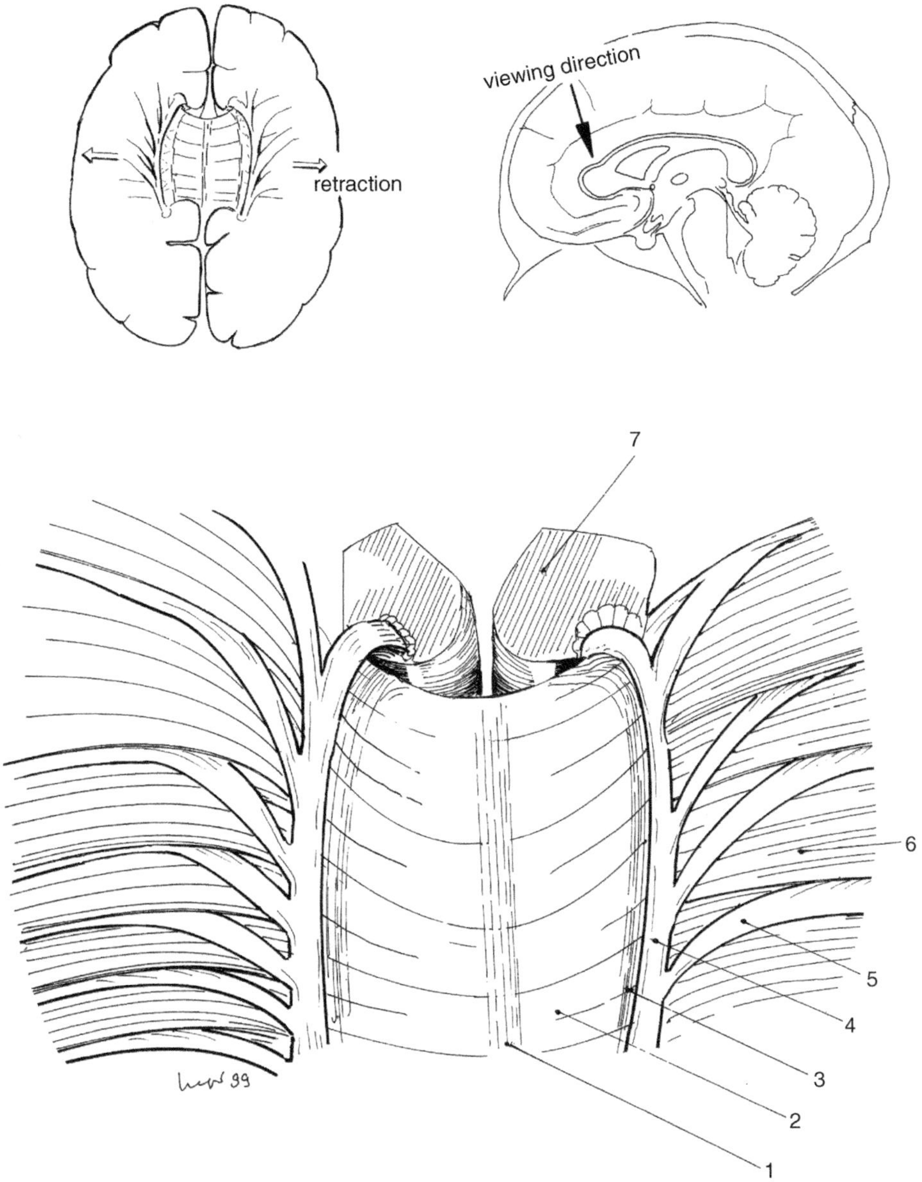

Fig. 2.29 Allocortex. *Cadaver fiber dissection* (Seeger 2000. Details are only to present by anatomical dissections. For fiber tracking, see Karaus et al. 2009). *1* Indusium griseum, striae mediales; *2* corpus callosum; *3* indusium griseum, striae laterales; *5* allocortical fiber bundles; *4* stria longitudinalis cinguli (allocortical part of the mesocortical gyrus cinguli); *6* neocortical fiber bundles, part of corpus callosum; *7* gyrus cinguli

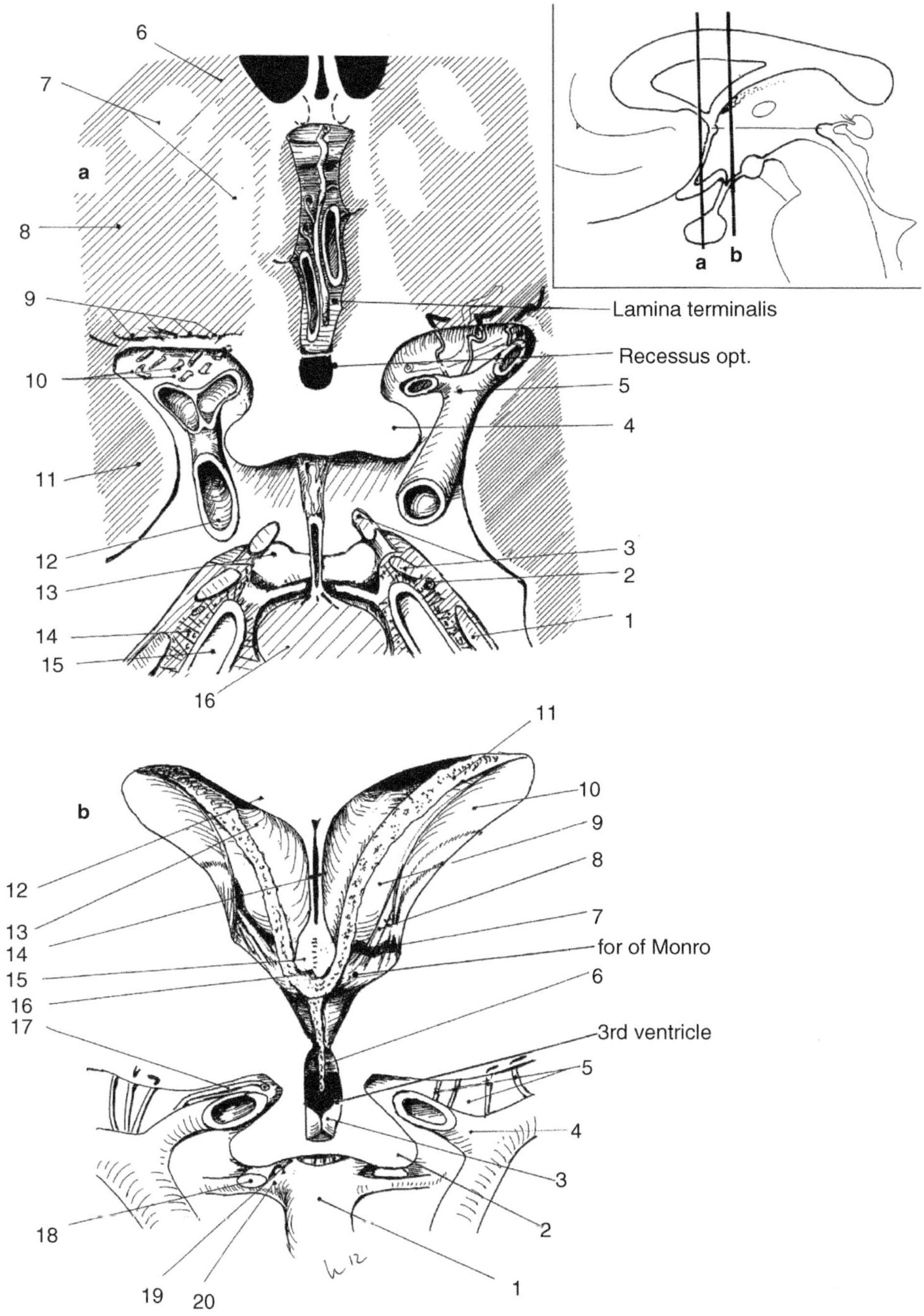

Fig. 2.30 Basal midline regions similar to neuronavigation. Some details are not sufficiently presented at imagings. **a** *1* N. ophthalmicus, *2* N. trochlearis, *3* N. oculomotorius, *4* chiasma, *5* carotid bifurcation, *6* nucleus caudatus (enclosing nucleus accumbens), *7* capsula interna, *8* putamen, *9* striate arteries, *10* Subst. perforata anterior, *11* amygdala, *12* A. carotis interna, *13* processus clinoideus posterior, *14* sinus cavernosus, *15* carotid siphon, *16* hypophysis. **b** *1* A. basilaris, *2* tractus opticus, *3* corpus mammillare, *4* carotid bifurcation, *5* perforating arteries, *6* commissura anterior, *7* V. thalamostriata, *8* stria terminalis, *9* lamina affixa thalami, *10* nucleus caudatus, *11* plexus choroideus, *12* corpus callosum, *13* corpus fornicis, *14* septum pellucidum, *15* columna fornicis, *16* tip of fissura transversa, *17* A. recurrens (Heubner), *18* N. oculomotorius, *19* A. cerebri posterior, *20* A. cerebelli superior

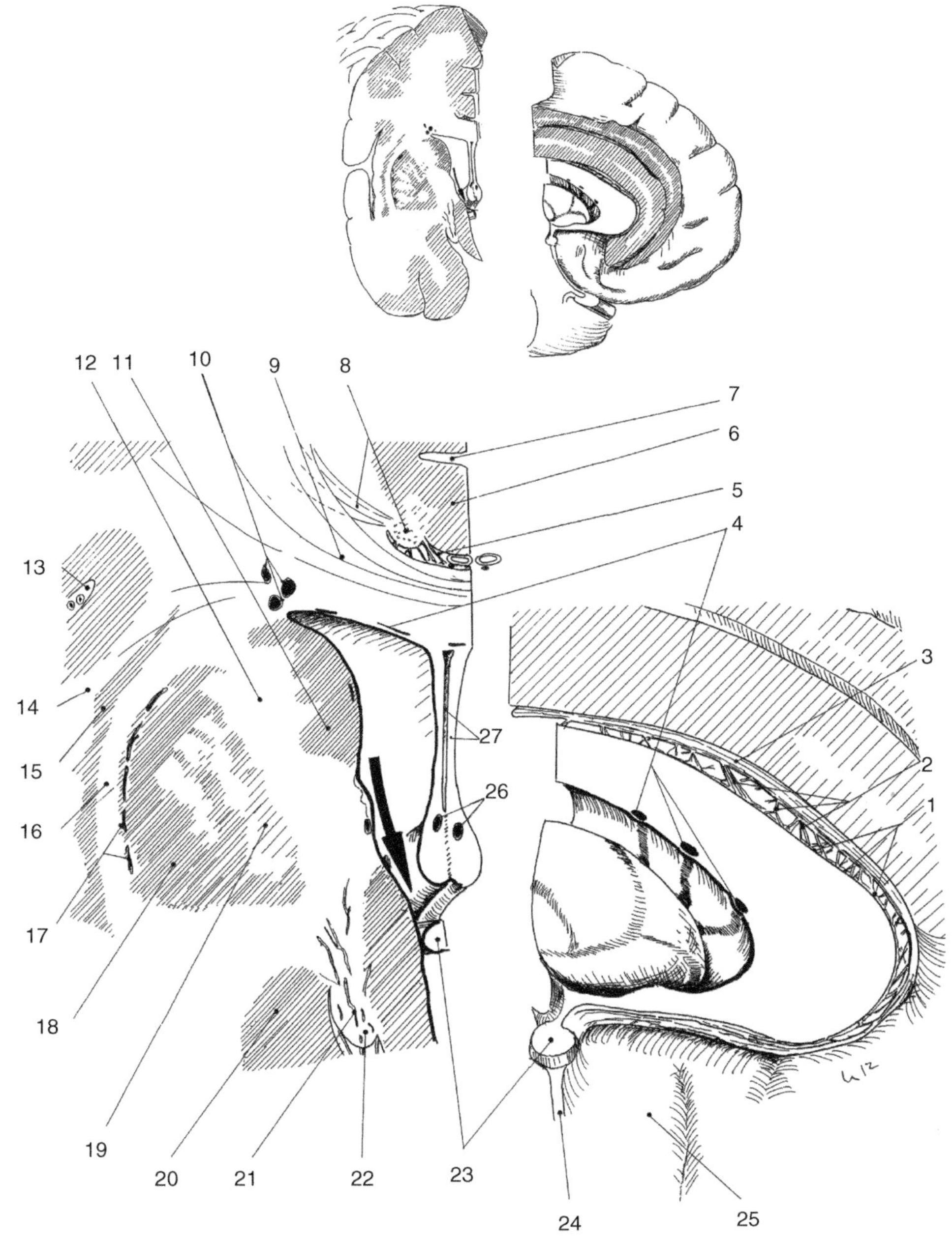

Fig. 2.31 Continuation of Fig. 2.30. *Stria longitudinalis cinguli must be preserved* at midline approaches. *1* Indusium griseum, *2* fiber connections between indusium griseum and stria longitudinalis cinguli, *3* stria longitudinalis cinguli, *4* transversal veins, *5* cisterna corporis callosi, *6* gyrus cinguli, *7* sulcus cinguli, *8* as 3, *9* fibers of corpus callosum, *10* Vv. longitudinales, *11* caput nuclei caudati, *12* knee of capsula interna, *13* sulcus superior insulae, *14* capsula extrema, *15* claustrum, *16* capsula externa, *17* Aa. striatae, *18* putamen, *19* globus pallidus, *20* amygdala, *21* as 17, *22* tractus opticus, *23* commissura anterior, *24* lamina terminalis, *25* area subcallosa. *Arrow*: foramen interventriculare (Monroi)

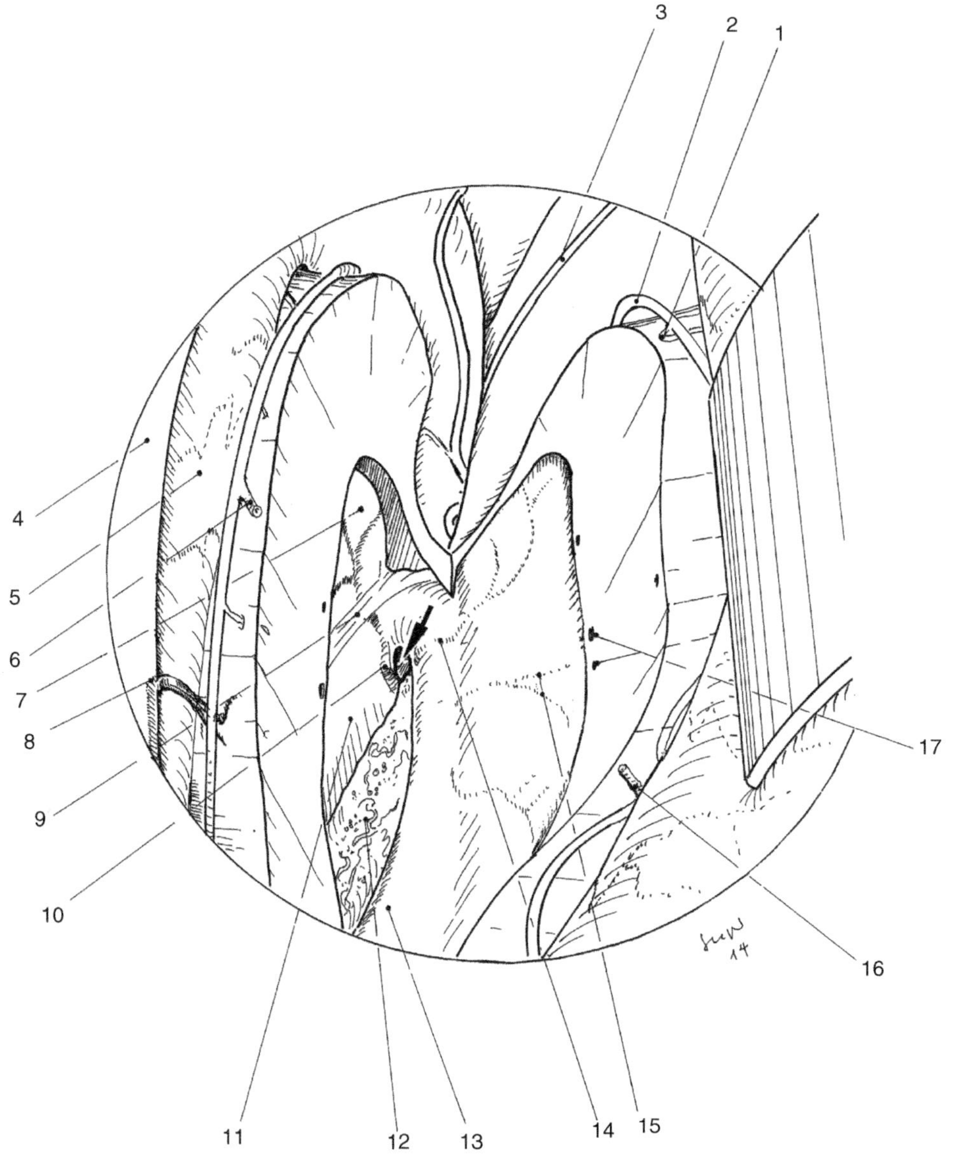

Fig. 2.32 Anatomical dissection similar to a surgical approach to the floor of the contralateral ventricle. *1* Atypical midline artery feeding stria longitudinalis cinguli; *2* A. pericallosa; *3* A. callosomarginalis; *4* falx; *5* base of gyrus cinguli (overlying stria longitudinalis); *6* midline crossing branch of A. pericallosa (common findings (Marino 1976)), cut at cadaver brain dissection; *7* caput nuclei caudati; *8* beginning of sinus sagittalis inferior; *9* stria terminalis; *10* V. thalamostriata; *11* thalamus, pars affixa; *12* plexus choroideus; *13* crus fornicis; *14* V. septi pellucidi; *15* medial septal veins; *16* as 6; *17* transversal dorsal ventricular veins. *Arrow*: foramen interventriculare (Monroi). 1, 6, and 16 noted at surgery

2.2.4 **Third Ventricle** (Figs. 2.33, 2.34, and 2.35)

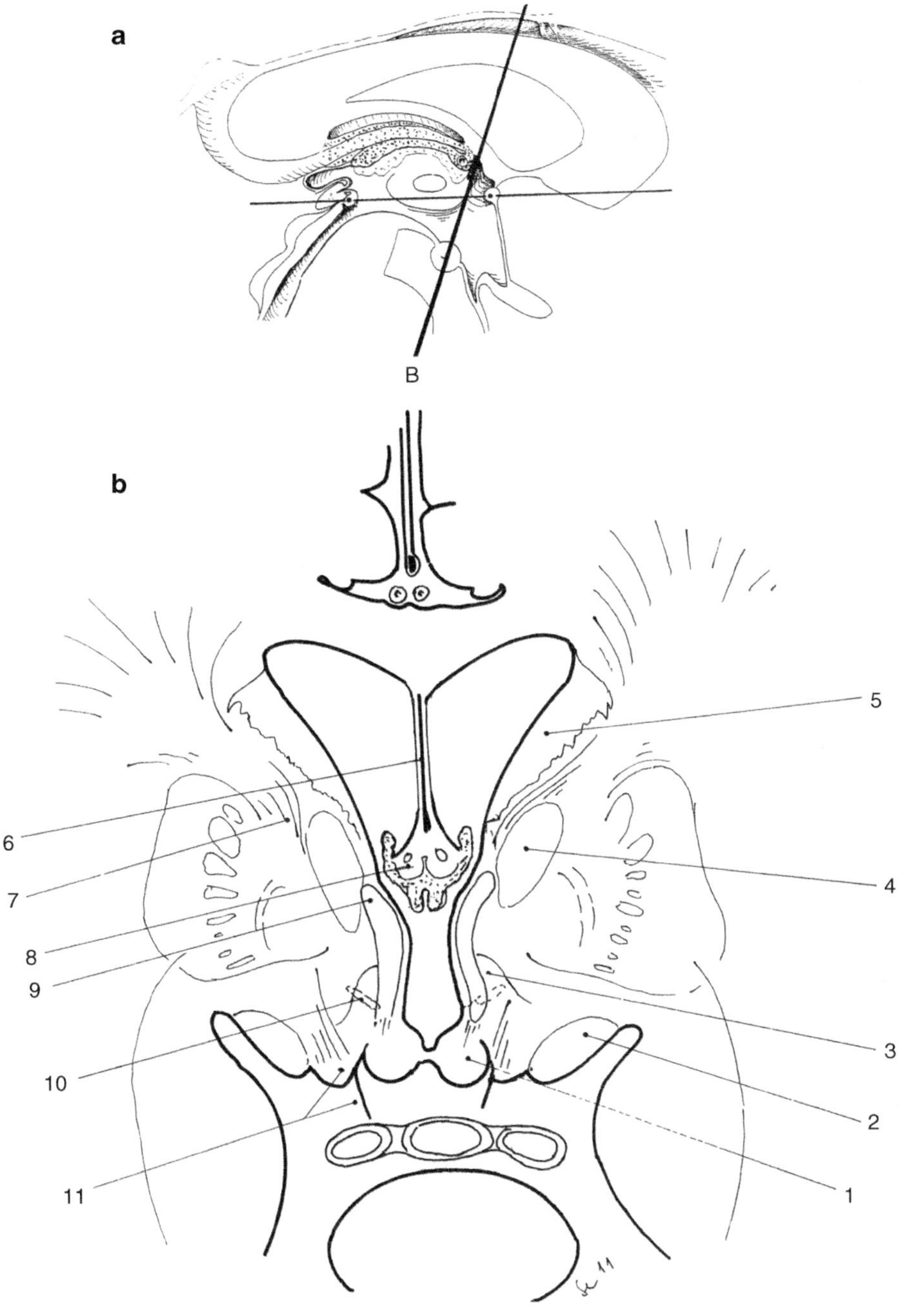

Fig. 2.33 Transforaminal endoscopic route to the third ventricle and basal cisterns. MRT level for neuronavigation. **a** Topogram, **b** anatomical drawing. *1* Corpus mammillare; *2* tractus opticus; *3* hypothalamus; *4* thalamus; *5* nucleus caudatus; *6* cavum septi pellucidi; *7* knee of capsula interna; *8* columna fornicis, pars libera; *9* pars tecta fornicis; *10* tractus mamillothalamicus, projection; *11* crus cerebri, tractus pyramidalis

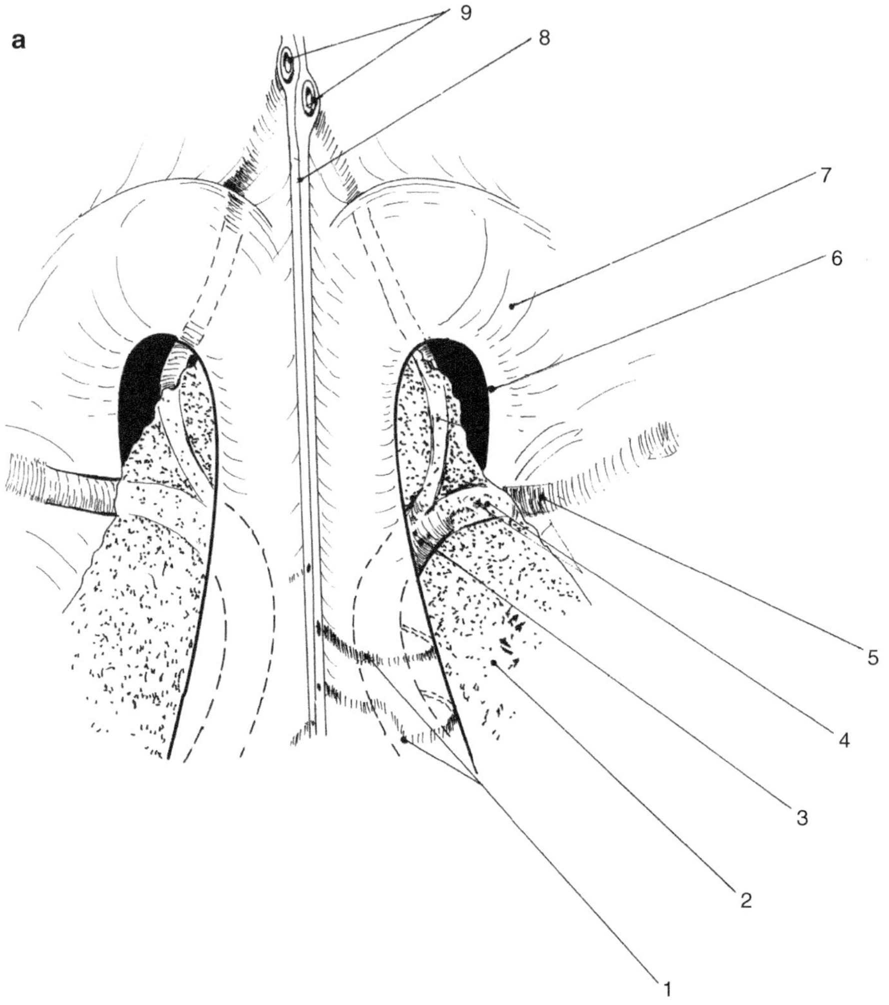

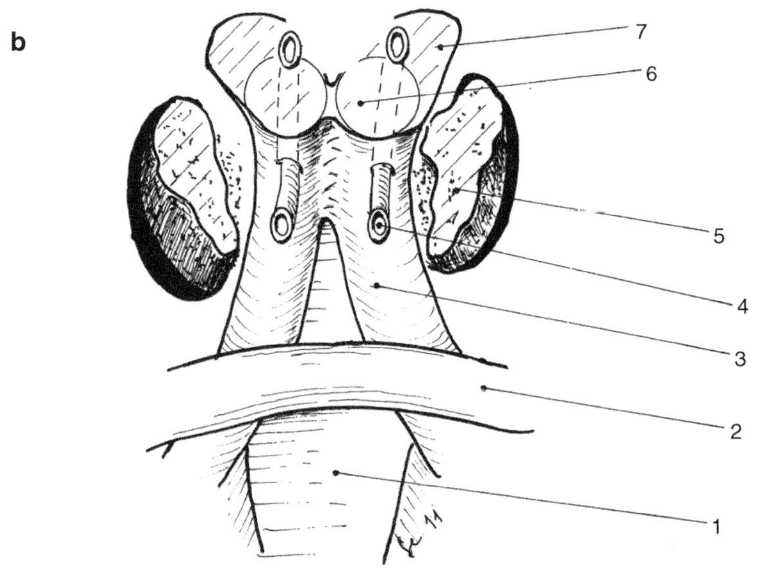

Fig. 2.34 For. interventriculare (Monroi). Anatomical dissection. Fine surgical details are not exact to define by imagings. **a** Dorsal viewing direction. **b** Basal viewing direction. **a** *1* Medial septal veins, *2* plexus choroideus, *3* V. cerebri interna, *4* angulus venosus, *5* V. thalamostriata, *6* For. interventriculare, *7* columna fornicis, *8* cavum septi pellucidi, *9* Vv. septi pellucidi. **b** *1* Lamina terminalis, *2* commissura anterior, *3* columna fornicis, *4* V. septi pellucidi, *5* plexus choroideus, *6* columna fornicis, *7* columna fornicis, gray-colored segment

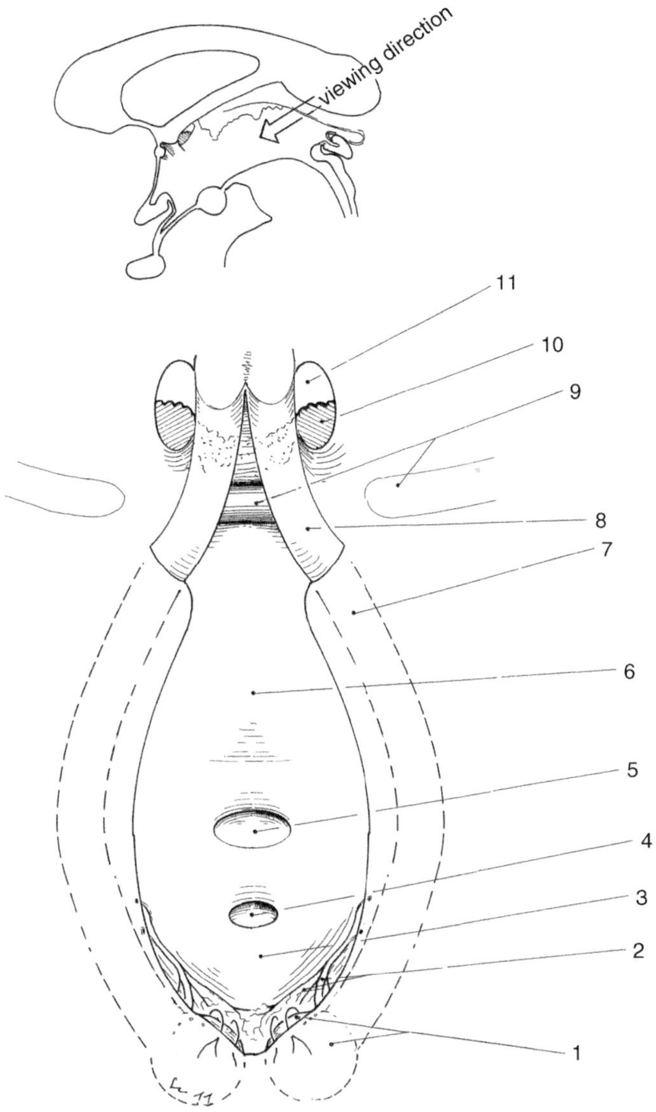

Fig. 2.35 Anterior basal relief of the third ventricle. Anatomical presentation similar to endoscopy. *1* Corpus mammillare; *2* vascular network subependymal cave at surgery; *3* floor of hypothalamus; *4* recessus infundibuli; *5* recessus opticus; *6* lamina terminalis; *7* pars tecta columnae fornicis, projection; *8* pars libera columna fornicis; *9* commissura anterior; *10* plexus choroideus; *11* foramen interventriculare. Plexus choroideus of the third ventricle eliminated. Fornices cut

2.2.5 Frontobasal and Frontotemporobasal Structures (Figs. 2.36, 2.37, 2.38, 2.39, 2.40, 2.41, 2.42, 2.43, 2.44, 2.45, 2.46, 2.47, 2.48, 2.49, 2.50, 2.51, 2.52, 2.53, 2.54, 2.55, and 2.56)

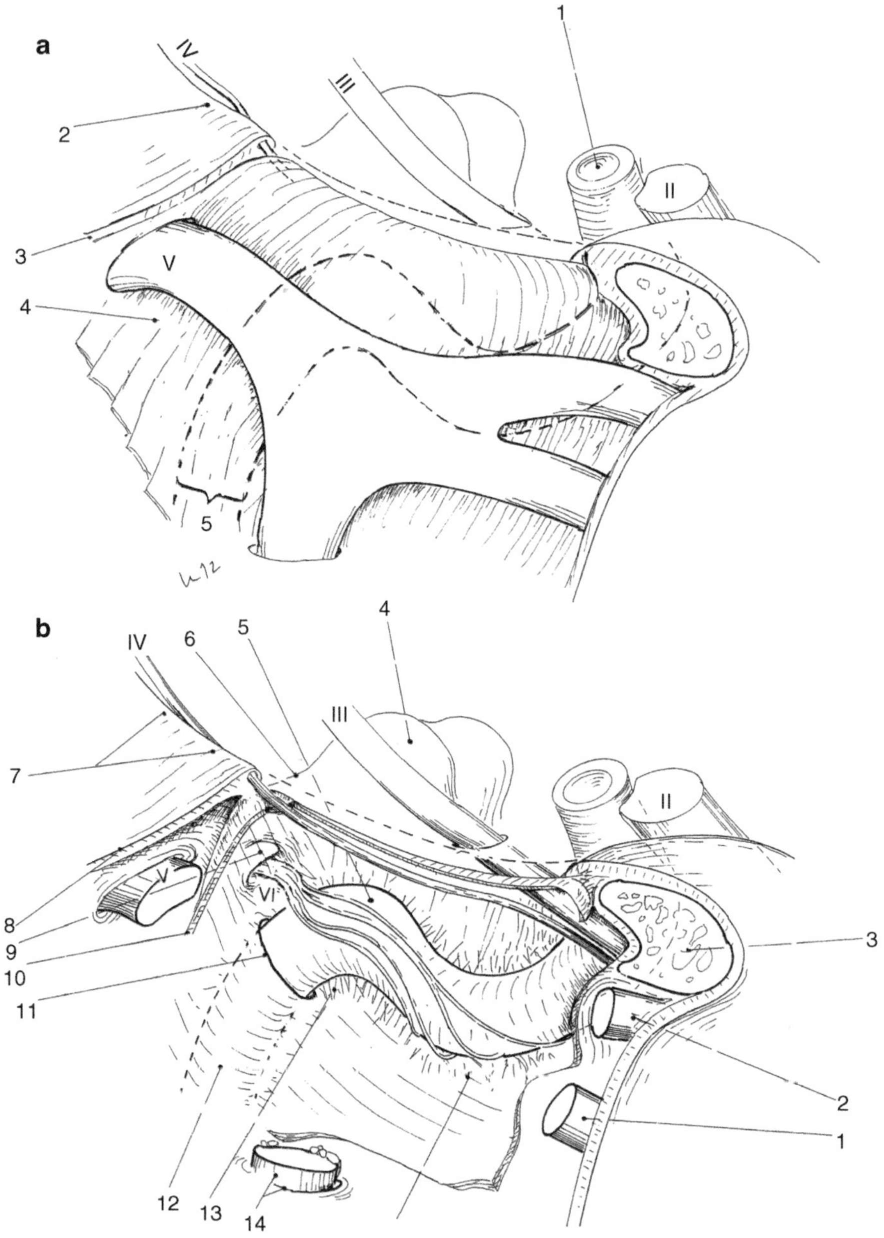

Fig. 2.36 Craniobasal structures. **a** Inner relief of dura and adjacent structures. **b** Bony relief and adjacent structures. *1* N. abducens; *2* hamulus petrosus, variable; *3* canalis caroticus, projection; *4* foramen lacerum, projection; *5* N. mandibularis and foramen ovale; *6* N. ophthalmicus; *7* N. oculomotorius and its dural penetration point, projection; *8* N. maxillaris at foramen rotundum; *9* A. carotis interna; *10* A. ophthalmica; *11* processus clinoideus posterior overlying the carotid siphon

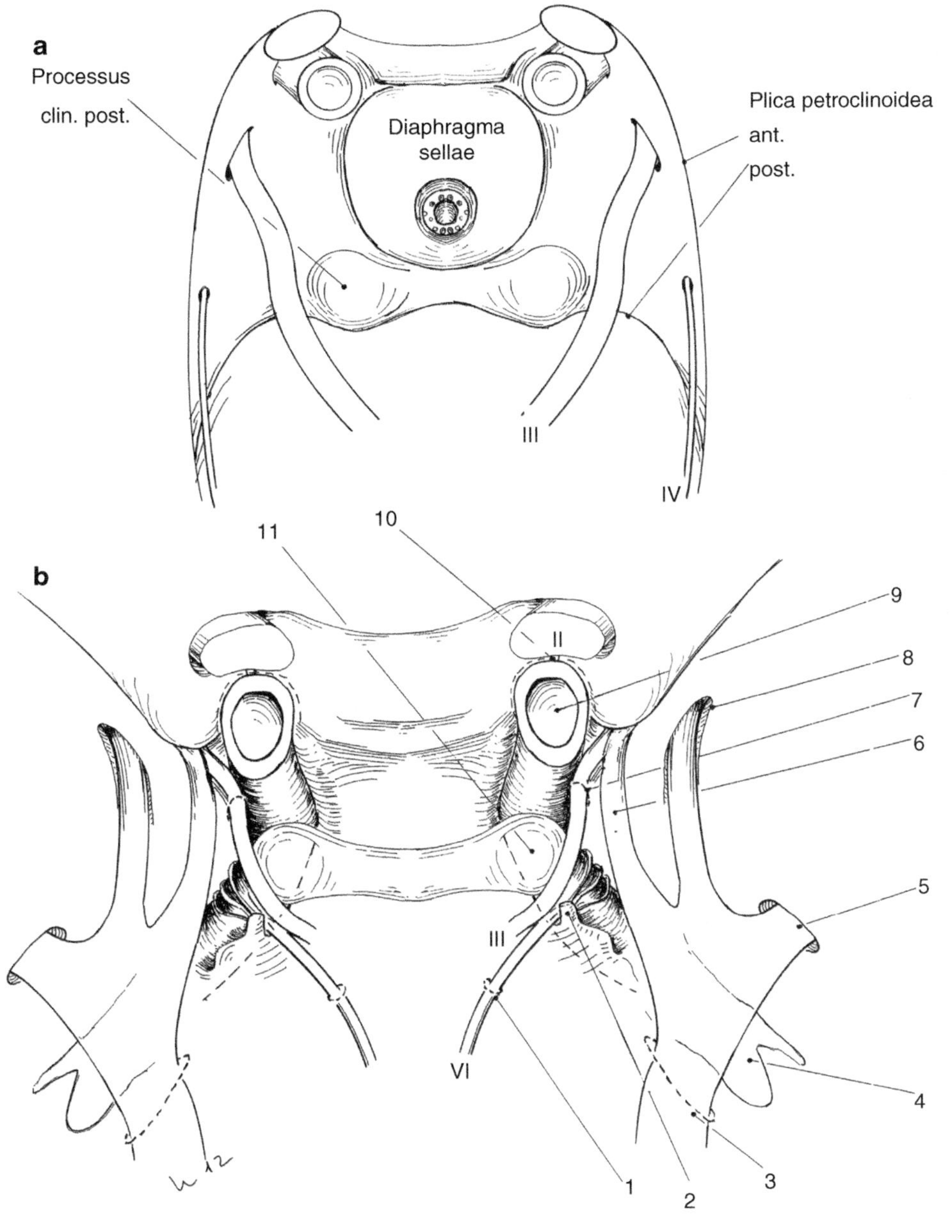

Fig. 2.37 Dural layers and its contents. **a** Inner layers of dura. **b** Middle and profound layers of dura. **a** *1* A. carotis interna; *2* tentorial edge; *3* tentorial fold, superficial layer incised and resected; *4* middle dural layers; *5* A. carotis interna, projection. **b** *1* N. maxillaris; *2* N. ophthalmicus; *3* processus clinoideus anterior, transectional plane; *4* processus clinoideus posterior; *5* curvatura posterior of the carotid siphon; *6* plica petroclinoidea posterior; *7* tentorial edge; *8* superficial tentorial layer incised and resected; *9* hamulus petrosus enclosing N. abducens, projection; *10* middle tentorial layers; *11* A. carotis interna transpassing apertura interna of canalis caroticus; *12* canalis caroticus of os petrosum, projection; *13* sinus cavernosus; *14* N. mandibularis transpassing foramen ovale; *15* middle segment of sinus cavernosus surrounding the variable carotid siphon II to VI cranial nerves

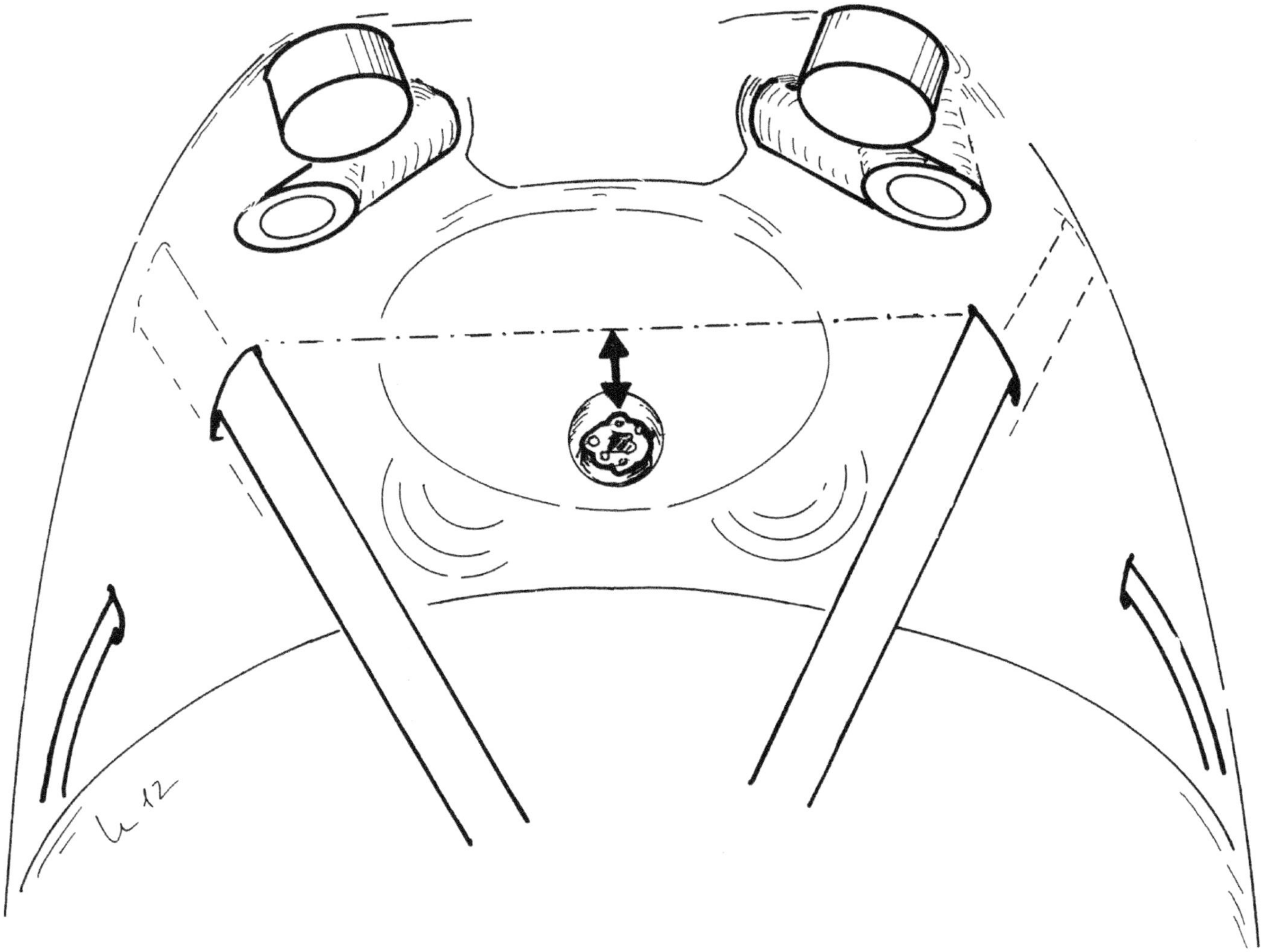

Fig. 2.38 Dural penetration points. N. oculomotorius: anterior from infundibulum (*arrows*). N. trochlearis: anterior from plica petroclinoidea posterior

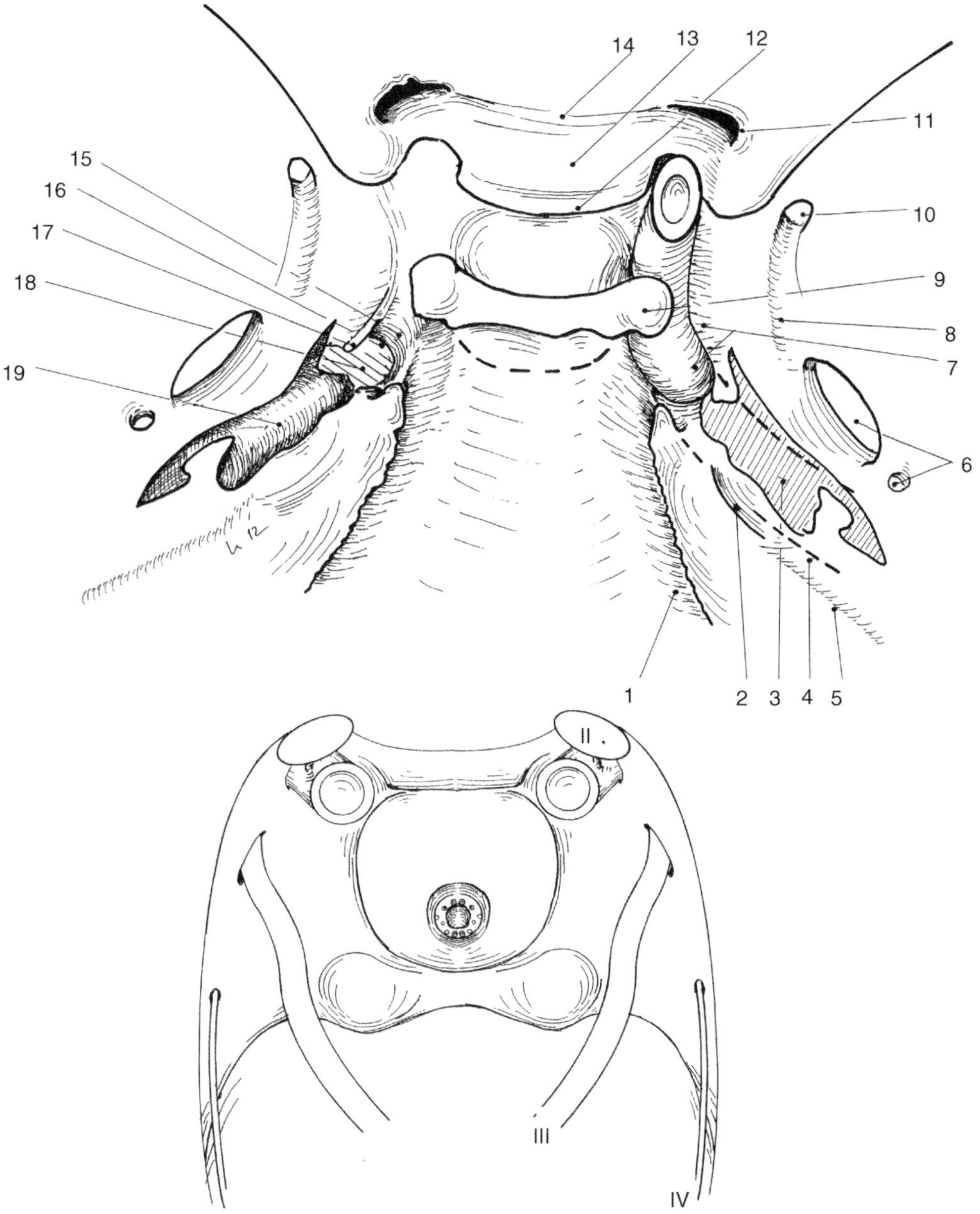

Fig. 2.39 *Comparison of the bony and dural relief* and its contents. *1* Sulcus petrosus inferior; *2* impressio trigemini; *3* foramen lacerum, wide variant hatched; *4* canalis caroticus, projection; *5* sulcus petrosus superior; *6* For. spinosum et ovale; *7* region of sinus cavernosus and A. carotis interna, projection; *8* bony sulcus at N. maxillaris; *9* processus clinoideus posterior; *10* canalis rotundus; *11* For. opticum; *12* tuberculum sellae; *13* sulcus chiasmatis presented especially at childhood; *14* interoptic line; *15* sulcus caroticus; *16* anterior margin of For. lacerum; *17* lingula sphenoidalis surrounding A. carotis interna; *18* For. lacerum occluded by synchondrosis, hatched; *19* floor of canalis caroticus, bony defect of its roof (common finding) II to IV cranial nerves

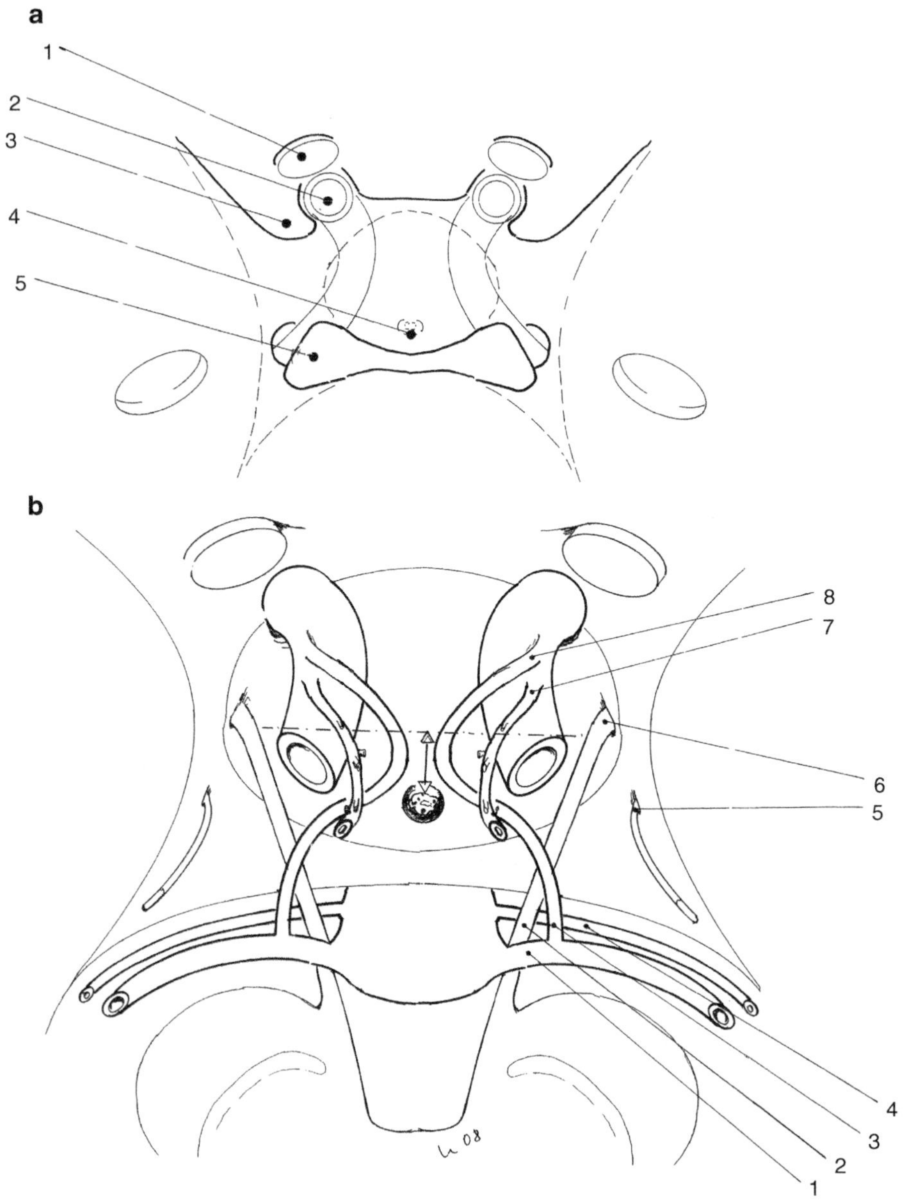

Fig. 2.40 Landmarks **a** and contents **b** of basal cisterns. **a** *1* N. opticus; *2* A. carotis interna; *3* processus clinoideus anterior; *4* infundibulum, projection; *5* processus clinoideus posterior. **b** *1* A. cerebri posterior, *2* N. oculomotorius, *3* A. communicans posterior, *4* A. cerebelli superior, *5* dural penetration point of N. trochlearis, *6* dural penetration point of N. oculomotorius, *7* A. chorioidea anterior, *8* A. communicans posterior

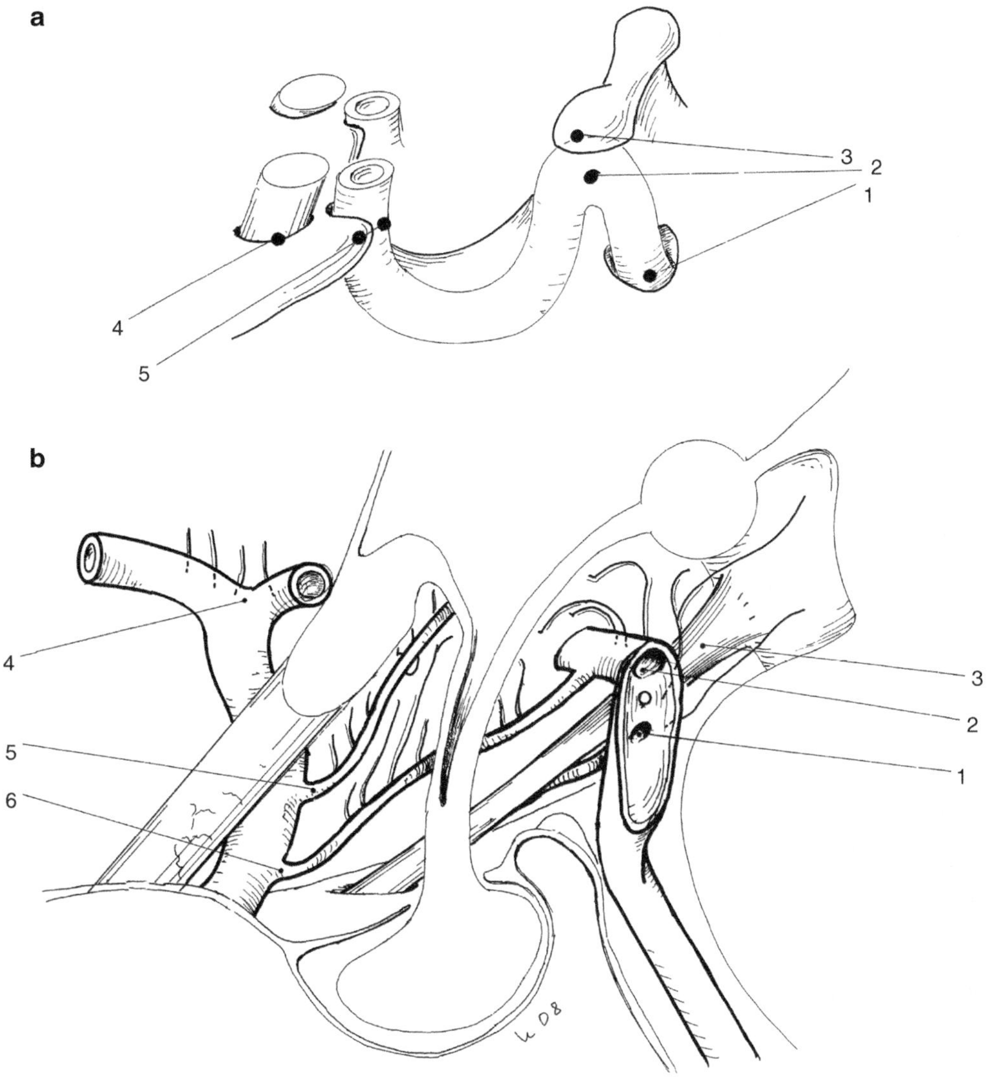

Fig. 2.41 Addendum for Fig. 2.40, lateral presentation. **a** *1* A. carotis interna; *2* as 1, curvatura posterior; *3* processus clinoideus posterior; *4* foramen opticum; *5* processus clinoideus anterior and curvatura anterior of the carotid siphon. **b** *1* A. cerebelli superior, mouth; *2* A. cerebri posterior, mouth; *3* N. oculomotorius; *4* bifurcation of A. carotis interna; *5* A. chorioidea anterior; *6* A. communicans posterior

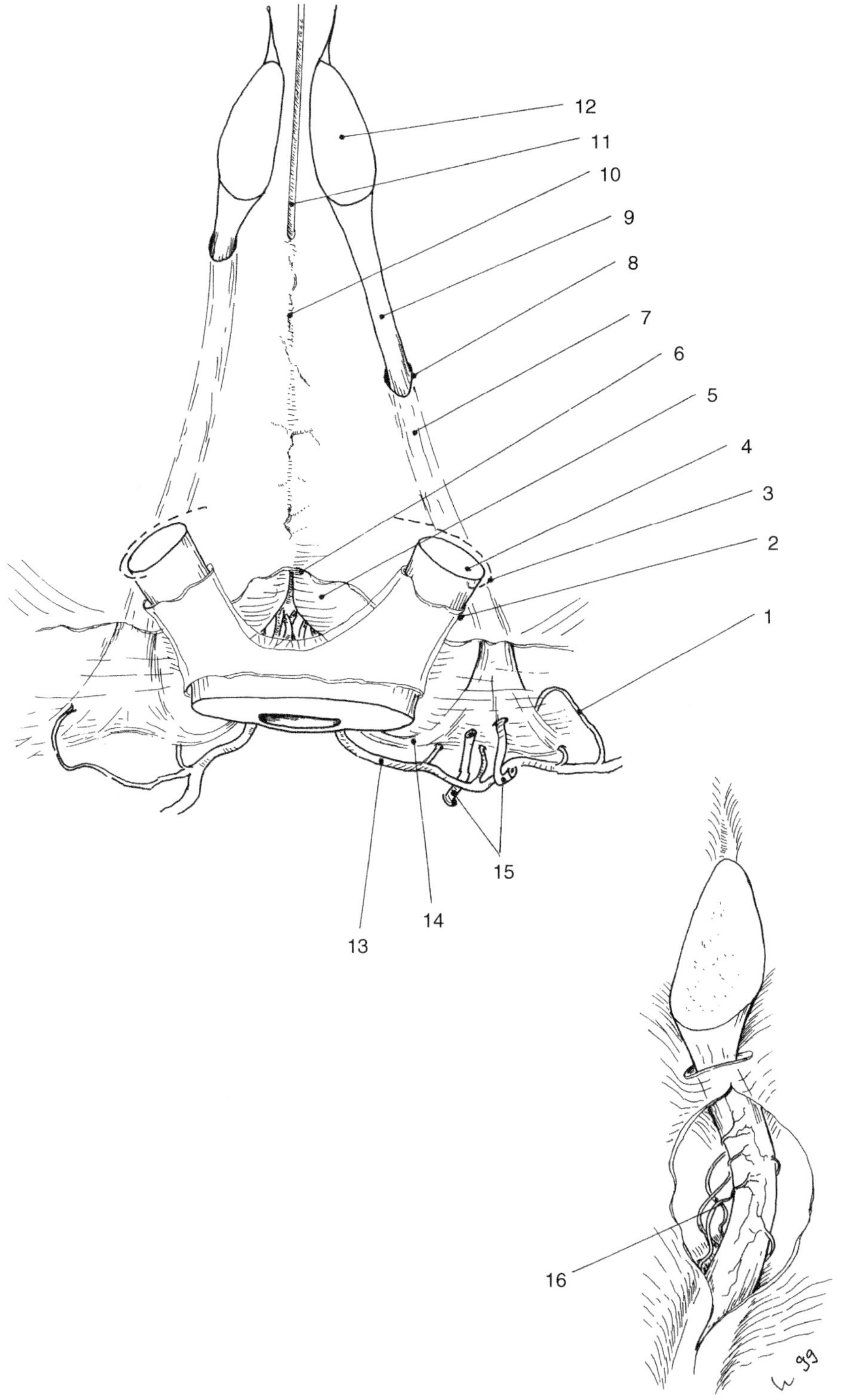

Fig. 2.42 Cistern of tractus olfactorius and trigonum olfactorium with arteries. N. olfactorius is the only cranial nerve which presents a long subdural route. This is favorable for the surgical elevation of lobus frontalis. Variants of the arachnoid penetration point and its bilateral asymmetries have to be considered. At surgical preservation of olfactory arteries, the posterior segment of gyrus rectus should be preserved (damaging of its allocortical part may be followed by psychological deficits (Hütter 1999, 2000)). *1* Olfactory branch(es) of Heubner's artery (Rhoton, schematized); *2* arachnoid wall of N. opticus; *3* foramen opticum, projection; *4* N. opticus; *5* gyrus rectus; *6* basal arachnoid membrane of cisterna laminae terminalis; *7* tractus olfactorius, intracisternal segment, and arachnoidea transparent; *8* as 7, arachnoid penetration; *9* subdural segment of tractus olfactorius; *10* sulcus medianus masked by thickened arachnoidea and vessels; *11* falx; *12* bulbus olfactorius; *13* A. recurrens (Heubneri); *14* trigonum olfactorium, striae mediales; *15* perforating arteries of trigonum olfactorium; *16* arteries at sulcus rectus feeding tractus olfactorius should be preserved (Applied by Prof. Samii, personal communication to the author)

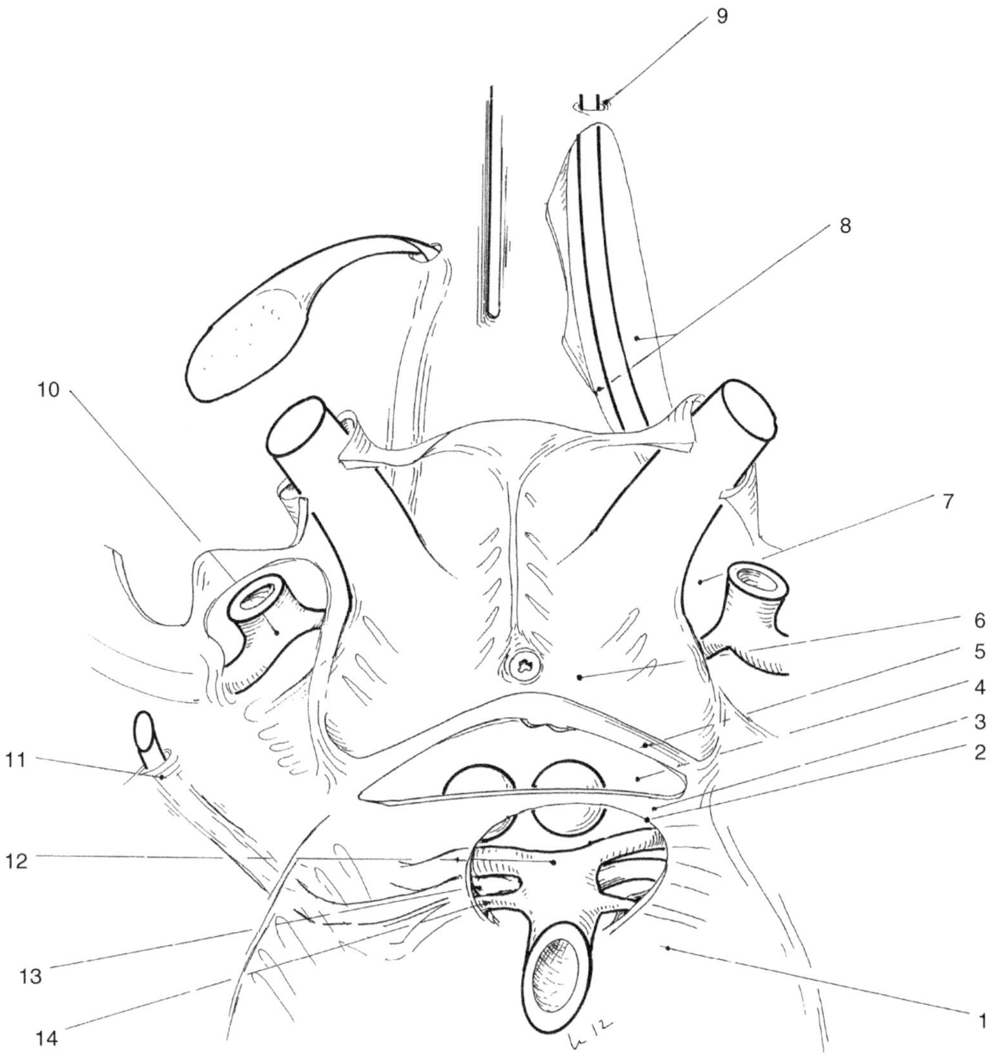

Fig. 2.43 *Liliequist's membrane* (for Liliequist's membrane and findings of Rhoton (2000), see Figs. 2.43, 2.46, and 3.32) *and adjacent structures. 1* Cisterna pontis medialis; *2* gap of pars mesencephalica of Liliequist's membrane; *3* wall of pars mesencephalica of Liliequist's membrane; *4* cisterna interpeduncularis; *5* wall of pars diencephalica of Liliequist's membrane; *6* cisterna optica; *7* cisterna carotica; *8* cisterna olfactoria; *9* tractus olfactorius penetrating arachnoidea; *10* carotid bifurcation; *11* arachnoid trabeculas thickened, enclosing N. oculomotorius; *12* bifurcation of A. basilaris; *13* N. oculomotorius; *14* A. cerebelli superior enclosed by pars mesencephalica of Liliequist's membrane

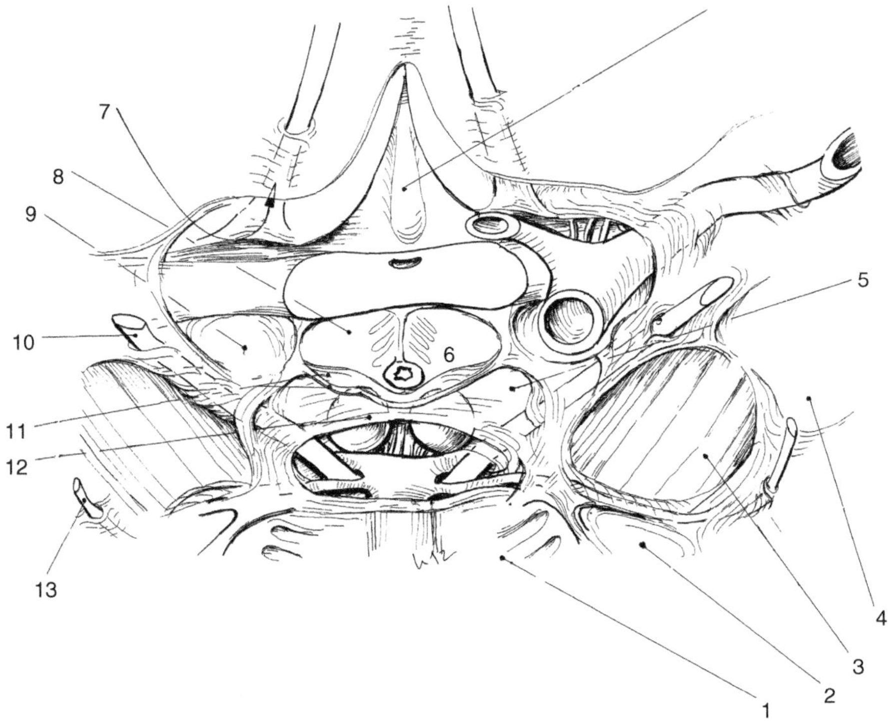

Fig. 2.44 Cisterna carotica and adjacent cisterns, details. Trabecular tunnels of A. cerebri media, adhesions of N. oculomotorius, and the gap of pars mesencephalica of Liliequist's membrane at bifurcation of A. basilaris. *1* Cisterna pontis medialis; *2* cisterna pontis lateralis; *3* cisterna cruralis; *4* cisterna ambiens; *5* cisterna interpeduncularis, pars superficialis; *6* cisterna laminae terminalis; *7* cisterna olfactoria; *8* cisterna chiasmatis; *9* cisterna carotica; *10* N. oculomotorius; *11* diencephalic portion of Liliequist's membrane; *12* mesencephalic portion of Liliequist's membrane; *13* N. trochlearis

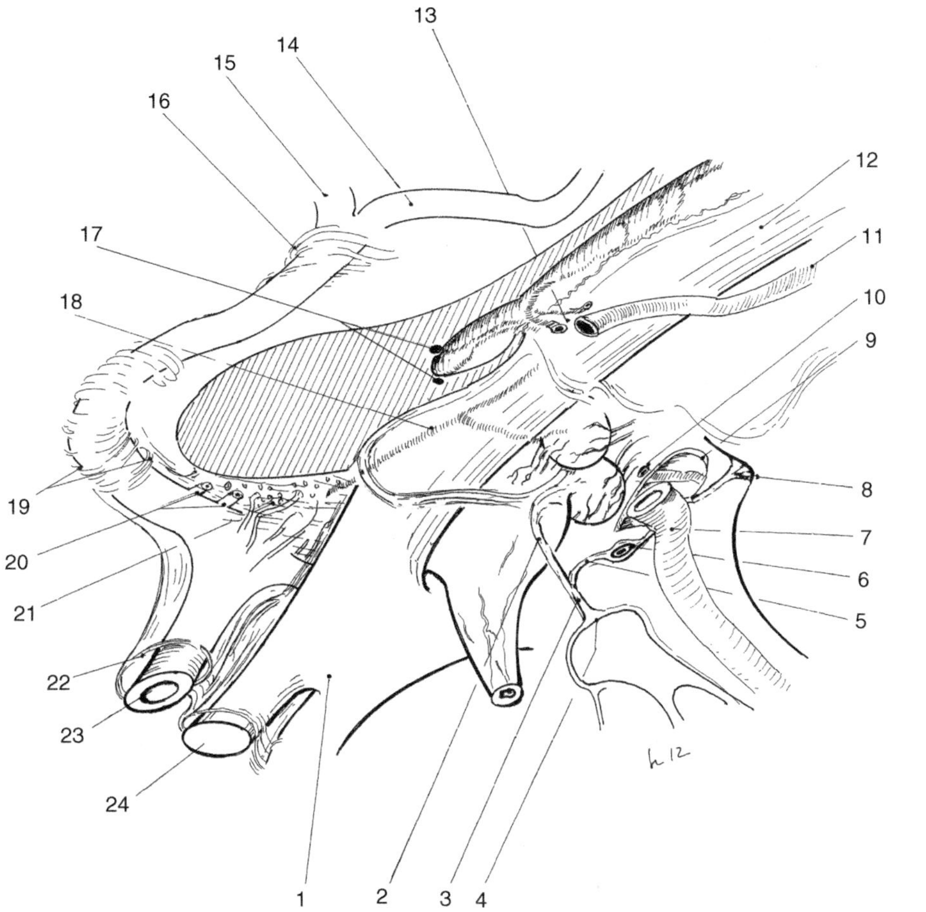

Fig. 2.45 Central position of A. carotis interna and its cistern between sulcus lateralis (sylvian cistern) and cisterna ambiens. A. cerebri media is the main structure of the sylvian cistern. A. cerebri posterior and tractus opticus are the main structures of cisterna ambiens. Between both cisterns is interposed cisterna carotica. All structures are surrounding the base of uncus (hatched: transectional plane). *1* Chiasma; *2* diencephalic segment of Liliequist's membrane; *3* as 2, common segment; *4* dorsum sellae, insertion of 3; *5* mesencephalic segment of Liliequist's membrane; *6* A. cerebelli superior, enclosed by 5; *7* A. basilaris; *8* insertion of the mesencephalic segment of Liliequist's membrane at the pontomesencencephalic rim; *9* arachnoid gap (Key and Retzius) containing bifurcation of A. basilaris; *10* V. interpeduncularis; *11* V. basalis (Rosenthal); *12* tractus opticus; *13* temporal choroid point; *14* A. cerebri media, distal trunk; *15* A. operculofrontalis; *16* arachnoid tunnel of A. cerebri media; *17* inferior tip vein (Huang and Wolf 1974b); 18 cisterna ambiens, tractus opticus, and one of its veins; *19* further arachnoid tunnel; *20* lateral striate arteries (Aa. lenticulostriatae); *21* substantia perforata anterior penetrated by 20; *22* cisterna carotica; *23* A. carotis interna; *24* N. opticus

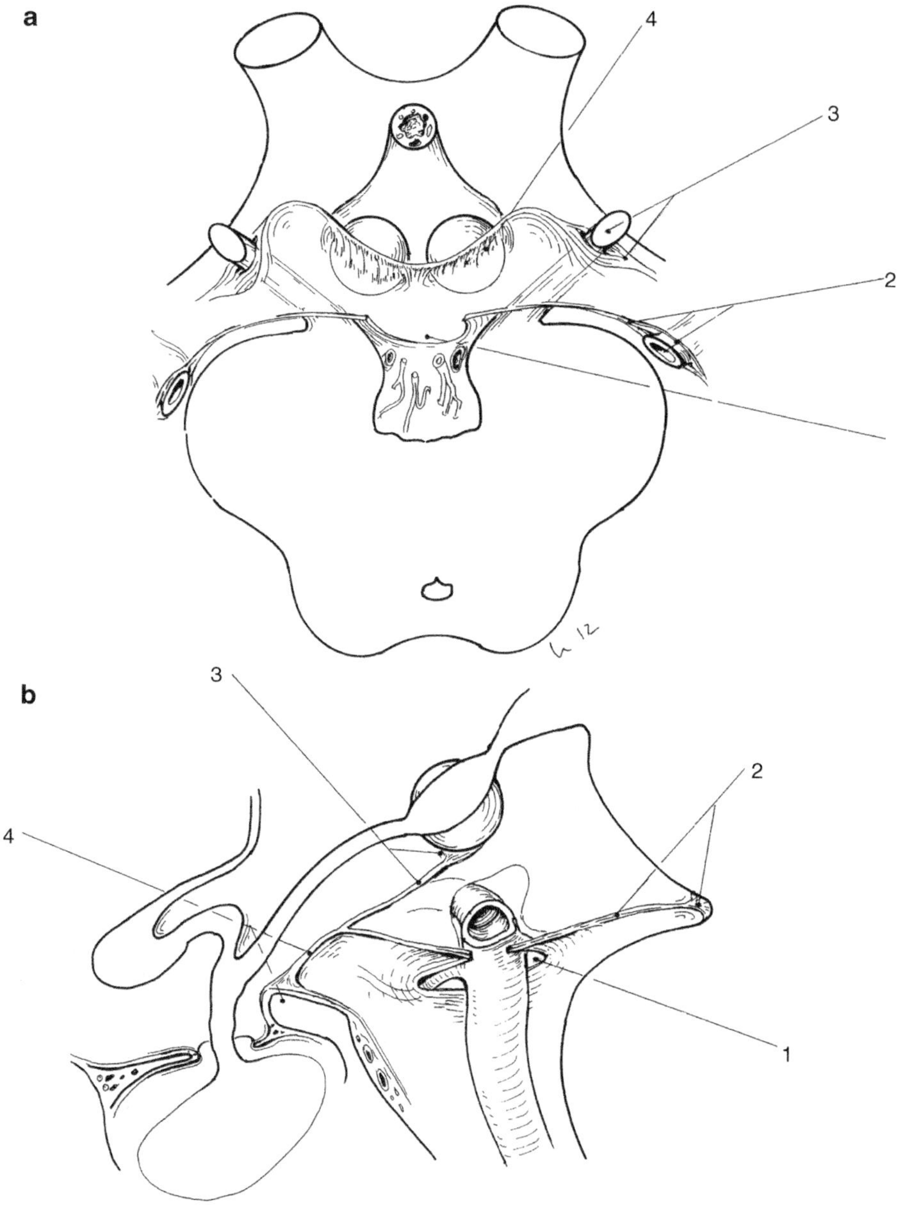

Fig. 2.46 Liliequist's membrane (drawing at the base of photographs of cadaver brain dissections of Rhoton 2000). Synopsis. **a** *1* Gap of pars mesencephalica, *2* pars mesencephalica, *3* N. oculomotorius and arachnoid adhesions, *4* pars diencephalica, insertion. **b** *1* Gap of pars mesencephalica, *2* pars mesencephalica, *3* pars diencephalica, *4* common part of Liliequist's membrane and its insertion at dorsum sellae

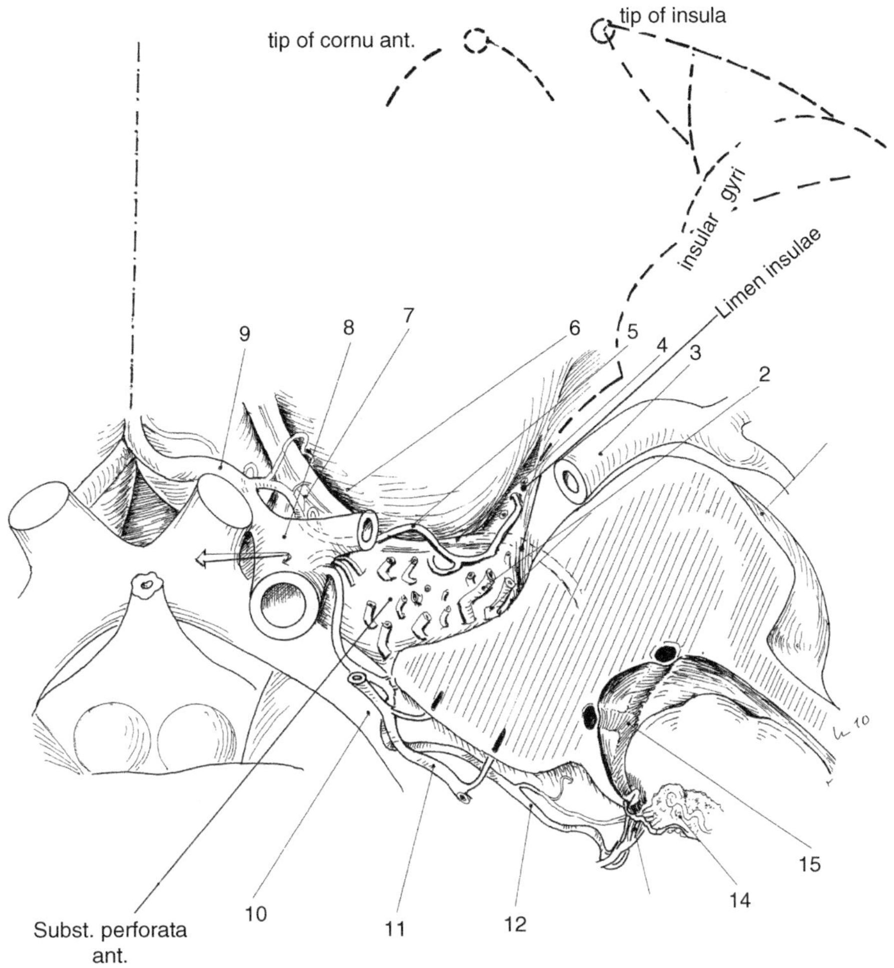

Fig. 2.47 Ground of basal cisterns and surrounding cerebrobasal structures. Conclusion. *1* Uncus, *2* lateral striate arteries (Aa. lenticulostriatae), *3* A. cerebri media, *4* striae laterales of trigonum olfactorium at limen insulae, *5* Heubner's artery, *6* sulcus rectus, *7* trigonum olfactorium, *8* bifurcation of A. carotis interna, *9* A1, *10* tractus opticus, *11* A. communicans posterior, *12* A. chorioidea anterior, *13* temporal choroid point, *14* plexus choroideus, *15* inferior tip vein, thick calibrated variant. *Dotted*: projection of surrounding structures

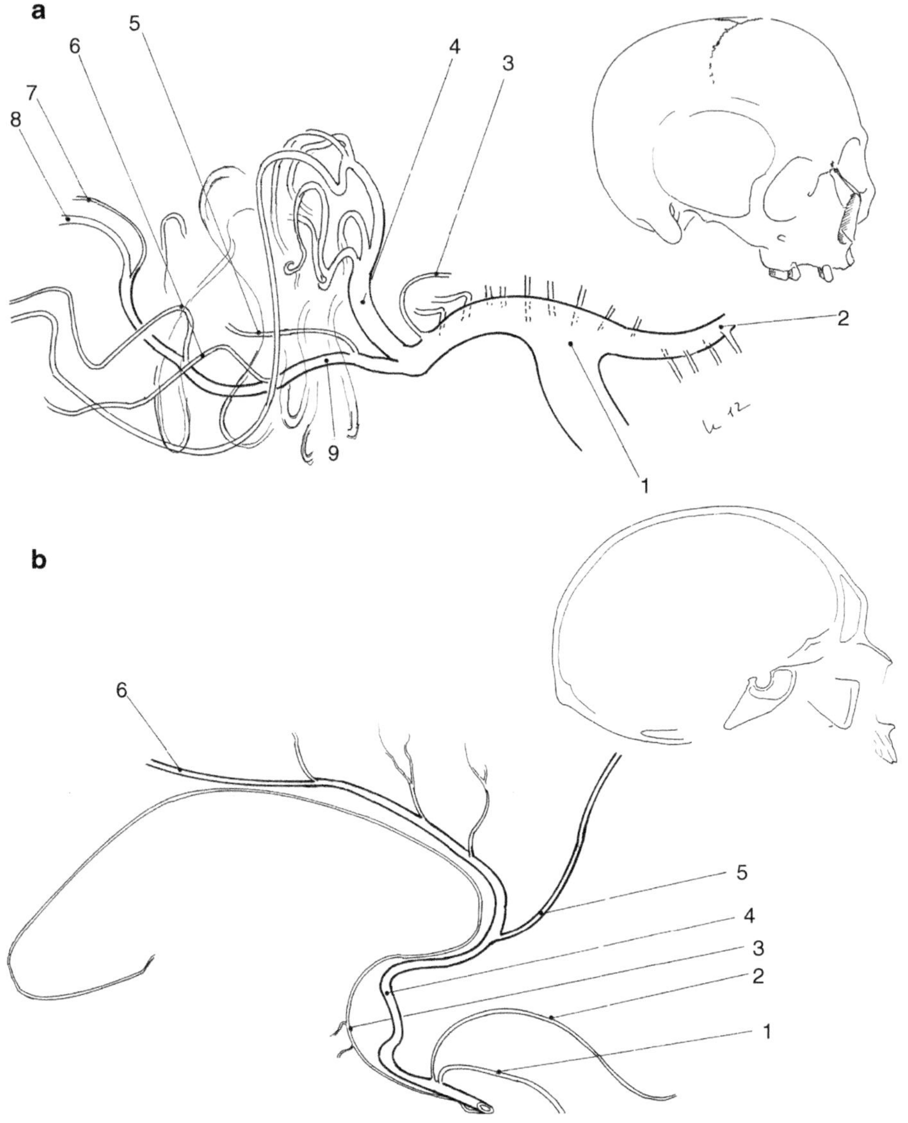

Fig. 2.48 A. carotis interna and its ramification. **a** A. cerebri media and its branches. **b** A. cerebri anterior and its branches. **a** *1* Carotid bifurcation, *2* A1, *3* frontobasal lateral artery, *4* A. operculofrontalis, *5* A. sulci centralis, *6* temporal branch(es) of A. cerebri media, *7* A. angularis, *8* A. temporooccipitalis, *9* A. cerebri media, distal branch. **b** *1* Frontobasal medial artery, *2* A. frontopolaris, *3* A. corporis callosi mediana, *4* A1, *5* A. callosomarginalis, *6* A. pericallosa. For A. communicans posterior and A. chorioidea anterior, see Fig. 2.50f Aa. hypophysealis inferior and superior omitted

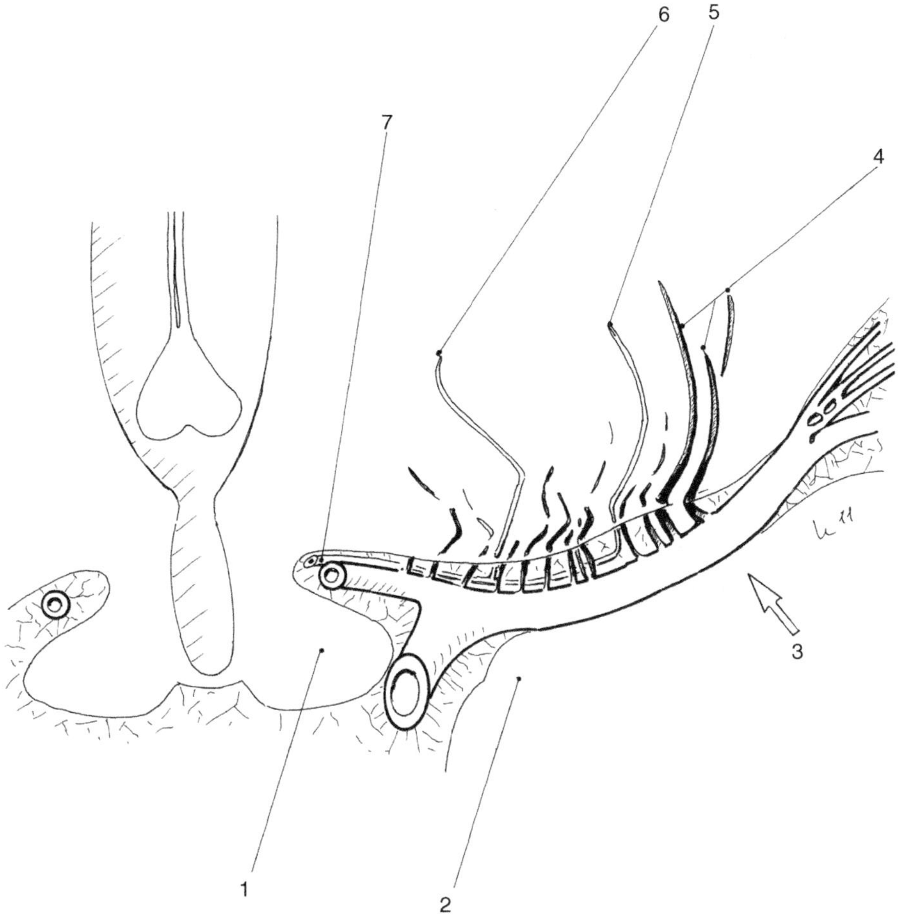

Fig. 2.49 Aa. striatae (these fine branches of cerebral arteries are less variable than large branches of circulus arteriosus (Willisi)). Branches of the carotid bifurcation and A. cerebri media. *1* Tractus opticus, *2* uncus, *3* limen insulae (*arrow*), *4* lateral striate arteries (Aa. lenticulostriatae), *5* lateral main branch of A. recurrens (Heubner), *6* medial main branch of A. recurrens (Heubner), *7* A. recurrens (Heubner)

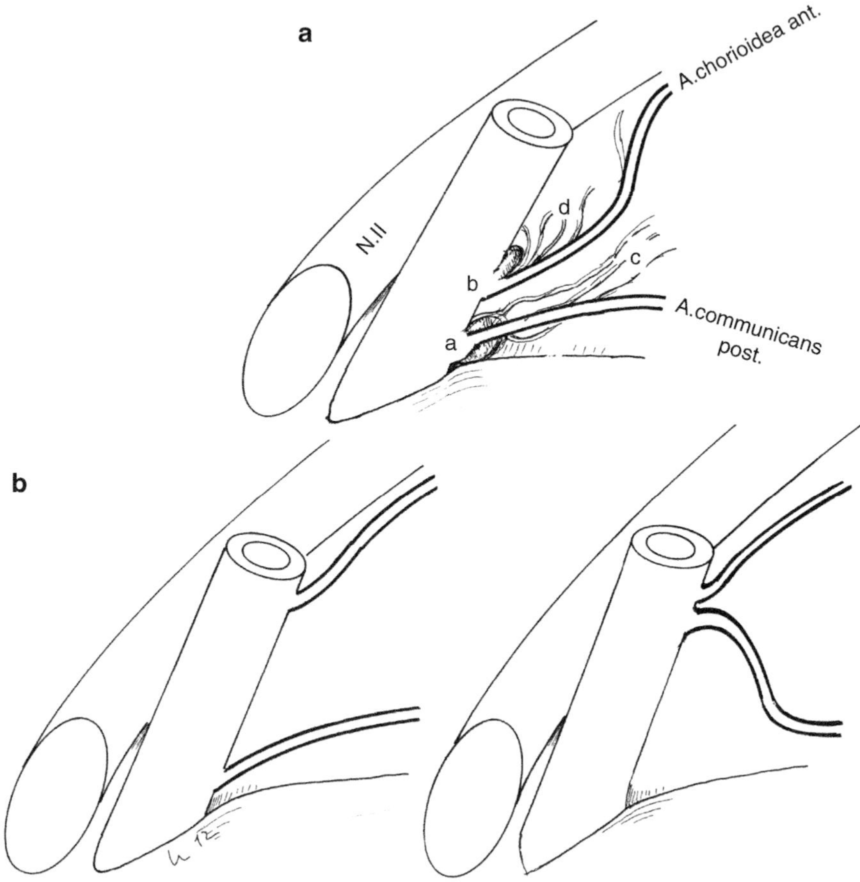

Fig. 2.50 Main proximal branches of A. carotis interna and perforating arteries. **a** A. communicans posterior and A. chorioidea anterior, common findings. Interruption of perforating branches is dangerous (Yaşargil). **b** Variants. **a** *a* and *b*: location of typical aneurysms. *c* and *d*: perforating arteries. **b** Examples of variants

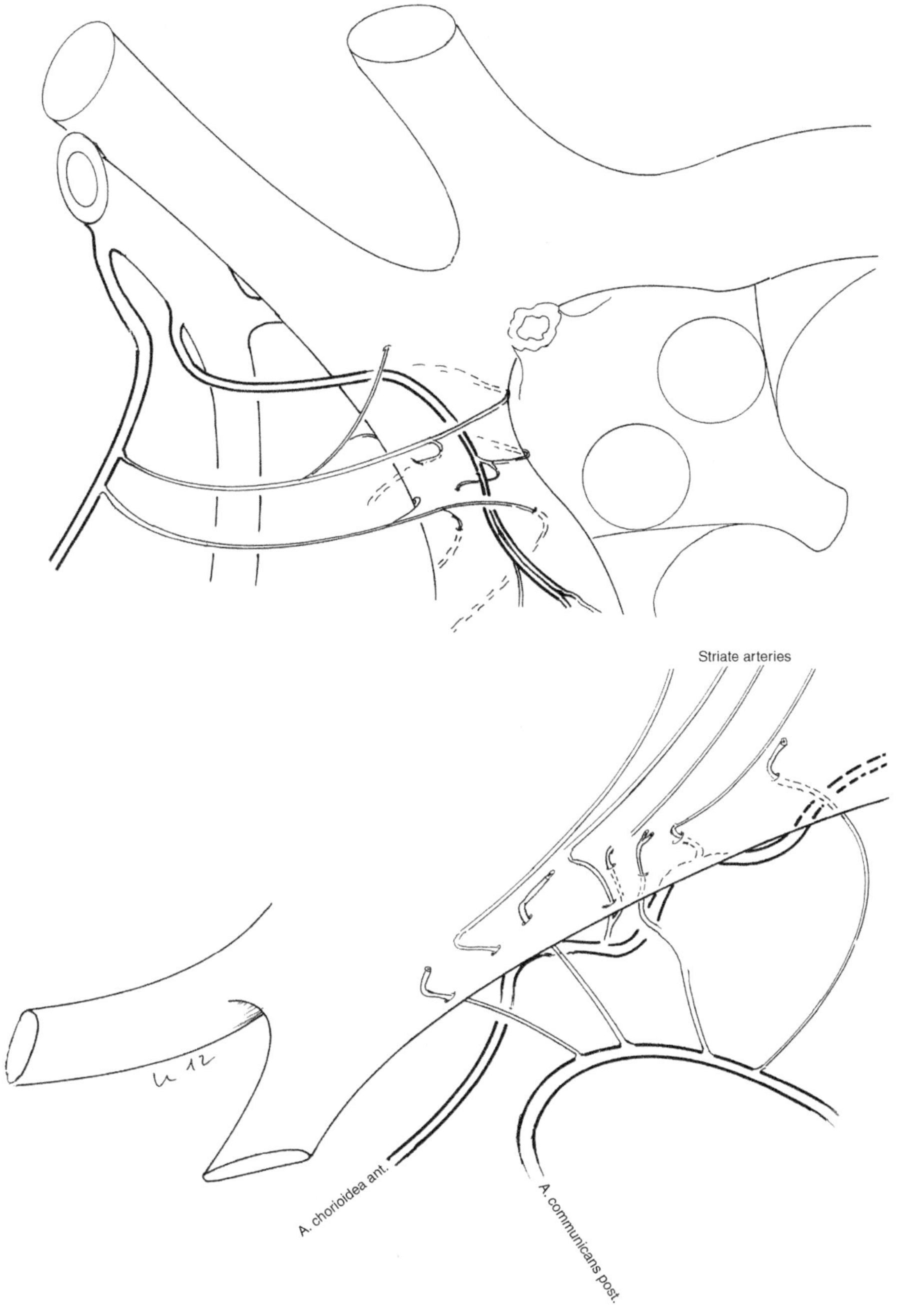

Fig. 2.51 Addendum for Fig. 2.50. Perforating branches penetrating tractus opticus and feeding basal ganglia. These branches are feeding not only tractus opticus. Most of them are feeding basal ganglia and the pyramidal tract after penetrating tractus opticus (Yaşargil)

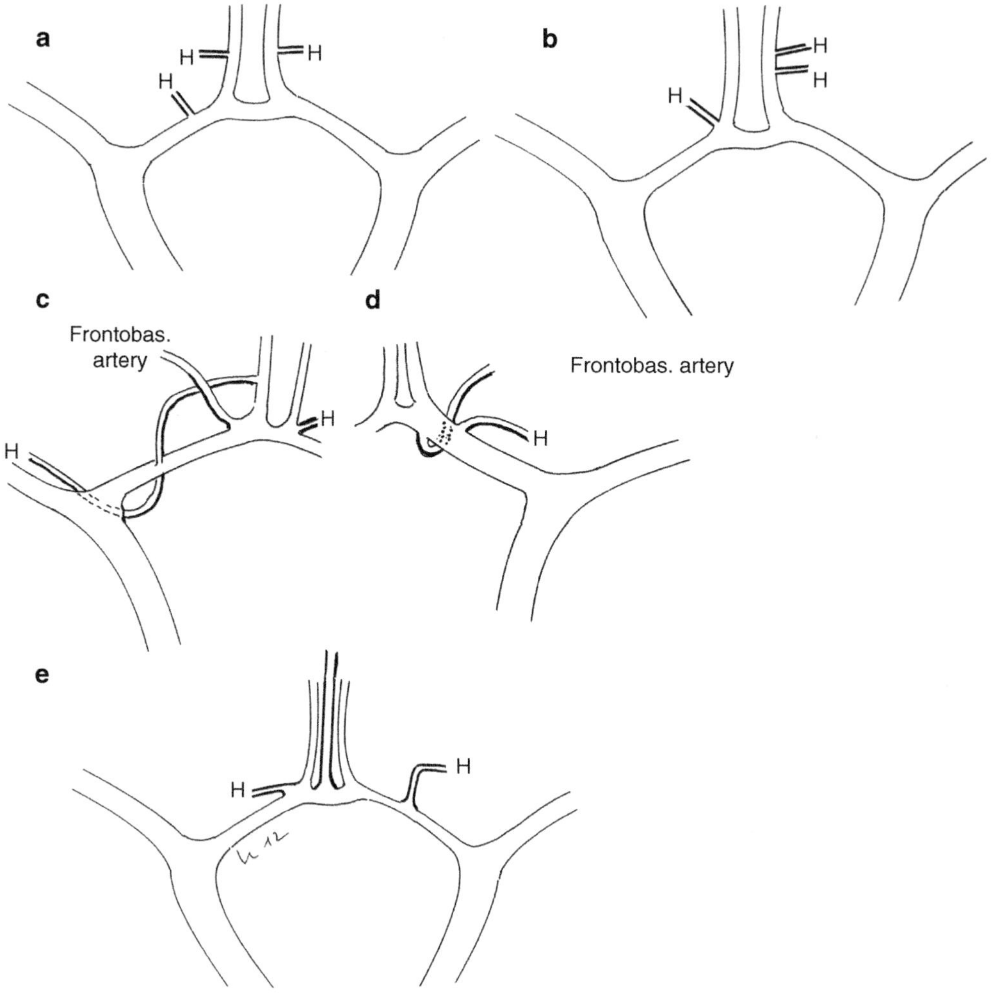

Fig. 2.52 Heubner's artery, variants of proximal branches (according to Kribs and Kleihues 1971, schematized). Perforating branches of Heubner's artery are feeding basal ganglia and pyramidal tract. At surgery, Heubner's artery could be mixed with other frontomedial arteries (Yaşargil). **a** Bilateral and right-sided variants. **b** Right sided, normal; left sided, dislocation and duplication. **c** and **d** Mixed with other arteries is possible. **e** Left sided: originating from A1, combined with a further variant (see Fig. 2.55f). *H* A. recurrens (Heubner)

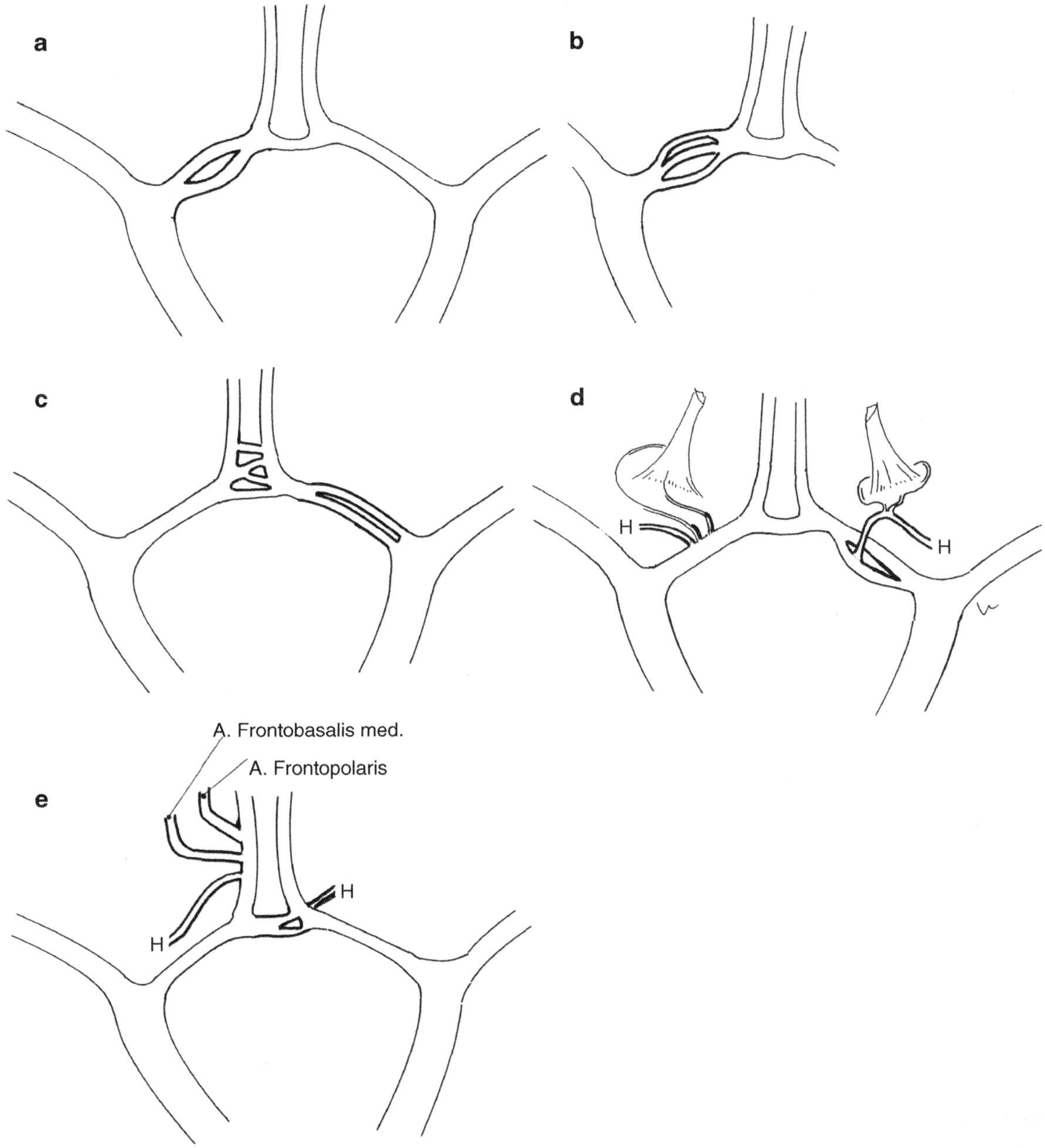

Fig. 2.53 Variability of A. cerebri anterior and of A. communicans anterior. Perforating branches of A. cerebri anterior and Heubner's artery present perforating branches which are feeding the basal ganglia, pyramidal tract, and adjacent structures. **a–c** v. Mitterwallner (1955). **d** Perlmutter and Rhoton (1976). **e** Yaşargil (1984b). *H* A. recurrens (Heubner)

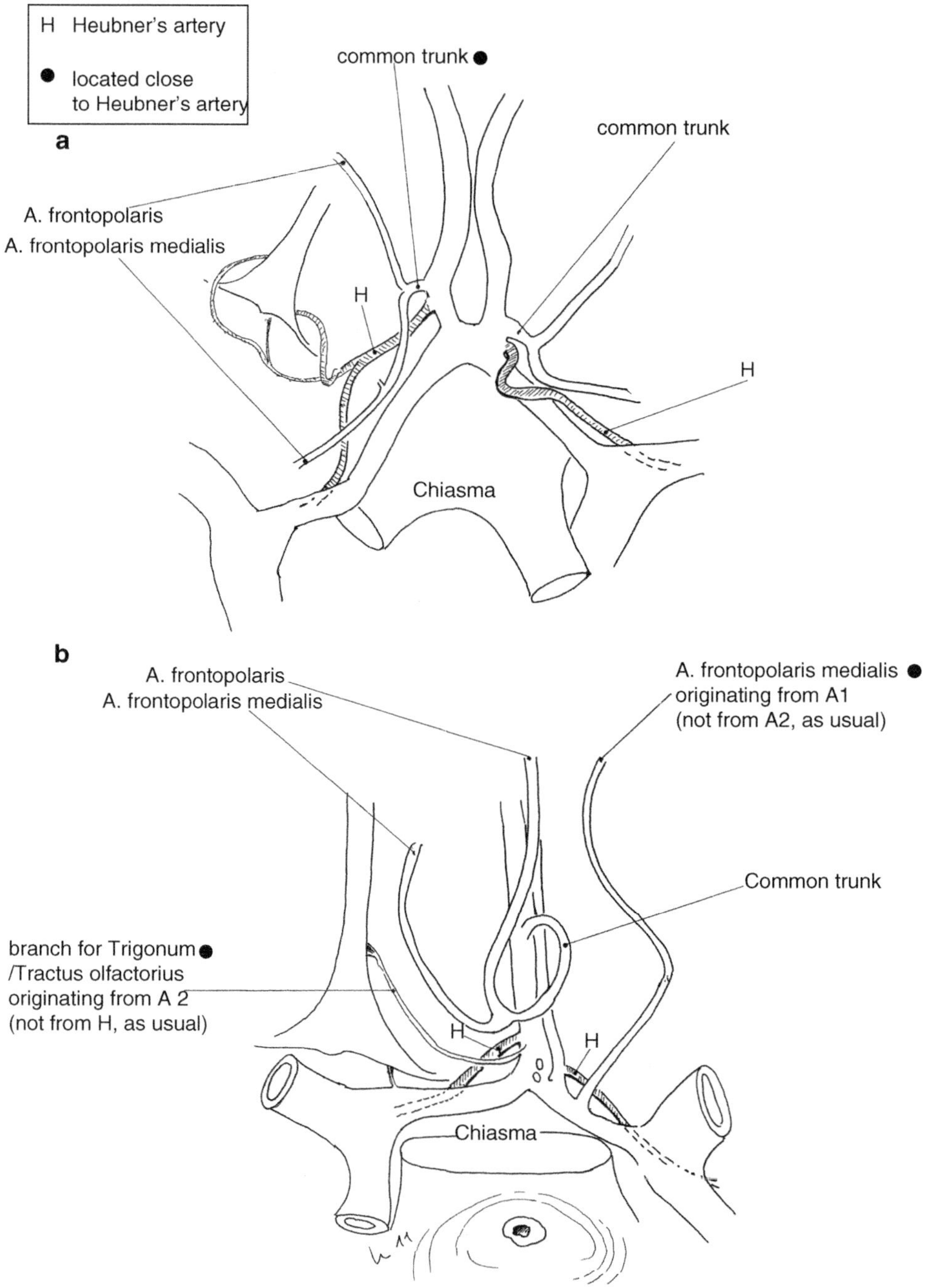

Fig. 2.54 Casts of variants, examples. Schematized

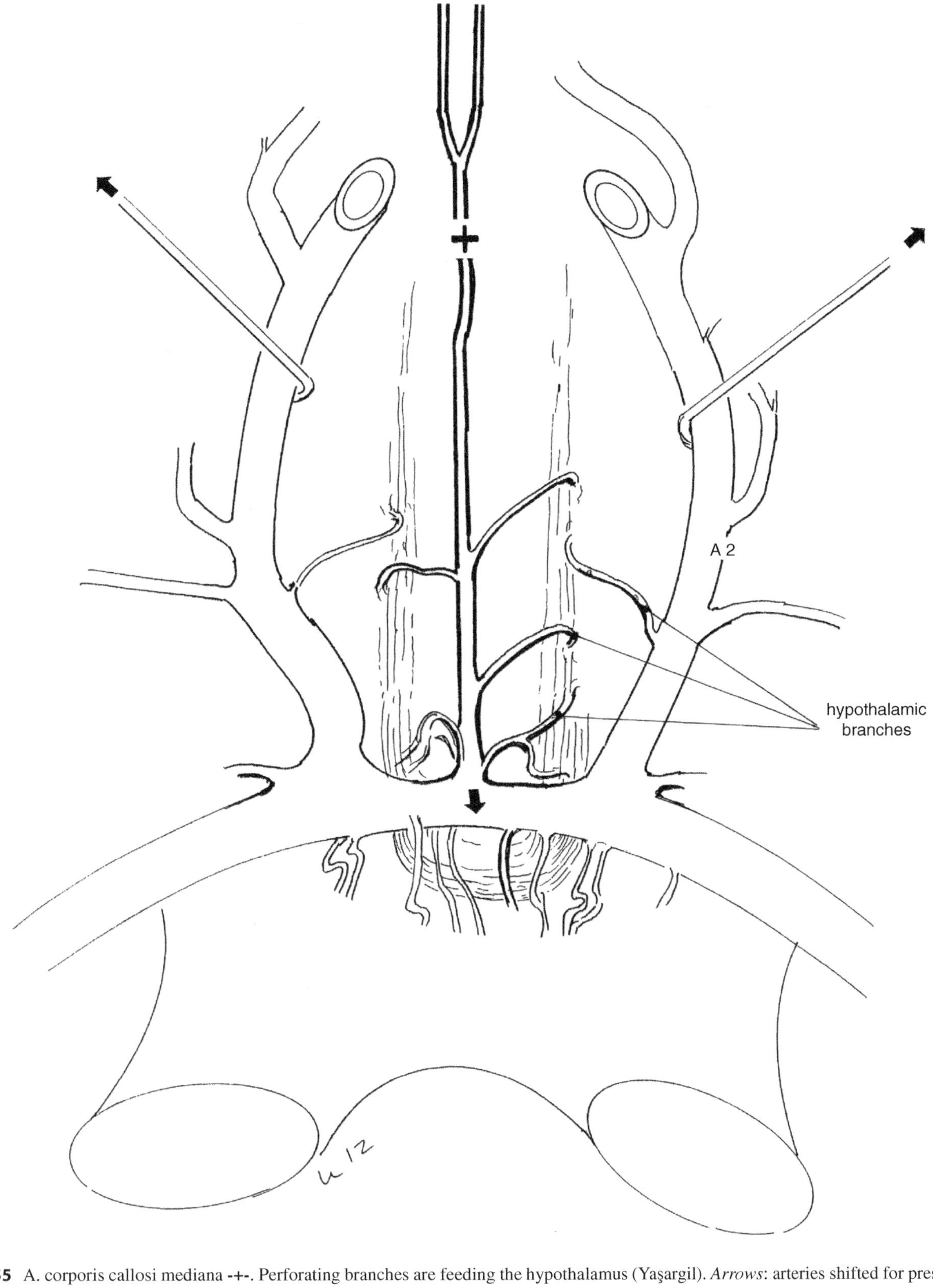

Fig. 2.55 A. corporis callosi mediana -+-. Perforating branches are feeding the hypothalamus (Yaşargil). *Arrows*: arteries shifted for presentation of perforating branches

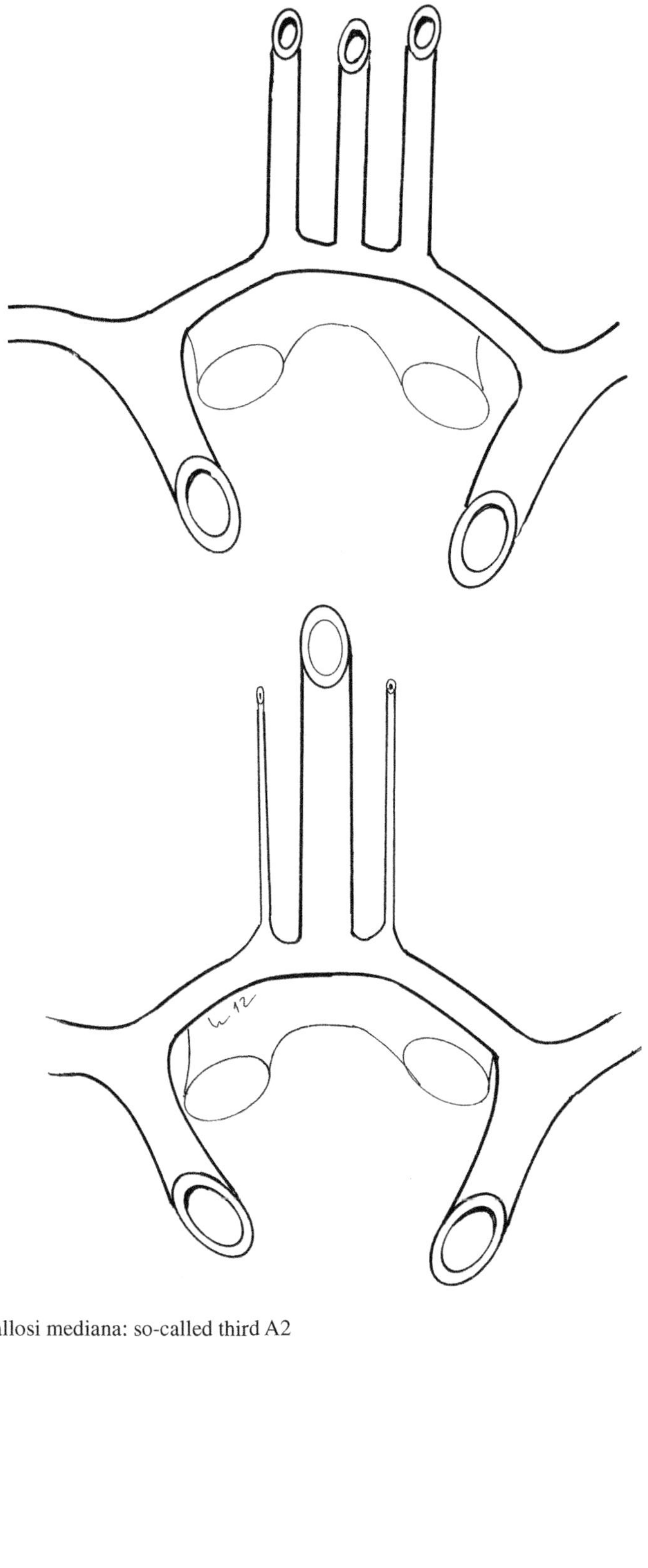

Fig. 2.56 Variants of A. corporis callosi mediana: so-called third A2

3 Special Surgical Aspects

3.1 Extradural Topography

3.1.1 Surgical Routes (Fig. 3.1)

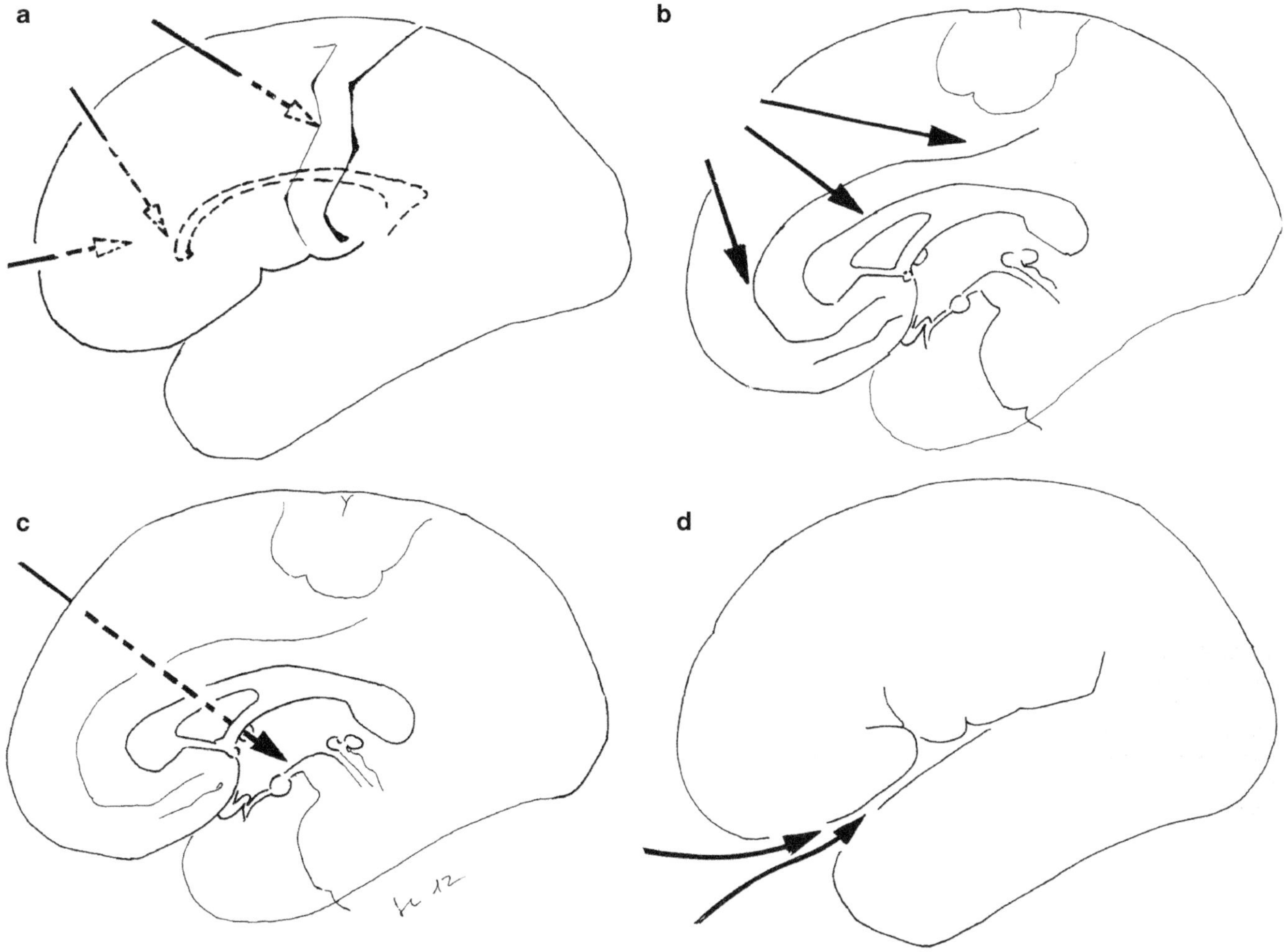

Fig. 3.1 Survey. (**a**) Frontolateral routes. (**b**) Midline routes. (**c**) Dorsolateral transforaminal routes to the third ventricle. (**d**) Pterional routes

W. Seeger, J. Zentner, *Anatomical Basis of Cranial Neurosurgery*, https://doi.org/10.1007/978-3-319-63597-2_3

3.1.2 Extracranial Surgical Aspects (Figs. 3.2 and 3.3)

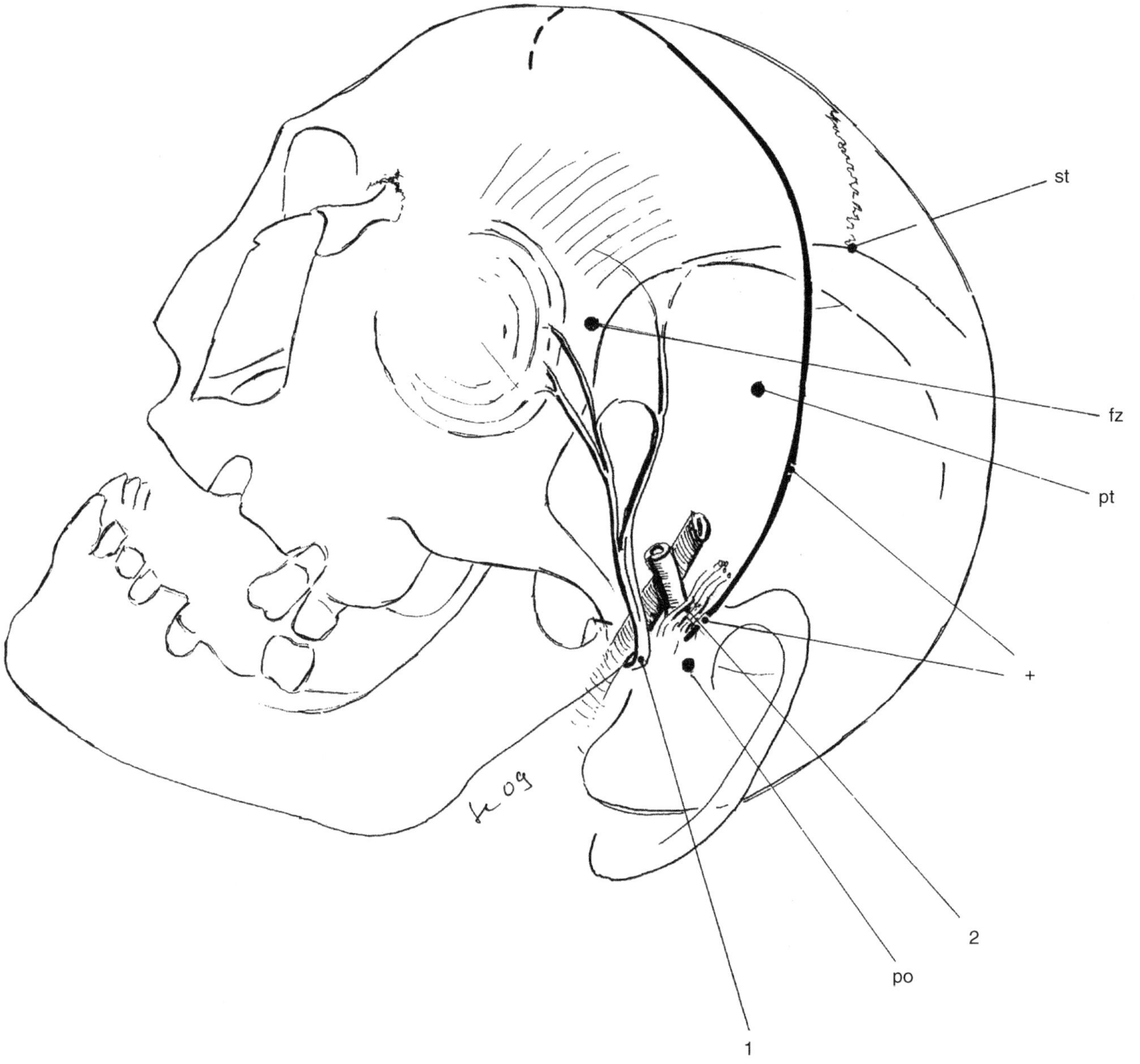

Fig. 3.2 Skin-galea incision -+-.Frontal branch of N. facialis and adjacent structures could be endangered. *fz* frontozygomatic point, *po* porion, *pt* pterion, *st* stephanion; *1* frontal ramification of N. facialis, *2* A. temporalis superficialis and N. auriculotemporalis

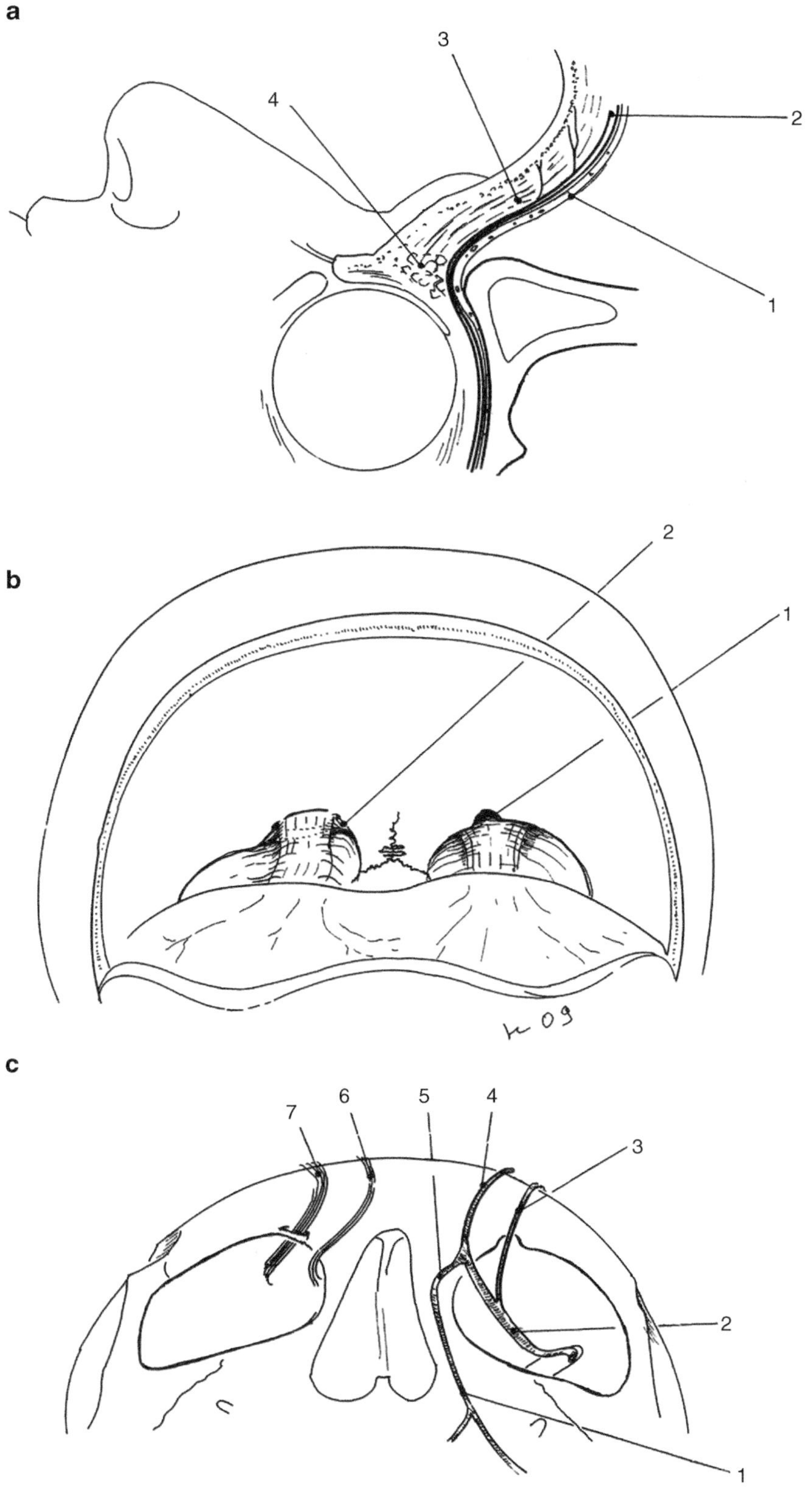

Fig. 3.3 Subperiosteal loosening (see Fig. 2.2b′) of skin-galea flap before performing craniotomy nerves and vessels of the supraorbital and frontomedial region should be preserved. Damage may be followed by anesthesia dolorosa. (**a**) Layers, (**b**) foramen supraorbitale right, incisura supraorbitalis left (common findings), (**c**) connections of facial veins to dural sinuses could be routes for intracranial phlebothrombosis. **a** *1* Periosteum, *2* nerves and vessels, *3* M. epicranius, frontal portion, *4* M. orbicularis oculi. (**b**) *1* At incisura supraorbitalis (common variant), vessels and nerves are easy to lose at surgical approach. *2* Foramen supraorbitale enclosing nerves and vessels. The inferior bony bridge of the foramen should be resected before loosening of nerves and vessels. (**c**) *1* V. facialis, *2* V. supraorbitalis, *3* as 2, *4* R. frontalis medialis, *5* V. angularis, *6* as 4, *7* as 3

3.1.3 Craniotomies (Figs. 3.4, 3.5, 3.6, 3.7, and 3.8)

a

b

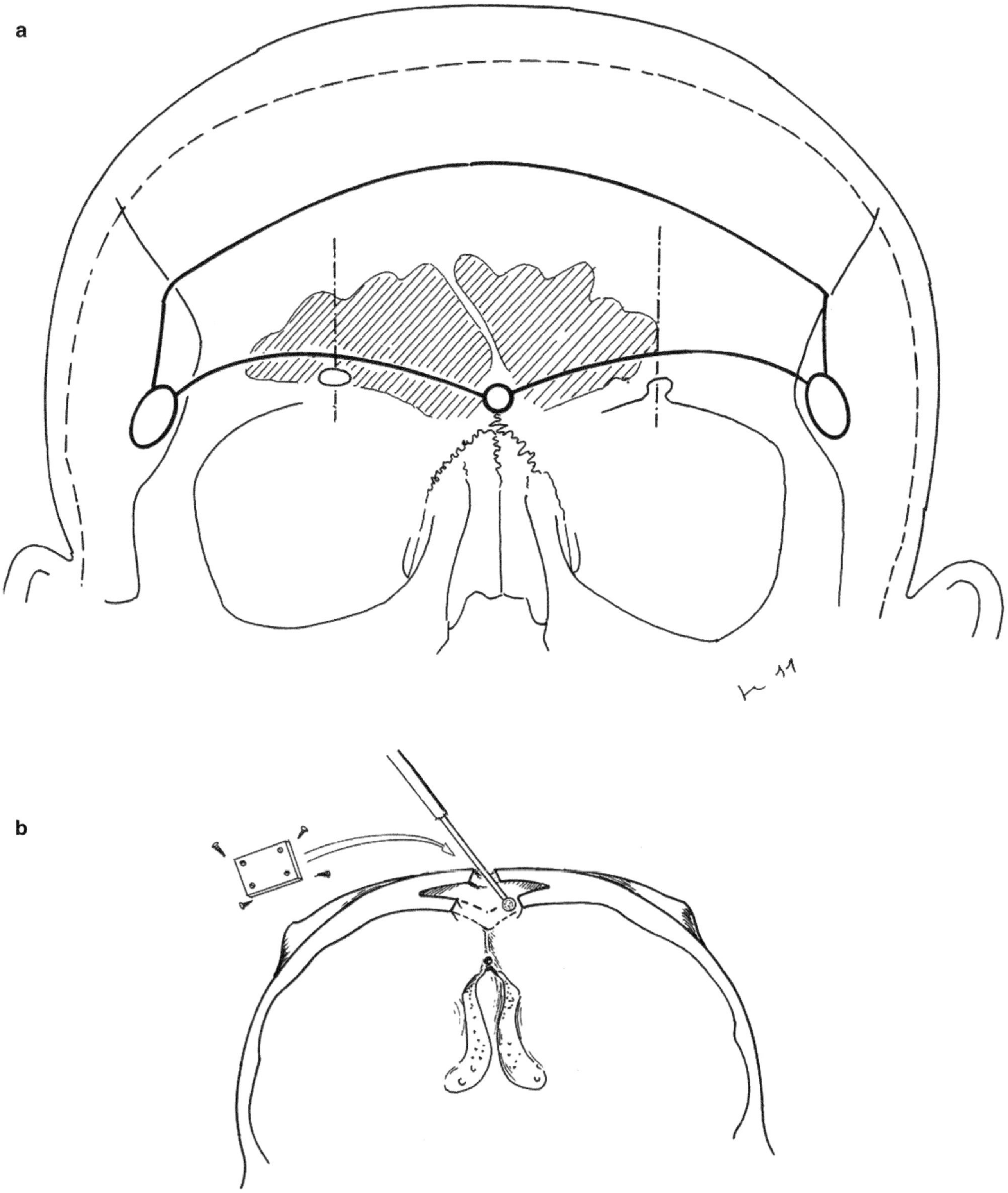

Fig. 3.4 Bifrontal craniotomy. Variability of crista frontalis should be noticed at preoperative axial CT. If it is configurated like a sharp crest, a burr hole may be placed at the midline for avoiding damage of sinus sagittalis superior and cerebral cortex. **a** Small burr hole, if other cosmetic techniques are not available. *Dotted*: skin incision. **b** Resection of the inner wall of the sinus frontalis and of crista frontalis

a

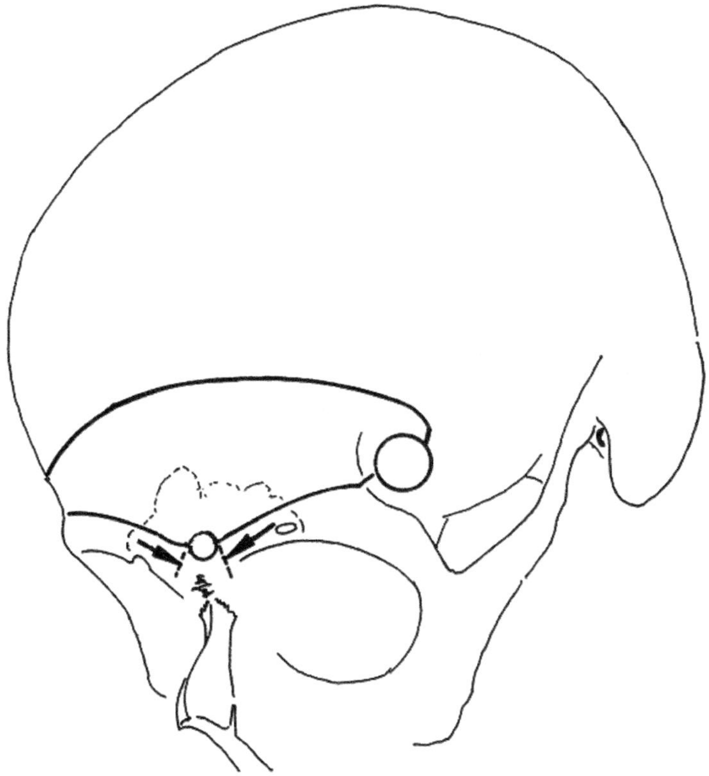

b

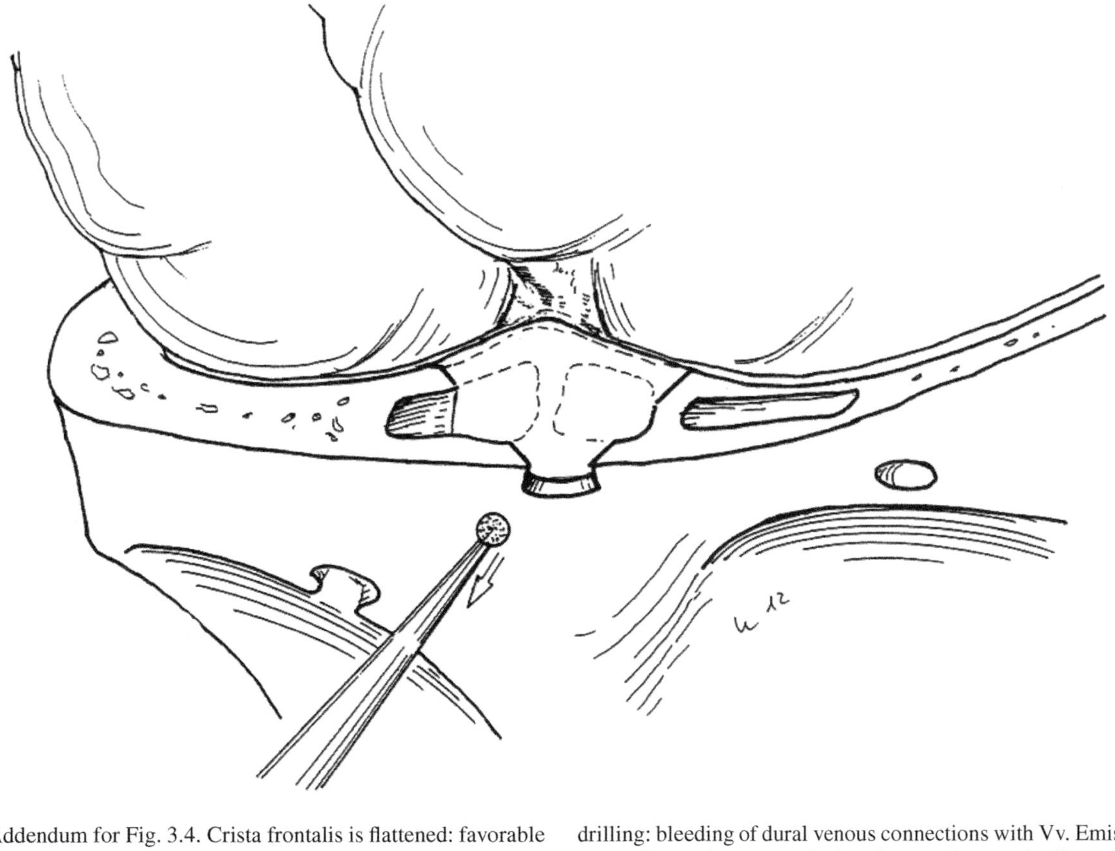

Fig. 3.5 Addendum for Fig. 3.4. Crista frontalis is flattened: favorable for craniotomy. **a** Nevertheless craniotomy may be hindered by dural adhesions (*arrows*). A Medial burr hole may be recommended. **b** After drilling: bleeding of dural venous connections with Vv. Emissariae may occur. Spare electrocoagulation close to sinus sagittalis superior

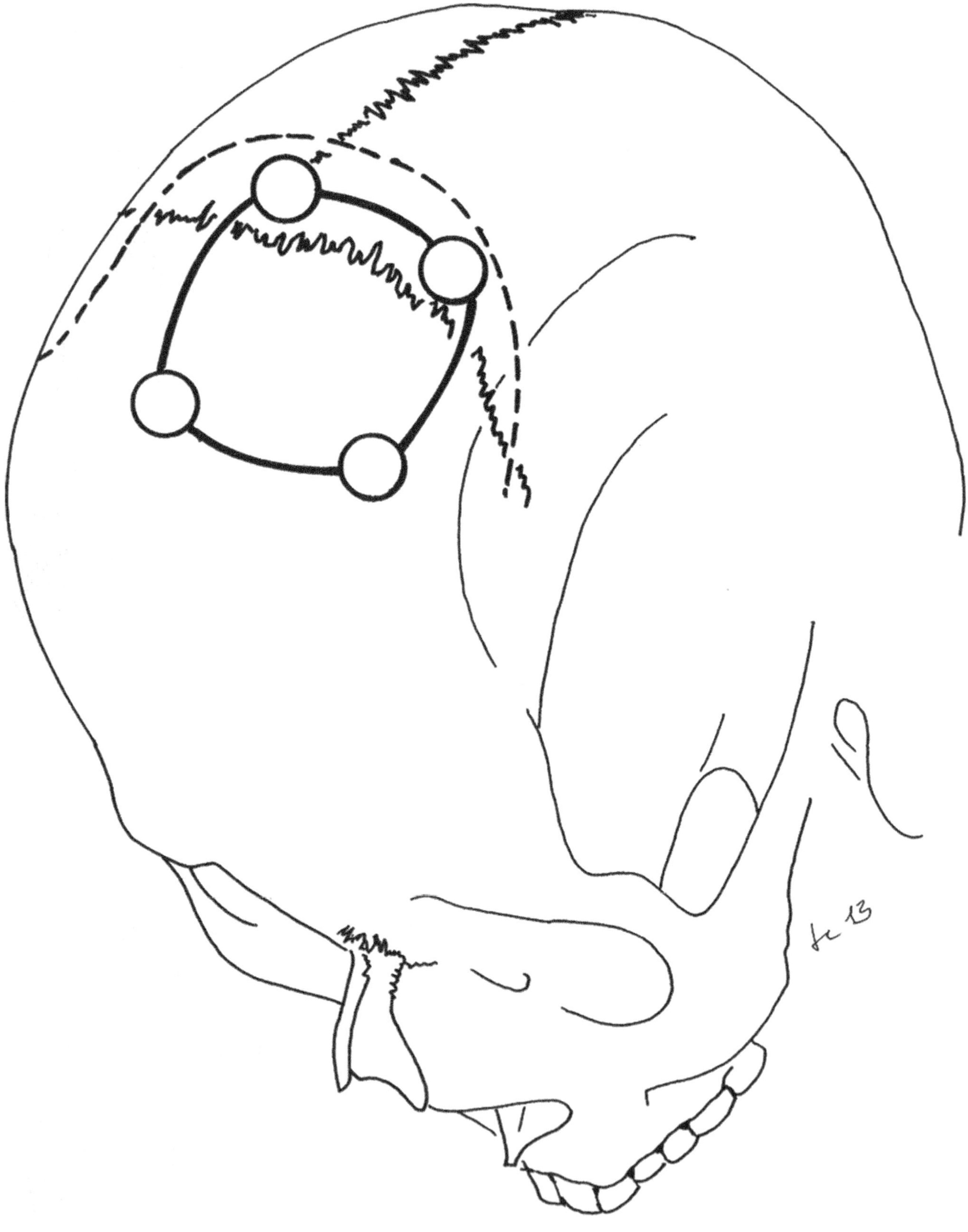

Fig. 3.6 Craniotomy for large lesions in the depth (This and the following craniotomies are applied by Prof. Zentner, Freiburg, 2013)

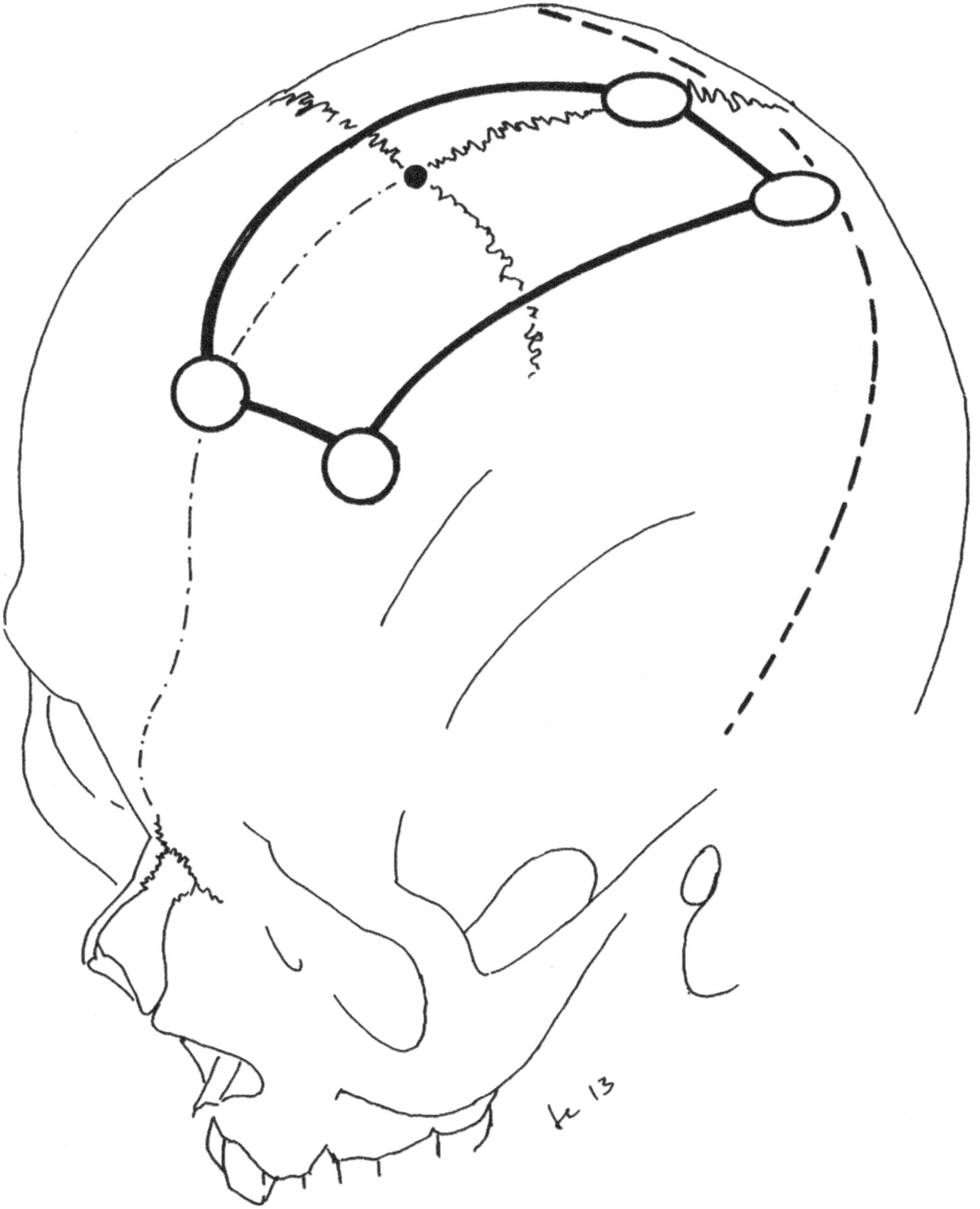

Figs. 3.7 Craniotomy for large lesions at the midline

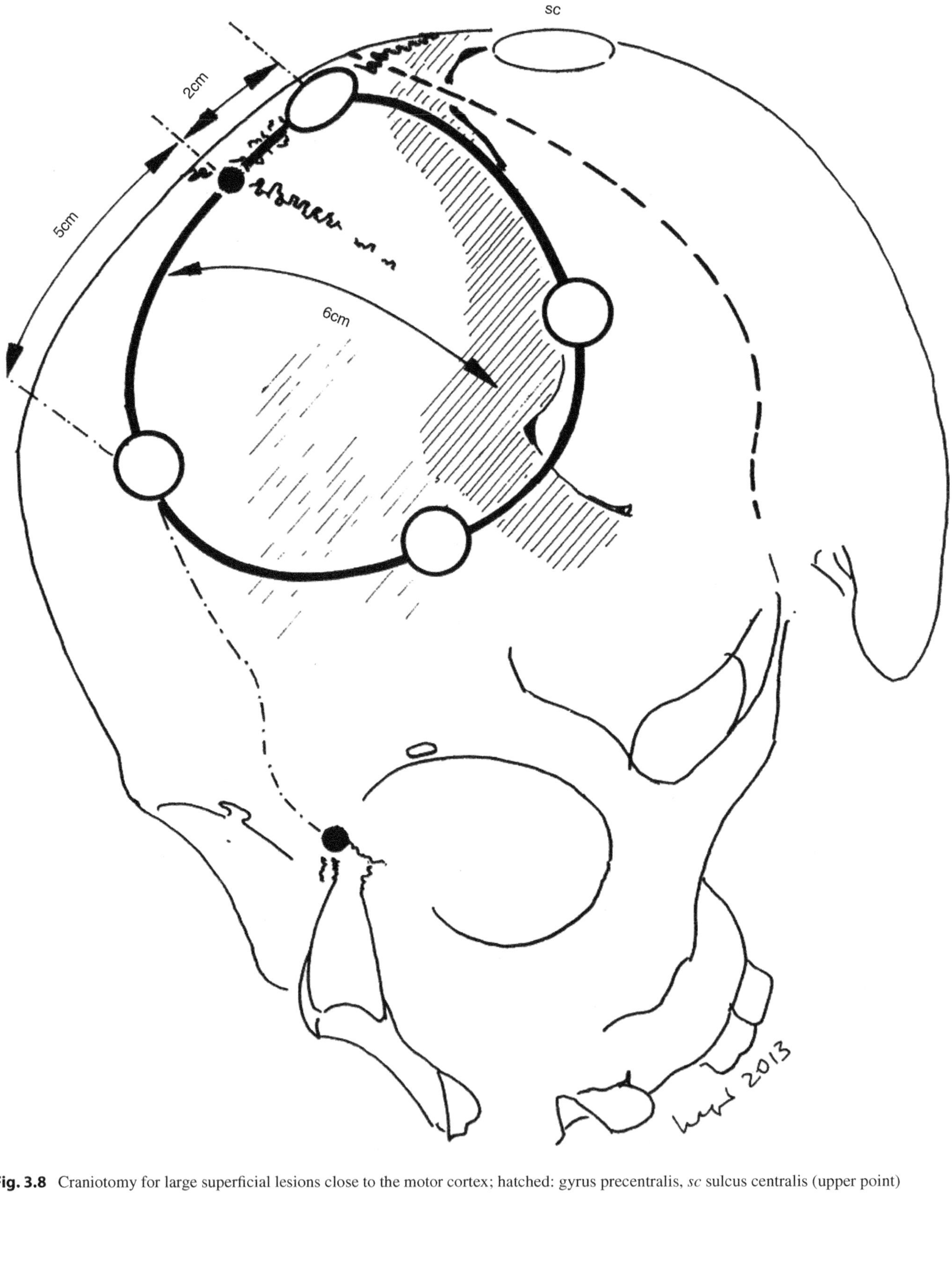

Fig. 3.8 Craniotomy for large superficial lesions close to the motor cortex; hatched: gyrus precentralis, *sc* sulcus centralis (upper point)

3.2 Intradural Topography

3.2.1 Topographical Aspects of Superficial and Deep Approaches (Figs. 3.9, 3.10, 3.11, 3.12, 3.13, 3.14, and 3.15)

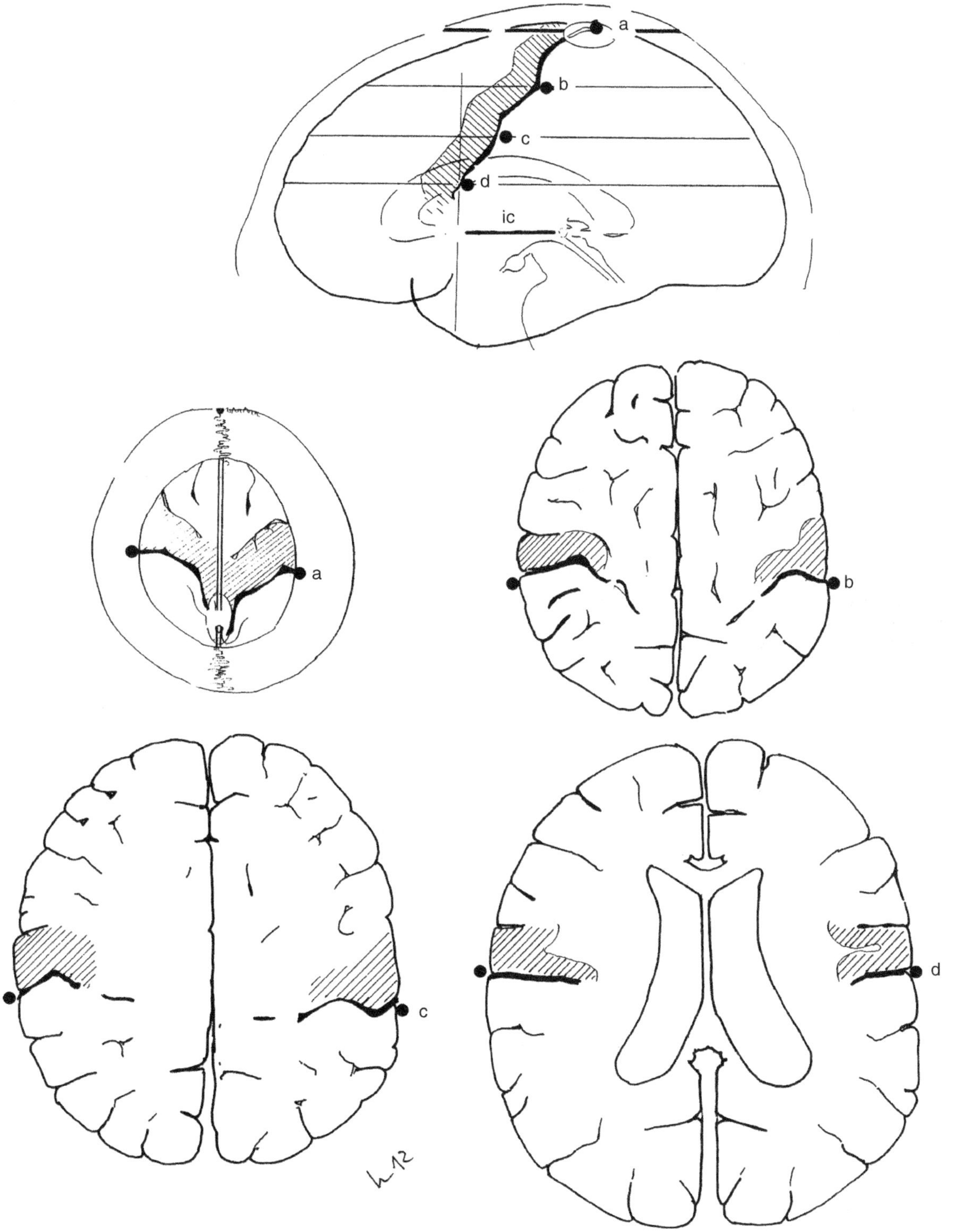

Fig. 3.9 Defining of sulcus centralis and gyrus precentralis by conventional axial MRT planes. *ic* intercommissural level; hatched: gyrus precentralis

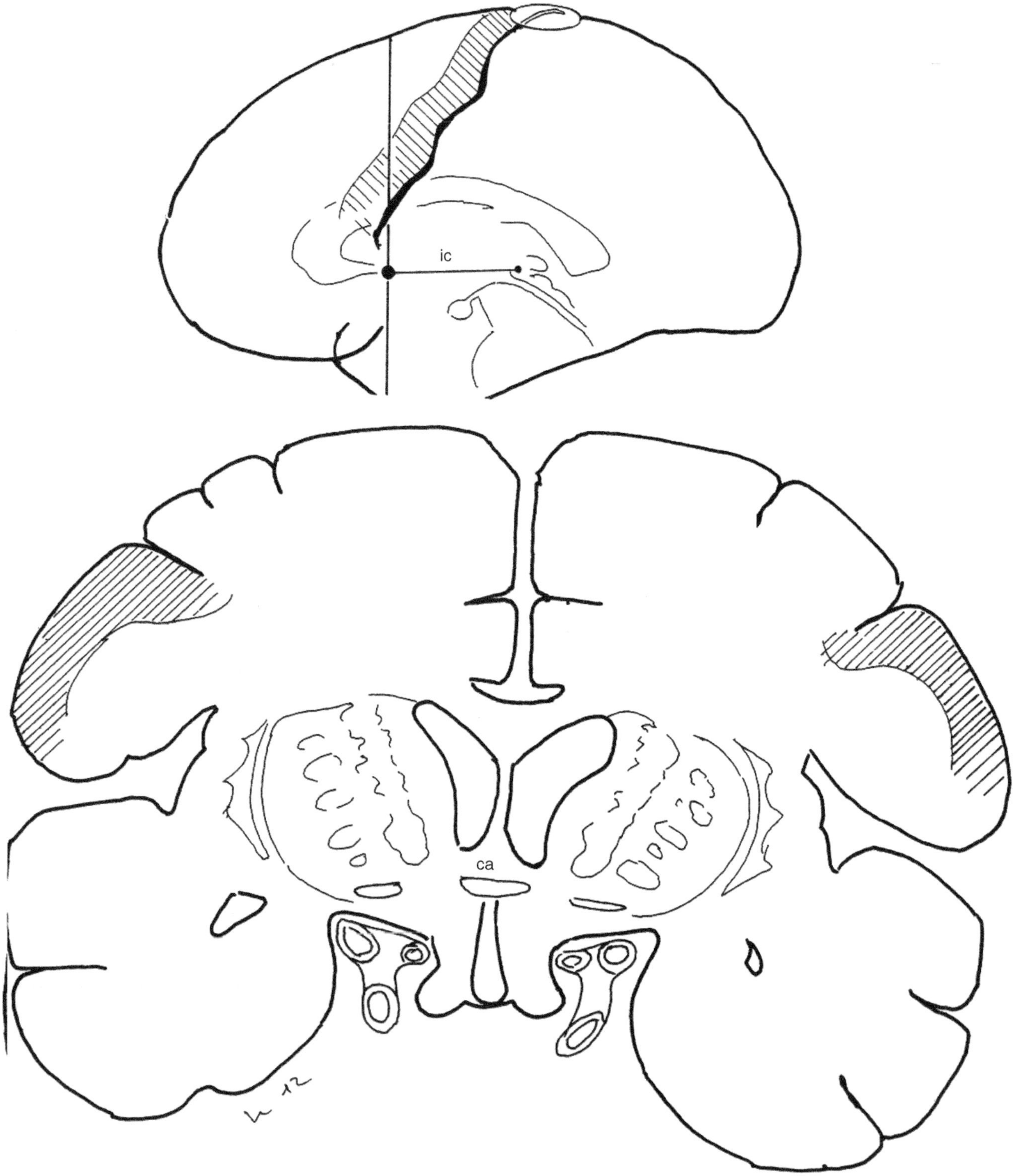

Fig. 3.10 Addendum for Fig. 3.9. Coronal MRT planes present only small segments of the motor cortex

a

b

c

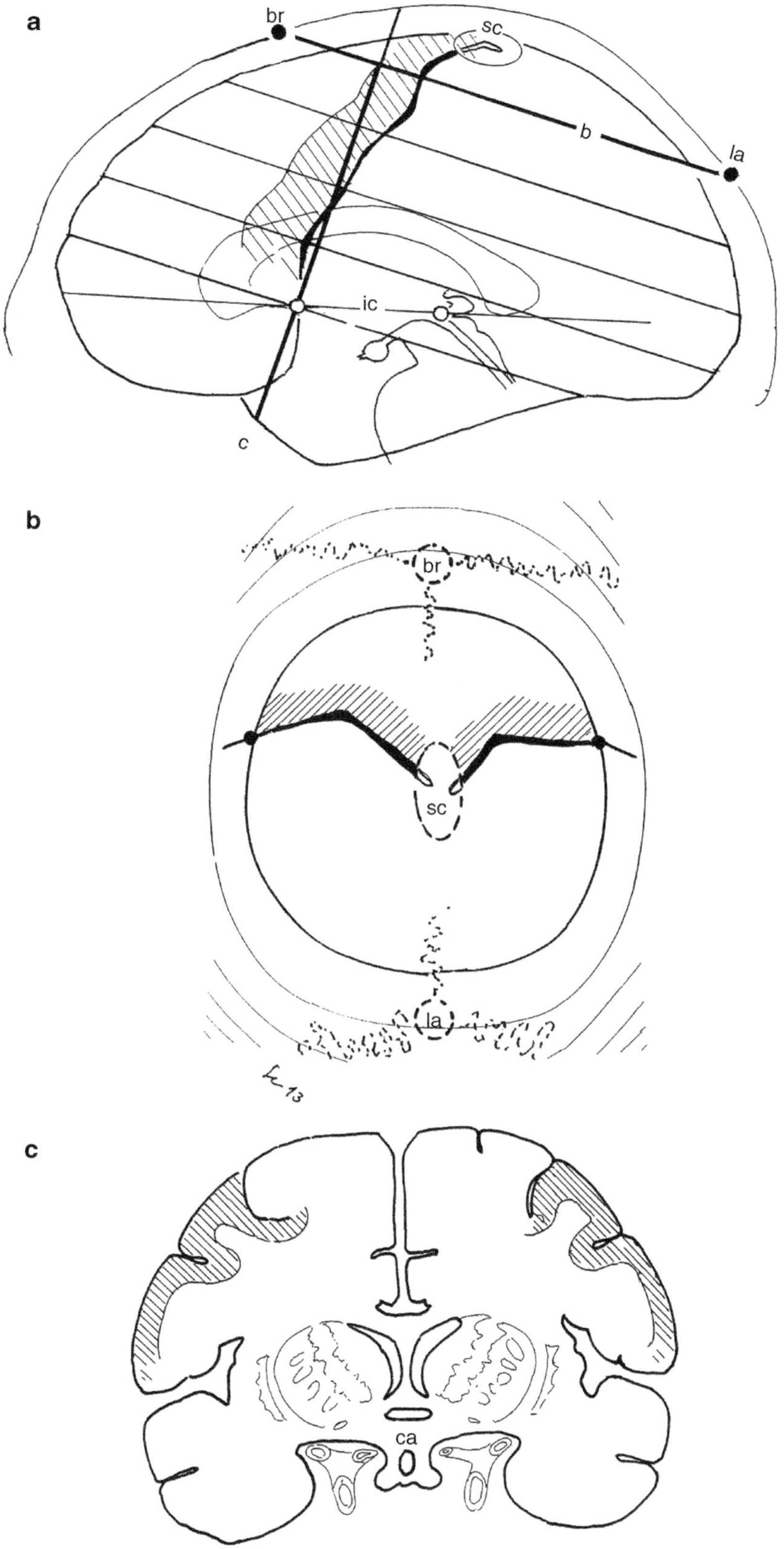

Fig. 3.11 Favorable defining of the motor area by using the Bregma-Lambda plane for CT. If conventional MRTs are unclear, then Bregma-Lambda method may be used before an exact neurophysiological defining will be performed. **a** Topogram; **b** level br-la; **c** level c; *br* bregma, *ic* intercommissural line, *la* lambda, *sc* sulcus centralis (upper point); hatched: gyrus precentralis

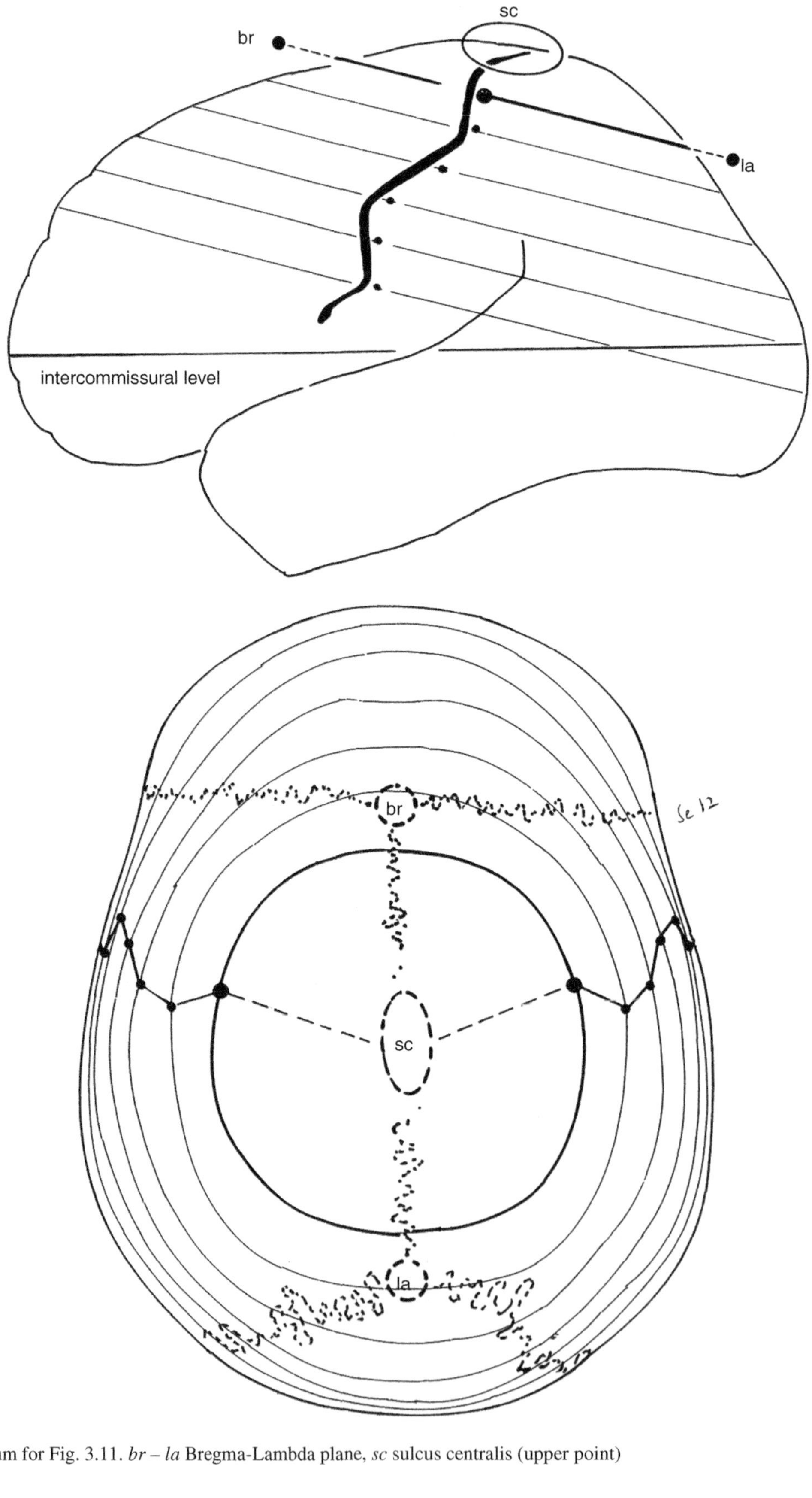

Fig. 3.12 Addendum for Fig. 3.11. *br – la* Bregma-Lambda plane, *sc* sulcus centralis (upper point)

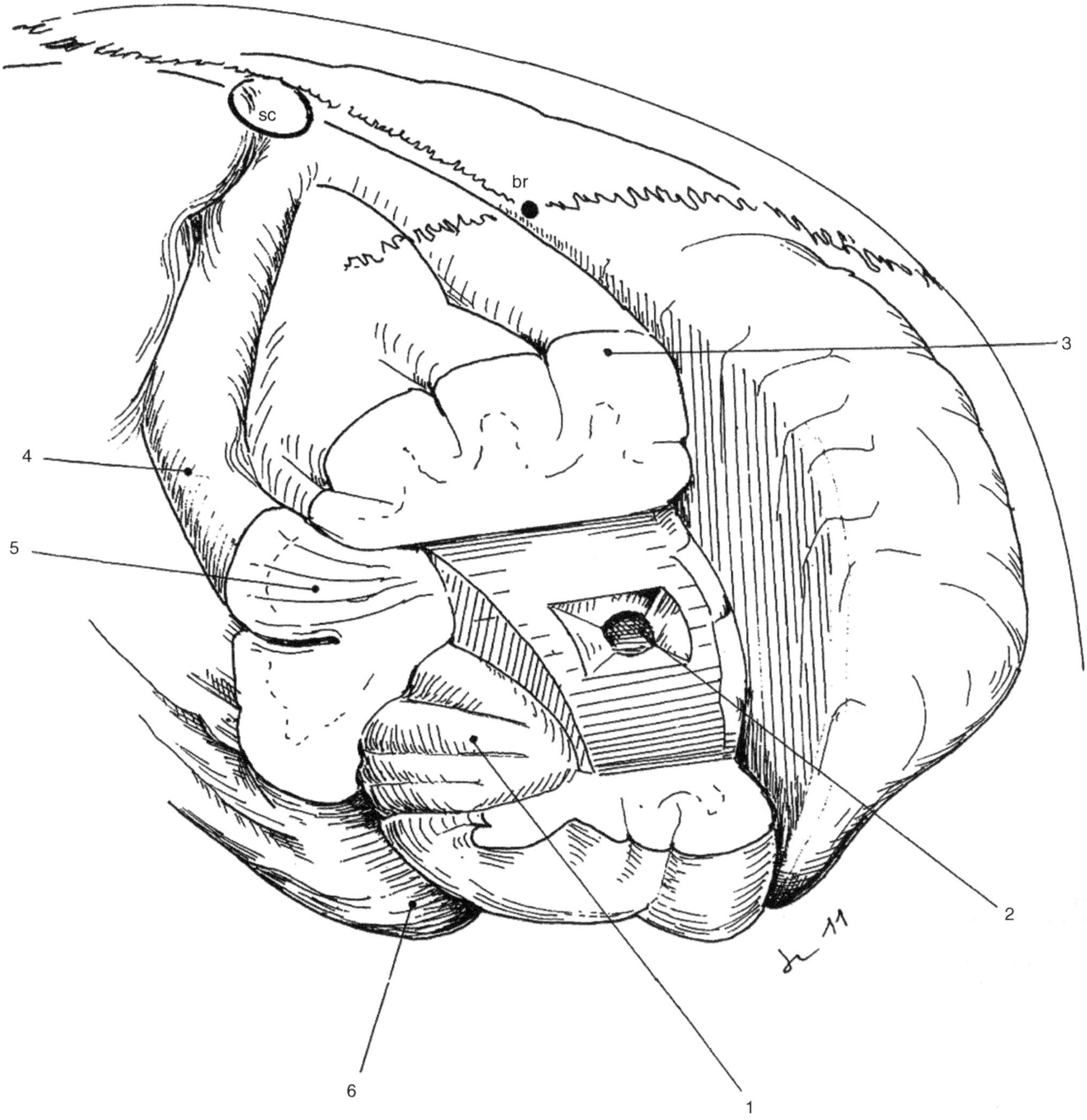

Fig. 3.13 Areas in the depth. Anatomical details similar to neuronavigation. Landmarks added. Survey. *br* bregma, *sc* sulcus centralis (upper point). *1* Insula; *2* tip of the ventricle; *3* first frontal gyrus; *4* gyrus precentralis; *5* tractus corticospinalis, part of corona radiata; *6* temporal pole

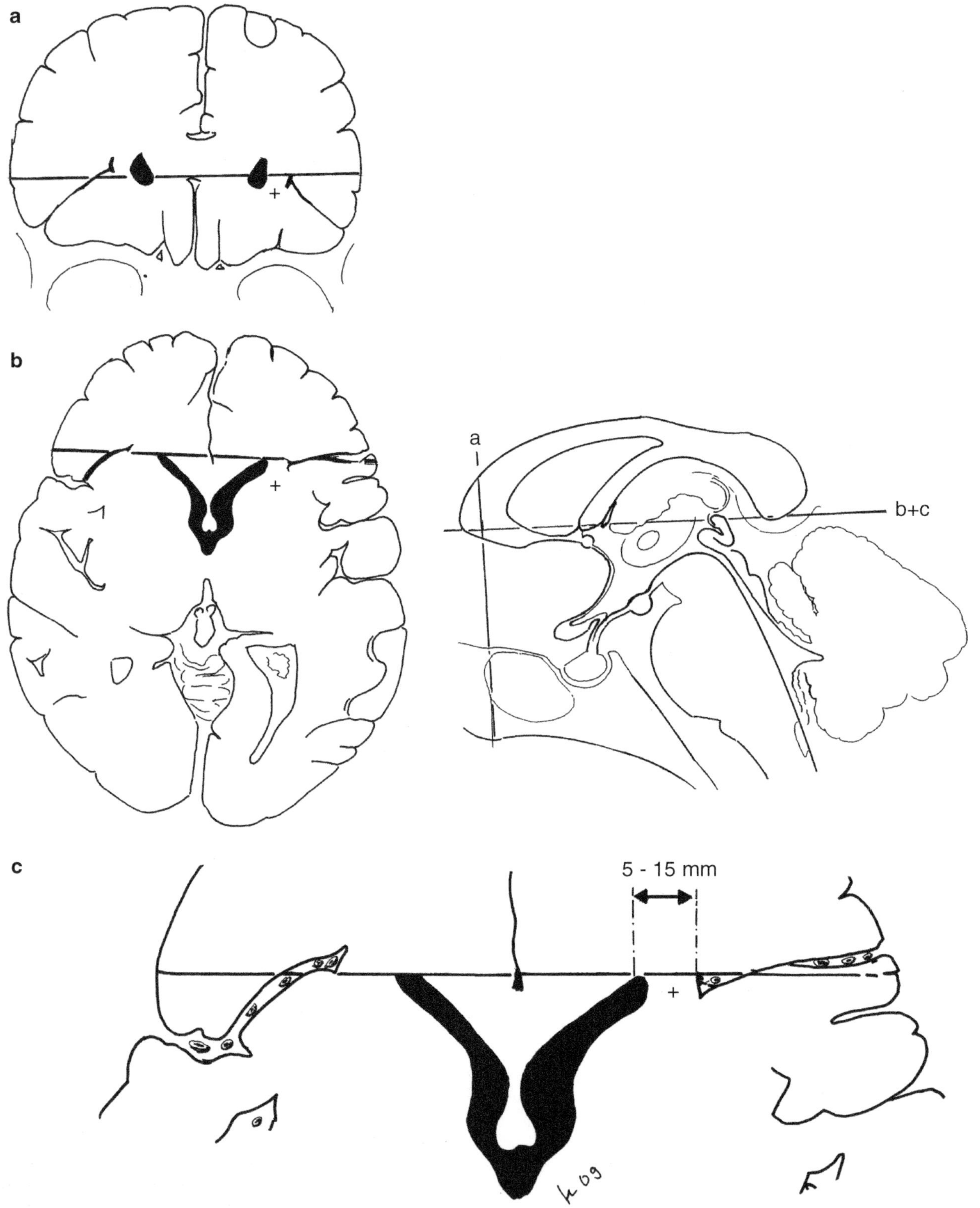

Fig. 3.14 Surgical target region between insula and cornu anterius -+-, example MRT **a** and **b** Topogram for **c** , **c** small lesion (e.g., cavernoma -+-) between the anterior sylvian area and tip of cornu, anterius

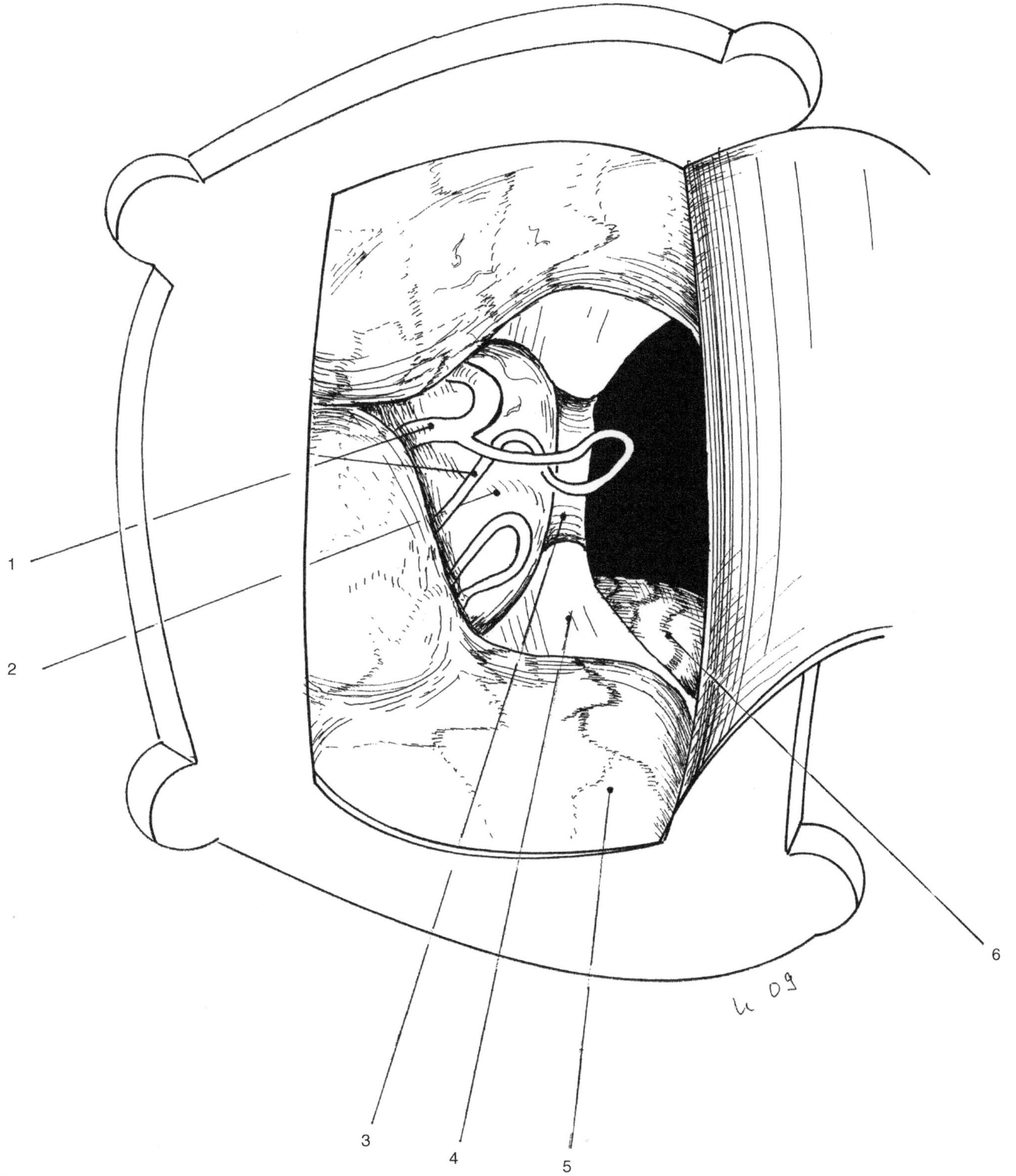

Fig. 3.15 Continuation of Fig. 3.14. Surgical topography: *1* sylvian vessels. Danger for intraventricular bleeding. *2* Insula, *3* cerebral defect after removal of a lesion, *4* white matter, *5* cortex, *6* caput nuclei caudati

3.2.2 Topographical Aspects of Midline Approaches (Figs. 3.16, 3.17, 3.18, 3.19, 3.20, 3.21, 3.22, 3.23, 3.24, and 3.25)

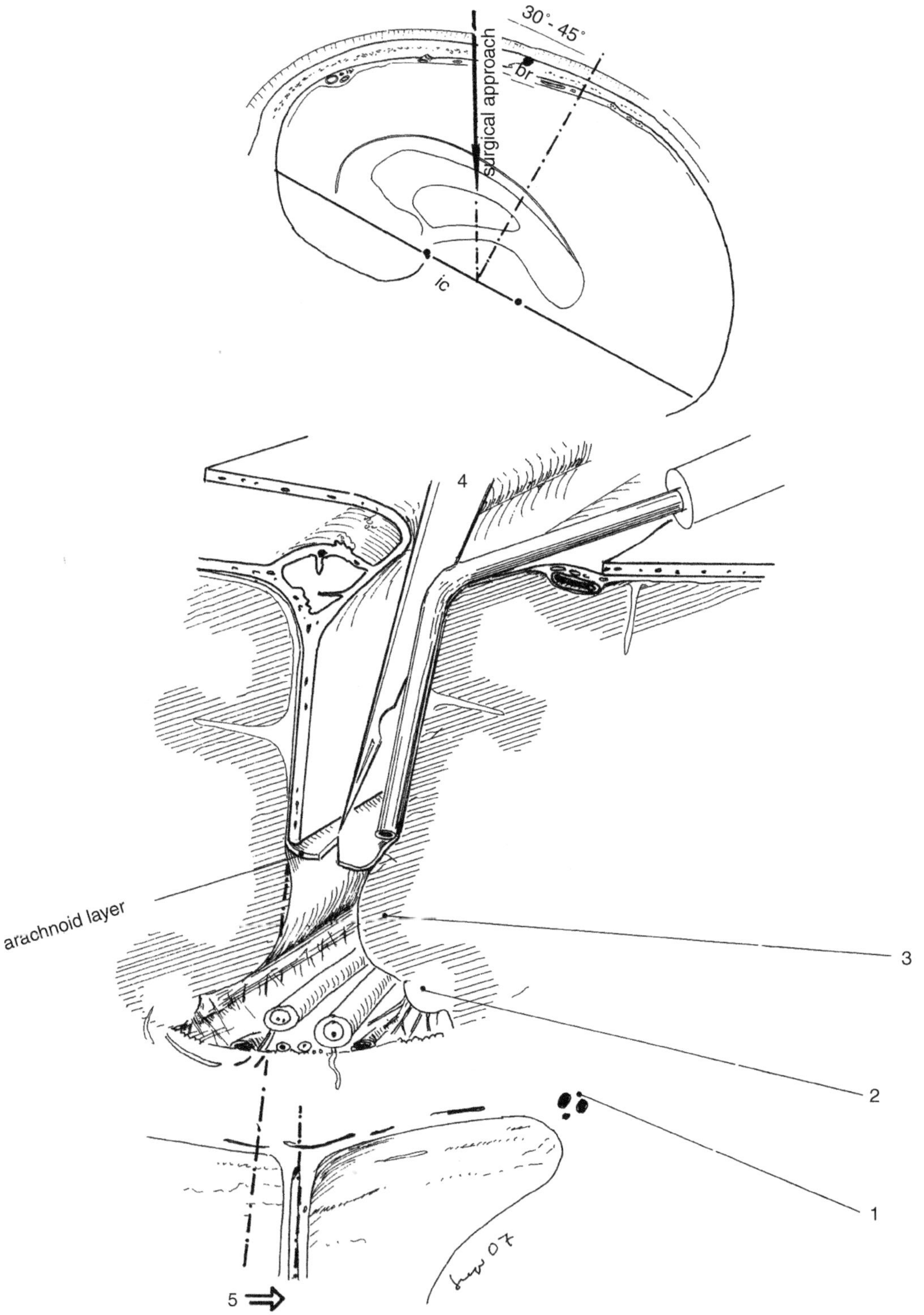

Fig. 3.16 Midline surgery. Principles. *br* bregma, *ic* intercommissural line; *1* Vv. longitudinales, *2* stria longitudinalis cinguli, *3* gyrus cinguli, *4* opening of cisterna corporis callosi, *5* brain shifting by instruments (*arrow*). Unexpected opening of the contralateral ventricle should be taken in consideration

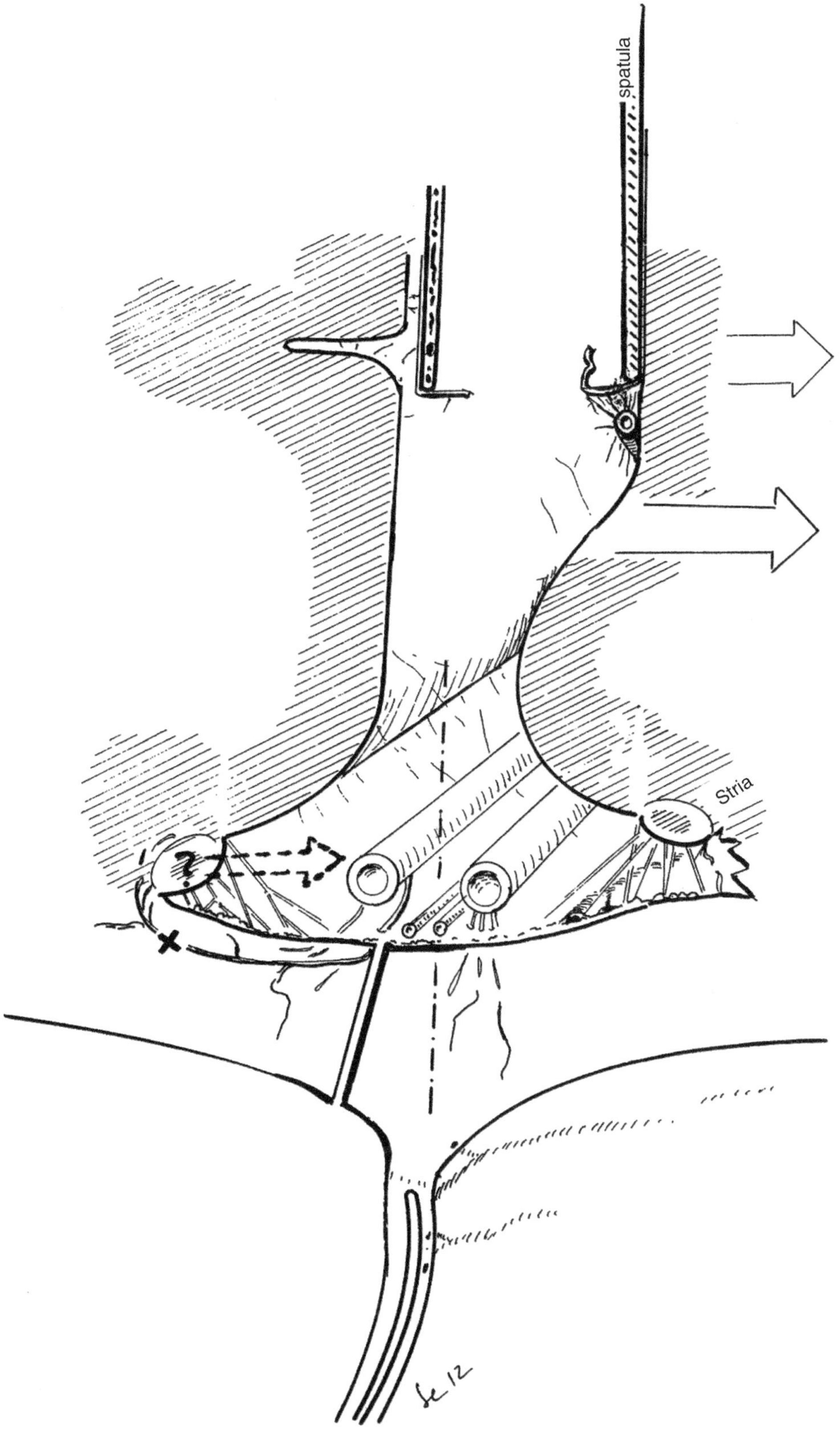

Fig. 3.17 Sectional enlargement of Fig. 3.16. Stria longitudinalis cinguli and its arteries –x- could be shifted to the midline area. Now stria and its arteries are endangered during splitting of the corpus callosum. Before splitting, shifting of stria could be diminished by loosening of arachnoid trabeculas at cisterna corporis callosi

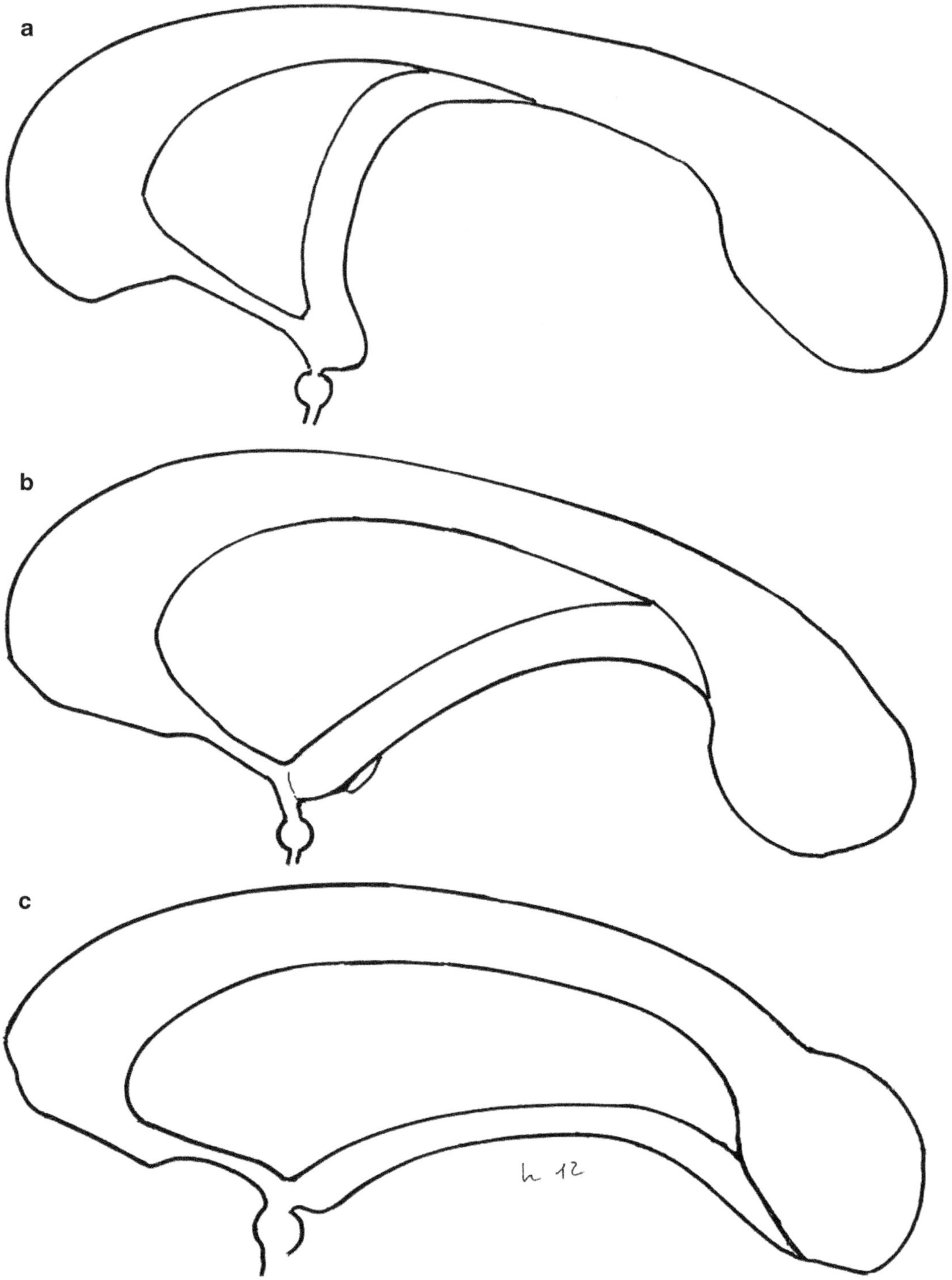

Fig. 3.18 Variants of Fornix (Lang and Ederer 1980) must be taken in consideration especially at callosotomy. **a** 4%, **b** 71%, **c** 19%

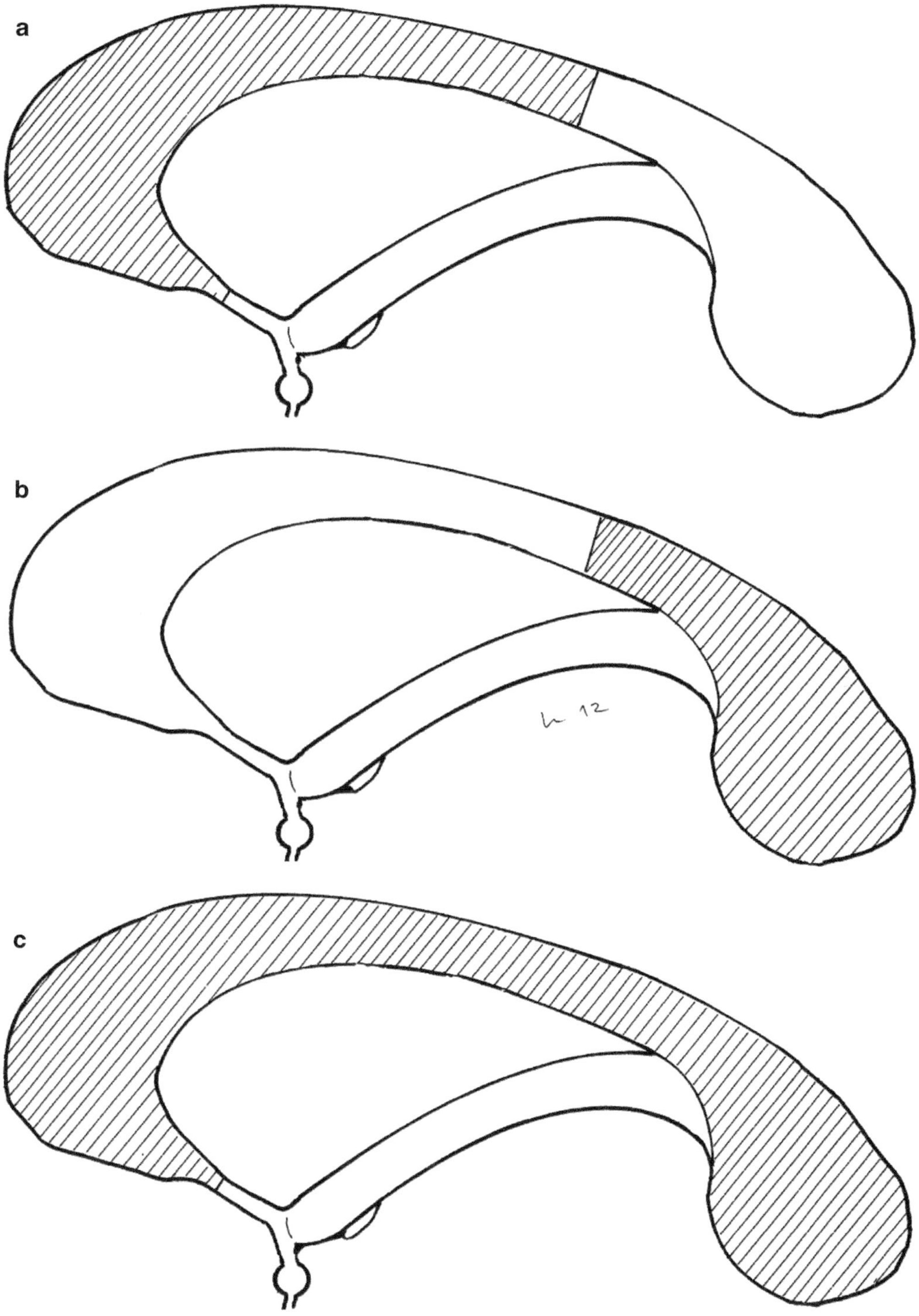

Fig. 3.19 Callosotomy (Zentner 1997, modified). Principles of surgery. **a** Anterior 2/3 callosotomy. **b** Posterior 1/3 callosotomy. **c** Complete callosotomy

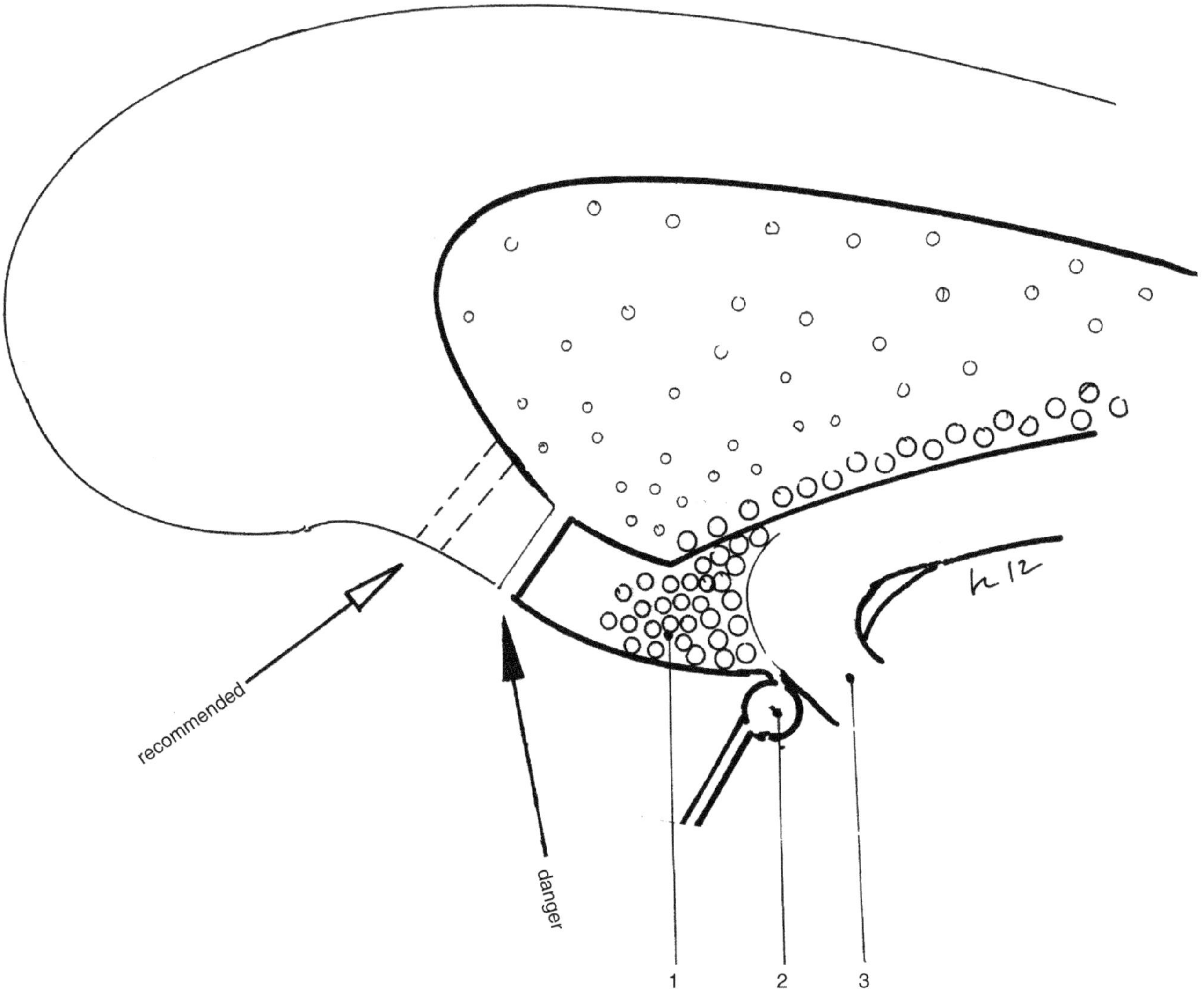

Fig. 3.20 Addendum for anterior callosotomy. Columnae fornicis are endangered, if rostrum corporis callosi is thickened. It could be mixed with genu corporis callosi (common variant). Fornices and rostrum corporis callosi can be defined using neuronavigation: *1* perifornical gray matter, part of fornix; *2* commissura anterior; *3* columna fornicis

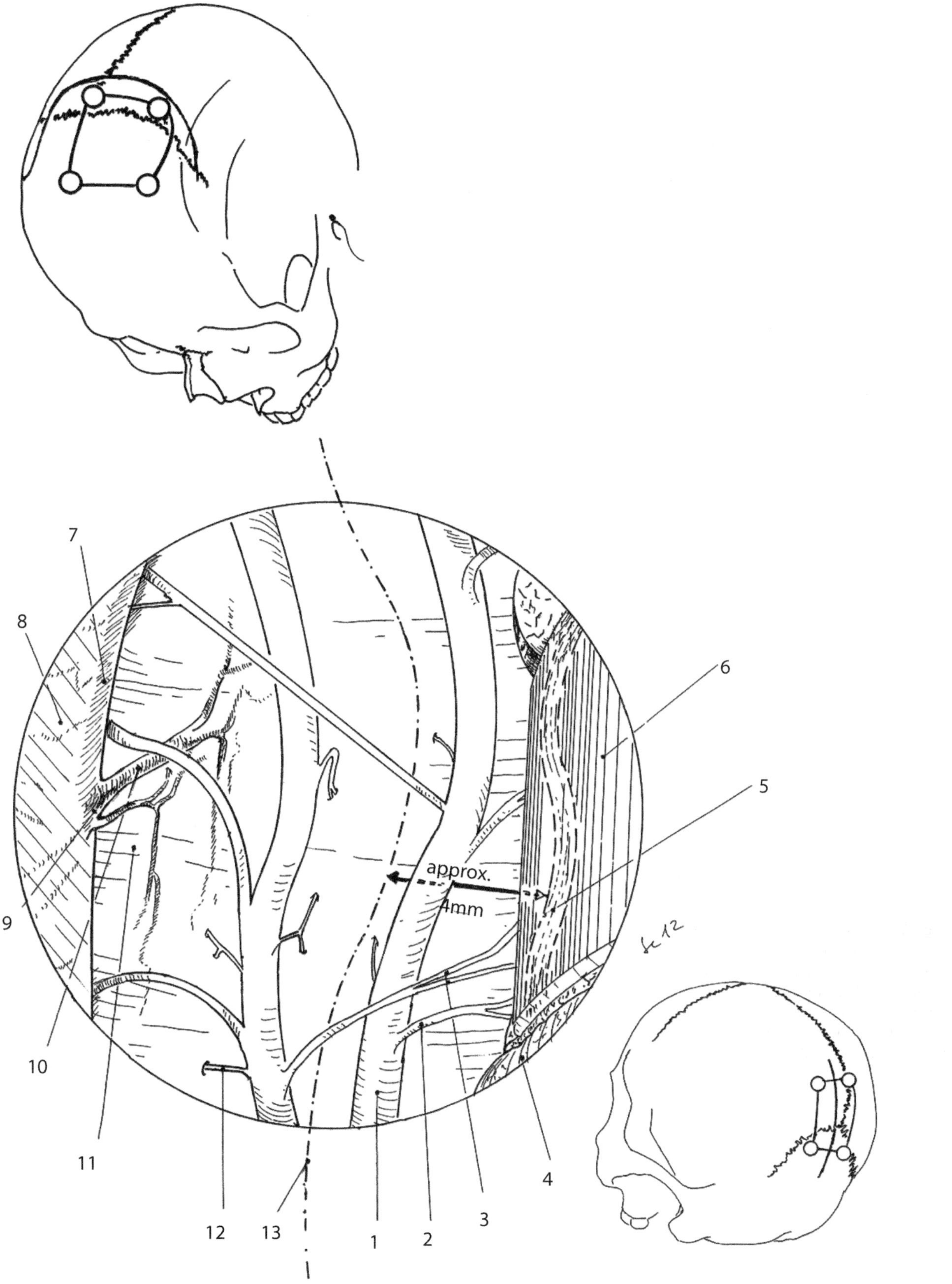

Fig. 3.21 Surgical topography of midline approach using a frontal, parietal, or occipital route. *1* A. pericallosa; *2* homolateral branch of 1; *3* midline crossing branch, spare; *4* gyrus cinguli; *5* stria longitudinalis cinguli, projection, spare; *6* spatula; *7* sinus sagittalis inferior; *8* falx; *9* begin of sinus sagittalis inferior; *10* bridging veins; *11* corpus callosum; *12* perforating arteries (After interruptions, large defects of corpus callosum could follow. Defect of stria longitudinalis is possible.) (After interruptions, large defects of corpus callosum could follow. Defect of stria longitudinalis is possible.) spare, if possible; *13* midline

Completion for Midline Approaches (Figs. 3.22, 3.23, 3.24, and 3.25)

Falx variants and bilateral arteries

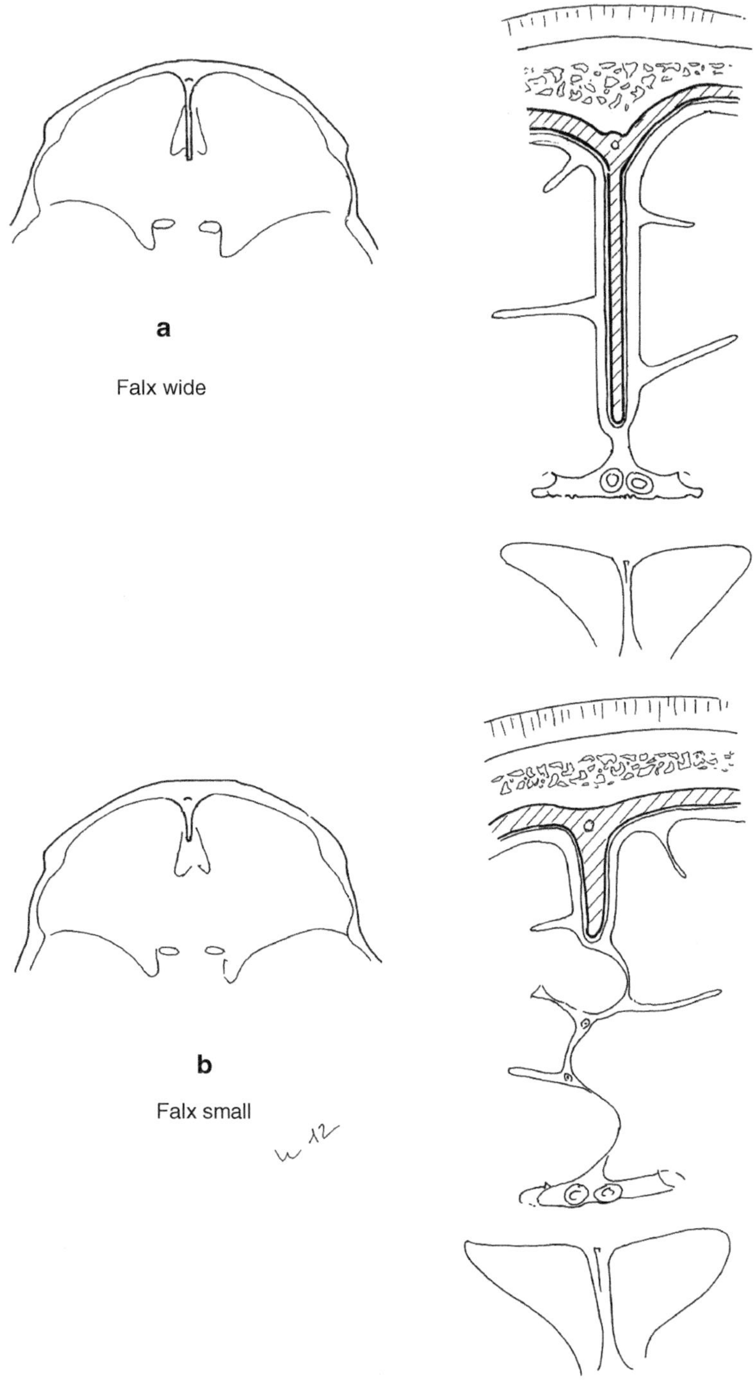

Fig. 3.22 Falx variants. **a** Favorable for subdural routes. **b** Less favorable

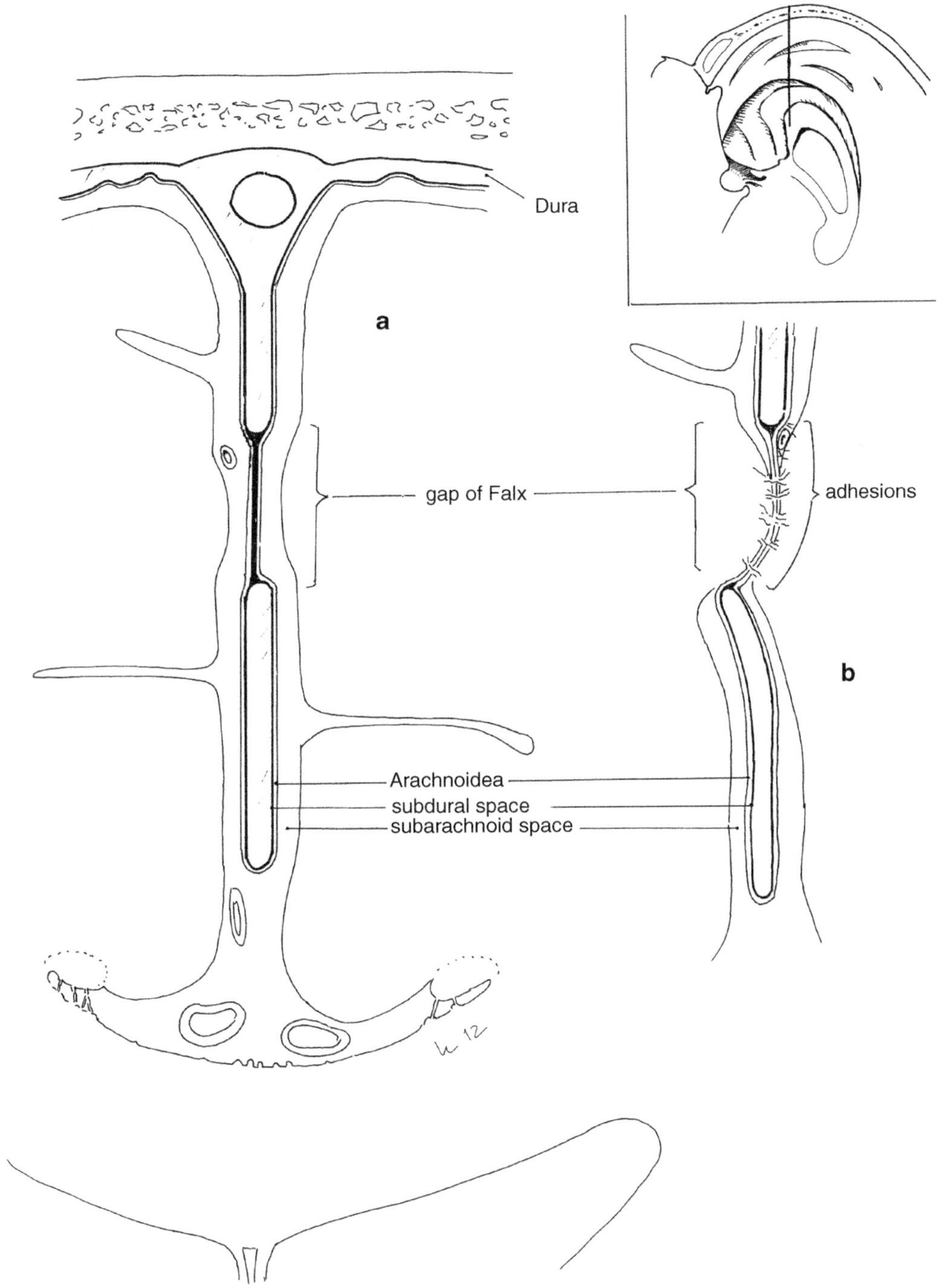

Fig. 3.23 Continuation of Fig. 3.22. Gaps of the frontal segment of falx. **a** Not problematic for subdural routes along falx. **b** Less favorable

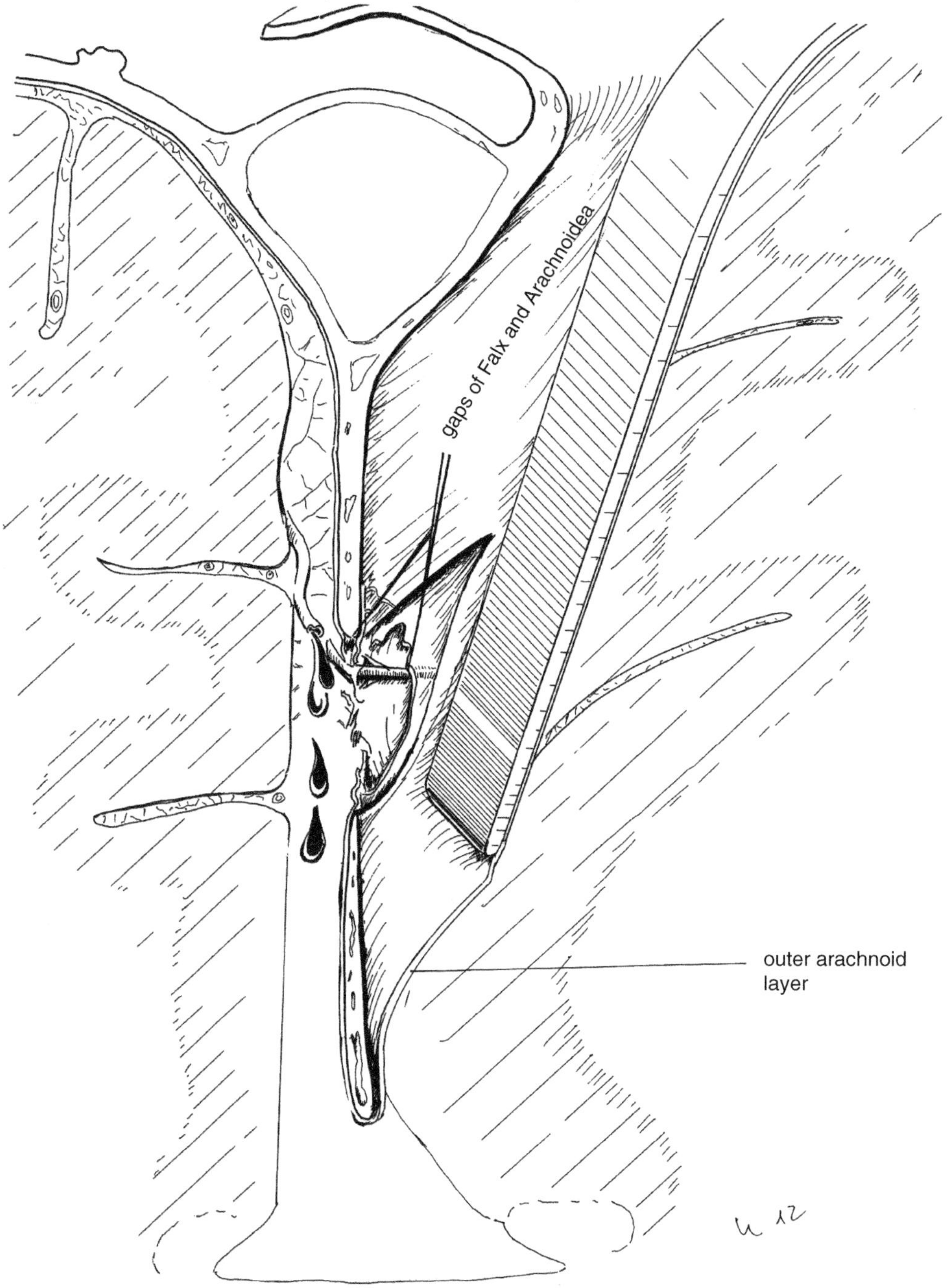

Fig. 3.24 Continuation of Fig. 3.23. Danger for bleeding of homo- or contralateral branches of A2

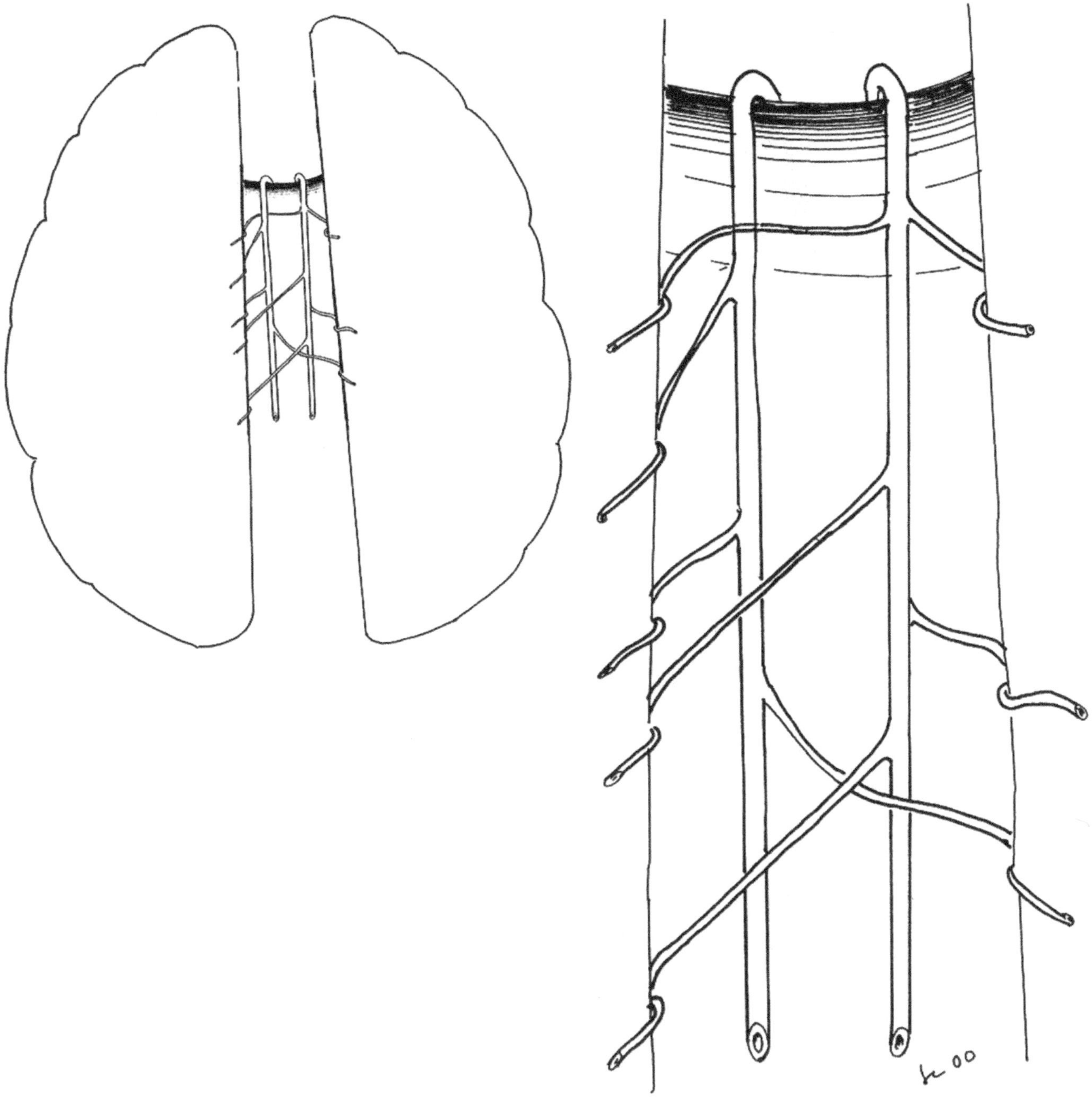

Fig. 3.25 Midline crossing arteries (according to anatomical brain dissections of Marino 1975) could hinder the surgical approach (for surgical topography, see Fig. 3.21). Danger for rupture or bleeding of crossing arteries at surgery

3.2.3 Topographical Aspects of Approaches to the Third Ventricle (Figs. 3.26, 3.27, 3.28, 3.29, 3.30, 3.31, 3.32, 3.33, and 3.34)

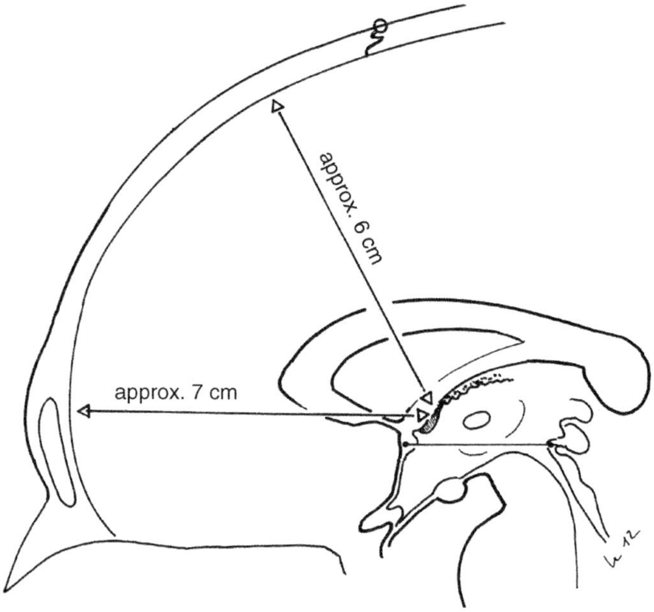

Fig. 3.26 Defining foramen interventriculare. Foramen interventriculare could be masked by plexus choroideus, e.g., now neuronavigation may be insufficient. It could be assisted by distance measurements and by phlebography (see Fig. 3.27)

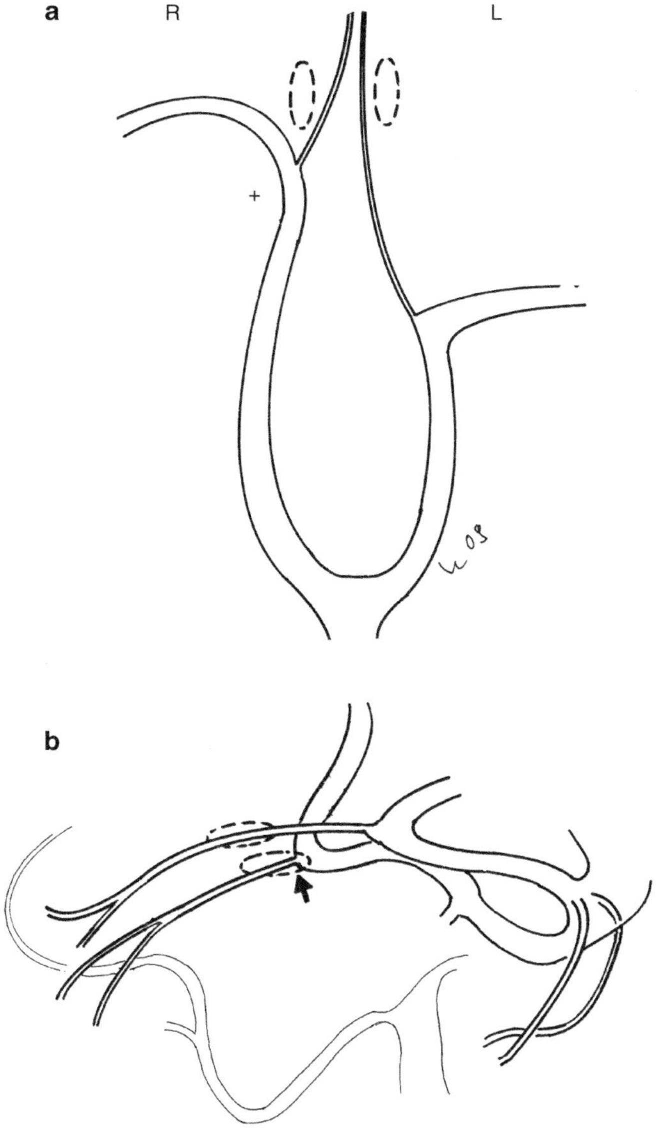

Fig. 3.27 Defining of foramen interventriculare using phlebography. **a** *Right*: usual location of angulus venosus -+-. *Left*: V. ventriculi lateralis directa (common variant). *Dotted*: foramina interventricularia. **b** Lateral projection for defining of the foramen. Here the common venous variant (V. ventriculi lateralis directa) must be taken in consideration, using an axial projection (see **a**). Angulus venosus is located in the posterior wall of foramen interventriculare (*arrow*), surrounded by plexus choroideus

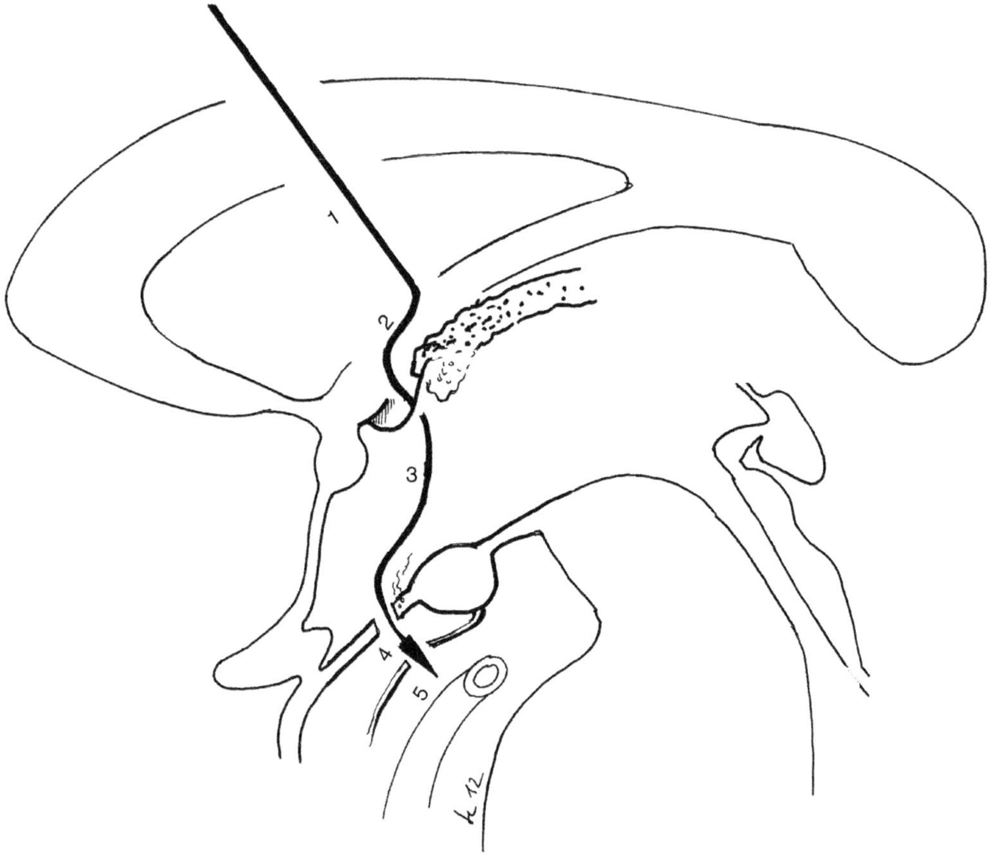

Fig. 3.28 Routes for ventriculocisternostomy. Principles: *1* transcerebral and midline approach, schematized; *2* presentation of plexus choroideus and of foramen interventriculare (Monroi); *3* presentation of the third ventricle, recessus infundibuli, and bulging of corpora mamillaria; *4* incision of the floor of the third ventricle; *5* incision of Liliequist's membrane and identification of A. basilaris or A. cerebri posterior

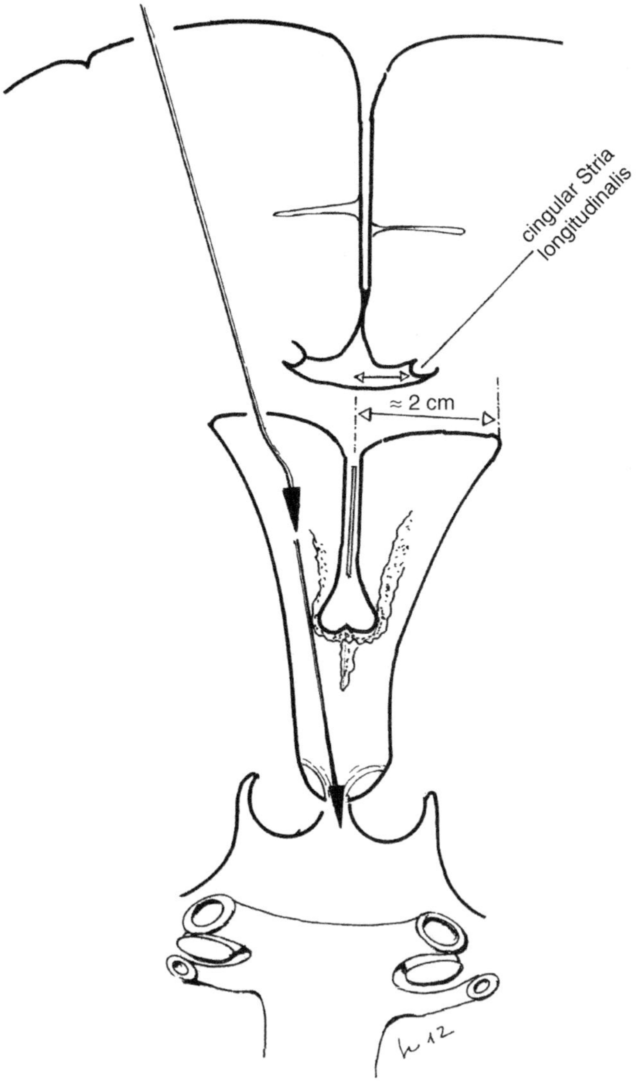

Fig. 3.29 *Avoiding lesions of stria longitudinalis cinguli and columna fornicis at ventriculocisternostomy* using neuronavigation. Lesions of the frontal segment of stria longitudinalis cinguli or lesions of columna fornicis could be followed by psychological deficits. The distance measurement between midline and stria longitudinalis cinguli is approx. 4 mm. Stria longitudinalis could be endangered by midline approaches but less endangered by neuronavigation-guided endoscopic approaches

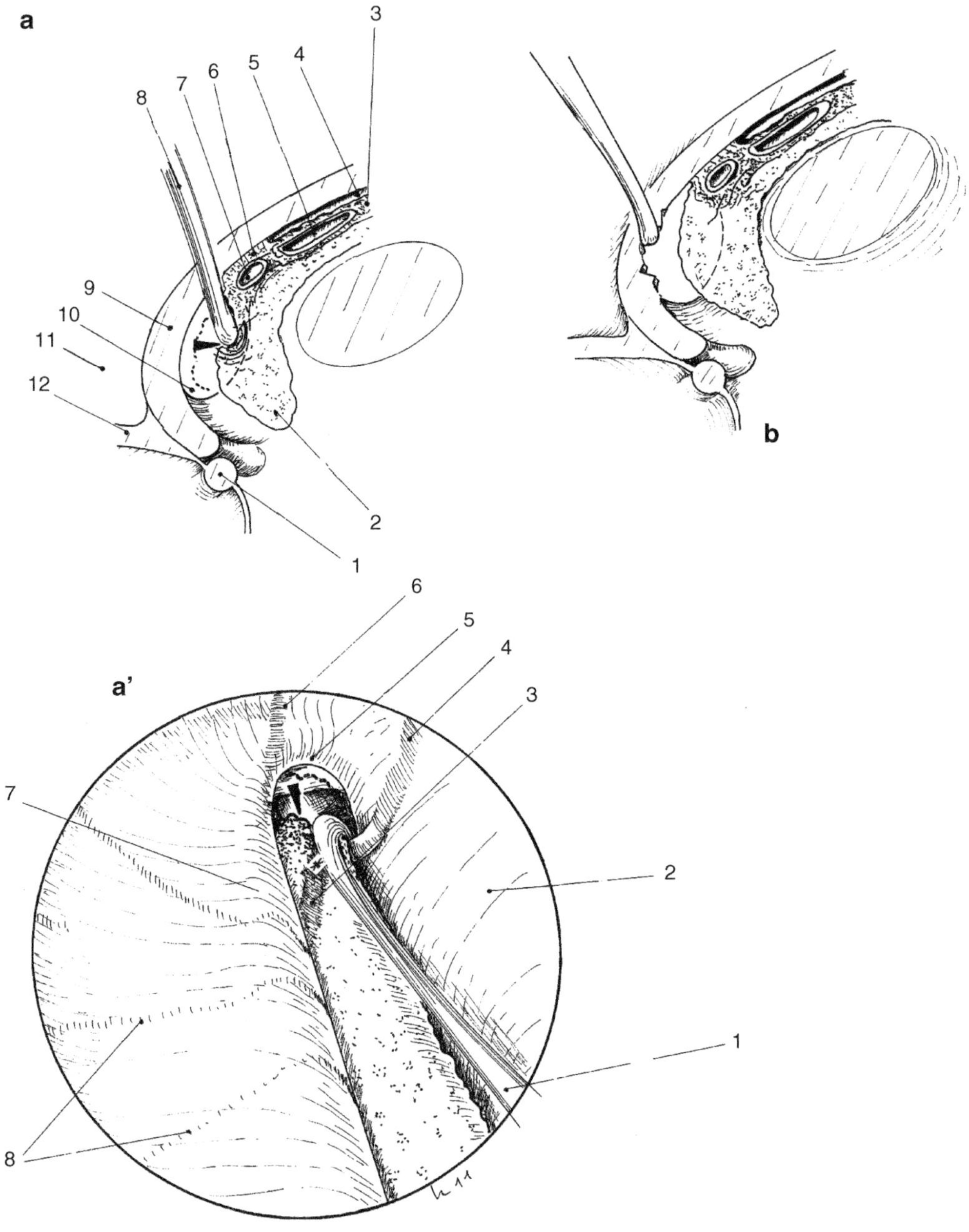

Fig. 3.30 *Avoiding lesions of columna fornicis* at transforaminal ventriculocisternostomy. Plexus shifting, if necessary (**a** and **a'**). Cave: bleeding. Shifting of fornix dangerous (**b**). **a** *1* Commissura anterior, *2* plexus choroideus, *3* velum interpositum, *4* fissura transversa, *5* V. cerebri interna, *6* posterior wall of For. interventriculare (plexus choroideus), *7* V. thalamostriata/V. cerebri interna (angulus venosus), *8* plexus and veins shifted aside (cave bleeding), *9* columna fornicis, *10* margin of For. interventriculare, *11* septum pellucidum, *12* rostrum corporis callosi. **a'** *1* Plexus choroideus shifted aside; *2* lamina affixa thalami; *3* V. cerebri interna; *4* V. thalamostriata; *5* columna fornicis, margin; *6* V. septi pellucidi; *7* columna-corpus fornicis; *8* medial septal veins. **b** Shifting of fornix cave

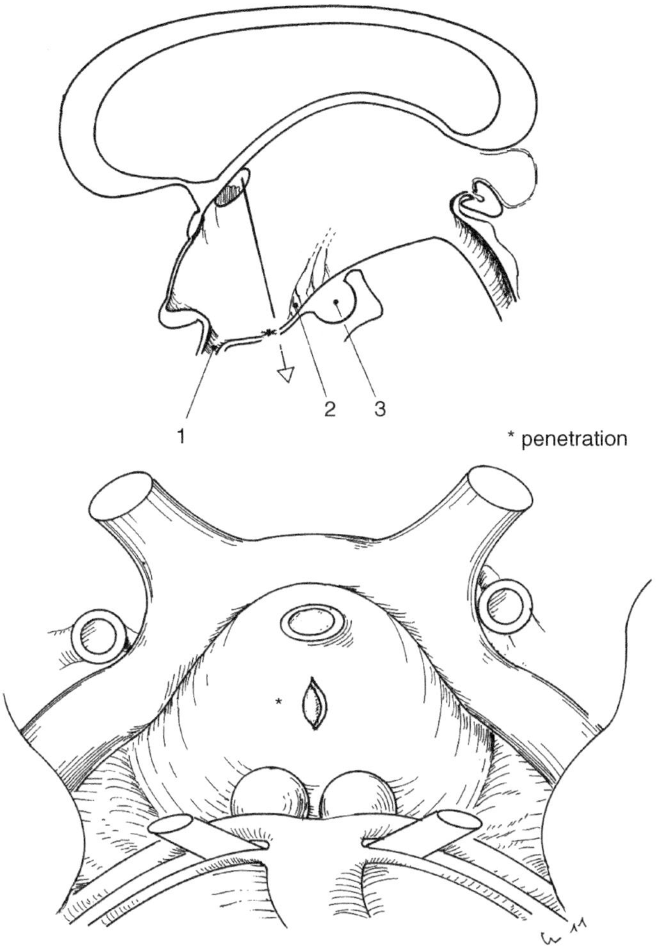

Fig. 3.31 *Avoiding lesions of corpora mamillaria and infundibulum* at ventriculocisternostomy. At hydrocephalus these regions are less endangered by surgery (for ventriculocisternostomy at nearly normal anatomical conditions, see Fig. 3.32). *1* Recessus infundibuli, *2* fine subependymal vessels, *3* corpus mammillare. *Arrow*: route + incision of the ventricular ground (area of the flattened tuber cinereum)

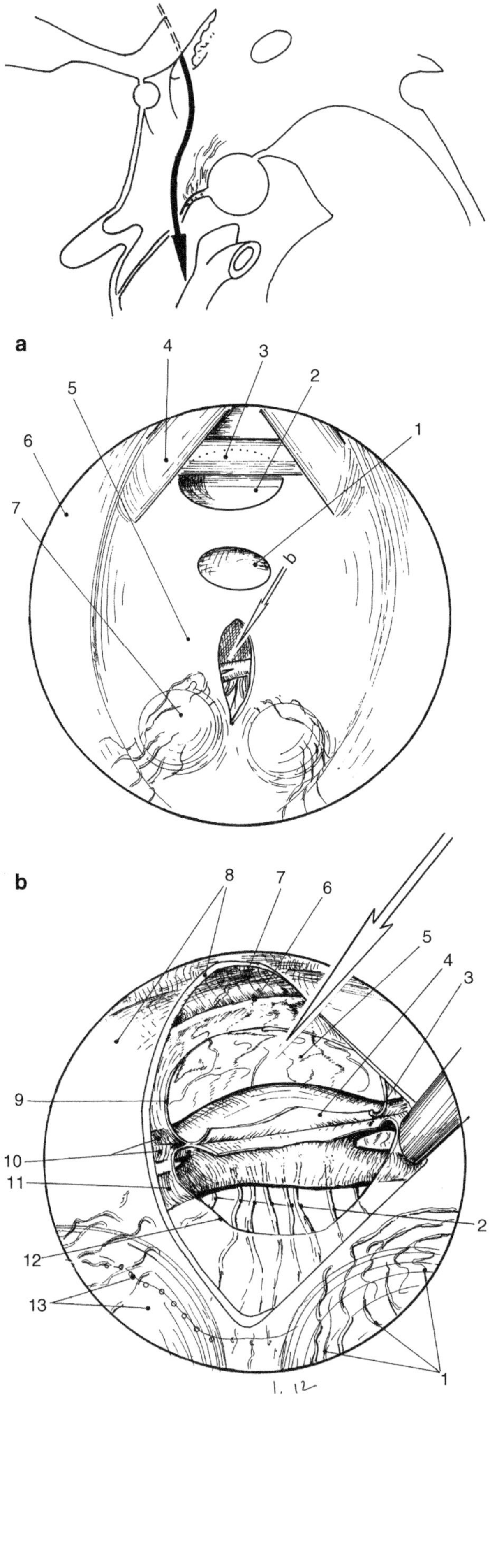
a
1
2
3
4
5
6
7
b
b
1
2
3
4
5
6
7
8
9
10
11
12
13

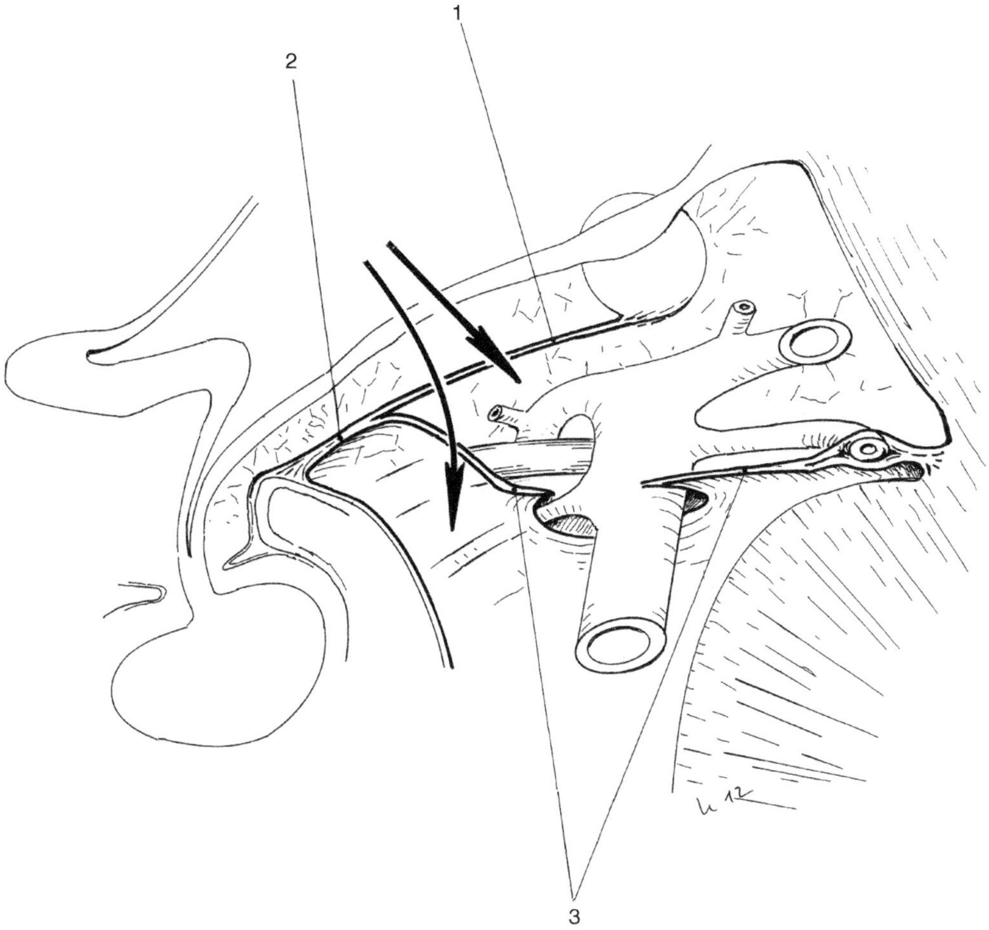

Fig. 3.33 Endoscopic penetration of Liliequist's membrane. *Superior arrow*: usual route. *Inferior arrow*: rare route (finding at endoscopic surgery, performed by Dr. Preuss and PD Dr. Nestler, Gießen, demonstrated to the author 2009). *1* diencephalic part of Liliequist's membrane, *2* common part, *3* mesencephalic part

Fig. 3.32 Ventriculocisternostomy at minimal hydrocephalus, for example (Endoscopic surgery performed by PD Dr. Nestler, Neurochir. Universitätsklinik Gießen. No deficits after surgery. Surgical and postsurgical demonstration to the author 2009), endoscopic topography. **a** Ground of the third ventricle, **b** basal cisterns. **a** *1* Recessus infundibuli, *2* recessus opticus, *3* commissura anterior, *4* columna fornicis, *5* hypothalamus, *6* thalamus, *7* corpus mammillare; **b** *1* fine subependymal vessels; *2* perforating arteries; *3* diencephalic part of Liliequist's membrane (incised); *4* mesencephalic part of Liliequist's membrane (incised); *5* pons; *6* common part of Liliequist's membrane, insertion at dorsum sellae; *7* dorsum sellae; *8* floor of the third ventricle (incised); *9* common part of Liliequist's membrane (incised); *10* A. cerebelli superior; *11* A. cerebri posterior; *12* diencephalic part of Liliequist's membrane (incised); *13* corpus mammillare, bulging the ground of the third ventricle insertion of the diencephalic part of Liliequist's membrane at corpus mammillare. *Black arrow*: surgical route. *Light arrow*: cisterna pontis medialis

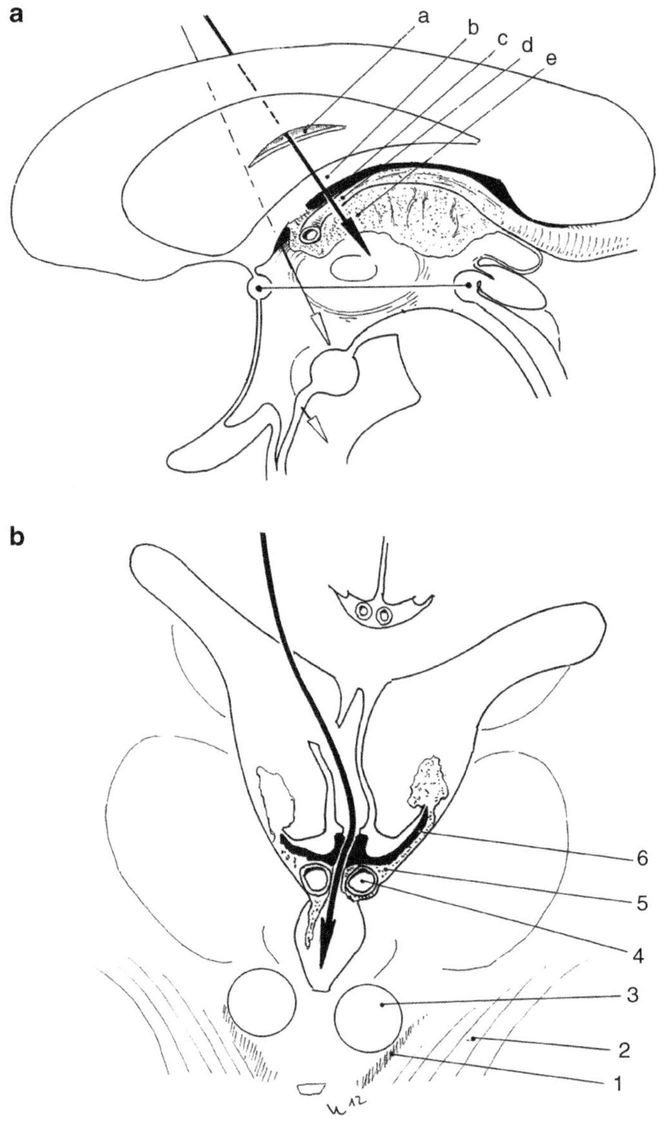

Fig. 3.34 Retroforaminal route and interfornical splitting (Appuzo et al. 1982; Appuzo 1987), principle. **a** Surgical route, sagittal presentation, topography. *Black arrow*, retroforaminal interfornical route; *light arrow*, transforaminal route. **b** Coronal presentation. **a** *a* Incision of the homolateral wall of septum pellucidum, *b* interfornical splitting, *c* crossing fissura transversa and splitting velum interpositum between Vv. cerebri internae and lateral to it, *d* shifting veins and shifting fine chorioid arteries aside, *e* shifting plexus choroideus of the third ventricle aside. **b** *1* Substantia nigra, *2* crus cerebri, *3* nucleus ruber, *4* V. cerebri interna, *5* velum interpositum, *6* fissura transversa

3.2.4 Topographical Aspects of Frontotemporal (Pterional) and Frontobasal Approaches (Figs. 3.35, 3.36, 3.37, 3.38, 3.39, 3.40, 3.41, 3.42, 3.43, 3.44, 3.45, 3.46, 3.47, 3.48, and 3.49)

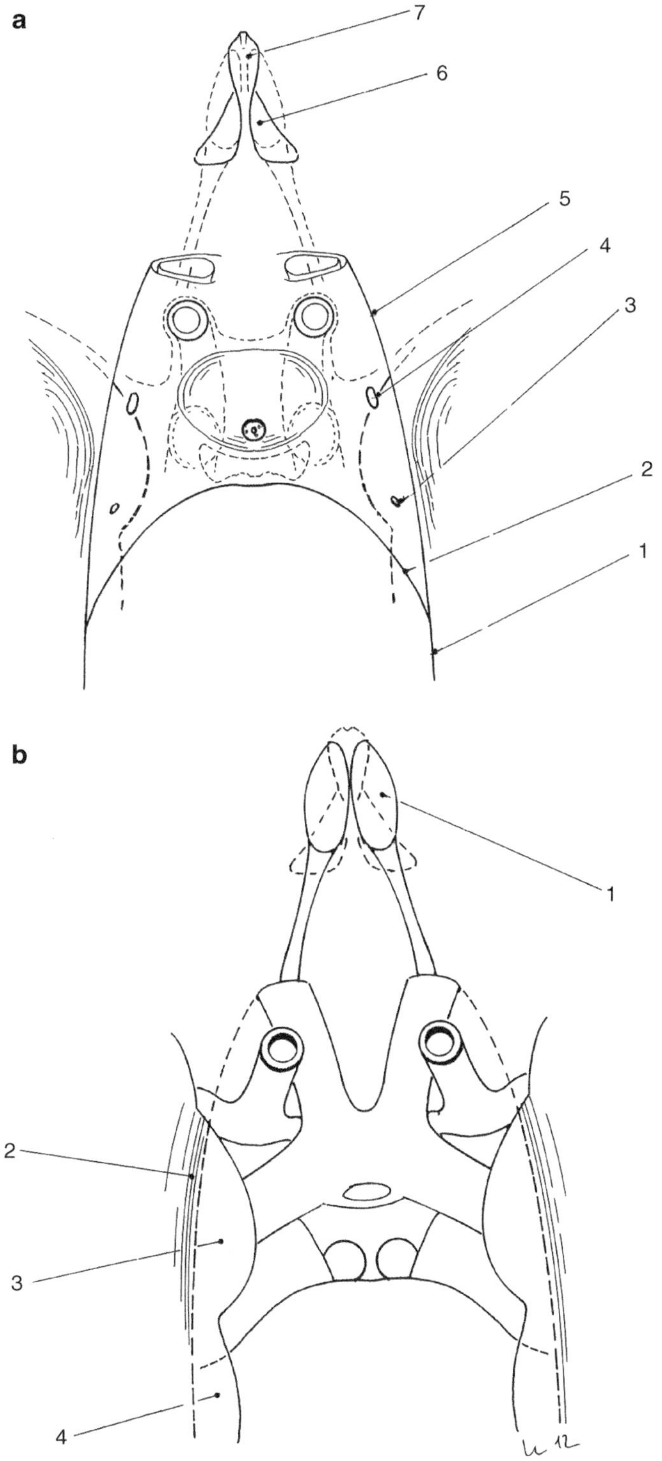

Fig. 3.35 Basal surgical target areas. **a** Ground of basal cisterns. Projections of uncus and N. olfactorius drawn in (*dotted*). **b** Dorsal roof of basal cisterns. Projection of tentorial fold and lamina cribrosa/crista galli drawn in (dotted). **a** *1* Tentorial edge, *2* plica petroclinoidea posterior, *3* dural penetration gap for N. trochlearis, *4* dural penetration for of N. oculomotorius, *5* plica petroclinoidea anterior, *6* lamina cribrosa (for preservation of bulbus olfactorius at surgery, see Figs. 2.11 and 2.42), *7* crista galli. **b** *1* Bulbus olfactorius overlying lamina cribrosa; *2* plica petroclinoidea anterior, projection; *3* uncus, cisternal segment; *4* gyrus parahippocampalis, cisternal segment

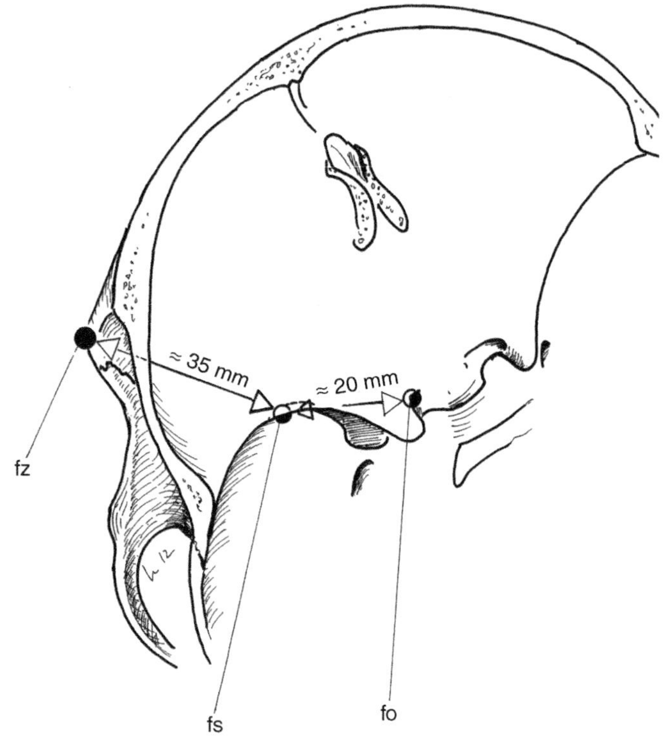

Fig. 3.36 *Defining N. opticus at the foramen opticum during surgery*, if foramen opticum cannot be identified, e.g., at basal tumors. *fo* For. opticum, *fs* lateral point of fissura orbitalis superior, *fz* frontozygomatic point

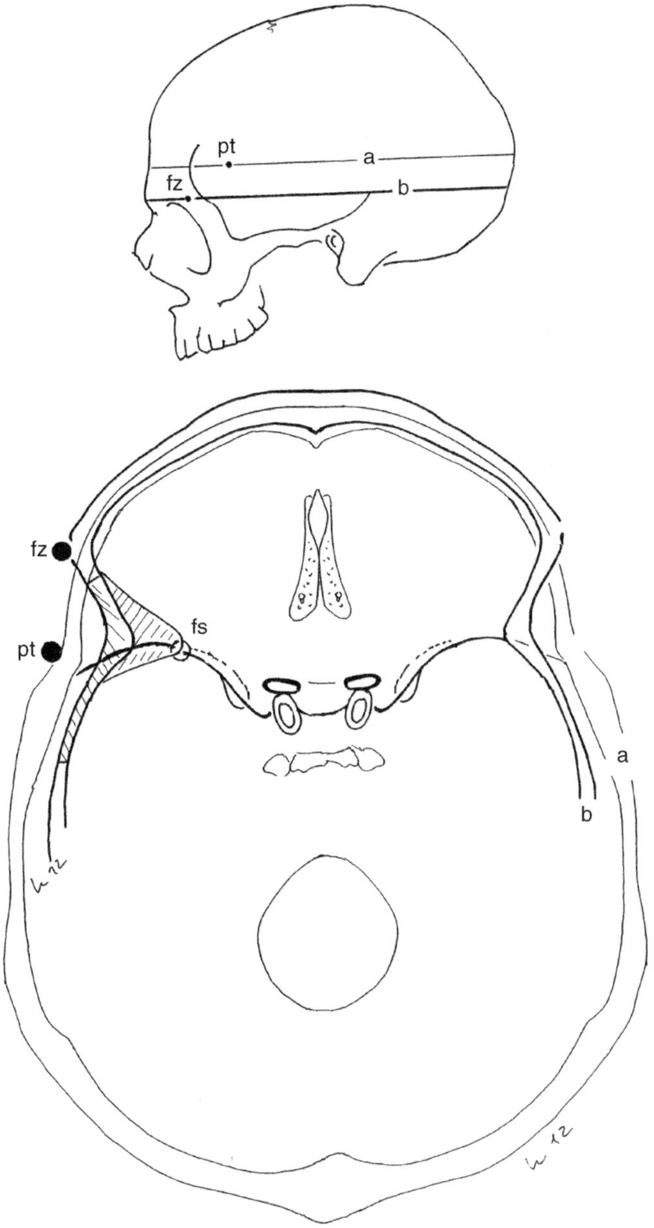

Fig. 3.37 Landmarks for craniotomy. Basal bony resection hatched; *fz* frontozygomatic point, *pt* pterion, *fs* lateral point of fissura orbitalis superior

Fig. 3.38 Pterional craniotomy and surrounding surgical topography. *Sketch dotted*: sylvian region

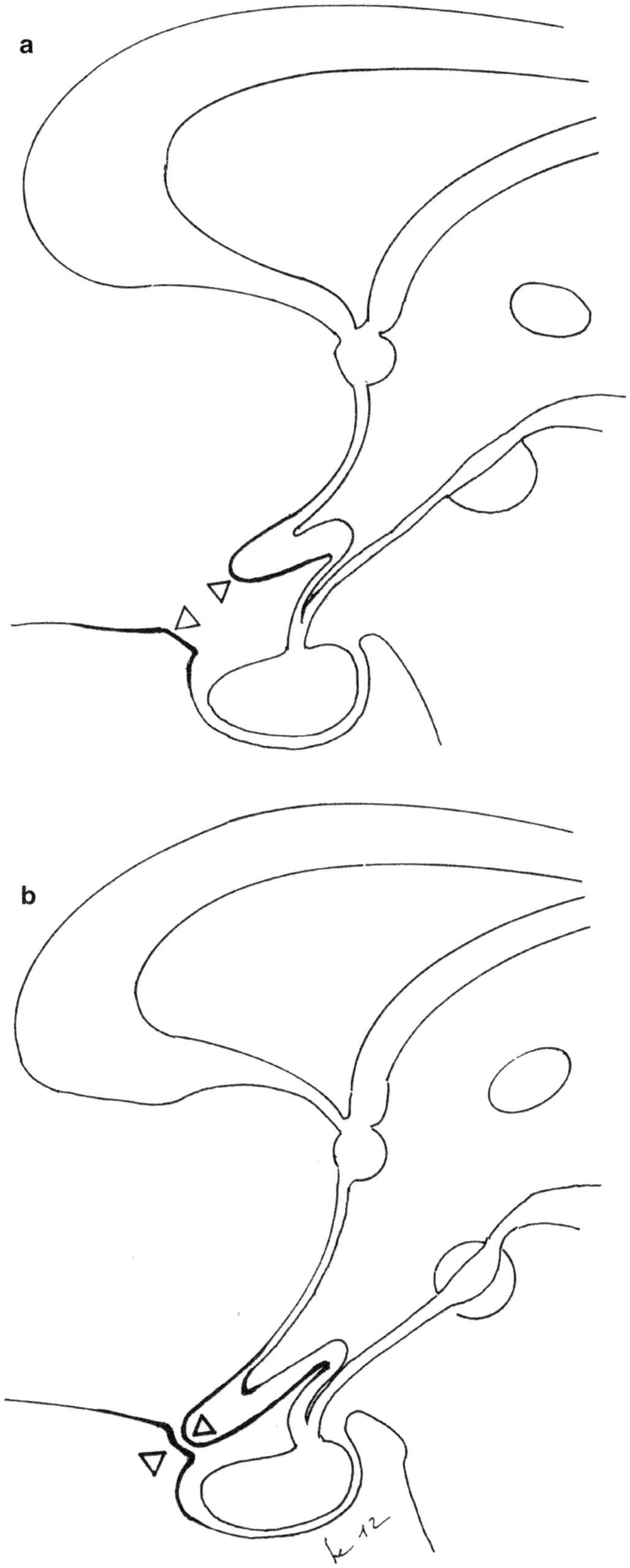

Fig. 3.39 Prefixed chiasma could hinder the interoptic suprasellar approach. Usual finding at childhood (combined with a marked sulcus chiasmatis at tuberculum sellae at childhood), variant at adults. **a** Usual findings at adults, **b** prefixed chiasma (Rhoton)

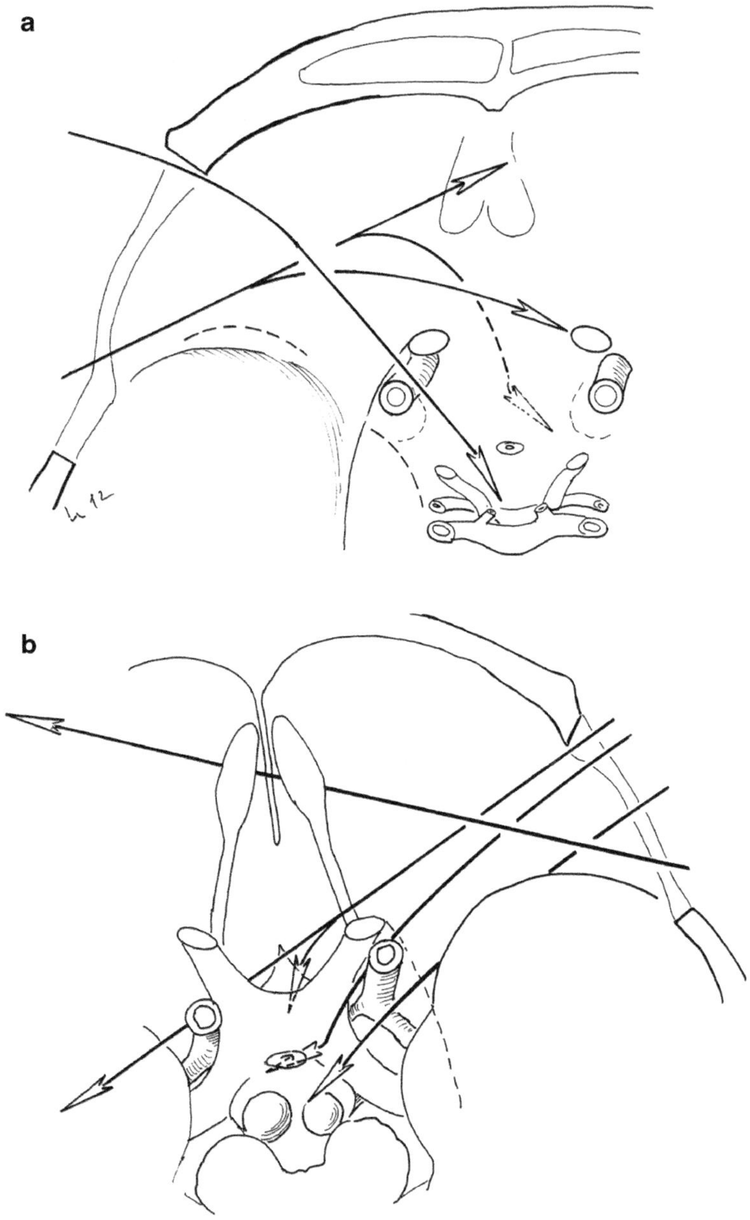

Fig. 3.40 *Pterional routes and target areas* according to Yaşargil (1984a, b) (personal demonstrations to the author since 1969 by Prof. Yaşargil). **a** Basal route and target area. **b** Dorsal route and target area

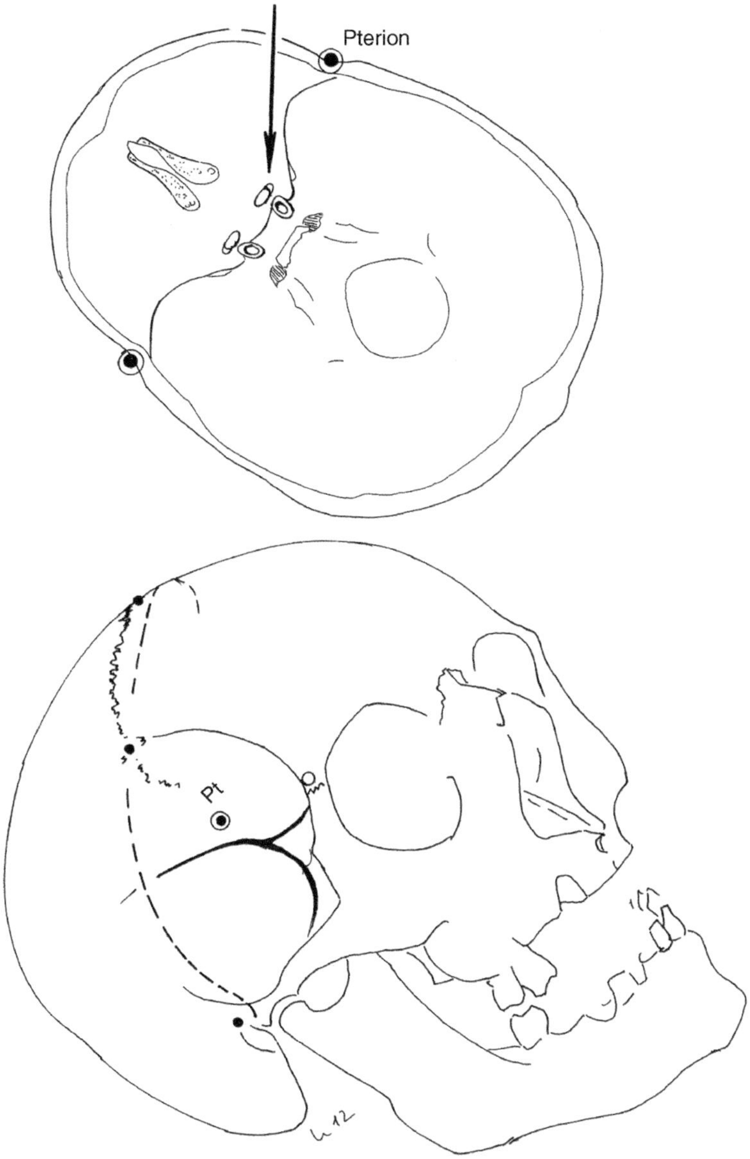

Fig. 3.41 Pterional surgical approach (*arrow*). Survey. Skin-galea incision (*dotted*). Sylvian area added. *pt* Pterion

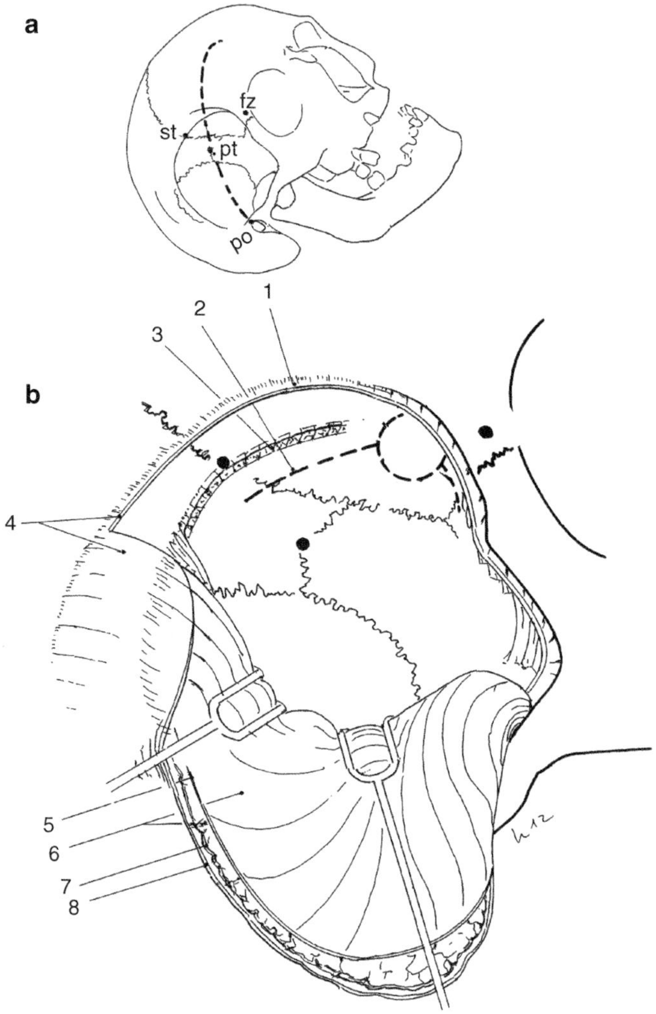

Fig. 3.42 Pterional extracranial surgical topography. **a** Skin incision (*dotted*). **b** Skin-galea-muscle-flap, burr hole at the beginning of craniotomy (projection). **a** *fz* frontozygomatic point, *po* porion, *pt* pterion, *st* stephanion. **b** *1* Linea temporalis superior, insertion of fascia temporalis; *2* craniotomy, projection; *3* linea temporalis inferior, insertion of M. temporalis; *4* fascia temporalis and M. temporalis cut and shifted; *5* periosteum; *6* M. temporalis; *7* aponeurosis temporalis and M. temporalis cut and shifted; *8* as 4

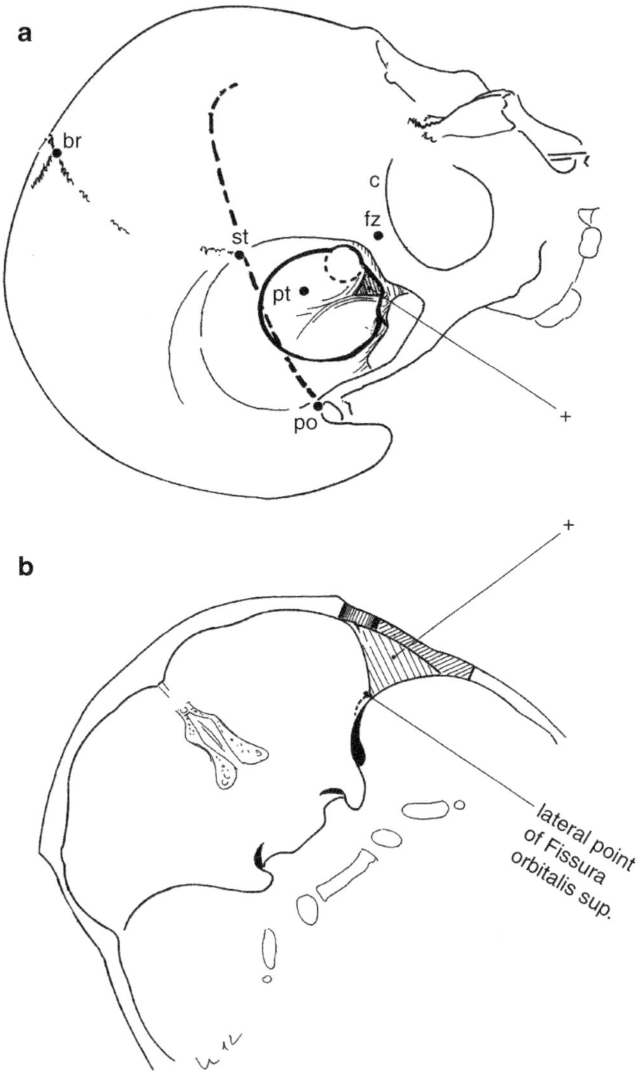

Fig. 3.43 Pterional craniotomy. **a** Lateral segment of craniotomy renders prominent (1st step); basal segment segment -+- hatched (second step). Dotted: skin incision. **b** Basal segment -+- as **a** *br* bregma, *fz* frontozygomatic point, *po* porion, *pt* pterion, *st* stephanion

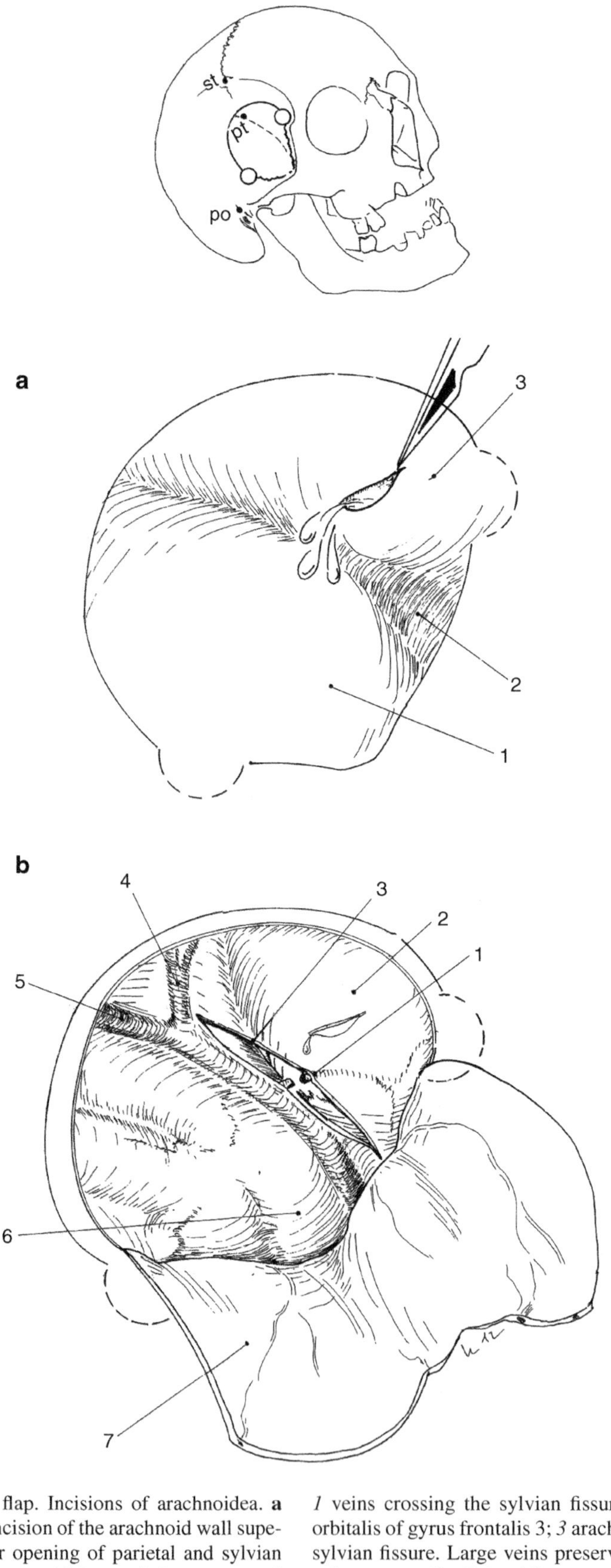

Fig. 3.44 After performing a dura flap. Incisions of arachnoidea. **a** Incision for CSF-decompression. **b** Incision of the arachnoid wall superior to V. temporalis superficialis for opening of parietal and sylvian CSF spaces. **a** *1* Outer temporal arachnoid layer, *2* frontotemporal thickening of the outer arachnoid layer, *3* frontal arachnoidea incised; **b** *1* veins crossing the sylvian fissure. Small veins interrupted; *2* pars orbitalis of gyrus frontalis 3; *3* arachnoidea incised; *4* veins crossing the sylvian fissure. Large veins preserved; *5* V. temporalis superficialis; *6* polus temporalis; *7* dural flap

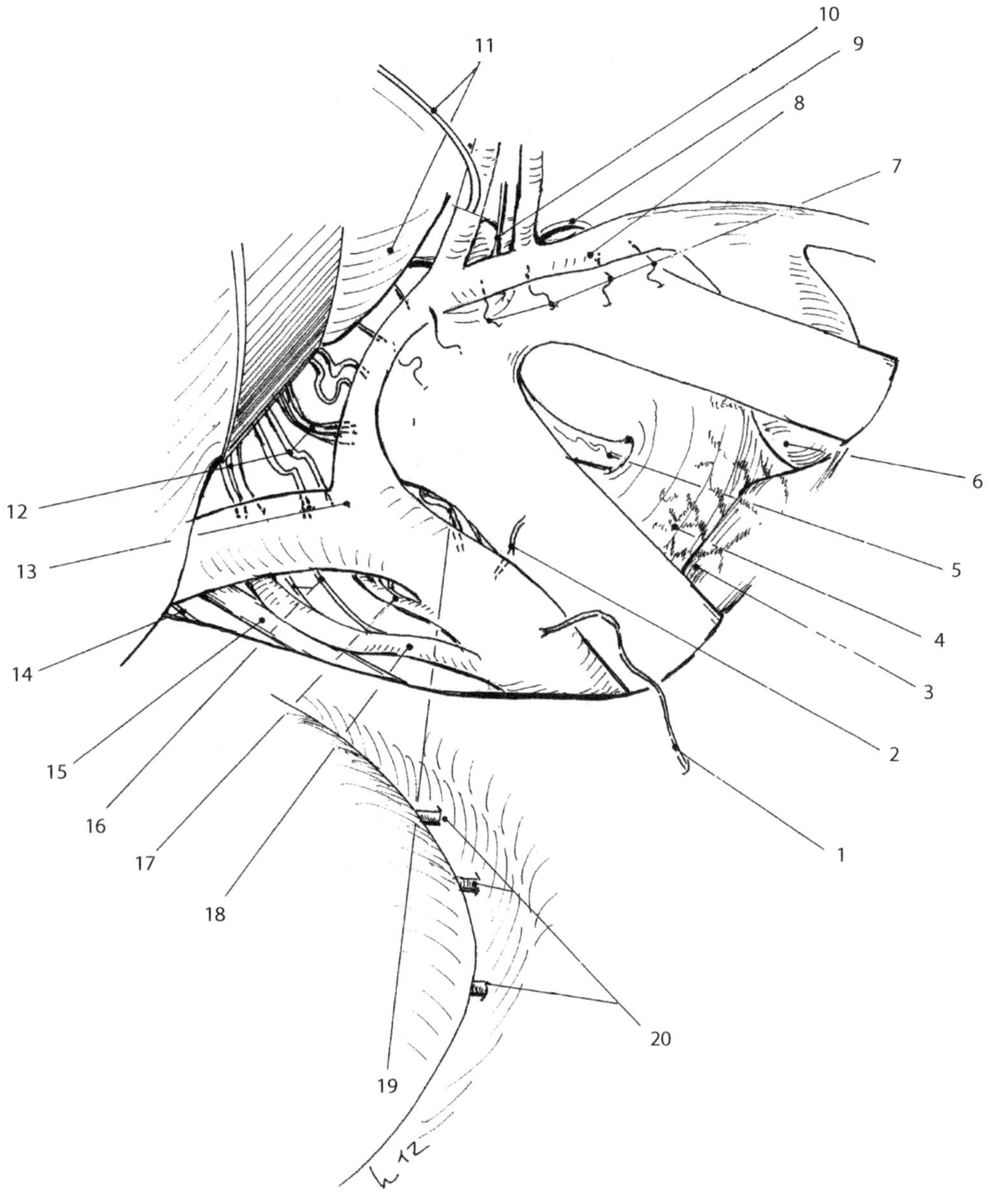

Fig. 3.45 After opening of the sylvian cistern. Main target area of the pterional microsurgical approach: *1* dural branch of A. carotis interna(Yaşargil), *2* branch for N. opticus, *3* tuberculum sellae, *4* diaphragma sellae enclosing sinus intercavernosus anterior, *5* infundibulum, *6* A. carotis interna, *7* perforating arteries, *8* A1, *9* Heubner's artery, *10* A. corporis callosi mediana, *11* mediobasal frontal artery at gyrus rectus, *12* perforating arteries of substantia perforata anterior, *13* carotid bifurcation, *14* N. trochlearis, *15* N. oculomotorius, *16* perforating arteries, *17* A. chorioidea anterior, *18* A. communicans posterior, *19* A. hypophysealis superior, *20* anterior temporal bridging veins

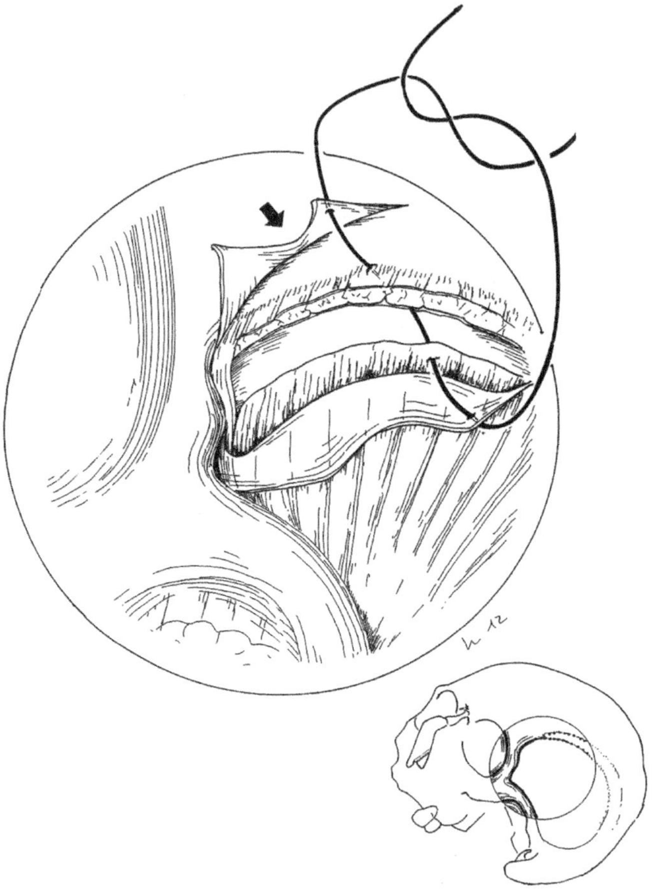

Fig. 3.46 Wound closure. Suturing of fascia and M. temporalis (applied by Prof. Zentner, Freiburg) avoiding hyperextension and rupture of these layers. *Arrow*: incision of fascia temporalis

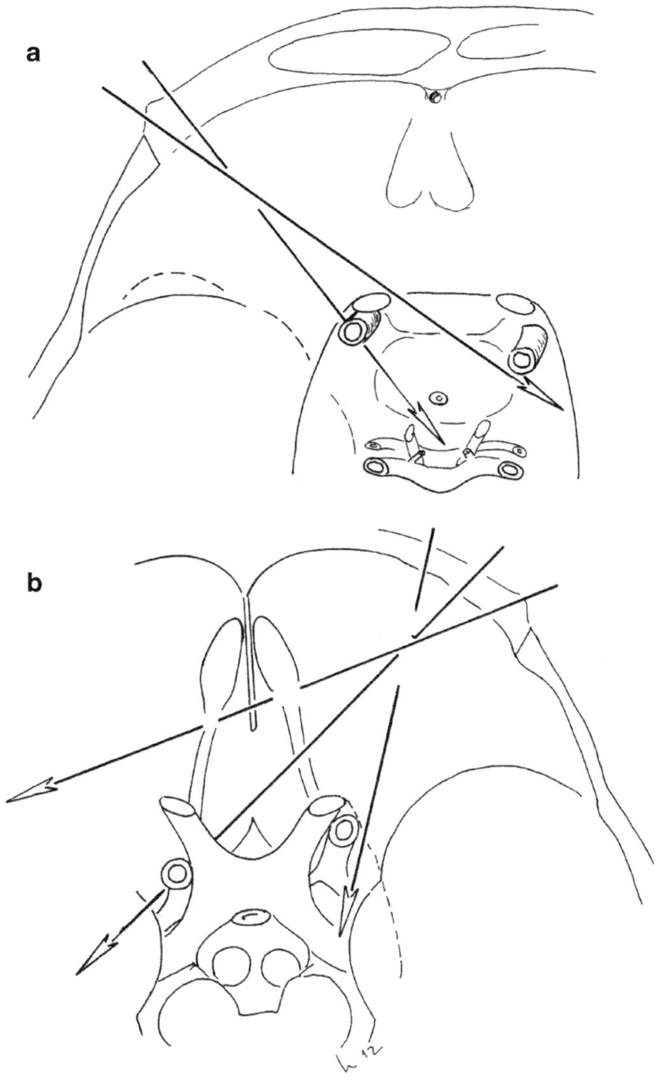

Fig. 3.47 Lateral frontobasal surgical routes and target areas. Opening of the sylvian cistern according to Dandy (1938) (modified by Samii for microsurgery, see Ramina R et al. 2008, pp. 92 ff), modified by Samii. **a** Basal route and target area. **b** Dorsal route and target area

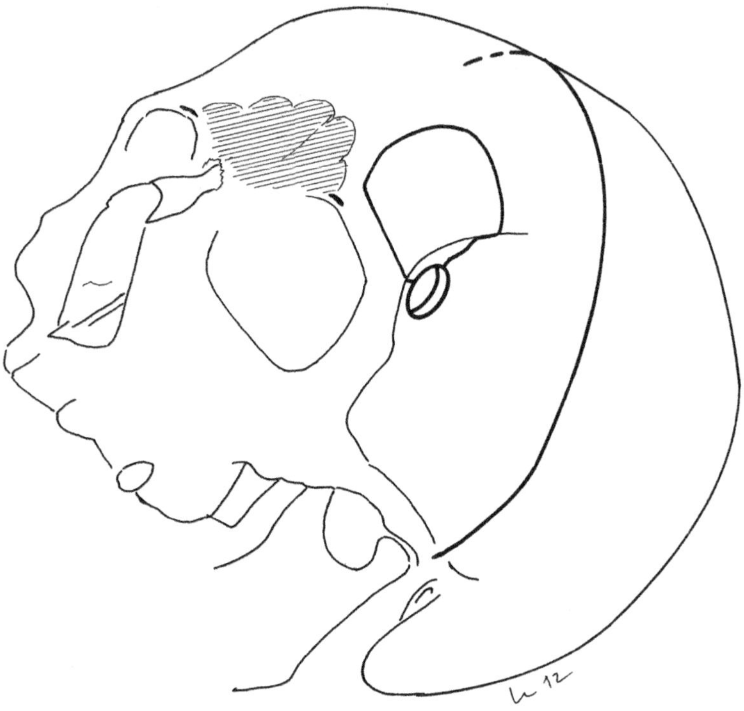

Fig. 3.48 Craniotomy according to Samii; skin incision mapped

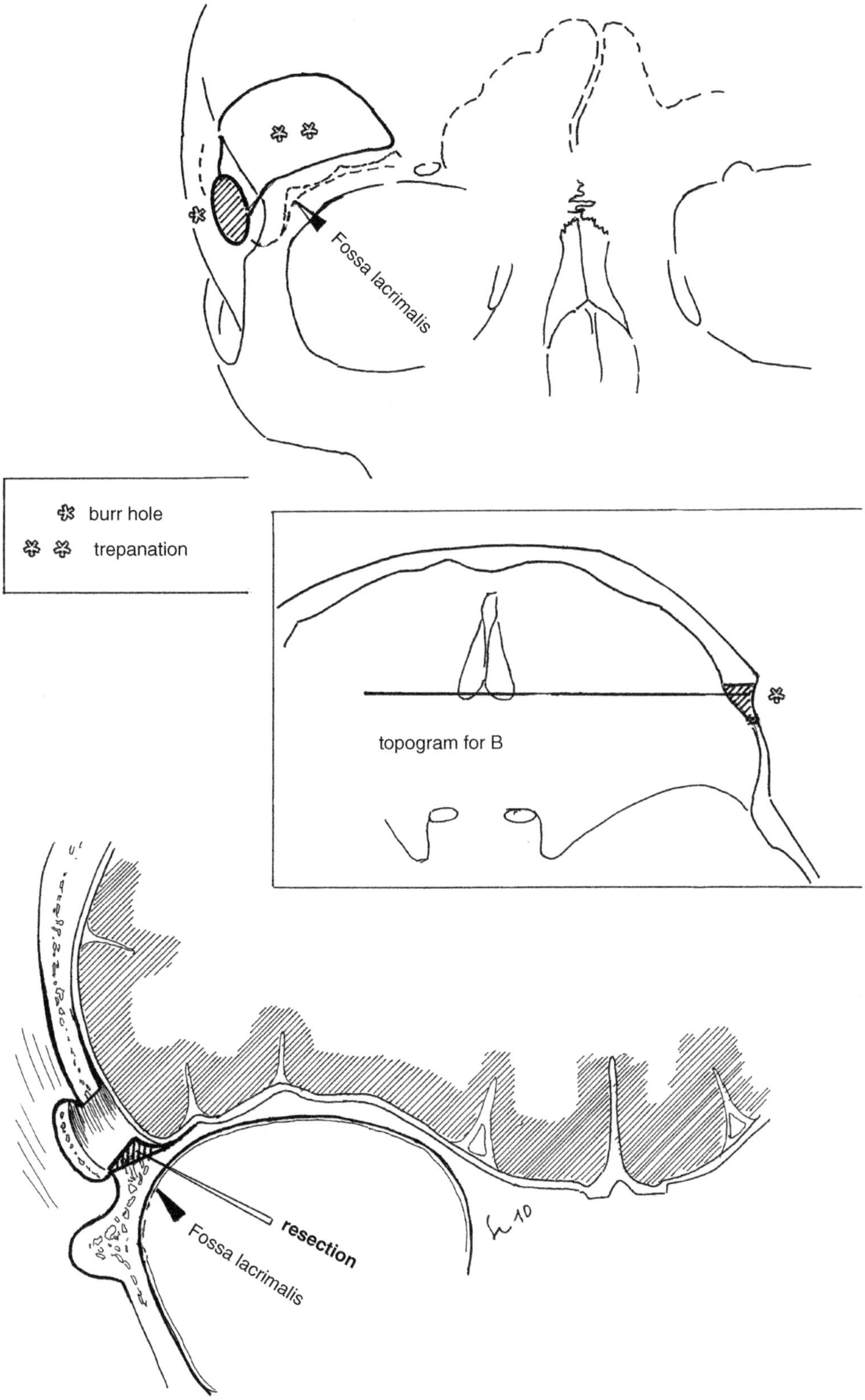

Fig. 3.49 As Fig. 3.48, details. Danger for opening of fossa lacrimalis and periorbita (According to Pia HW, 1957, chief of the author 1957–1975)

4 Completion for Chapters 1 to 3: Topographical Areas of Rhinoliquorrhea

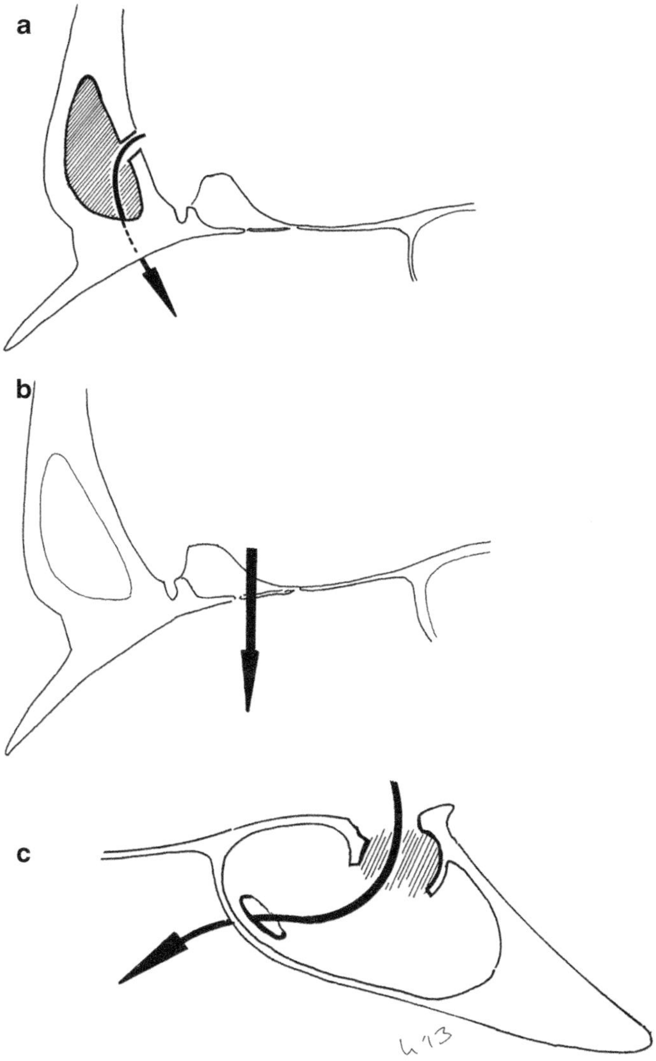

Fig. 4.1 Rhinoliquorrhea by craniobasal defects (*arrows*). **a** Defect of the wall of sinus frontalis and adjacent layers. Note: ductus nasofrontalis could be duplicated at both sides. This should be considered at surgical repair of CSF fistulas. **b** Defect of lamina cribrosa and adjacent layers. **c** After traumatic defects, e.g., and after transnasal surgery at the region of sella

W. Seeger, J. Zentner, *Anatomical Basis of Cranial Neurosurgery*, https://doi.org/10.1007/978-3-319-63597-2_4

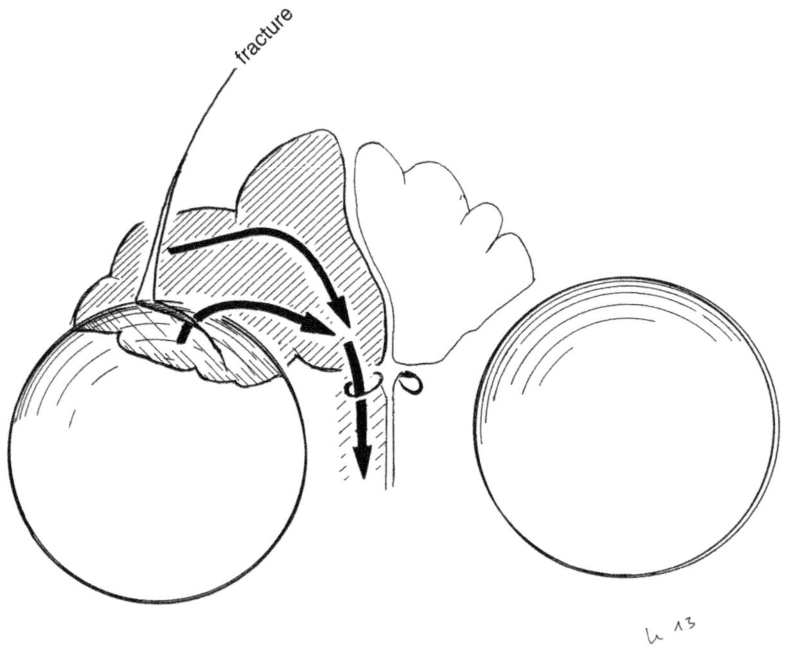

Fig. 4.2 Defect of the orbital wall of sinus frontalis after trauma. *Arrows*: rhinoliquorrhea. * common widening of sinus frontalis

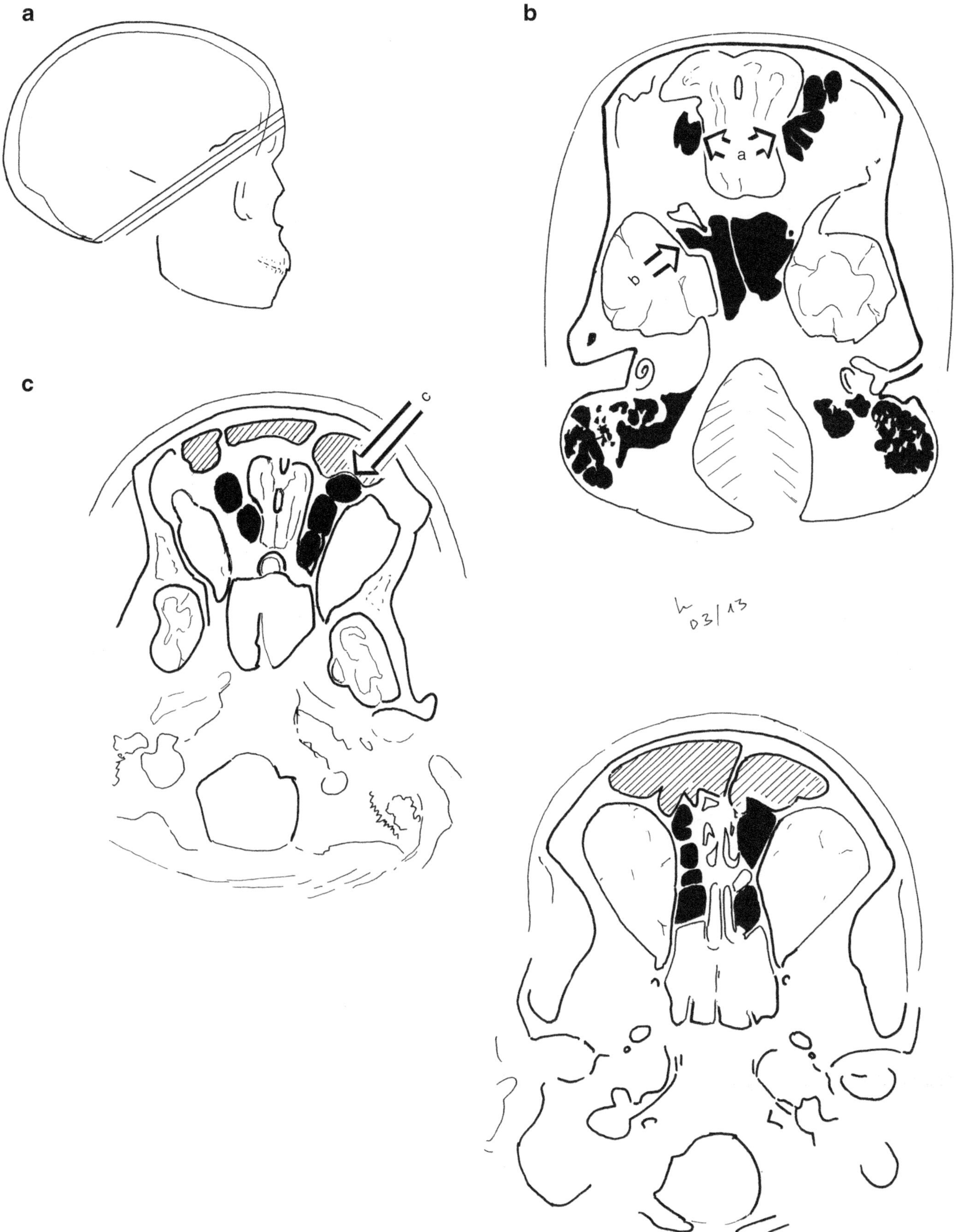

Fig. 4.3 *Widenings of paranasal sinuses* may suggest rhinoliquorrhea after trauma. An example of rare widenings of paranasal sinuses was found at the computer tomogram of a normal child after a cranial trauma. CSF fistula had to be excluded. Typical and atypical hypothetic routes of rhinoliquorrhea had to be taken in consideration: **a** widenings and protruding of anterior ethmoid cells into the orbital roof, **b** widening and protruding of sinus sphenoidalis into processus clinoideus anterior, **c** as a

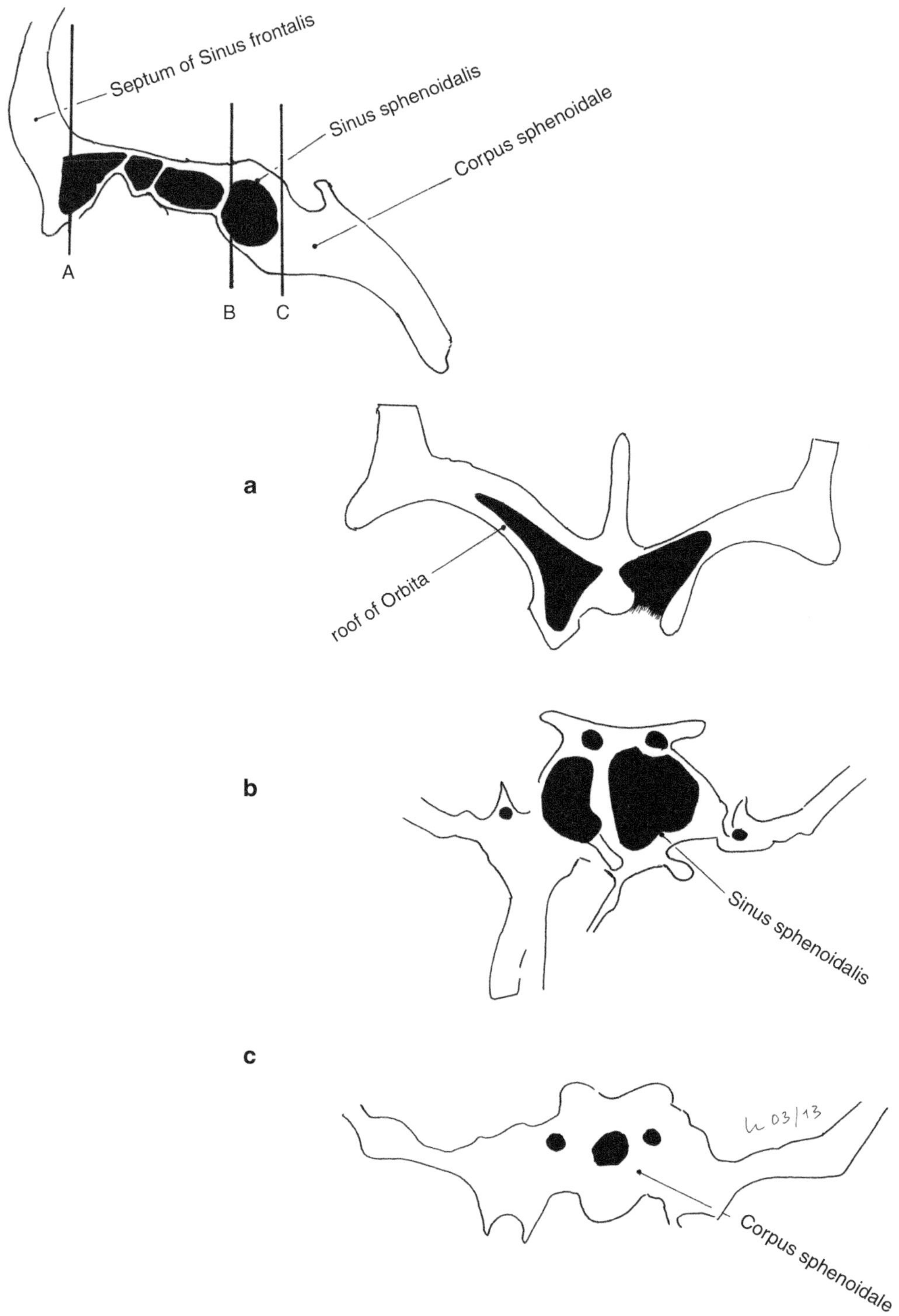

Fig. 4.4 Further example of atypical paranasal sinuses. Sinus frontalis widened, sinus sphenoidalis narrow. CT of a cadaver skull dissection. **a** Sinus frontalis protruded into the roof of orbita. Common variant **b.** Common finding. **c** The posterior region of corpus sphenoidale is not pneumatized. Usual finding at childhood (pneumatization of sinus sphenoidalis begins usually at the 4th year), rare at adults

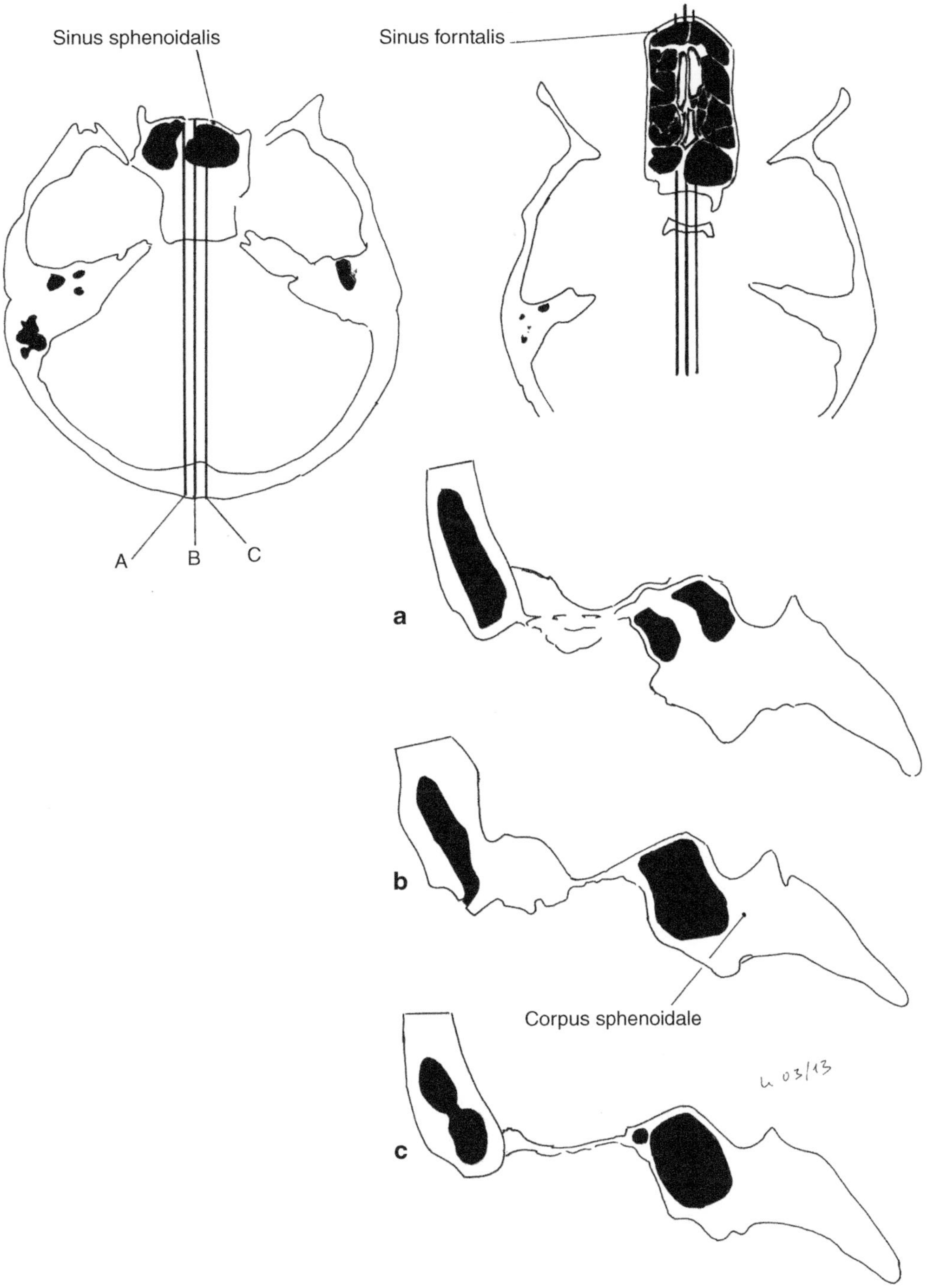

Fig. 4.5 Continuation of Fig. 4.4

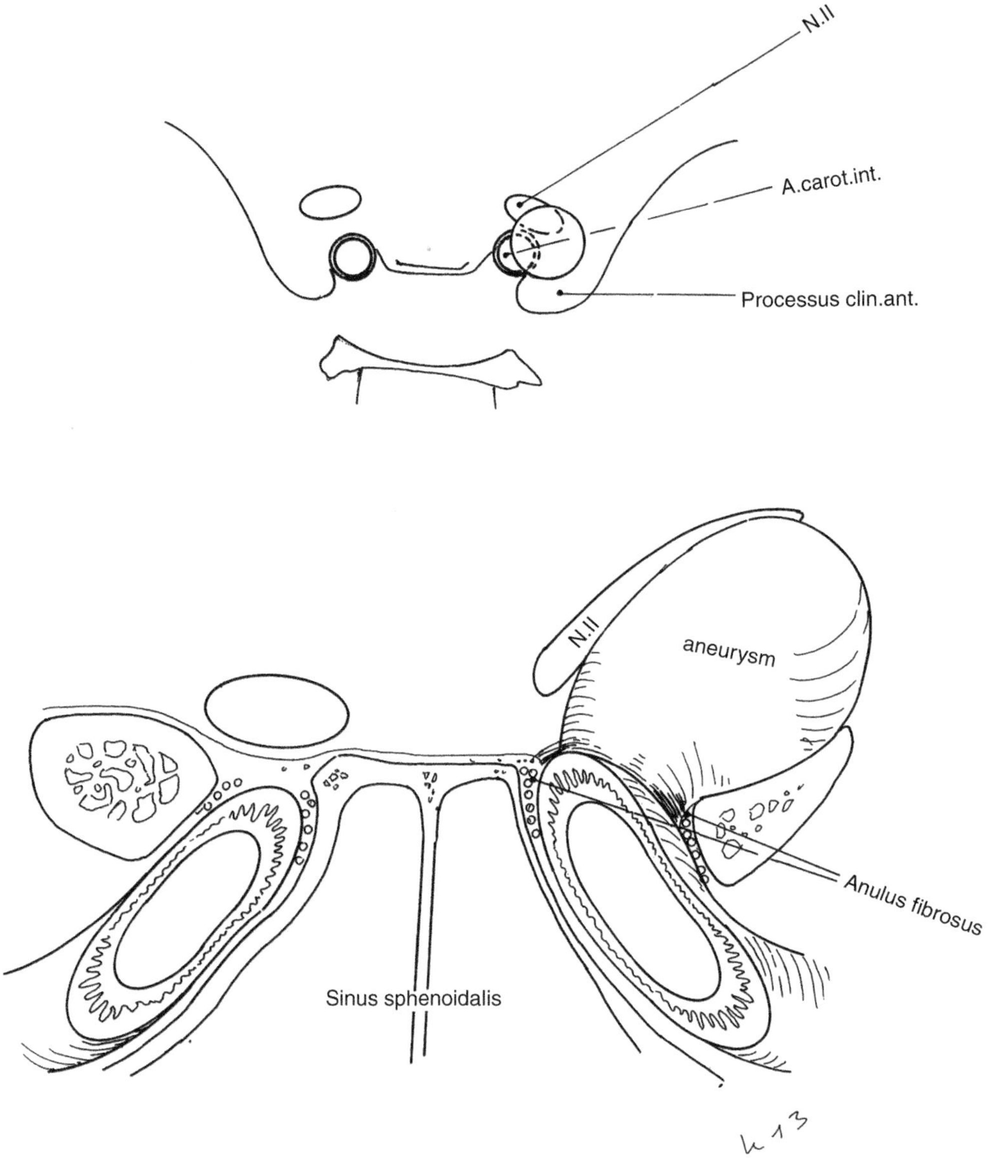

Fig. 4.6 Danger for rhinoliquorrhea after resection of processus clinoideus anterior. Aneurysm at curvatura anterior of the carotid siphon. Findings before surgery

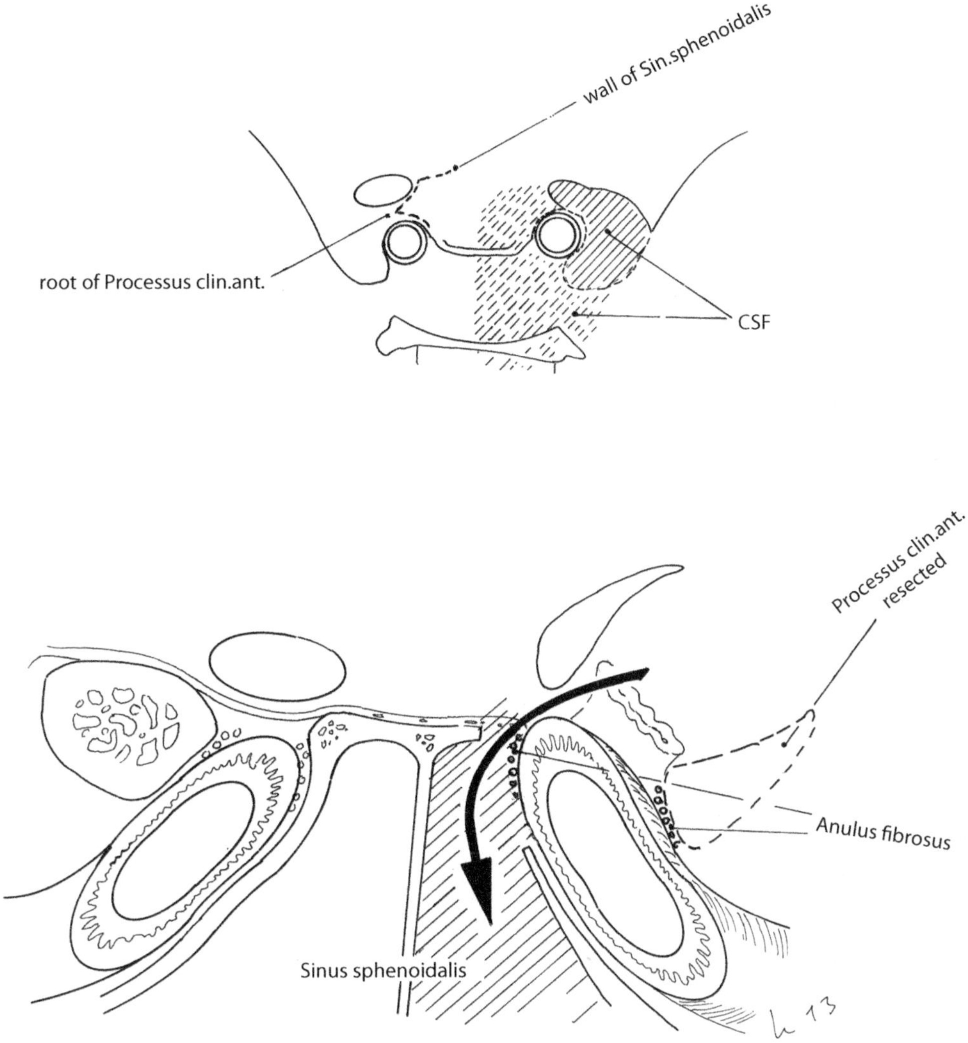

Fig. 4.7 After resection of processus clinoideus anterior: aneurysm eliminated (*upper illustration*). Liquorrhoea transpassing sinus sphenoidalis may happen (*arrow*), if the bony defect is not sufficiently closed at surgery (For an ethmoidal route of rhinoliquorrhea at pneumatization of processus clinoideus ant, see Fig. 4.9)

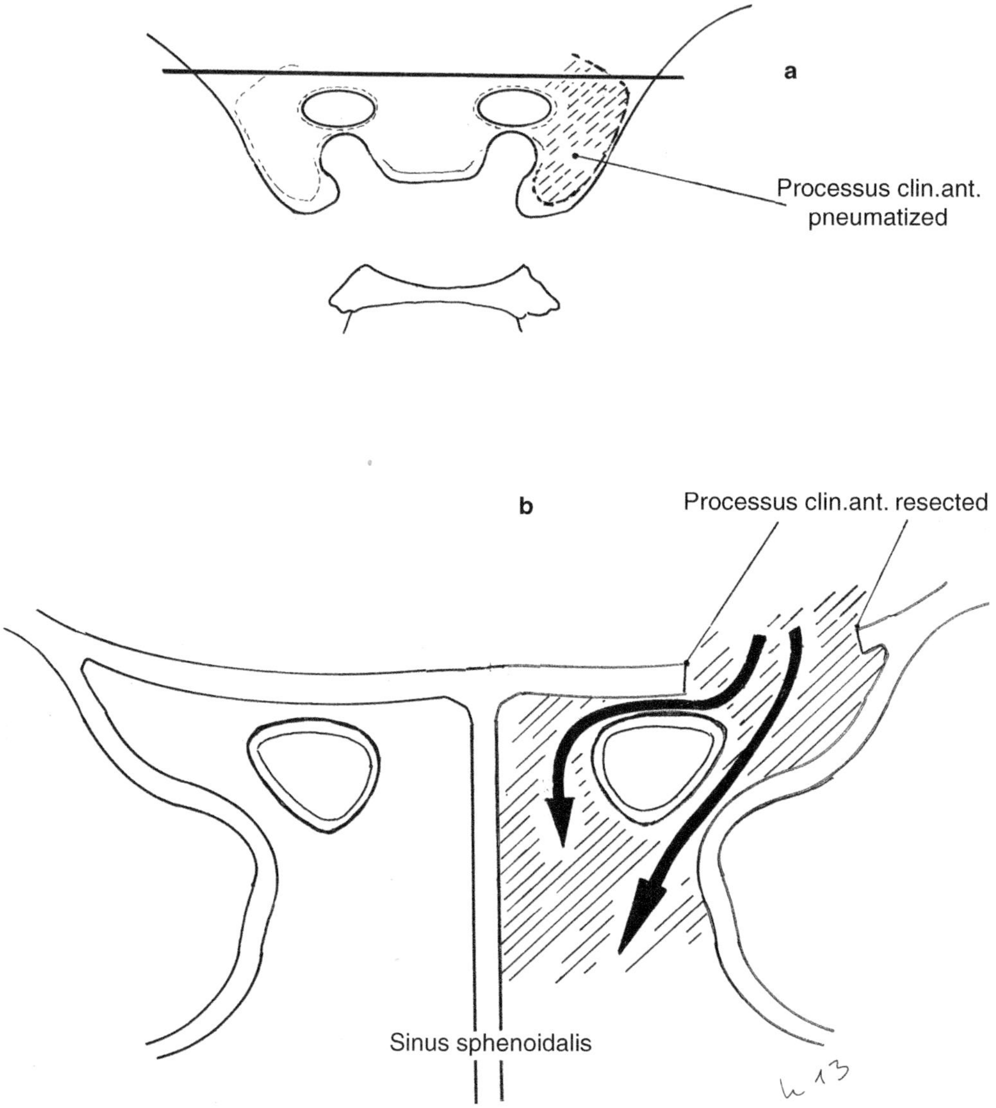

Fig. 4.8 **a** Pneumatization of processus clinoideus anterior. All paranasal sinuses are extremely widened. Canalis opticus surrounded by sinus sphenoidalis and a pneumatized processus clinoideus anterior. Cadaver skull dissection. **b** Hypothetical routes of rhinoliquorrhea superior and inferior from canalis n. optici after trauma or surgery, e.g., (*arrows*)

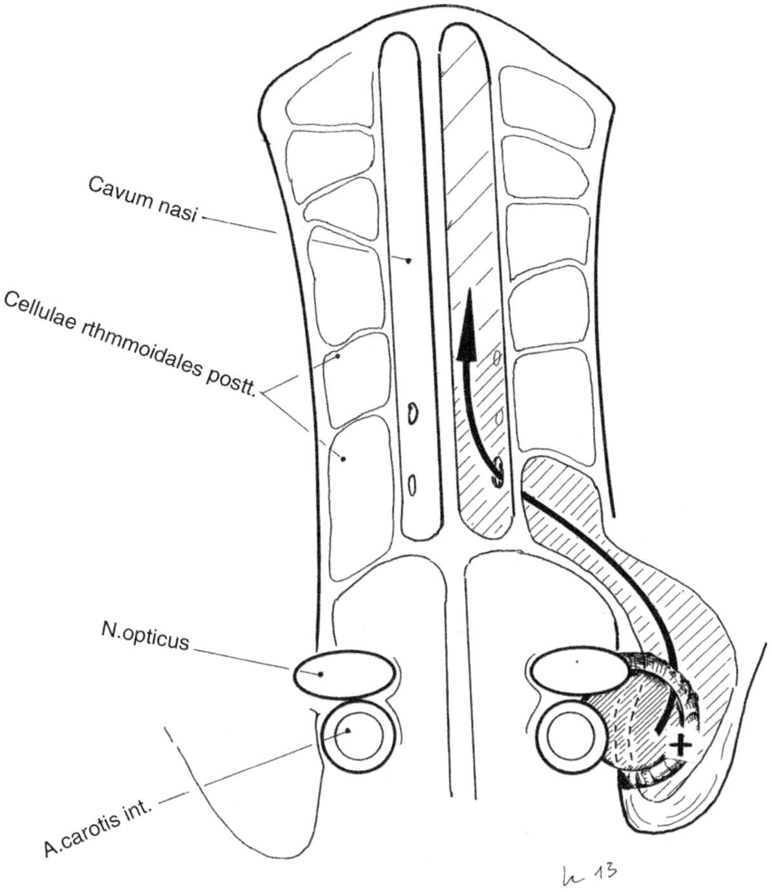

Fig. 4.9 Pneumatization of Processus clinoideus anterior. Processus clinoideus anterior pneumatized by cellulae ethmoidales posteriores. Anatomical dissection of a rare variant (Lang 1981). See Fig. 4.3. Hypothetical defect after surgery -+-. Rhinoliquorrhea (*arrow*) could transpass cellulae ethmoidales posteriores or sinus sphenoidalis or both

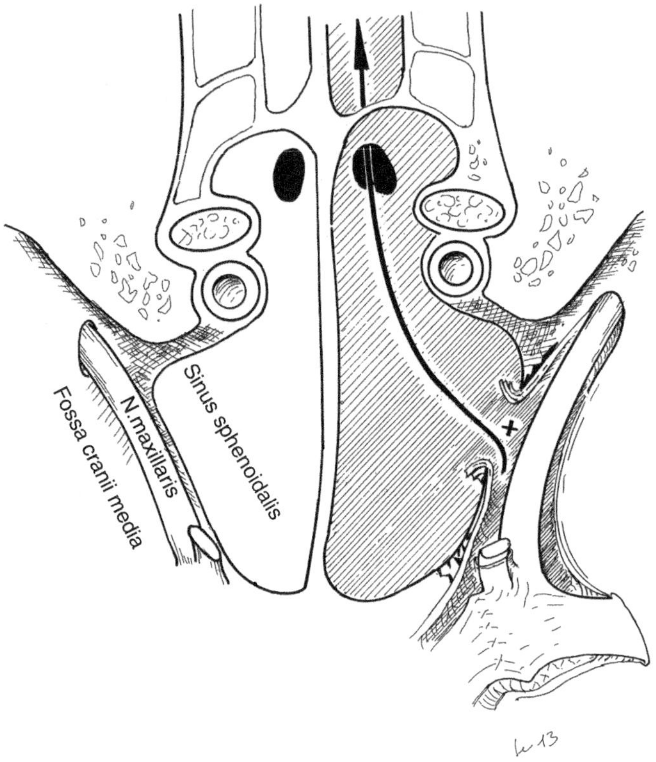

Fig. 4.10 Rhinoliquorrhea (*arrow*) by a posttraumatic defect of the wall of sinus sphenoidalis and of canalis rotundus. Widening of the paranasal sinuses with rhinoliquorrhea (three experiences of the author) + defect of the wall of sinus sphenoidalis

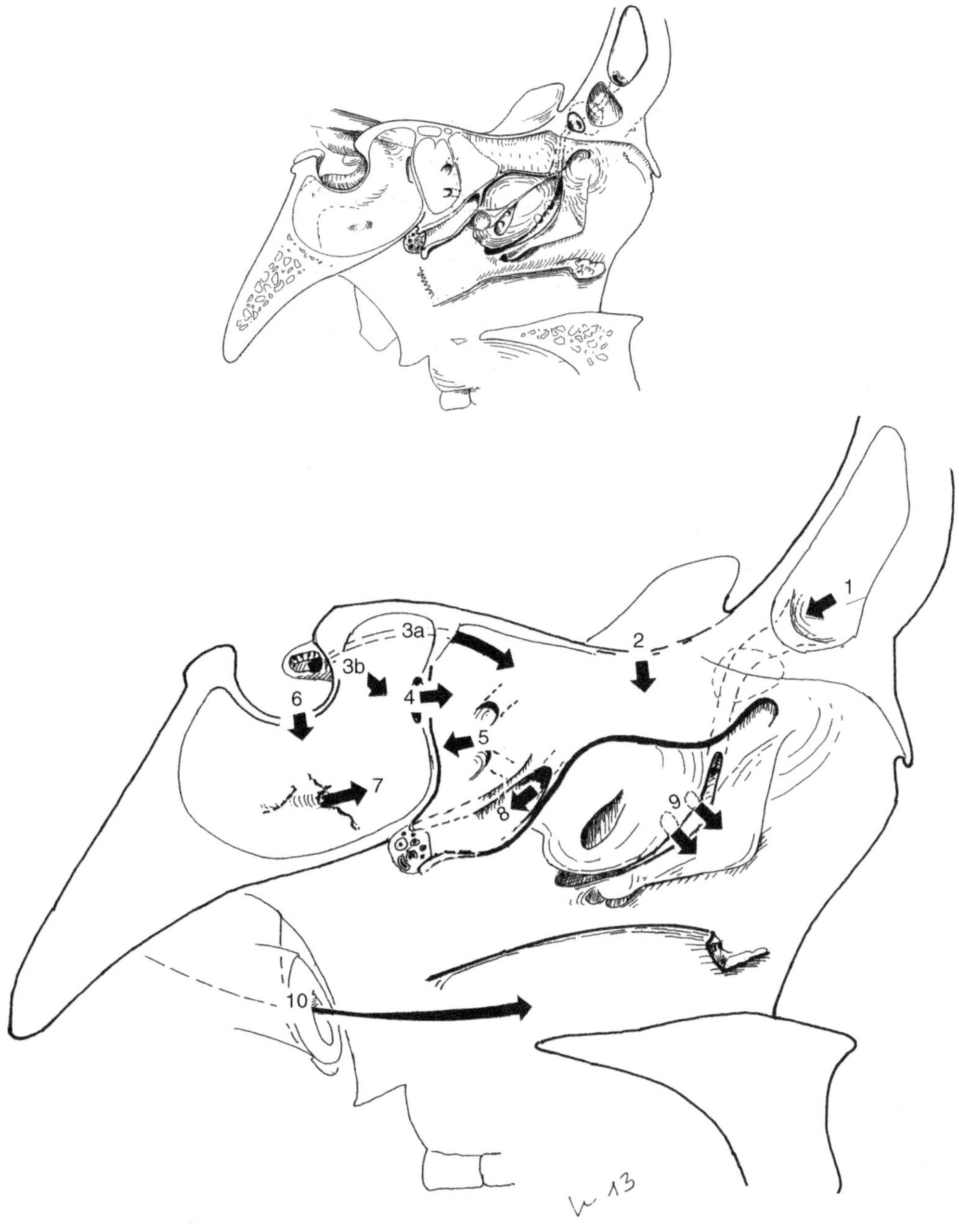

Fig. 4.11 Rhinoliquorrhea. Conclusion. *Arrows*: routes of rhinoliquorrhea. *1* Ductus nasofrontalis; *2* lamina cribrosa; *3a* processus clinoideus anterior and cellulae ethmoidales posteriores; *3b* processus clinoideus anterior and sinus sphenoidalis; *4* apertura sphenoidalis; *5* cellulae ethmoidales posteriores; *6* sella; *7* canalis rotundus, CSF space of N. maxillaris; *8* cellulae ethmoidales posteriores with inconstant connections to bulla ethmoidalis; *9* cellulae ethmoidales anteriores and infundibulum ethmoidale; *10* cellulae mastoideae, cavum tympanicum, and tuba

Part II

Temporal Region

1.1 Introduction

This introduction to Part II contains a short summary of each chapter.

1.2 Chapter 5 Diagnostic Base

The main aspect of imagings is avoiding of distortions at MRTs of the temporal region. This will be arrived using conventional presentations. Exact presentations without distortions are possible using hippocampaxial MRTs as illustrated here. After this, hippocampaxial projections of lobus temporalis are compared with conventional temporal MRTs of the skull and brain. This may be helpful for understanding errors at using conventional MRTs.

1.3 Chapter 6 Anatomical Base of Surgery

The extracranial fronto-temporo-basal (pterional) region was presented at the frontal region. Now are additional presented temporal layers and landmarks for defining of the upper limit of lobus temporalis (sulcus lateralis Sylvii). Intradural presented in this chapter are allocortex and neocortex of the temporal lobe. Temporal allocortex (limbic cortex) is a combination of amygdala with hippocampus and gyrus dentatus (area dentata) and with gyrus parahippocampalis and fimbria fornicis. These structures could be eliminated by amygdalohippocampectomy without damaging neocortex. Neocortical temporal areas are Wernicke's angularis area and other neocortical areas of the dominant and nondominant hemisphere. Unilateral elimination of the acustic areas (e.g., Heschl's gyrus) would not be followed by functional deficits according to the bilateral connections of the acustic system at pons and mesencephalon.

1.4 Chapter 7 Special Surgical Aspects

Presented are the anatomical base of resections of the temporal lobe, keyhole resections, amygdalohippocampectomy, and other temporal surgical approaches. Preconditions for amygdalohippocampectomy are surgical routes to cornu inferius of the lateral ventricle before performing amygdalohippocampectomy. There are three surgical types of routes: the transylvian route, the infratemporal route, and transcortical routes. In this book are presented Yaşargil's transylvian route and the infratentorial route. The infratemporal approach can be used for amygdalohippocampectomy and for other basal target areas. Precondition for an infratemporal route is the exact defining of the basal limit of craniotomy as basal as possible. Labyrinth, antrum mastoideum, knee of sinus transversus, and articulatio mandibularis must be preserved. The ground of fovea mandibularis is located close to fossa cranii media.

Diagnostic Base

5

5.1 Landmarks (Figs. 5.1 and 5.2)

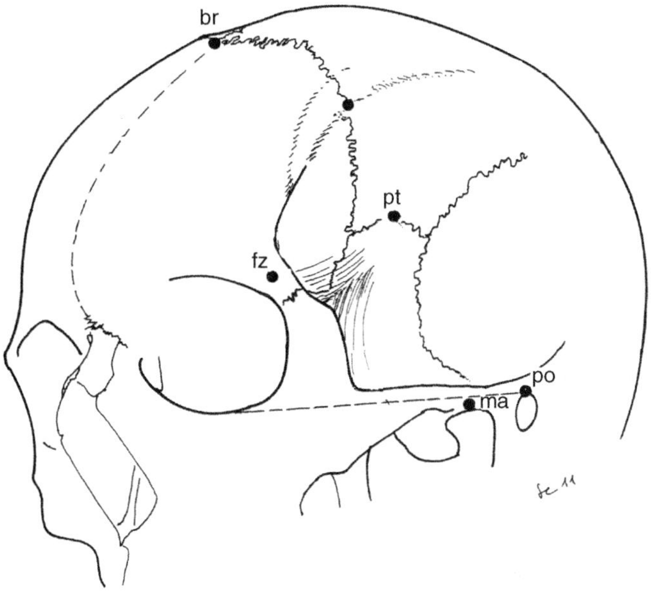

Fig. 5.1 Temporal and surrounding landmarks of the skull for planning surgery. *br* bregma, *fz* frontozygomatic point, *ma* fossa mandibularis, *po* porion, *pt* pterion, *st* stephanion; *dotted*: orbitomeatal line

W. Seeger, J. Zentner, *Anatomical Basis of Cranial Neurosurgery*, https://doi.org/10.1007/978-3-319-63597-2_5

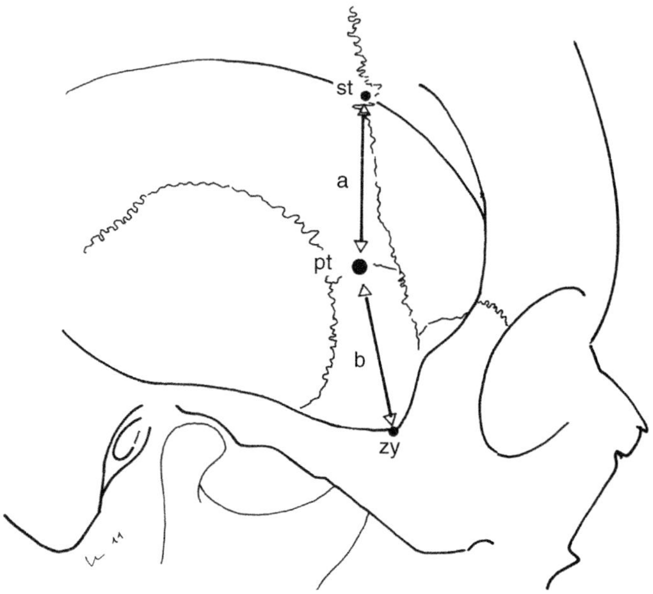

Fig. 5.2 Defining of pterion for craniotomy pterion is an anthropological frontotemporal landmark (Yaşargil's landmark for fronotemporal craniotomy) at the upper point of sutura sphenoparietalis (at adults often not recognizable). Approximate defining, if anatomical defining at the outside of the skull is unclear. Distance measurement a similar to *b*. *pt* pterion, *st* stephanion, *zy* angle of os zygomaticum

5.2 **Radiology** (Figs. 5.3, 5.4, 5.5, and 5.6)

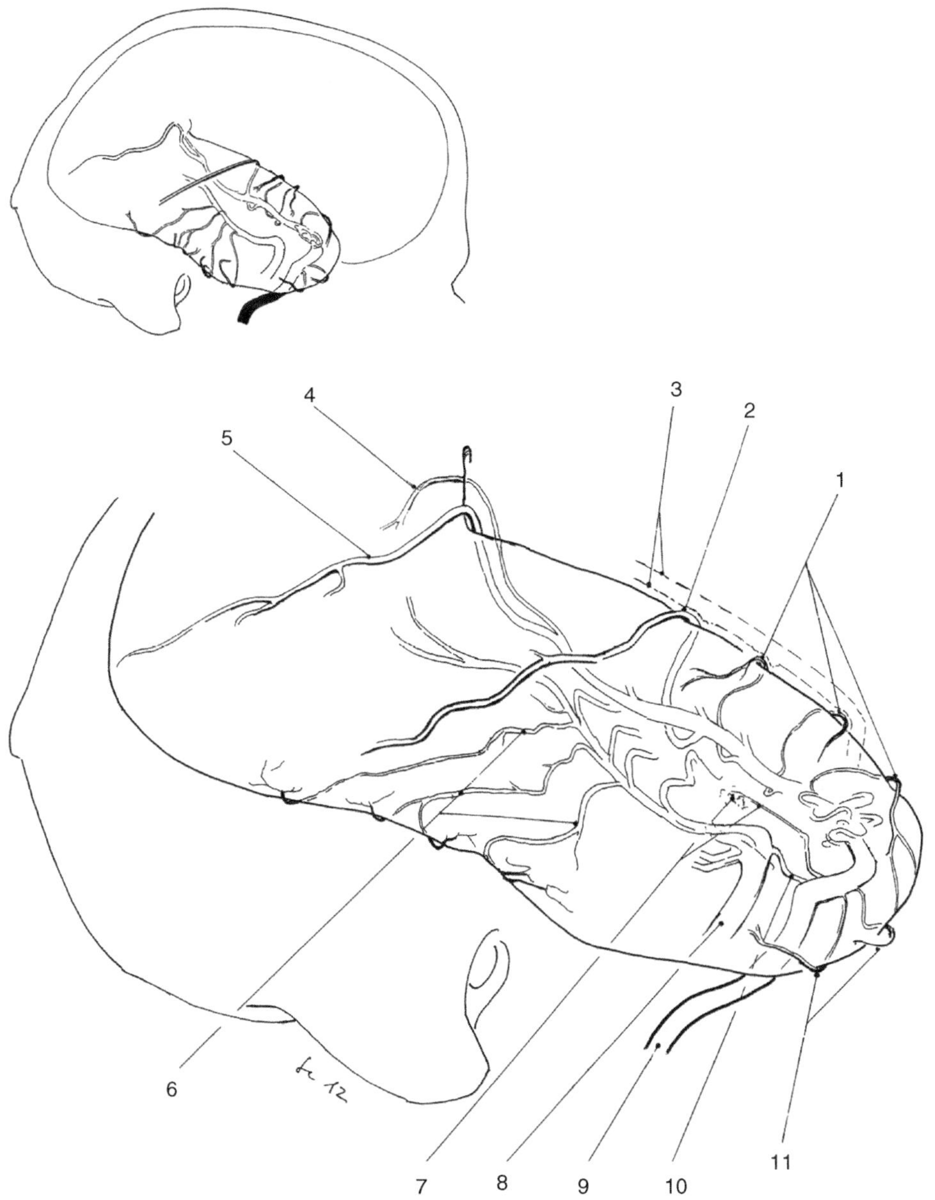

Fig. 5.3 Arteries of lobus temporalis, schematized transparent drawing of structures similar to imagings. Landmarks added for defining of arteries assisting imagings, if necessary: *1* temporal branches of A. cerebri media (and A. carotis interna); *2* A. temporalis posterior; *3* projection of V. temporalis superficialis; *4* A. angularis; *5* A. temporooccipitalis; *6* temporobasal branches of A. cerebri posterior, projections; *7* A. chorioidea anterior close to the temporal chorioid point, marking the anterior, segment of cornu inferius, projection; *8* A. basilaris, projection; *9* A. carotis interna; *10* A. communicans posterior, projection; *11* temporal branches of A. carotis interna (and A. cerebri media)

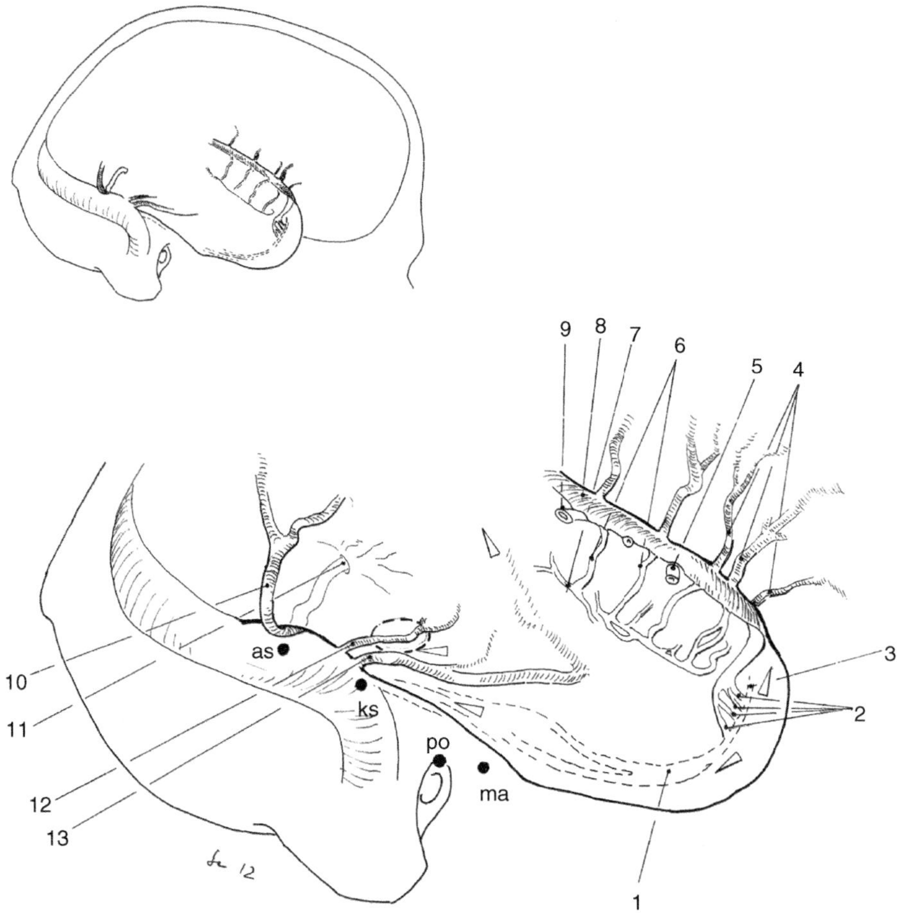

Fig. 5.4 Continuation of Fig. 5.3. *Veins*: as asterion; *ea* eminentia arcuata, projection; *ks* knee of sinus transversus; *ma* fossa mandibularis; *po* porion. *1* Sinus paracavernosus, projection; *2* temporopolar bridging veins, projection; *3* sinus alae minoris; *4* frontoparietal veins; *5* one of Rr. temporales of A. cerebri media; *6* sylvian venous connections to 7, projections; *7* V.(Vv.) temporalis profunda projection; *8* V. temporalis superficialis; *9* A. temporalis posterior undercrossing 8; *10* venous variant; *11* as 10, bridging connection to tentorium, projection; *12* typical small vein close to V. Labbé (Scheremet 1V. Labbé (must be preserved at surgery (Cambria 1980)), *Arrows*: directions of blood flow

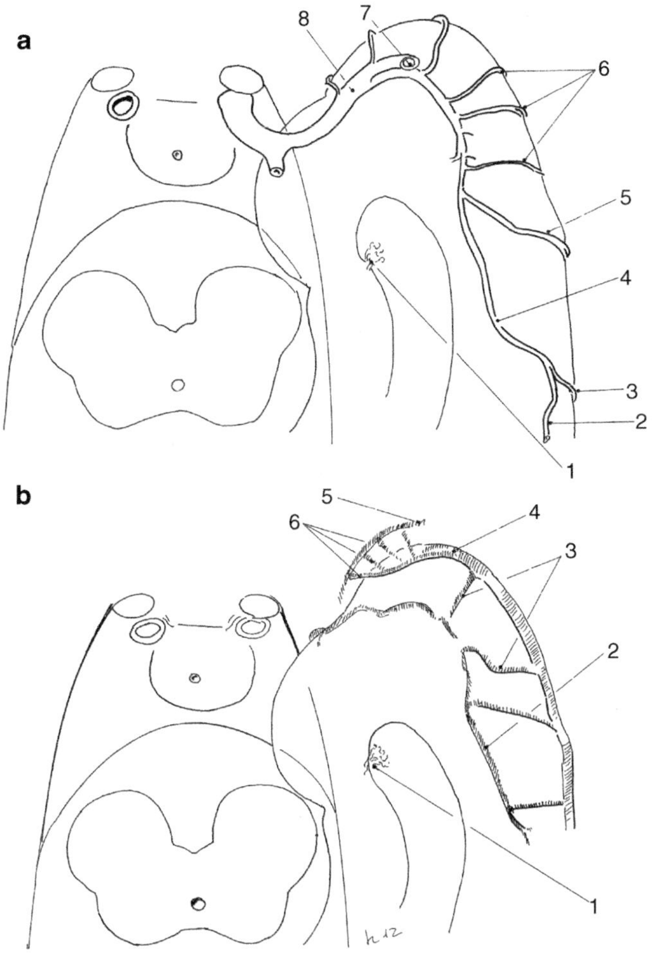

Fig. 5.5 Vessels at the temporal surface of sulcus lat. (Sylvii), schematized. **a** Arteries, **b** veins. **a** *1* Temporal chorioid point; *2* A. temporooccipitalis; *3* A. angularis; *4*. A. cerebri media, distal trunk; *5* A. temporalis posterior; *6* Rr. temporales; *7* A. operculofrontalis; *8* A. cerebri media, main trunk; **b** *1* temporal chorioid point, *2* V.(Vv.) temporalis profunda, *3* connections of 2 and 4, *4* V. temporalis superficialis, *5* sinus alae minoris – sinus sphenoparietalis, *6* temporopolar bridging veins (at pterional surgical approaches not involved)

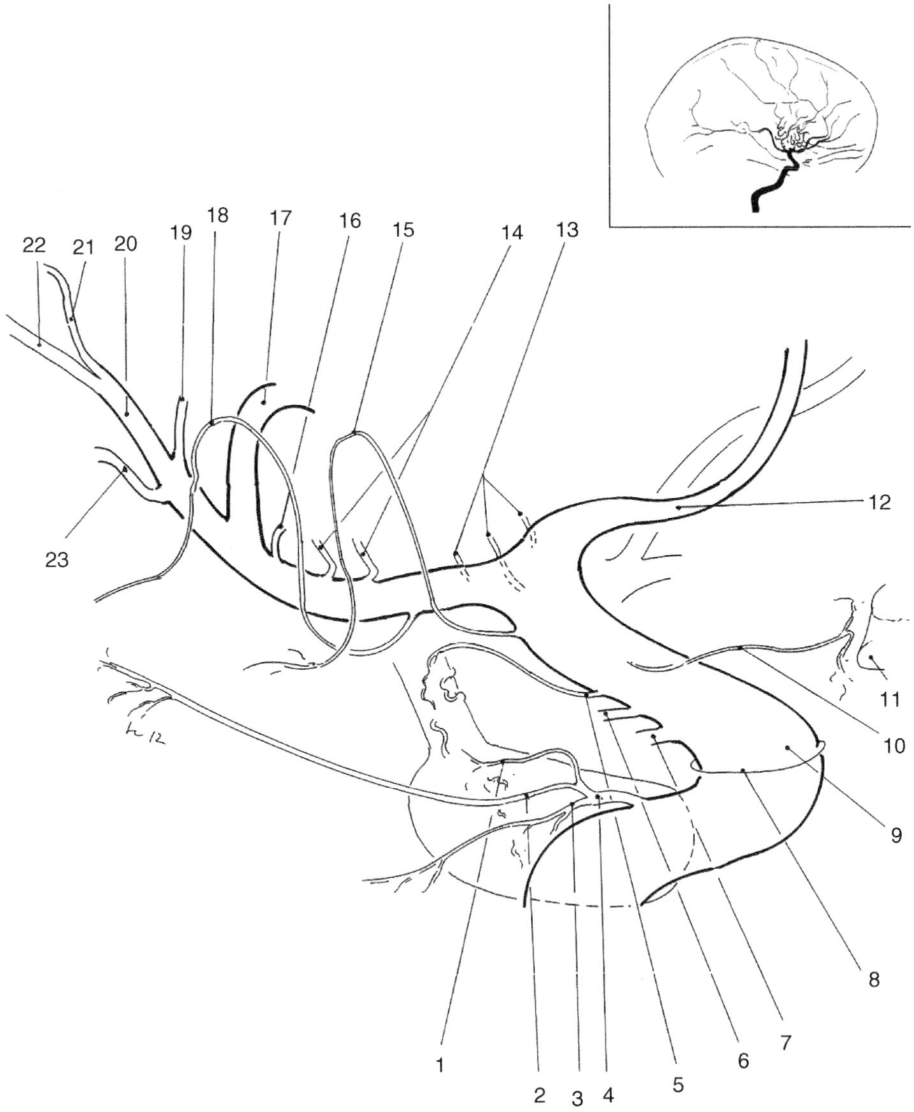

Fig. 5.6 Branches of A. carotis int., schematized. *1* A. hypophysealis inferior; *2* A. tentorii (Bernasconi); *3* meningeal branch; *4* truncus meningohypophysealis; *5* A. hypophysealis superior; *6* A. chorioidea anterior; *7* A. communicans posterior; *8* A. carotis interna, dural penetration; *9* carotid siphon, curvatura anterior; *10* dural branch; *11* region of tuberculum sellae; *12* A1; *13* medial striate arteries; *14* lateral striate arteries (Aa. lenticulostriate); *15* one of numerous parietal arteries; *16* A. sulci centralis; *17* A. operculofrontalis; *18* as 15; *19* as 15; *20* A. cerebri media, distal trunk; *21* A. angularis; *22* A. temporooccipitalis; *23* A. temporalis posterior

5.3 Further Imaging (Figs. 5.7, 5.8, 5.9, 5.10, 5.11, and 5.12)

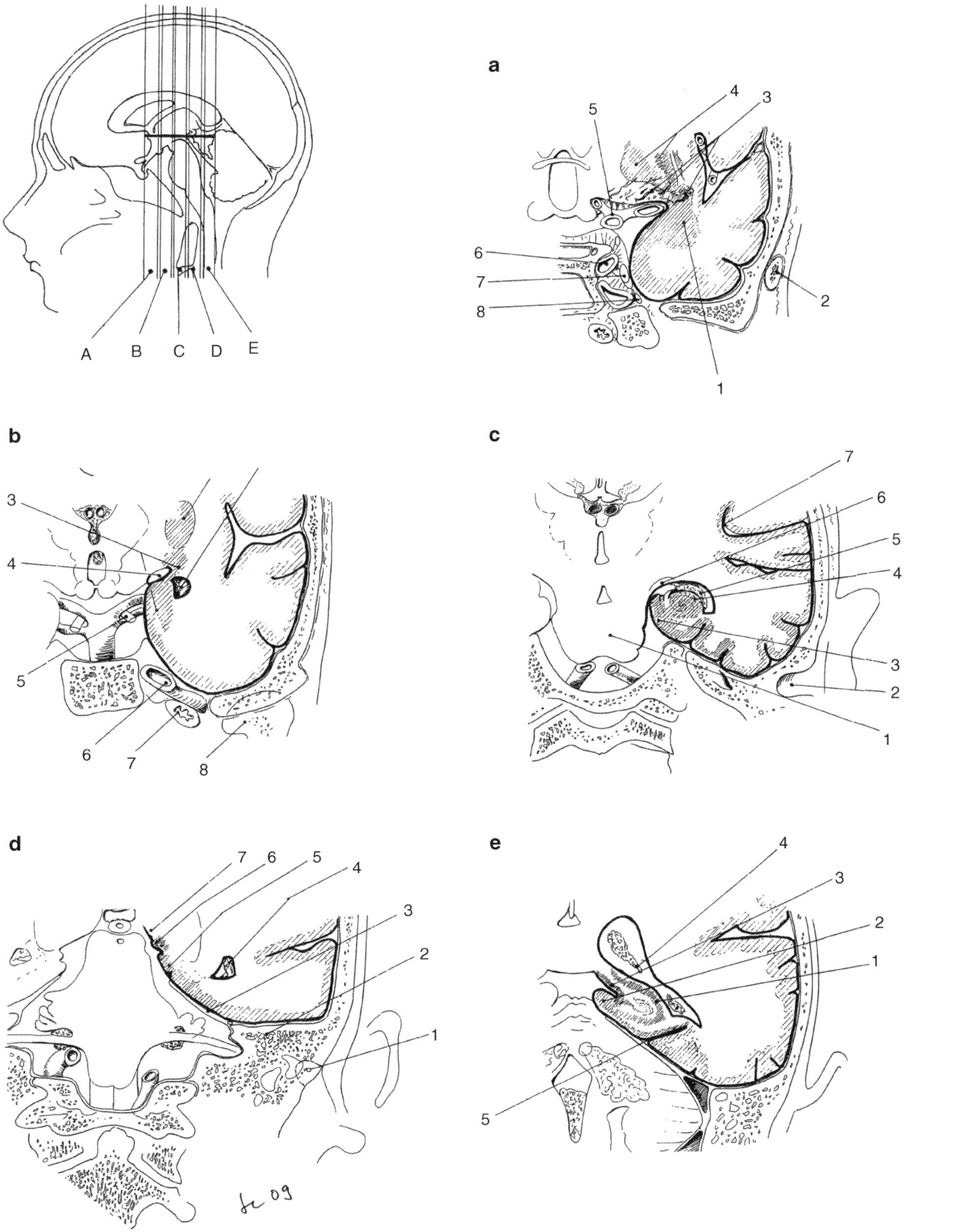

Fig. 5.7 Conventional MRTs. Coronal slices (**a–e**) are performed vertical to the intercommissural line and not performed right-angled to the axis of the temporal lobe. Now temporal presentations are distorted. **a** *1* Amygdala, *2* arcus zygomaticus, *3* carotid cistern, *4* gangliae basales, *5* carotid bifurcation, *6* carotid siphon (curvatura posterior), *7* N. ophthalmicus, *8* N. abducens; **b** *1* tip of cornu inferius, *2* gangliae basales, *3* amygdala, *4* tractus opticus, *5* N. oculomotorius, *6* A. carotis interna, *7* tuba auditiva, *8* capitulum mandibulae; **c** *1* pons, *2* meatus acusticus externus, *3* gyrus parahippocampalis, *4* hippocampus, *5* plexus choroideus, *6* tractus opticus, *7* sulcus lateralis (sylvian fissure); **d** *1* porus acusticus externus (projection), *2* os petrosum, *3* tentorium, *4* cornu inferius, *5* gyrus parahippocampalis, *6* area dentata, *7* isthmus gyri cinguli; **e** *1* hippocampus, *2* gyrus parahippocampalis, *3* area dentata, *4* atrium

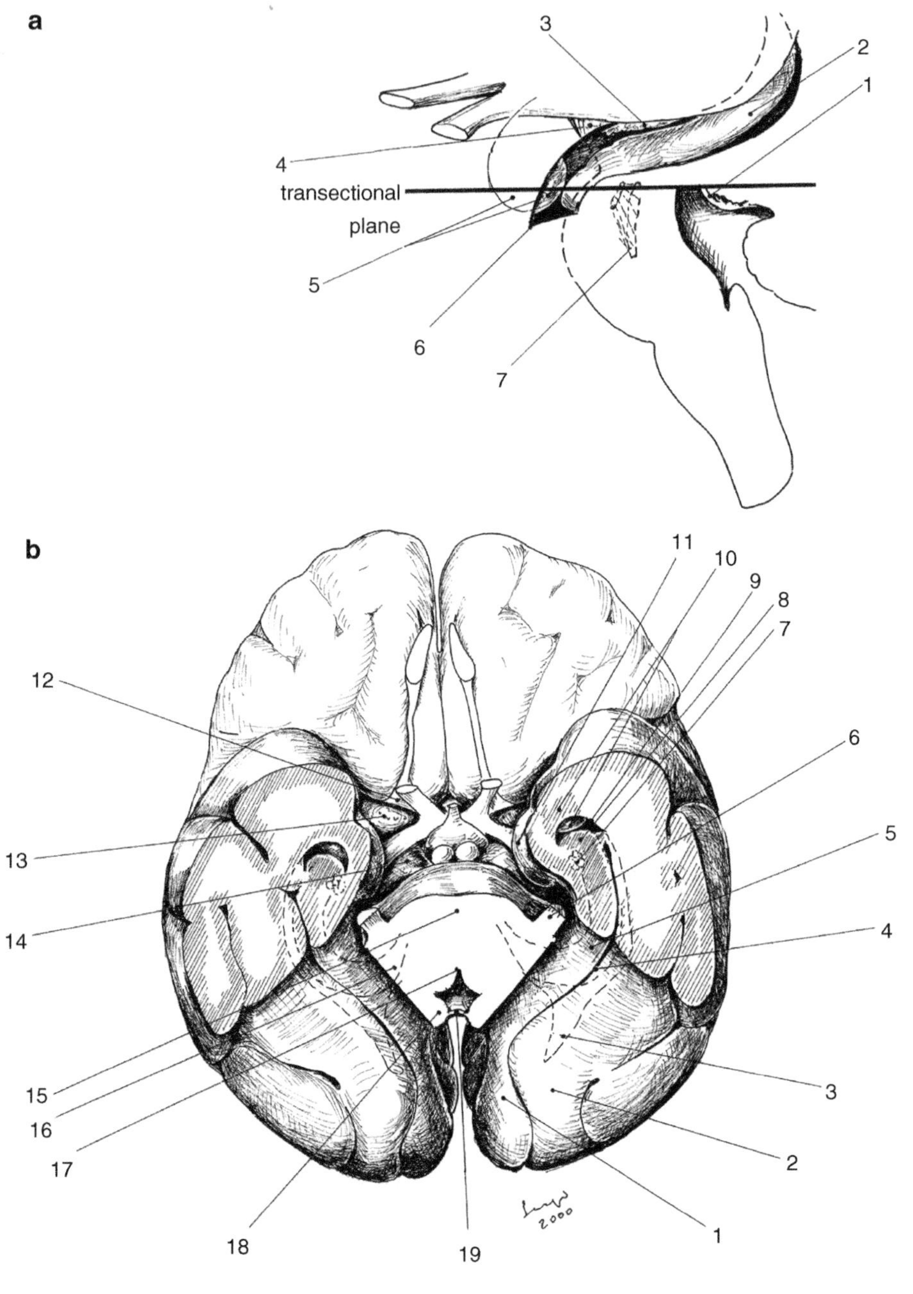

Fig. 5.8 Cadaver brain dissections. Plane **a** is not performed temporoaxial. **a** Topogram, **b** Axial transection. **a** *1* Velum medullare anterius, *2* atrium, *3* plexus choroideus, *4* crus cerebri, *5* uncus, *6* tip of cornu inferius, *7* N. trigeminus, projection. **b** *1* Gyrus occipitotemporalis medius, *2* gyrus occipitotemporalis lateralis, *3* cornu posterius (projection), *4* sulcus collateralis, *5* gyrus parahippocampalis, *6* N. trigeminus, *7* region of the chorioid point, *8* pes hippocampi, *9* tip of cornu inferius, *10* amygdala, *11* uncus, *12* trigonum olfactorium, *13* substantia perforata anterior, *14* crus cerebri, *15* pons, *16* atrium (projection), *17* sulcus medianus, *18* pedunculus cerebellaris superior, *19* velum medullare anterius

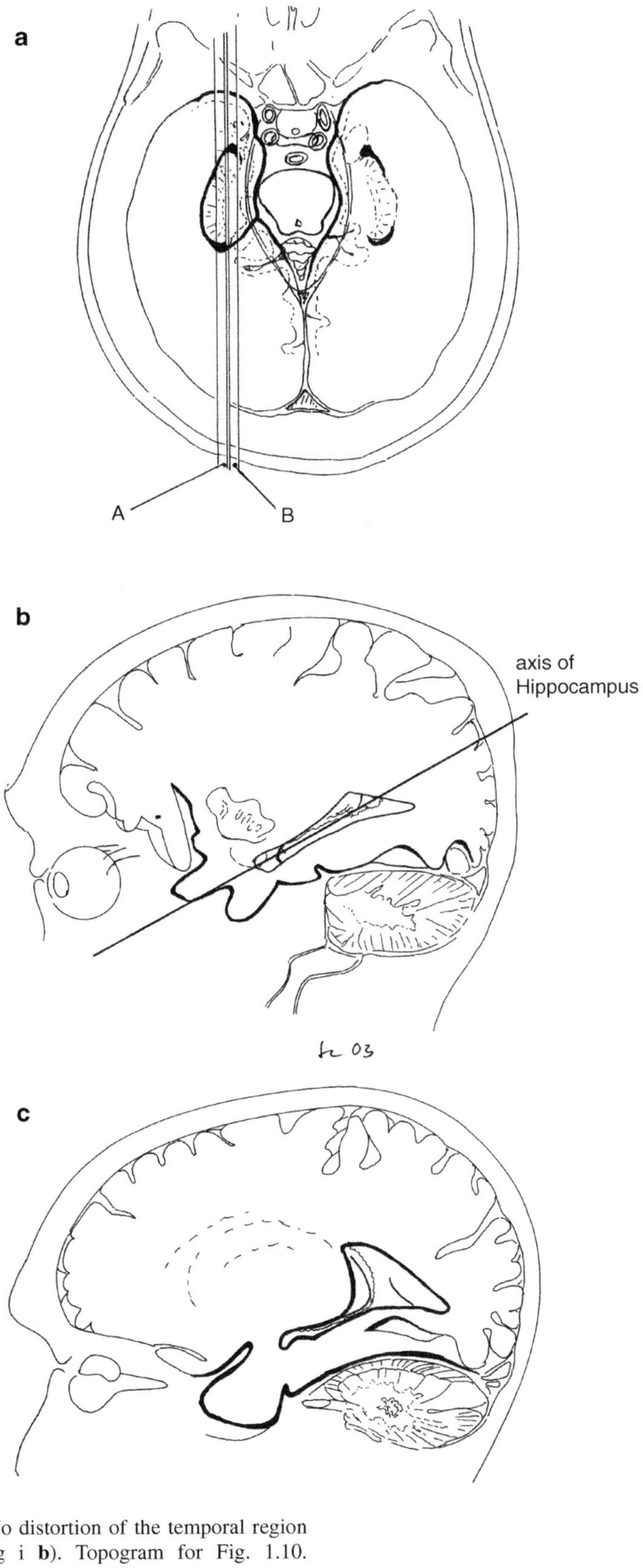

Fig. 5.9 Hippocampaxial MRT. No distortion of the temporal region (used by Prof. Zentner, Freiburg i **b**). Topogram for Fig. 1.10. **a** Hippocampaxial presentation, **b** ventriculoaxial presentation, if **a** is not available

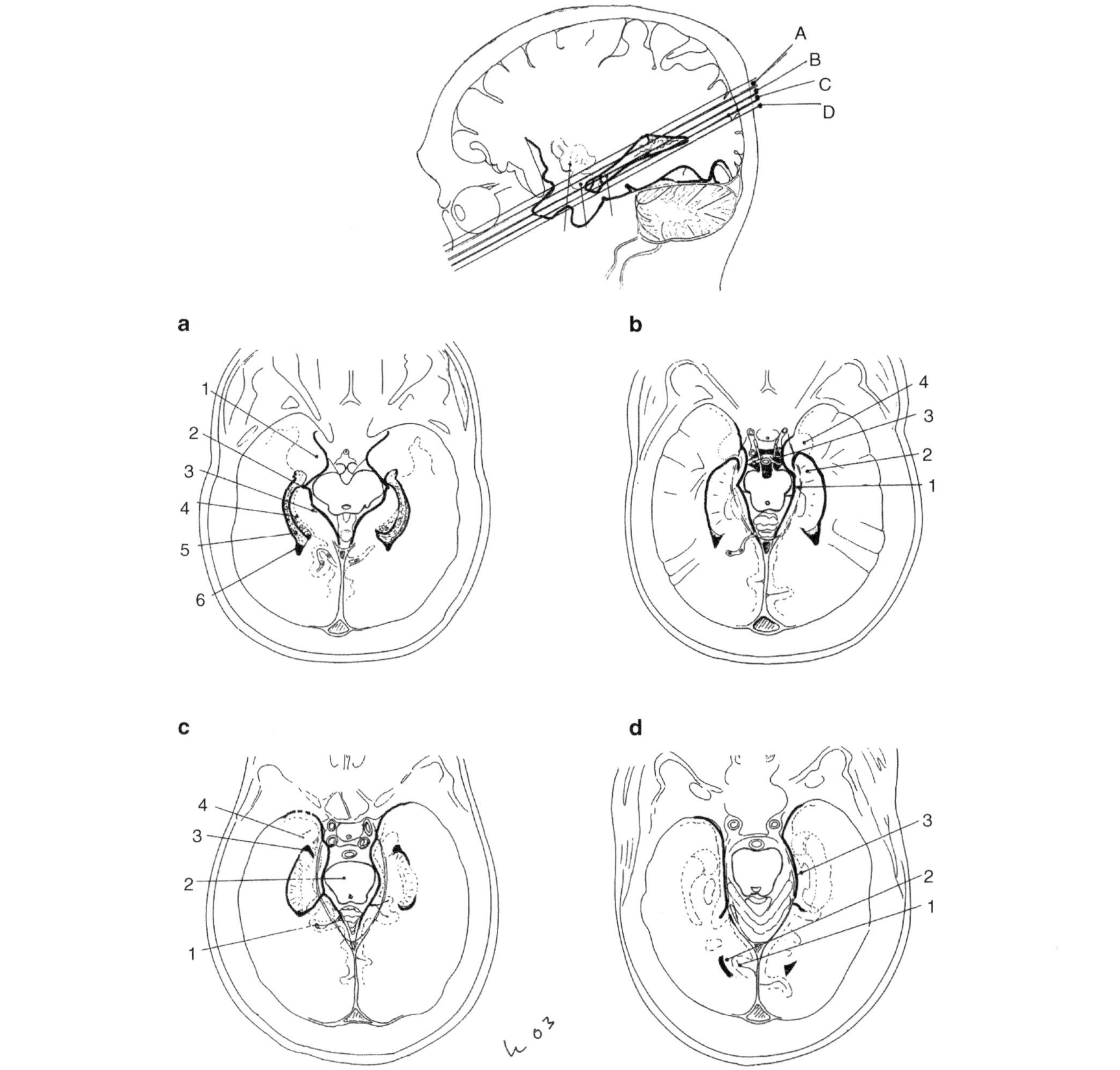

Fig. 5.10 Continuation of Fig. 5.9. **a** *1* Uncus, *2* chorioid point, *3* gyrus parahippocampalis, *4* hippocampus, *5* atrium, *6* cornu posterius; **b** *1* tentorial fold, *2* pes hippocampi, *3* N. oculomotorius and bifurcation of A. basilaris, *4* amygdala; **c** *1* sulcus calcarinus and sulcus parietooccipitalis, typical fusion rostral, *2* pons, *3* tip of cornu inferius, *4* uncus; **d** *1* sulcus calcarinus, *2* calcar avis, *3* margin of gyrus parahippocampalis

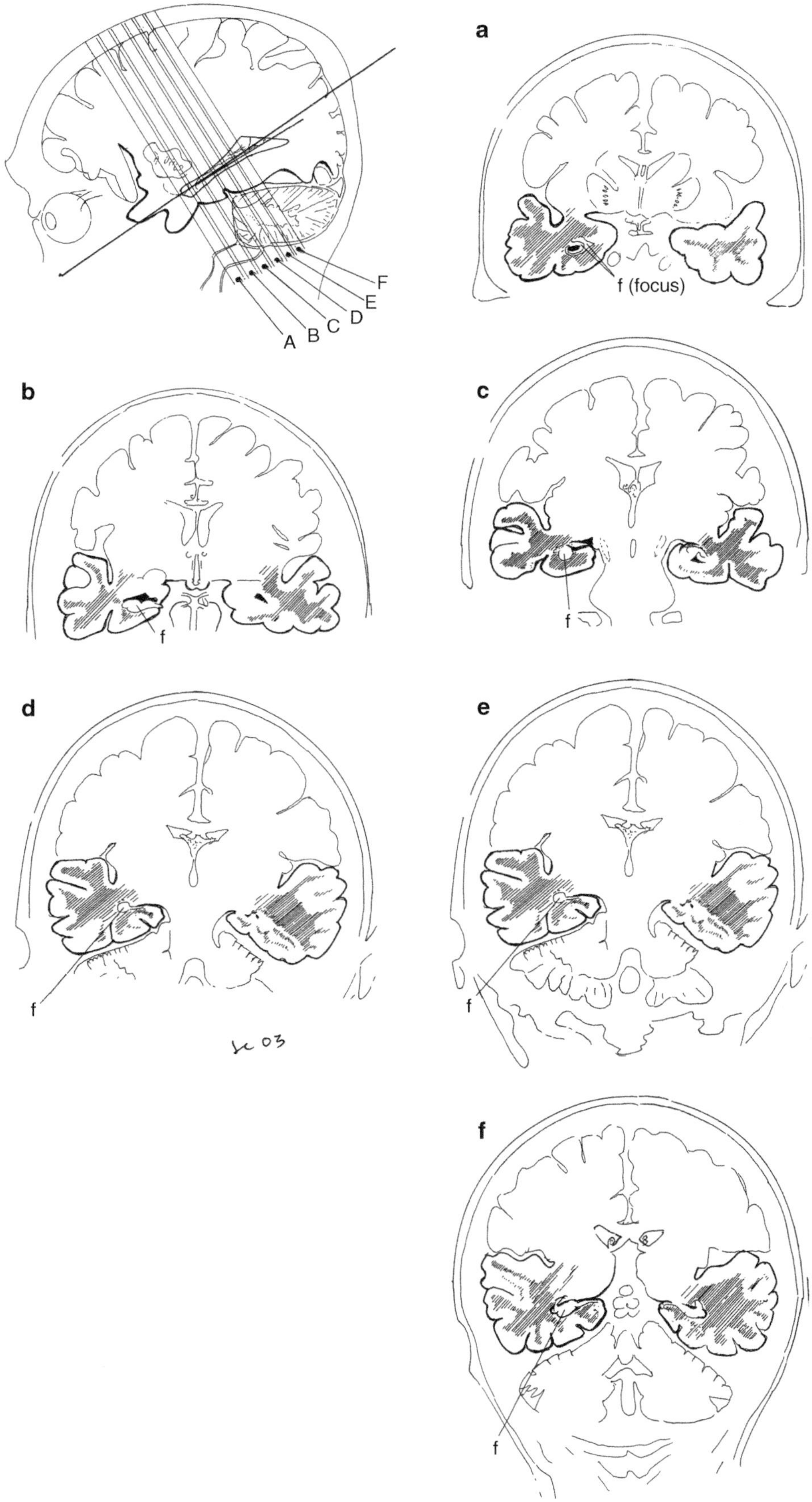

Fig. 5.11 Example of the exact location of an epilepsy focus (Experience of Prof. Zentner, Freiburg) –*f*– by hippocampaxial presentation. Frontal planes of MRT vertical to the axis of hippocampus (**a**–**e**). **a** Chiasma, **b** corpora mammillaria, **c** and **d** region of focus –*f* –, **e** tectum

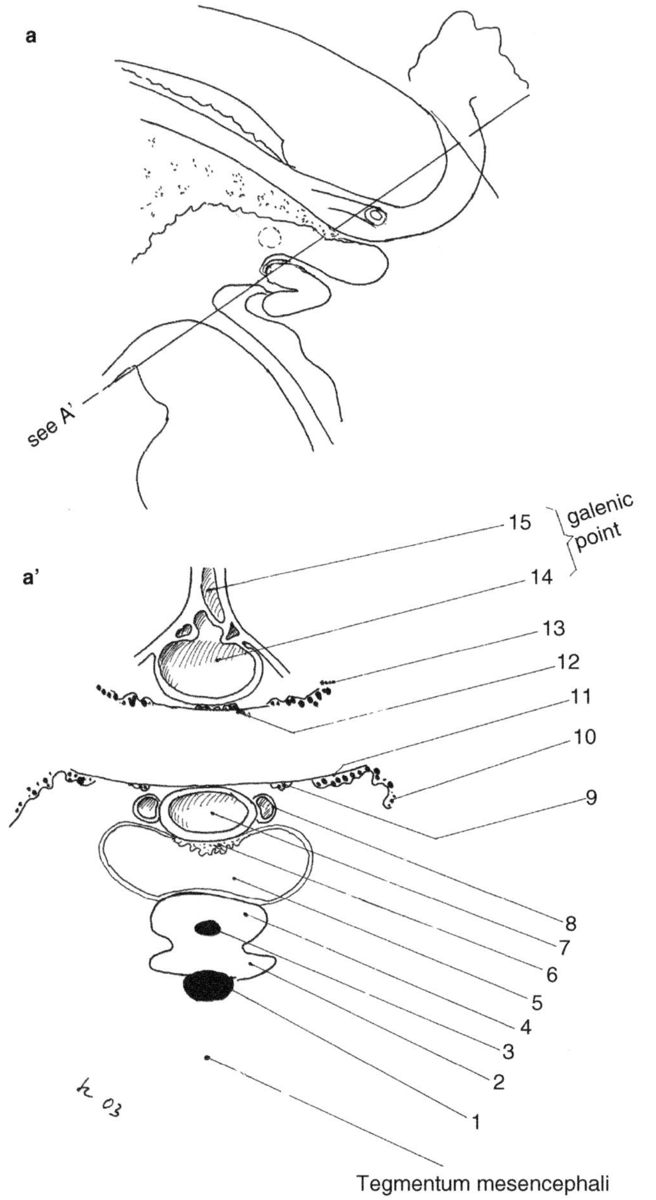

Fig. 5.12 Area between hippocampus/gyrus dentatus and striae longitudinales cinguli/indusium griseum (Figs. 2.19 and 2.37). MRT schematized. Bilateral lesions must be avoided. *1* Aquaeductus; *2* commissura posterior; *3* recessus pinealis; *4* corpus pineale; *6* plexus choroideus; *5* recessus suprapinealis; *7* V. magna (Galeni); *8* V. basalis (Rosenthal); *9* allocortex, beginning of striae mediales of indusium griseum; *10* as 9, beginning of stria longitudinalis cinguli and of striae laterales of indusium griseum; *11* as 10; *12* indusium griseum, striae mediales; *13* as 12, striae laterales and stria longitudinalis cinguli; *14* V. magna (Galeni); *15* sinus rectus

6 Anatomical Base of Surgery

6.1 Extradural Topography (Figs. 6.1, 6.2, and 6.3)

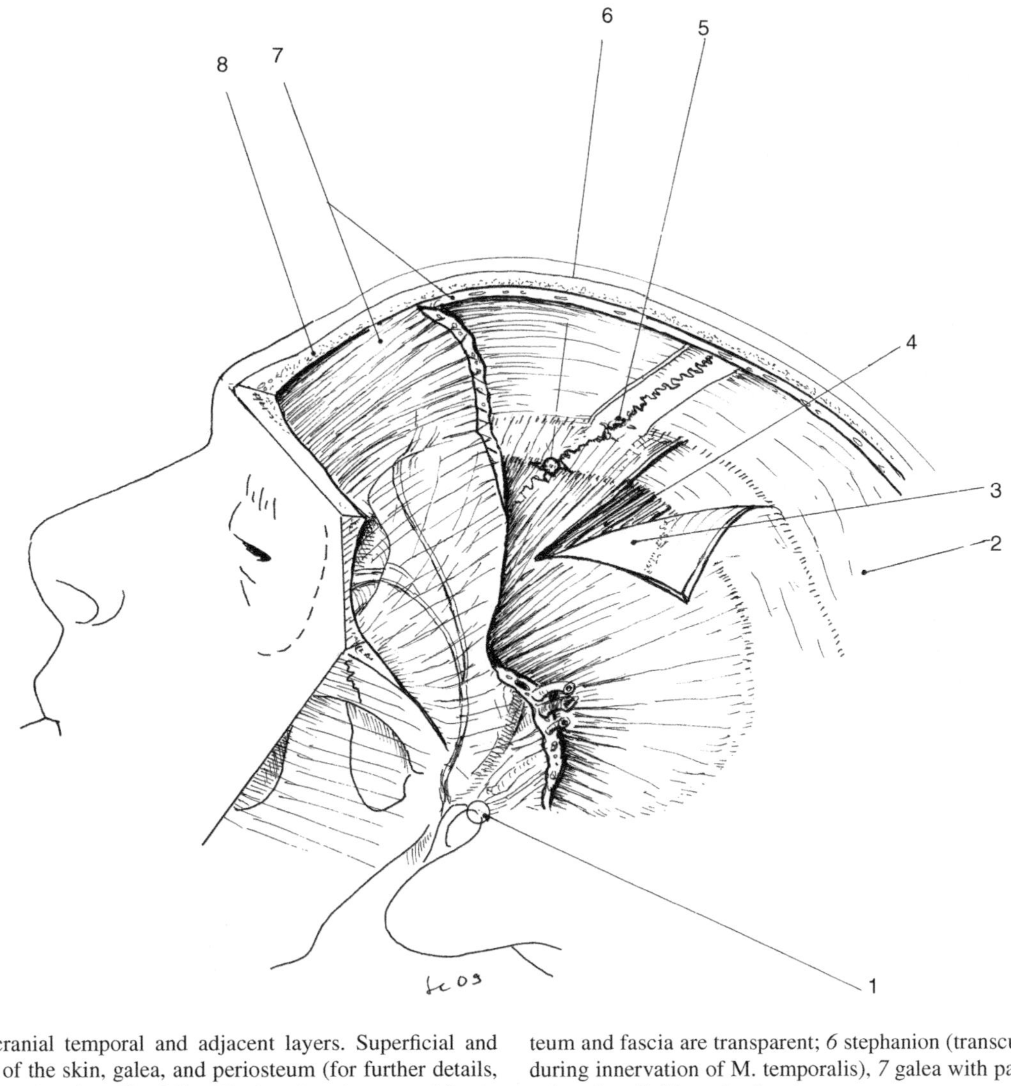

Fig. 6.1 Extracranial temporal and adjacent layers. Superficial and profound layers of the skin, galea, and periosteum (for further details, see volume 1, frontal region, Fig. 2.3). *1* Porion; *2* periosteum; *3* fascia temporalis; *4* M. temporalis; *5* sutura coronalis recognizable, if periosteum and fascia are transparent; *6* stephanion (transcutaneous palpable during innervation of M. temporalis), *7* galea with pars frontalis of M. epicranius; *8* skin and subcutis

W. Seeger, J. Zentner, *Anatomical Basis of Cranial Neurosurgery*, https://doi.org/10.1007/978-3-319-63597-2_6

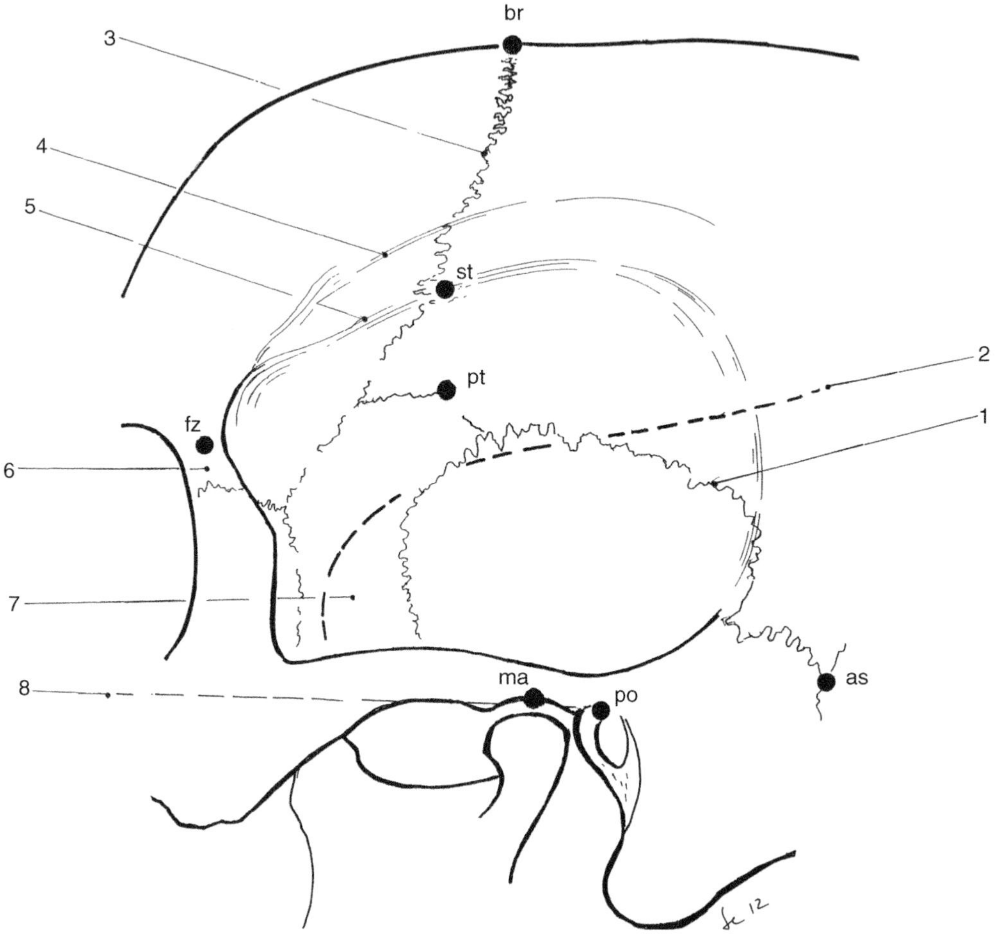

Fig. 6.2 Outside landmarks of the cranial convexity frontotemporal. Cadaver skull dissection. The outline of sulcus temporalis (Sylvii) (*dotted*) is located close to sutura squamosa. The basal region of the lobus temporalis is located close to fossa mandibularis, porion, and asterion (see Fig. 2.25). These relationships must be considered at temporal craniotomies for avoiding errors. *as* asterion, *br* bregma, *fz* frontozygomatic point, *ma* fossa mandibularis, *po* porion, *pt* pterion, *st* stephanion. *1* Sutura squamosa, variable; *2* sulcus lateralis (Sylvii), projection; *3* sutura coronalis; *4* linea temporalis superior; *5* linea temporalis inferior; *6* processus zygomaticus; *7* ala major; *8* orbitomeatal line

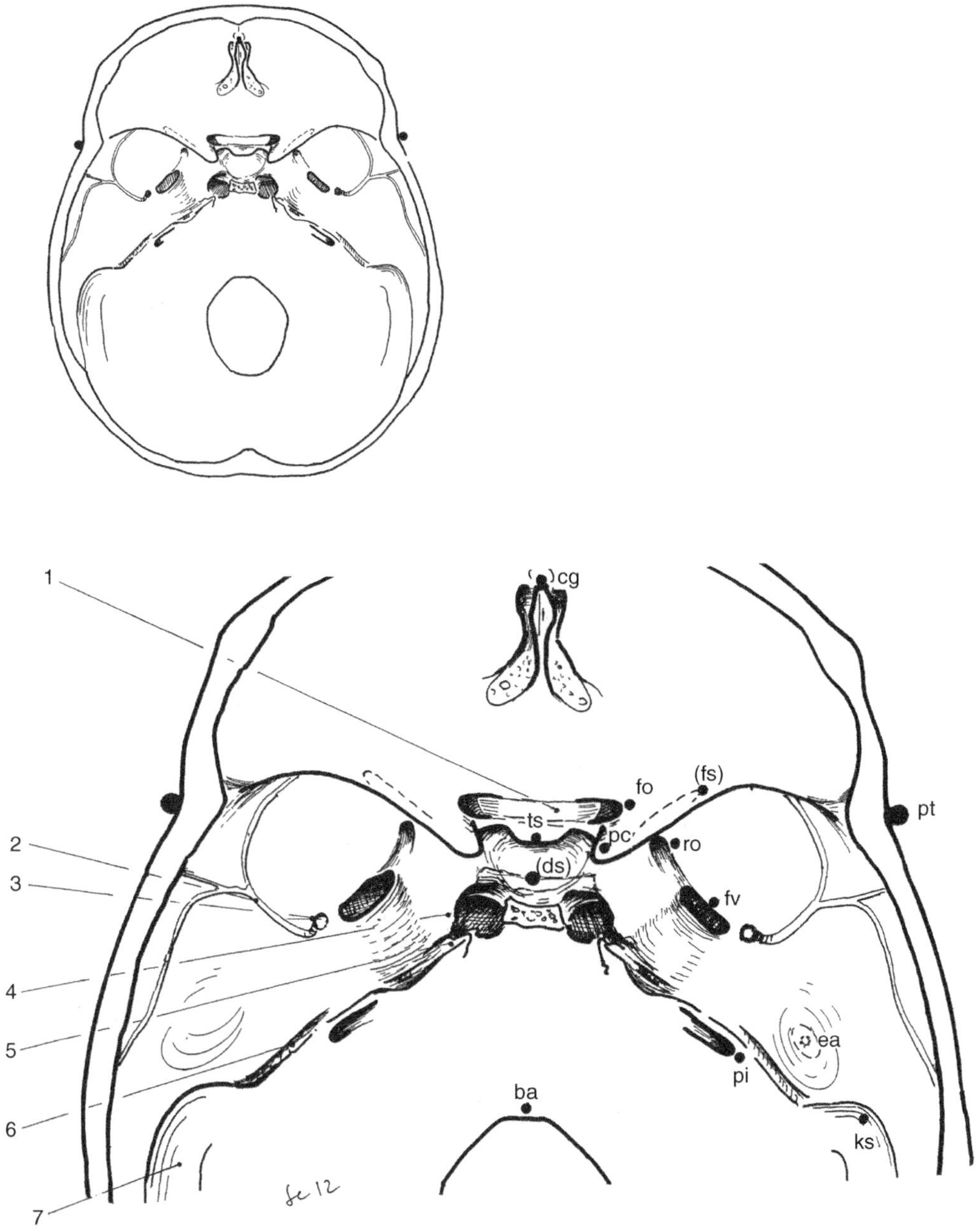

Fig. 6.3 Inside landmarks craniobasal. Cadaver skull dissection: foramen spinosum (enclosing A. meningea media) is located close to foramen ovale (enclosing N. mandibularis). This should be considered at surgical evacuations of temporal epidural hematomas and at punctures of foramen ovale. Apertura interna of canalis caroticus is located at the posterior margin of foramen lacerum. Eminentia arcuata (enclosing canalis semicircularis labyrinthi) is a landmark for sinus petrosus superior (overlying sulcus petrosus sup.) and the palpable superior margin of os petrosum. At this area, an important bundle of basal bridging veins medial from V. Labbé is located. Cave during surgical elevation of lobus temporalis. *ba* basion; *cg* crista galli et For. caecum; *ds* dorsum sellae (defect); *ea* eminentia arcuata; *fo* foramen opticus; *fs* fissura orbitalis superior, lateral point, projection; *fv* foramen ovale; *ks* knee of sulcus transversus; *pc* processus clinoideus anterior overlying the medial segment of fissura orbitalis superior projection; *pi* porus acusticus internus; *pt* pterion; *ro* canalis rotundus; *ts* tuberculum sellae. *1* Sulcus chiasmatis, *2* sulcus of A. meningea media, *3* For. spinosum, *4* For. lacerum, *5* apex ossis petrosi, *6* sulcus petrosus superior, *7* sulcus transversus

6.2 Intradural Topography

6.2.1 Survey (Figs. 6.4, 6.5, 6.6, 6.7, 6.8, 6.9, 6.10, 6.11, and 6.12)

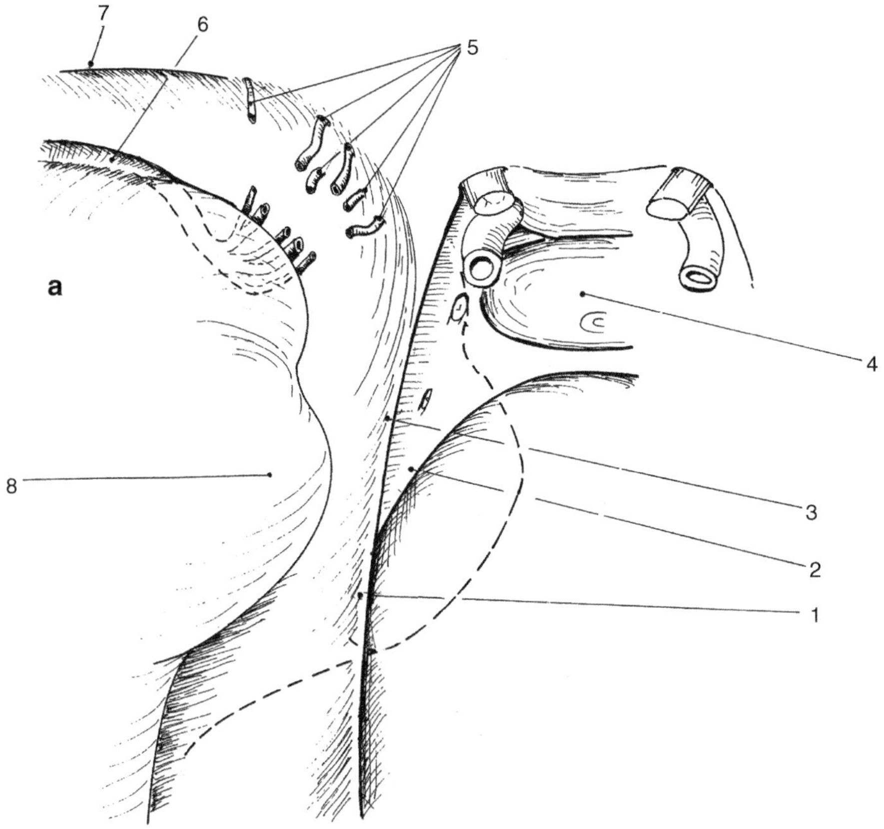

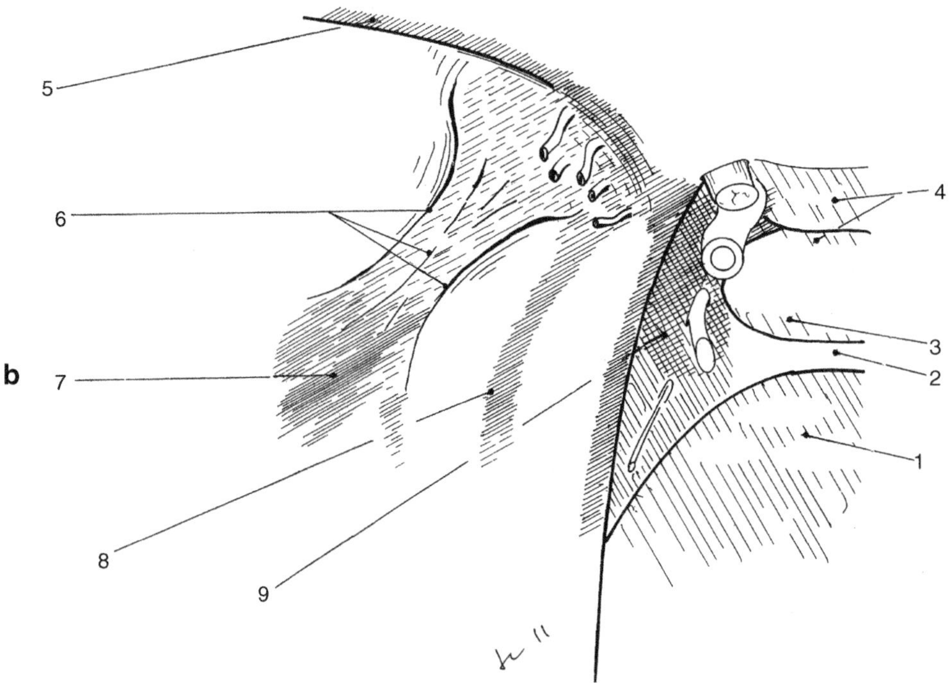

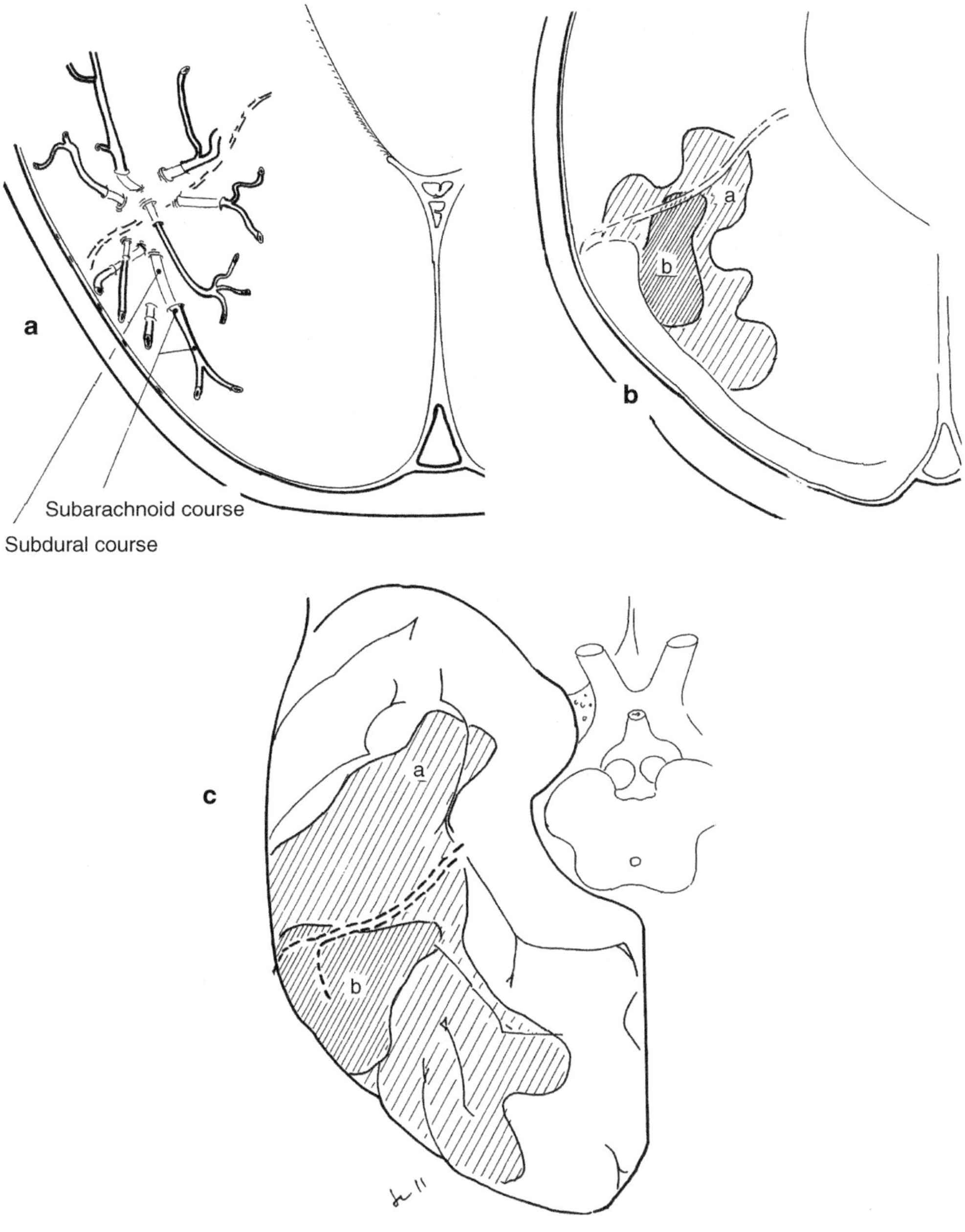

Fig. 6.5 Posterior temporobasal medial bridging veins (Scheremet 1984). Interruption could be followed by encephalomalacia similar to interruption of V. Labbé. **a** Leptomeningeal relationships of veins, **b** density of its dural exit points; *a* low density of dural penetration points, *b* high density, **c** density of its cerebral exit points; **a** low density, **b** high density

Fig. 6.4 Anterior basal veins and sinuses. **a** At pterional surgical approaches, an interruption of temporopolar bridging veins for elevation of lobus frontalis is not necessary. **b** Connections of temporobasal bridging veins with basal sinuses. The dural envelopments of sinus cavernosus and sinus paracavernosus are thickened and intransparent. **a** *1* tentorial fold, *2* plica petroclinoidea posterior, *3* plica petroclinoidea anterior, *4* diaphragma sellae, *5* temporopolar bridging veins, *6* V. temporalis superficialis, *7* dural fold of ala minor, *8* uncus; **b** *1* dural venous plexus of clivus; *2* dura of dorsum sellae, hypovascularized; *3* sinus intercavernosus posterior; *4* sinus intercavernosus anterior; *5* sinus alae minoris; *6* sinus sphenopetrosus, enclosed by a thickened dural fold; *7* as 6, often transparent; *8* sinus paracavernosus (Lang); *9* sinus cavernosus, often transparent

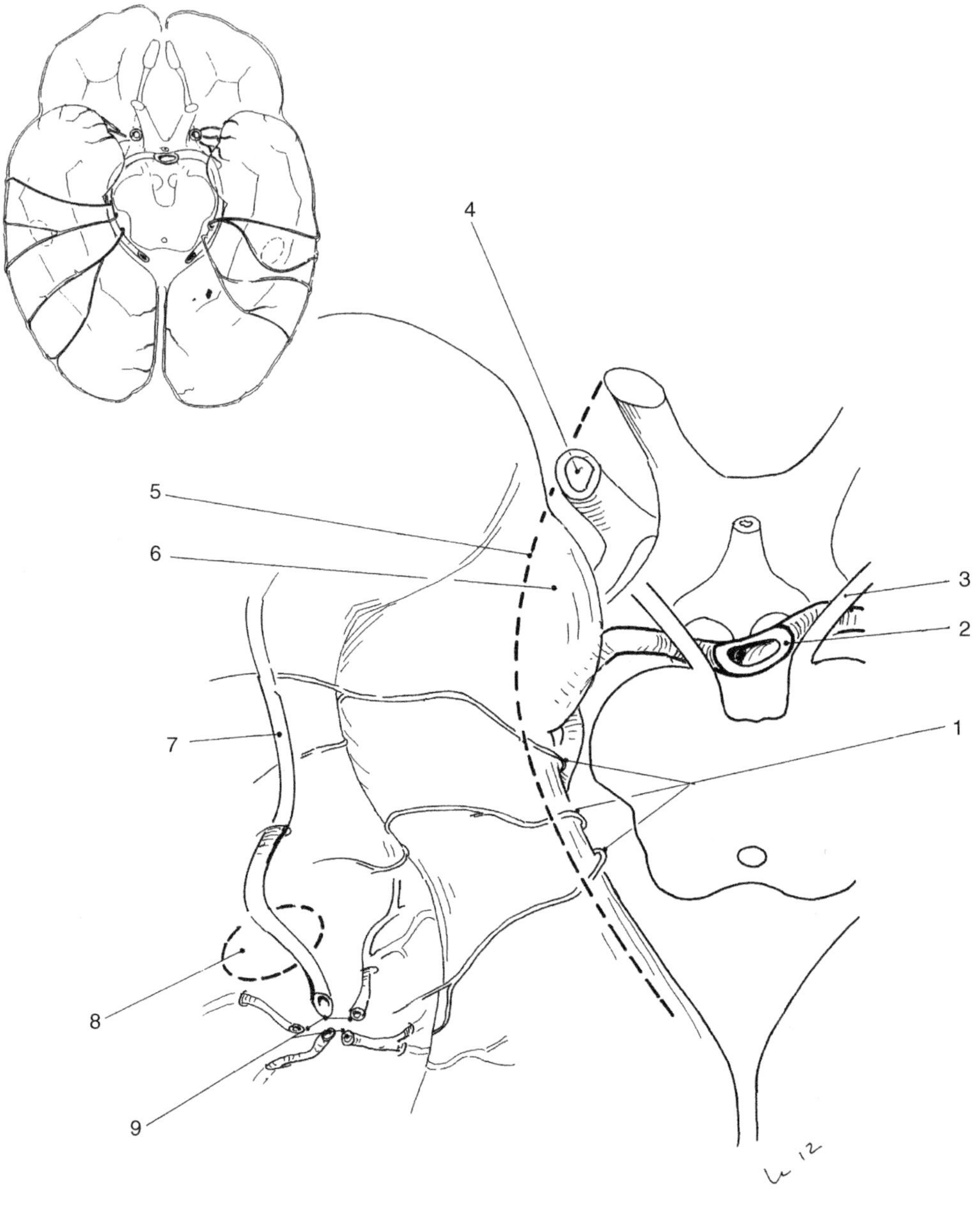

Fig. 6.6 Temporal basal surgical target areas and adjacent medial structures. Eminentia arcuata (projection) is a landmark for medial basal bridging veins. *1* Temporobasal branches of A. cerebri posterior, *2* A. basilaris, *3* N. oculomotorius, *4* A. carotis interna, *5* plica petroclinoidea anterior (projection), *6* uncus, *7* basal bridging vein (subdural course), *8* eminentia arcuata (projection), *9* basal bridging veins, subdural course

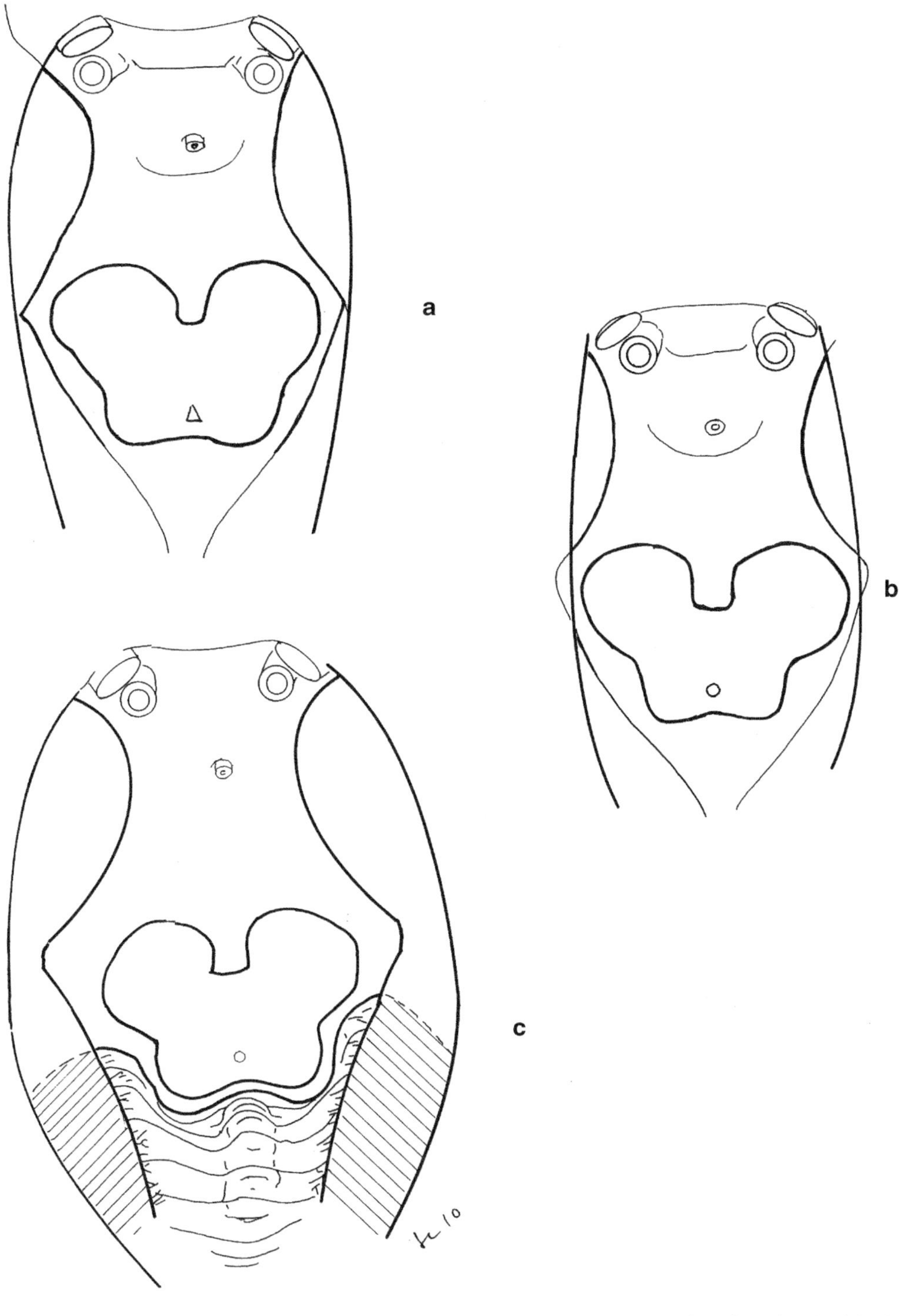

Fig. 6.7 Tentorial notch, variants. Surgical aspects (Experiences of the author). **a** Tentorial notch, usual finding. **b** Narrow variant. Low tolerance for brain shifting at space occupying lesions or surgical elevation of lobus temporalis. **c** Wide variant. Arachnoid adhesions between cerebellum (hatched) and gyrus parahippocampalis could hinder surgical elevation of lobus temporalis

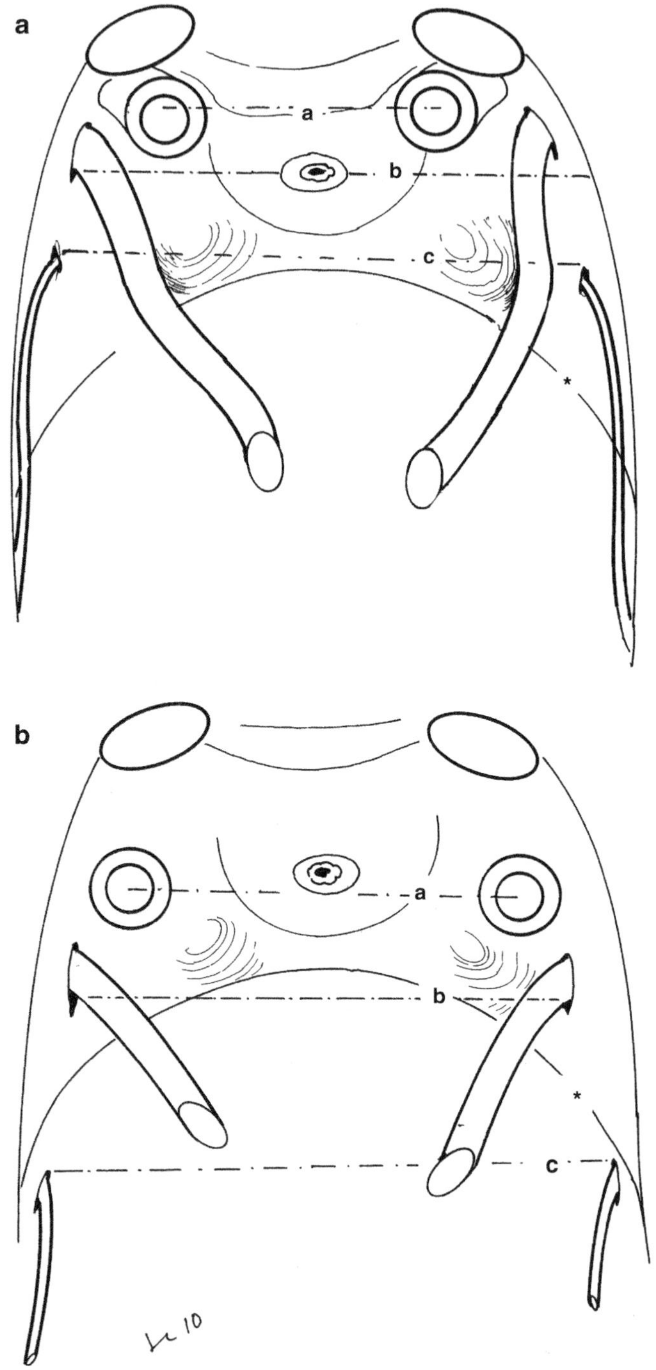

Fig. 6.8 Suprasellar dura penetrating structures (Lang 1975). **a** Usual findings: *a* transversal level of A. carotis interna, *b* dural exit points of N. oculomotorius are located at the transversal level of infundibulum or anterior to it, *c* dural exit point of N. trochlearis is located anterior from plica petroclinoidea posterior. **b** Rare variant. A. carotis interna is retropositioned. This may be favorable for pterional surgical approaches to medial target areas. * Plica petroclinoidea posterior

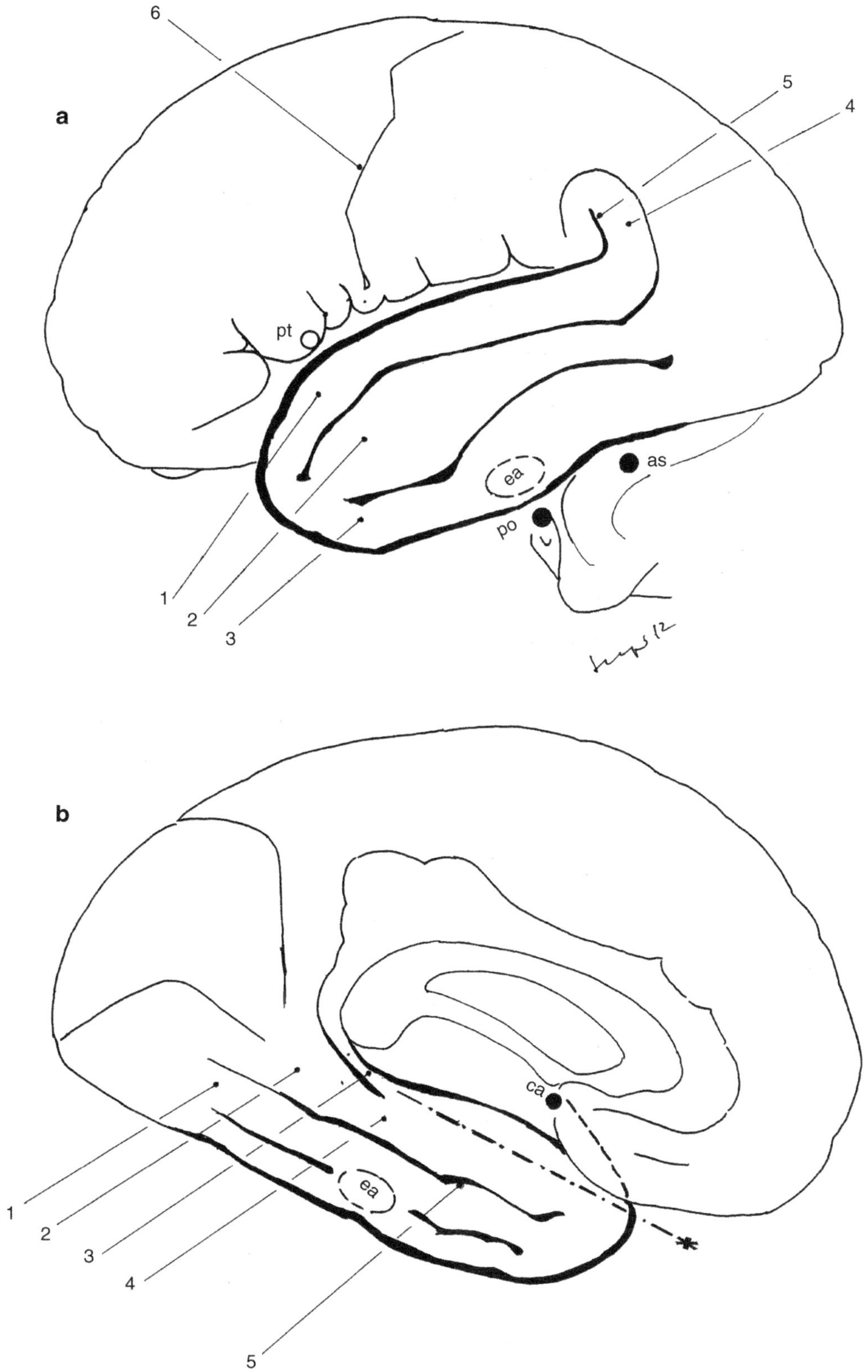

Fig. 6.9 Lateral temporal region. Temporal cortex renders prominent. Bony landmarks added for assisting neuronavigation, if necessary. Small distance measurement between porion and lobus temporalis. Small distance measurement between asterion (sinus transversus) and lobus temporalis. **a** *as* asterion; *ea* eminentia arcuata, projection; *po* orion; *pt* pterion, projection. *1* First temporal gyrus, *2 seco*nd temporal gyrus, *3* third temporal gyrus, *4* gyrus angularis, *5* pars ascendens of sulcus lateralis (Sylvii), *6* sulcus centralis. **b** *ca* commissura anterior, *ea* eminentia arcuata, projection. *1* Gyrus occipitotemporalis lateralis, *2* gyrus occipitotemporalis medialis, *3* isthmus gyri cinguli, *4* gyrus parahippocampalis, *5* sulcus collateralis, * axis of hippocampus

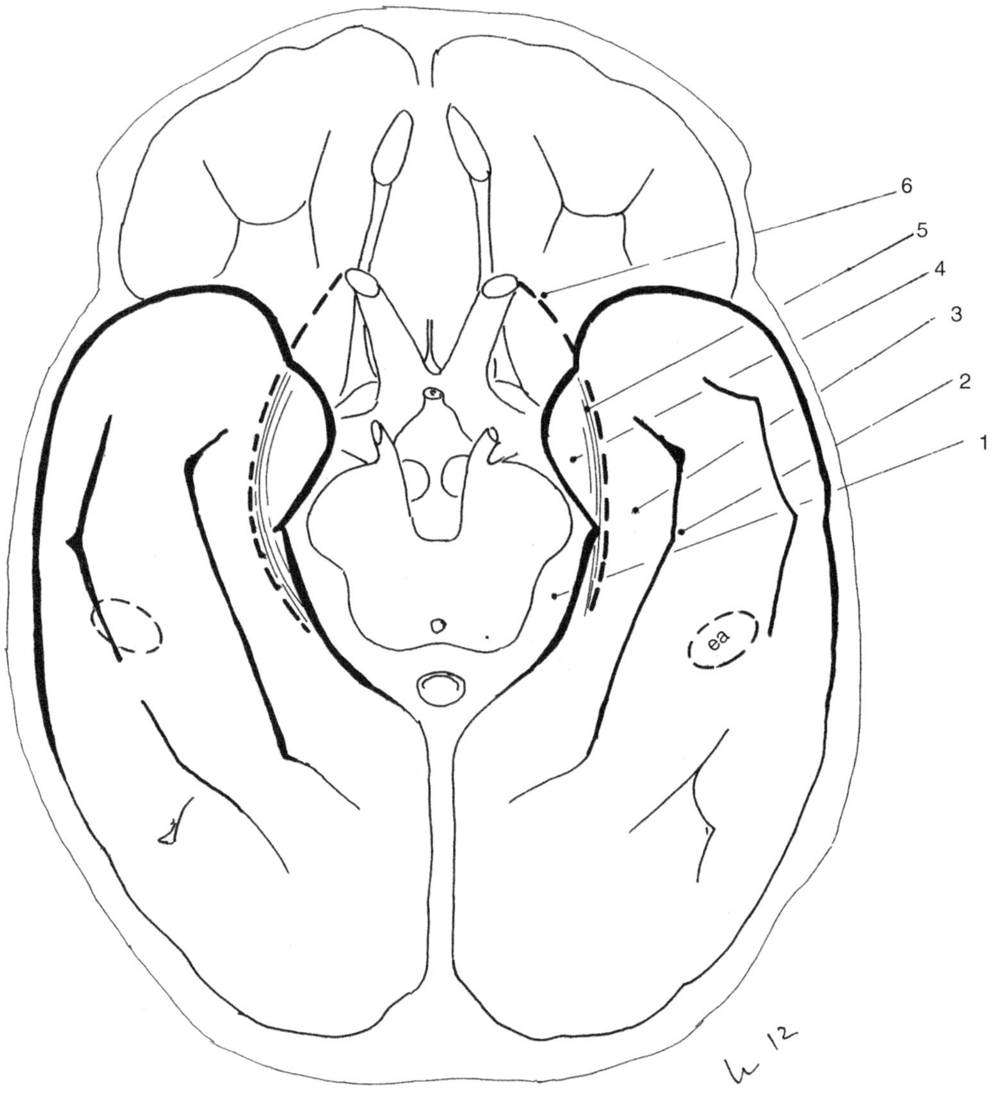

Fig. 6.10 Basal temporal region. *ea* eminentia arcuata, projection. *1* Cisterna ambiens, *2* sulcus collateralis, *3* gyrus parahippocampalis, *4* uncus, *5* impression of uncus (by plica petroclinoidea anterior), *6* plica petroclinoidea anterior, projection

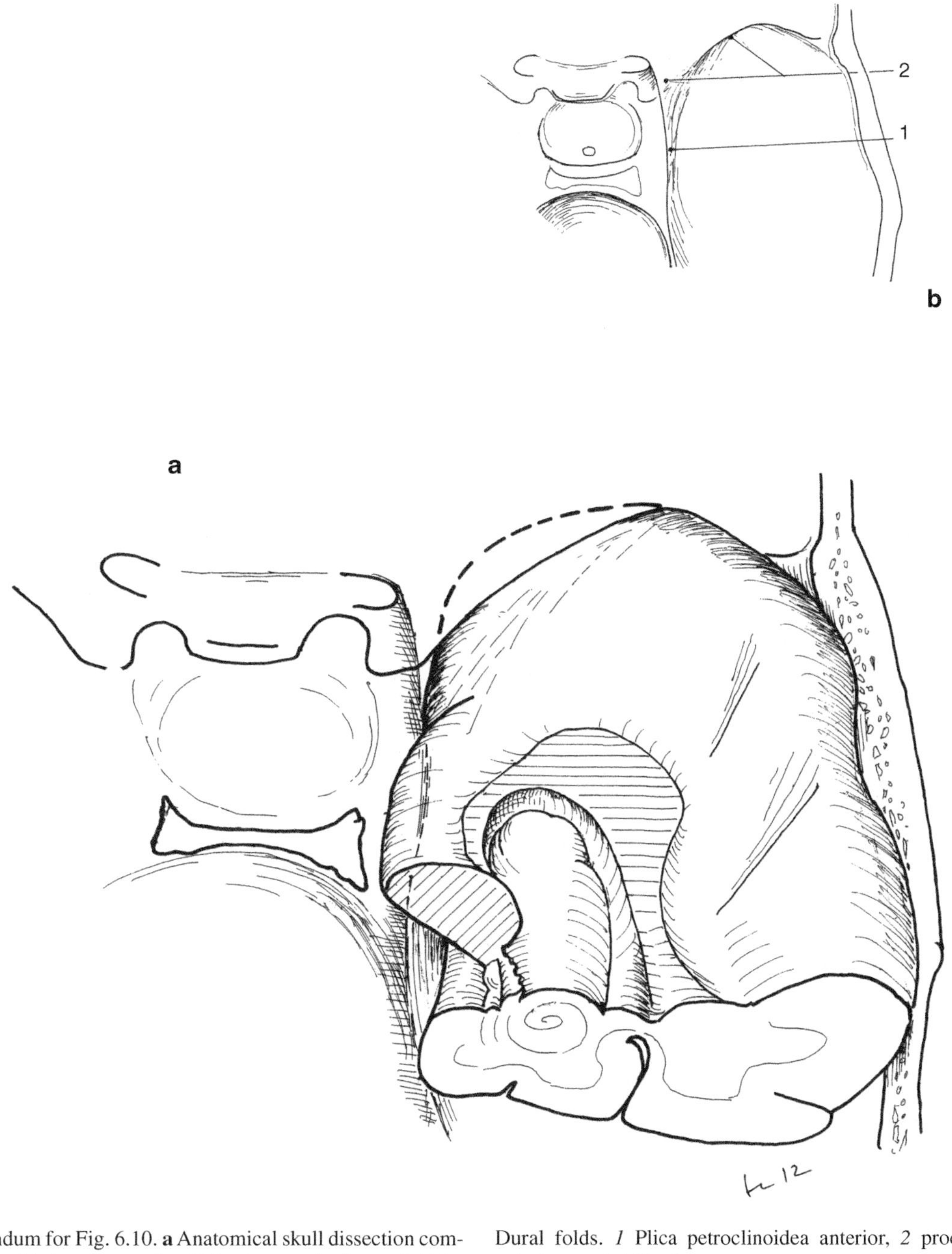

Fig. 6.11 Addendum for Fig. 6.10. **a** Anatomical skull dissection combined with a cast of lobus temporalis. Polus temporalis is overlaid by ala minor. Uncus is located dorsal from plica petroclinoidea anterior. **b** Dural folds. *1* Plica petroclinoidea anterior, *2* processus clinoideus anterior overlaid by *1*

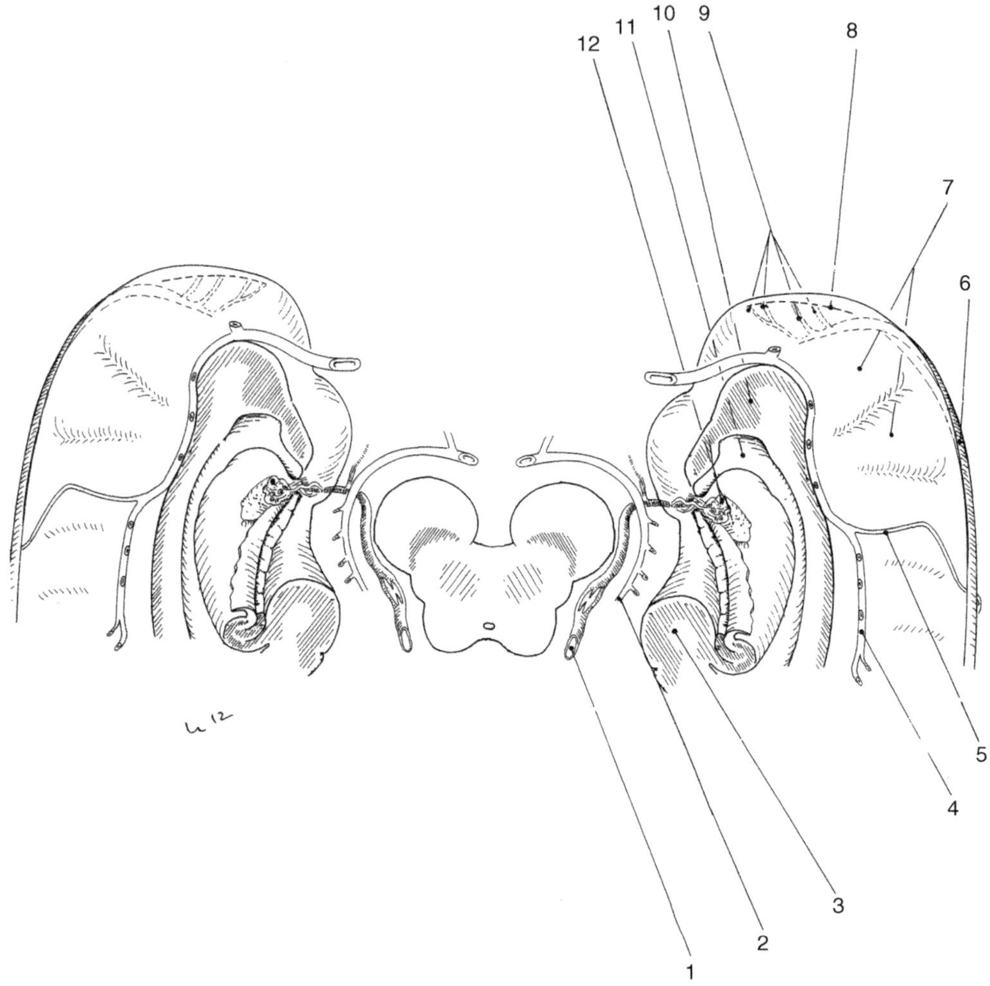

Fig. 6.12 Dorsal temporal (sylvian) region with adjacent structures. *1* V. basalis (Rosenthal); *2* A. cerebri posterior; *3* gyrus parahippocampalis; *4* A. cerebri media; *5* A. temporalis posterior; *6* V. temp. superficialis; *7* Heschl's gyri (compartment of the acoustic cortex); *8* ala minor, projection; *9* temporopolar bridging veins, projection; *10* uncus, transectional plane; *11* tip of cornu inferius; *12* temporal chorioid point

6.2.2 **Lateral Structures** (Figs. 6.13 and 6.14)

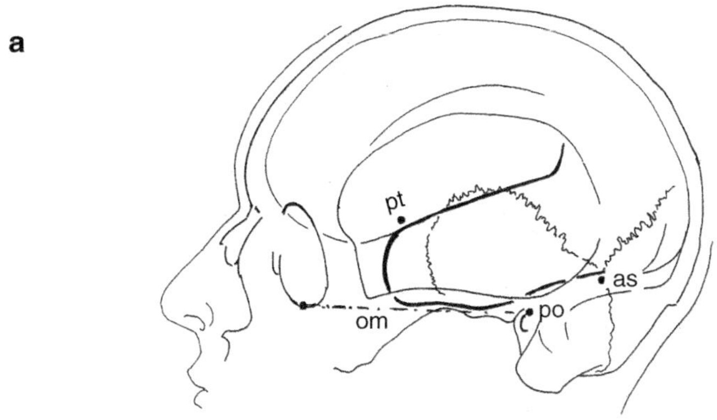

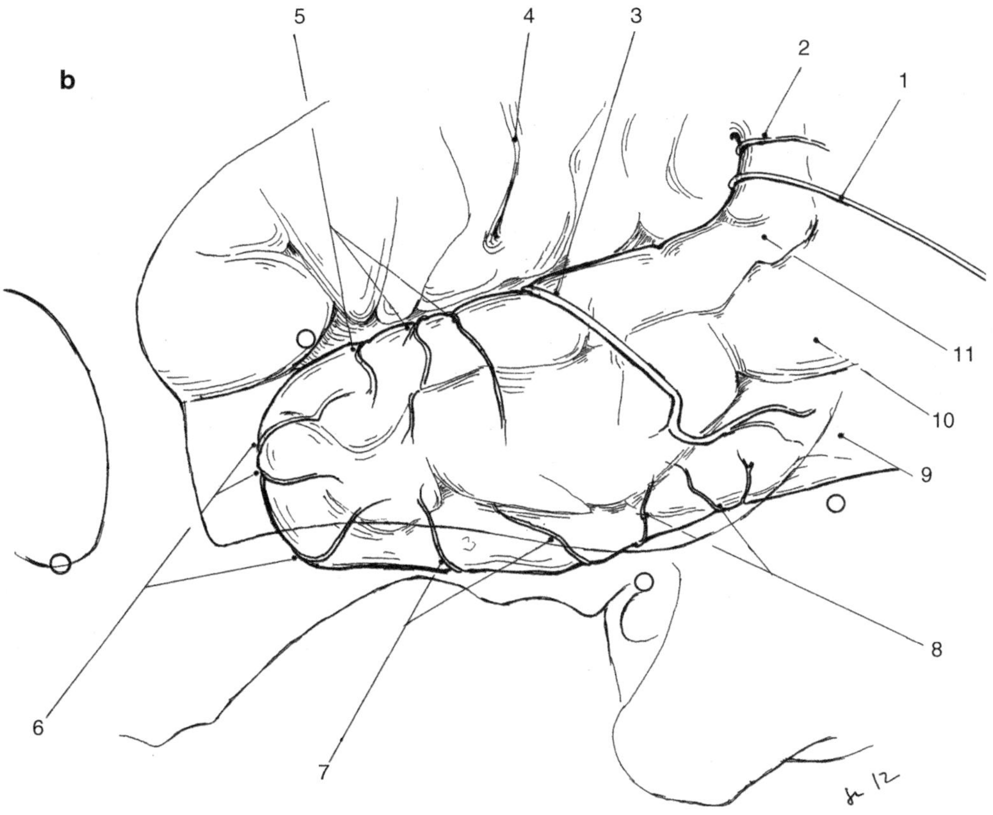

Fig. 6.13 Superficial temporal arteries, schematized. Landmarks added for assisting imagings, if necessary. **a** Bony landmarks, **b** arteries. **a** *as* asterion, *om* orbitomeatal line, *po* porion, *pt* pterion. **b** *1* A. temporooccipitalis, *2* A. angularis, *3* A. temporalis posterior, *4* sulcus centralis, *5* temporal branches of A. cerebri media, *6* temporal branches of A. carotis interna, *7* temporobasal branches of A. cerebri posterior, *8* as 7, *9* third temporal gyrus, *10* second temporal gyrus, *11* pars angularis of the first temporal gyrus (Wernicke). *Light circles*: landmarks as **a**

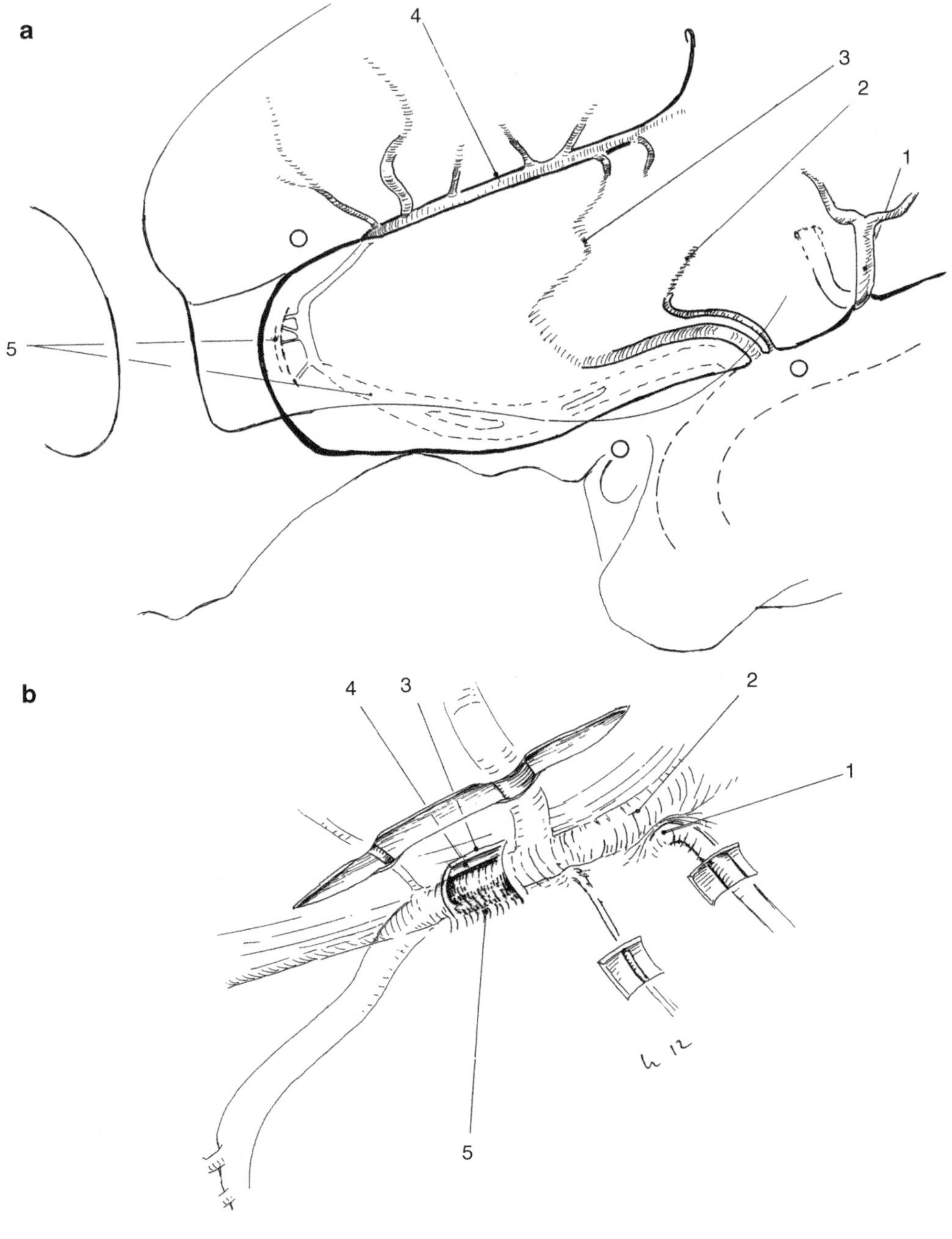

Fig. 6.14 *V. temporalis superficialis* and further temporal veins. **a** V. temporalis superficialis and other cortical veins. **b** Opening of the sylvian cistern. Arachnoid incision is performed superior from V. temporalis superficialis. Now brain damage close to temporal arachnoid adhesions can be avoided. **a** *1* Temporooccipital bridging vein to tentorium, variant (Lang), *2* veins of Labbé, *3* variable parietal veins, *4* V. temporalis superficialis, *5* temporopolar bridging veins are connected with sinus paracavernosus, projection. *Light circles*: landmarks as Fig. 1.2. **b** *1* Temporal arteries underlying V. temporalis superficialis, *2* V. temporalis superficialis, *3* superficial arachnoid layer of the parietal cortex, *4* parietal CSF space, *5* arachnoid adhesions

6.2.3 **Sylvian Structures** (Figs. 6.15, 6.16, 6.17, and 6.18)

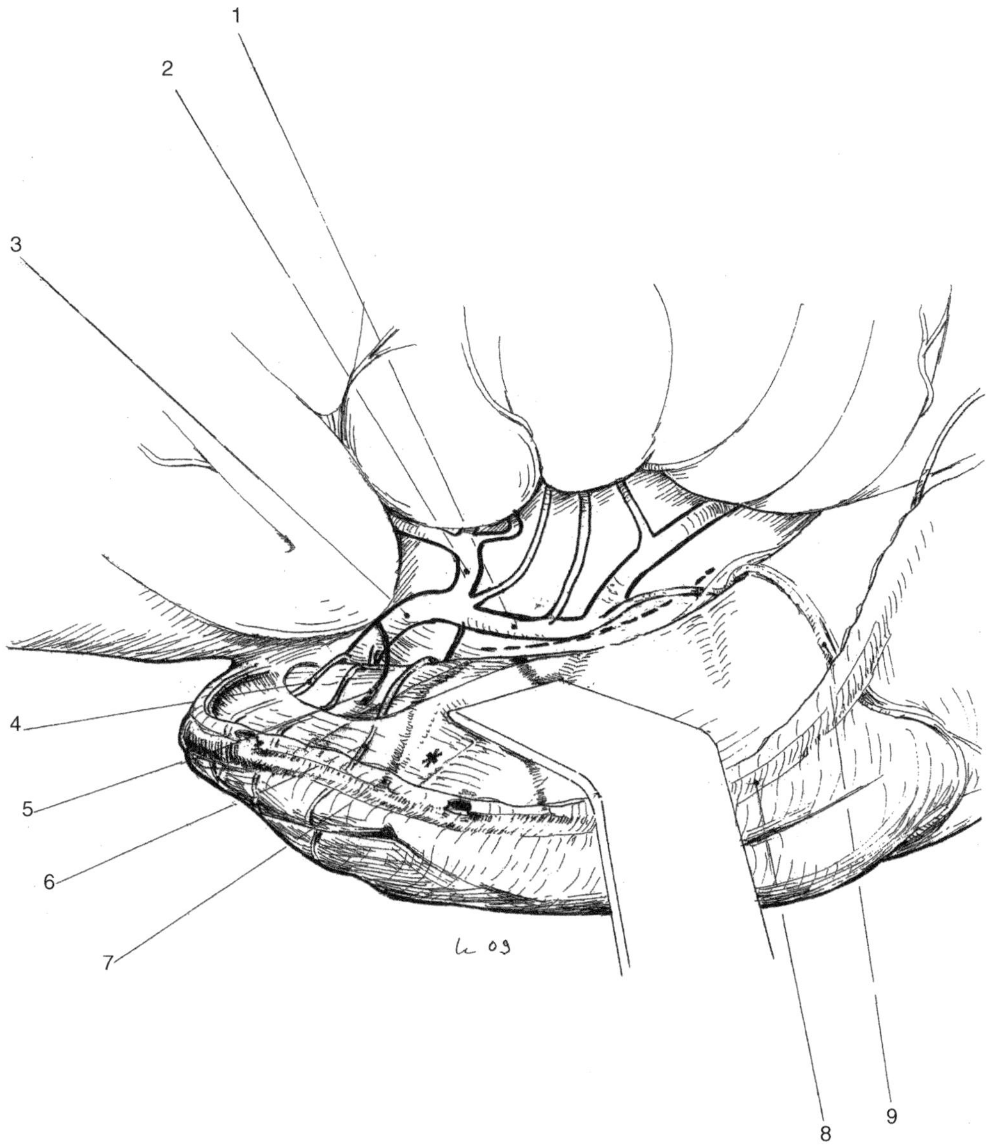

Fig. 6.15 Sylvian arteries and arachnoidea, schematized. Temporal arteries are undercrossing the temporodorsal arachnoid layer which is retracted by spatula. Parietal arteries remain in the so-called sylvian cistern. Surgical aspects: incision of sulcus inferior insulae (*dotted*) for opening of cornu inferius, e.g., for amygdalohippocampectomy. *1* A. cerebri media, distal trunk; *2* A. operculofrontalis; *3* A. cerebri media, main trunk; *4* one of numerous temporopolar branches of A. cerebri media; *5* a small frontotemporal vein crossing sulcus lateralis (sylvian cistern) should be interrupted. At surgical opening of the sylvian cistern; *6* as 4; *7* connection(s) of superficial with profound sylvian veins; *8* V. temporalis superficialis; *9* A. temporalis posterior, *: Heschl's gyri. *Dotted*: incision of sulcus inferior insulae

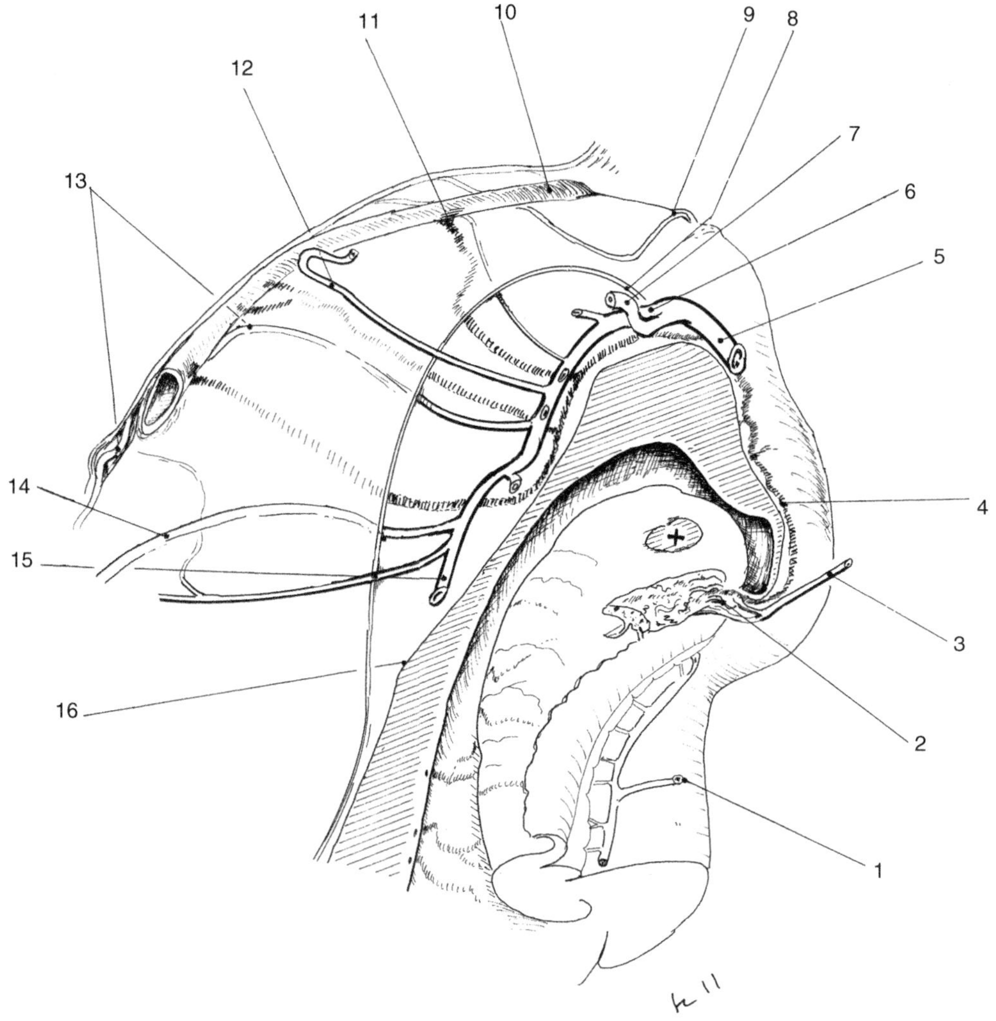

Fig. 6.16 Temporodorsal sylvian region, cornu inferius, and vessels. *1* A. hippocampi; *2* temporal chorioid point; *3* A. chorioidea anterior; *4* V.(Vv.) temporalis profunda (ae); *5* A. cerebri media, main trunk; *6* distal trunk of A. cerebri media; *7* A. operculofrontalis; *8* arachnoid wall between sylvian cistern and temporodorsal subarachnoid space. Note positions of temporal and parietal vessels to it; *9* temporopolar branch of A. carotis interna; *10* V. temporalis superficialis; *11* connection of 10 with Vv. temporales profundi; *12* one of the parietal branches of A. cerebri media; *13* temporal branches of A. cerebri media; *14* A. temporalis posterior; *15* A. angularis et A. temporooccipitalis; *16* sulcus inferior insulae, + adhesion of hippocampus with the roof of cornu inf., (common finding). It may be transected at surgery, e.g., amygdalohippocampectomy, if necessary

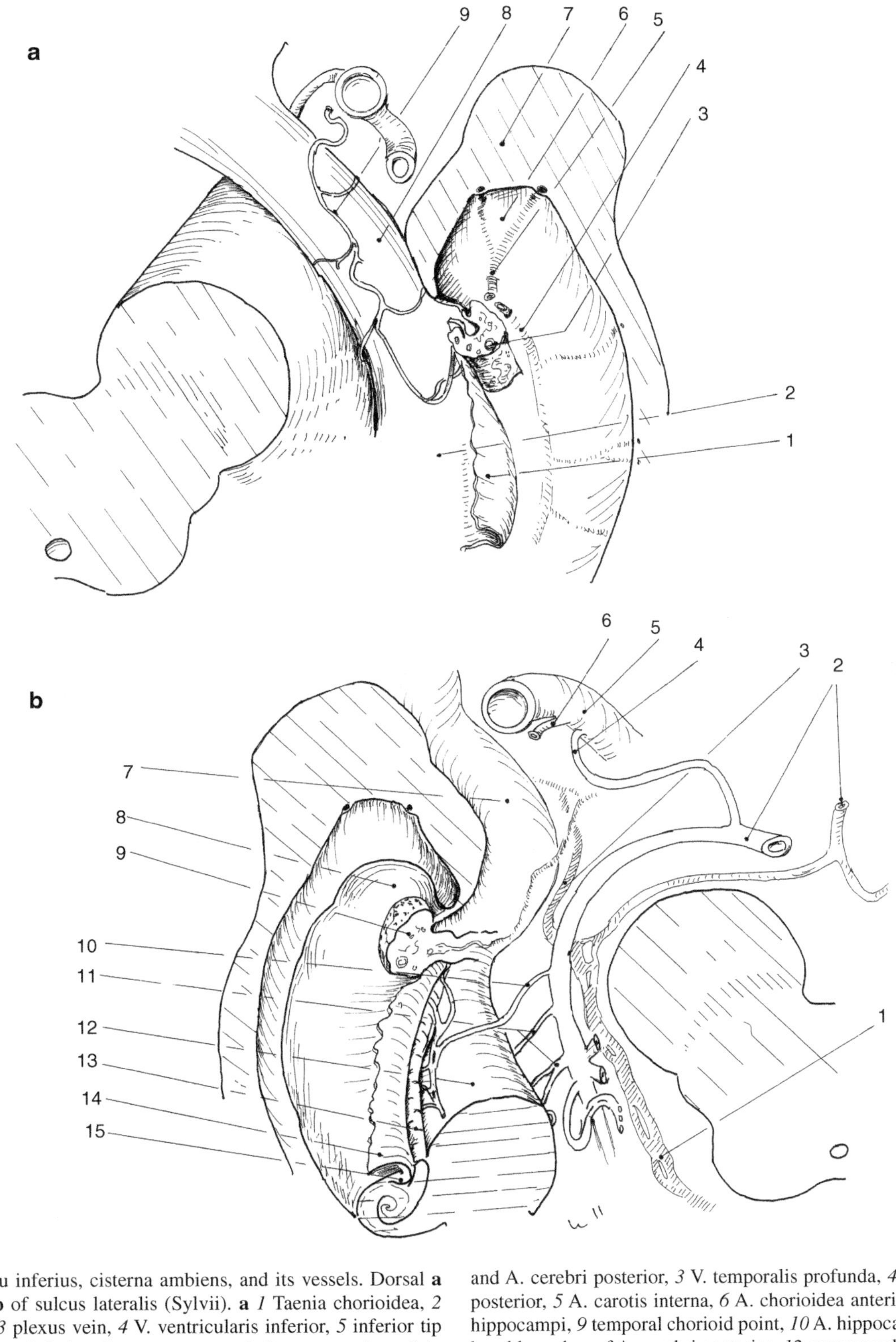

Fig. 6.17 Cornu inferius, cisterna ambiens, and its vessels. Dorsal **a** and basal wall **b** of sulcus lateralis (Sylvii). **a** *1* Taenia chorioidea, *2* tractus opticus, *3* plexus vein, *4* V. ventricularis inferior, *5* inferior tip vein, *6* tip of cornu inferius, *7* uncus transected, *8* tractus opticus, *9* A. chorioidea anterior; **b** *1* V. basalis (Rosenthal), *2* V. interpeduncularis and A. cerebri posterior, *3* V. temporalis profunda, *4* A. communicans posterior, *5* A. carotis interna, *6* A. chorioidea anterior, *7* uncus, *8* pes hippocampi, *9* temporal chorioid point, *10* A. hippocampi, *11* temporobasal branches of A. cerebri posterior, *12* gyrus parahippocampalis, *13* gyrus dentatus, *14* taenia fornicis, *15* fimbria fornicis

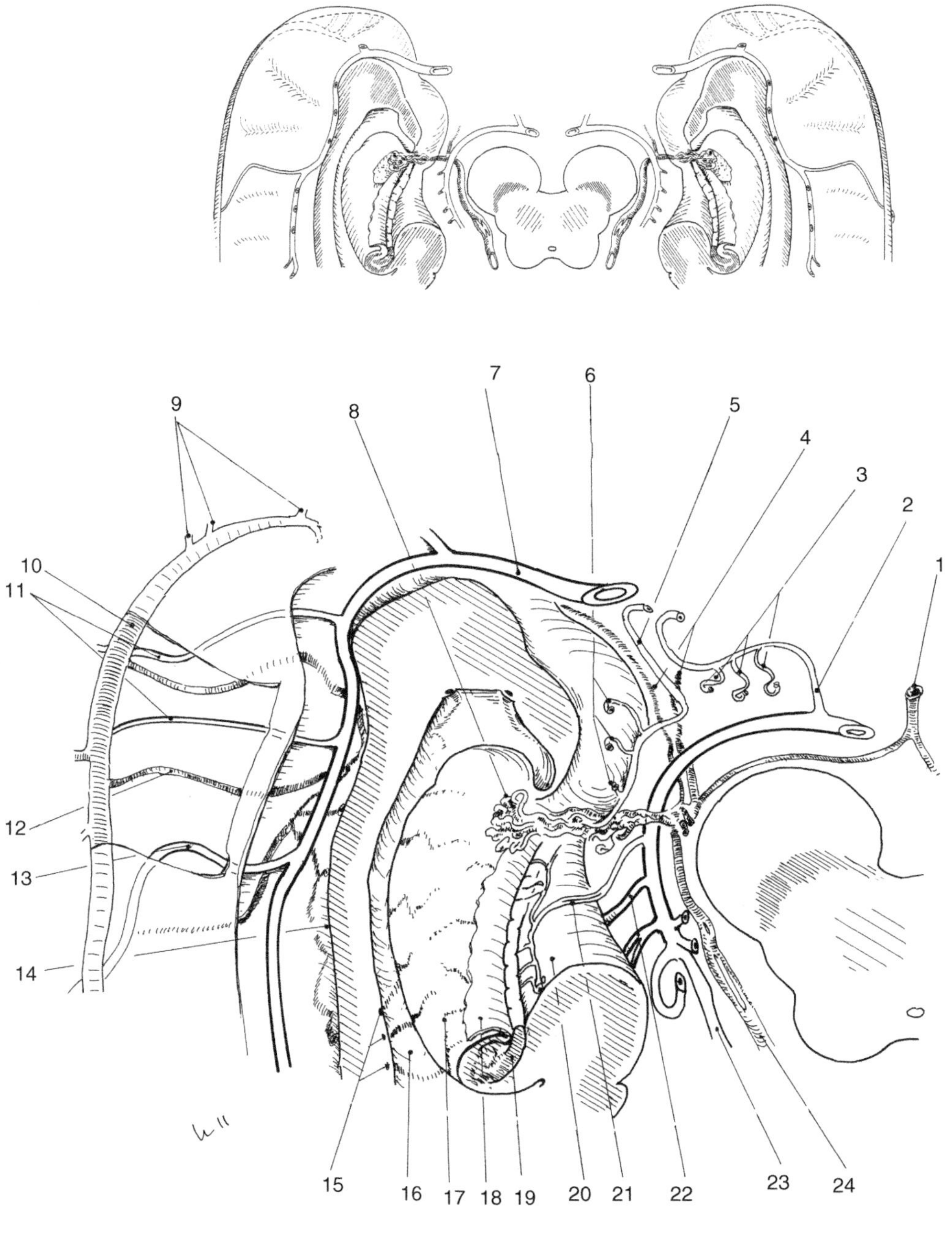

Fig. 6.18 Conclusion. Minimized survey of Fig. 6.12 added. *1* V. interpeduncularis, *2* A. communicans posterior, *3* perforating arteries, *4* Vv. temporales profundi, *5* A. chorioidea anterior, *6* perforating arteries, *7* A. cerebri media, *8* temporal chorioid point, *9* bridging veins, *10* V. temporalis superficialis, *11* temporal branches of A. cerebri media, *12* anastomosis between superficial and profound sylvian veins, *13* A. temporalis posterior, *14* sulcus inferior insulae, *15* lateral atrial veins, *16* atrium, *17* hippocampus, *18* fimbria fornicis, *19* gyrus dentatus, *20* gyrus parahippocampalis, 21 A. hippocampi, *22* temporobasal branches of A. cerebri posterior, *23* A. cerebri posterior, *24* V. basalis (Rosenthal)

6.2.4 Temporal Allocortex (Figs. 6.19, 6.20, 6.21, 6.22, 6.23, 6.24, 6.25, and 6.26)

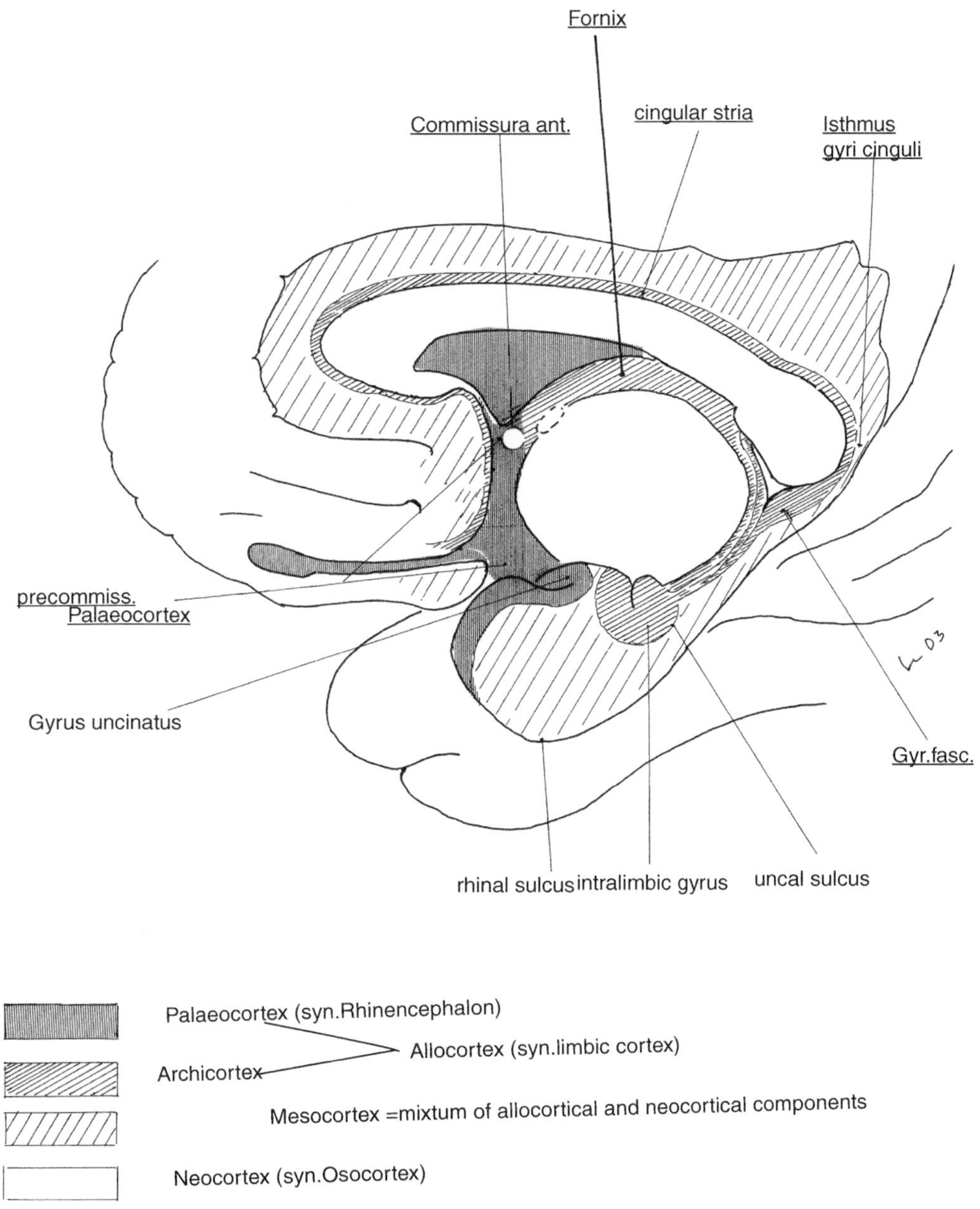

Fig. 6.19 Allocortex («limbic» system) (schematized). Survey. Unilateral surgical eliminations of the temporal component of allocortex are performed, e.g., for therapy of epilepsy. Bilateral lesions are followed by severe psychological deficits. Extratemporal allocortical structures are frontal and parietooccipital striae longitudinales and indusium griseum of corpus callosum, area subcallosa, posterior segment of gyrus rectus, further tractus et striae olfactoriae and bulbus olfactorius et fila olfactoria

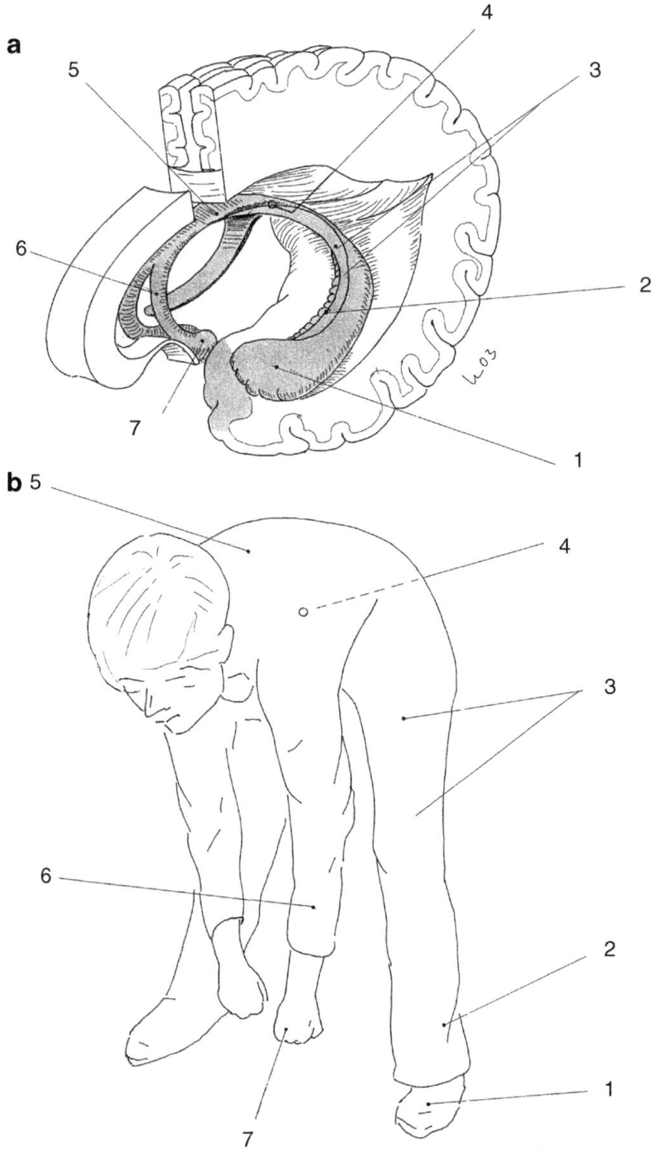

Fig. 6.20 Addendum for Fig. 6.19 (Feneis in Kahle (2001)). **a** *1* Pes hippocampi, *2* sulcus fimbriodentatus, *3* fimbria fornicis, *4* crus fornicis, *5* corpus fornicis, *6* columna fornicis, *7* corpus mammillare. **b** Demonstration during a lecture for anatomy of fornix and temporal allocortex (Prof. Feneis, Anatomisches Institut der Universität Tübingen (Germany), teaching 1950. Recollection of the author)

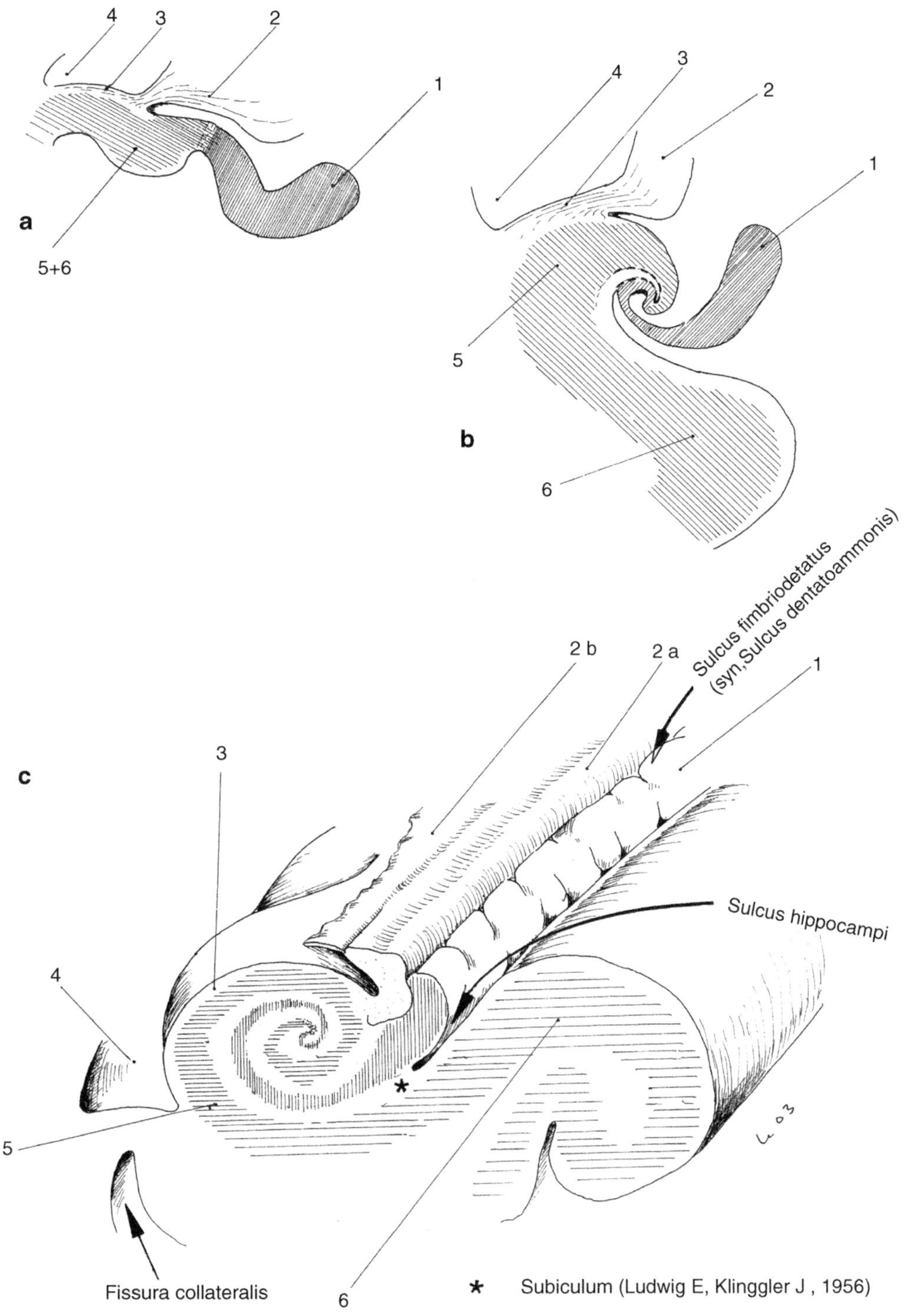

Fig. 6.21 Temporal allocortex. **a** and **b** Development (According to Duvernoy 1988, pp 35 and 93, schematized). **c** Allocortex of adults. **a** and **b:** *1* gyrus dentatus, *2* fornix, *3* alveus, *4* cornu inferius, *5* hippocampus, *6* gyrus parahippocampalis. **c** *1* Gyrus dentatus, *2a* fimbria fornicis, *2b* taenia fornicis, *3* alveus, *4* cornu inferius, *5* hippocampus, *6* gyrus parahippocampalis

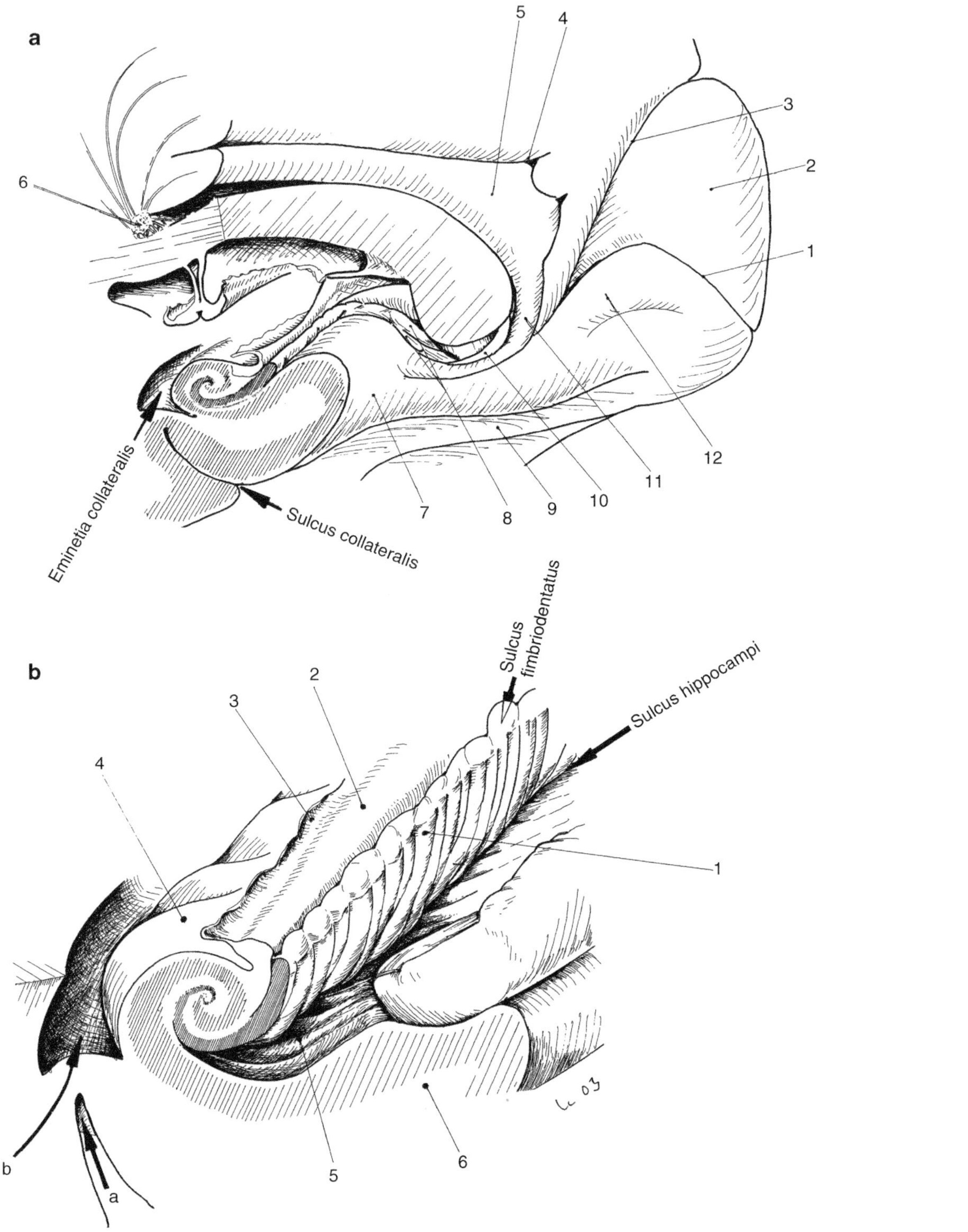

Fig. 6.22 Allocortex and surrounding structures. **a** Survey. **b** Details after digital loosening of area dentata from gyrus parahippocampalis. Numerous fine fiber connections were interrupted. Area dentata: hippocampus plus gyrus dentatus. **a** *1* Sulcus calcarinus, *2* cuneus, *3* sulcus parietooccipitalis, *4* sulcus cinguli, *5* gyrus cinguli, *6* stria longitudinalis cinguli, *7* gyrus parahippocampalis, *8* gyrus dentatus-fasciolaris, *9* gyrus occipitotemporalis lateralis, *10* tail of hippocampus located at the outer temporal cortex, *11* isthmus gyri cinguli, *12* gyrus occipitotemporalis medialis; **b** *1* gyrus dentatus, *2* fimbria fornicis, *3* taenia fornicis, *4* alveus, *5* digital loosening of area dentata from gyrus parahippocampalis, *6* gyrus parahippocampalis; *a* ground of sulcus collateralis, *b* ground of cornu inferius

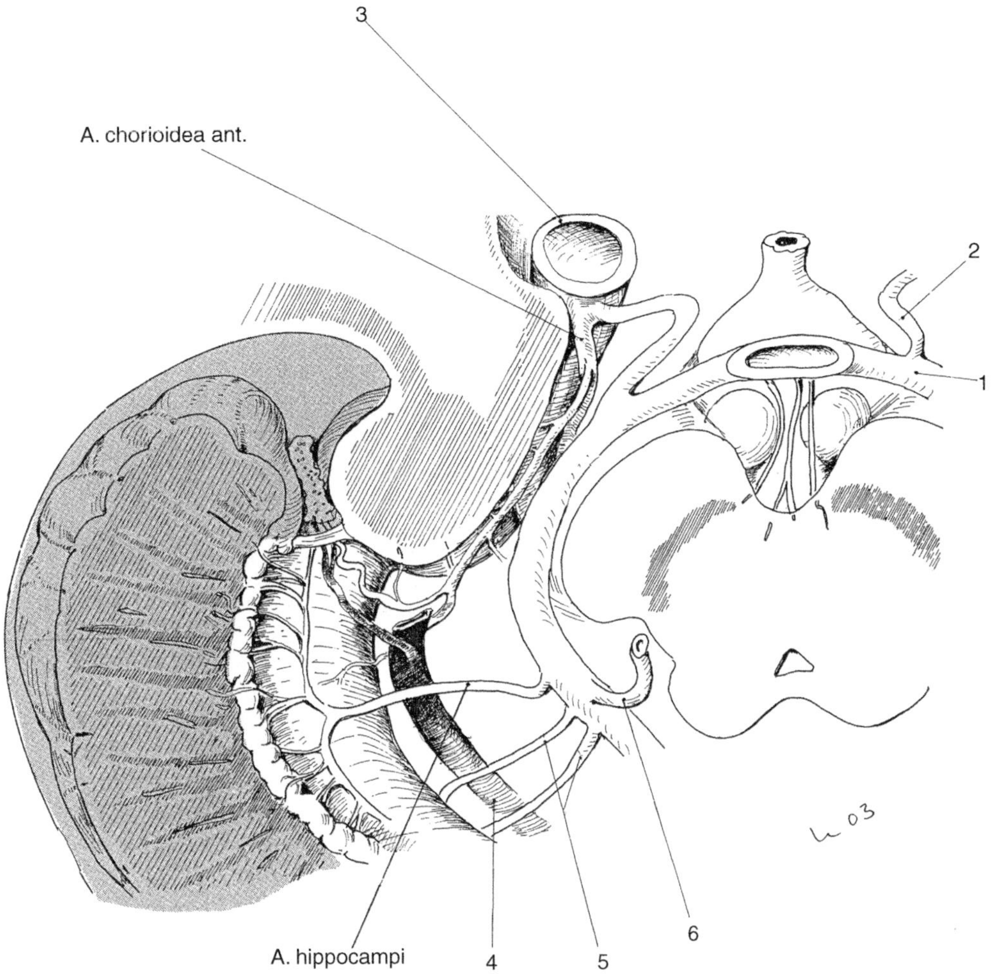

Fig. 6.23 Feeding arteries of the temporal allocortex. A. chorioidea ant., branches feeding amygdala and temporal chorioid point A. hippocampi feeding area dentata and gyrus parahippocampalis. *1* A. cerebri posterior, *2* A. communicans posterior, *3* A. carotis interna, *4* V. basalis (Rosenthal), *5* temporobasal branches of A. cerebri posterior, *6* A. chorioidea posterior lateralis

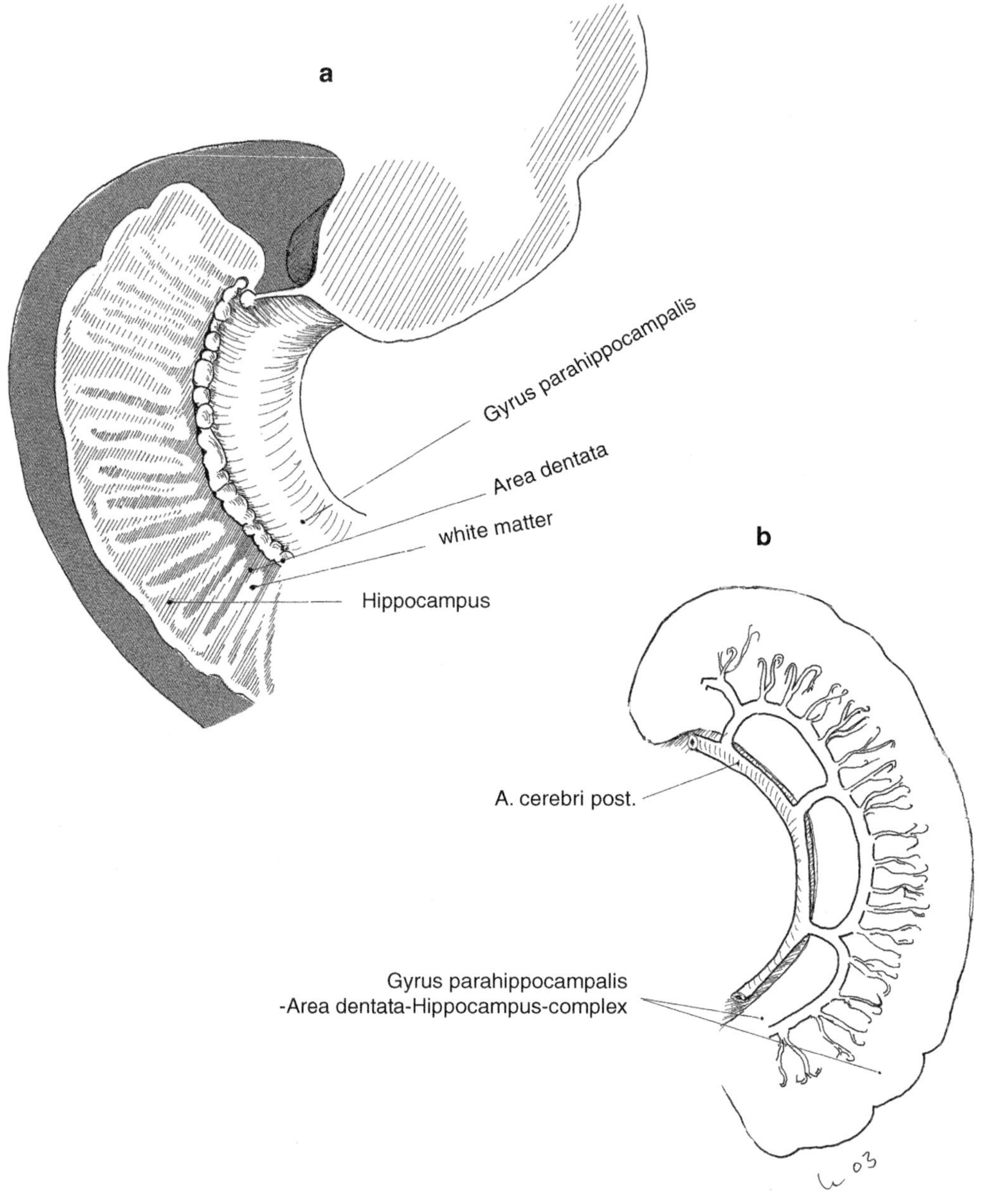

Fig. 6.24 Feeding arteries of area dentata (hippocampus-gyrus-dentatus complex). **a** Transectional plane of the hippocampus, **b** ramification of A. hippocampi (According to Stephan 1975, pp 115 and 119)

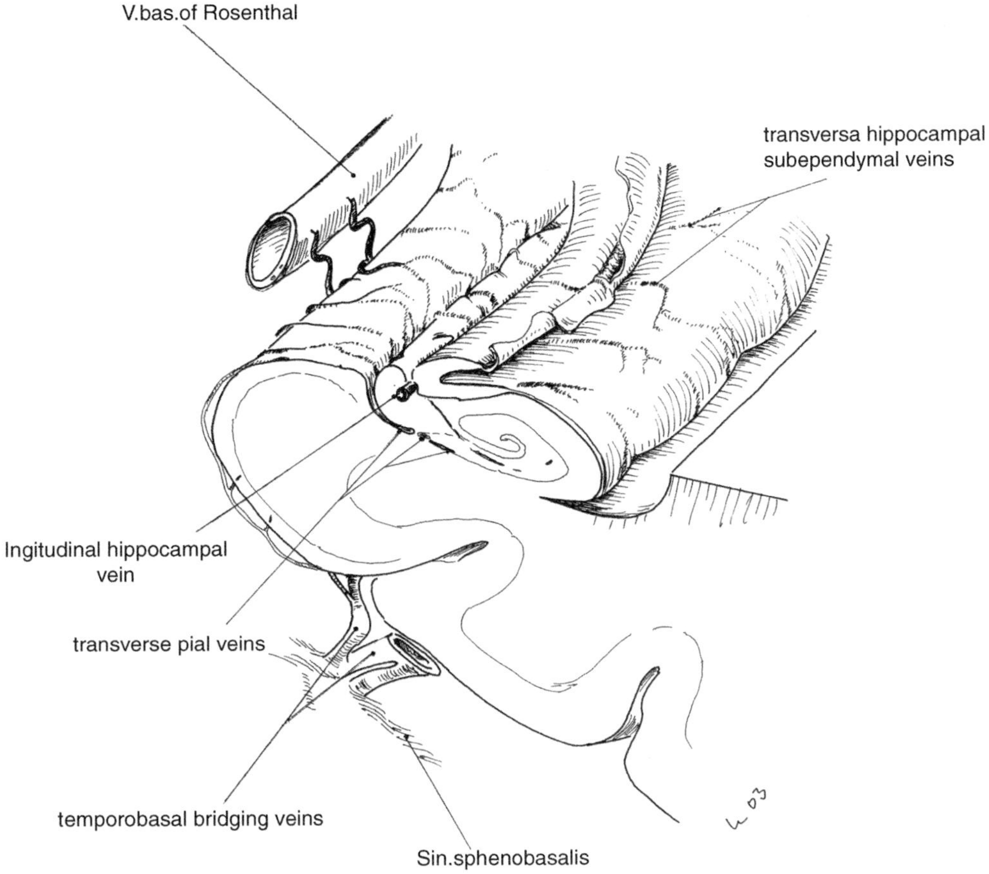

Fig. 6.25 Draining veins of the temporal allocortex

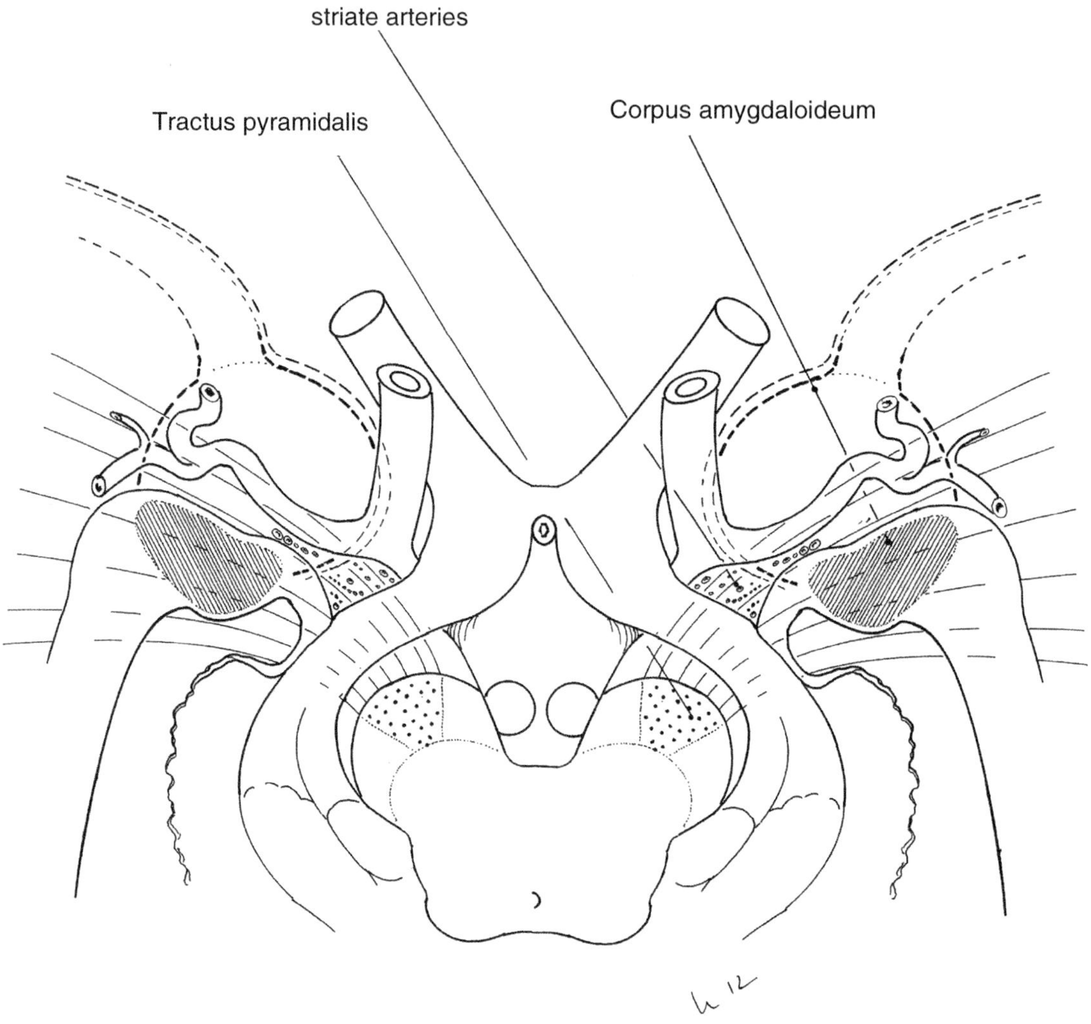

Fig. 6.26 Corpus amygdaloideum (amygdala) and surrounding structures. Essential adjacent structures must be preserved at amygdalohippocampectomy

7 Special Surgical Aspects

7.1 Extradural Surgery (Figs. 7.1 and 7.2)

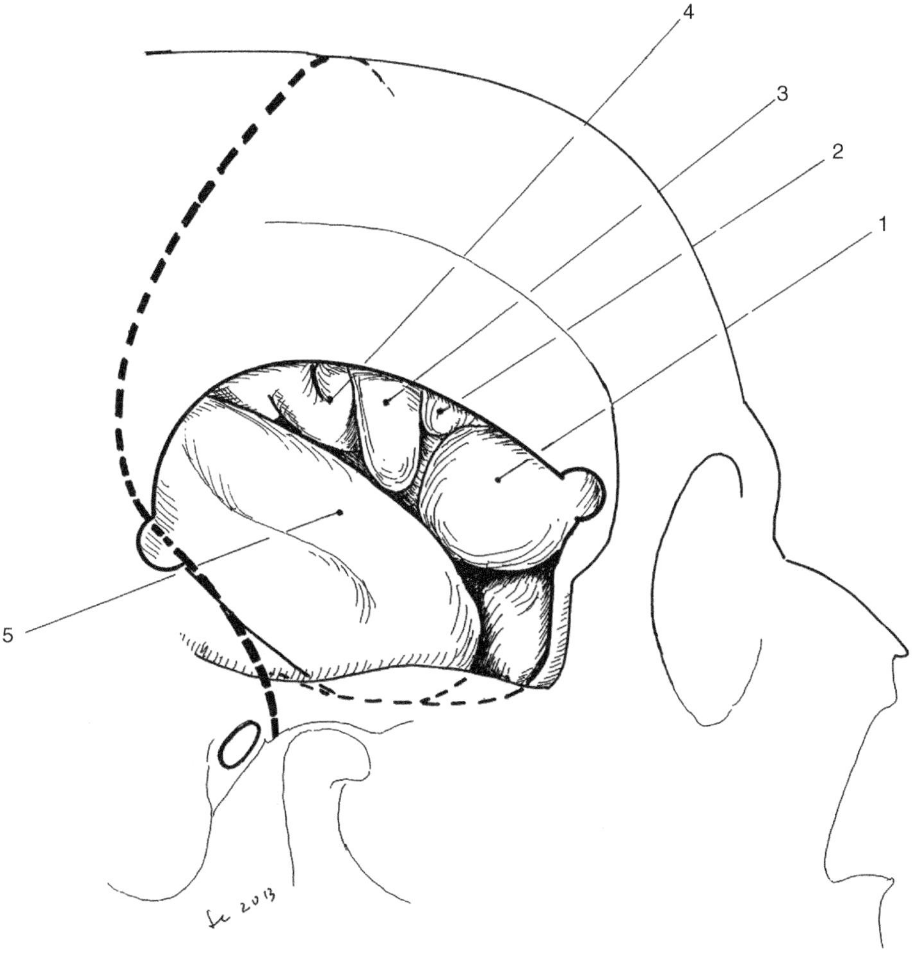

Fig. 7.1 Pterional craniotomy widened for amygdalohippocampectomy (Used by Prof. Zentner, Freiburg 2014). Skin-galea incision dotted. Cerebral relief drawn in *1* third frontal gyrus, pars orbitalis; *2* as 1, pars triangularis; *3* as 1, pars opercularis; *4* gyrus precentralis; *5* first temporal gyrus

W. Seeger, J. Zentner, *Anatomical Basis of Cranial Neurosurgery*, https://doi.org/10.1007/978-3-319-63597-2_7

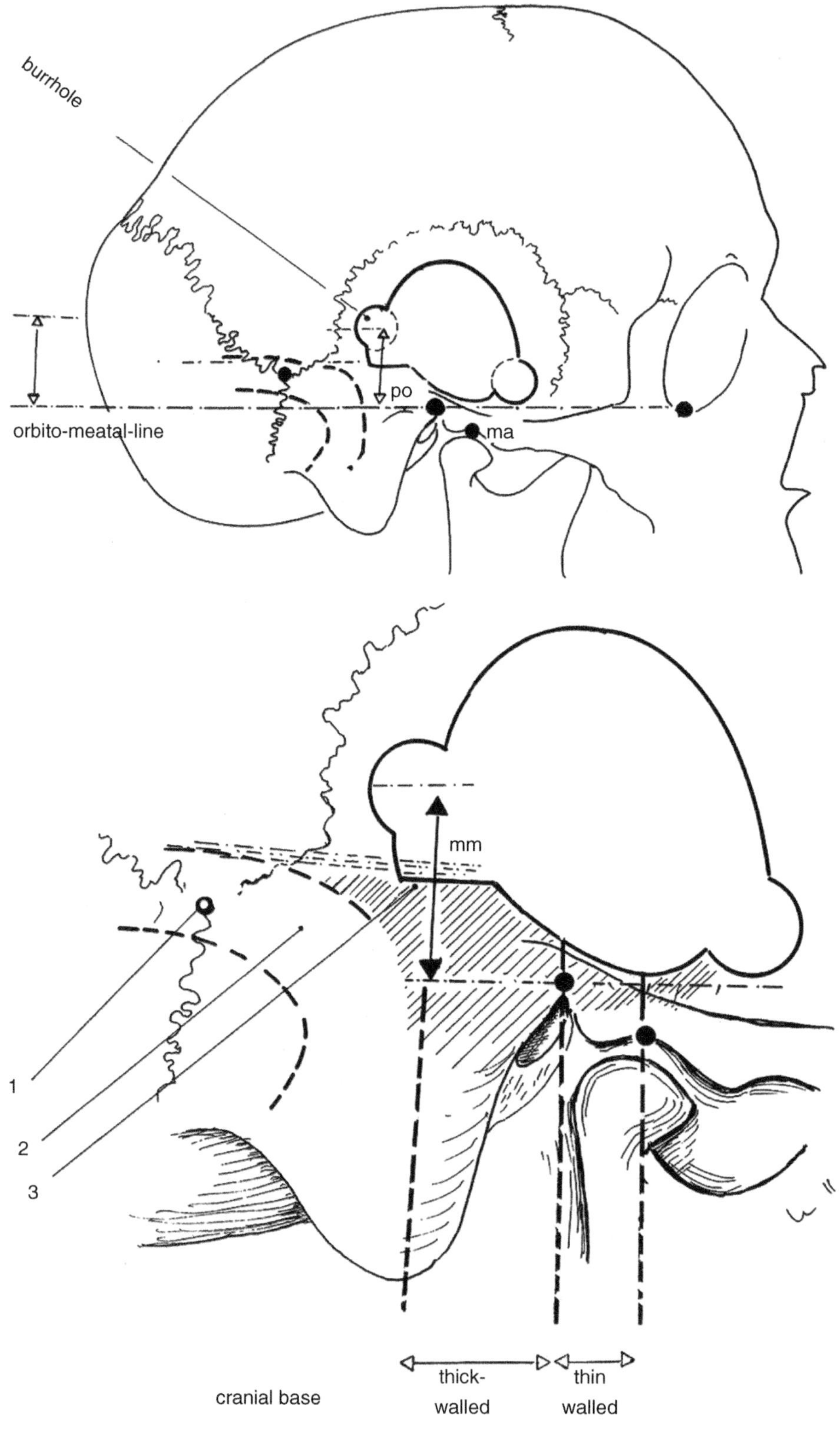

Fig. 7.2 Craniotomy for temporobasal approaches. Craniotomy should be performed as basally as possible for avoiding cerebral or craniobasal damage. Endangered are lobus temporalis after too far dorsally performed craniotomy. Endangered are antrum mastoideum of cavum tympanicum, labyrinth, and sinus transversus after a too far basally performed craniotomy. Pneumatic cells must be considered. Note the small distance measurement between fossa mandibularis and fossa cranii media. *ma* fossa mandibularis, *po* porion; *1* asterion, *2* knee of sinus transversus, *3* lateral insertion of os petrosum at squama temporalis, close to antrum tympanicum hatched: area of widened mastoid cells

7.2 Intradural Surgery

7.2.1 Standard Resections of Lobus Temporalis (Figs. 7.3, 7.4, and 7.5)

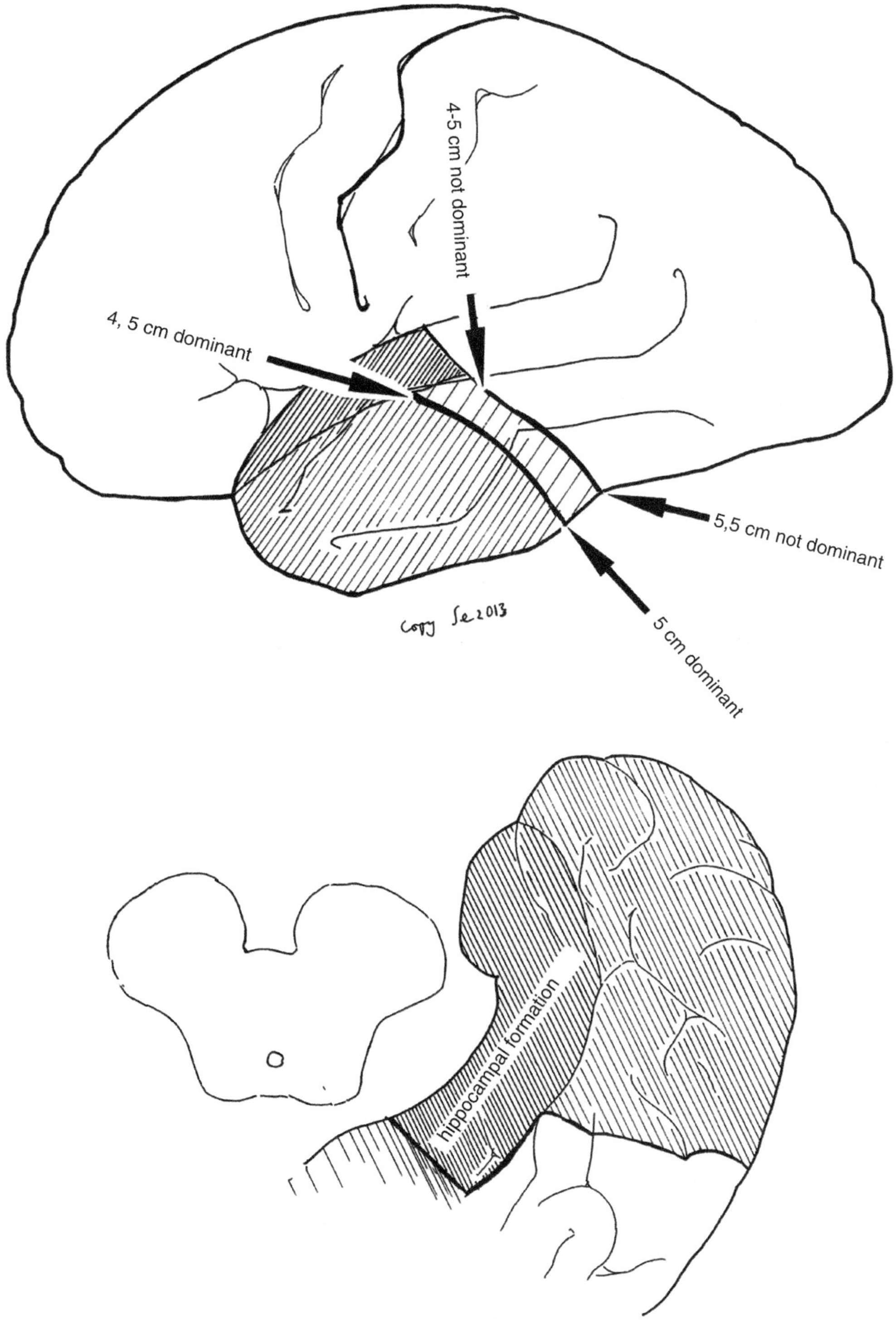

Fig. 7.3 Resection of the temporal lobe at the dominant and nondominant hemisphere with and without amygdalohippocampectomy (According to drawings of Zentner 1997, modified presentation)

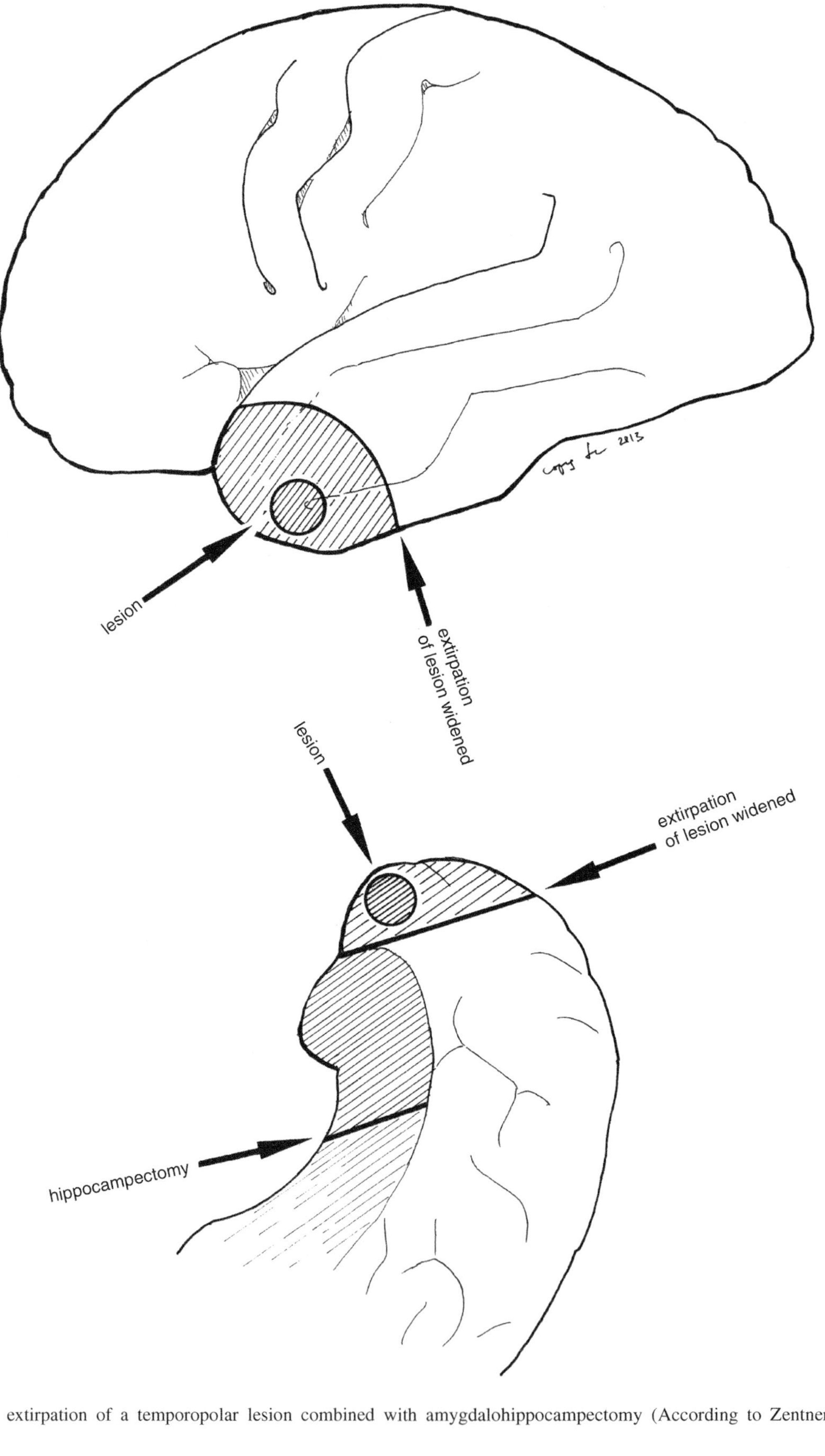

Fig. 7.4 Keyhole extirpation of a temporopolar lesion combined with amygdalohippocampectomy (According to Zentner 1997, modified presentation)

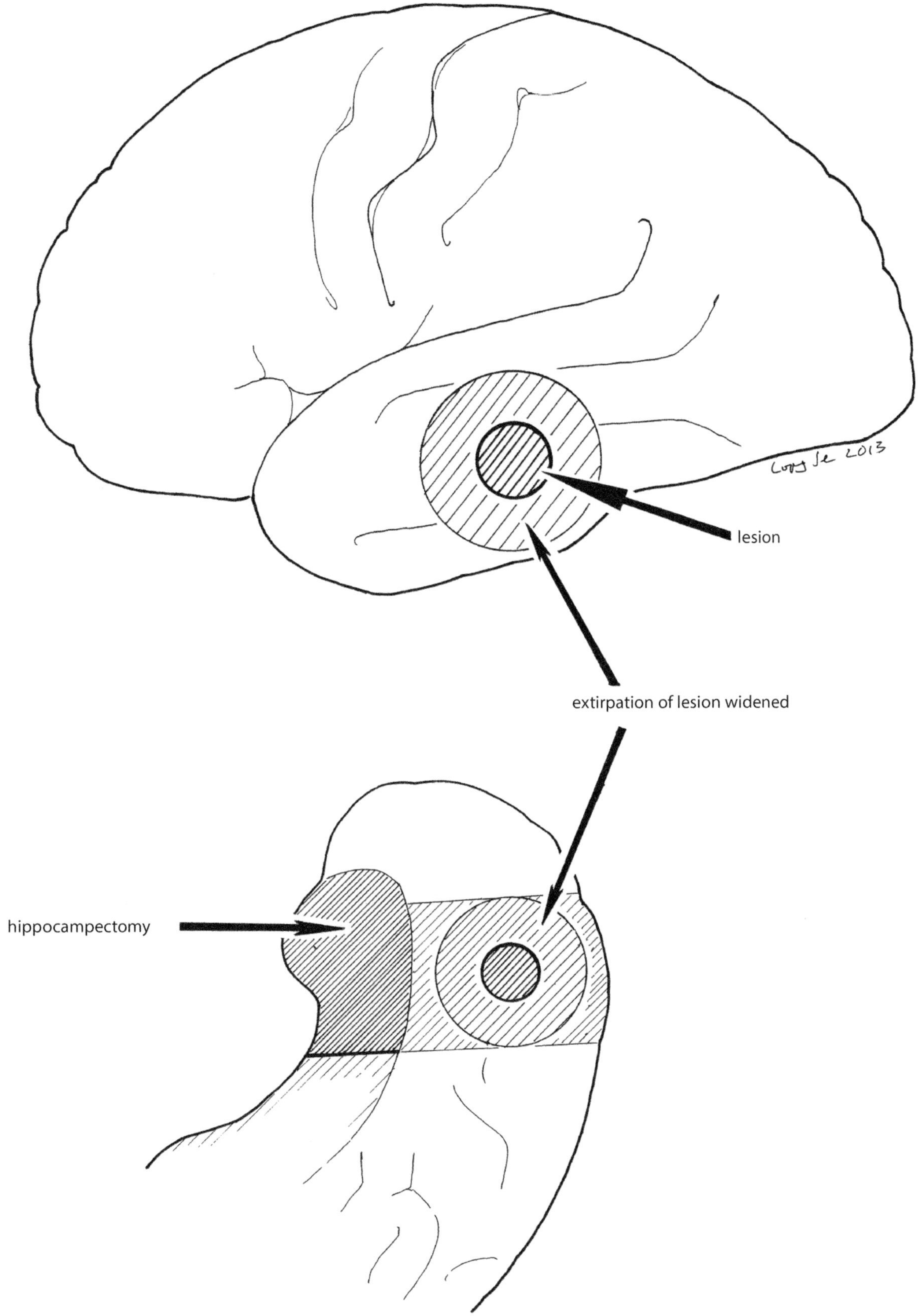

Fig. 7.5 Keyhole extirpation of a central temporal lesion combined with amygdalohippocampectomy (According to Zentner 1997, modified presentation)

7.2.2 **Amygdalohippocampectomy** (Figs. 7.6, 7.7, 7.8, 7.9, 7.10, 7.11, 7.12, 7.13, 7.14, 7.15, 7.16, and 7.17)

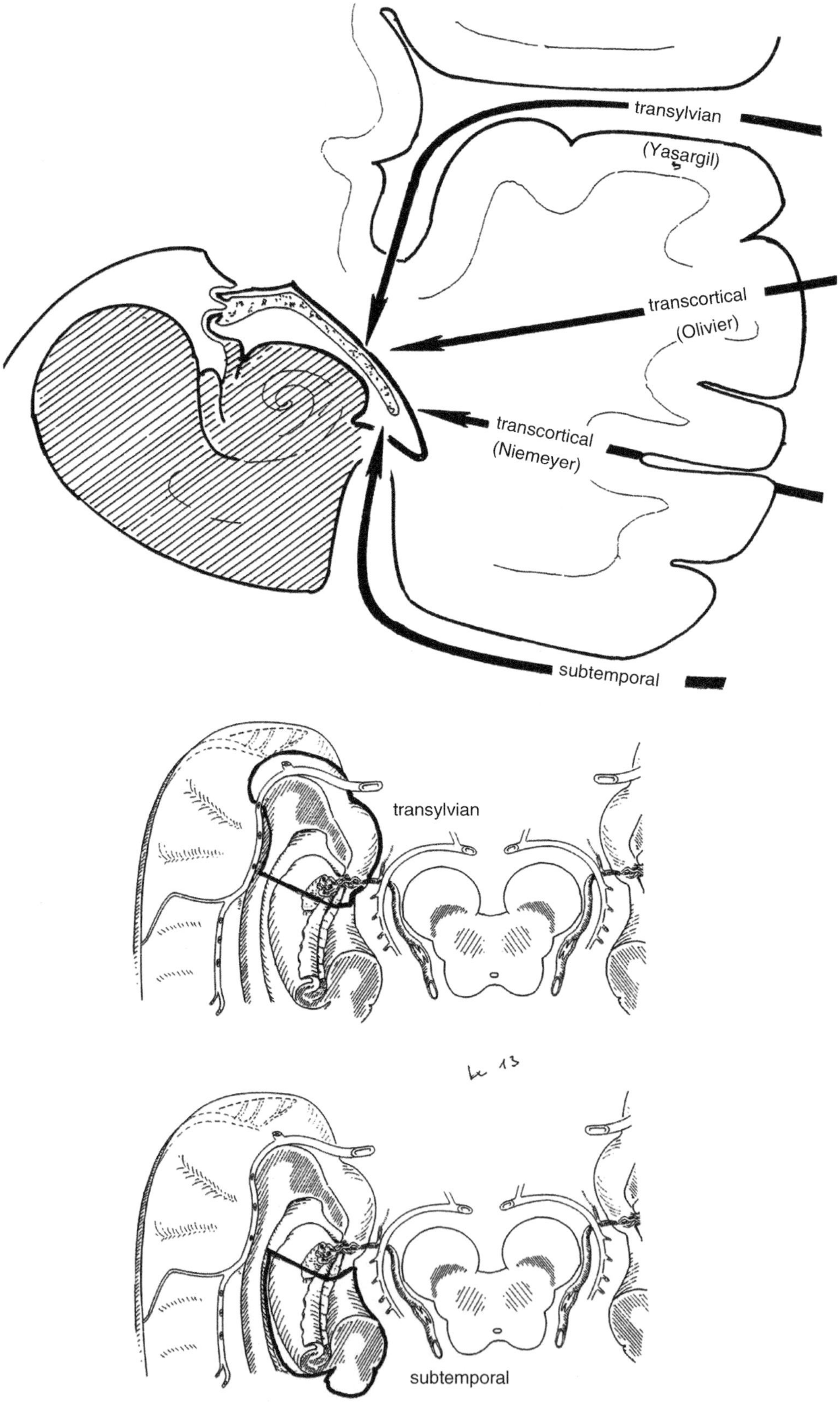

Fig. 7.6 Routes and target regions. Survey. Routes (*arrow*) and target areas (hatched) could be widened to a complete amygdalohippocampectomy (Prof. Zentner, personal communication to the author 2014)

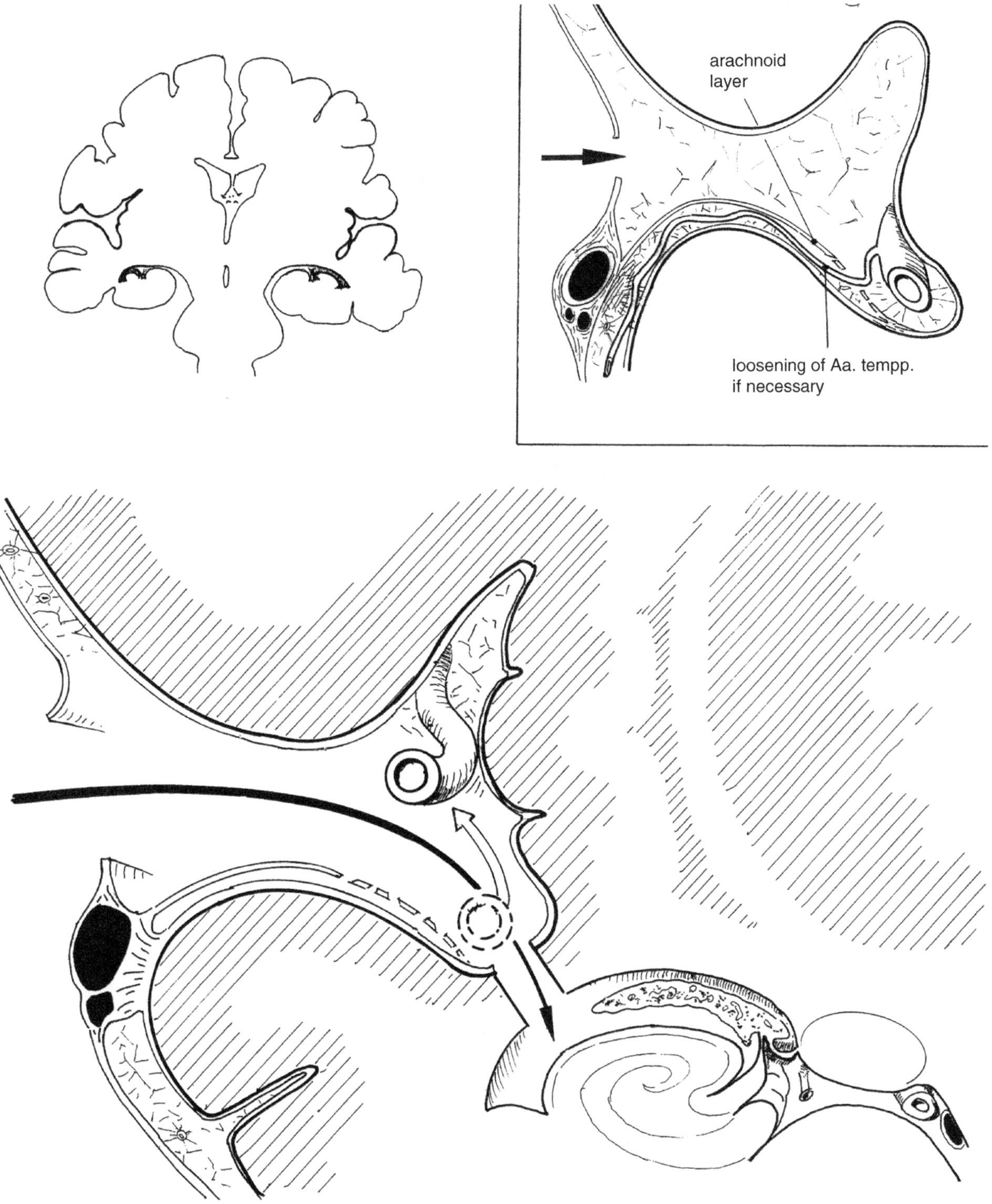

Fig. 7.7 Transsylvian route to cornu inferius for amygdalohippocampectomy (*black arrows*) according to Yaşargil et al. (1985). If Sulcus inferior insulae is overlaid by A. cerebri media, now this artery should be shifted to a superior direction (*light arrow*). One or two small perforating arteries could be interrupted, if necessary (Yaşargil). Sulcus inferior insulae must be identified, before it will be incised for opening cornu inferius. Basal ganglia, pyramidal tract, and perforating vessels could be endangered. If cornu inferius is presented, amygdalohippocampectomy can be performed

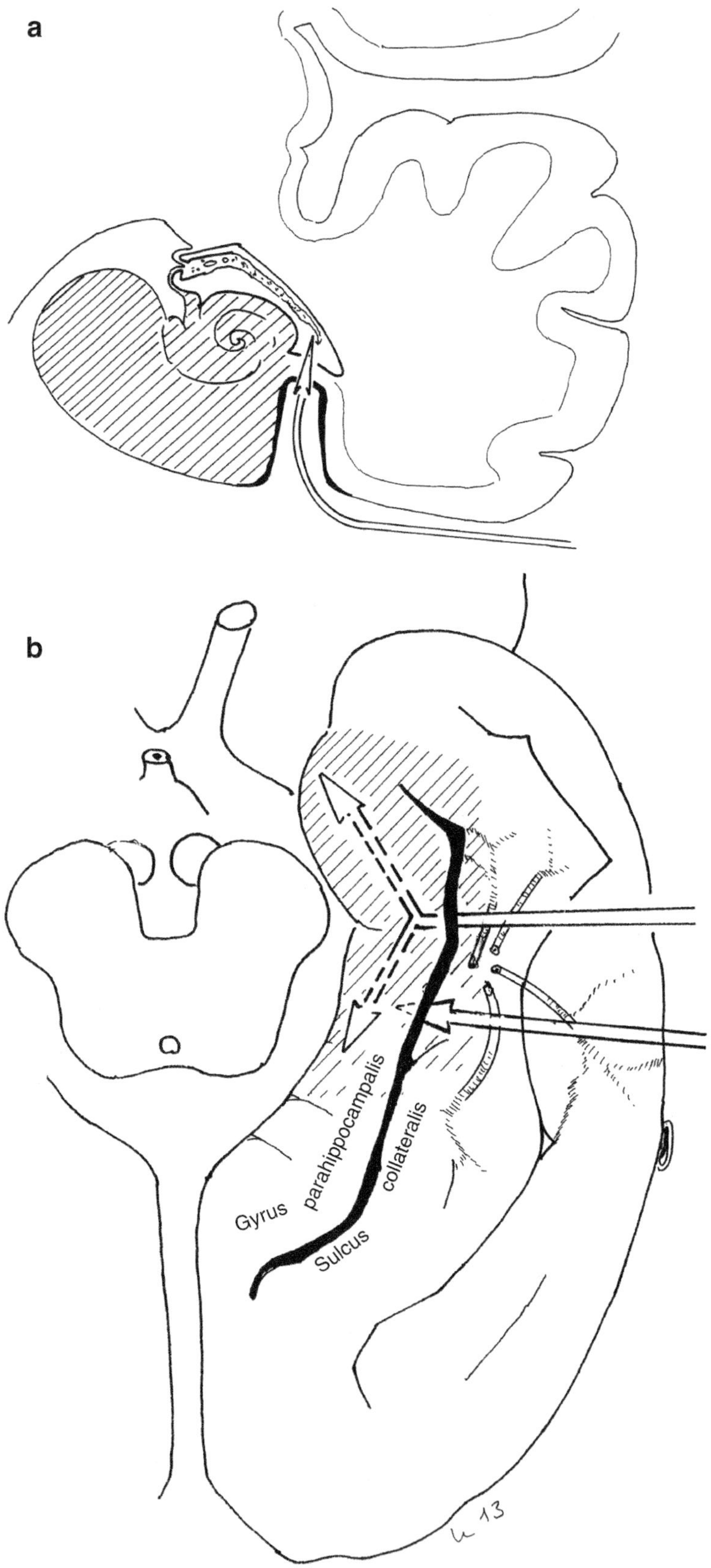

Fig. 7.8 Subtemporal route to cornu inferius for amygdalohippocampectomy (*light arrows*) (Applied by Prof. Zentner, Freiburg i B 2013) transcrossing sulcus collateralis and cornu inferius (**a** and **b**)

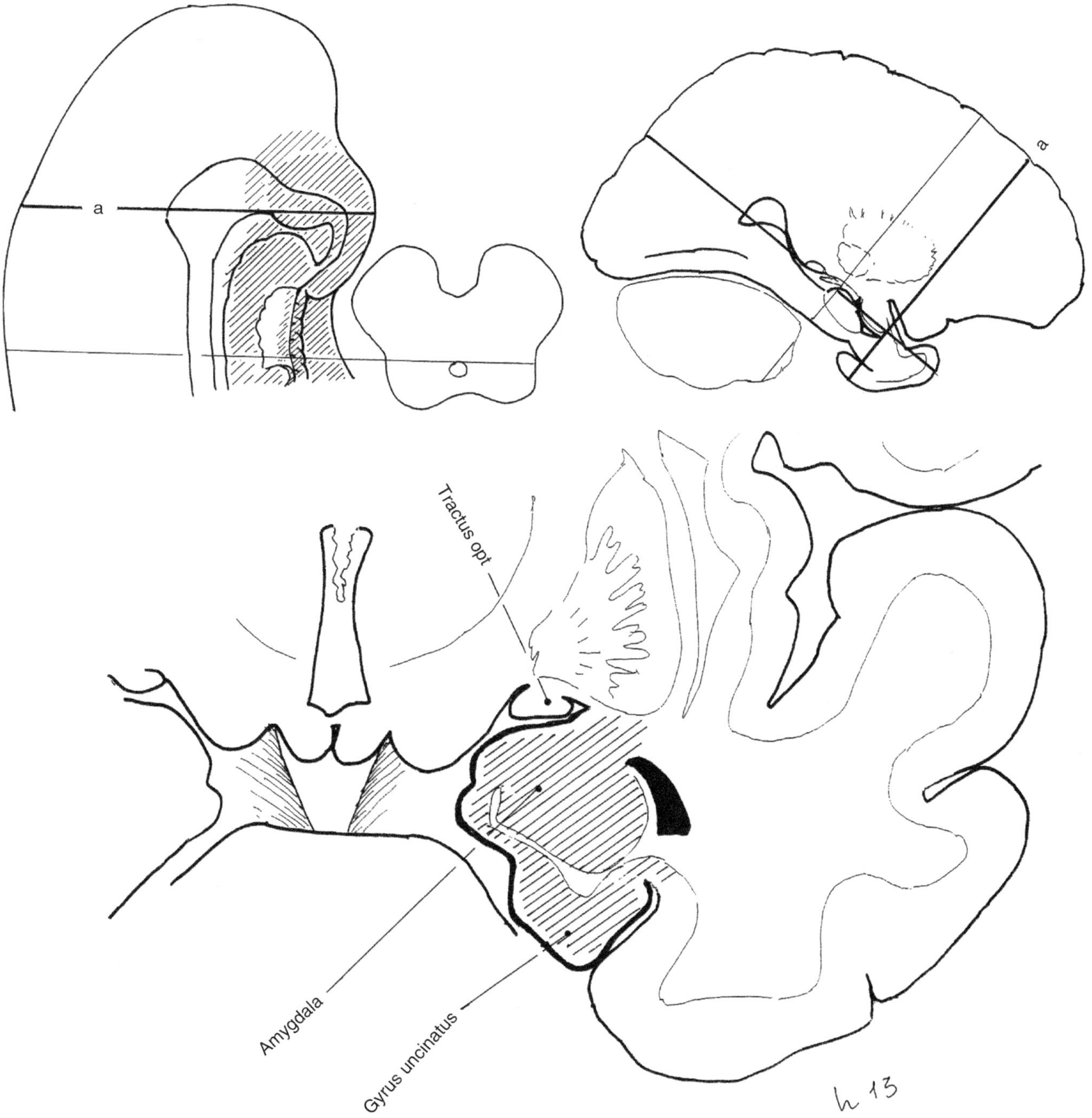

Fig. 7.9 Preservation of tractus opticus. Area of amygdalohippocampectomy hatched

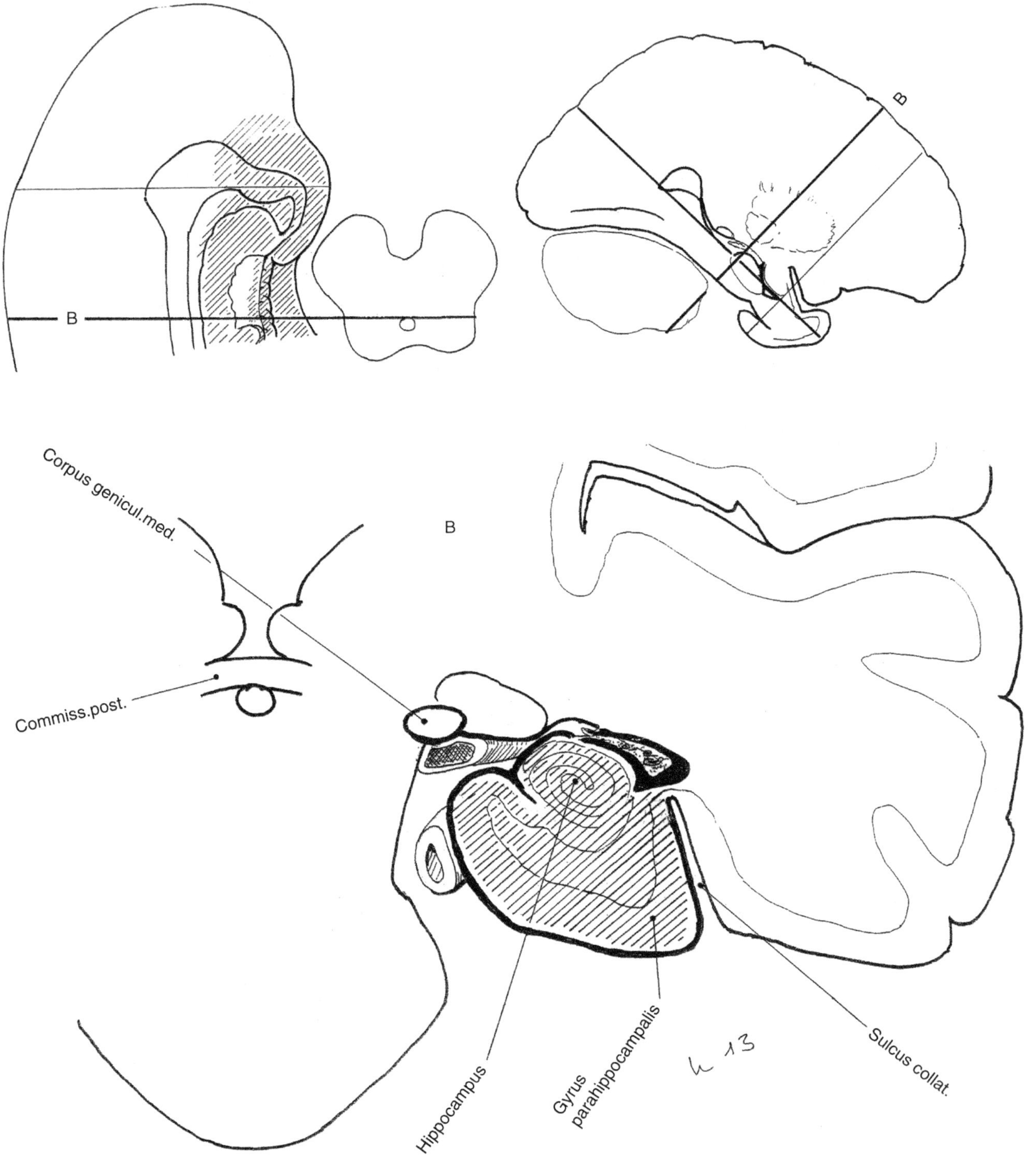

Fig. 7.10 Continuation of Fig. 7.9. Posterior region

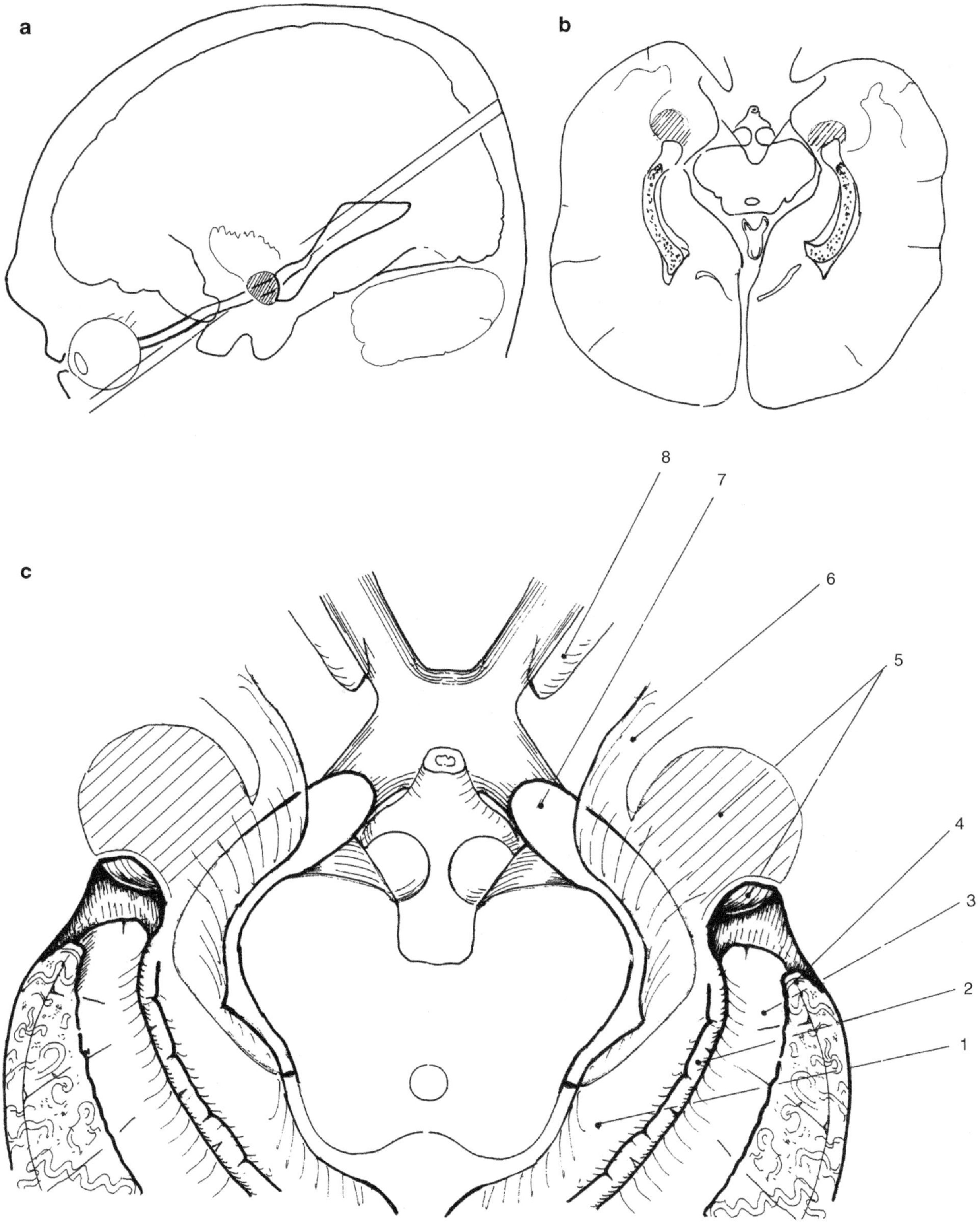

Fig. 7.11 Continuation of Fig. 7.10. Anterior superior limit area of amygdalohippocampectomy. **a** and **b** Topogram. **c** Drawing similar to neuronavigation. *1* gyrus parahippocampalis, *2* gyrus dentatus, *3* pes hippocampi, *4* tip of plexus choriodeus (landmark for surgery) *5* amygdala, *6* gyrus uncinatus, *7* tractus opticus, *8* processus clinoideus anterior

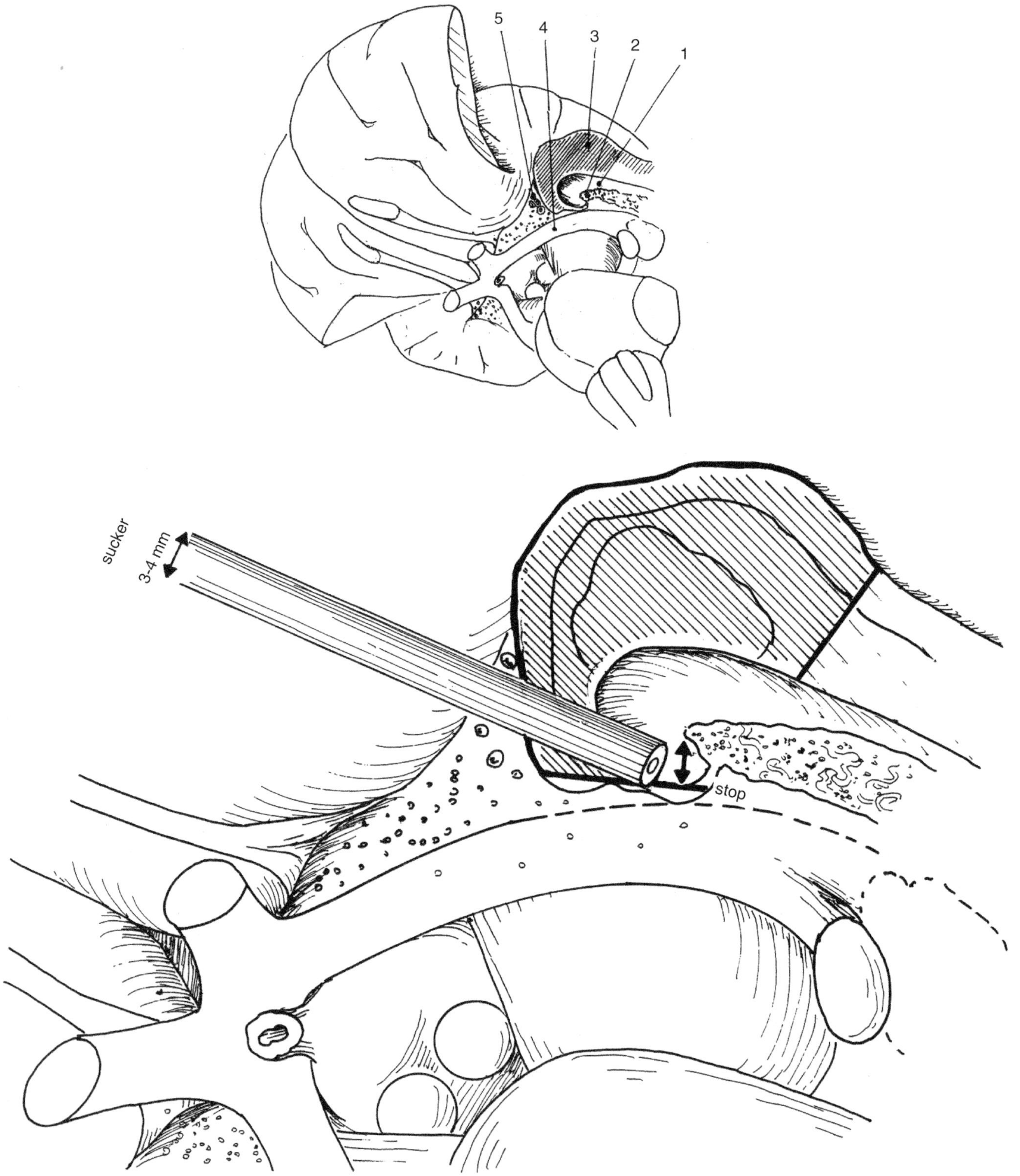

Fig. 7.12 Preservation of tractus opticus and striate arteries during amygdalohippocampectomy (Applied by Prof. Zentner, Freiburg 2013). Technical aspects. *1* cornu inferius; *2* tip of plexus choriodeus; *3* uncus, transected; *4* tractus opticus; *5* striate arteries

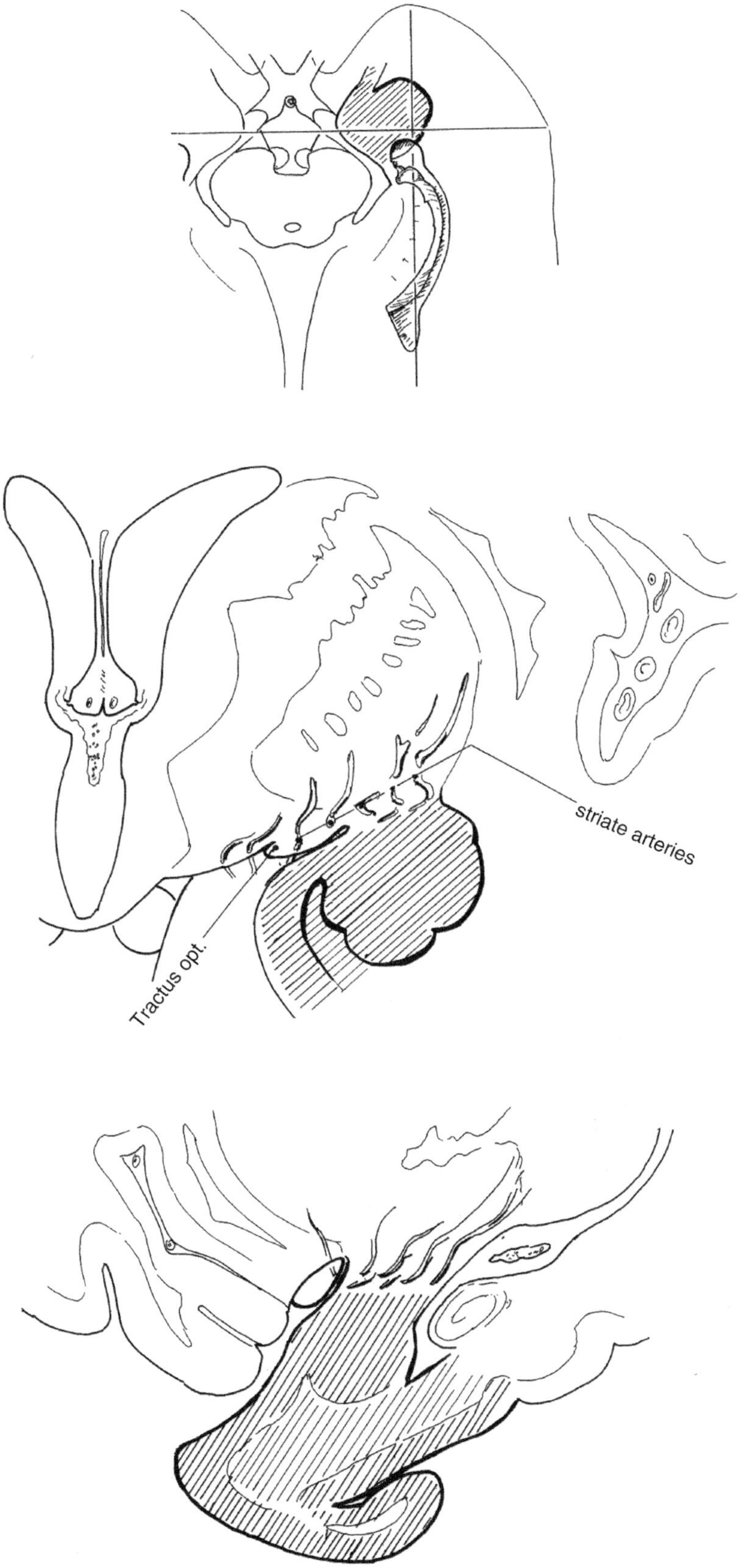

Fig. 7.13 Preservation of striate arteries, gangliae basales et tractus pyramidalis at surgery (According to anatomical dissections and casts of Hussein et al. 1987). Hatched: area of resection at surgery

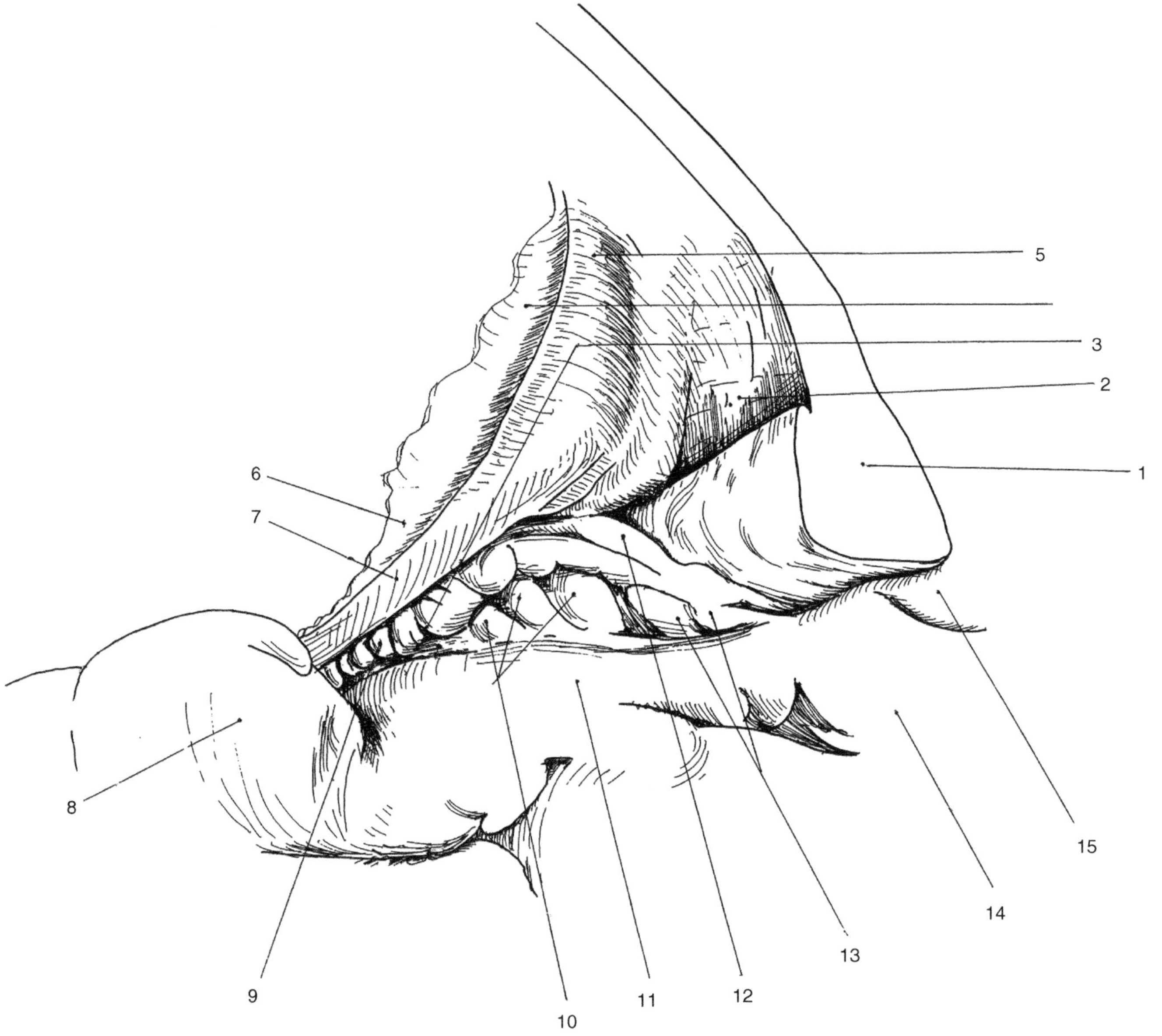

Fig. 7.14 Posterior area of amygdalohippocampectomy. Cave columna et crus fornicis. Anatomical aspects. *1* splenium corporis callosi, *2* commissura fornicis, *3* sulcus hippocampi, *4* taenia fornicis, *5* crus fornicis, *6* as 4, *7* fimbria fornicis, *8* uncus, *9* gyrus dentatus, *10* Retzius' gyri, *11* gyrus parahippocampalis, *12* tail of hippocampus, *13* gyrus fasciolaris, *14* gyrus occipitotemporalis medialis, *15* isthmus gyri cinguli

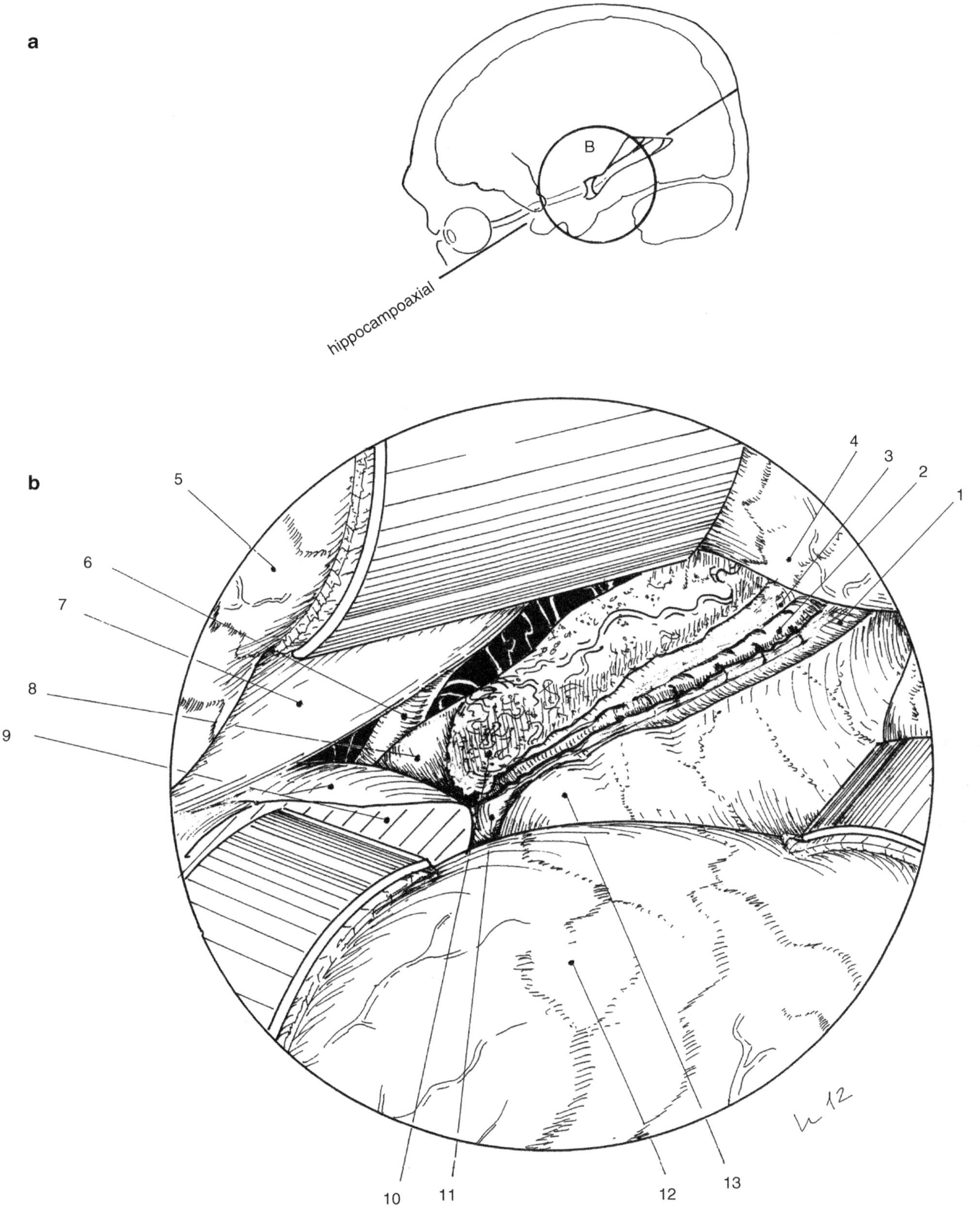

Fig. 7.15 Middle region of amygdalohippocampectomy. Cadaver brain dissection. Microsurgical aspects. *1* fimbria and taenia fornicis, taenia split, plexus choriodeus shifted; *2* gyrus dentatus after loosening and shifting of plexus choriodeus; *3* gyrus parahippocampalis after loosening and shifting of plexus choriodeus; *4* lobus parietalis; *5* lobus frontalis; *6* A. cerebri posterior overlying by the transparent taenia chorioidea; *7* tractus opticus; *8* gyrus parahippocampalis after shifting of plexus choriodeus, taenia chorioidea transparent; *9* uncus residual, transectional plane; *10* plexus choriodeus shifted, taenia chorioidea transparent; *11* tip of cornu inferius; *12* lobus temporalis; *13* hippocampus

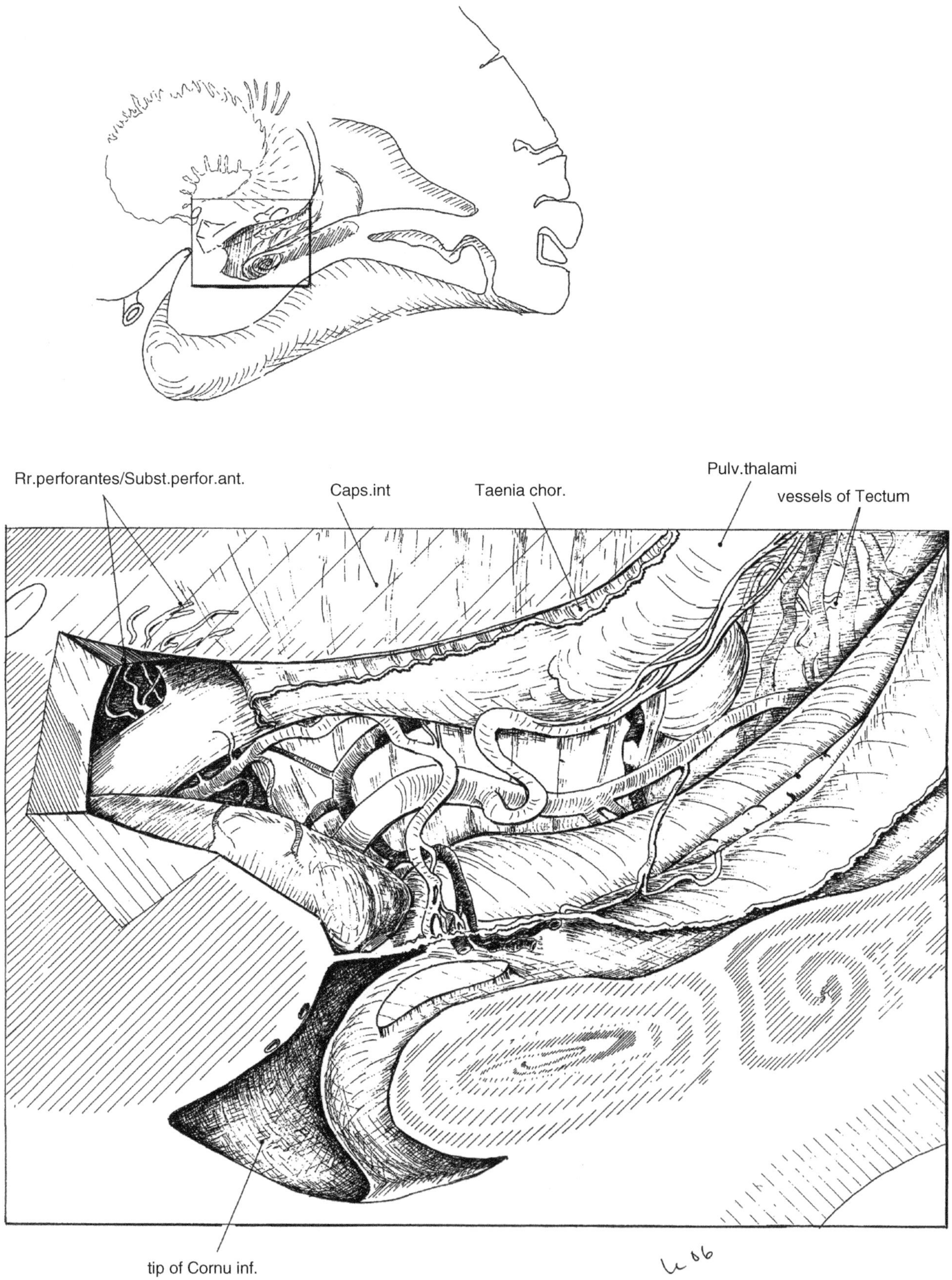

Fig. 7.16 Continuation of Fig. 7.15. Anterior and middle target region of amygdalohippocampectomy

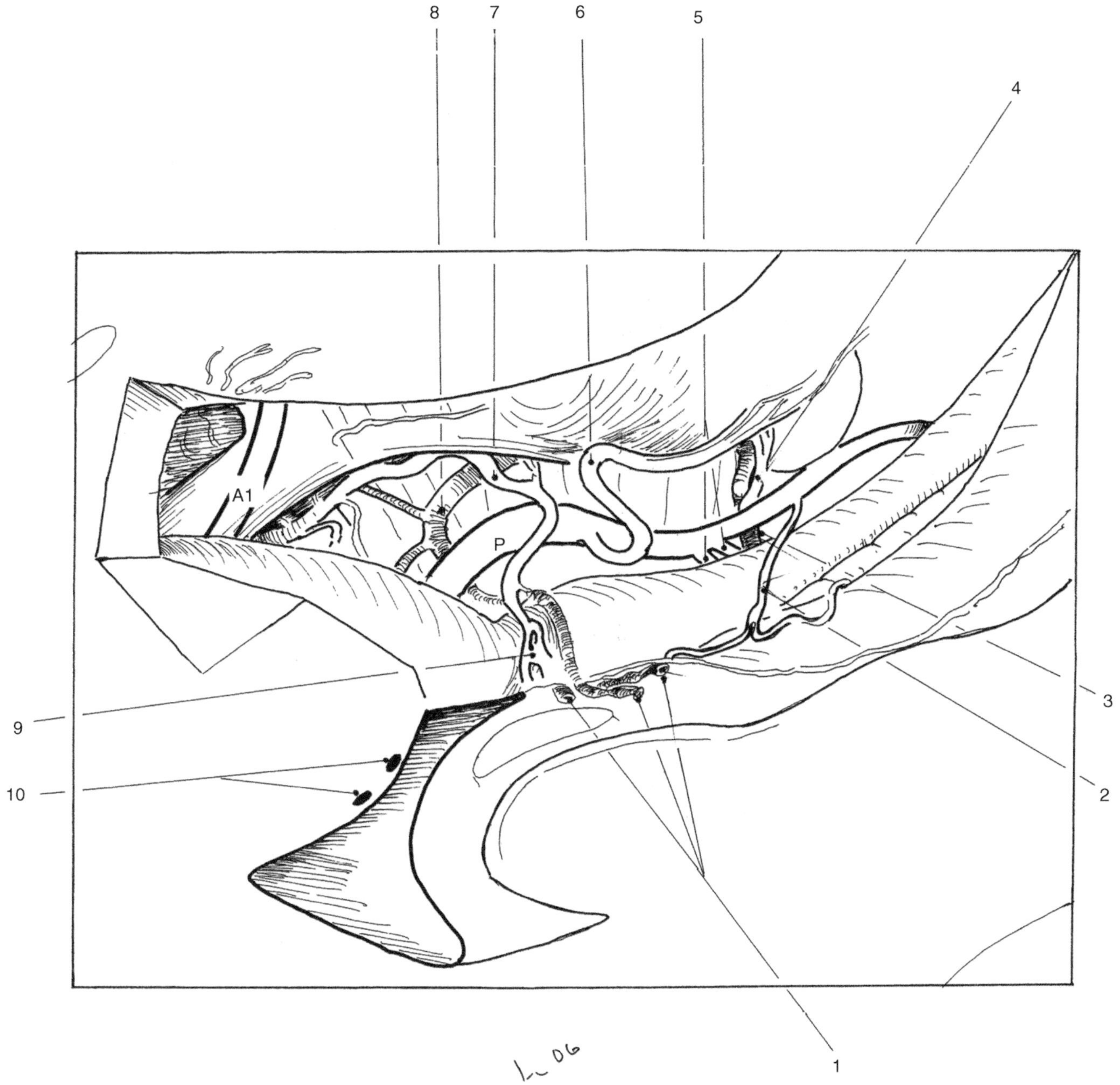

Fig. 7.17 As Fig. 7.16. Vessels: *1* temporal chorioid point, *2* A. hippocampi, *3* V. mesencephalica lateralis, *4* A. thalamogeniculata, *5* temporobasal branches of A. cerebri posterior, *6* knee of A. choriodeus posterior lateralis marking corpus geniculatum laterale, *7* A. chorioidea anterior, *8* V. basalis (Rosenthal), *9* branches of A. chorioidea anterior at the temporal chorioid point, *10* inferior tip vein; A1 A. cerebri anterior, *P* A. cerebri posterior

7.2.3 Subtemporal Surgical Approaches to Basal Cisterns (Figs. 7.18, 7.19, and 7.20)

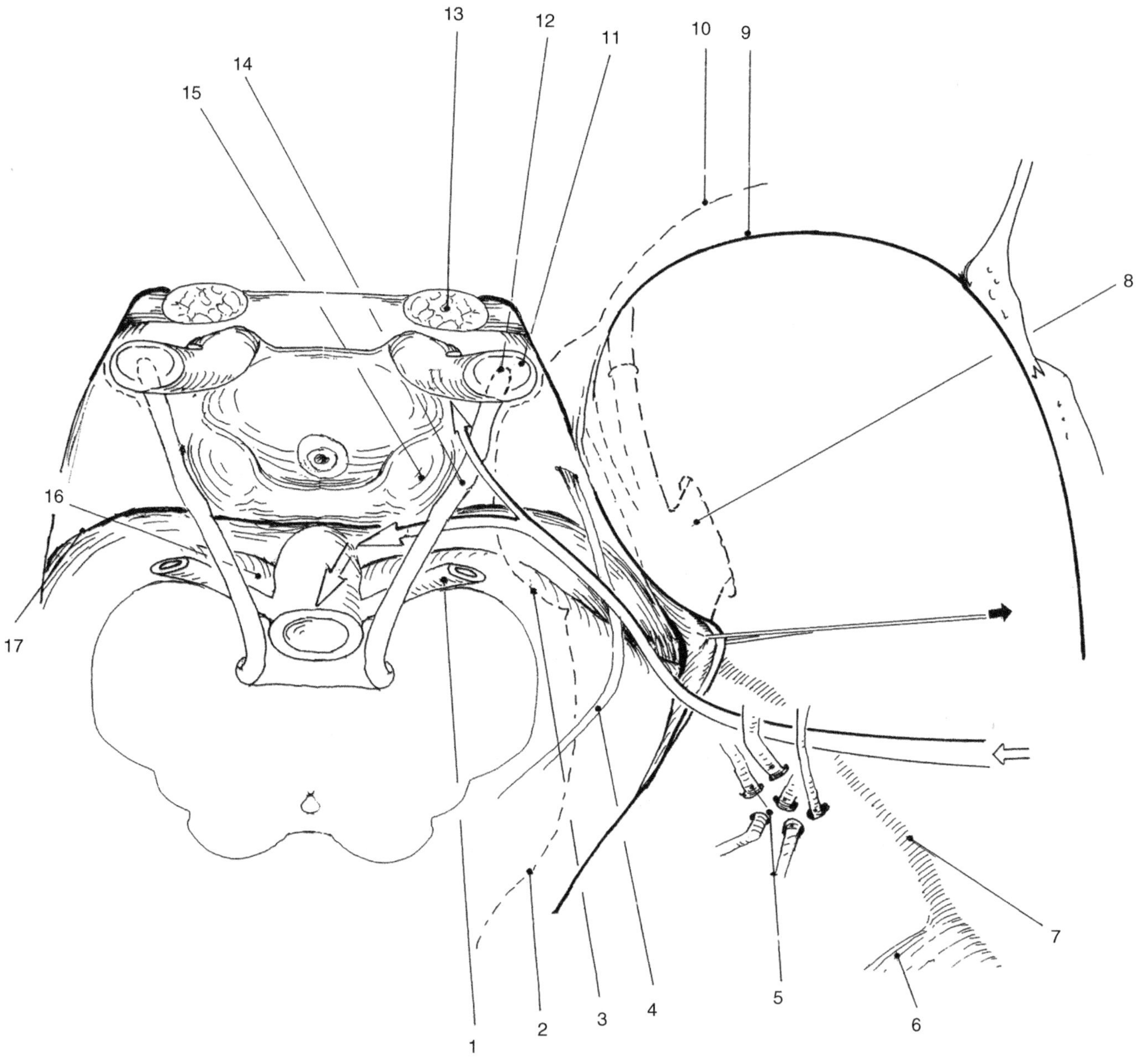

Fig. 7.18 Subtemporal route according to Dandy, modified by Drake *Black arrow*: retraction of tentorium. *Light arrow*: surgical route. *1* A. cerebelli superior; *2* gyrus parahippocampalis, projection; *3* uncus, projection; *4* N. trochlearis; *5* temporobasal bridging veins; *6* knee of sinus transversus; *7* sinus petrosus superior, dura sometimes transparent; *8* N. mandibularis, projection; *9* ala minor, *10* polus temporalis, projection; *11* A. carotis interna, *12* N. oculomotorius, projection; *13* N. opticus; *14* N. oculomotorius; *15* processus clinoideus posterior; *16* pons; *17* plica petroclinoidea posterior

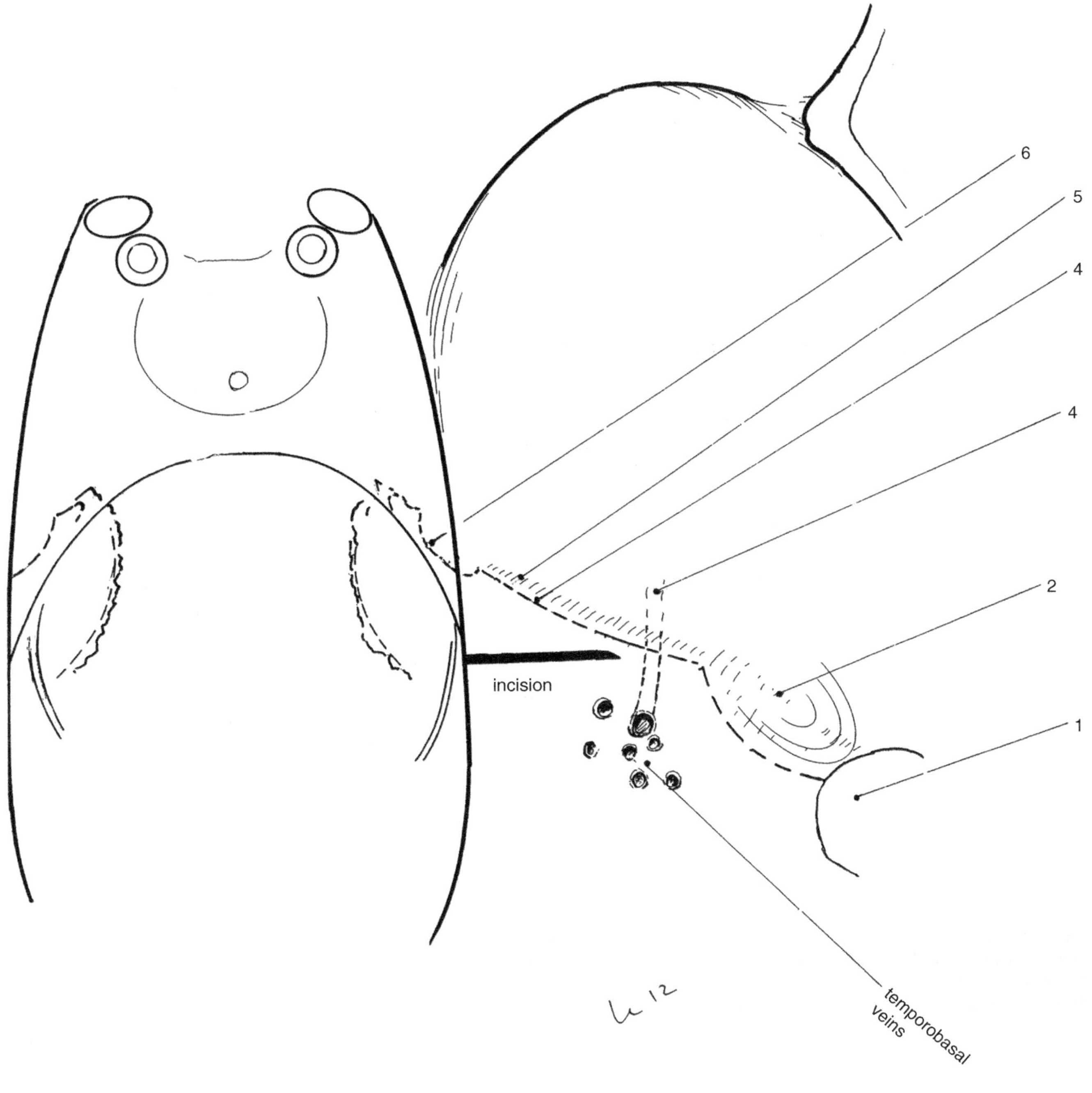

Fig. 7.19 Addendum for Fig. 7.18. *Incision of tentorium* according to Dandy, if Drake's tentorial retraction is not applicable. *1* knee of sinus transversus, *2* eminentia arcuata, *3* temporobasal bridging vein (s), *4* margo superior ossis petrosi, *5* sinus petrosus superior, *6* incisura trigeminalis. *Dotted*: projections

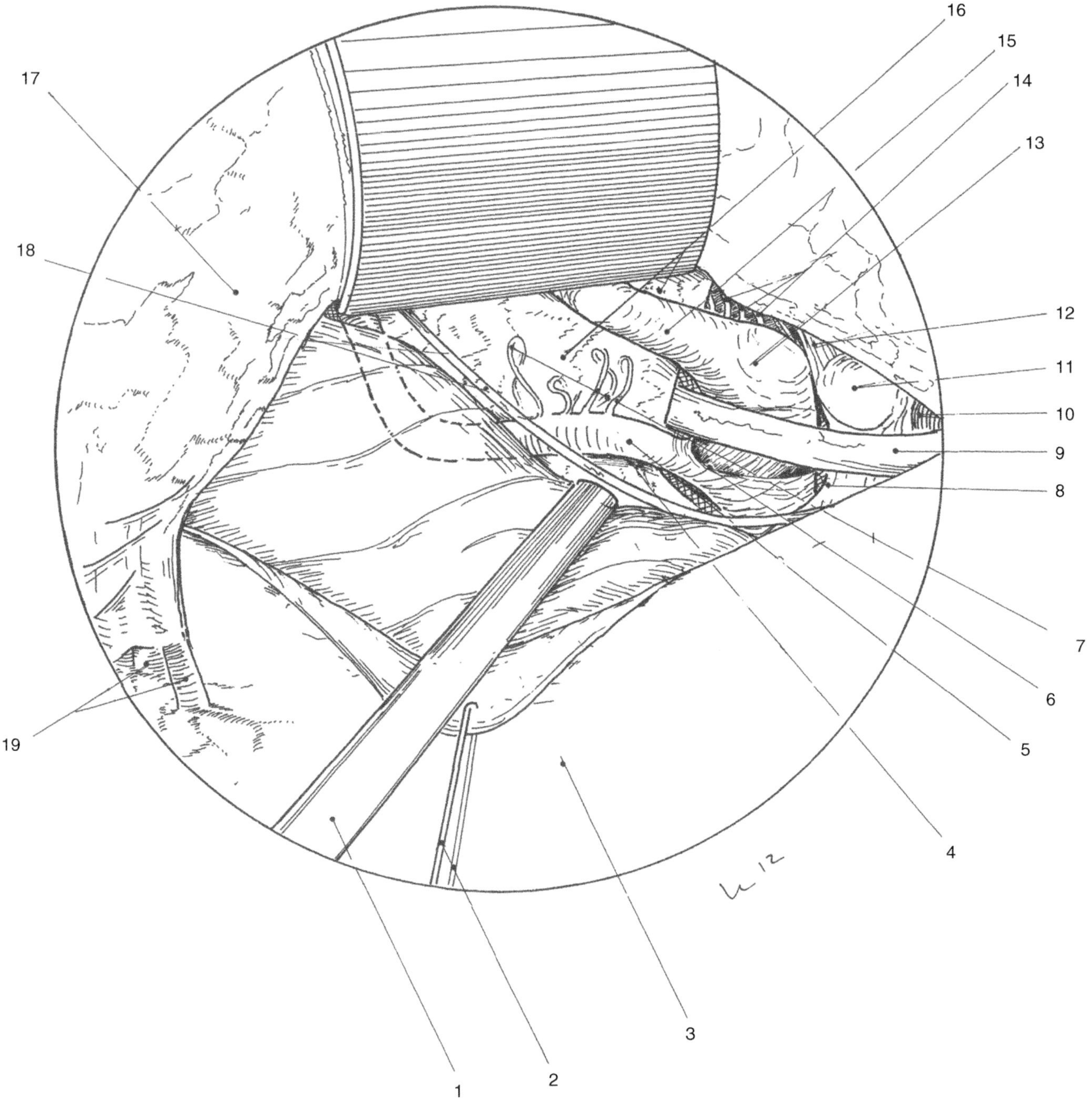

Fig. 7.20 Surgical topography *1* retraction of lobulus quadrangularis cerebelli; *2* Drake's retraction of tentorium without tentorial incision; *3* tentorium; *4* pontomesencephalic rim; *5* A. cerebelli superior; *6* Liliequist's membrane, pars mesencephalica, margin of its gap; *7* perforating arteries; *8* plica petroclinoidea posterior; *9* N. oculomotorius; *10* dura; *11* processus clinoideus posterior; *12* dorsum sellae; *13* A. basilaris; *14* perforating arteries; *15* A. cerebri posterior; *16* crus cerebri; *17* third temporal gyrus; *18* N. trochlearis; *19* temporobasal bridging veins

Part III

Parietal Region

1.1 Introduction

This introduction to Part III contains a short summary of each chapter.

1.2 Chapter 8 Diagnostic Base

Landmarks may assist neuronavigation for exact defining of small target areas, if necessary.

1.3 Chapter 9 Anatomical Base of Surgery

Asymmetries of midline structures, incongruence of os parietale and lobus parietalis, and defining of sinus sagittalis superior should be regarded for avoiding errors at surgery:

- Asymmetries of sutura sagittalis, bregma and sinus sagittalis superior are often not exact to define by imagings.
- Especially the parietal segment of sinus sagittalis superior is variable, its lateral limit is often not to define by imagings or during surgery. It may be mixed up with lacunae.
- The anterior limit of lobus parietalis is identic with sulcus centralis. This region is located far distant from os frontale, overlaid by the middle area of os parietale.
- The fine cerebral midline structures in the depth, especially fissura transversa and adjacent structures, are often not exact to define by imagings.

1.4 Chapter 10 Special Surgical Aspects

- Interhemispheric surgical approaches may be hindered by bridging veins and by unclear defined lateral limits of the sinus.
- At approaches to the third ventricle, the middle segment of fornix could be splitted by opening of cavum septi pellucidi. Its posterior segment can be splitted by opening of commissura fornicis between crura fornicis. These areas are variable and well to define using imagings.
- At transylvian approaches, it should be considered the short distance measurement between the middle segment of sulcus superior insulae and tractus pyramidalis.

8 Diagnostic Base

8.1 Landmarks (Figs. 8.1 and 8.2)

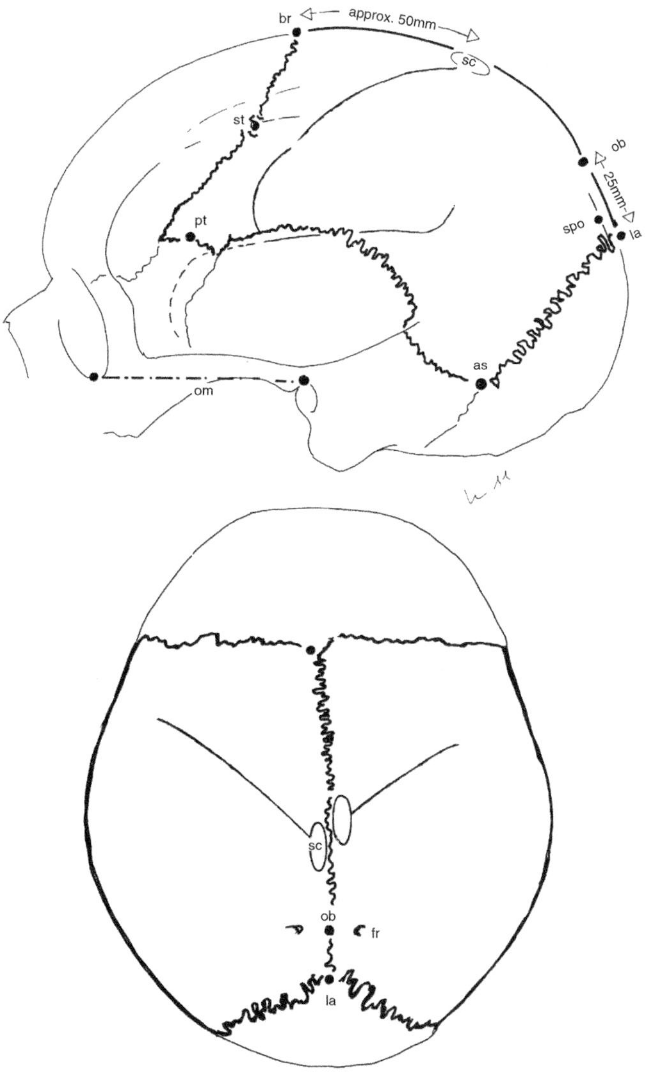

Fig. 8.1 Parietal landmarks. Sulcus centralis added, schematized: *as* asterion, *br bregma*, *fr* foramina parietalia for defining of obelion (more precise than lambda), *la* lambda, *ob* obelion, *om* orbitomeatal line, *pt* pterion, *sc* sulcus centralis (for details see Fig. 3.10 ff), *spo* sulcus parietooccipitalis, upper point, *st* stephanion

W. Seeger, J. Zentner, *Anatomical Basis of Cranial Neurosurgery*, https://doi.org/10.1007/978-3-319-63597-2_8

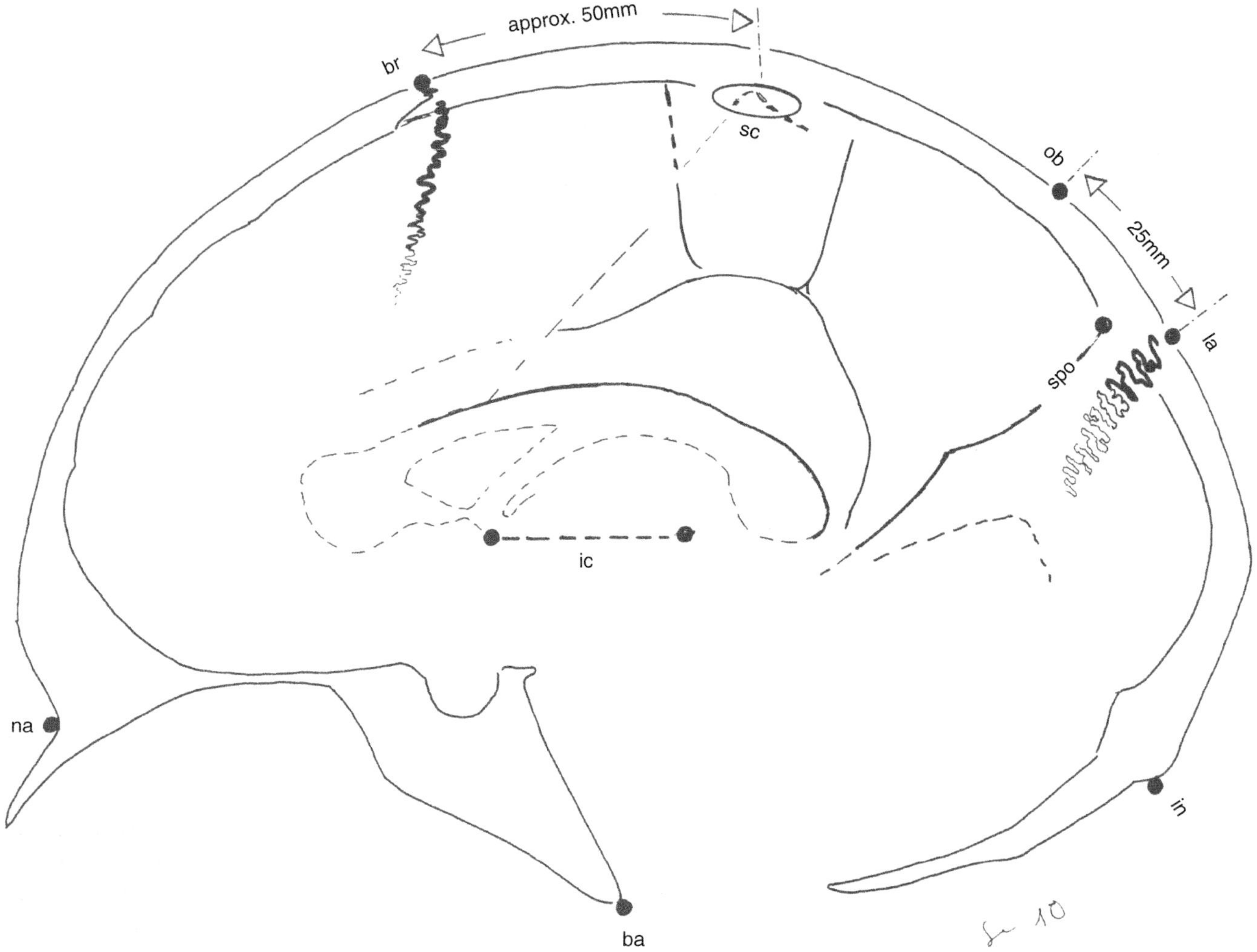

Fig. 8.2 Continuation of Fig. 8.1. Parietal medial sulci added: *ba* basion, *br* bregma (anterior limit of os parietale), *ic* intercommissural line, *in* inion, *la* lambda, *na* nasion, *ob* obelion, *sc* sulcus centralis (anterior limit of lobus parietalis) , *spo* sulcus parietooccipitalis

8.2 Radiology (Figs. 8.3, 8.4, 8.5, and 8.6)

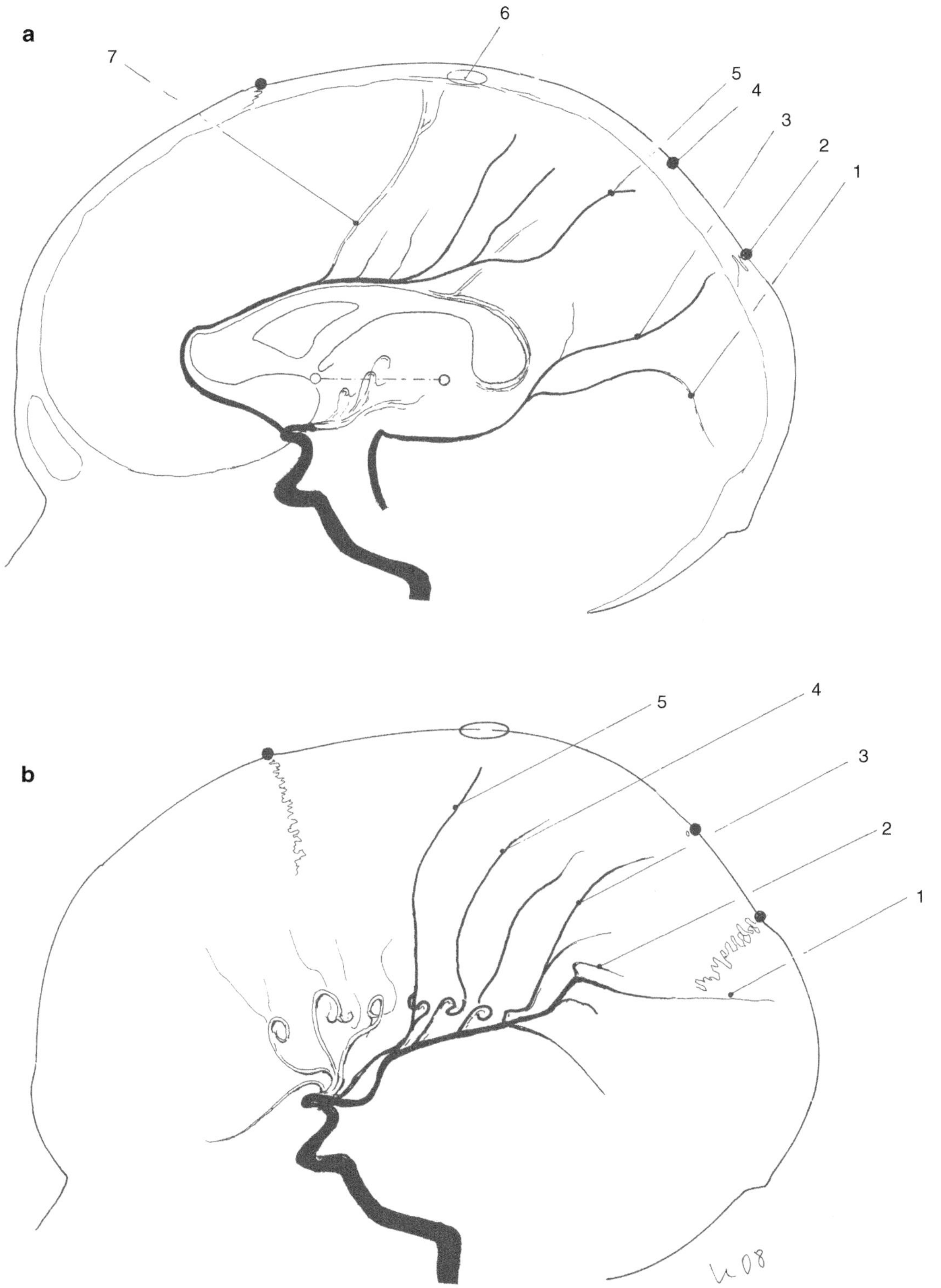

Fig. 8.3 Medial (**a**) and lateral arteries (**b**). Survey, schematized. (**a**) *1* A. calcarina, *2* lambda, *3* A. parietooccipitalis, *4* obelion, *5* A. praecunealis, *6* sulcus centralis, *7* A. paracentralis; (**b**) *1* A. temporooccipitalis, *2* A. angularis, *3* A. parietalis posterior (posteriores), *4* A. parietalis anterior (anteriores), *5* A. sulci centralis

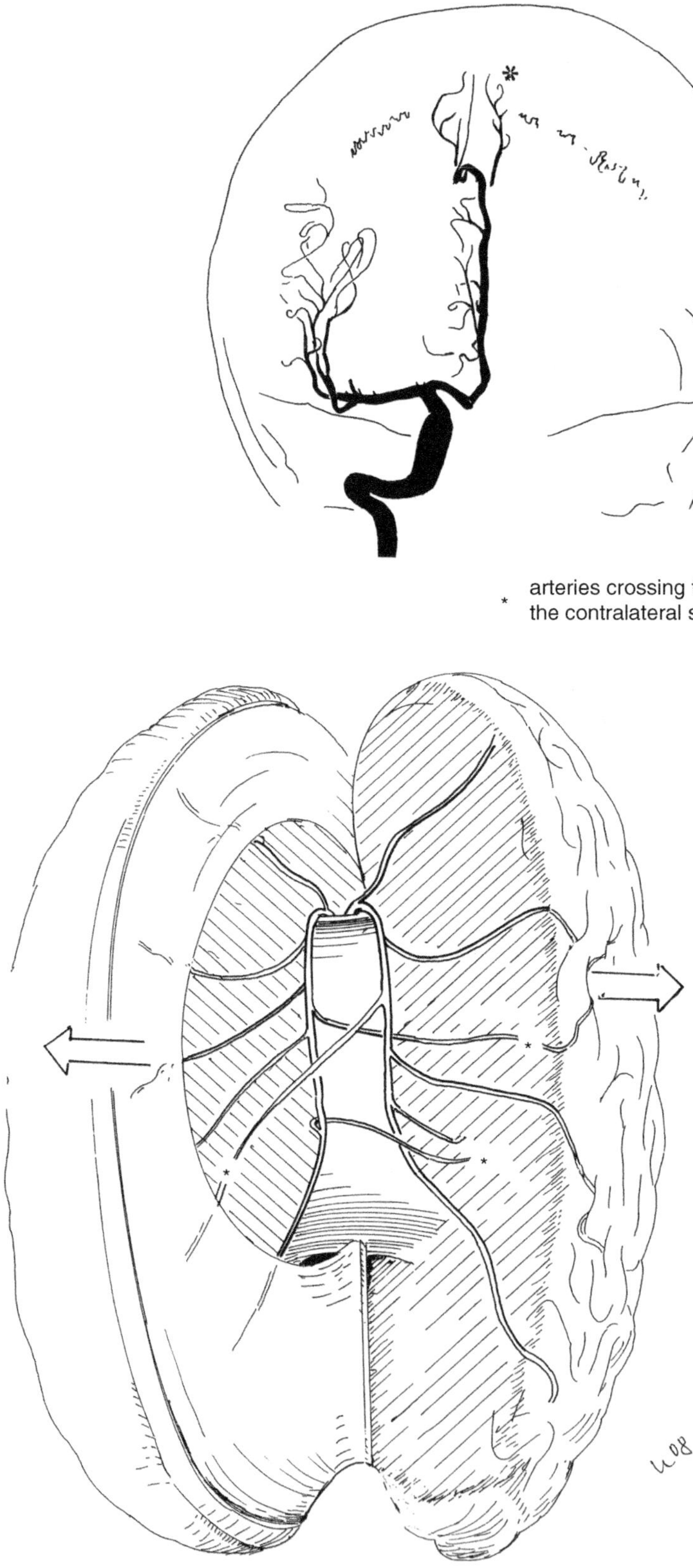

Fig. 8.4 Midline crossing branches of A. cerebri ant. (According to Marino 1976) Note contralateral branches of A2 (during contralateral digital compression of A. carotis communis). Homolateral interruption of A1 could be followed by bihemispheric encephalomalacia. *Left*, falx, and dura *right*, cortex

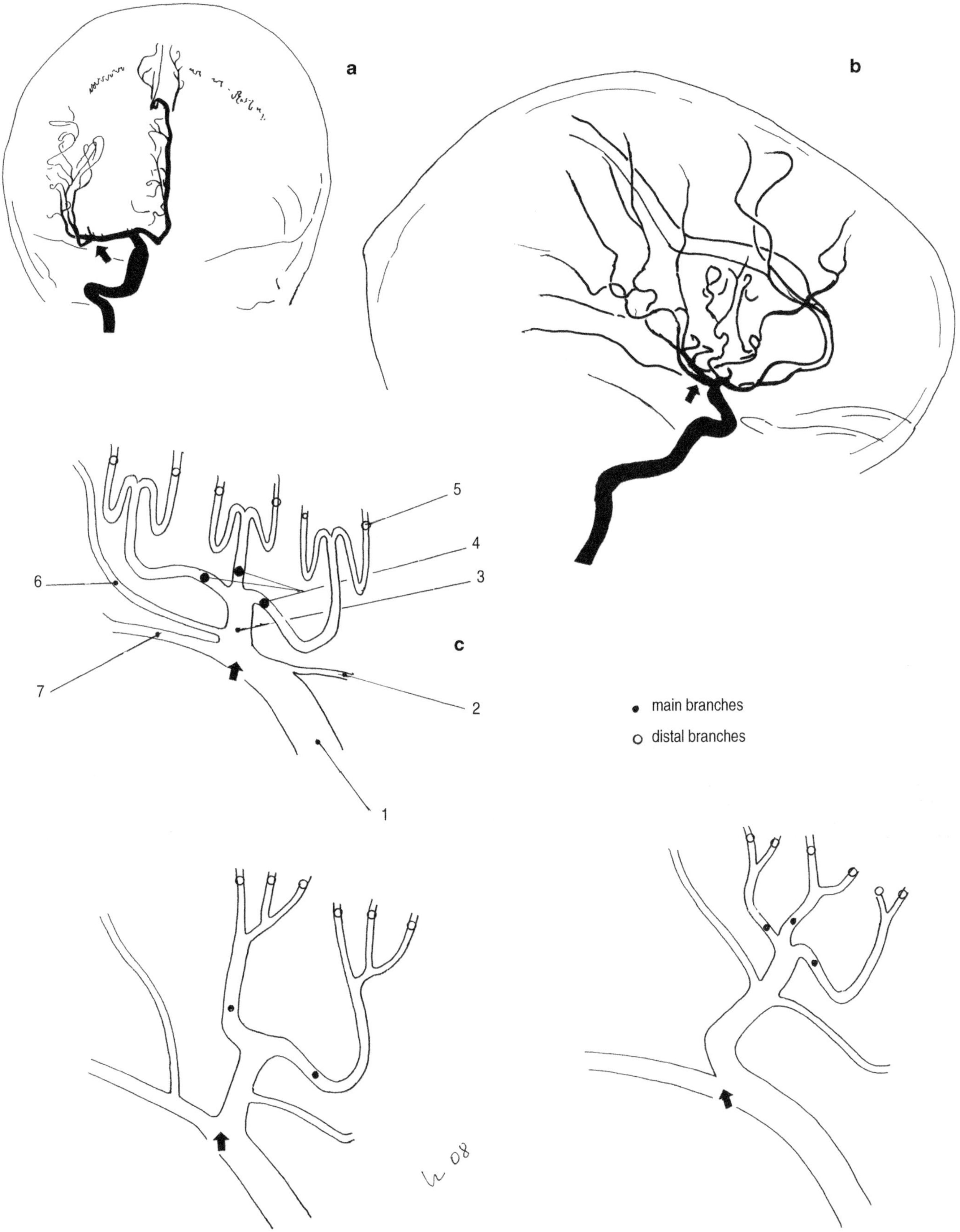

Fig. 8.5 A. cerebri media, trifurcation, schematized. (**a**) and (**b**) Angiogram. *Arrows*: main division of A. cerebri media. (**c**) Variability (sketches on the base of Ring 1969). *1* A. cerebri media, main trunk, *2* lateral basal branch, *3* A. operculofrontalis, *4* and *5* branches, *6* A. sulci centralis, *7* A. cerebri media, distal trunk

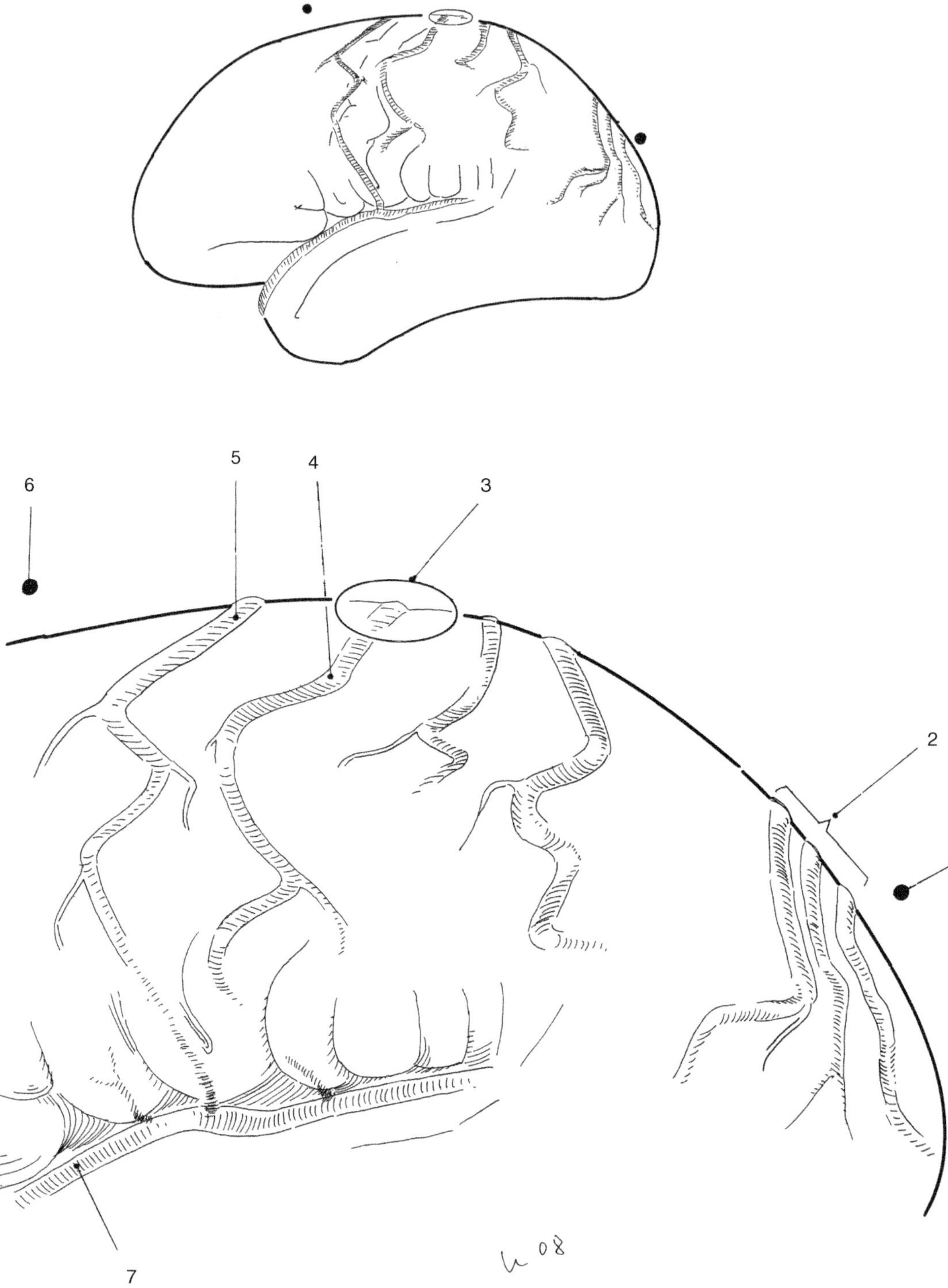

Fig. 8.6 Lateral veins (*sketches*). Landmarks added for assisting neuronavigation, if necessary. *1* Lambda, *2* parietooccipital bridging veins, *3* sulcus centralis, *4* V. centralis (Rolandi), *5* vein of Trolard (upper anastomotic vein, variable), *6* bregma, *7* V. temporalis superficialis

8.3 **Further Imaging** (Figs. 8.7, 8.8, 8.9, 8.10, and 8.11)

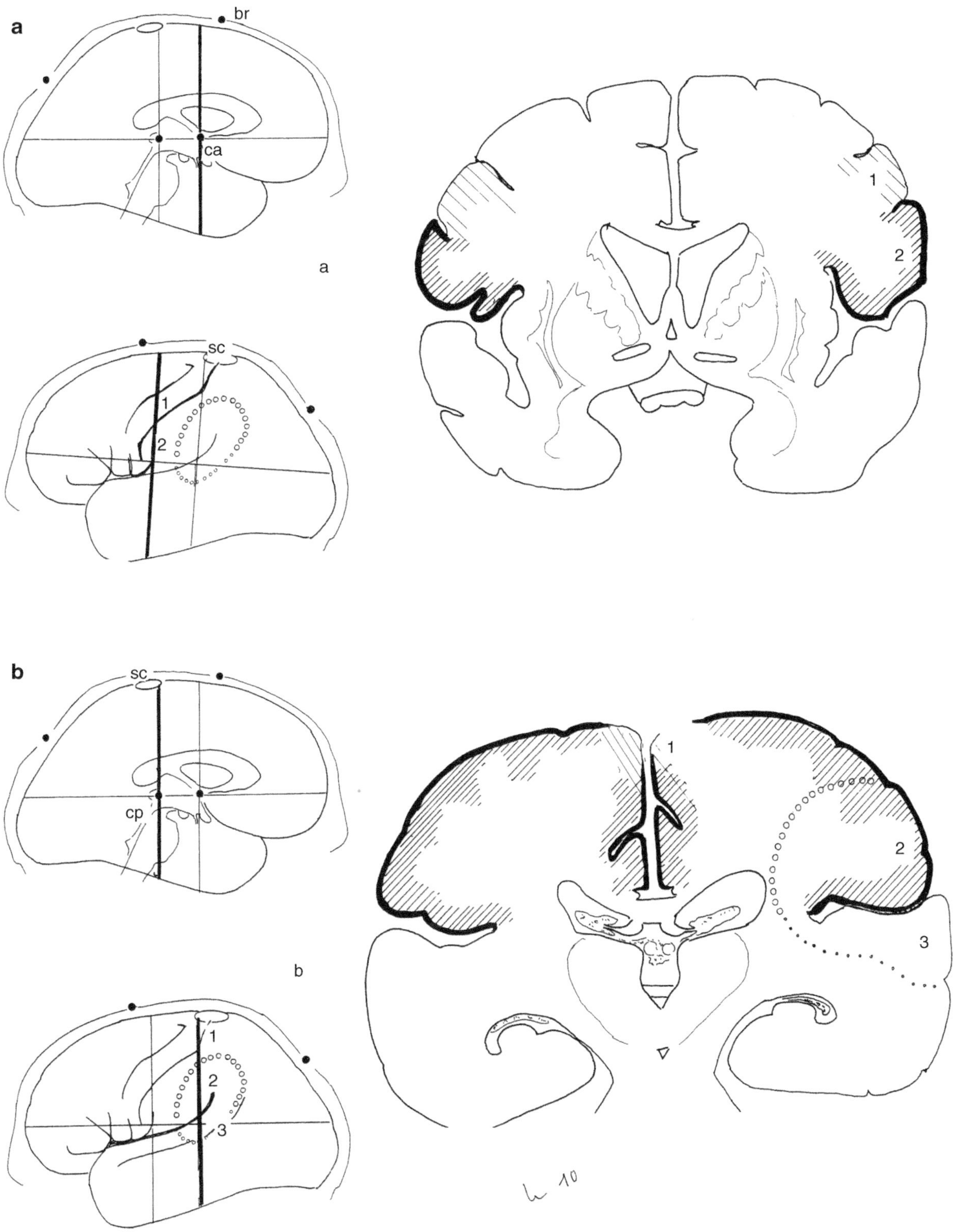

Fig. 8.7 *MRT*, coronal planes. Parietal cortex hatched and rendered prominent. Landmarks added for assisting imagings, if necessary: (**a**) *br* bregma, *ca* commissura anterior, *1* gyrus praecentralis, *2* gyrus postcentralis; (**b**) *cp* commissura posterior, *sc* sulcus centralis, *1* lobulus paracentralis, *2* gyrus angularis, *3* gyrus temp.1

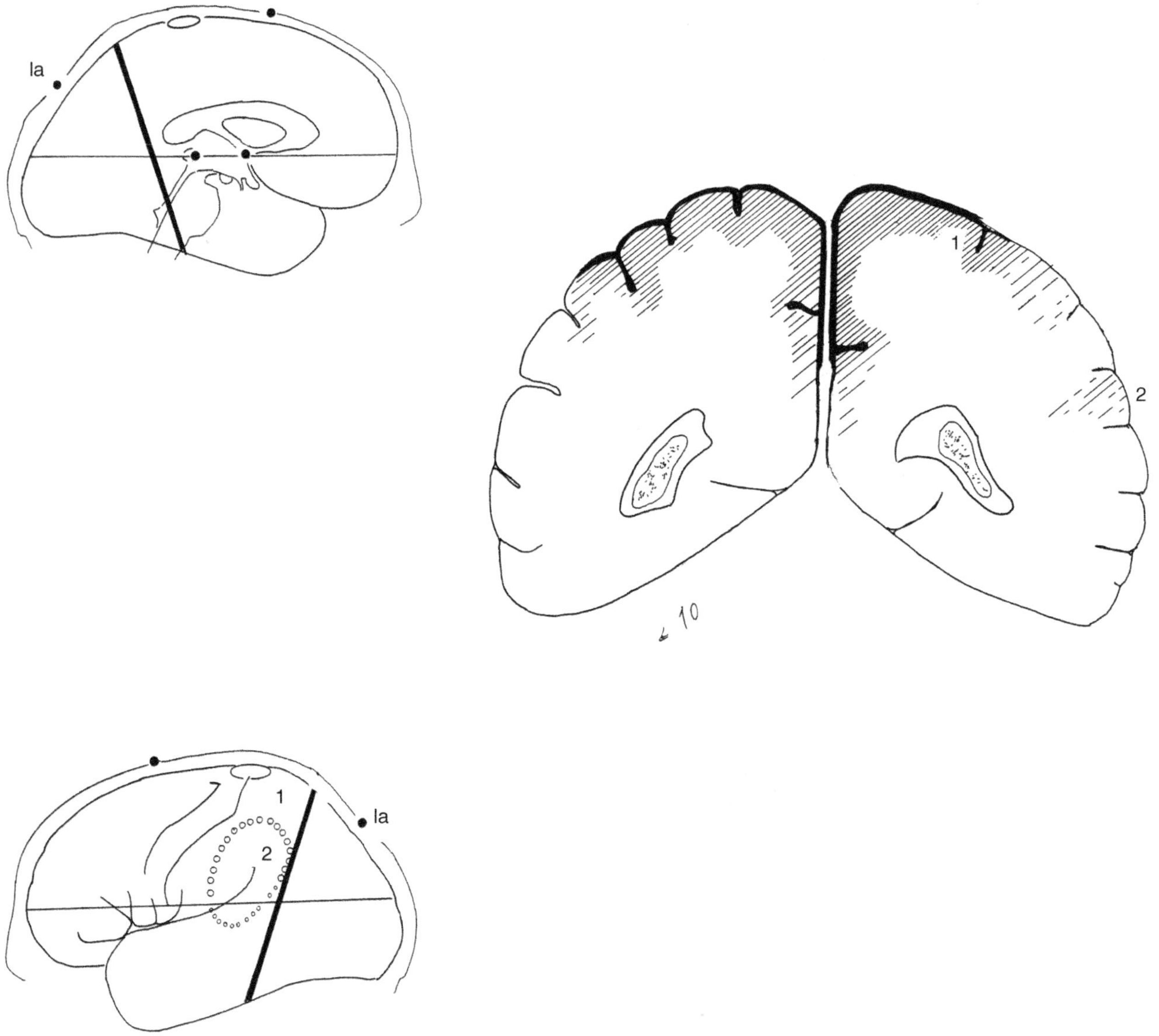

Fig. 8.8 Continuation of Fig. 8.7. *la* lambda *1* dorsal parietal cortex, *2* cortex of the angularis area

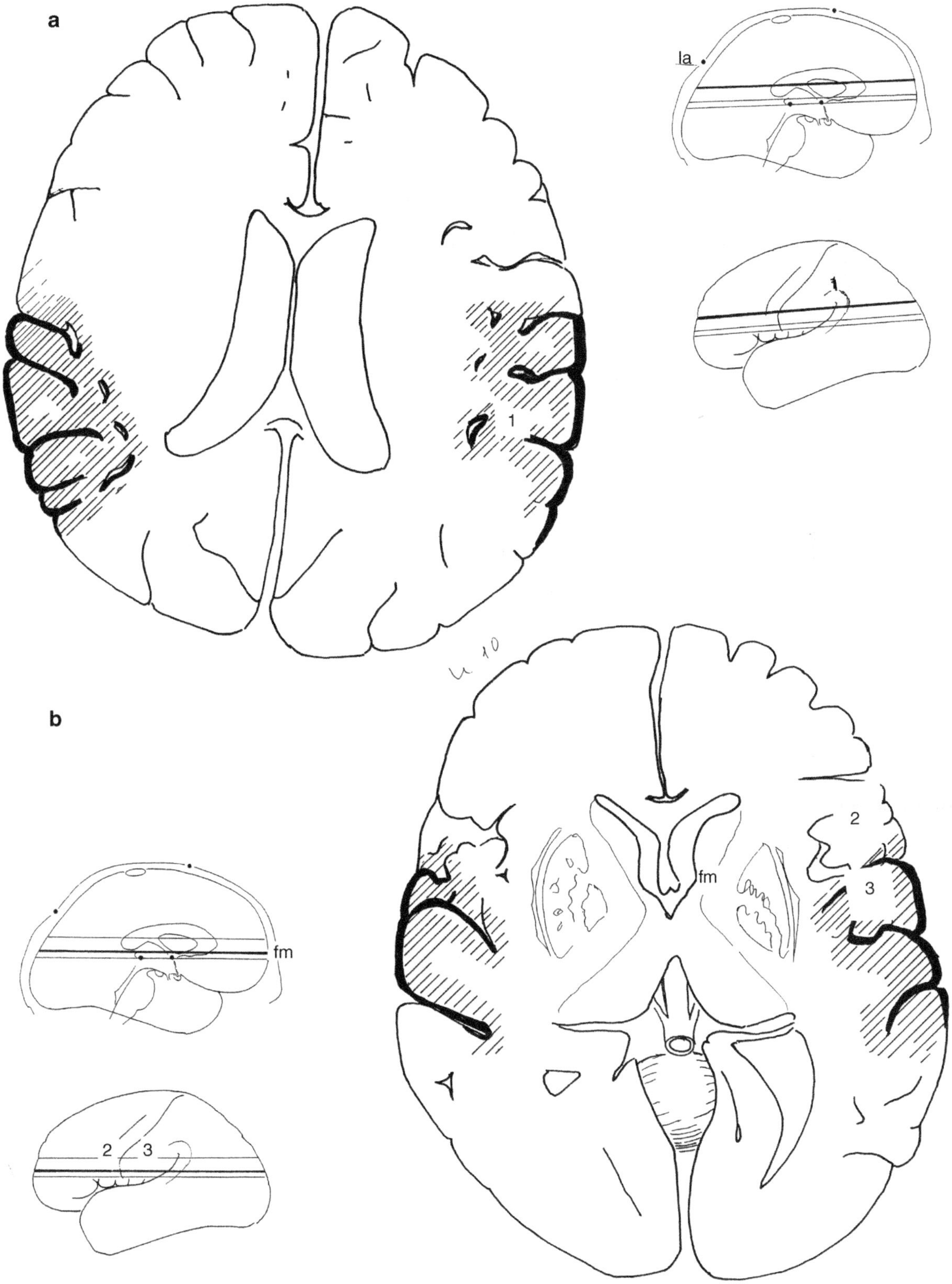

Fig. 8.9 Continuation of Fig. 8.8. (**a**) *la* lambda *1* angularis area. (**b**) *fm* foramen interventriculare (Monroi), *2* gyrus praecentralis, *3* gyrus postcentralis

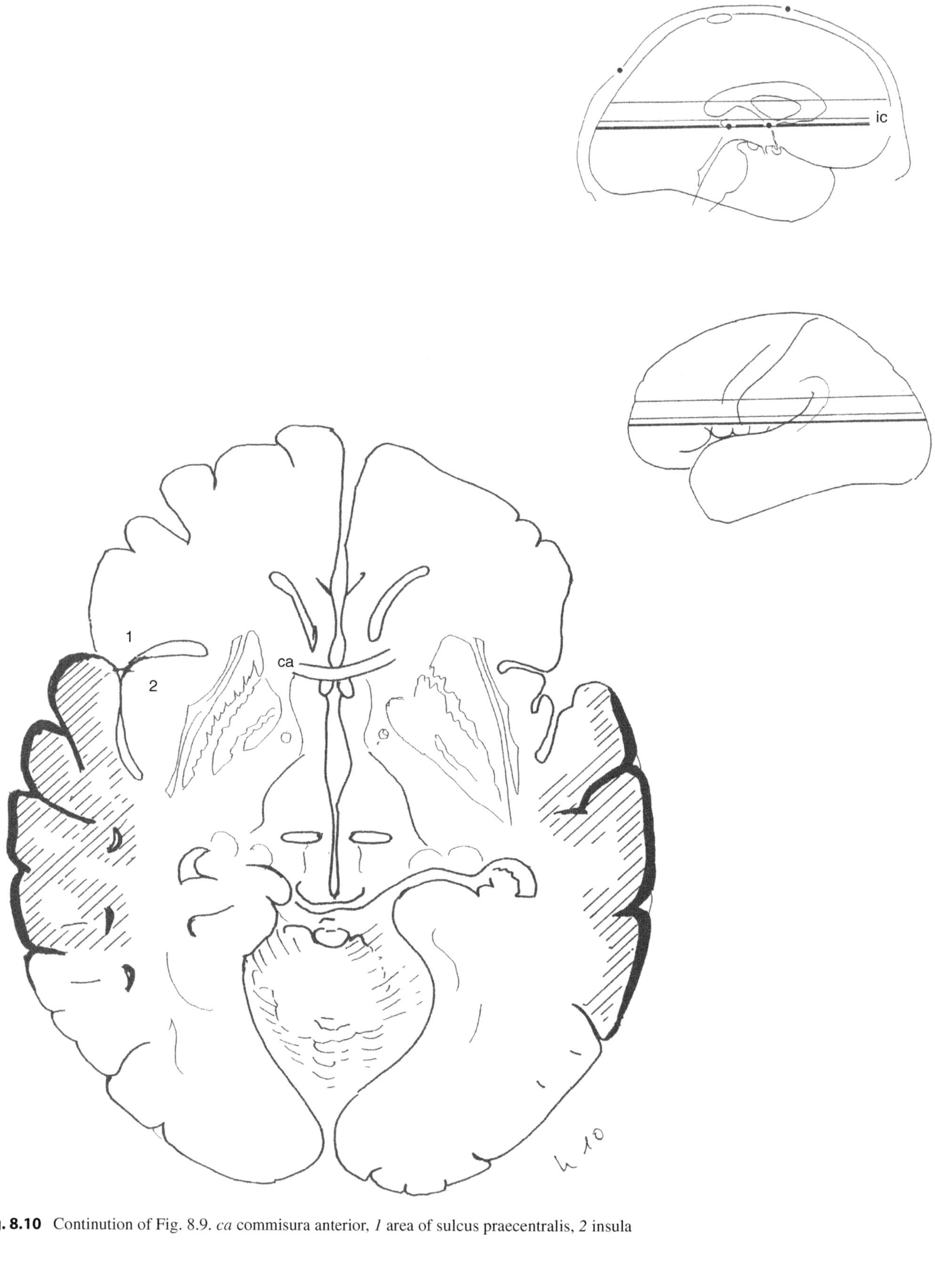

Fig. 8.10 Continution of Fig. 8.9. *ca* commisura anterior, *1* area of sulcus praecentralis, *2* insula

a

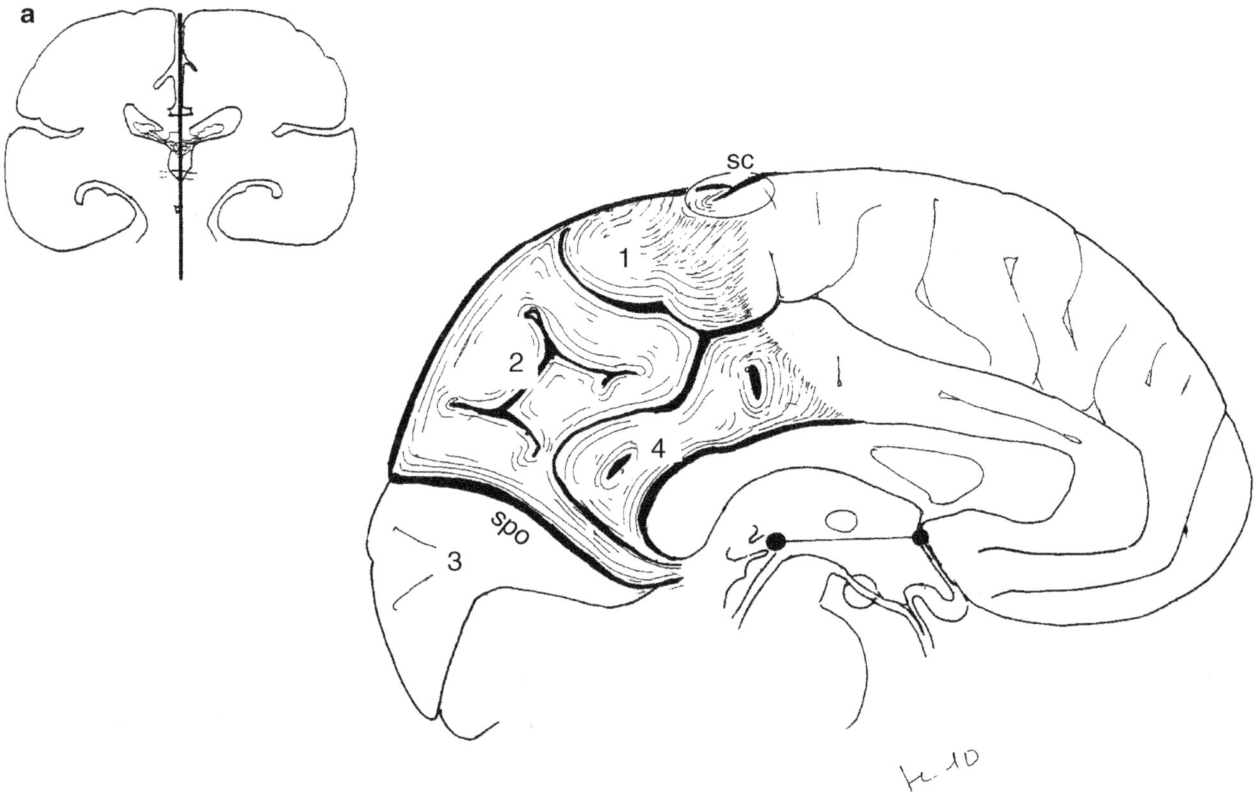

b

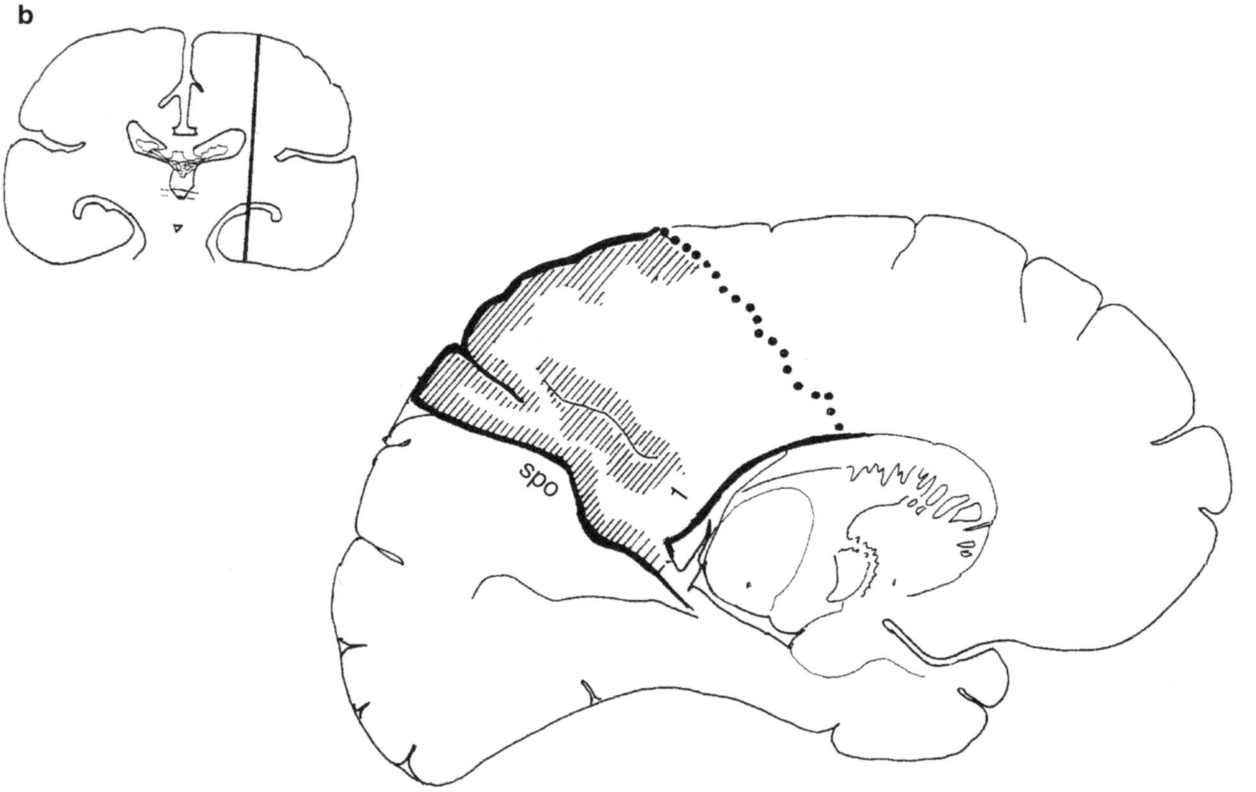

Fig. 8.11 Continuation of Fig. 8.10. (**a**) The complete medial cortex can be presented at one sagittal slice in contrast to lateral presentations of the cortex. The medial relief of the cortex is less variable than the lateral relief. (**b**) The ground of sulcus parietooccipitalis is located close to calcar avis of the ventricle: (**a**) *sc* sulcus centralis, *spo* sulcus parietooccipitalis, *1* lobulus paracentralis, postcentral segment, *2* lobulus praecunealis, *3* cuneus, *4* gyrus cinguli; (**b**) *spo* sulcus parietooccipitalis, 1 dorsal wall of atrium

9 Anatomical Base of Surgery

9.1 Extradural Topography

9.1.1 Extracranial Layers (Fig. 9.1)

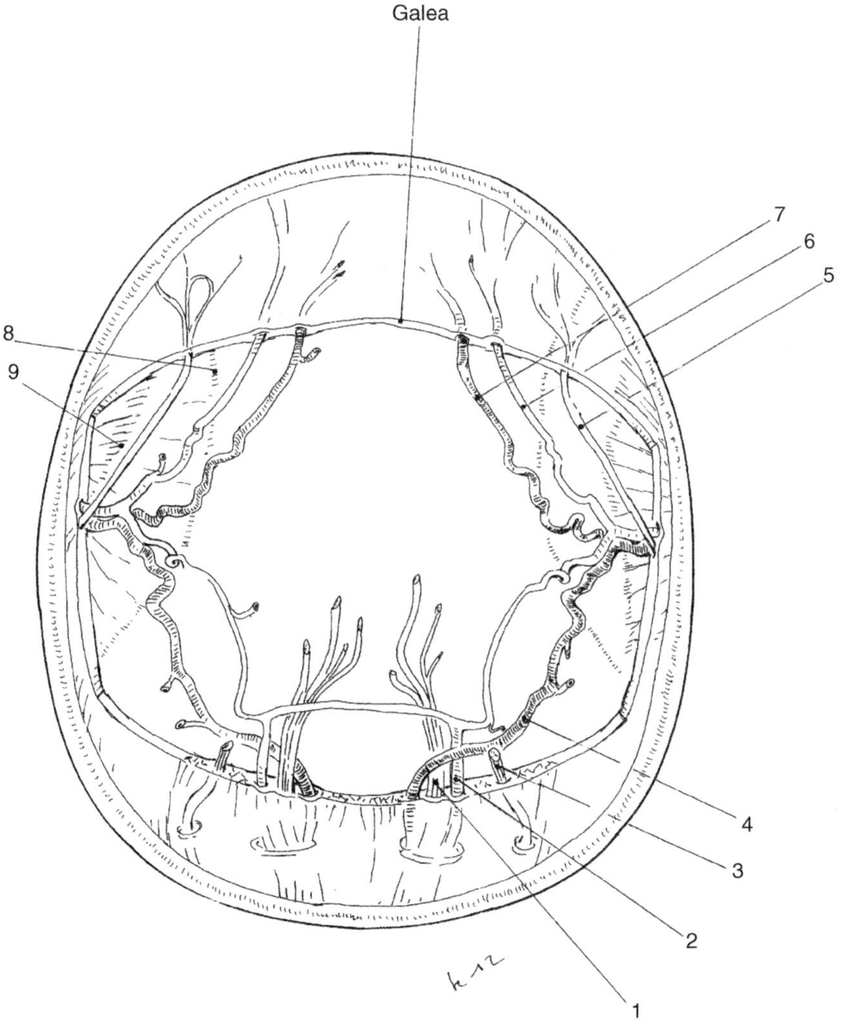

Fig. 9.1 Vessels and nerves of galea aponeurotica (galea) resected. Its contents are preserved: *1* N. occipitalis major, *2* A. occipitalis and its main anastomotic branches, *3* N. occipitalis minor, *4* anastomotic vein(s), *5* N. facialis, frontal ramification, *6* A. temporalis superficialis, frontal branch. Anastomotic branches are supraorbital arteries: *7* V. temporalis superficialis, *8* linea temporalis superior, insertion of fascia temporalis, *9* linea temporalis inferior, insertion of M. temporalis

W. Seeger, J. Zentner, *Anatomical Basis of Cranial Neurosurgery*, https://doi.org/10.1007/978-3-319-63597-2_9

9.1.2 **Skull** (Figs. 9.2 and 9.3)

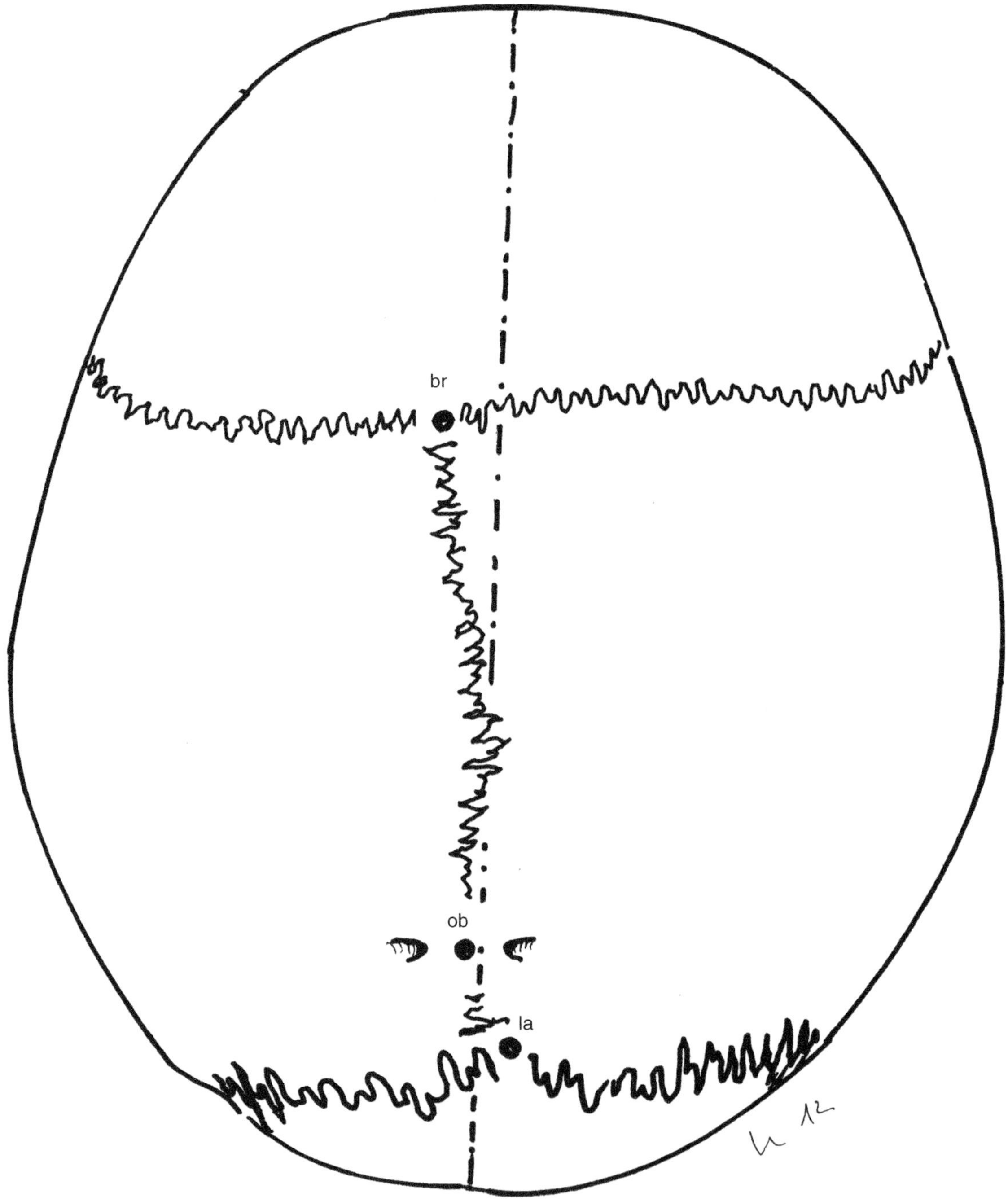

Fig. 9.2 Incongruity of sutura sagittalis and midline. Cadaver skull dissection, common finding. This must be considered at surgical midline approaches: *br* bregma, *la* lambda, *ob* obelion

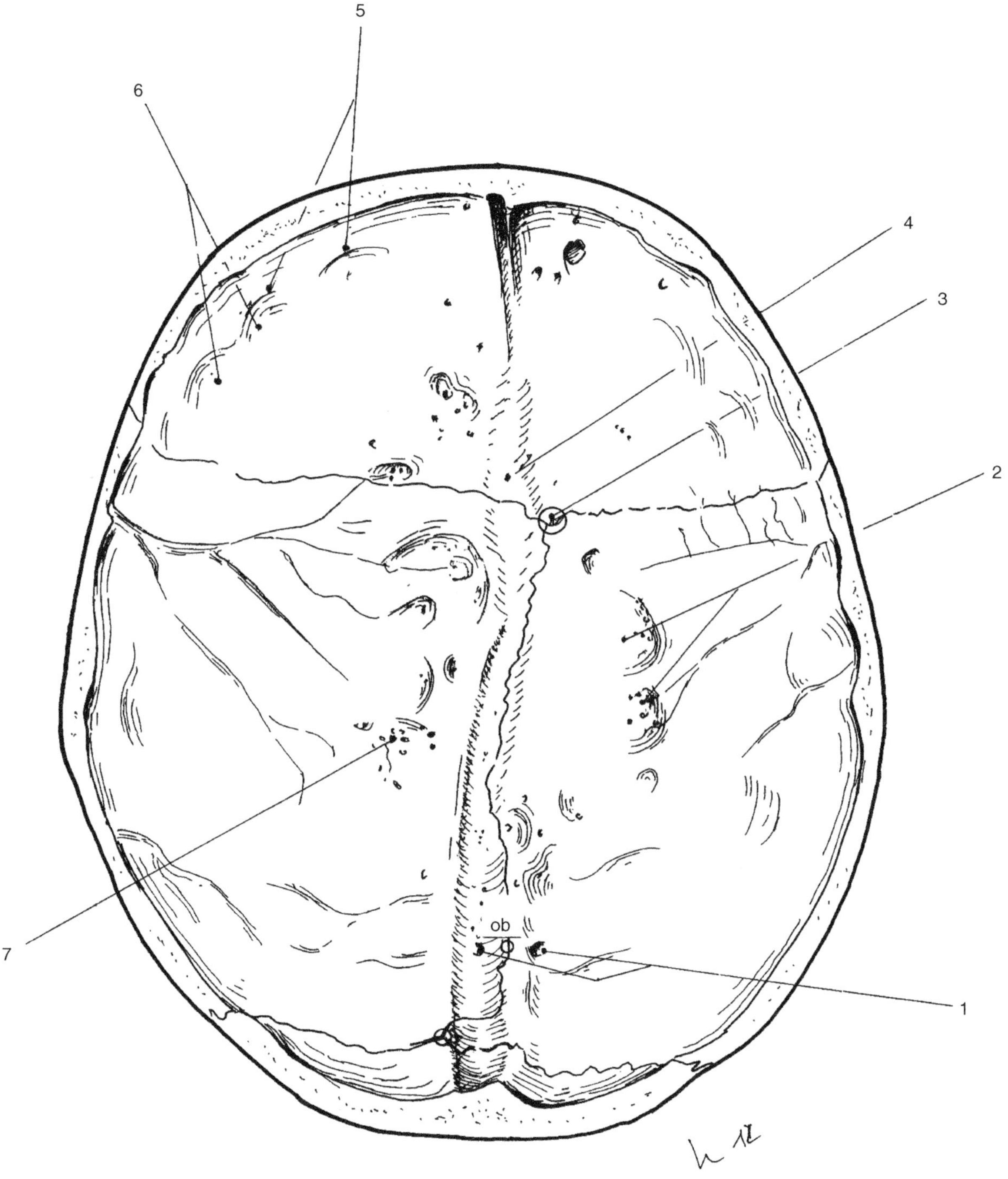

Fig. 9.3 Continuation of Fig. 9.2. Inside asymmetry of sulcus sagittalis (bed of sinus sagittalis superior) according to sutura sagittalis. *ob* obelion, *1* foramina parietalia, *2* foveolae granulares, enclosing Pacchioni's granulations and some lacunae, *3* sutures at the inside not congruent to sutures at the outside (bregma, recognizable at axial CTs), *4* sulcus sagittalis, *5* juga cerebralia, *6* impressiones gyrorum, *7* as *2*

9.1.3 Sinus Sagittalis Superior (Figs. 9.4, 9.5, and 9.6)

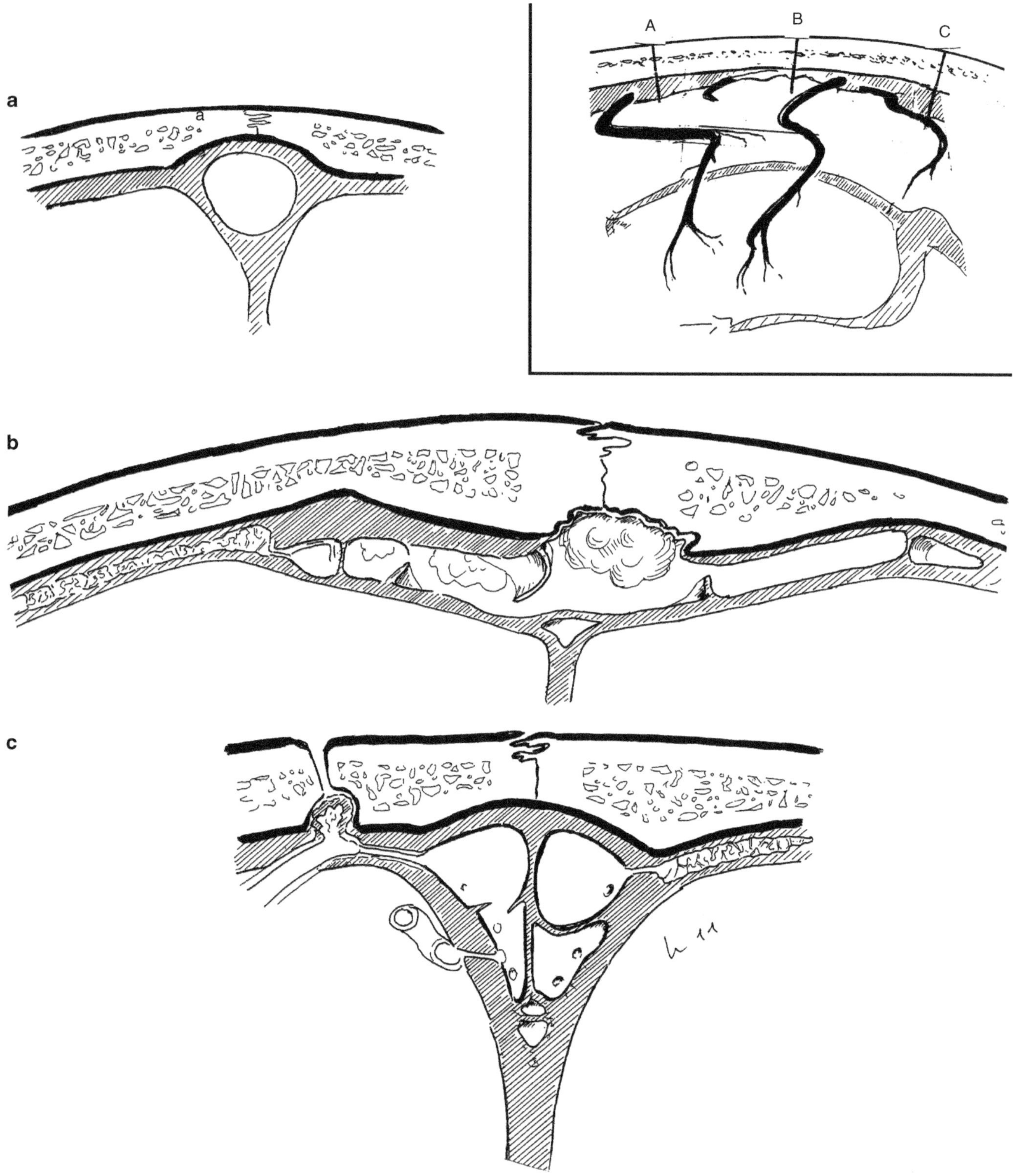

Fig. 9.4 Sinus sagittalis sup and sulcus sagittalis. Common findings. Sketch for topogram added: **a** frontal segment, **b** parietooccipital segment, **c** occipital segment. The lateral limits of sinus sagittalis superior could be unclear at imagings and during surgery. Sinus could be mixed up with lacunae

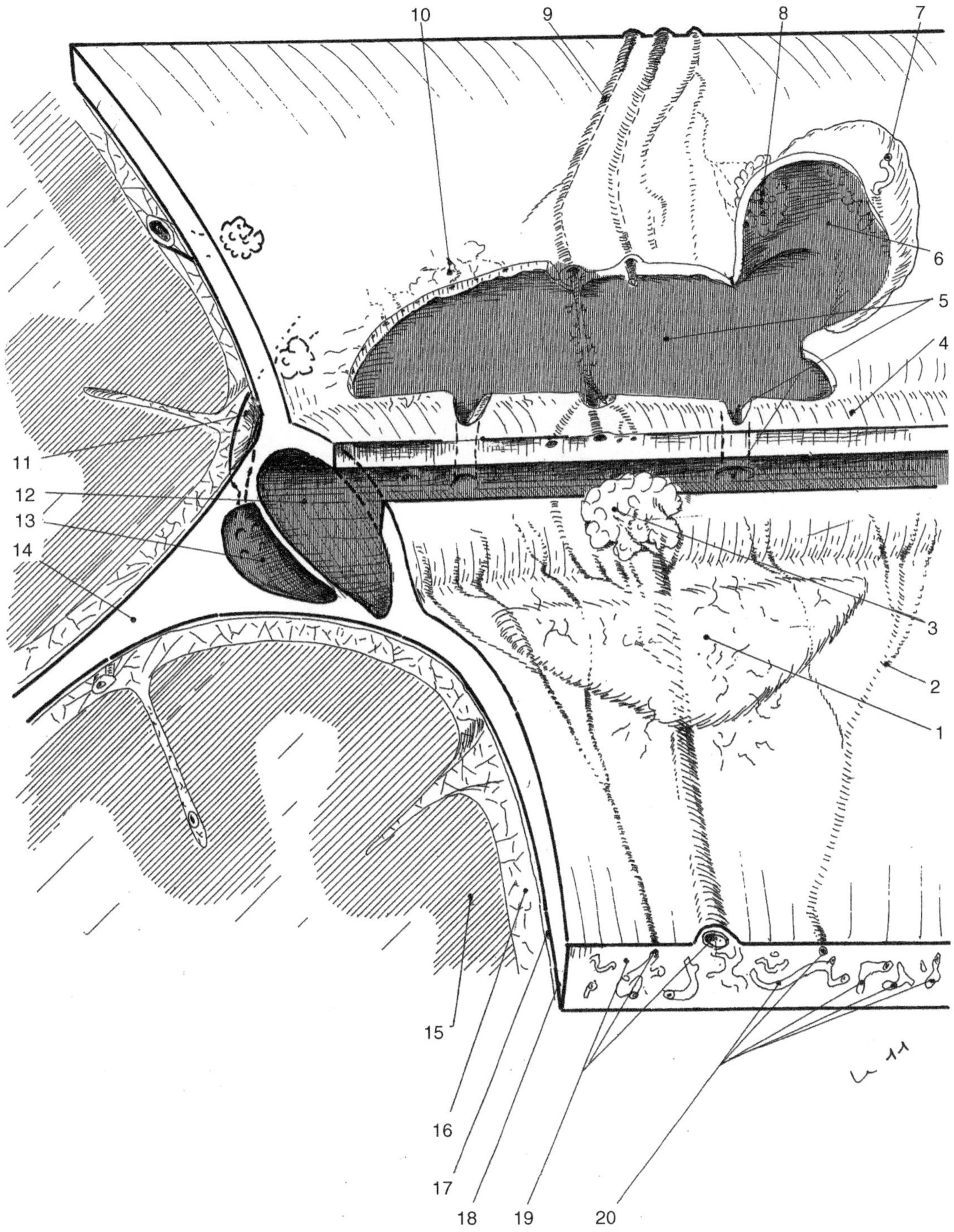

Fig. 9.5 Sinus sagittalis superior, dural veins, and lacunae, schematized. Fine venous network penetrates all dural structures close to the sinuses. Venous network and adjacent bridging veins could be dilated close to extracerebral tumors, e.g., meningiomas. The widened veins could substitute the compressed sinus and bridging veins. This must be considered at surgery. A transversal incomplete or complete horizontal septum is dividing the sinus in a superficial and a usually smaller deep channel. Lacunae are connected by surgery without endangering cerebral circulation. Bridging veins are not connected with lacunae (Browder and Kaplan 1976), but it could be located at the deep wall of lacunae. The mouth of bridging veins is usually connected with the inferior sulcus of the deep channel of the sinus (Lang 1979, 1991). *1* Lacuna; *2* meningeal arteries and veins; *3* granulationes arachnoidales (Pacchioni); *4* sinus wall; *5* lacuna and mouth at the sinus wall; *6* as 5, bulging to tabula int. containing granulationes arachnoidales; *7* vessel(s) of granulationes arachnoidales; *8* granulationes arachnoidales at the inside of lacunae; *9* meningeal veins; *10* fine venous network penetrating all dural structures;*11* bridging vein; *12* superficial part of the sinus, often larger than the inner part; *13* deep part of the sinus, *14* falx; *15* cortex; *16* subarachnoid space; *17* outer arachnoid layer; *18* subdural space; *19* dural veins; *20* dural arteries

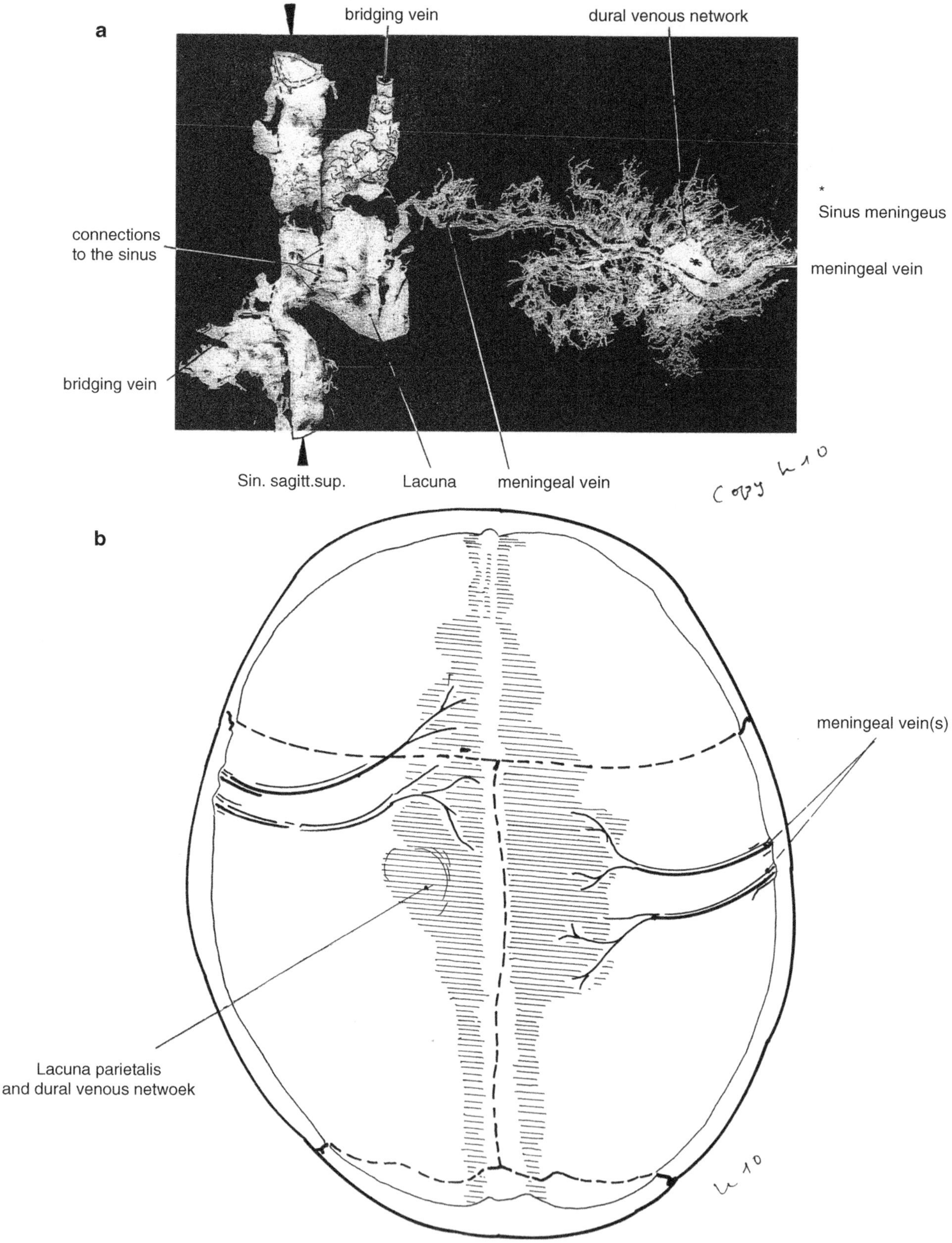

Fig. 9.6 Lacunae, dural veins, and venous network adjacent to the sinuses. (**a**) Cast (According to Browder and Kaplan 1976, p 20, Indian ink copy). (**b**) Schematized vascular bony sulci at a cadaver skull dissection. Hatched: area of numerous foveolae arachnoidales and bony sulci for vessels

9.1.4 Dural Sinuses and Its Intradural Connections (Figs. 9.7, 9.8, and 9.9)

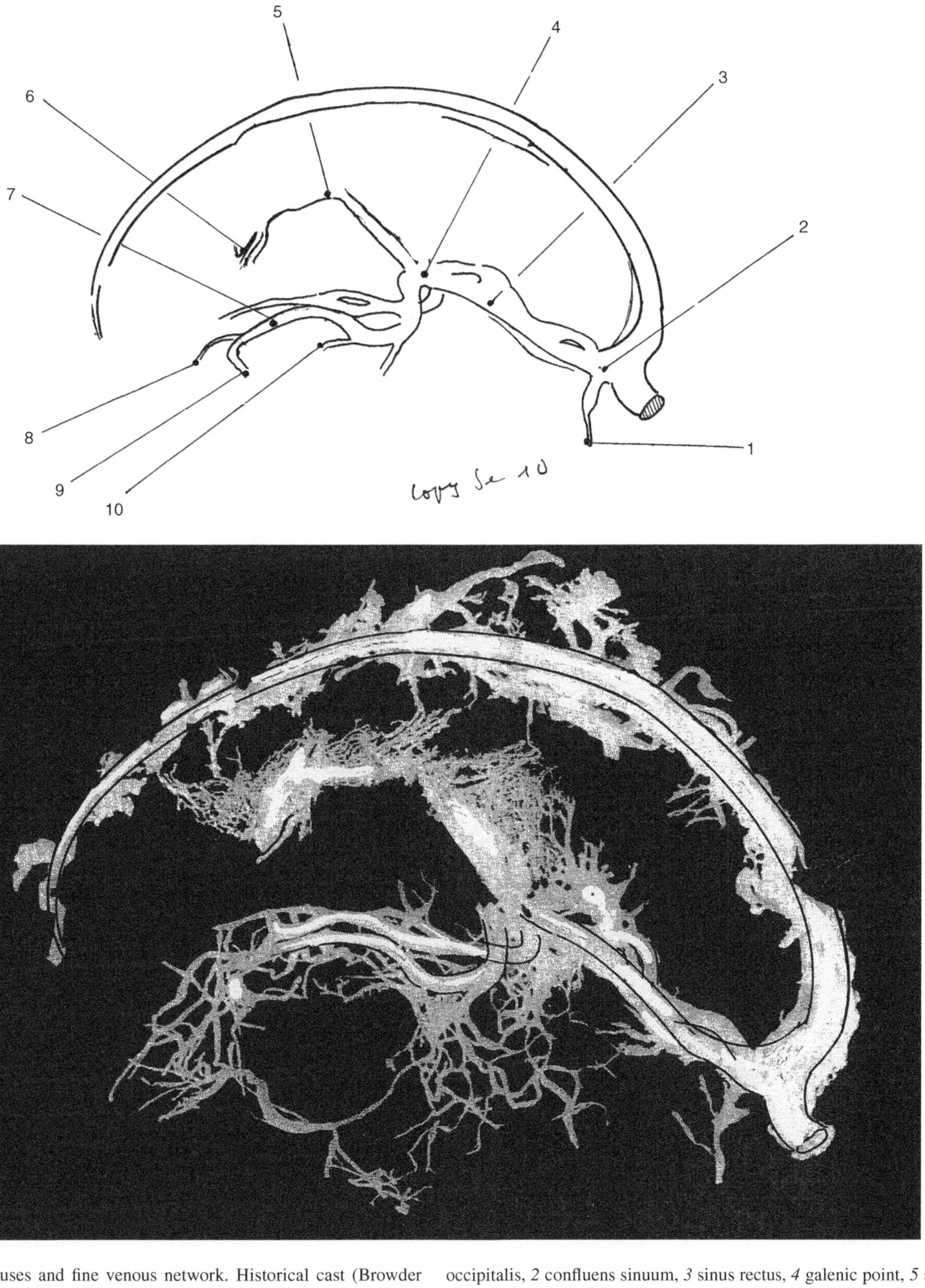

Fig. 9.7 Sinuses and fine venous network. Historical cast (Browder and Kaplan (1976), p 96, Indian ink copy). Similar to modern MRT shapes of sinuses rendered prominent. Simplified copy added. *1* Sinus occipitalis, *2* confluens sinuum, *3* sinus rectus, *4* galenic point, *5* sinus sagittalis inferior, *6* bridging veins of 5, *7* v. cerebri interna, *8* v. septi pellucidi, *9* v. thalamostriata, *10* thalamic vein (inconstant)

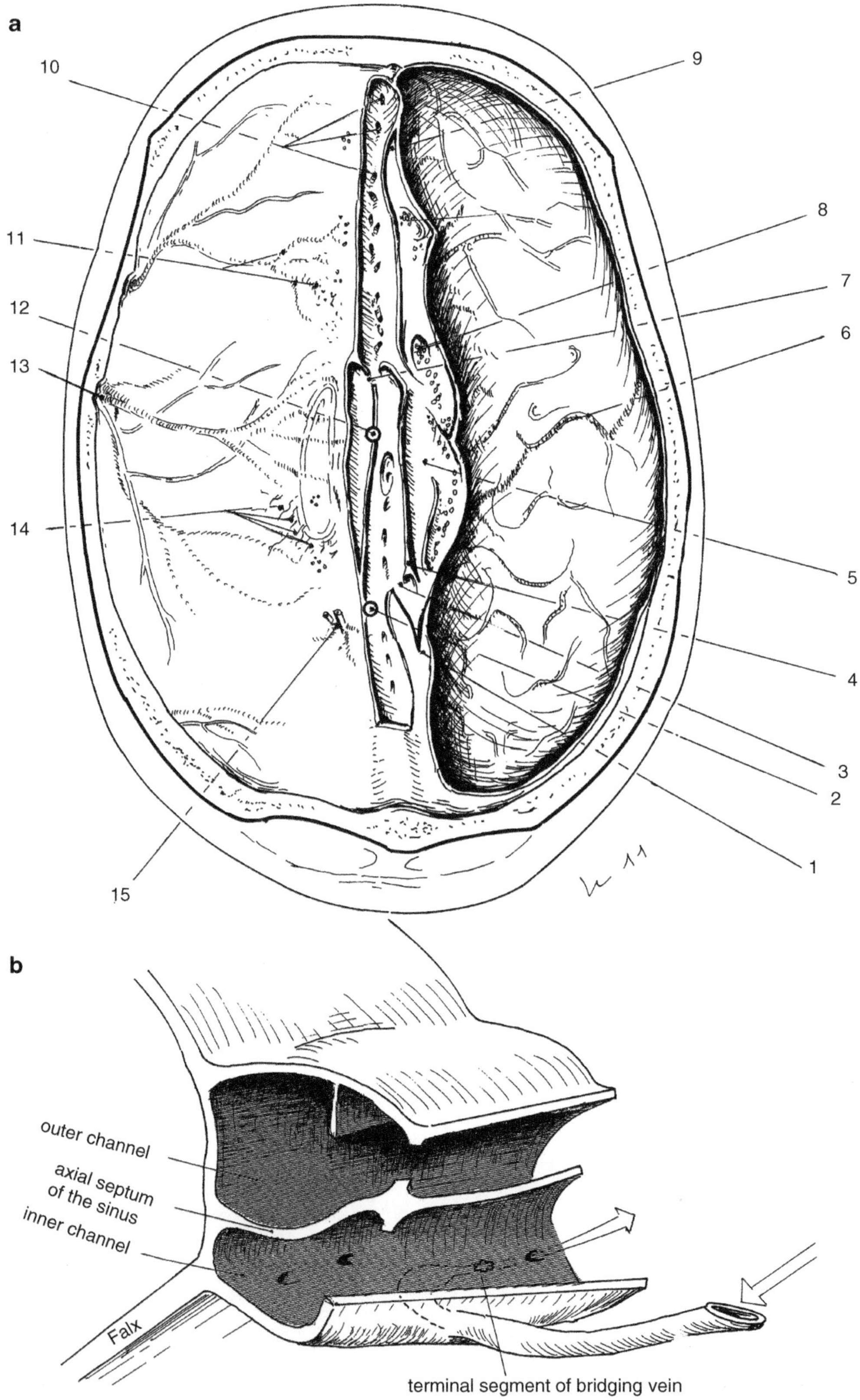

Fig. 9.8 Sinus sagittalis superior. (**a**) Inside of the sinus (Lang 1979, p 49, schematized). (**b**) Architecture of septa. *Arrows*: blood stream. Note mouth of bridging veins at sinus ground. *1* Lacuna, posterior region; *2* obelion, projection; *3* mouth of lacuna; *4* right-sided lateral wall of the sinus; *5* lacuna, parietalis (common finding), long type; *6* V. centralis (Rolandi); *7* sinus wall medial from lacuna; *8* typical groups of granulationes arachnoidales at the inside of lacunae; *9* lacuna, anterior limit; *10* usual location of the mouth of bridging veins at the ground of the sinus; *11* fine vascular network; *12* precentral region; *13* sinus sphenoparietalis; *14* fine meningeal arteries and fine venous network; *15* Vv. emissariae of the foramen parietale Fig 9.9

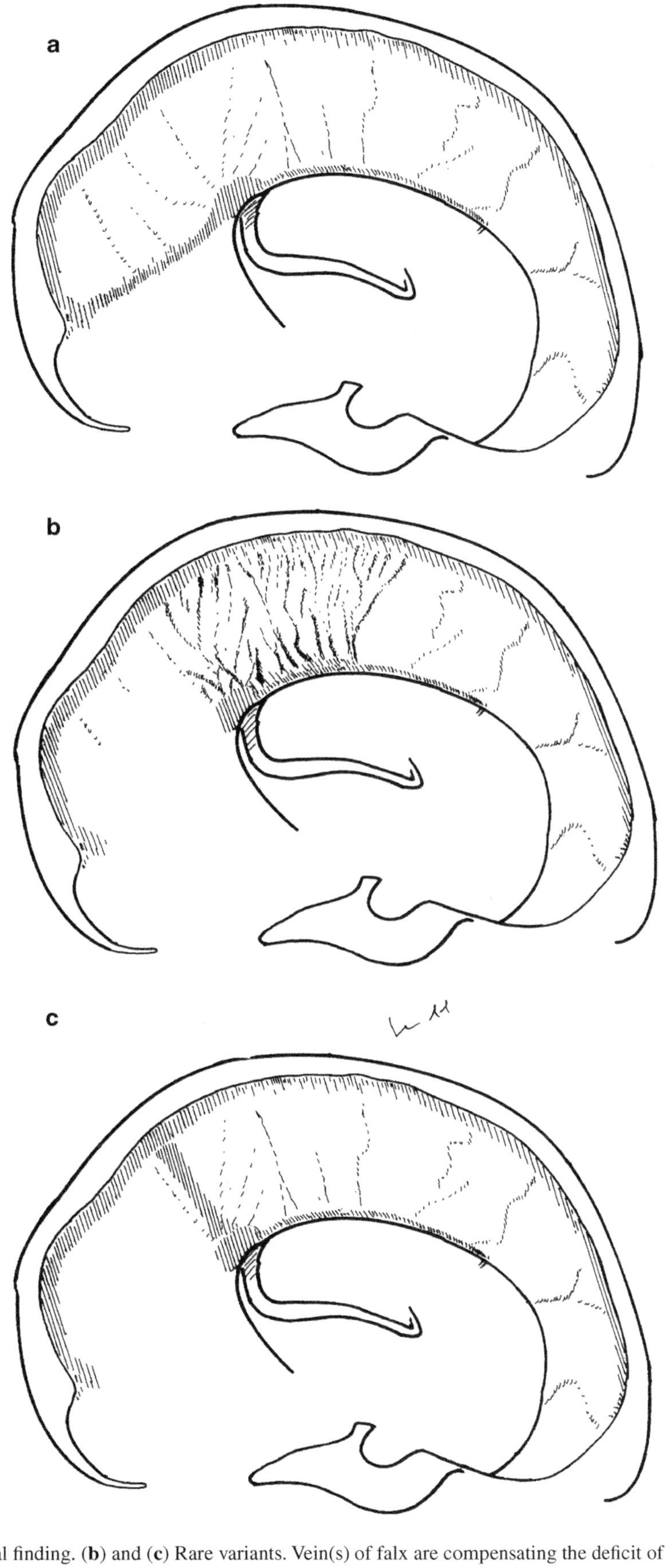

Fig. 9.9 Sinus rectus. (**a**) Usual finding. (**b**) and (**c**) Rare variants. Vein(s) of falx are compensating the deficit of sinus rectus

9.2 Intradural Topography

9.2.1 Survey (Figs. 9.10, 9.11, 9.12, 9.13, 9.14, 9.15, and 9.16)

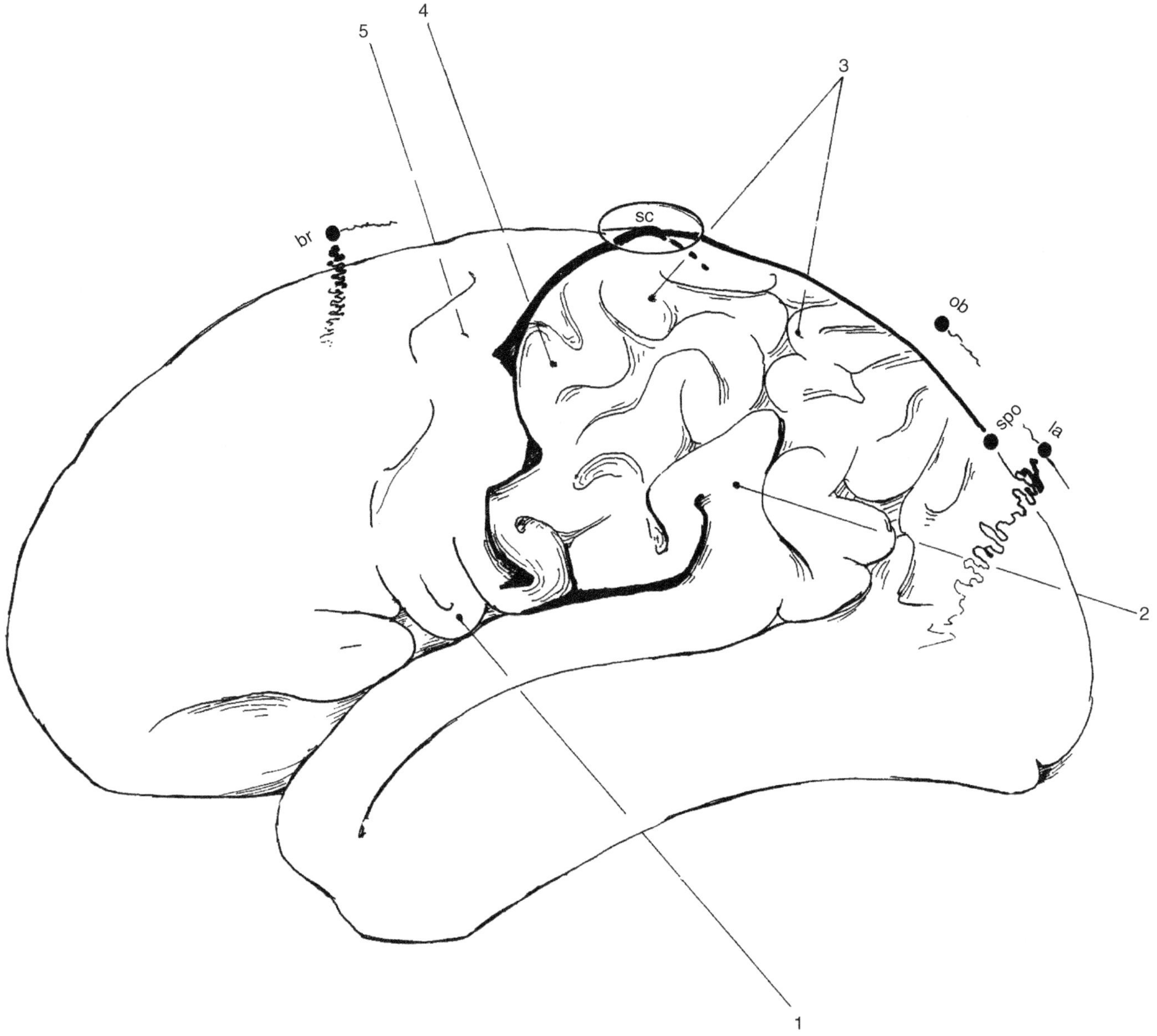

Fig. 9.10 Cortex of lobus parietalis. Numerous cortical functions of lobus parietalis may be endangered by surgical manipulations at the parietal cortex (Fig. 3.14), more endangered than at surgery of lobus frontalis, temporalis, or occipitalis. The ground of cortical sulci is extended into the depth of centrum semiovale (Figs. 3.20 and 3.22). Surgical routes are possible along sulci without opening its pial walls in the depth (Yaşargil). An exact topographical location of lesions during surgery is necessary and could be defined and carefully approached using neuronavigation and supported by bony landmarks, if necessary. This is illustrated here. But small lesions in the depth could be problematic, if stereotactic methods are contraindicated (e.g., for some cavernomas). Solely neuronavigation may not be sufficient for topographical definings of small lesions. The problem is brain shifting after dura opening with flow up CSF. Recommended is using a superficial vein for landmark, e.g., a ramification or loop of a vein. Distance measurements as short as possible between the venous configuration, a sulcus, and the lesion in the depth make possible an exact defining of the lesion during surgery. This is possible with or without using a sector scan (experience of the author).Arteries are not recommended for landmarks at transulcal surgical routes, because arteries are pulsing in the CSF space and are not fixed at the brain by pia mater like superficial cortical veins. *br* bregma, *la* lambda, *ob* obelion, *sc* sulcus centralis, *spo* sulcus parieto-occipitalis. *1* Gyrus frontalis, 3 pars opercularis; *2* gyrus angularis; *3* lobulus parietalis superior and inferior, high variability; *4* gyrus postcentralis; *5* gyrus praecentralis

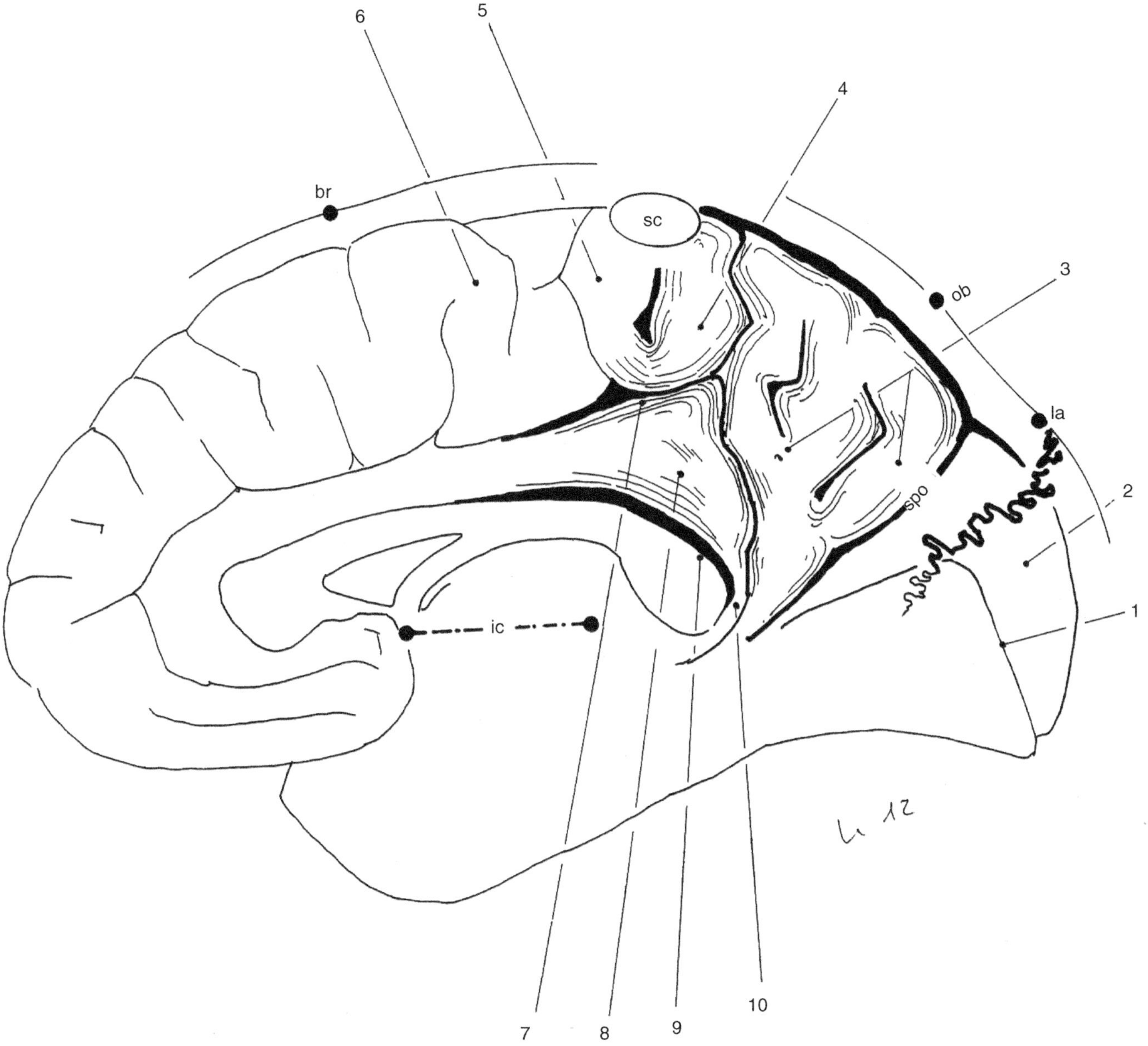

Fig. 9.11 Continuation of Fig. 9.10. Well to define is the medial parietal cortex, in contrast to the lateral cortex. br bregma, ic intercommissural line, la lambda, *ob* obelion, *sc* sulcus centralis, *spo* sulcus parietooccipitalis. *1* Sulcus calcarinus, *2* cuneus, *3* praecuneus, *4* lobulus paracentralis, parietal segment, *5* as 4, frontal segment, *6* gyrus front. 1, *7* sulcus cinguli, *8* gyrus cinguli, *9* sulcus corporis callosi, *10* isthmus gyri cinguli

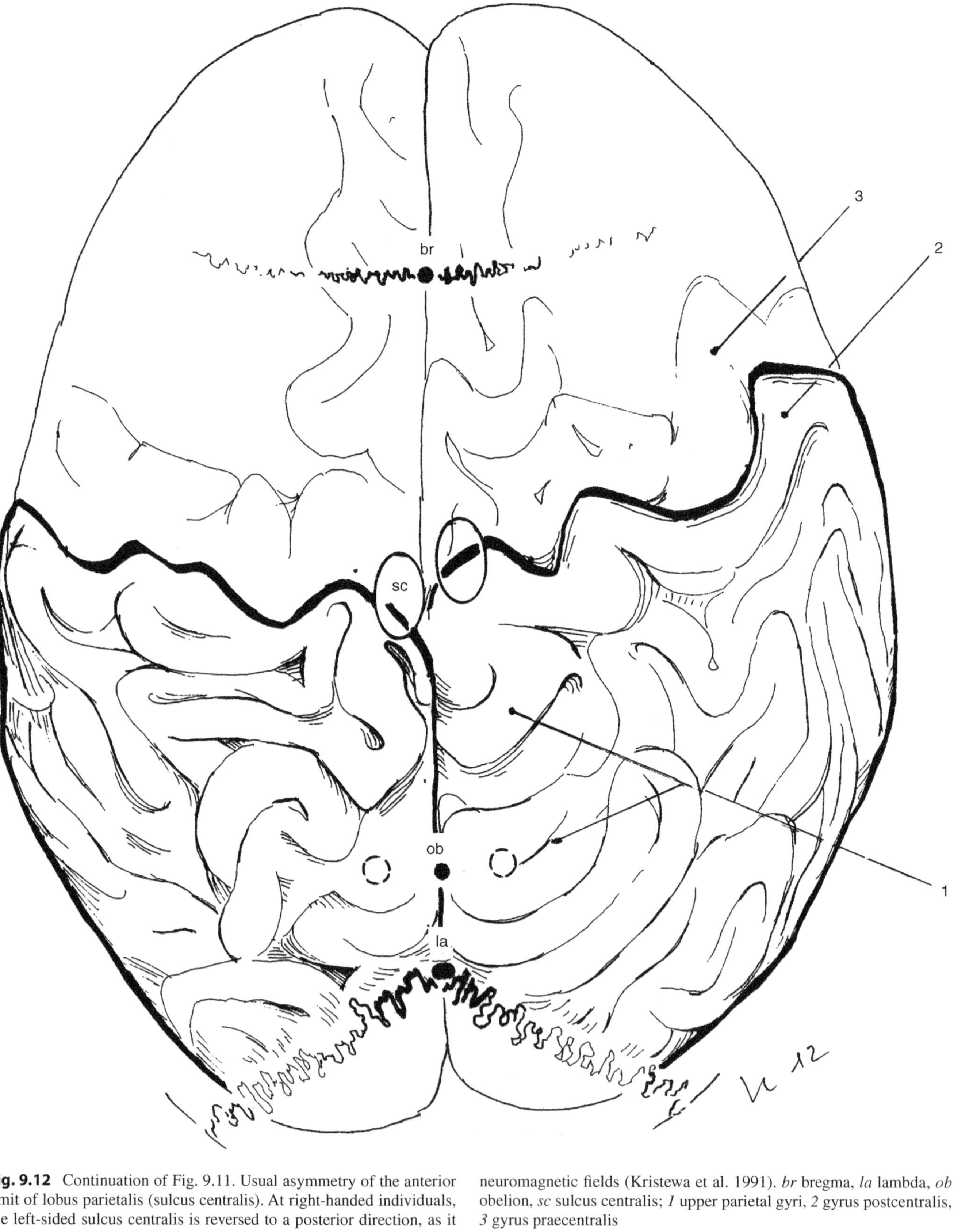

Fig. 9.12 Continuation of Fig. 9.11. Usual asymmetry of the anterior limit of lobus parietalis (sulcus centralis). At right-handed individuals, the left-sided sulcus centralis is reversed to a posterior direction, as it was presented by Bischoff 1868 and is presented today by imagings and neuromagnetic fields (Kristewa et al. 1991). *br* bregma, *la* lambda, *ob* obelion, *sc* sulcus centralis; *1* upper parietal gyri, *2* gyrus postcentralis, *3* gyrus praecentralis

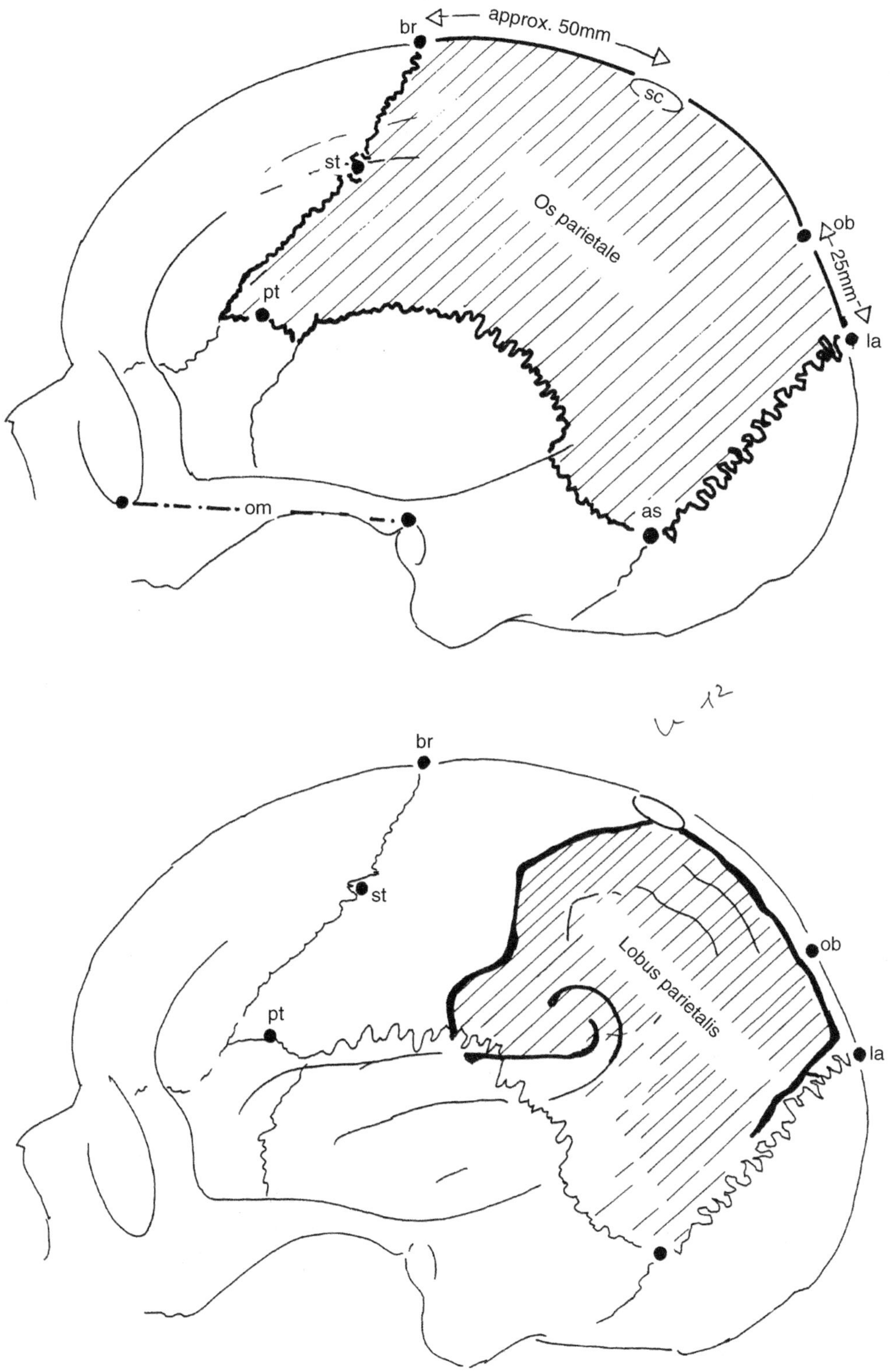

Fig. 9.13 Comparison of lobus parietalis with os parietale. For avoiding topographical errors at craniotomy. *as* asterion, *br* bregma, *la* lambda, *ob* obelion, *om* orbitomeatal line, *pt* pterion, *sc* sulcus centralis, *st* stephanion

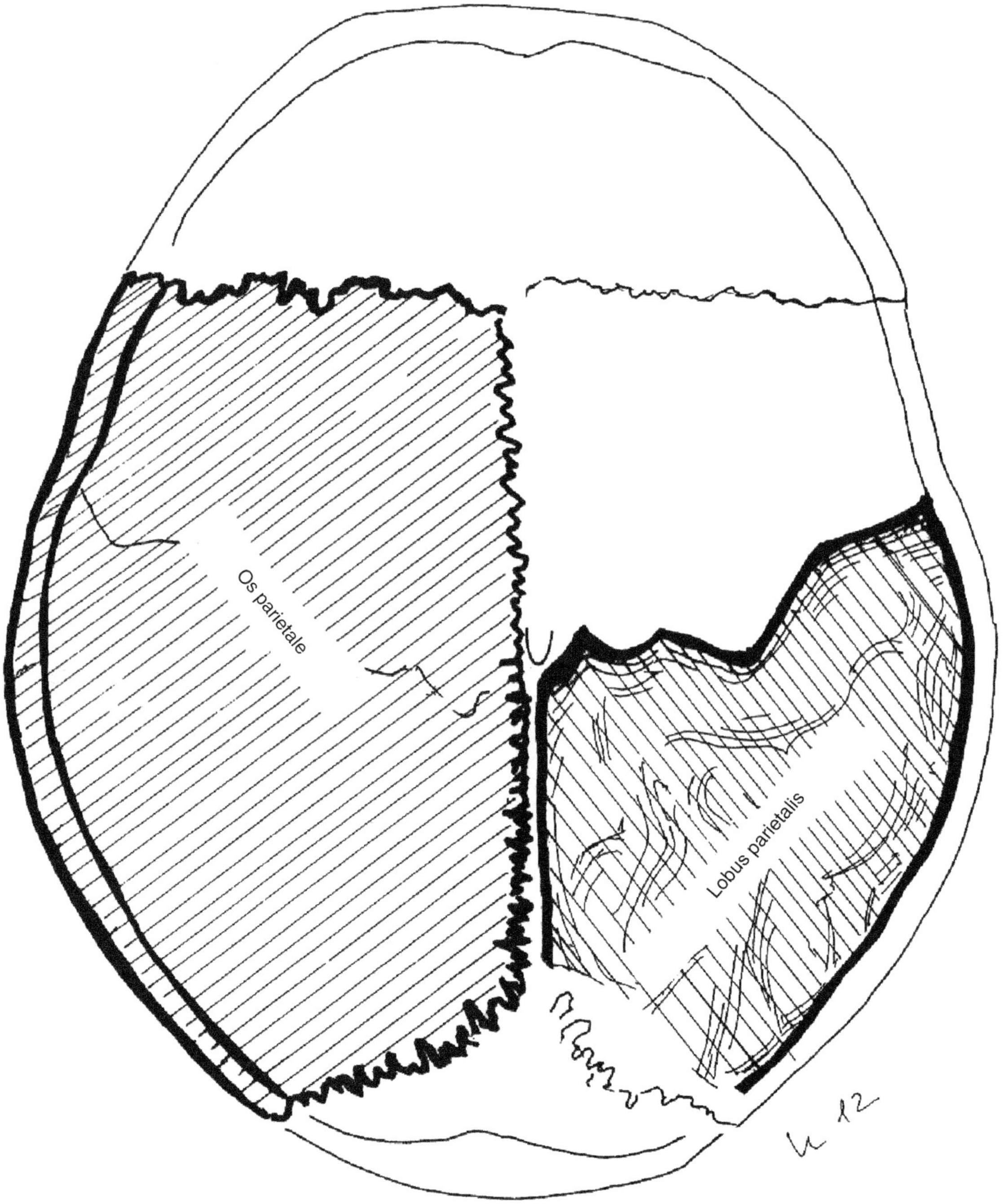

Fig. 9.14 Continuation of Fig. 9.13

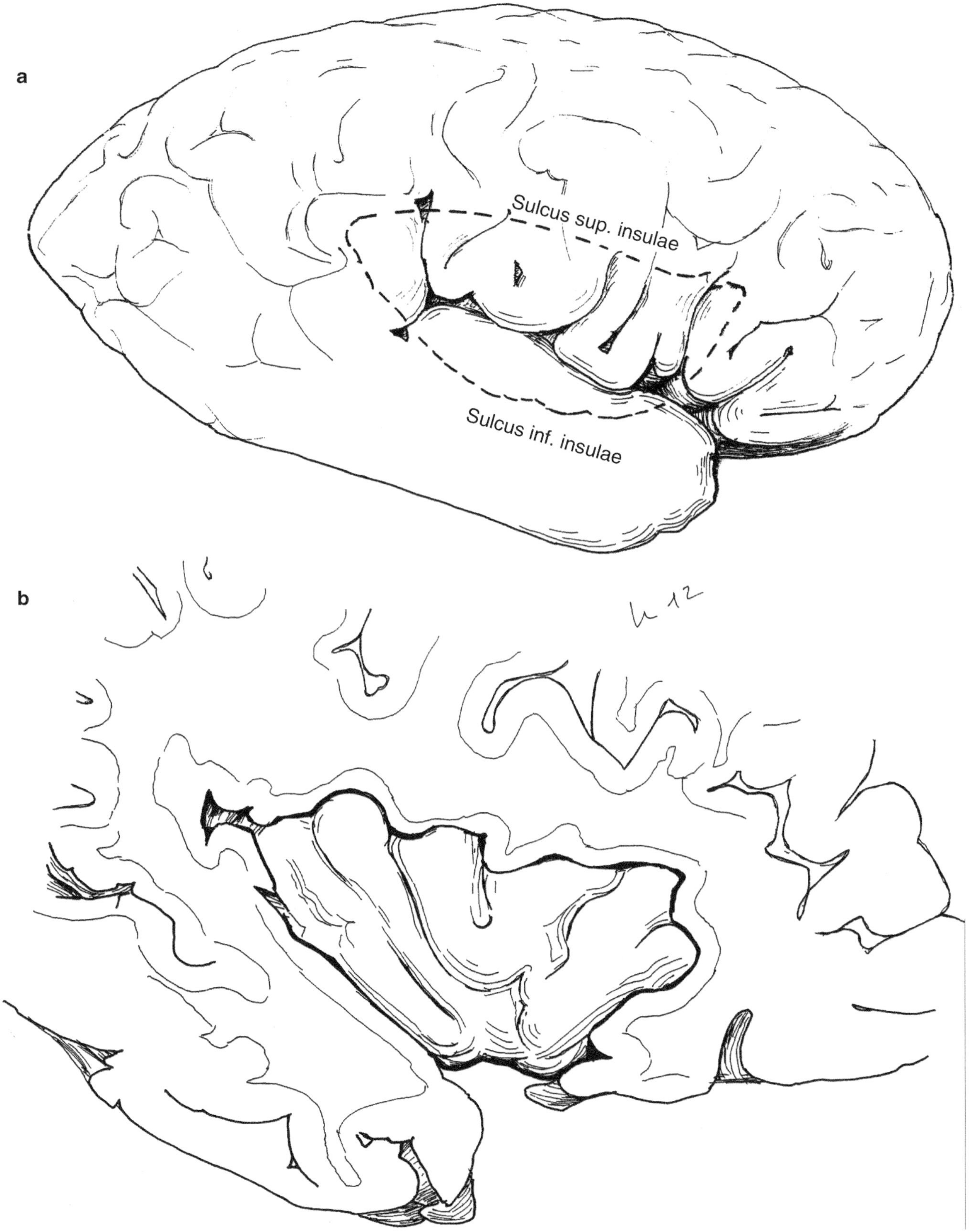

Fig. 9.15 Lateral parietal region and insula. (**a**) Insula, projection. (**b**) Relief of insula similar to neuronavigation

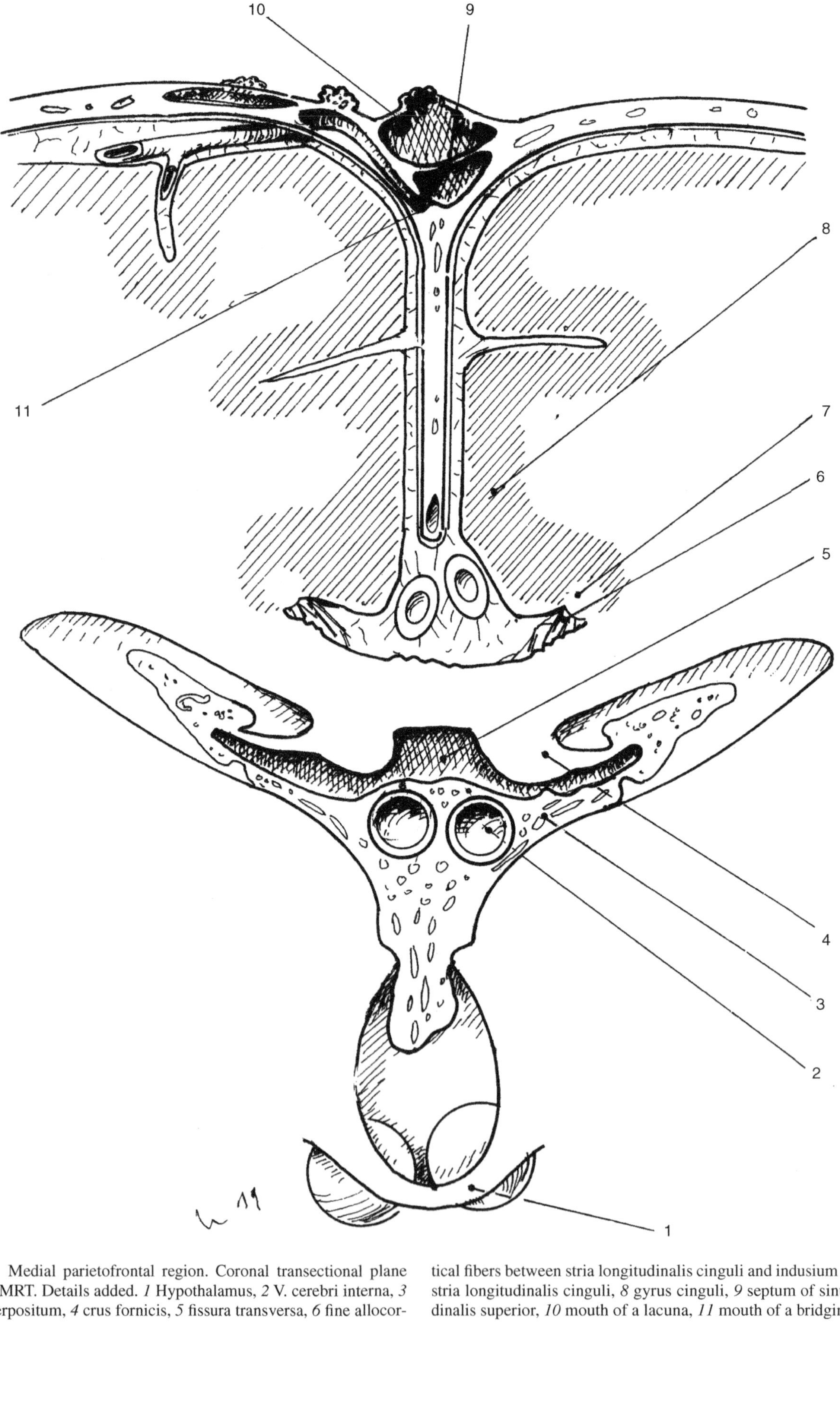

Fig. 9.16 Medial parietofrontal region. Coronal transectional plane similar to MRT. Details added. *1* Hypothalamus, *2* V. cerebri interna, *3* velum interpositum, *4* crus fornicis, *5* fissura transversa, *6* fine allocortical fibers between stria longitudinalis cinguli and indusium griseum, *7* stria longitudinalis cinguli, *8* gyrus cinguli, *9* septum of sinus longitudinalis superior, *10* mouth of a lacuna, *11* mouth of a bridging vein

9.2.2 Lateral Structures (Figs. 9.17, 9.18, 9.19, 9.20, 9.21, and 9.22)

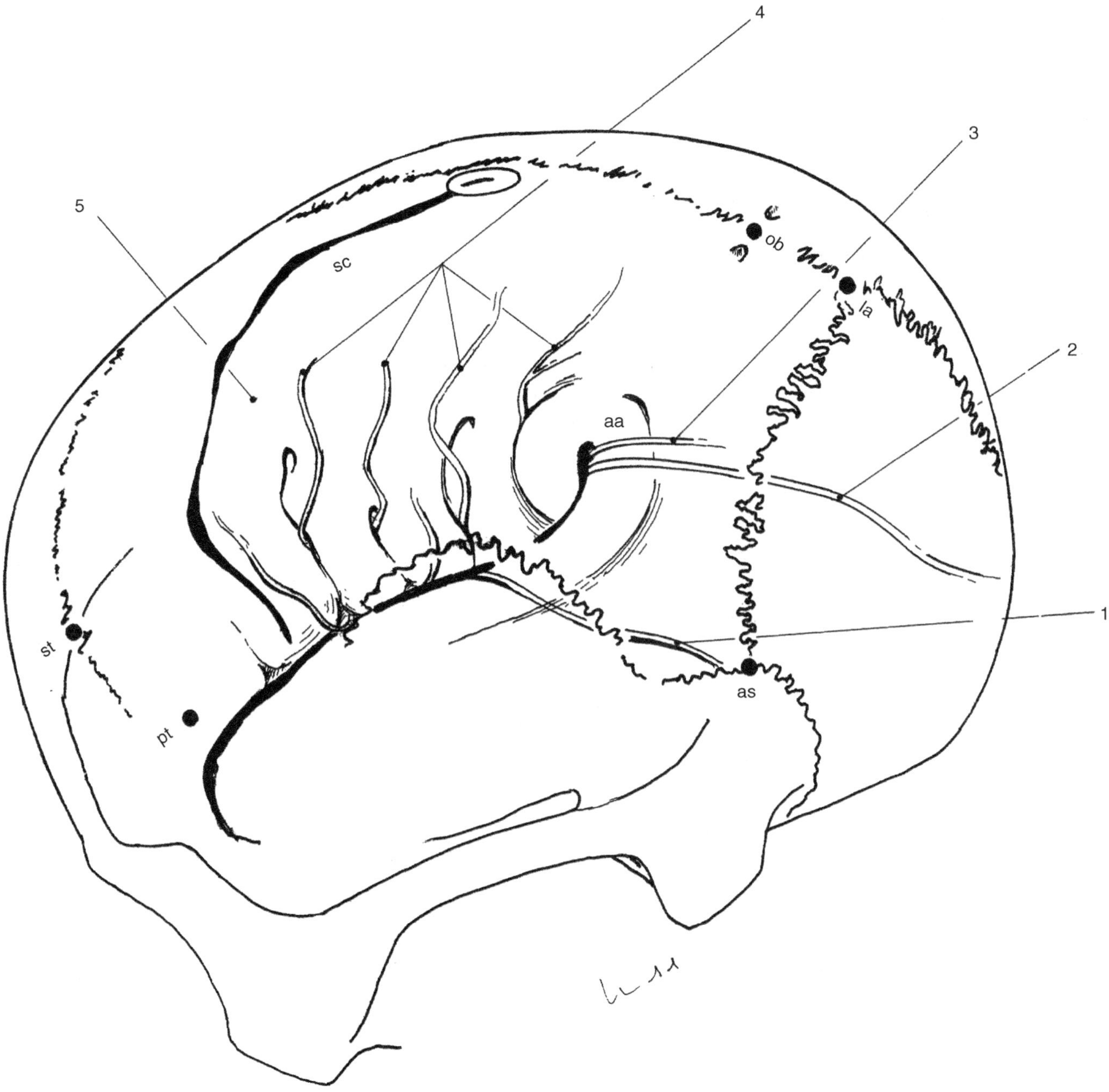

Fig. 9.17 Superficial cortical arteries, schematized. Similar to neuronavigation, assisted by landmarks, if necessary. *aa* angularis area *as* asterion, *la* lambda, *ob* obelion, *pt* pterion, *sc* sulcus centralis, *st* stephanion. *1* A. temporalis posterior, *2* A. temporooccipitalis, *3* A. angularis, *4* Rr. parietales of A. cerebri media, *5* gyrus postcentralis

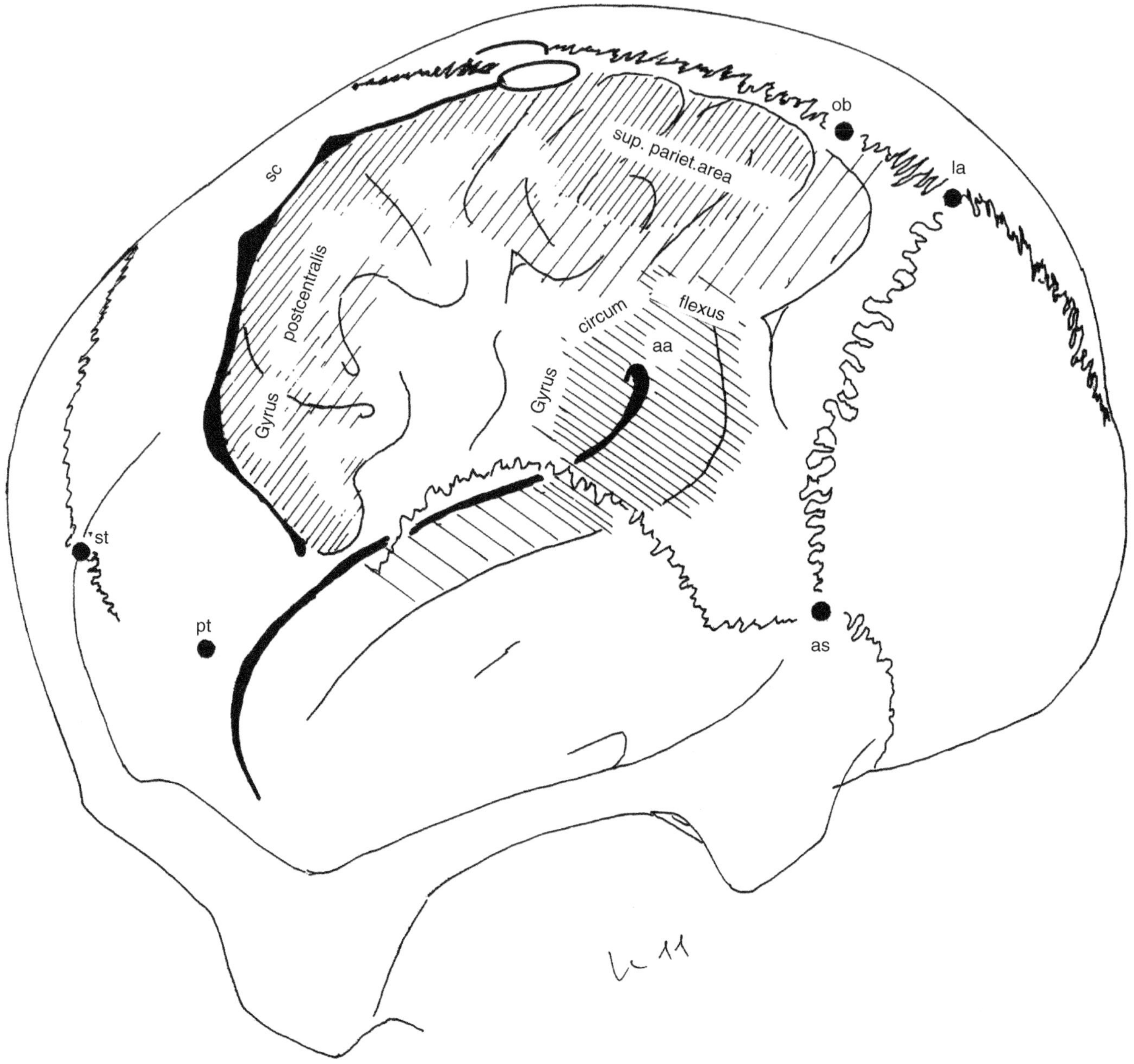

Fig. 9.18 Lateral cortex, schematized. Essential regions hatched. High variability of anatomical areas and its functions. It may be defined using neurophysiological methods and neuronavigation, supported by landmarks, if necessary. *aa* angularis area, *as* asterion, *la* lambda, *ob* obelion, *pt* pterion, *sc* sulcus centralis, *st* stephanion

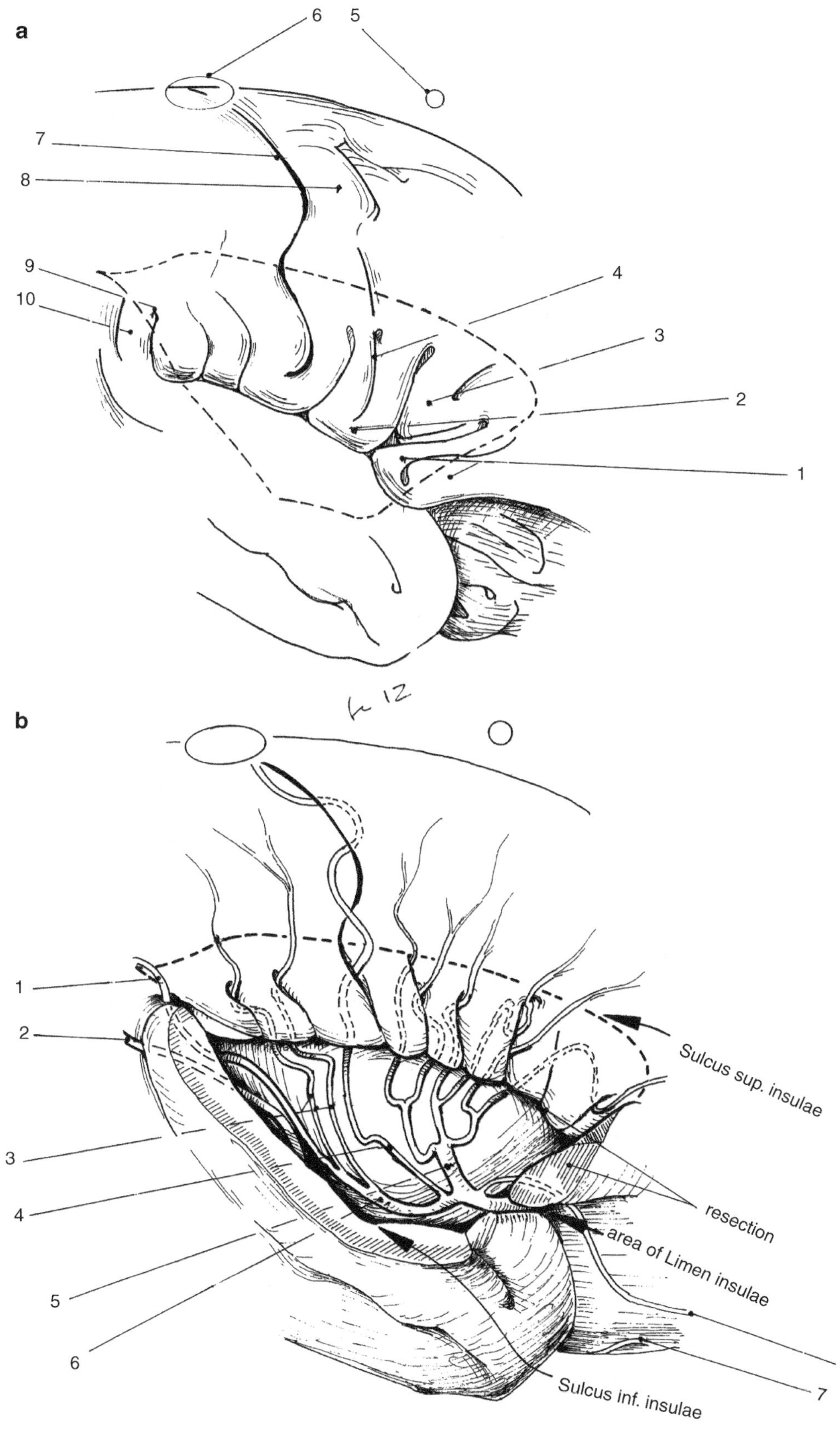

Fig. 9.19 Sylvian segments of A. cerebri media. These branches are running close to sulcus superior insulae. The sulcus is located close to corona radiata, especially its tractus pyramidalis. (**a**) Survey with projection of sulcus superior insulae (*dotted*), (**b**) fronto-parietooccipital branches of A. cerebri media. (**a**) *1* Third frontal gyrus, pars orbitalis; *2* as 1, pars opercularis; *3* as 1, pars triangularis; *4* sulcus praecentralis; *5* bregma; *6* sulcus centralis; *7* as 6; *8* gyrus praecentralis; *9* pars ascendens of sulcus lat. (sylvii); *10* gyrus angularis. (**b**) *1* A. angularis, *2* A. temporooccipitalis, *3* parietal branches of A. cerebri media, *4* A. sulci centralis, *5* A. operculofrontalis, *6* A. cerebri media, distal trunk

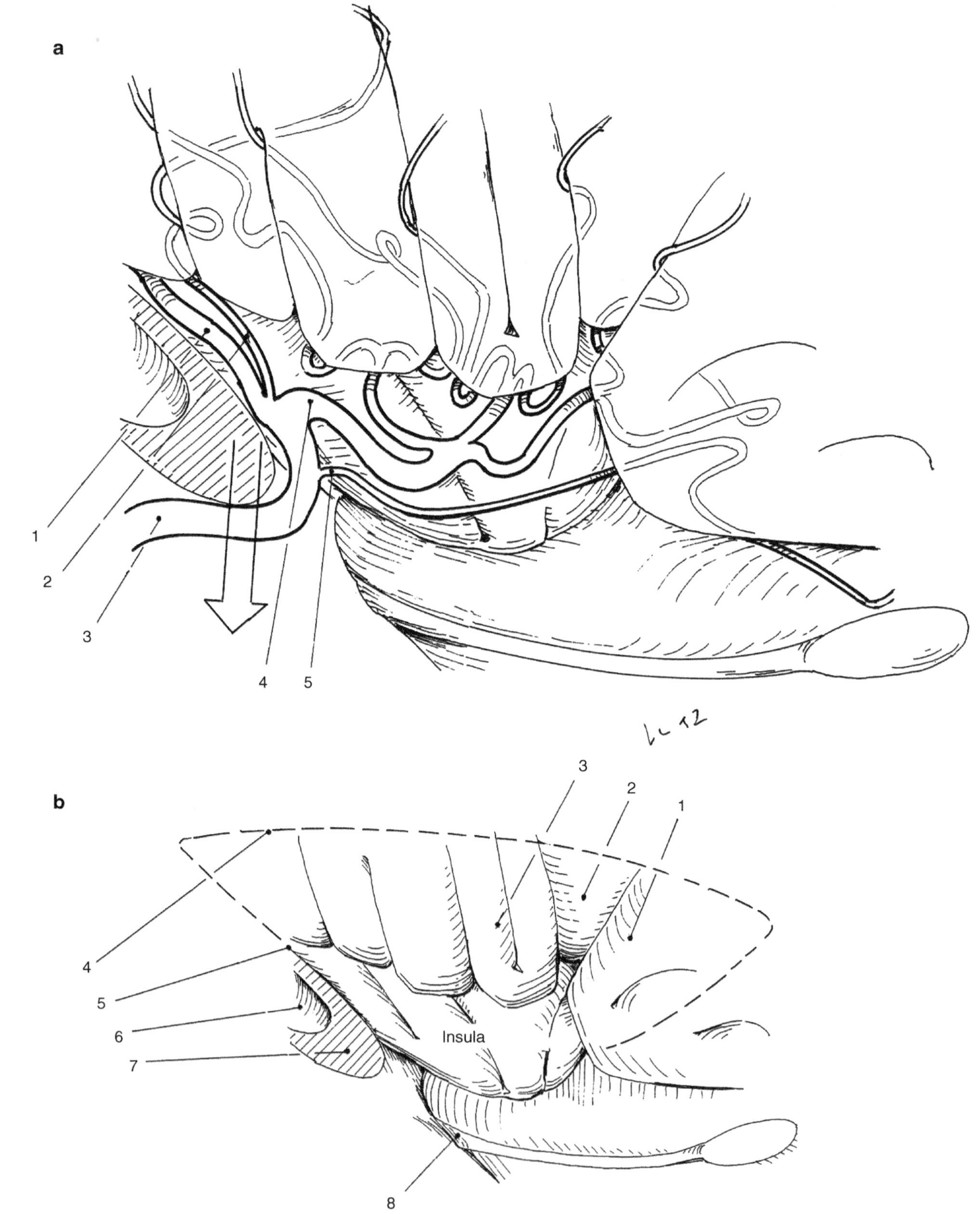

Fig. 9.20 Addendum for Fig. 9.19. **a** Sectional magnification. (**b**) Relief of Cortex. (**a**) *1* A. cerebri media, distal trunk, *2* A. sulci centralis, *3* A. cerebri media, common trunk, *4* A. operculofrontalis, *5* lateral frontobasal artery. Uncus hatched, its residual shifted (arrow). (**b**) *1* Third frontal gyrus, pars orbitalis; *2* as 1, pars triangularis; *3* as, pars opercularis; *4* sulcus superior insulae, projection; *5* sulcus inferior insulae, projection; *6* tip of cornu inferius; *7* uncus, transected; *8* trigonum olfactorium

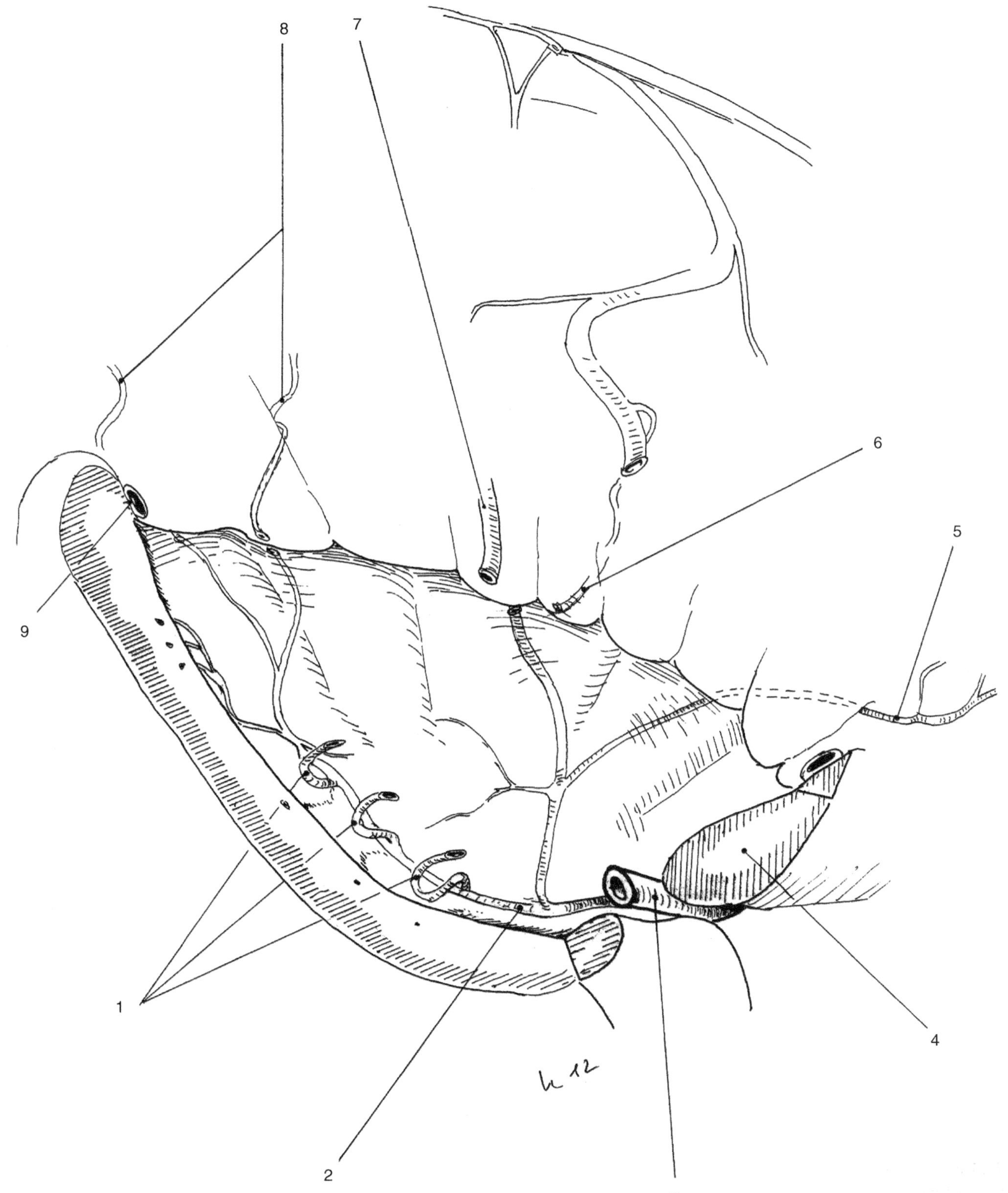

Fig. 9.21 Veins of insula. Sketch. Note the connections between V. temporalis profunda et superficialis *1*. Superficial cortical veins are feeding V. temporalis superficialis. Rare are anastomotic branches with bridging veins to sinus sagittalis sup. *6*. *1* anastomoses between superficial and profound sylvian veins, cut; *2* V. temporalis profunda; high variability of diameter, number, and course; *3* A. cerebri media overlapping *2*; *4* third frontal gyrus, pars orbitalis; *5* frontopolar vein(s); *6* anastomotic vein; *7* thick calibrated parietotemporal vein, cave; *8* small parietal vein draining to superficial or profound sylvian veins; *9* V. temporalis superficialis

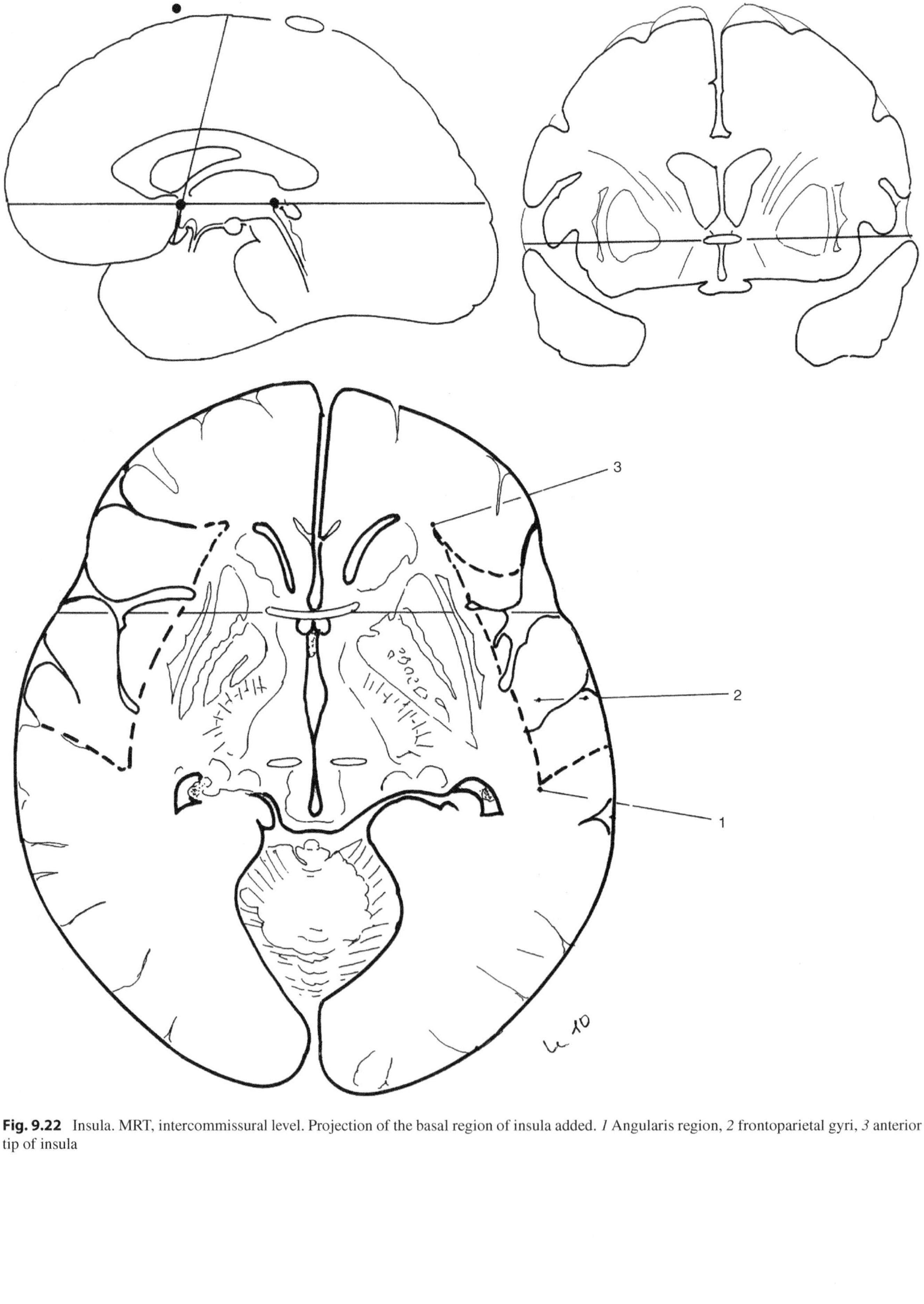

Fig. 9.22 Insula. MRT, intercommissural level. Projection of the basal region of insula added. *1* Angularis region, *2* frontoparietal gyri, *3* anterior tip of insula

9.2.3 **Midline Structures** (Figs. 9.23, 9.24, 9.25, and 9.26)

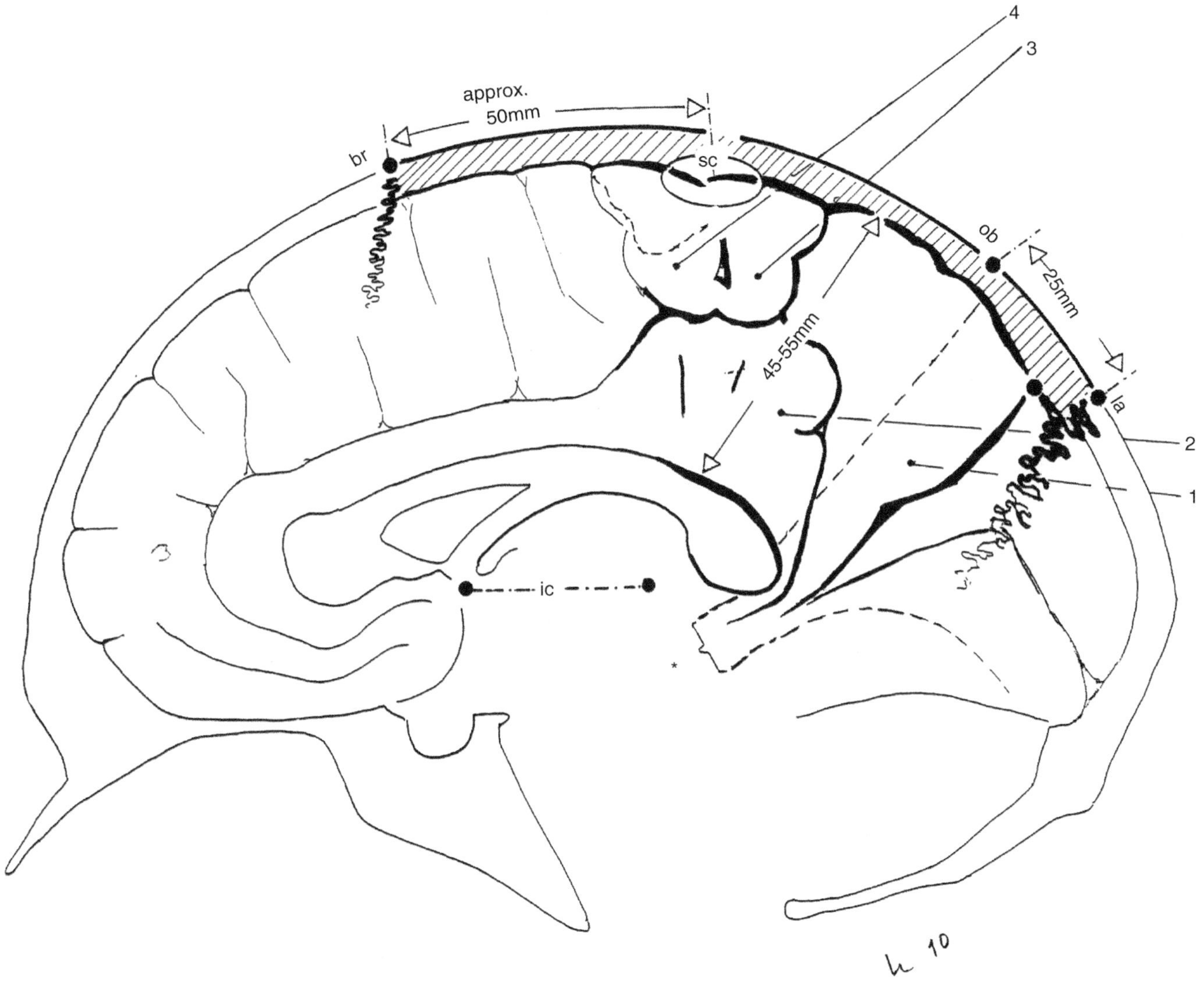

Fig. 9.23 Medial cortex. MRT. Parietal cortex and suturae rendered, os parietale hatched. Landmarks and distance measurements added. *br* bregma, *ic* intercommissural line, *la* lambda, *ob* obelion, *sc* sulcus centralis. *1* Praecuneus, *2* gyrus cinguli, *3* lobulus paracentralis, parietal segment, *4* as 3, frontal segment

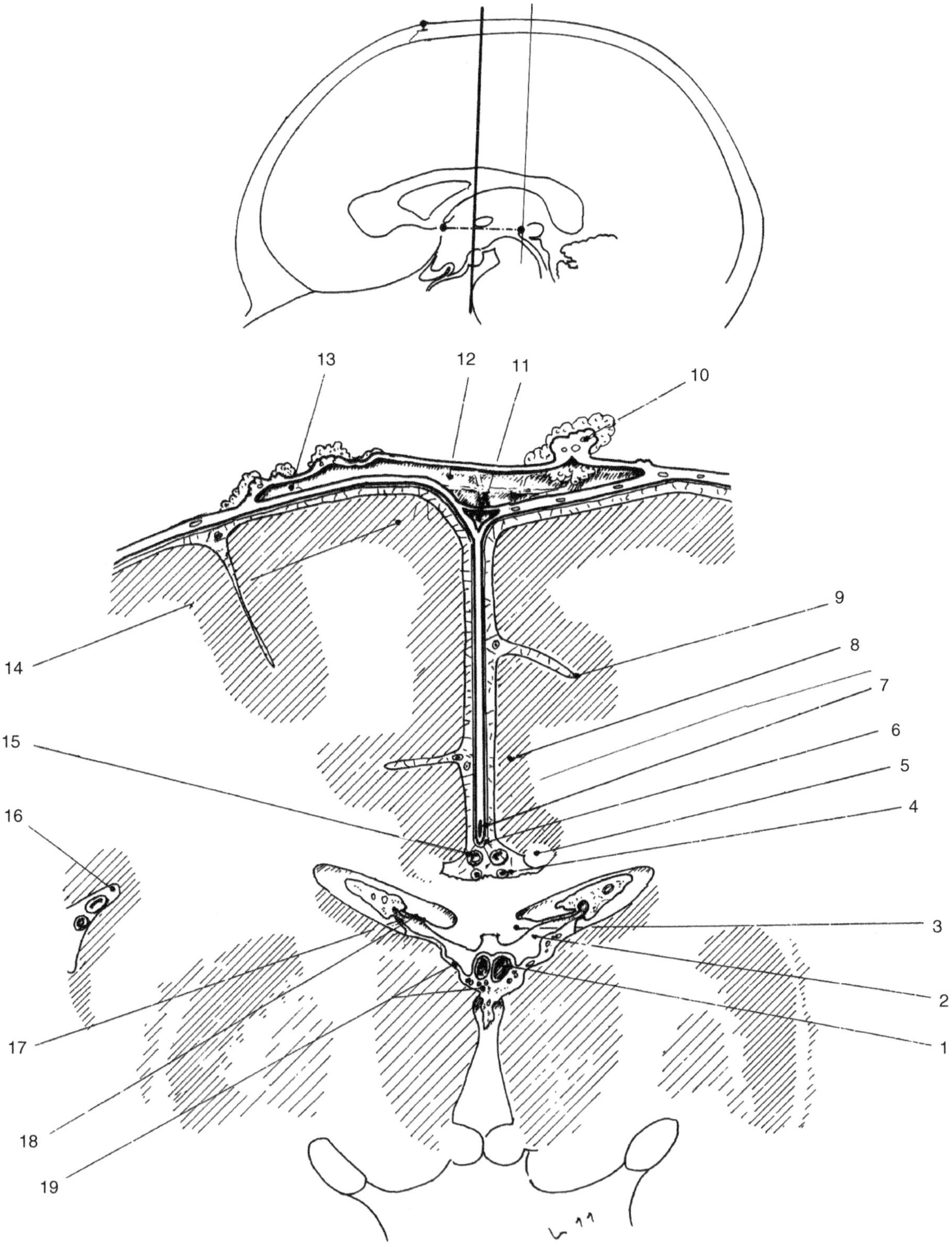

Fig. 9.24 Anterior medial parietal region. MRT, plane of corpora mamillaria, details added. *1* V. cerebri interna; *2* fissura transversa; *3* crus fornicis; *4* A. corporis callosi mediana; *5* stria longitudinalis cinguli; *6* edge of falx; *7* sinus sagittalis inferior; *8* gyrus cinguli; *9* sulcus cinguli; *10* granulationes arachnoidales (Pacchioni); *11* sinus sagittalis superior, deep located compartment (channel); *12* as 11, outer compartment; *13* mouth of lacuna; *14* first frontal gyrusm; *15* A. pericallosa; *16* sulcus superior insulae; *17* taenia chorioidea; *18* taenia fornicis; *19* velum interpositum

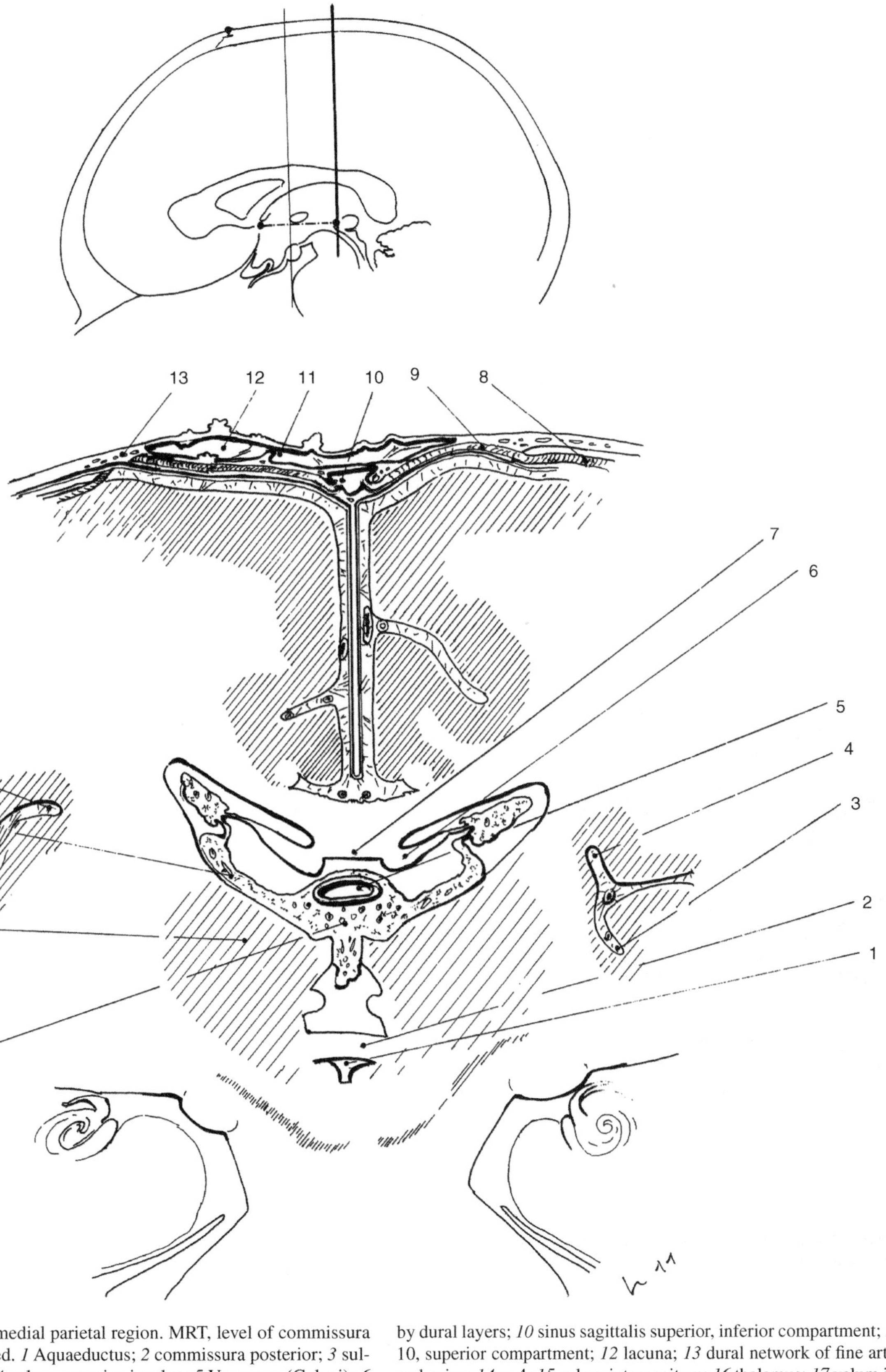

Fig. 9.25 Posterior medial parietal region. MRT, level of commissura posterior, details added. *1* Aquaeductus; *2* commissura posterior; *3* sulcus inferior insulae; *4* sulcus superior insulae; *5* V. magna (Galeni); *6* crus fornicis; *7* commissura fornicis; *8* bridging vein; *9* as 8, enclosed by dural layers; *10* sinus sagittalis superior, inferior compartment; *11* as 10, superior compartment; *12* lacuna; *13* dural network of fine arteries and veins; *14* as 4; *15* velum interpositum; *16* thalamus; *17* velum interpositum and plexus chorioideus of the third ventricle

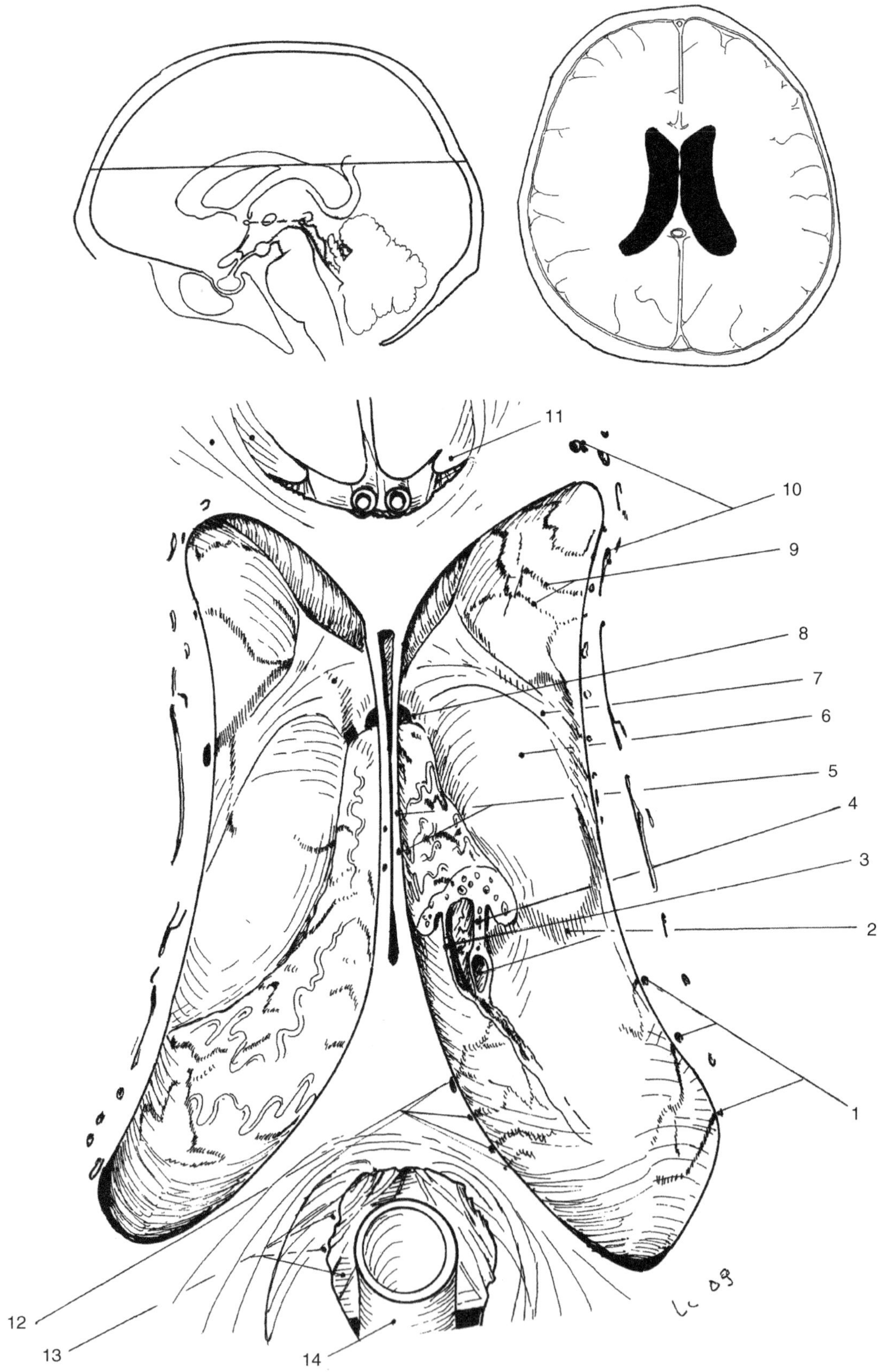

Fig. 9.26 Floor of ventriculus lateralis. Similar to neuronavigation, details added. *1* Lateral atrial veins; *2* V. ventriculi lat. directa, common variant; *3* taenia fornicis; *4* taenia chorioidea; *5* medial septal veins; *6* lamina affixa thalami; *7* stria terminalis; *8* columna fornicis; *9* Vv. nuclei caudati; *10* longitudinal veins along the lateral rim of the ventricle; *11* stria longitudinalis cinguli; *12* medial atrial veins; *13* stria longitudinalis cinguli; *14* V.magna (Galeni)

10 Special Surgical Aspects

10.1 Extradural Surgery (Figs. 10.1 and 10.2)

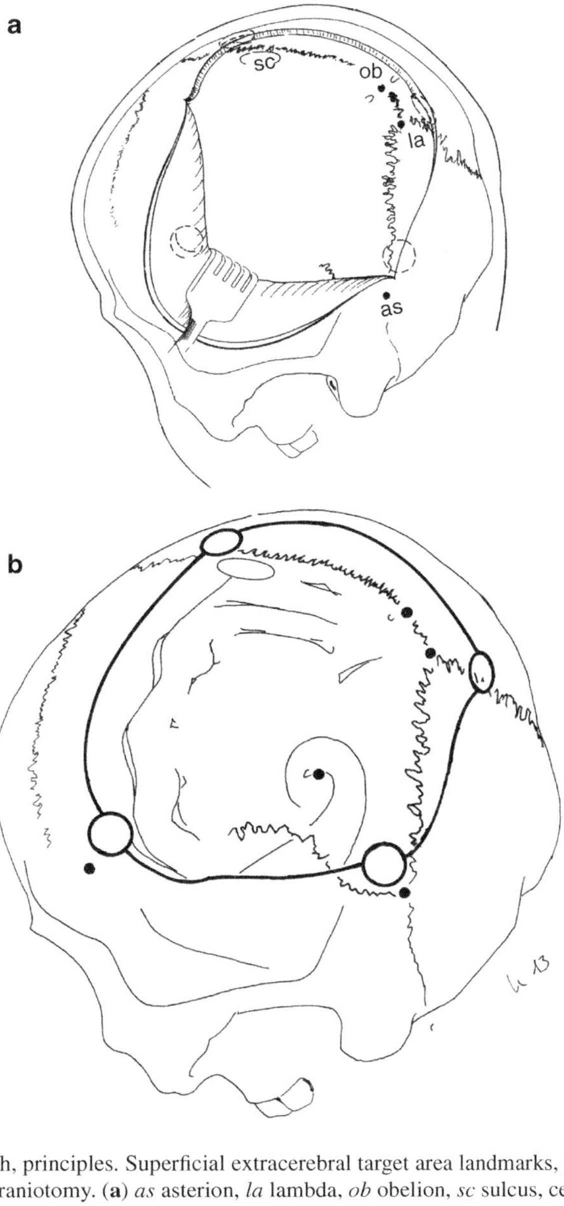

Fig. 10.1 Large parietal surgical approach, principles. Superficial extracerebral target area landmarks, pars ascendens of sulcus lat. (sylvii), and sulcus centralis. (**a**) Skin-galea-flap. (**b**) Craniotomy. (**a**) *as* asterion, *la* lambda, *ob* obelion, *sc* sulcus, centralis. (**b**) Landmarks as (**a**)

W. Seeger, J. Zentner, *Anatomical Basis of Cranial Neurosurgery*, https://doi.org/10.1007/978-3-319-63597-2_10

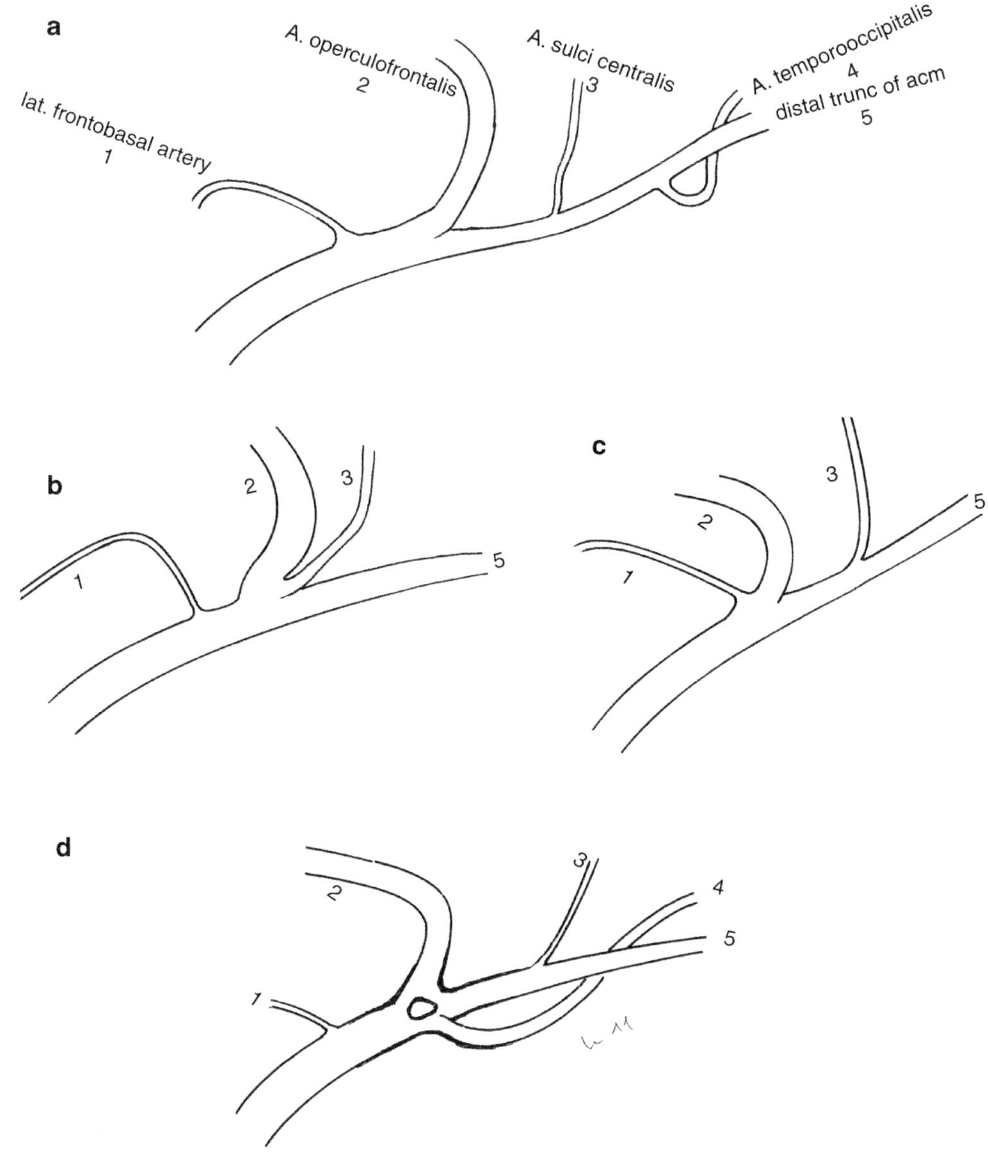

Fig. 10.4 Region of the so-called trifurcation, common findings (Yaşargil 1985, personal communication with W. Seeger). (**a**) Bifurcation of A. cerebri media. (**b**)–(**d**) Different types. These findings should be considered especially at microsurgical and interventional eliminations of aneurysms, AVMs, and meningiomas

Fig. 10.5 Interfornical approach to the third ventricle (Apuzzo et al. 1982). Steps of surgery, steps A to E are the principles of approaches which are presented at Figs. 3.30, 3.31, 3.32, 3.33, and 3.34. (*A*) Subdural route. (*B*) Arachnoid incision for opening of cisterna corporis callosi. (*C*) Incision of corpus callosum for opening of the lateral ventricle. Cave opening of the contralateral ventricle. Cave transversal veins of the ventricular roof at the usual longitudinal incision of corpus callosum. Transversal incision of corpus callosum would endanger stria longitudinalis cinguli. (*D*) Method of Apuzzo: longitudinal splitting of septum pellucidum (cave septal veins) and interfornical splitting. (*E*) Splitting velum interpositum along Vv. cerebri internae and between the numerous fine medial branches of Aa. choroideae posteriores before opening the third ventricle

10.2.2 Surgical Aspects of Midline Approaches (Figs. 10.5, 10.6, 10.7, 10.8, 10.9, and 10.10)

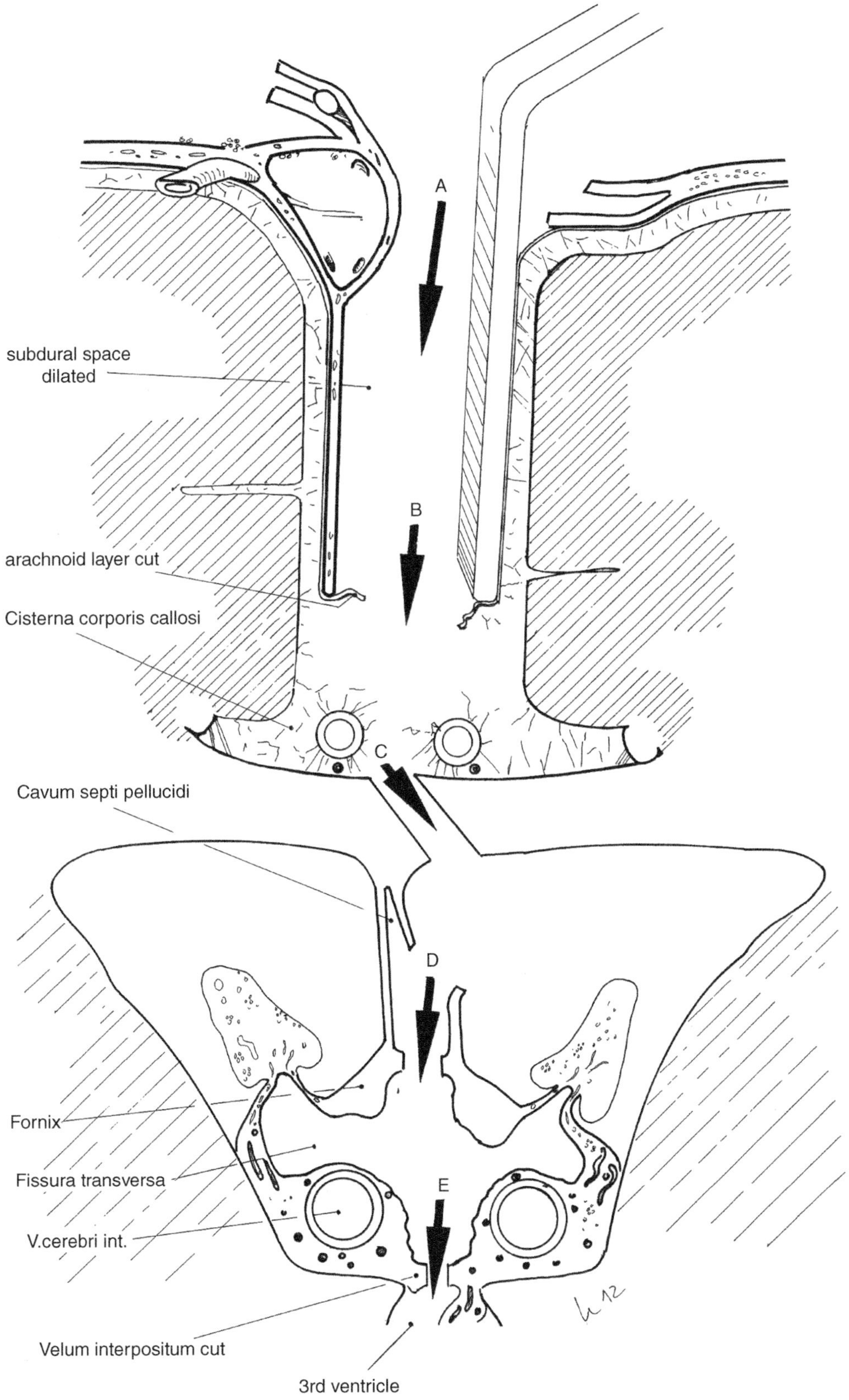

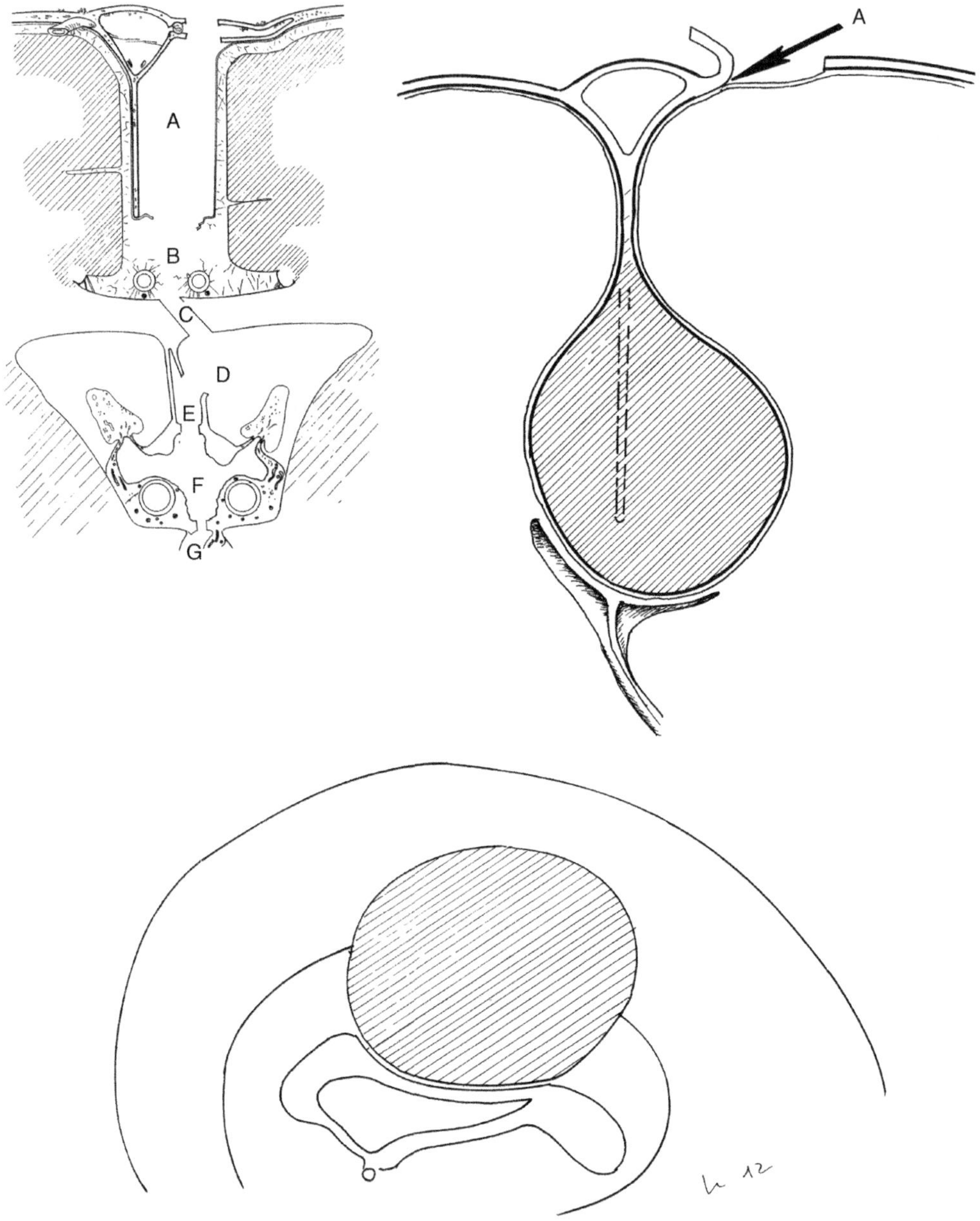

Fig. 10.6 Falx meningioma, example surgical approach according to step A of Fig. 10.5

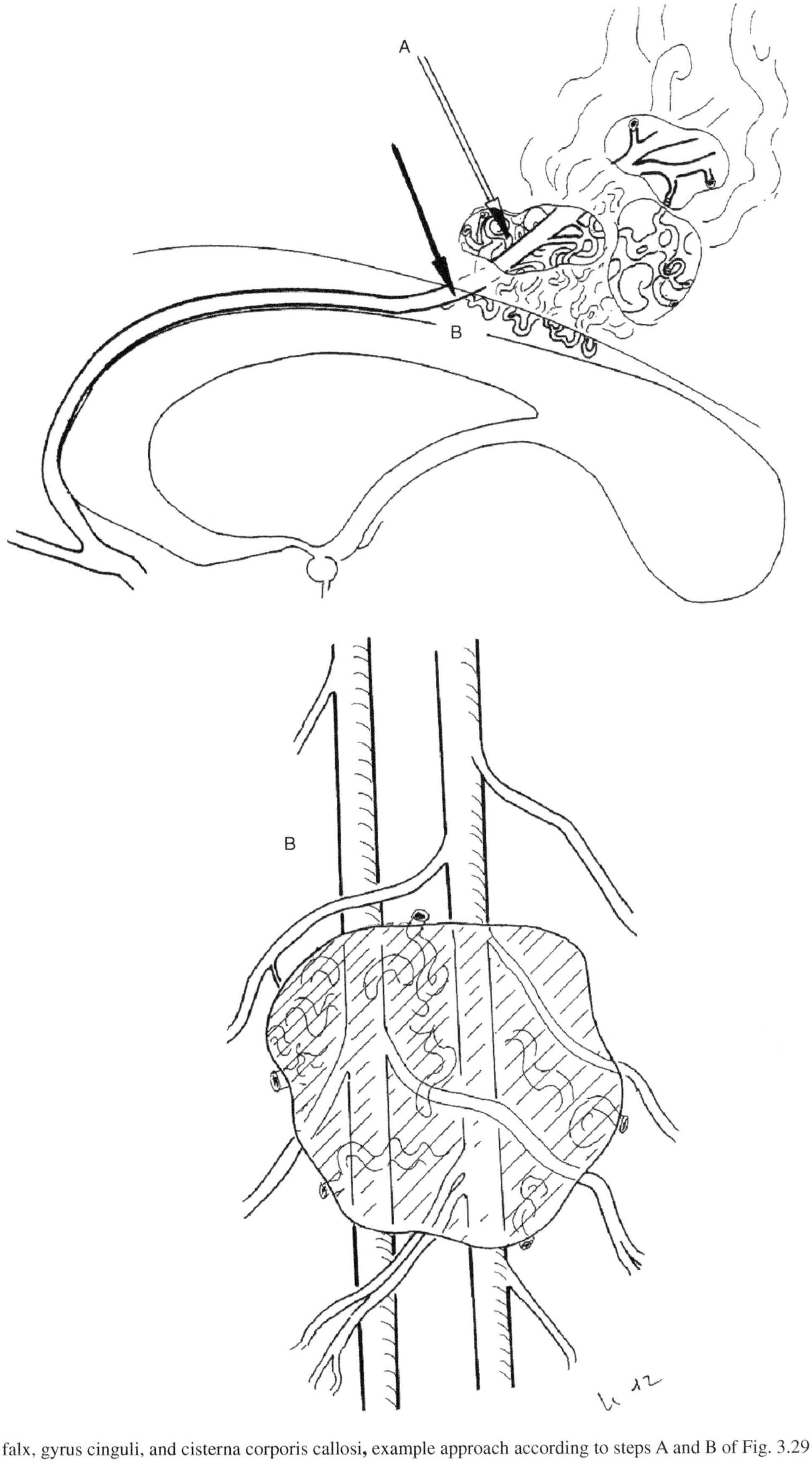

Fig. 10.7 AVM of falx, gyrus cinguli, and cisterna corporis callosi, example approach according to steps A and B of Fig. 3.29

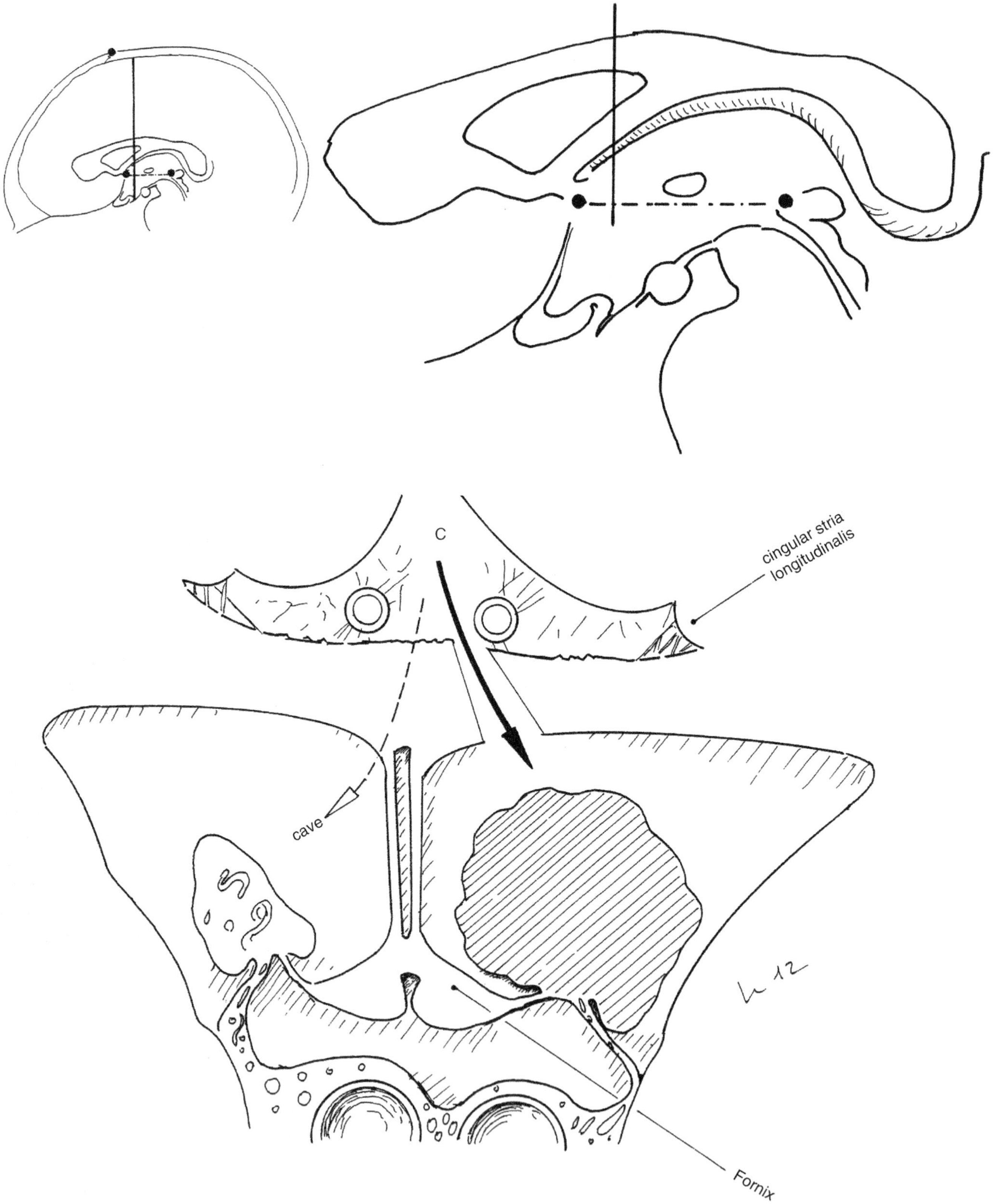

Fig. 10.8 Tumor of plexus choroideus of the lateral ventricle, if a transcortical neuronavigatory-guided approach is not indicated. Approach according to steps A to C of Fig. 3.29

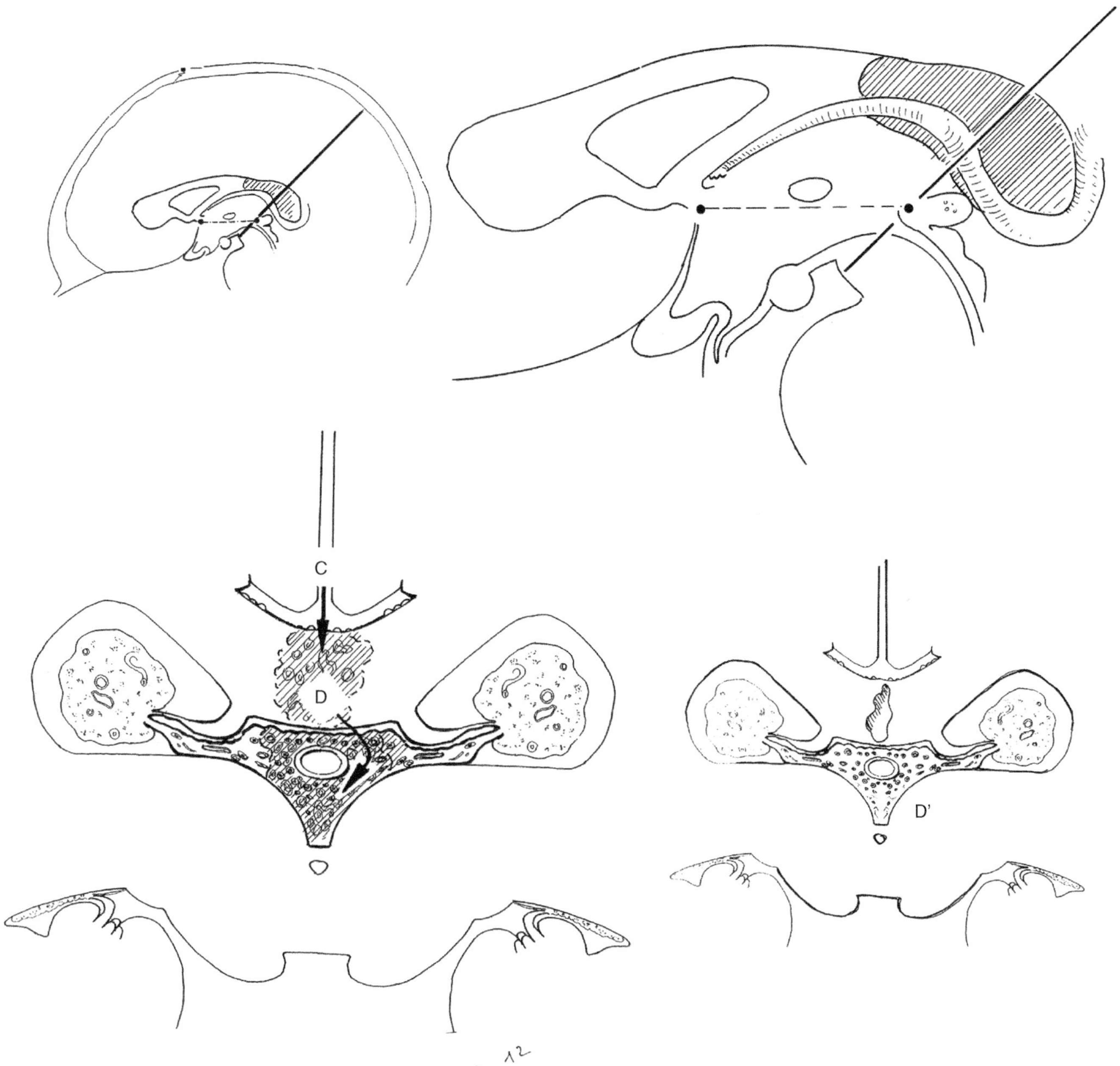

Fig. 10.9 AVM of corpus callosum and its cistern, of Fissura transversa, and of velum interpositum according to Yaşargil et al. 1992. Surgical approach according to steps A to D. This AVM is draining to the galenic vein. After a complete elimination of the AVM, the light-colored galenic vein changes to its normal venous color (D')

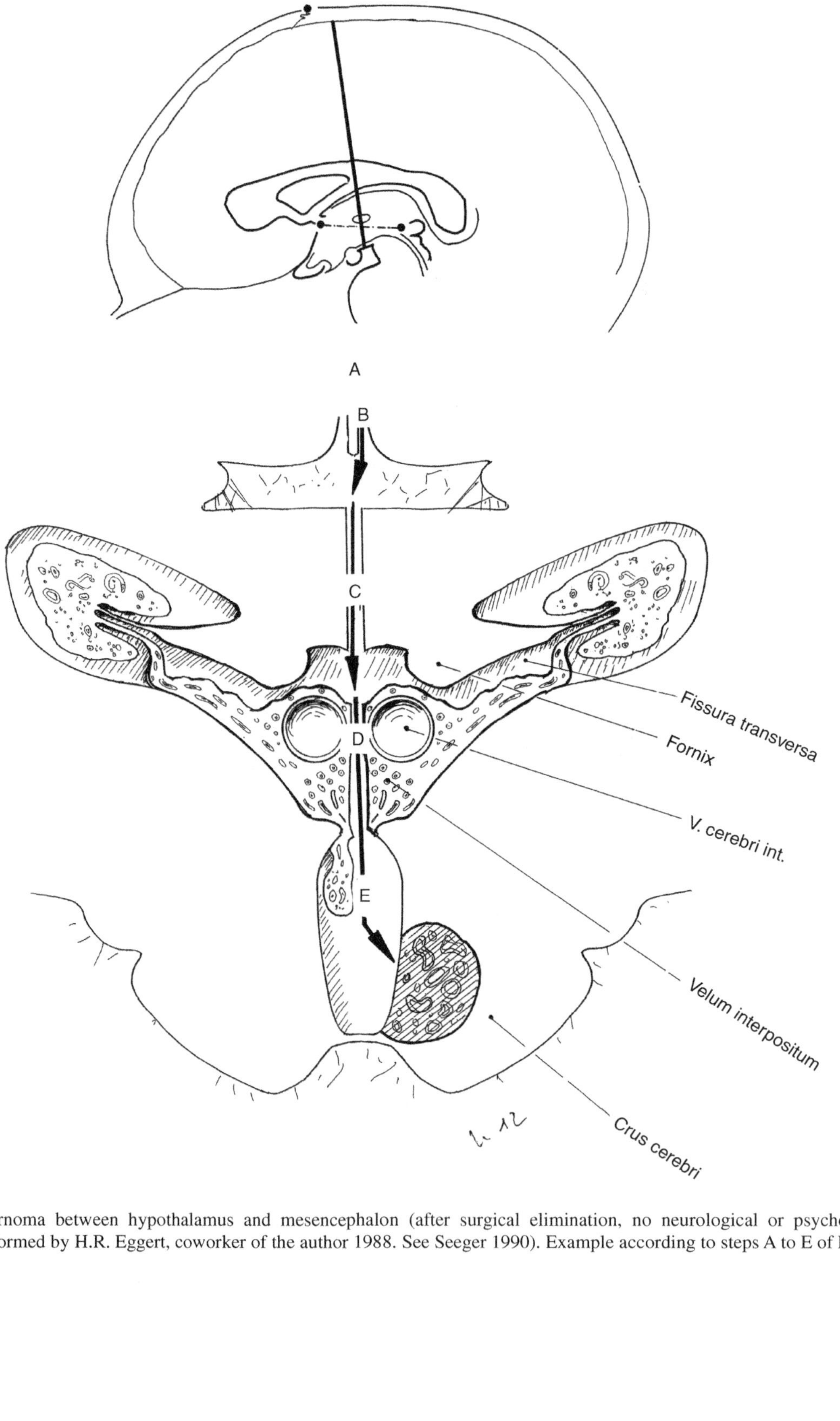

Fig. 10.10 Cavernoma between hypothalamus and mesencephalon (after surgical elimination, no neurological or psychological deficits. Microsurgery performed by H.R. Eggert, coworker of the author 1988. See Seeger 1990). Example according to steps A to E of Fig. 10.5

Part IV

Occipital Region

1.1 Introduction

This introduction to Part IV contains a short summary of each chapter.

1.2 Chapter 11 Diagnostic Base

The anatomical and functional limits of lobus occipitalis are unclear. The cortical area optica begins at dorsal precuneus, between sutura parietooccipitalis and sulcus calcarinus. Ventrobasal complex begins at the posterior area of gyrus parahippocampalis, anterior from isthmus gyri cinguli. The area angularis is located between lobus temporalis, parietalis, and occipitalis. Anatomical configuration and location of its functions are variable.

1.3 Chapter 12 Anatomical Base for Surgery

- The lumen of confluens sinuum presents less lacunae than other sinus segments. Trabeculae are rare or were missing, but septa and its topographical relationships with other sinuses are variable.
- Radiatio optica is better to present by fiber tracking than by all other methods but not the surrounding fibers which presents much fiber crossings. Fiber crossing are hindering fiber tracking.
- Cave bilateral lesions of the preoccipital segments of striae longitudinalis cinguli and its interruption would be followed by irreversible psychological deficits.
- Cave occipitobasal anterior veins of lobus occipitalis, if no resection of lobus is planned. Danger for encephalomalacia. The vein is a part of the inner cerebral veins close to the midline. Cave.

1.4 Chapter 13 Special Surgical Aspects

- At skin-galea incision, cave anastomotic branches of A. occipitalis
- Cave the numerous V. emissaria close to confluens sinuum
- Danger for intraventricular bleeding of A. calcarina or A. parietooccipitalis and its branches, which are presented in the depth of Sulci, close to cornu posterius.
- At puncture of cornu posterius and fornices and structures at cisterna tecti and ambiens could be endangered.

11 Diagnostic Base

11.1 Landmarks (Figs. 11.1 and 11.2)

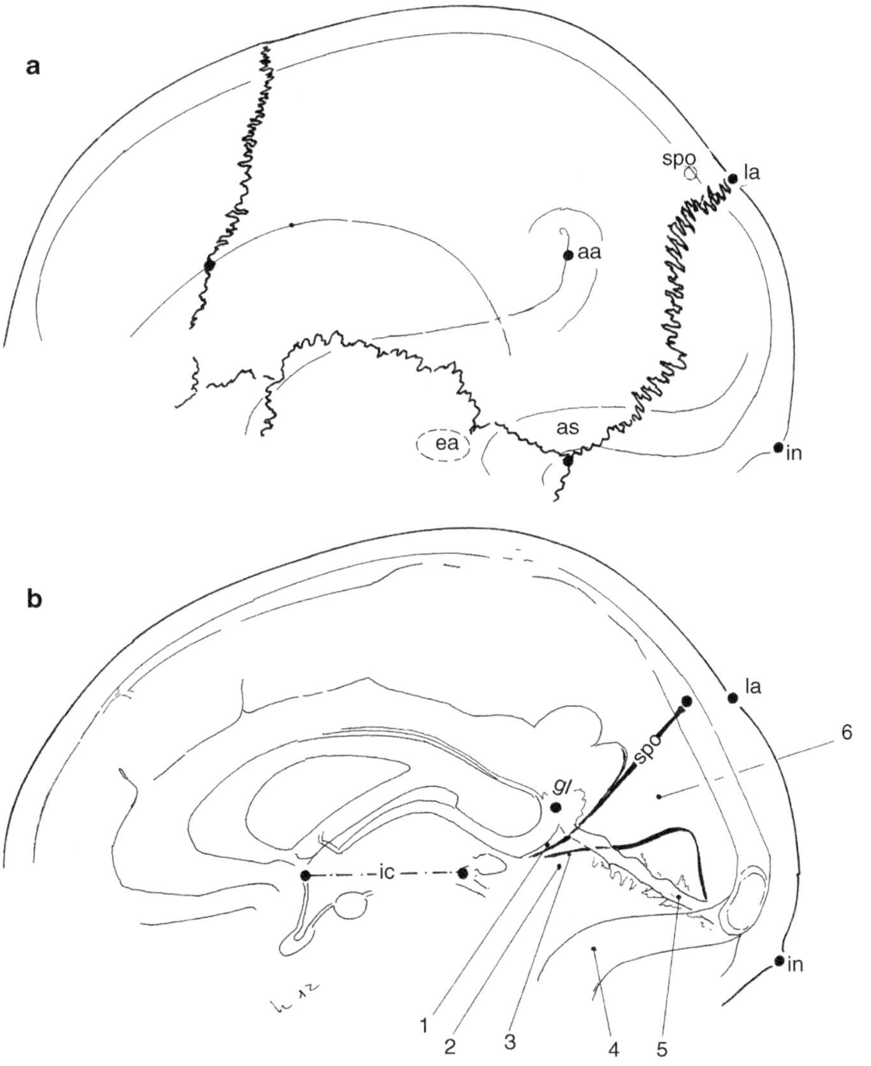

Fig. 11.1 Extra- and intracranial landmarks for assisting of imagings. (**a**) *aa* angularis area, *as* asterion, *ea* eminentia arcuata, projection, *in* inion, *la* lambda, *spo* sulcus parietooccipitalis, projection; (**b**) *gl* galenic point, *ic* intercommissural line, *in* inion, *la* lambda, *spo* sulcus parieto-occipitalis. *1* Isthmus gyri cinguli, *2* area of gyrus occipitotemporalis medialis, *3* sulcus calcarinus, *4* knee of sinus transversus, *5* sinus rectus, *6* cuneus

W. Seeger, J. Zentner, *Anatomical Basis of Cranial Neurosurgery*, https://doi.org/10.1007/978-3-319-63597-2_11

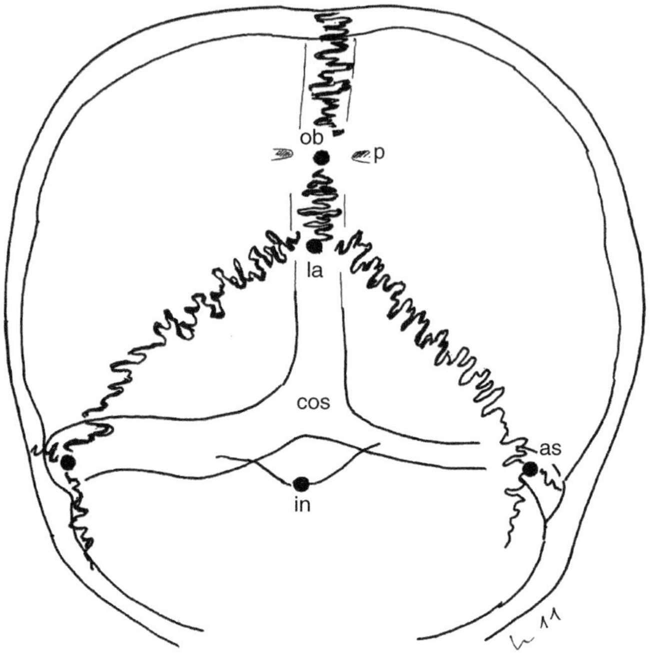

Fig. 11.2 Continuation of Fig. 11.7. *as* asterion, *cos* confluens sinuum, projection, *in* inion, *la* lambda, *ob* obelion, *p* for parietale

11.2 Radiology (Figs. 11.3, 11.4, and 11.5)

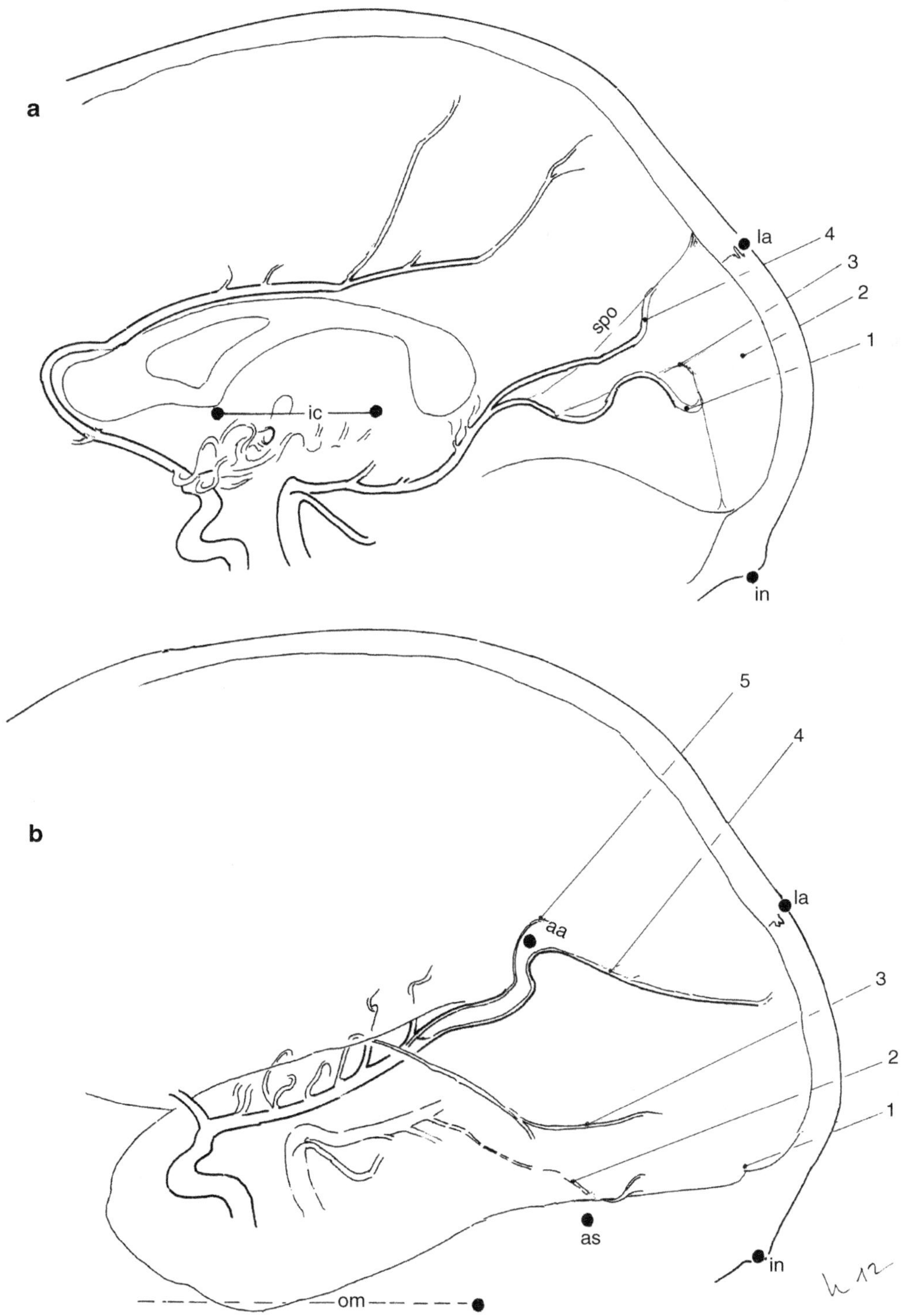

Fig. 11.3 Arteries, skull, and brain, schematized. Landmarks added for assisting imagings, if necessary. (**a**) Medial arteries. (**b**) Lateral arteries. (**a**) *ic* intercommissural line, *in* inion, *la* lambda, *spo* sulcus parietooccipitalis. *1* A. calcarina, *2* cuneus, *3* sulcus calcarinus, *4* A. parietooccipitalis. (**b**). *aa* angularis area, *as* asterion, *in* inion, *la* lambda, *om* orbitomeatal line. *1* Sulcus calcarinus, *2* temporobasal branch(es) of A. cerebri posterior, projection, *3* A. temporalis posterior, *4* A. temporooccipitalis, *5* A. angularis

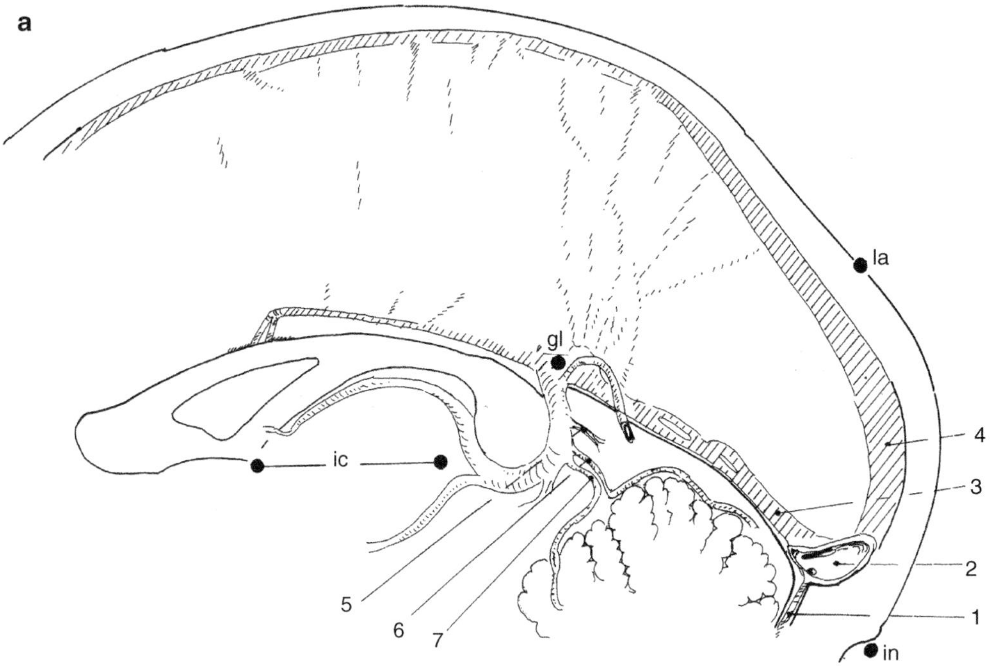

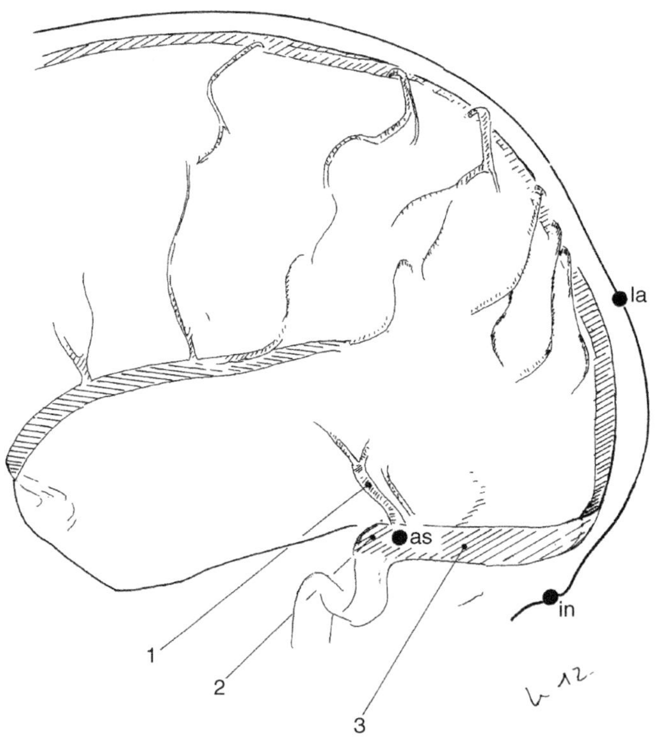

Fig. 11.4 Continuation of Fig. 11.3. Veins medial (**a**) and lateral (**a**) *gl* galenic point, *ic* intercommissural level, *in* inion, *la* lambda. *1* sinus occipitalis, *2* confluens sinuum, *3* sinus rectus, *4* sinus sagittalis sup., *5* occipitomedial bilateral vein(s), *6* V. supraculminalis, *7* V. cerebellaris precentralis. (**b**) *as* asterion, *in* inion, *la* lambda. *1* Vein of labbé, *2* knee of sinus transversus, *3* sinus transversus

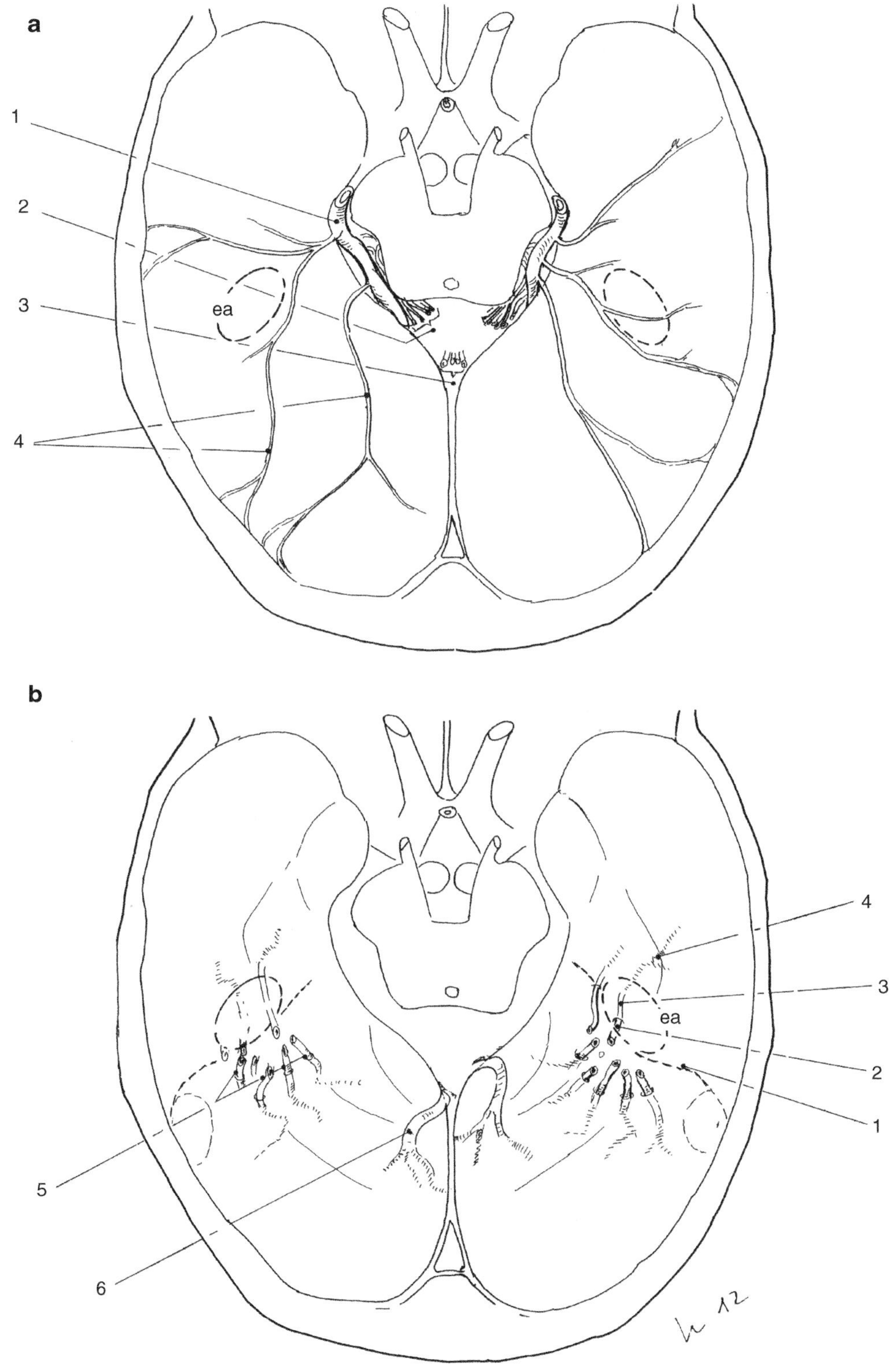

Fig. 11.5 Basal arteries (**a**) and veins (**b**). Eminentia arcuata of os petrosum is located close to sinus petrosus superior et margo superior ossis petrosi. It is a landmark for defining temporo-occipital basal bridging veins during surgery. Interruption of temporobasal arteries and of bridging veins at this area (V. labbé included) could be followed by neurological deficits. Differentiation of the subarachnoid and subdural segment of basal medial veins could be necessary for its preservation during surgery. (**a**) *ea* eminentia arcuata, projection. *1* A. cerebri posterior, *2* branches of A. choroideae posterior, *3* mesencephalic vessels, *4* posterior basal branches of A. cerebri posterior. (**b**) *1* Margo superior ossis petrosi, projection; *2* temporobasal vein, subdural segment, variable; *3* as 2, subarachnoid segment, variable; *4* as 2, subpial segment, variable; *5* as 2 to 4; *6* occipitomedial basal veins, draining to the galenic system

11.3 Further Imaging (Figs. 11.6, 11.7, 11.8, 11.9, and 11.10)

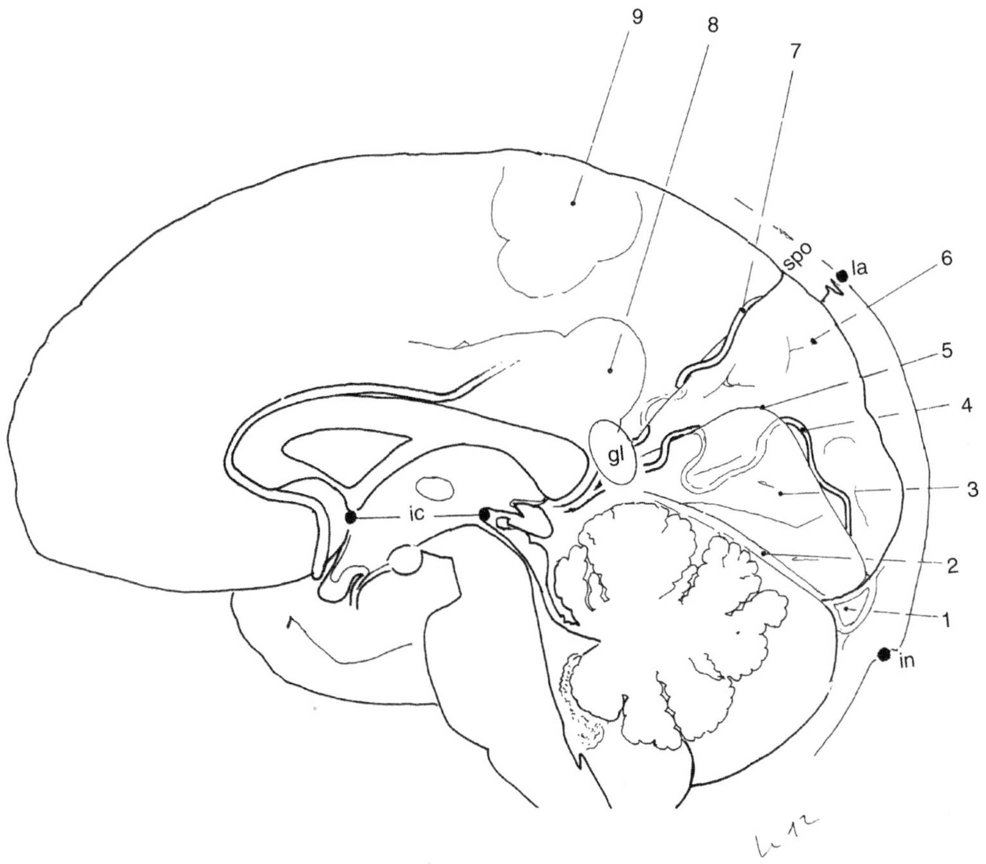

Fig. 11.6 *Neuronavigation*, modified. Landmarks added for assisting neuronavigation, if necessary. *gl* galenic region, *ic* intercommissural line, *in* inion, *spo* sulcus parietooccipitalis. *1* confluens sinuum, *2* sinus rectus, *3* cuneus, *4* A. calcarina, *5* sulcus calcarinus, *6* praecuneus, *7* A. parietooccipitalis, *8* gyrus cinguli, *9* lobulus paracentralis

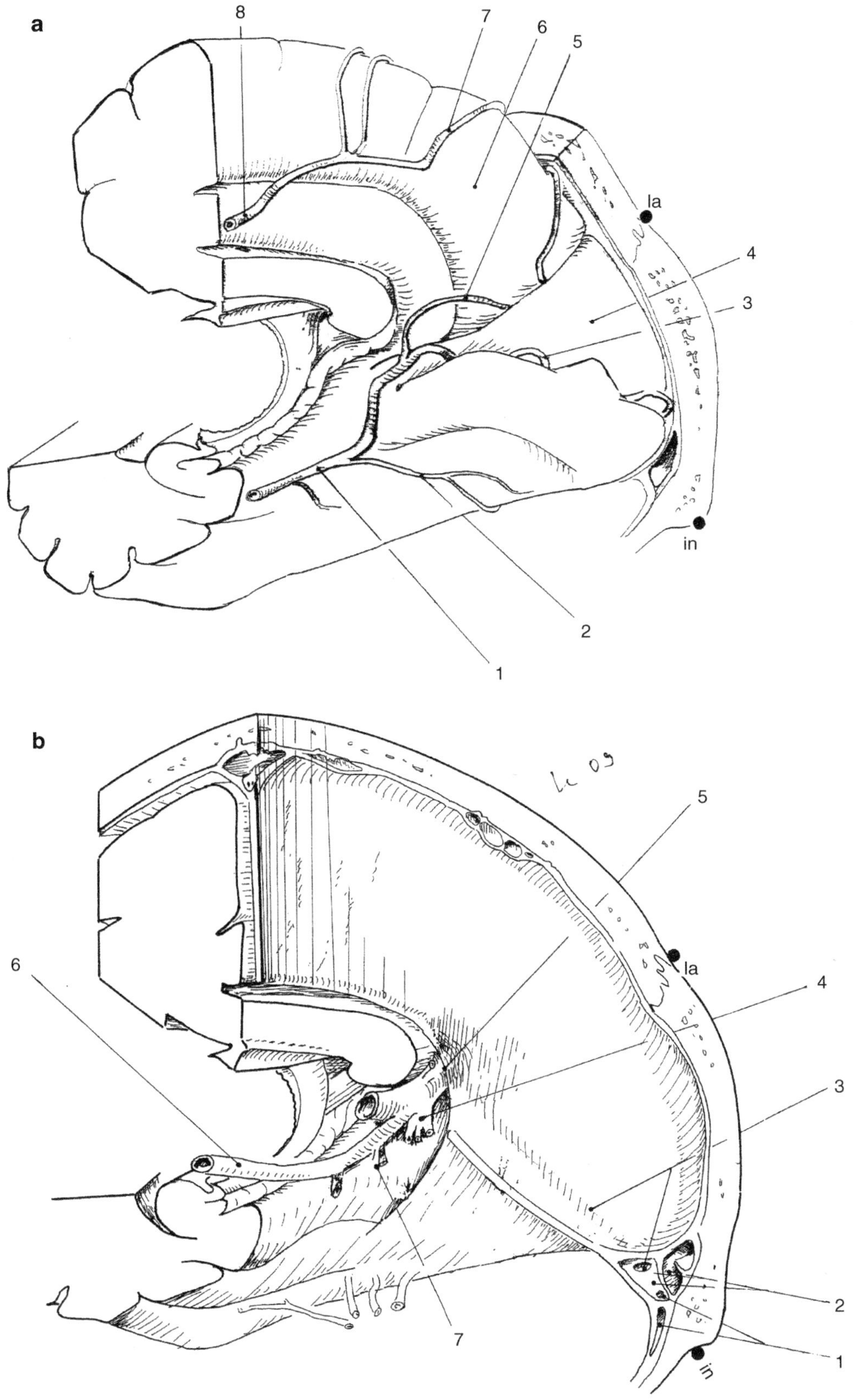

Fig. 11.7 Continuation of Fig. 11.6. Intra- and extradural occipitoparietal structures. (**a**) Landmarks, arteries, and cerebral relief. (**b**) Falx, tentorium, and galenic area. (**a**) *in* inion, *la* lambda. *1* A. cerebri posterior, *2* posterior basal branch of A. cerebri posterior, *3* A. calcarina, *4* cuneus, *5* A. parietooccipitalis, *6* praecuneus, *7* A. praecunealis, *8* A. pericallosa. B. in inion, la lambda. *1* sinus occipitalis, *2* confluens sinuum, *3* sinus rectus, *4* mediobasal anterior veins of lobus occipitalis, *5* ampulla of V. magna (Galeni), *6* V. basalis (Rosenthal), *7* V. mesencephalic lat

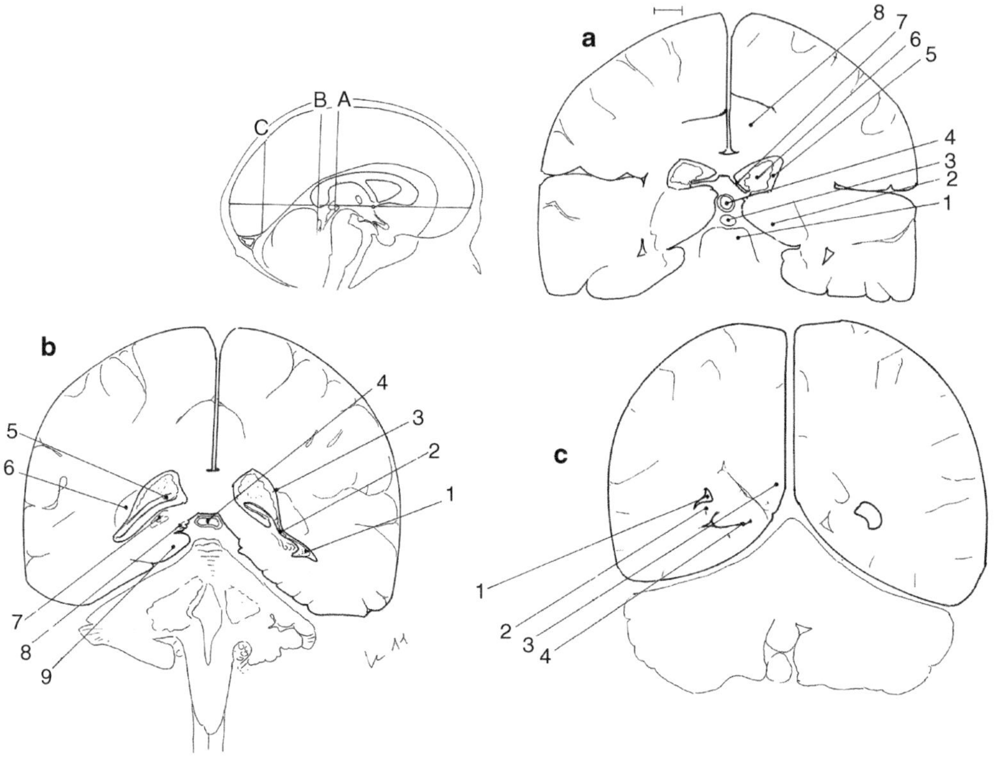

Fig. 11.8 Occipitoparietal MRT. (**a**) Plane at corpus pineale. (**b**) Area of splenium. (**c**) Tip of cornu posterius (length of cornu post. is variable). (**a**) *1* Tectum, *2* thalamus, *3* corpus pineale, *4* V. magna (Galeni), *5* atrium, *6* plexus choroideus, *7* crus fornicis, *8* gyrus cinguli. (**b**) *1* Cornu inferius; *2* as 1, narrow segment; *3* atrium; *4* V. magna (Galeni), *5* crus fornicis; *6* pulvinar thalami; *7* tail of hippocampus; *8* area dentata; *9* gyrus parahippocampalis. (**c**) *1* Tip of cornu posterius, *2* calcar avis, *3* precuneus, *4* sulcus calcarinus

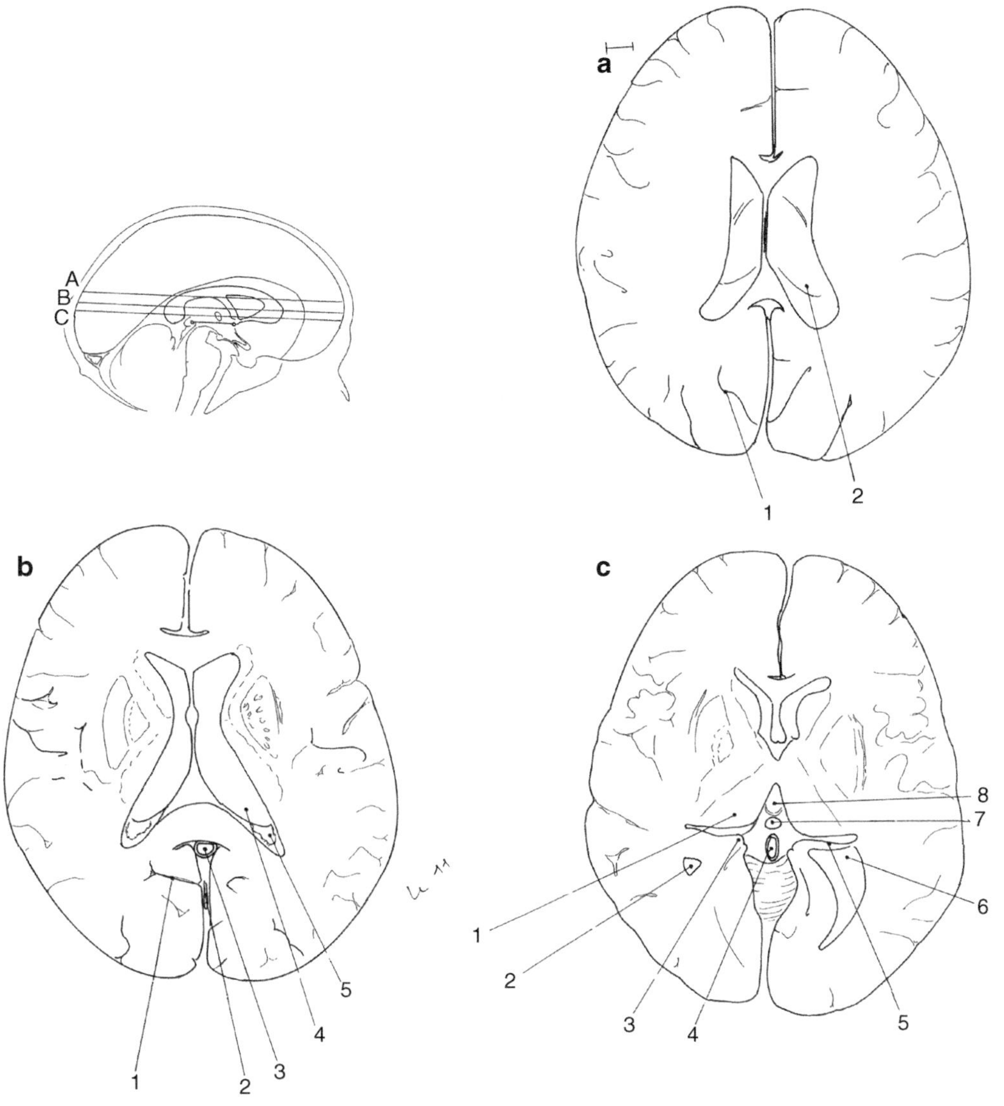

Fig. 11.9 Continuation of Fig. 11.8. (**a**) Transsectional plane at the dorsal ventricular area. (**b**) Middle region of splenium. (**c**) Transsectional plane of for. interventriculare (Monroi). (**a**) *1* Sulcus parietooccipitalis, *2* thalamus. (**b**) *1*. Sulcus parietooccipitalis, *2* sinus rectus, *3* V. magna (Galeni), *4* thalamus, *5* plexus choroideus. (**c**) *1* Thalamus, *2* tip of cornu posterius, *3* area dentata, *4* V. magna (Galeni), *5* crus fornicis, *6* atrium, *7* corpus pineale, *8* third ventricle

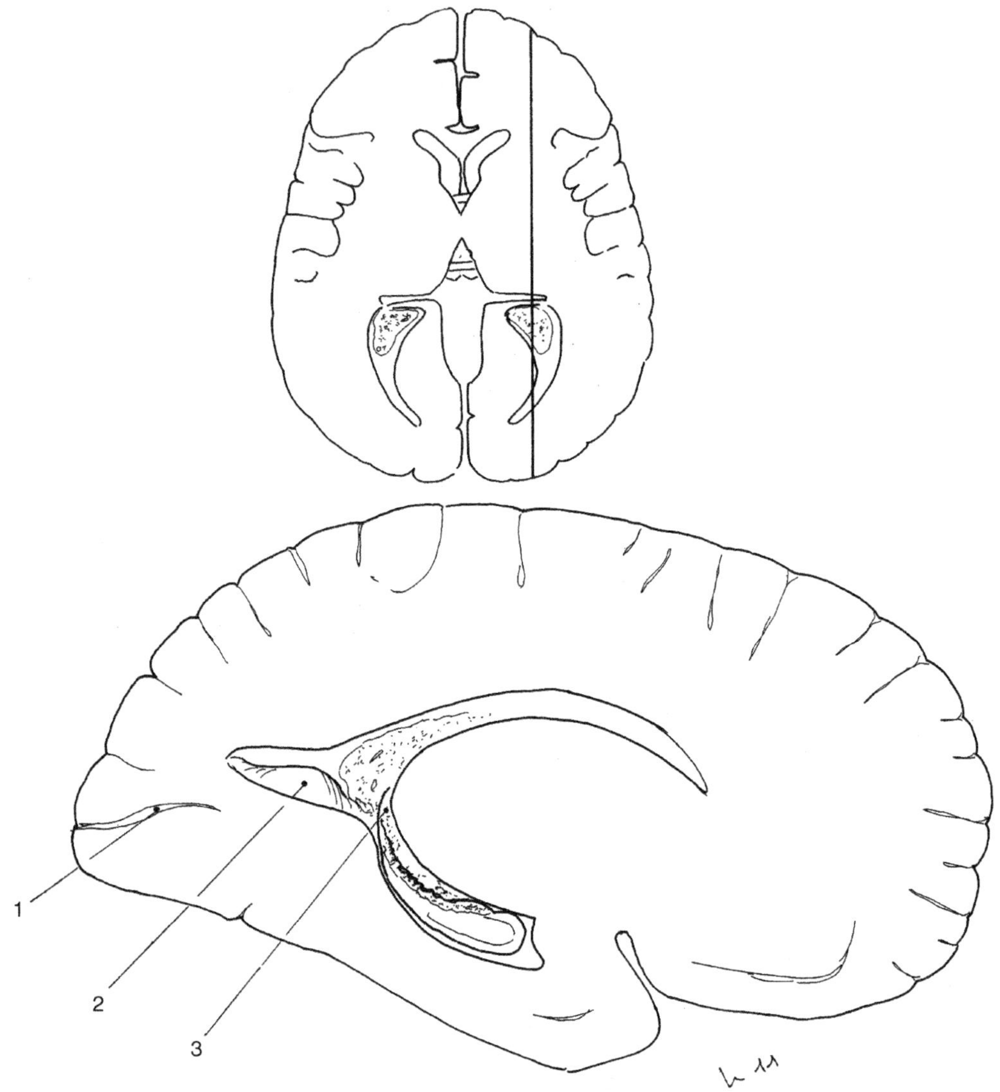

Fig. 11.10 Continuation of Fig. 11.9. *1* Sulcus calcarinus, *2* calcar avis, *3* narrow ventricular segment between the atrium and cornu inferius

Anatomical Base of Surgery

12

12.1 Extradural Topography (Figs. 12.1, 12.2, 12.3, 12.4, 12.5, 12.6, 12.7, 12.8, and 12.9)

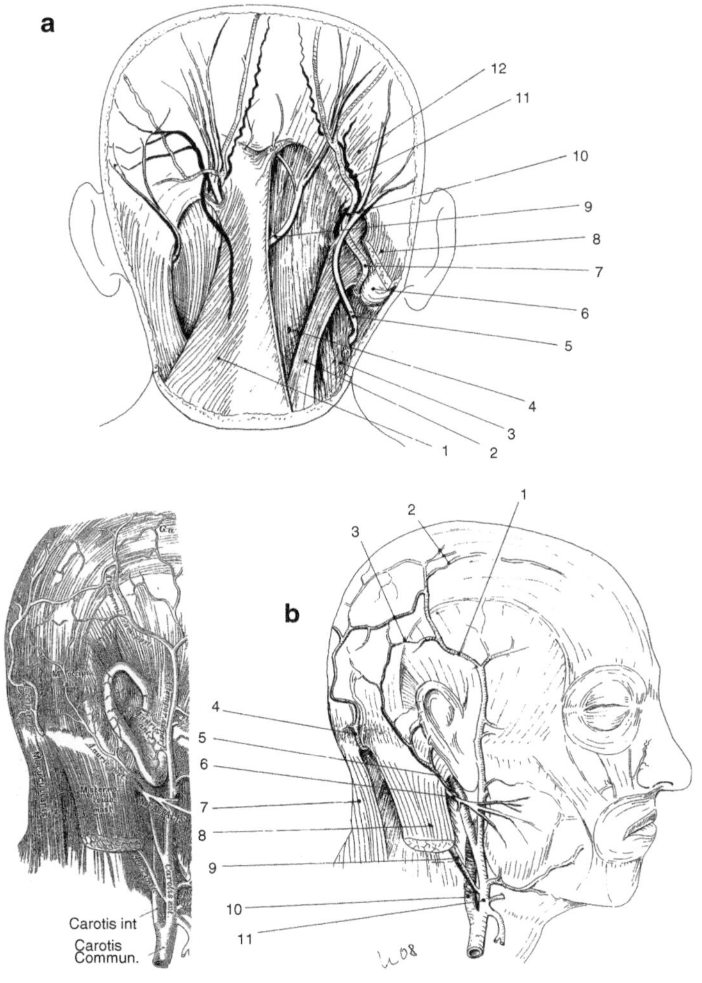

Fig. 12.1 Skin, galea aponeurotica, and periosteum. Wound healing could be endangered after interruptions of anastomotic connections of A. occipitalis with A. temporalis superficialis. Galea nerves spare. **a** Vessels and nerves. **b** Historical addendum and copy of it (Heitzmann 1875, VI, p 182 (anatomical Atlas with woodcuts according to drawings of Prof. Heitzmann, Wien)). Arteries rendered. **a** *1* M. trapezius; *2* M. longissimus capitis; *3* M. levator scapulae; *4* M. splenius capitis; *5* N. occipitalis minor; *6* processus mastoideus; *7* A. occipitalis; *8* M. sternocleidomastoideus; *9* N. occipitalis major; *10* gap of fascia nuchae, containing N. occipitalis minor and A. et V. occipitalis; *11* V. occipitalis, *12* M. occipitofrontalis, pars occipitalis. **b** *1* A. temporalis superficialis; 2 and 3 anastomotic branches; *4* A. occipitalis; *5* A. retroauricularis close to For. stylomastoideum and N. facialis, cave at surgery; *6* N. facialis; *7* M. trapezius; *8* M. sternocleidomastoideus; *9* A. occipitalis; *10* A. carotis interna; *11* A. carotis externa

W. Seeger, J. Zentner, *Anatomical Basis of Cranial Neurosurgery*, https://doi.org/10.1007/978-3-319-63597-2_12

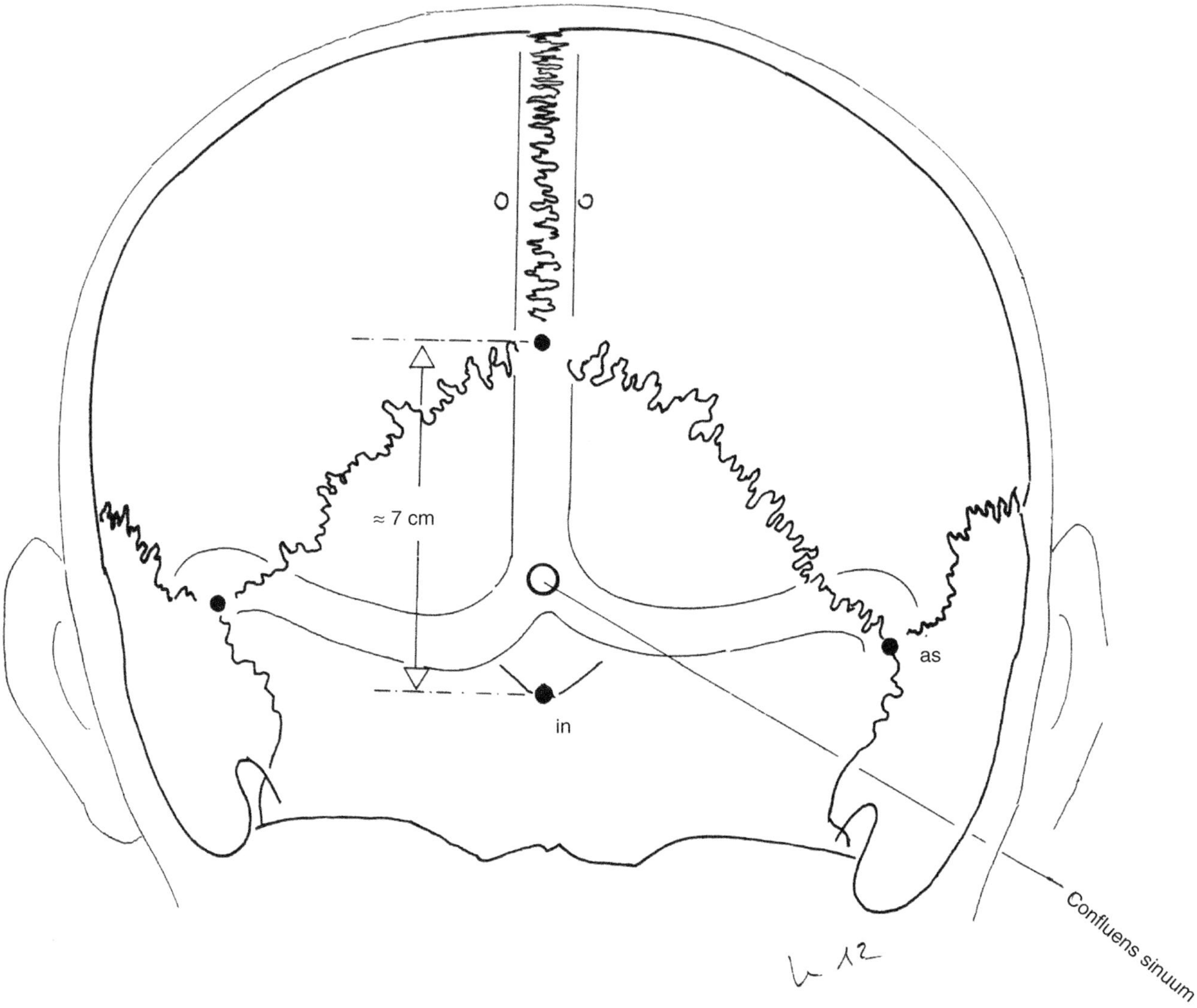

Fig. 12.2 Planum occipitale et planum nuchale. Projection of sinuses. Landmarks and measurement for planning craniotomies and punctures of cornu posterius. Projection of sinus sagittalis superior. Confluens sinuum et sinus transversus added. *as* asterion, *in* inion, *la* lambda

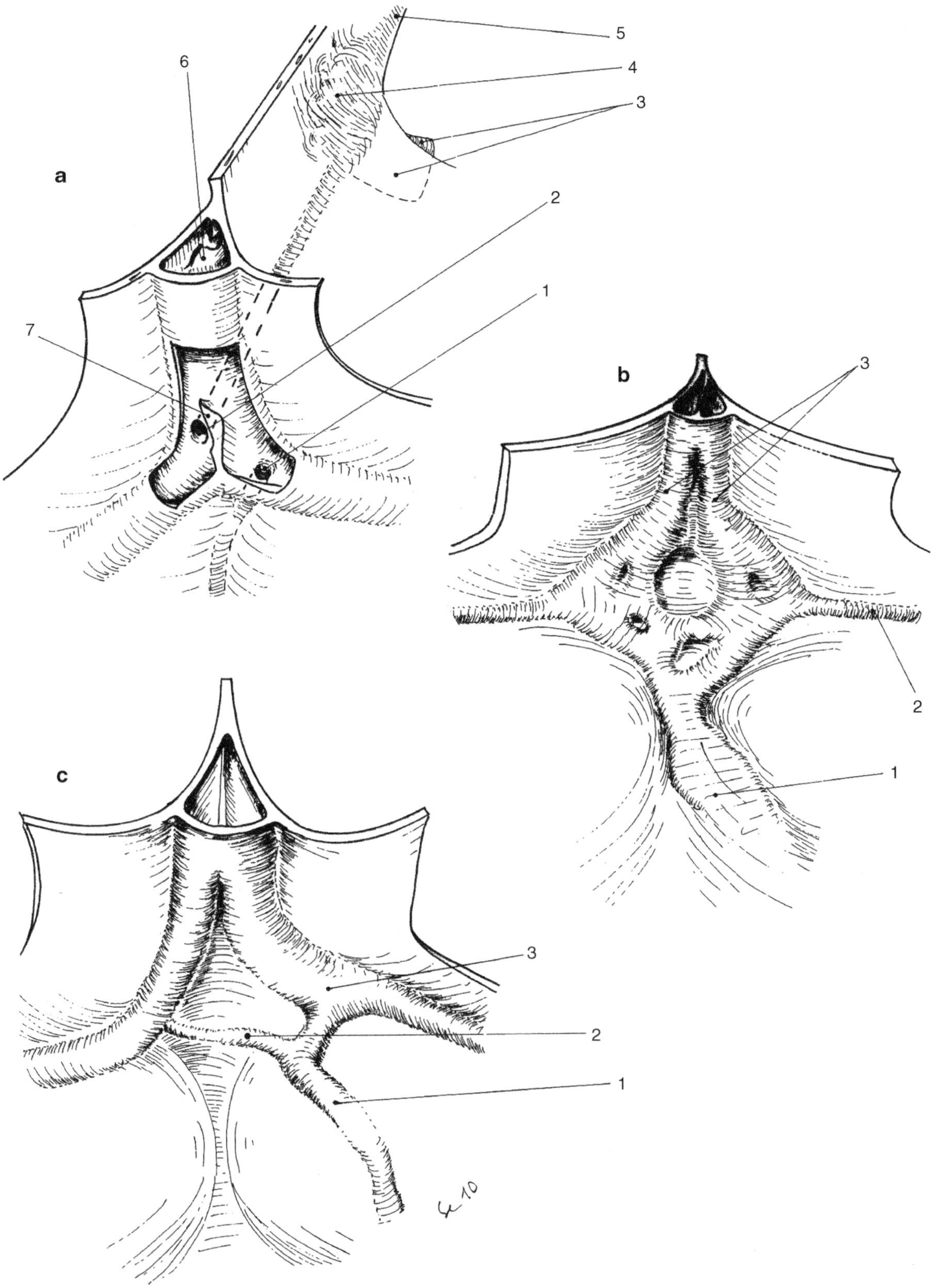

Fig. 12.3 Confluens sinuum (According to Browning 1953 and to Browder and Kaplan 1976, p73). **a** Common findings **b** and **c** variants. **a** *1* Sinus occipitalis, unilateral mouth at confluens, *2* sinus rectus, contralateral mouth at confluens, *3* galenic region of falx and tentorium, *4* venous network, *5* sinus sagittalis inferior, *6* sagittal septum of confluens sinuum, *7* incomplete septum. **b** *1* Sinus occipitalis wide, *2* sinus transversus hypoplastic, *3* duplication of sinus sagittalis superior. **c** *1* Sinus occipitalis wide and deviated, *2* sinus occipitalis wide and connected with sinus transversi, *3* sinus transversus medial wide, lateral connected with sinus occipitalis

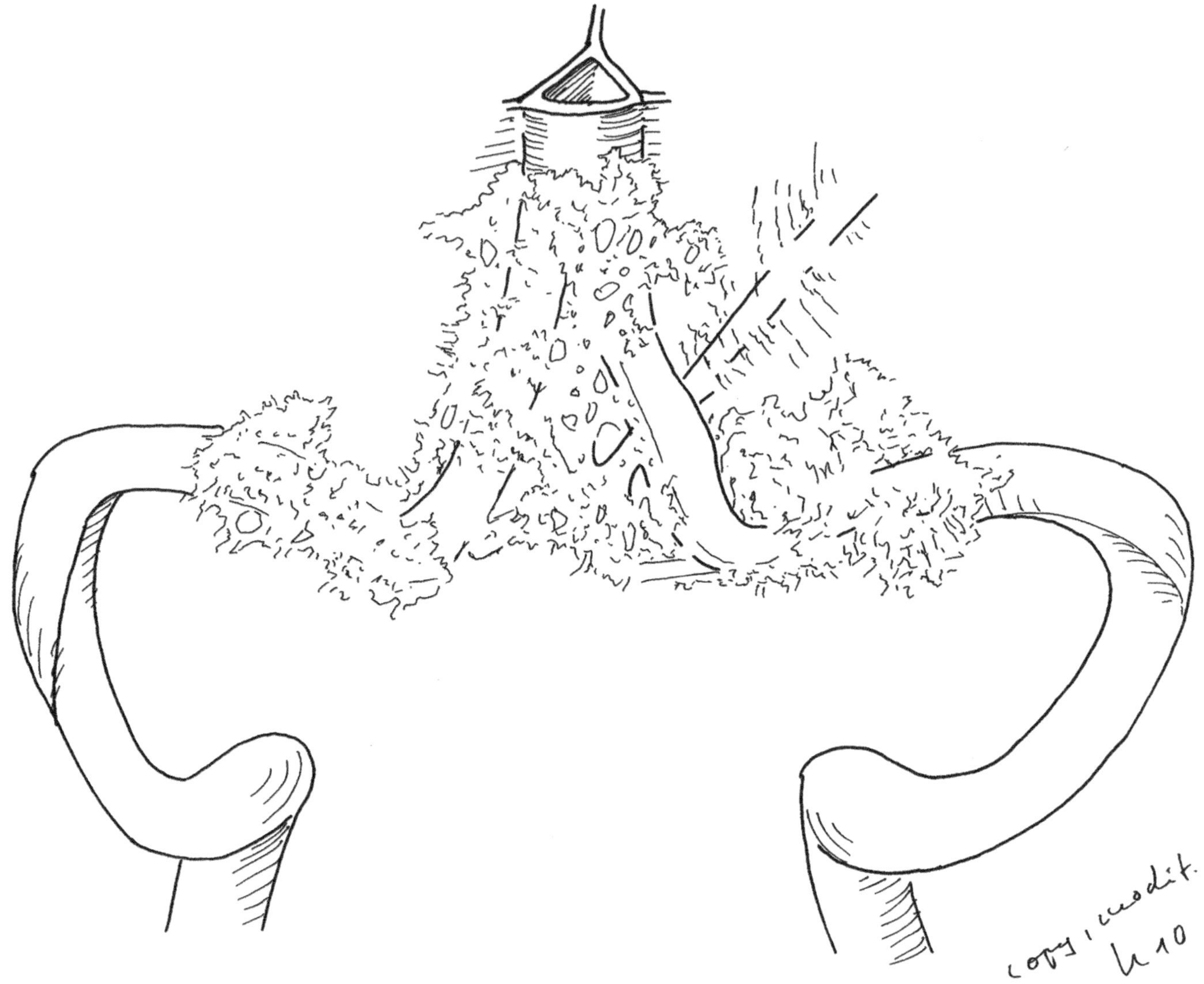

Fig. 12.4 Confluens sinuum and venous network (Cast of Browder and Kaplan 1976, p 70. Indian ink copy). Venous network is surrounding all intracranial sinuses, especially the region of confluens. At obturation of confluens (e.g., by tumors), the network of veins may be widened, compensating the occluded confluence. Now the surgical interruption of venous network of dilated veins could be dangerous

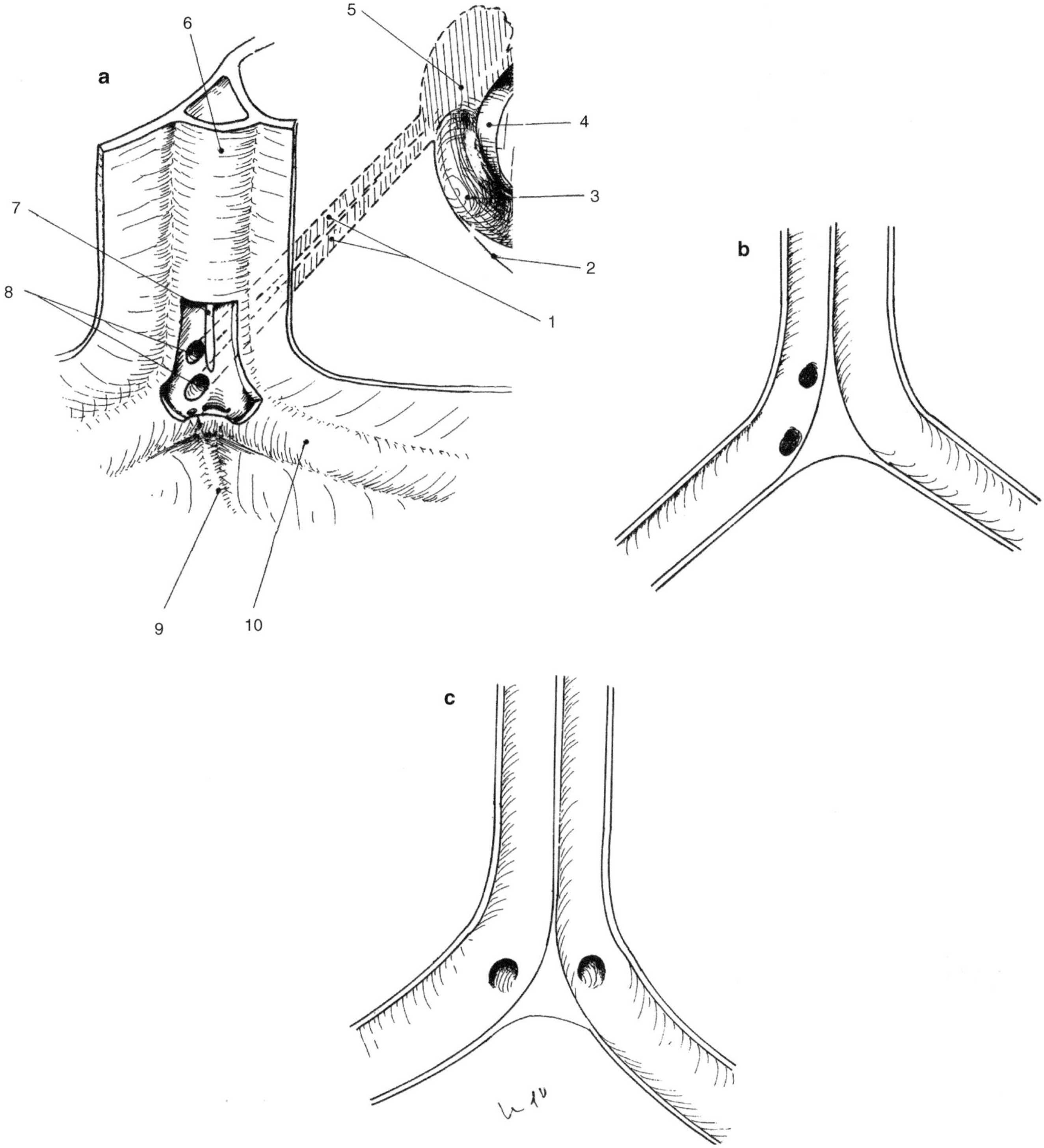

Fig. 12.5 Confluens sinuum and variants of sinus rectus. **a** Duplicated mouth of sinus rectus and incomplete septum of sinus sagittalis superior. **b** Duplicated mouth unilateral and complete septum of sinus sagittalis superior/confluens sinuum. **c** Duplicated mouth of sinus rectus and complete septum of sinus sagittalis superior. These variants should be taken in consideration at surgery, e.g., at meningiomas *1* sinus rectus duplicated, projection; *2* margin of tentorium; *3* V. magna (Galeni); *4* splenium corporis callosi; *5* galenic point; *6* sinus sagittalis superior; *7* septum, fenestrated at confluens; *8* duplicated mouth of sinus rectus; *9* sinus occipitalis; *10* sinus transversus

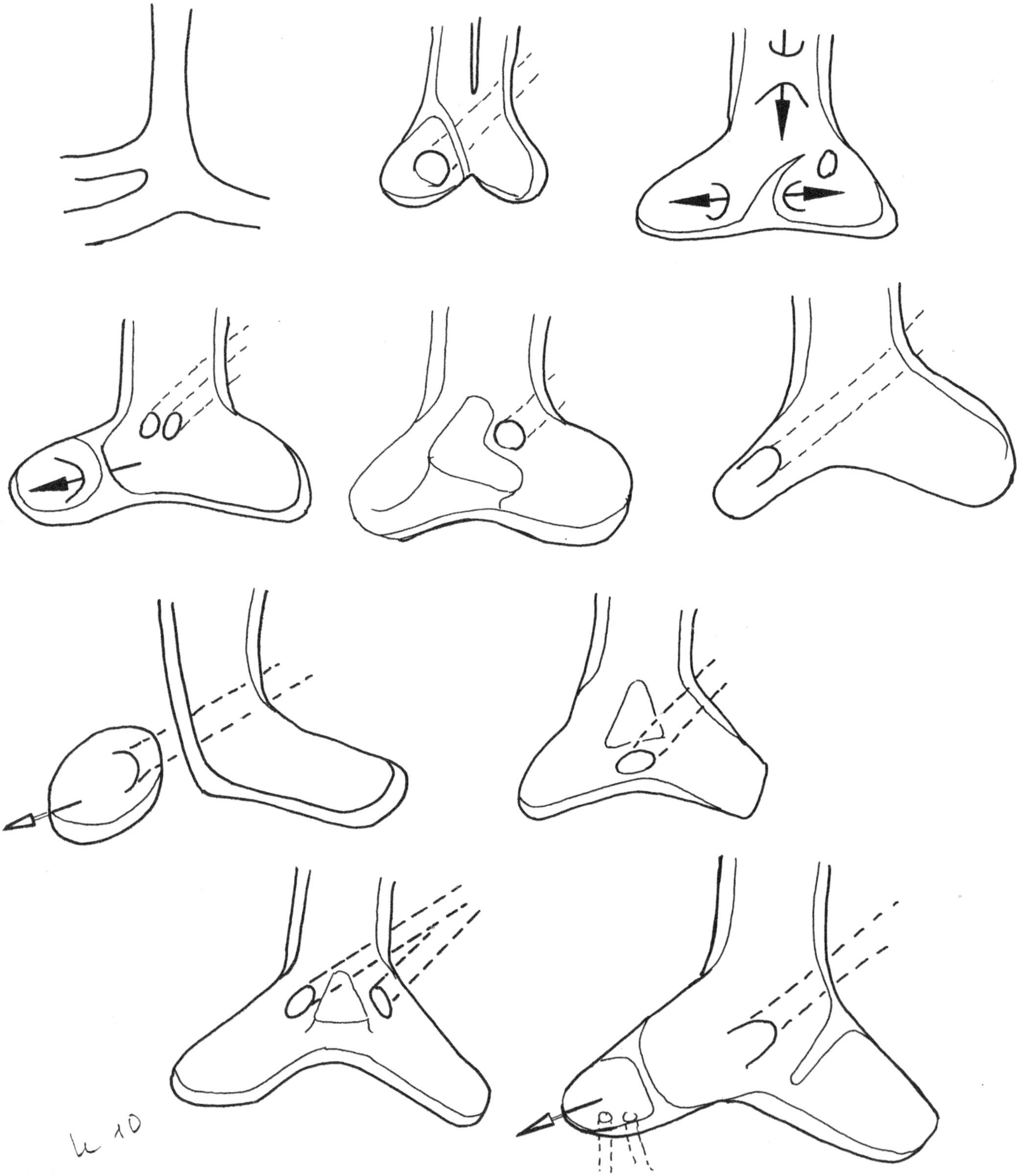

Fig. 12.6 Addendum for Fig. 12.4. Rare variants (According to Browning 1953, modified)

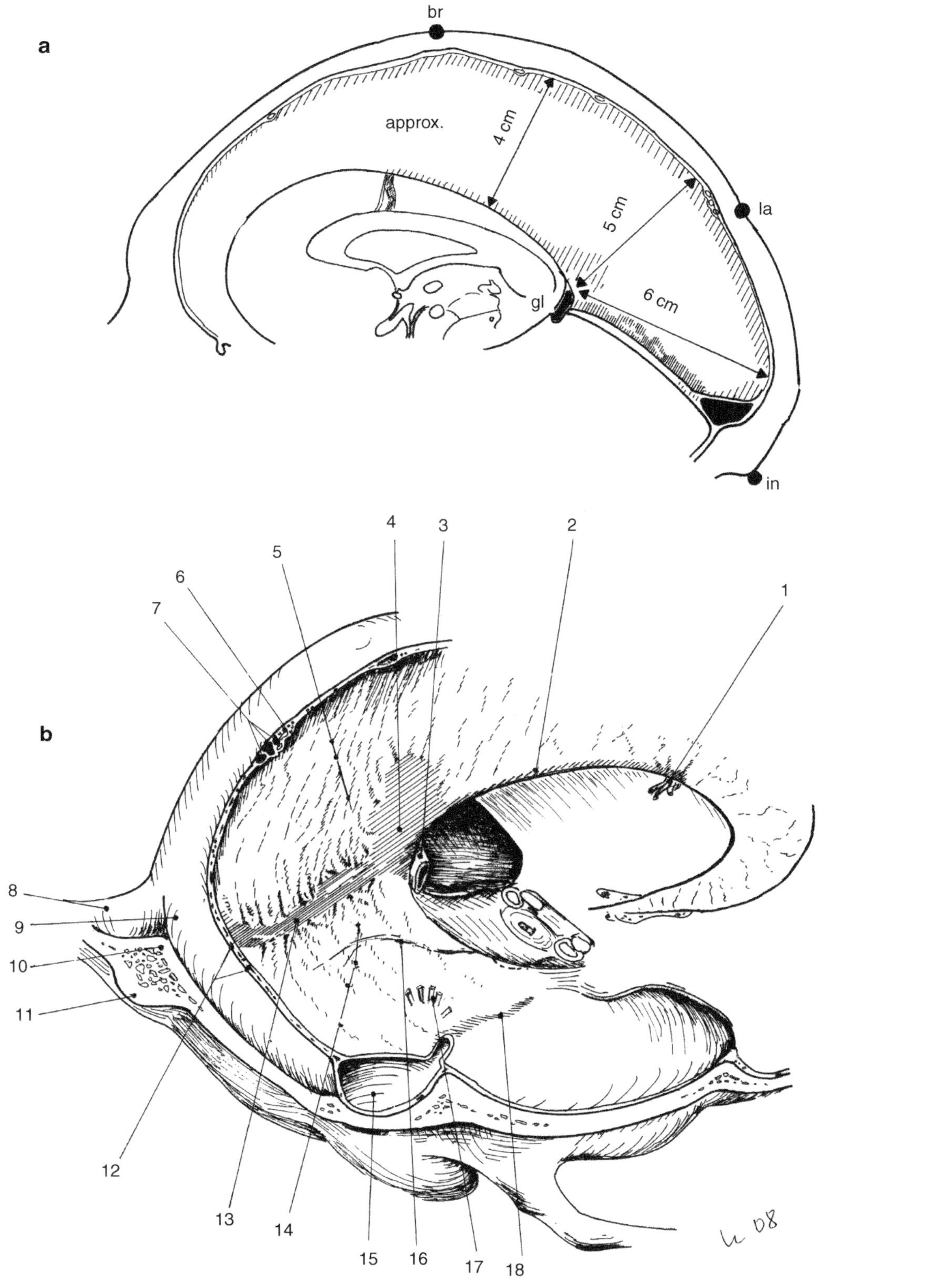

Fig. 12.7 Falx and tentorium landmarks and measurements for planning occipital surgical approaches. **a** Sinuses of falx. **b** Sinuses of falx and tentorium. **a** *br* Bregma, *gl* galenic point, *in* inion, *la* lambda. **b** *1* Bridging veins, *2* sinus sagittalis inferior, *3* V. magna (Galeni), *4* network of veins close to the galenic point, *5* numerous veins close and distant to the galenic point, *6* granulationes arachnoidales (Pacchioni), *7* bridging veins, *8* sinus transversus, *9* confluens sinuum, *10* protuberantia occipitalis interna, *11* protuberantia occipitalis externa (inion), *12* venous network, *13* sinus rectus, *14* tentorium, *15* knee of sinus transversus, *16* A. tentorii (Bernasconi), *17* bridging veins, *18* sinus petrosus superior

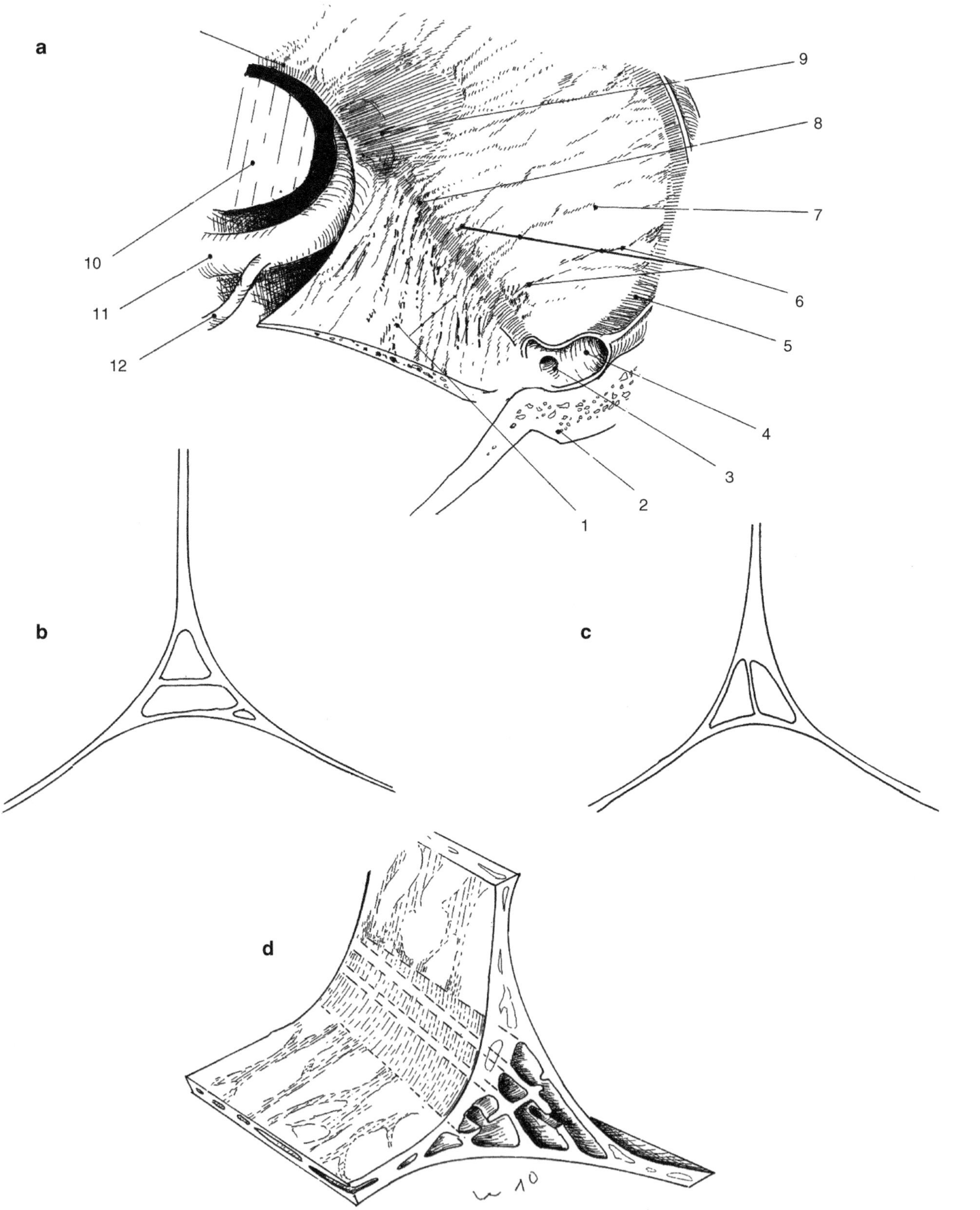

Fig. 12.8 Sinus rectus according to Browning 1953, schematized. **a** Usual presentations. **b** Platform-like septum. **c** Sagittal septum. **d** Multiple septa. At bleeding during surgery spare electrocoagulation, if possible. Suturing should be preferred. Danger for sinus thrombosis. **a** *1* Tentorial veins, high density; *2* inion; *3* mouth of sinus rectus; *4* confluens sinuum; *5* sinus sagittalis superior; *6* veins of falx, high density; *7* falx; *8* sinus rectus; *9* galenic point; *10* splenium corporis callosi; *11* V. magna (Galeni); *12* V. basalis (Rosenthal). **b** The platform-like septum could be multiple fenestrated. **c** The sagittal septum could be multiple fenestrated. **d** Common finding

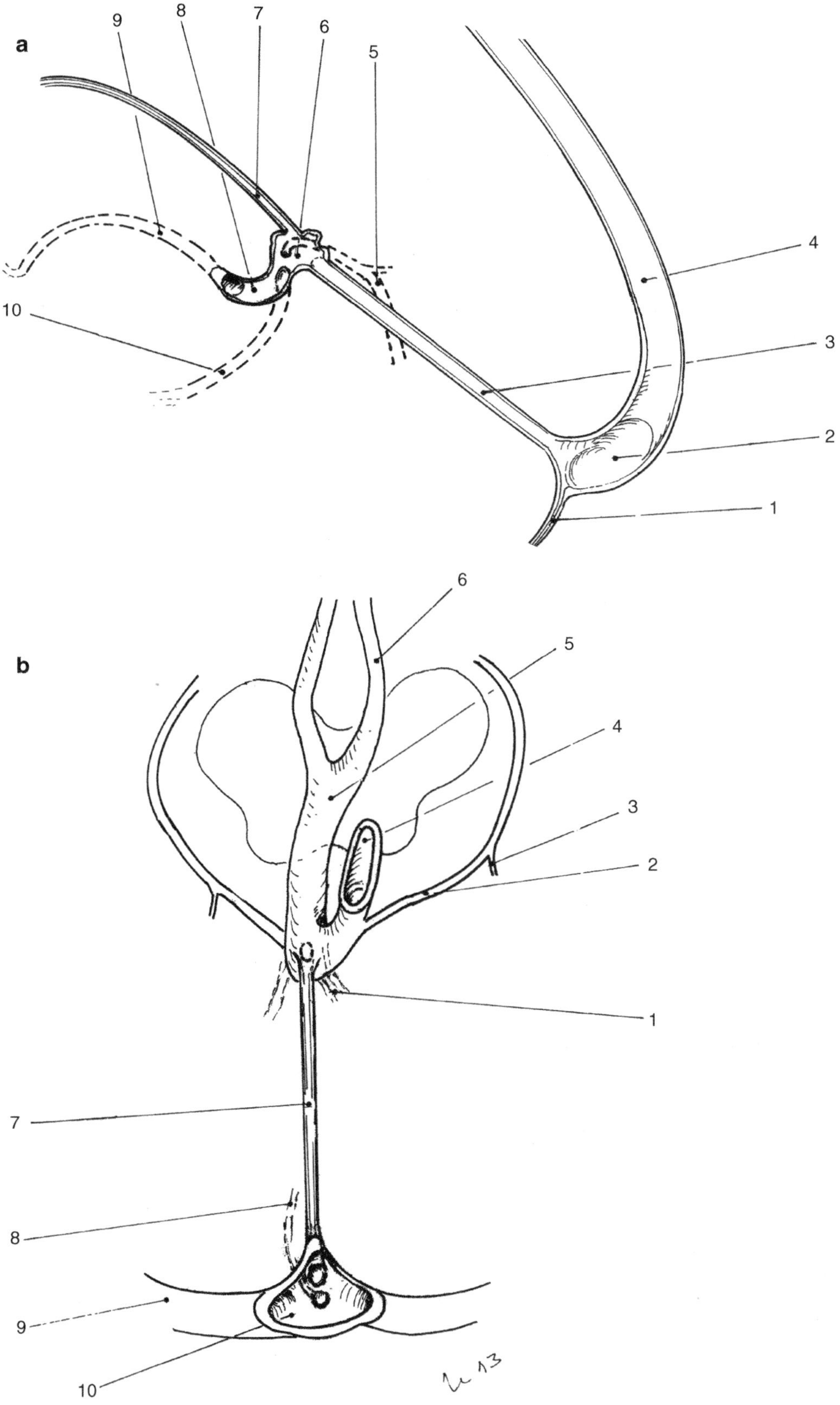

Fig. 12.9 Galenic system conclusion. **a** *1* Sinus occipitalis; *2* confluens sinuum; *3* sinus rectus; *4* sinus sagittalis superior; *5* occipitobasal medial veins of lobus occipitalis, projection; *6* galenic point; *7* sinus sagittalis inferior; *8* V. magna (Galeni); *9* V. cerebri interna; *10* V. basalis (Rosenthal). **b** *1* Occipitobasal veins of lobus occipitalis, projection; *2* V. basalis (Rosenthal); *3* V. mesencephalica lateralis; *4* galenic point; *5* junction of Vv. cerebri internae; *6* V. cerebri interna; *7* sinus rectus; *8* sinus occipitalis, projection; *9* sinus transversus; *10* confluens sinuum

12.2 Intradural Topography

12.2.1 Survey (Figs. 12.10, 12.11, 12.12, and 12.13)

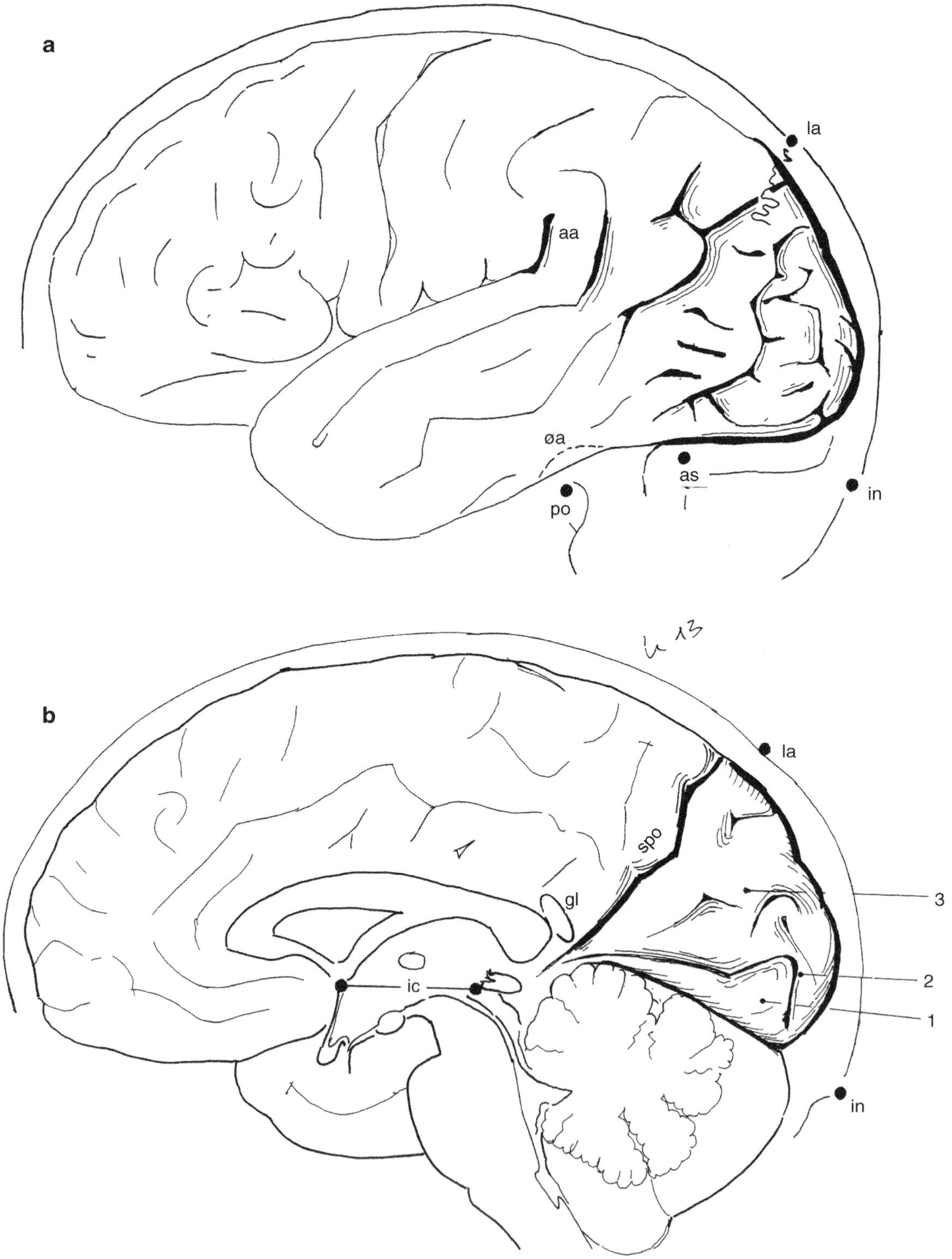

Fig. 12.10 Occipital lobe lateral **a** and medial cortex **b**. Landmarks added for assisting imagings, if necessary. **a** *aa* angularis region, *as* asterion, *ea* eminentia arcuata, projection, *in* inion, *la* lambda, *po* porion. **b** *gl* galenic region, *ic* intercommissural line, *in* inion, *la* lambda, *spo* sulcus parietooccipitalis. *1* Gyrus occipitotemporalis medialis, *2* sulcus calcarinus, *3* cuneus

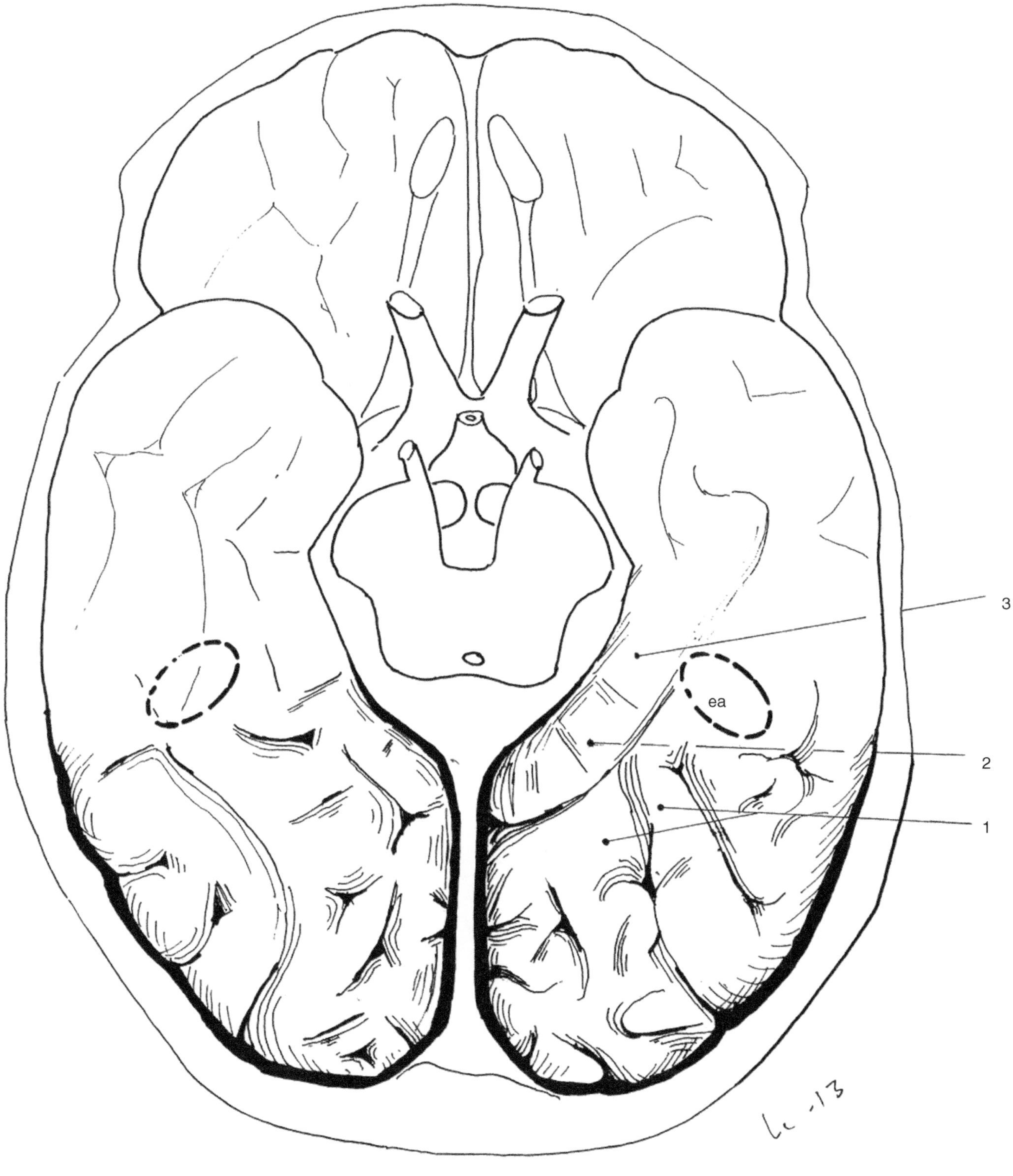

Fig. 12.11 Addendum for Fig. 12.10. *ea* eminentia arcuata, projection. *1* Gyrus occipitotemporalis lateralis, *2* gyrus occipitotemporalis medialis, *3* gyrus parahippocampalis

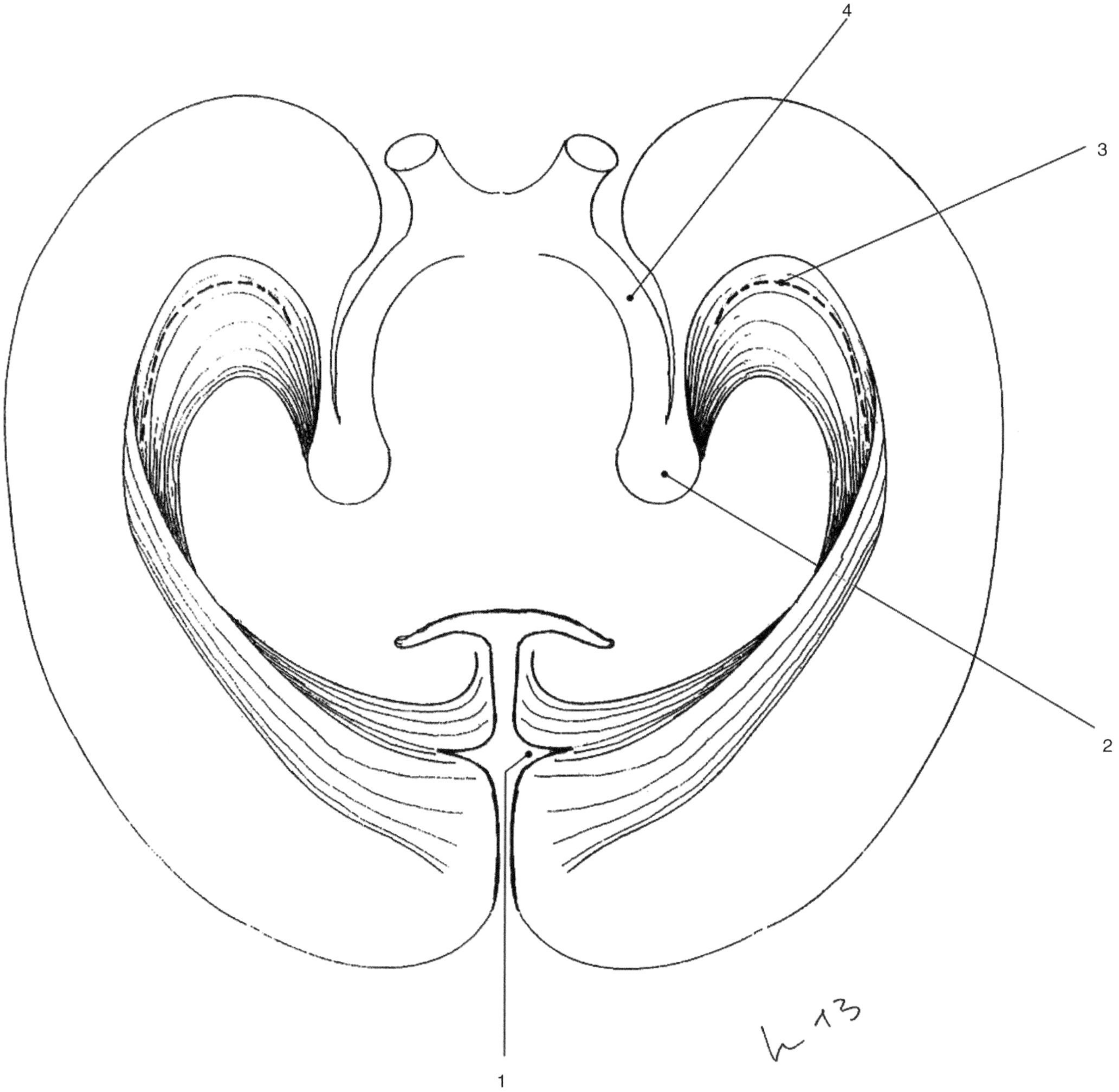

Fig. 12.12 Radiatio optica schematized. *1* Sulcus calcarinus; *2* corpus geniculatum laterale; *3* genu temporale of radiatio optica, projection; *4* tractus opticus

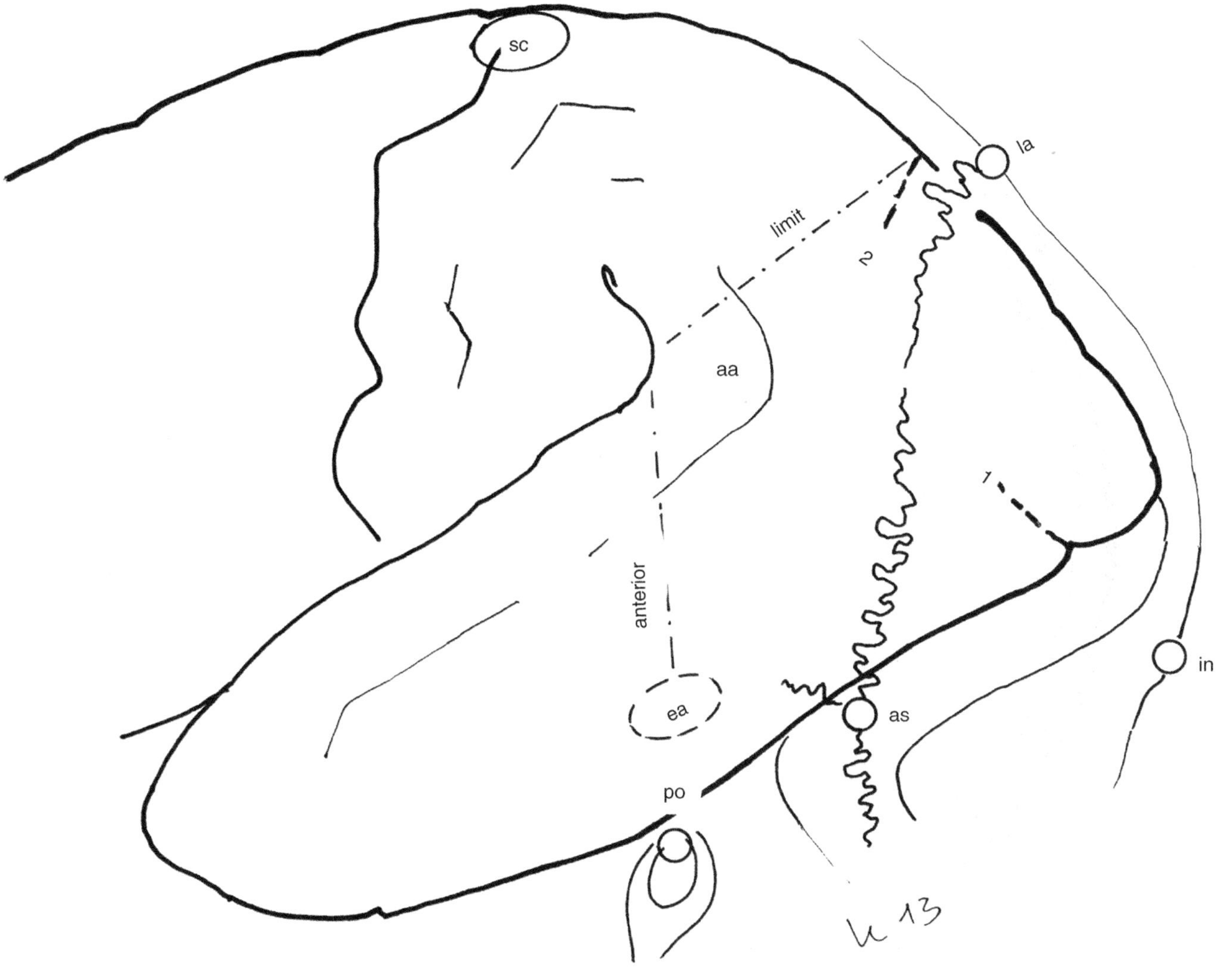

Fig. 12.13 Topography for planning lateral surgical approaches. *aa* angularis region, *as* asterion, *ea* eminentia arcuata, projection, *in* inion, *la* lambda, *po* porion, *sc* sulcus centralis. *1* region of sulcus calcarinus, projection; *2* region of sulcus parietooccipitalis, projection

12.2.2 Superficial and Profound Structures (Figs. 12.14, 12.15, 12.16, 12.17, 12.18, 12.19, 12.20, 12.21, 12.22, and 12.23)

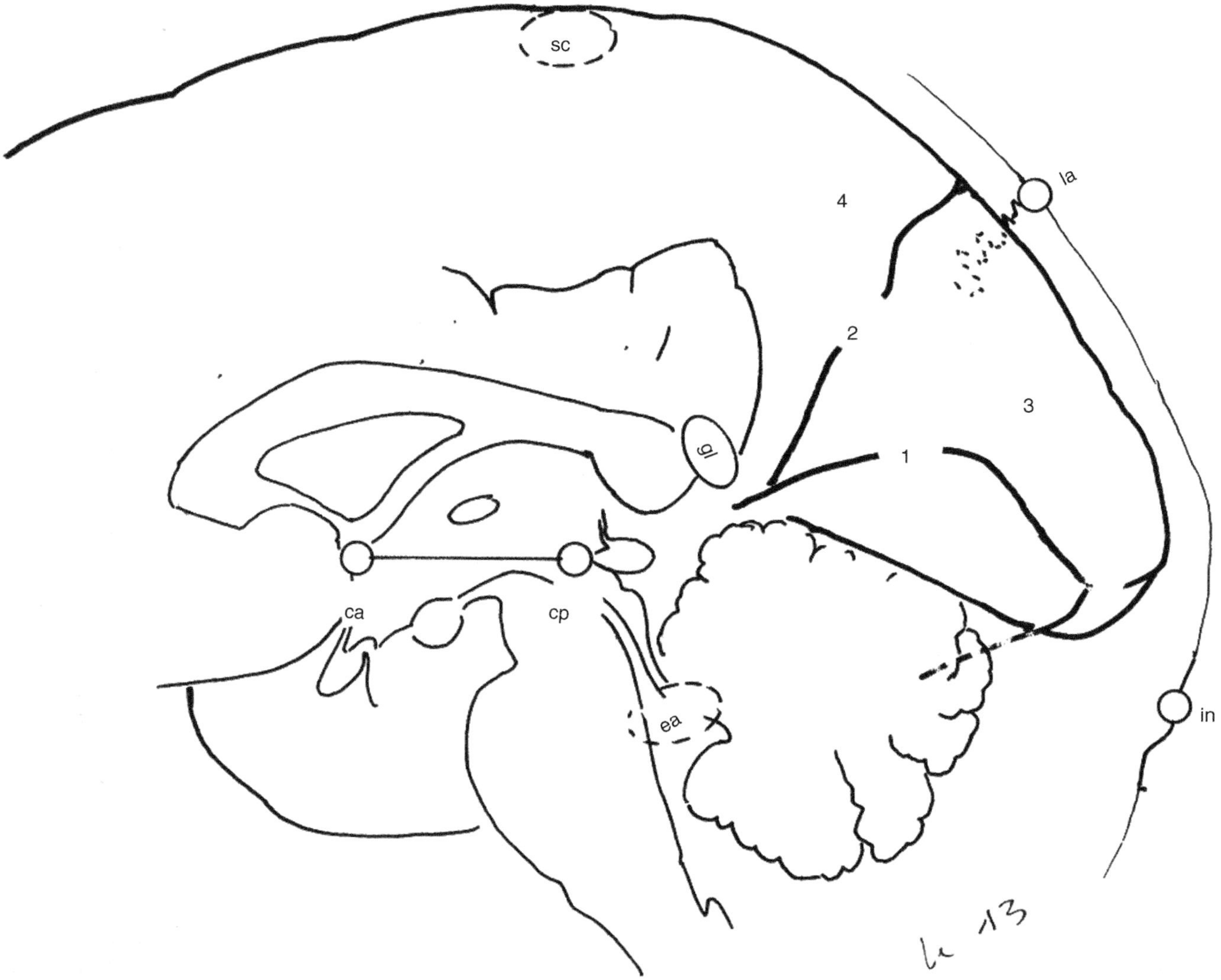

Fig. 12.14 Topography for planning medial surgical approaches. Well-defined relief of gyri in contrast to the varable lateral relief (see Fig. 12.13). *ca* commissura anterior, *cp* commissura posterio,r *ea* eminentia arcuata, projection, *gl* galenic point, *in* inion, *la* lambda, *sc* sulcus centralis, projection. *1* sulcus calcarinus, *2* sulcus parietooccipitalis, *3* cuneus, *4* precuneus

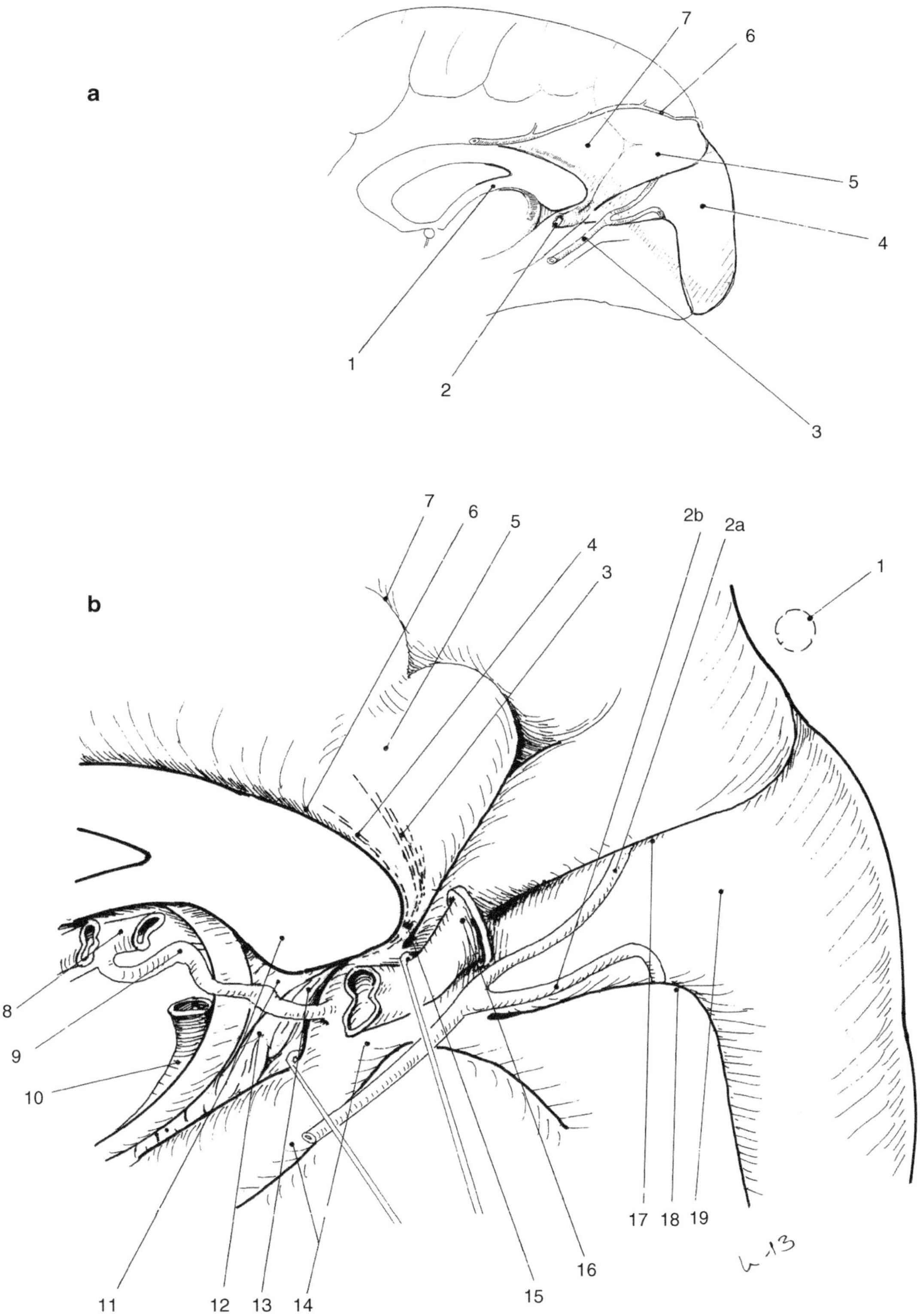

Fig. 12.15 Profound medial occipital and adjacent structures survey **a** and details **b. a** *1* Crus fornicis, *2* V. magna (Galeni), *3* A. cerebri posterior, *4* cuneus, *5* precuneus, *6* A. parietooccipitalis, *7* gyrus cinguli. **b** *1* Lambda, projection; *2a* A. parietooccipitalis; *2b* A. calcarina; *3* stria longitudinalis cinguli, projection; *4* stria medialis et lateralis of indusium griseum, projection; *5* gyrus cinguli, covering 3 and 4; *6* sulcus corporis callosi; *7* sulcus cinguli; *8* V. cerebri interna; *9* medial vein of lobus occipitalis; *10* V. basalis (Rosenthal); *11* splenium corporis callosi; *12* area dentata; *13* begin of indusium griseum and stria longitudinalis cinguli; *14* gyrus occipitotemporalis medialis and the beginning of gyrus parahippocampalis; *15* isthmus gyri cinguli and the beginning of gyrus parahippocampalis; *16* V. magna (Galeni); *17* sulcus parietooccipitalis; *18* sulcus calcarinus; *19* cuneus

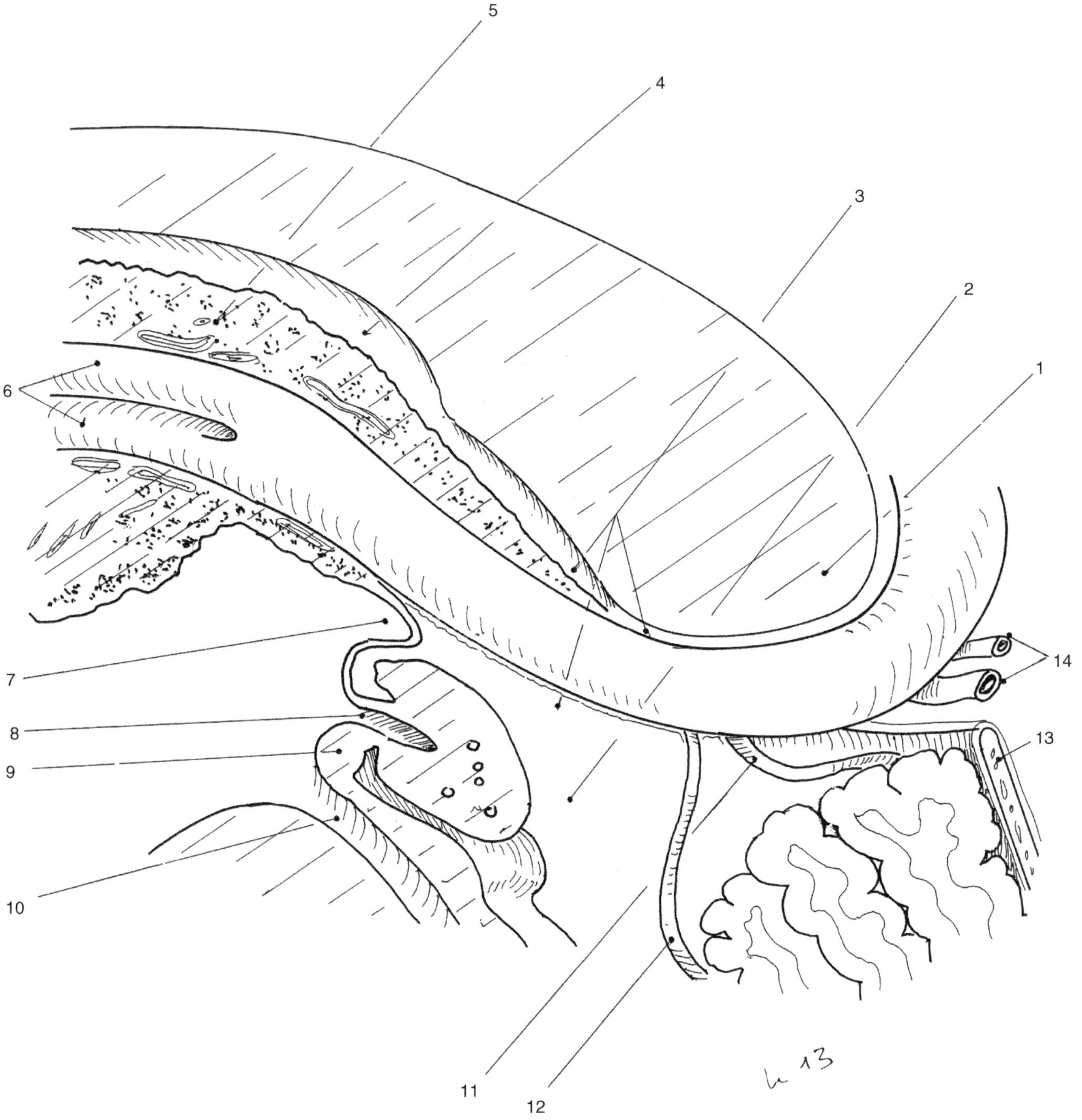

Fig. 12.16 Continuation of Fig. 12.15. Galenic area *1* splenium corporis callosi, *2* cisterna tecti, *3* galenic cistern, *4* fissura transversa cerebri, *5* velum interpositum, *6* Vv. cerebri internae, *7* recessus suprapinealis, *8* recessus pinealis, *9* commissura posterior, *10* aquaeductus, *11* V. supraculminalis, *12* V. cerebellaris precentralis, 13 tentorium, cut, *14* anterior mediobasal veins of lobus occipitalis

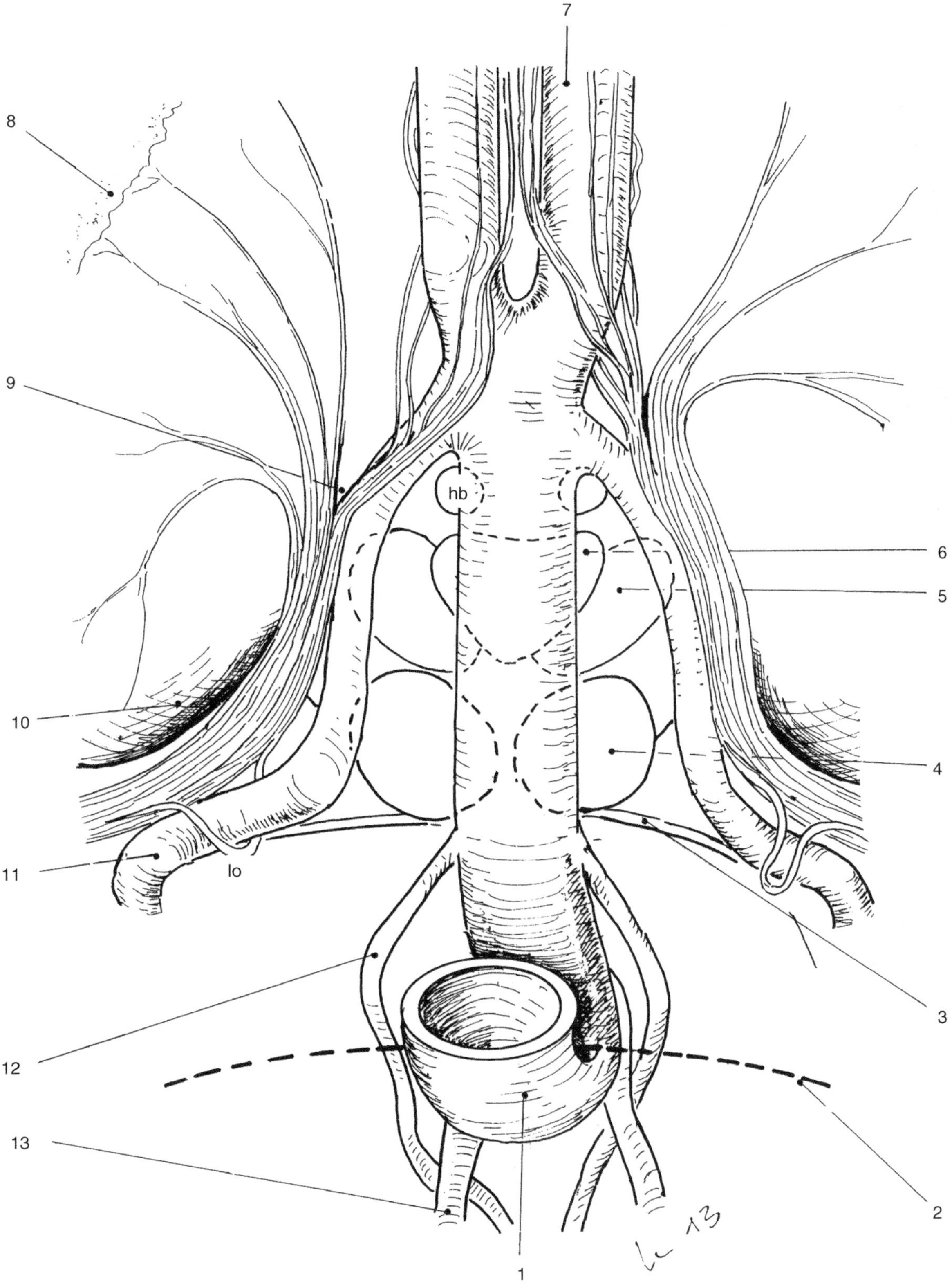

Fig. 12.17 Continuation of Fig. 12.16. Dorsal viewing direction. Habenulae are landmarks at imagings and during surgery for defining of divisions of Aa. choroideae posteriores laterales and mediales. Baumgartner's loop is a landmark during surgery for defining of pulvinar thalami. *1* V. magna (Galeni); *2* splenium, projection; *3* N. trochlearis; *4* colliculus inferior; *5* colliculus superior; *6* corpus pineale; *7* V. cerebri interna; *8* plexus choroideus, margin; *9* branches of A. chorioidea posterior, division (Plets 1969); *10* pulvinar thalami; *11* V. basalis (Rosenthal); *12* V. cerebellaris precentralis; *13* V. supraculminalis. *hb* habenula, *lo* Baumgartner's loop

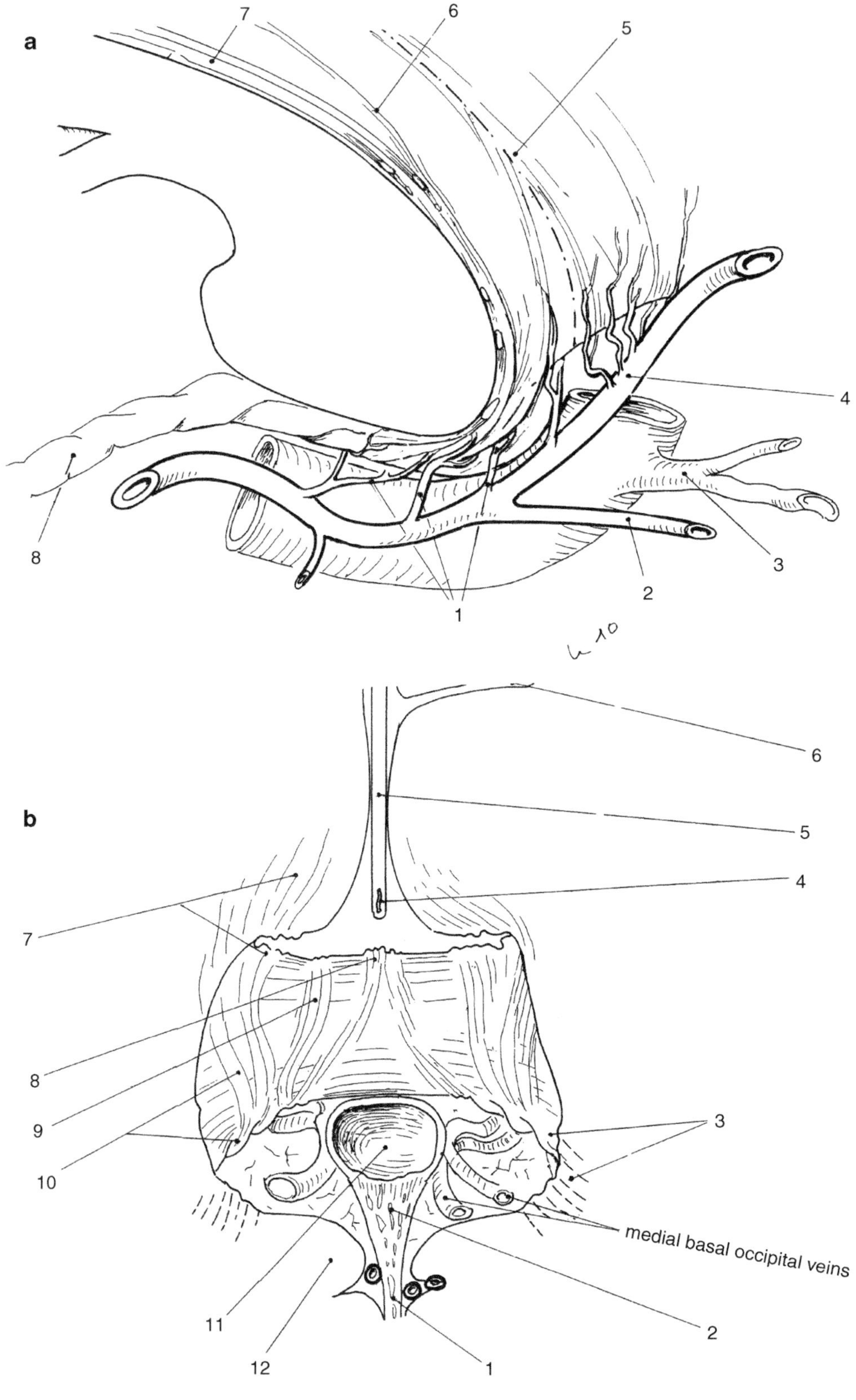

Fig. 12.18 Adjacent structures of splenium corporis callosi. Lateral **a** and posterior **b** viewing directions. Splenium is surrounded by essential vessels and cortex (optic and allocortical segments) at the narrow space between falx and tentorium. During surgery, a surgical shifting of the brain is hindered by the connection of V. magna (Galeni) with falx-tentorium. **a** *1* Branches of A. cerebri posterior feeding area dentata. Cave at surgery *2* A. calcarina, *3* mediobasal vein(s) of lobus occipitalis, *4* A. parietooccipitalis, *5* striae mediales of indusium griseum, marking the midline region, *6* striae laterales of indusium griseum, *7* stria longitudinalis cinguli, *8* area dentata. **b** *1* Falx; *2* high density of veins; *3* area dentata; *4* sinus sagittalis inferior; *5* falx; *6* sulcus cinguli; *7* gyrus cinguli, penetrated by fiber bundles of stria longitudinalis cinguli and fiber bundles of corpus callosum; *8* striae mediales of indusium griseum; *9* striae laterales of indusium griseum; *10* posterior begin of stria longitudinalis cinguli; *11* V. magna (Galeni); *12* isthmus gyri cinguli

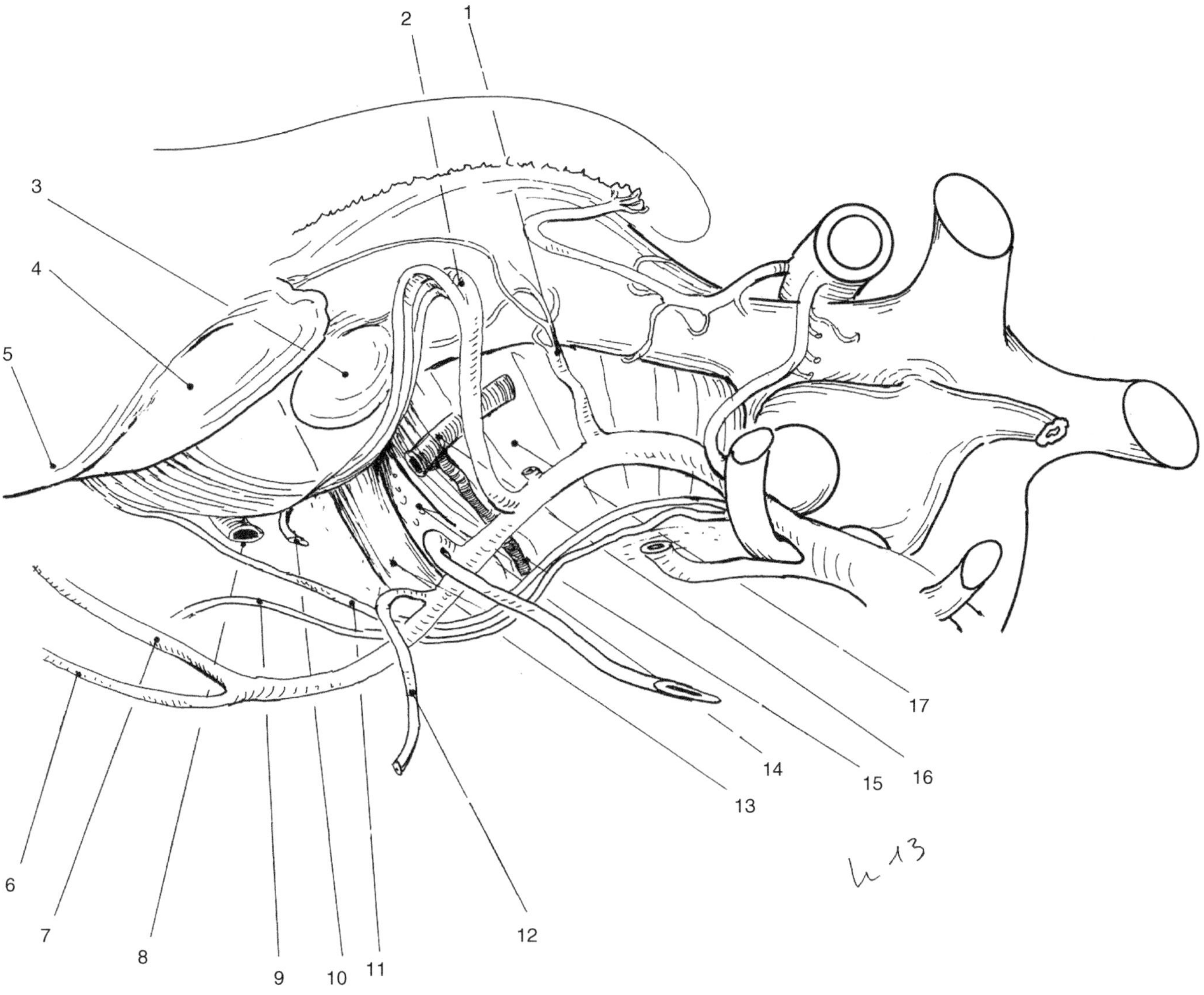

Fig. 12.19 Arteries of cisterna ambiens and adjacent structures. Numerous fine branches are feeding the midbrain, tractus opticus, the upper pons, and basal ganglia. It is originating from A. communicans posterior, A. chorioidea anterior, A. cerebri posterior, and its branches. These could hinder surgical approaches. Further hindering is the deep location of cisterna ambiens. Surgical routes along these natural CSF spaces present only small sections of cisterna ambiens. More favorable routes could present transcerebral endoscopic approaches. *1* A. thalamogeniculata, *2* loop of A. chorioidea posterior or its main ramification, *3* corpus geniculatum mediale, *4* corpus geniculatum laterale, *5* pulvinar thalami, *6* A. calcarina, *7* A. parietooccipitalis, *8* V. basalis (Rosenthal), *9* A. tecti, *10* N. trochlearis, *11* A. chorioidea posterior medialis, *12* temporo-occipital basal branch of A. cerebri posterior, *13* pedunculus cerebellaris superior, *14* trigonum lemnisci hypervascularized, inconstant finding, *15* V. mesencephalic lateralis connected with V. petrosa and sinus petrosus superior, *16* V. basalis (Rosenthal), *17* crus cerebri

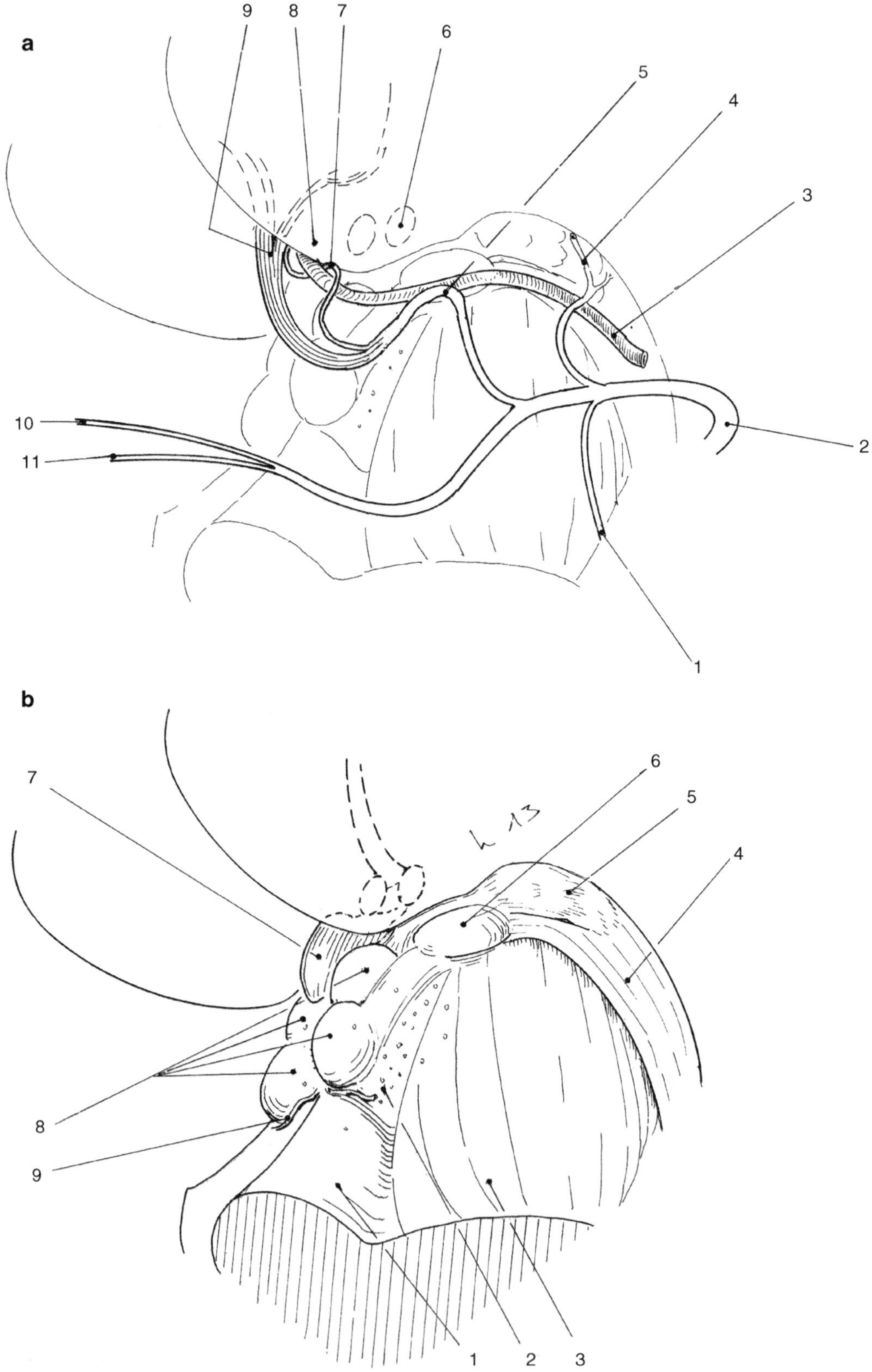

Fig. 12.20 Cisterna ambiens and cisterna tecti. Synopsis. **a** Vessels. **b** Mesencephalon. **a** *1* Anterior branch of temporobasal arteries, *2* A. cerebri posterior, *3* V. basalis (Rosenthal), *4* A. thalamogeniculata, *5* loop of A. chorioidea posterior lateralis, *6* habenula, projection, *7* Baumgartner's loop, *8* pulvinar thalami, *9* division of branches of A. chorioidea post. lateralis, *10* A. parietooccipitalis, *11* A. calcarina. **b** *1* Pedunculus cerebellaris superior, *2* trigonum lemnisci, *3* crus cerebri, *4* tractus opticus, *5* corpus geniculatum laterale, *6* corpus geniculatum mediale, *7* corpus pineale, *8* tectum (colliculi superiores et inferiores), *9* N. trochlearis

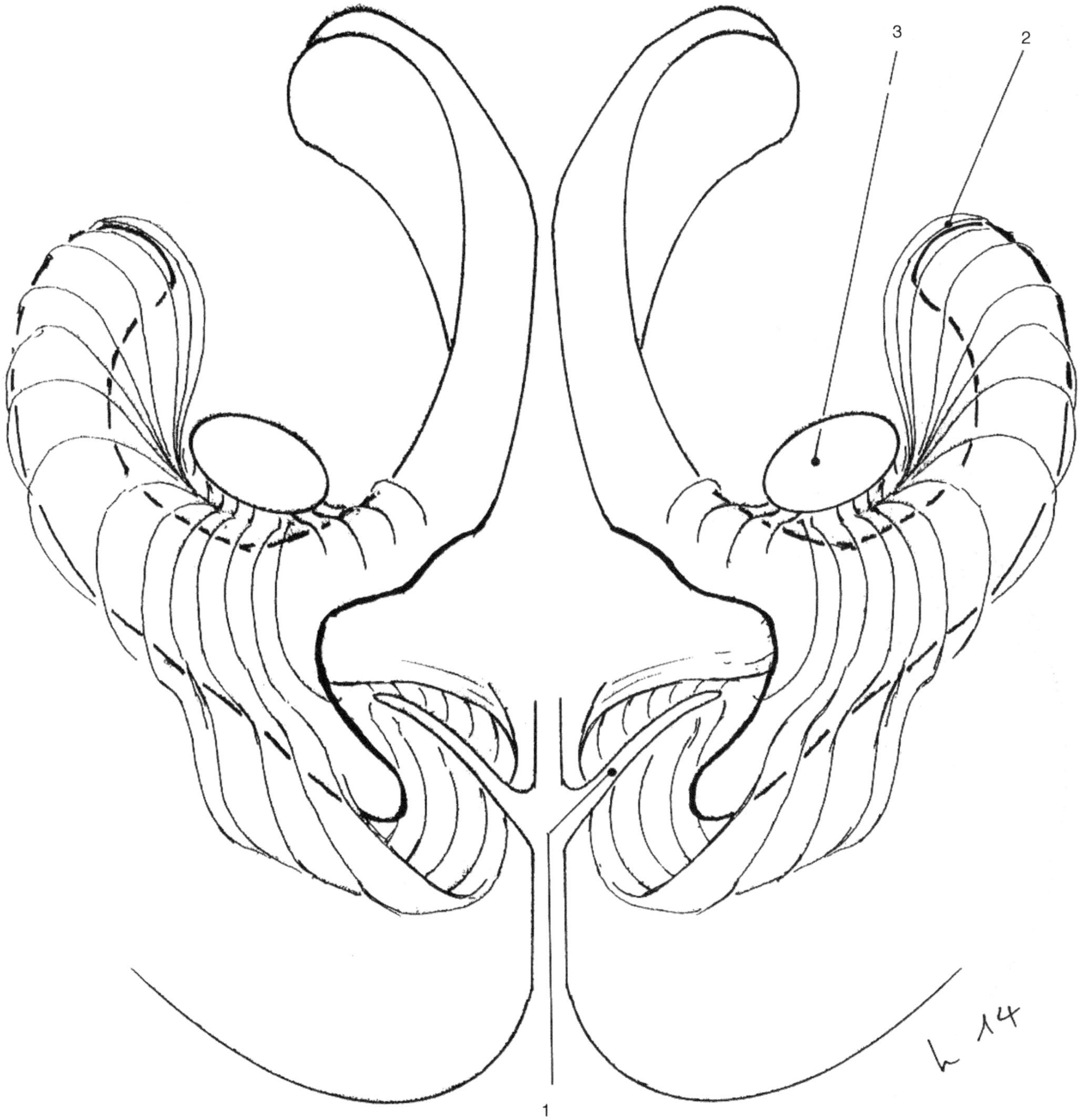

Fig. 12.21 Radiatio optica, schematized. *1* Sulcus calcarinus, *2* tip of cornu inf. et genu temporale of radiatio optica, *3* corpus geniculatum laterale

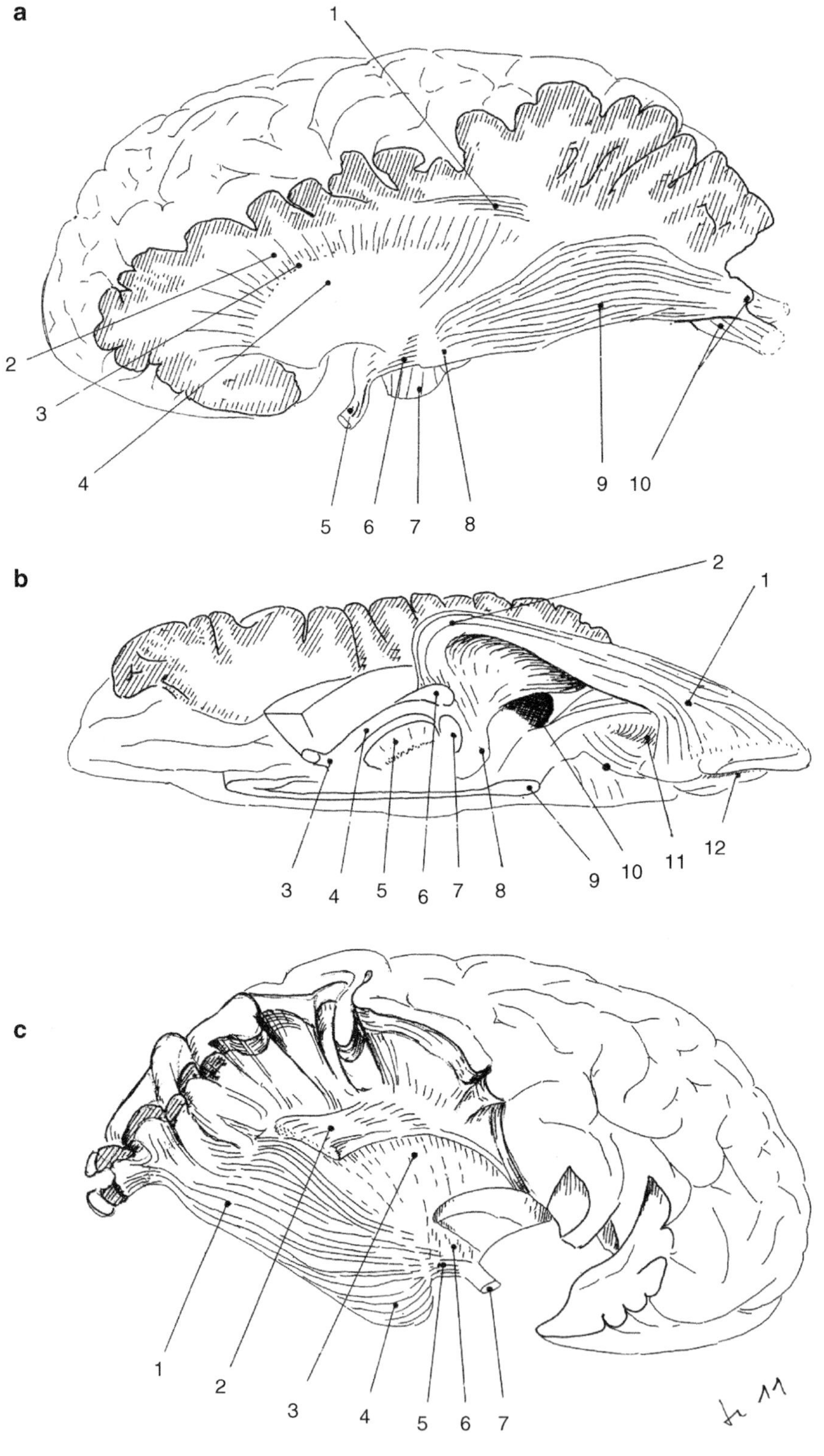

Fig. 12.22 Radiatio optica and adjacent structures. Anatomical fiber dissections according to Ludwig and Klingler 1956. **a** Lateral viewing direction. **b** Basal viewing direction. **c** Lateral posterior viewing direction. **a** *1* Fasciculus longitudinalis superior, *2* corona radiata, *3* capsula interna, *4* putamen, *5* N. opticus, *6* tractus opticus, *7* crus cerebri, *8* corpus geniculatum laterale, *9* radiatio optica, *10* genu occipitale of radiatio optica. **b** *1* Genu occipitale of radiatio optica, *2* genu temporale of radiatio optica, *3* chiasma, *4* tractus opticus, *5* crus cerebri, *6* corpus geniculatum laterale, *7* corpus geniculatum mediale, *8* pulvinar thalami, *9* splenium corporis callosi, *10* atrium, *11* U-fibers close to sulcus parietooccipitalis, *12* sulcus calcarinus. **c** *1* Radiatio optica, *2* fasciculus longitudinalis superior, *3* corona radiata, *4* genu temporale of radiatio optica, *5* tractus opticus, *6* capsula interna, *7* N. opticus

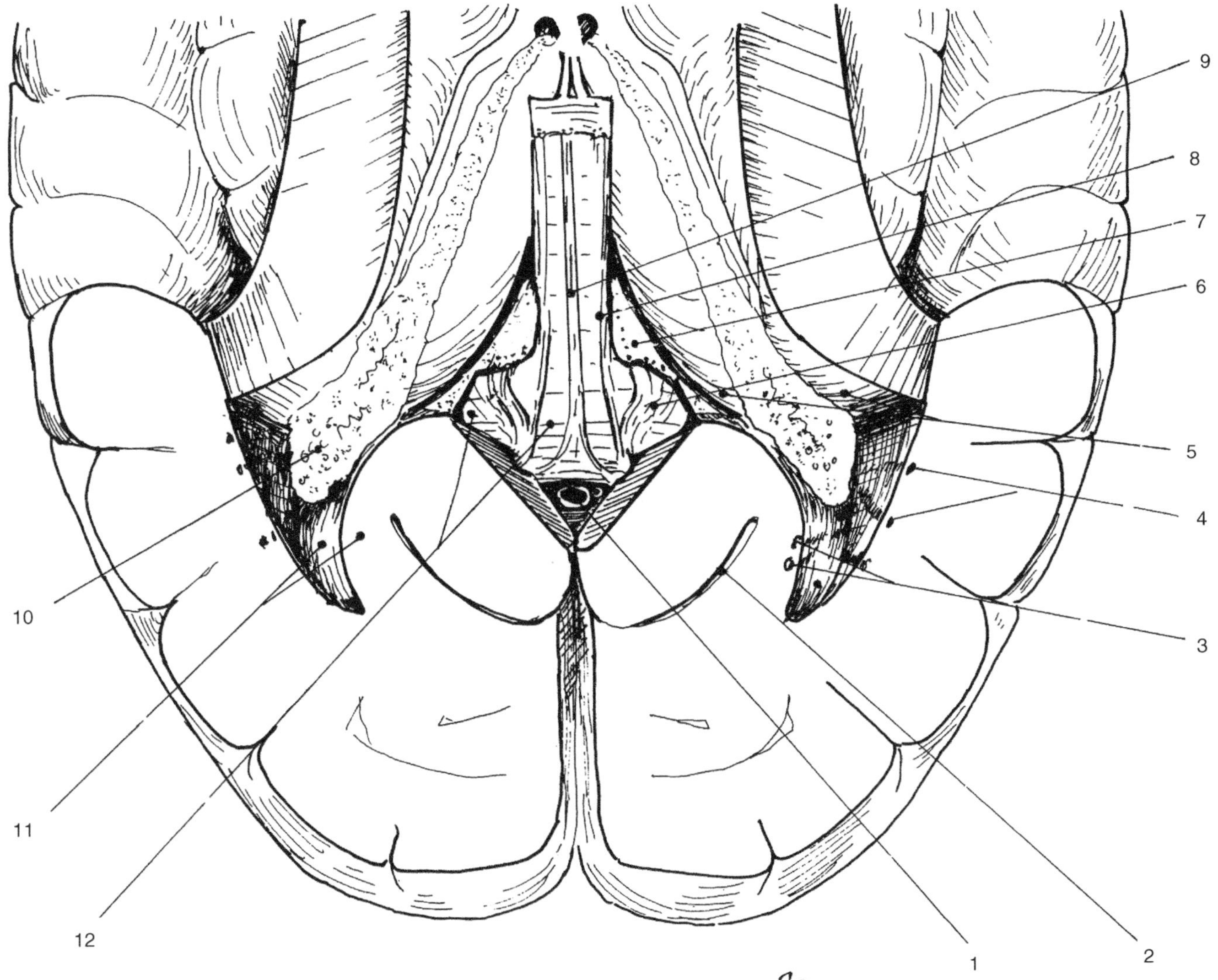

Fig. 12.23 Occipital lobe. Cadaver brain dissection. Conclusion. *1* V. magna (Galeni), *2* sulcus calcarinus, *3* medial atrial veins, *4* lateral atrial veins, *5* pulvinar thalami, *6* area striata of the optic cortex residual, *7* gyrus cinguli, residual, *8* striae laterales of indusium griseum, *9* striae mediales of indusium griseum, *10* plexus choroideus, *11* calcar avis, *12* splenium corporis callosi

Special Surgical Aspects 13

13.1 Extradural Topography

13.1.1 Skin-Galea-Incision (Fig. 13.1)

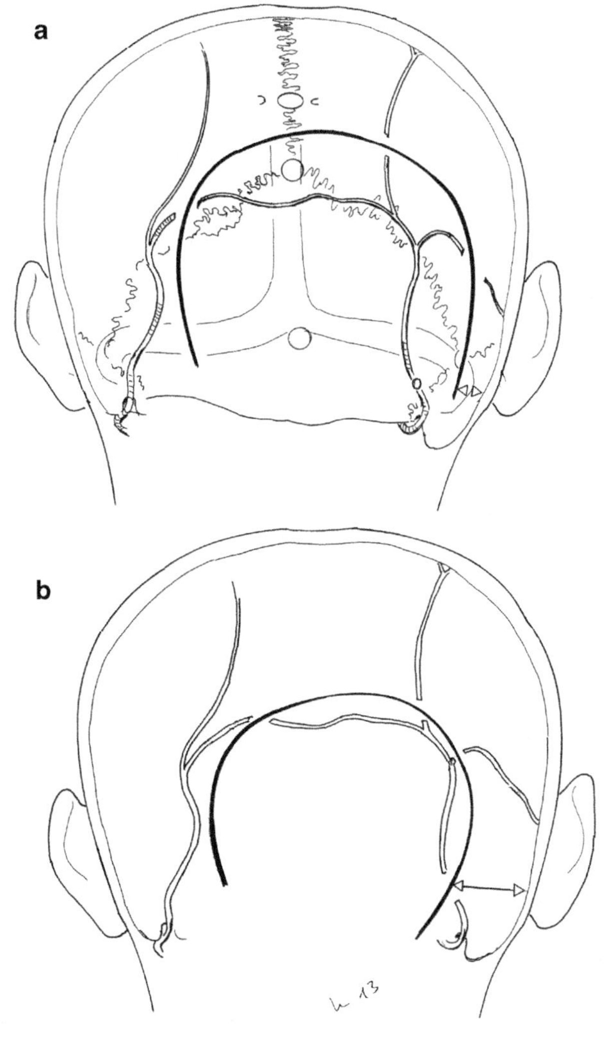

Fig. 13.1 Preservation of anastomotic arteries at skin-galea-incision. **a** Sufficient unilateral preservation of the right-sided A. occipitalis. **b** Danger for necrosis of the skin after surgery. Circles: landmarks

W. Seeger, J. Zentner, *Anatomical Basis of Cranial Neurosurgery*, https://doi.org/10.1007/978-3-319-63597-2_13

13.1.2 Craniotomy (Fig. 13.2)

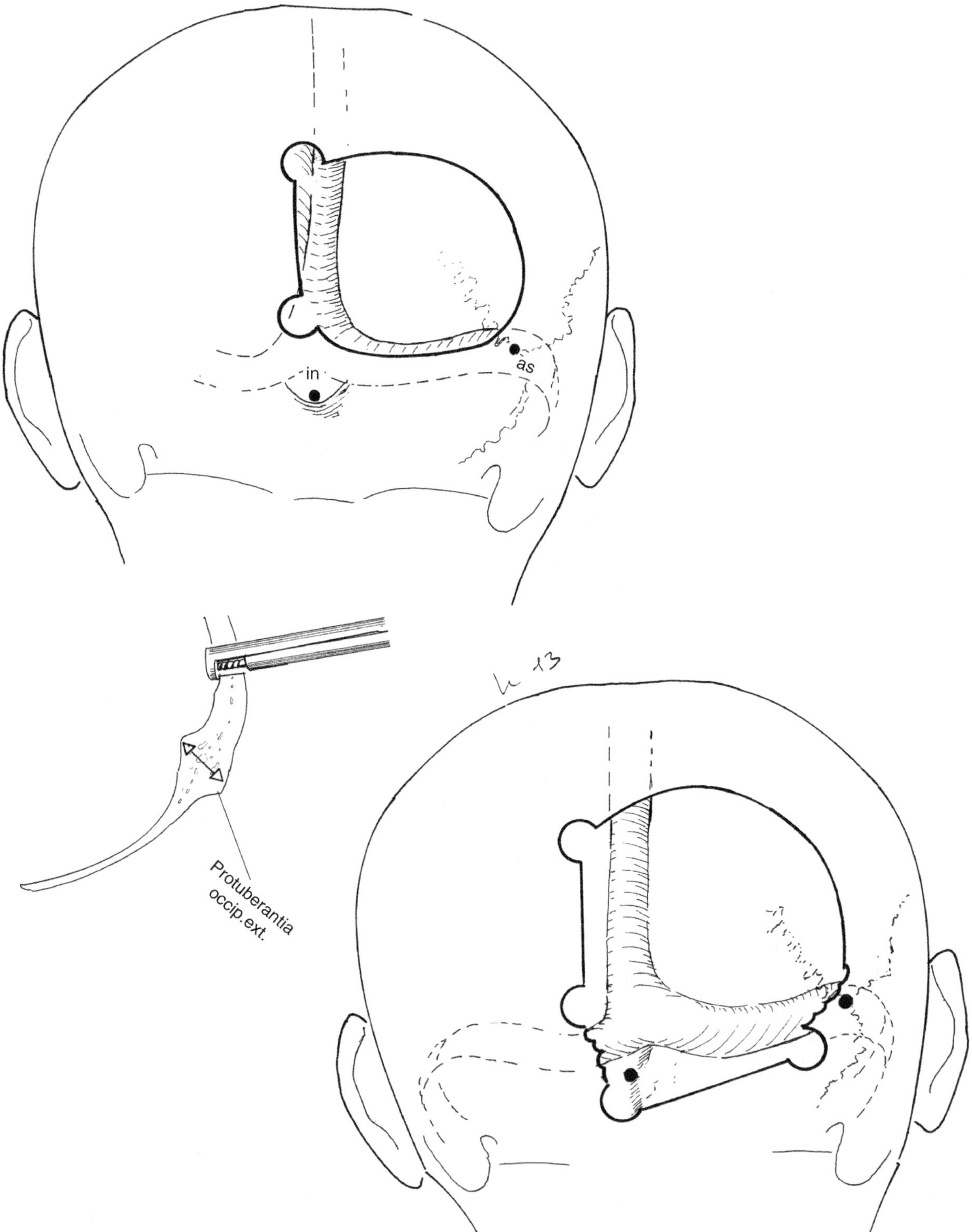

Fig. 13.2 Widening of craniotomy. This may be recommended, if the thickened bone and emissariae of the region of inion and confluens sinuum are hindering craniotomy. *as* asterion, *in* inion

13.2 Intradural

13.2.1 Intraventricular Bleeding (Fig. 13.3)

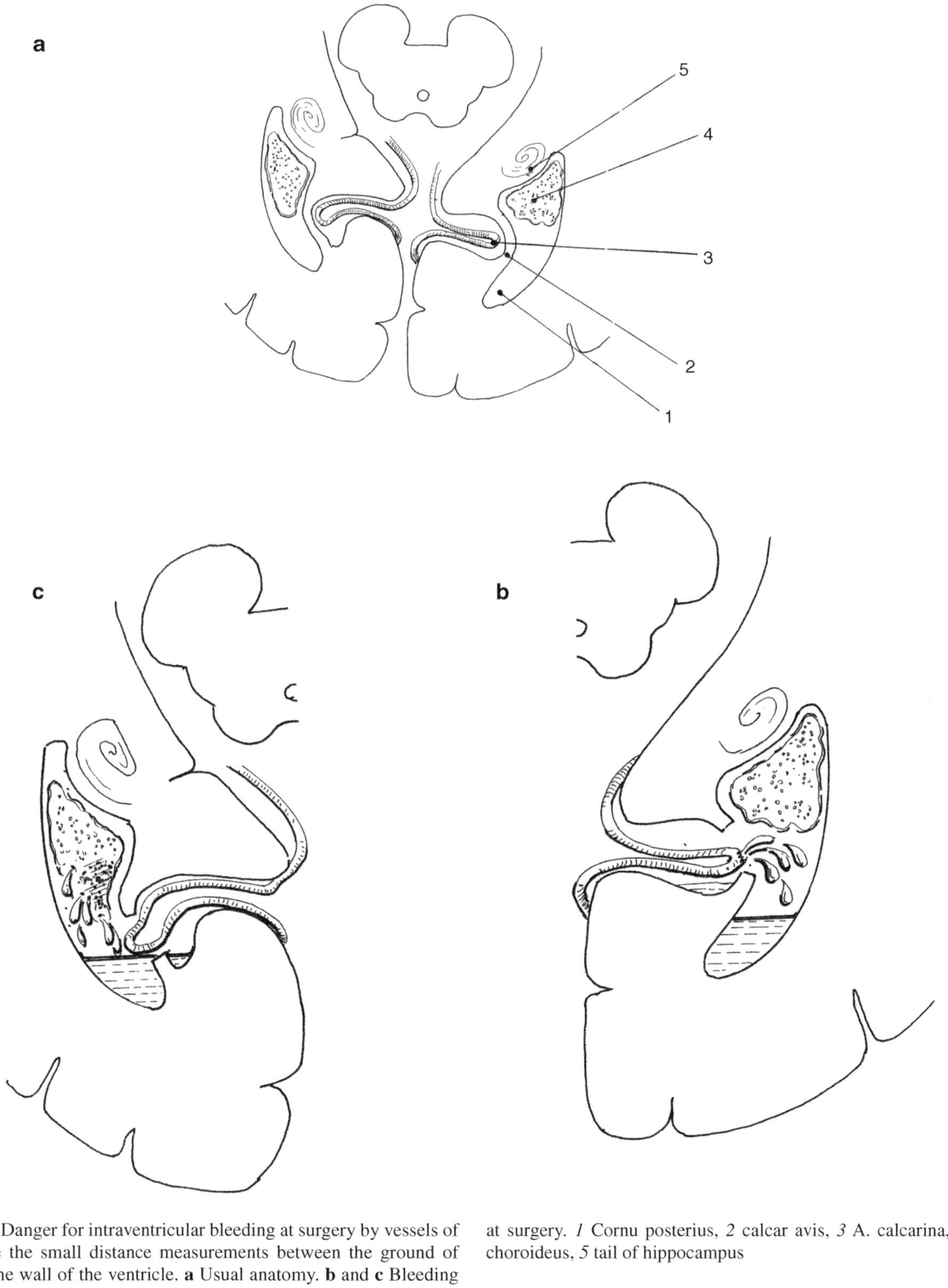

Fig. 13.3 Danger for intraventricular bleeding at surgery by vessels of sulci. Note the small distance measurements between the ground of sulci and the wall of the ventricle. **a** Usual anatomy. **b** and **c** Bleeding at surgery. *1* Cornu posterius, *2* calcar avis, *3* A. calcarina, *4* plexus choroideus, *5* tail of hippocampus

13.2.2 Surgical Topography (Fig. 13.4)

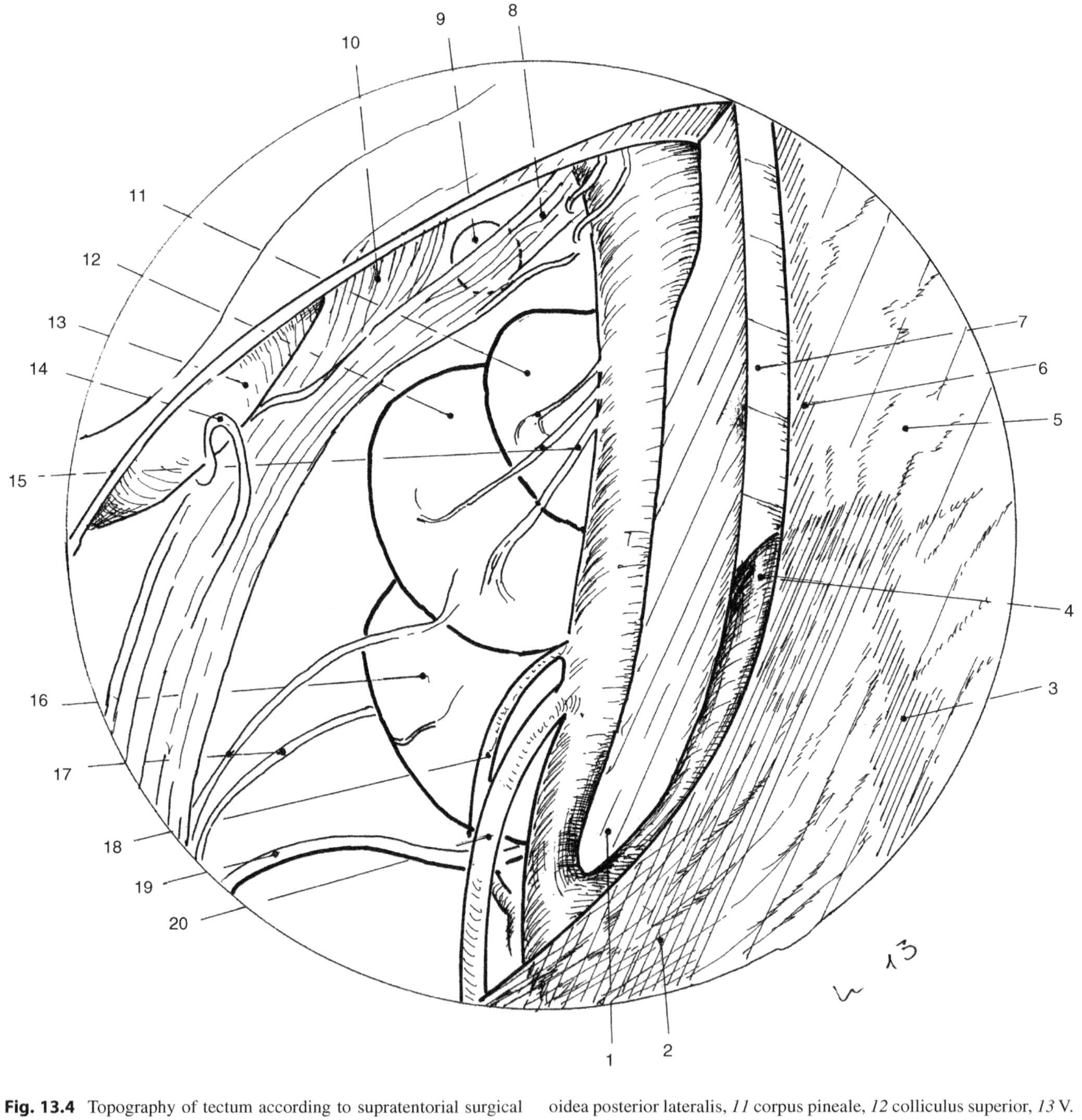

Fig. 13.4 Topography of tectum according to supratentorial surgical approaches cadaver brain dissection. *1* Splenium corporis callosi, *2* tentorium, *3* sinus rectus, *4* ampulla of V. magna (Galeni), *5* falx, *6* sinus sagittalis inferior, *7* corpus callosum, *8* medial ramification of A. chorioidea posterior lateralis, *9* habenula, *10* lateral ramification of A. chorioidea posterior lateralis, *11* corpus pineale, *12* colliculus superior, *13* V. basalis (Rosenthal), *14* Baumgartner's loop, *15* Vv. tecti, *16* colliculus inferior, *17* Aa. tecti, *18* V. cerebellaris precentralis, *19* N. trochlearis, *20* V. supraculminalis

13.2.3 Puncture of Cornu Post (Figs. 13.5, 13.6, and 13.7)

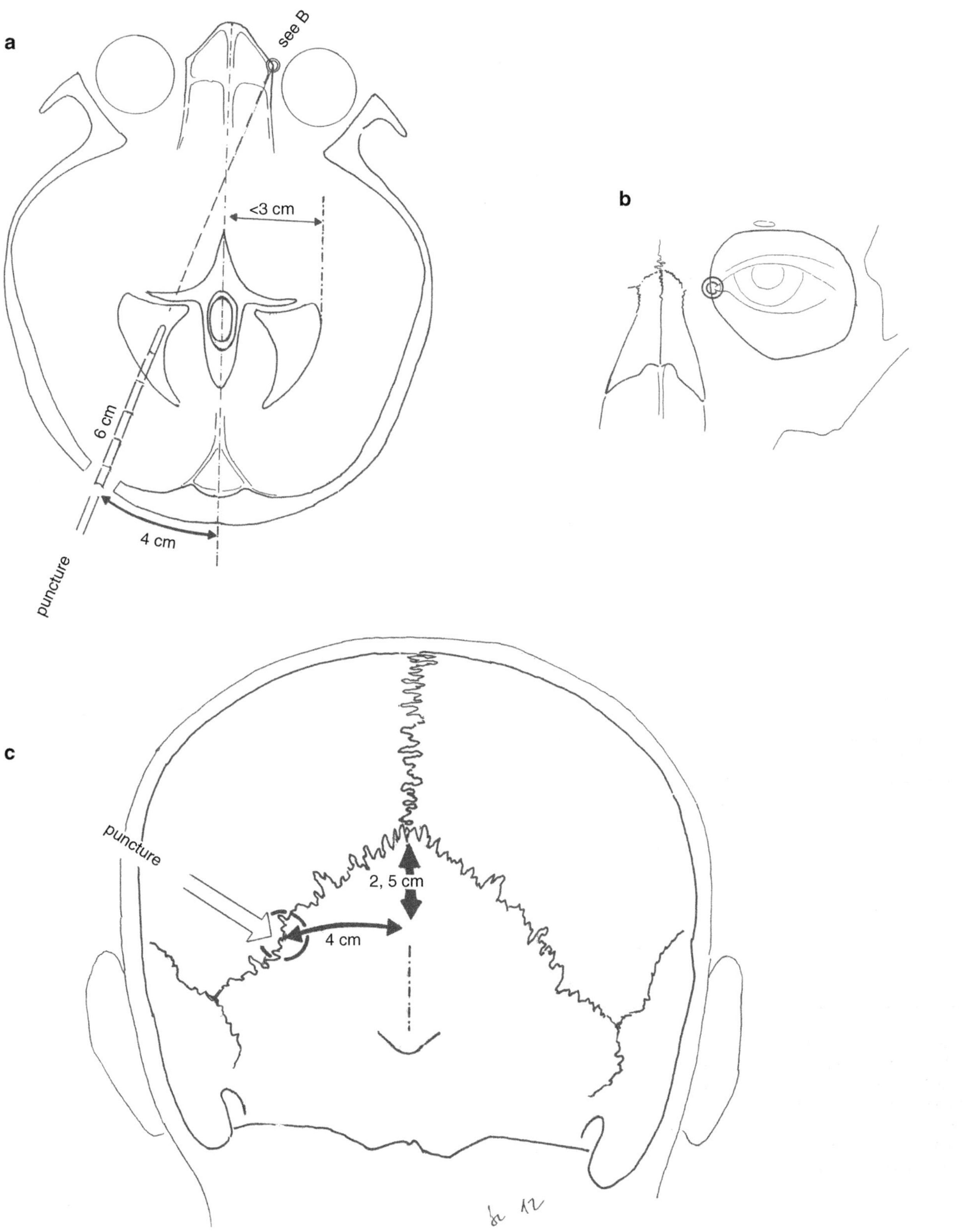

Fig. 13.5 Puncture of cornu posterius for assisting neuronavigation, if necessary. **a** and **b** Torkildsen method, modified. **c** Approximate measurements

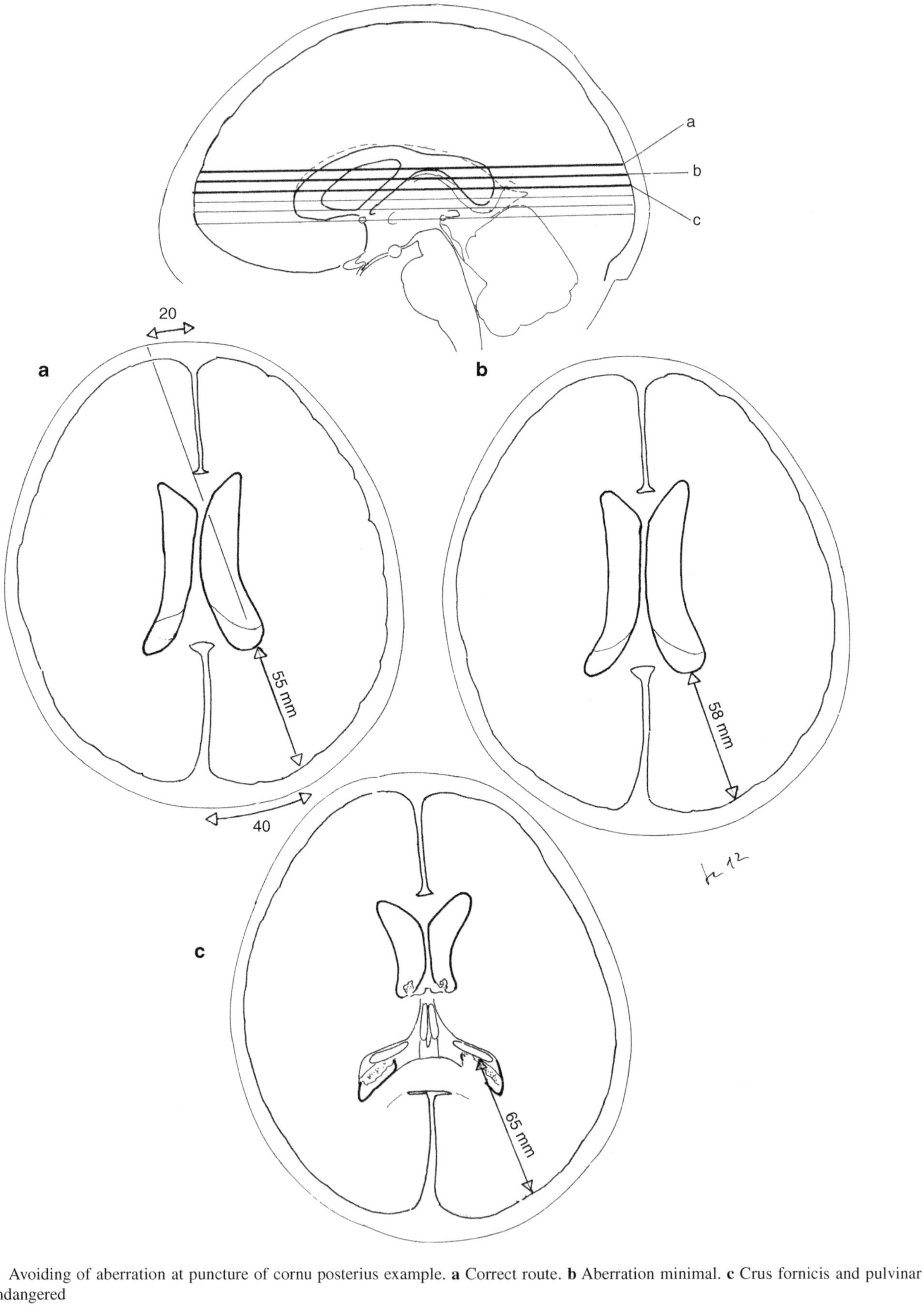

Fig. 13.6 Avoiding of aberration at puncture of cornu posterius example. **a** Correct route. **b** Aberration minimal. **c** Crus fornicis and pulvinar thalami endangered

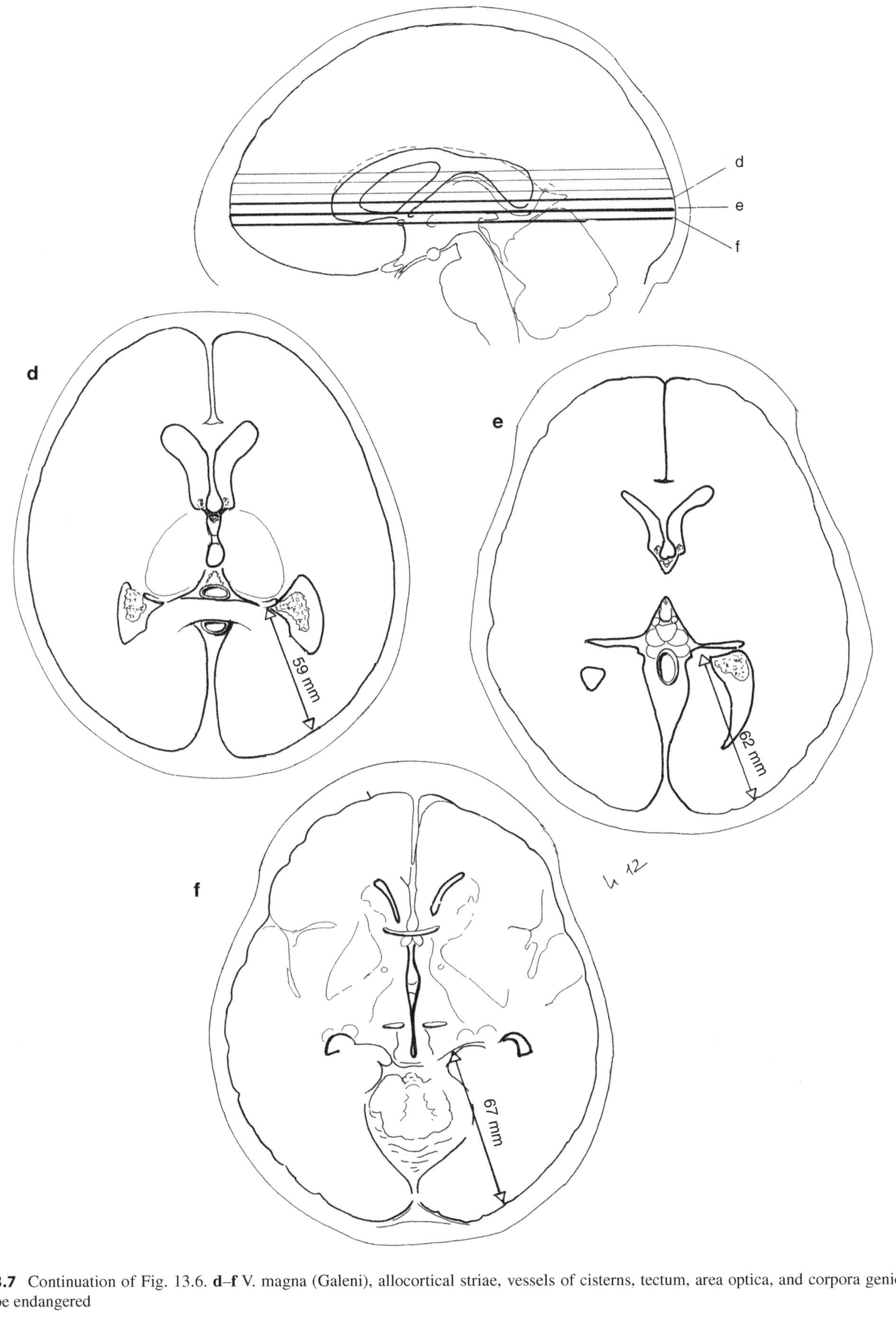

Fig. 13.7 Continuation of Fig. 13.6. **d–f** V. magna (Galeni), allocortical striae, vessels of cisterns, tectum, area optica, and corpora geniculata could be endangered

Part V

Infratentorial Region

1.1 Introduction

This introduction to Part V contains a short summary of each chapter.

1.2 Chapter 14 Diagnostic Base

- *Extra- and intracranial landmarks* can support planning surgery, because it could be helpful for the topographic orientation, if imagings are unclear. The schematized illustrations of vessels could be used as introduction for a better understanding of the high variability of vessels which is presented later on.
- A typical example of introduction for surgery is the intracerebellar surgical approach. Wide areas of *cerebellum* can be resected without following neurological deficits. It motivates neurosurgeons to start infratentorial surgery by performing intracerebellar approaches. But problematic is the deep located area. The high density of fibers of all three cerebellar pedunculi close to nucleus dentatus and the fourth ventricle presents a region which must be preserved at surgery. It can be preserved regarding the topographical relationships of the cerebellar sulci to the essential cerebellar structures in the depth. Anatomical dissections and a series of MRTs may help to define the problematic area by neuronavigation before and during surgery, as presented in this book.

1.3 Chapter 15 Anatomical Base of Surgery

- *Extracranial and cranial structures*: Muscles, nerves, and vessels of the neck are detailed presented for preservation at surgery, if possible. These aspects have to be considered for avoiding problems of wound healing, development of neuromas after transection of nerves, liquorrhea, and infections.
- *Dural structures*, which have to be considered, are sinuses, its connecting veins, and its variants. Illustrations may be helpful for avoiding bleeding, air embolism, and encephalomalacia by surgery.
- *Cisterna cerebellomedullaris* and its contents are presented according to the surgical approach to the fourth ventricle. The cistern is filled out by the bilateral vermis-tonsillar-complex. The ground of cistern is built by velum medullare posterius. This is the posterior wall of the fourth ventricle. The velum presents a hypo/avascular thin-walled compartment («nidus avis») and a plexus-choroideus compartment. PICA and branches of it are interposed between vermis-tonsillae and velum medullare posterius. PICA presents a variable loop close to the midline. The surgical route to the fourth ventricle is located between tonsilla and nodulus vermis. The fourth ventricle is opened after incision of velum medullare posterius.

- At surgical dilation of this opening, tractus flocculonodularis must be preserved for avoiding deficits of the optokinetic nystagmus, as presented.
- *Cisterna pontocerebellaris and surrounding cisterns* are presented as introduction for training neurosurgeons according to Yaşargil's creation of cisternal microsurgery. Further illustrated is the high variability of vessels, especially arteries, e.g., according to elimination of aneurysms and AVMs.
- *Intrarhombencephalic target areas for surgery,* especially circumscript lesions (e.g., cavernomas) are sometimes approached by microsurgical methods. It must be considered the microanatomy of nuclei and fibers of the brain stem. Favorable for surgery is the minimal variability of nuclei and fibers. Even less variable are the fine intracerebral branches of perforating arteries. An example for using a favorable surgical approach is the microsurgical elimination of lesions at the voluminous pedunculus cerebellaris inferior. But nucleus and tractus solitarius and other components of N. vagus are underlying the peduncle and must be preserved. This example demonstrates that it is helpful to present a specific introduction for microanatomy of nuclei and fibers of the brain stem, as given here. After this is added an illustration of the neuronal control of respiration at the dorsal pontomedullary area, according to surgical approaches to the fourth ventricle. Further illustrated is the chemosensory control of respiration at the ventral surface of the medullopontine area according to the transcondylar premedullary surgical approach. Perforating arteries of the brain stem are added, combined with essential nuclei and fiber systems.
- Presented are *landmarks at the superficial relief of the brain stem and cerebellum for orientation during surgical approaches.* These landmarks are polus inferior and superior olivae, ligamentum denticulatum, pontomedullary rim, obex, fastigium, frenulum tecti, cranial nerves, colliculi tecti, fossa interpeduncularis (corpora mamillaria), lobulus centralis cerebelli, ala lobuli centralis, lingula cerebelli, fissura horizontalis cerebelli, flocculus, uvula vermis, and nodulus vermis.

1.4 Chapter 16 Special Surgical Aspects

Anatomical base of special surgical approaches:

- *Tonsillo-nodular surgical approach to the fourth ventricle.* Preservation of paleocerebellar functions (optokinetic nystagmus) in contrast to splitting vermis.
- *Cerebellopontine surgical approach according to Yaşargil.* Principles of basal cisternal approaches with preservation of arachnoid walls of cisternae basales and its contents according to Yaşargil[1].
- *Transcondylar approaches to the premedullary cisterns.* Basal approach for craniospinal lesions ventral from medulla and pons (Seeger 1978).
- *Supracerebellar approaches to cisterna tecti.* F. Krause's approach, Yaşargil's modification for microsurgery with widening of the approach to adjacent CSF spaces.

[1]These principles are applicable for all basal cisterns (with modifications).

14 Diagnostic Base

14.1 Landmarks (Figs. 14.1 and 14.2)

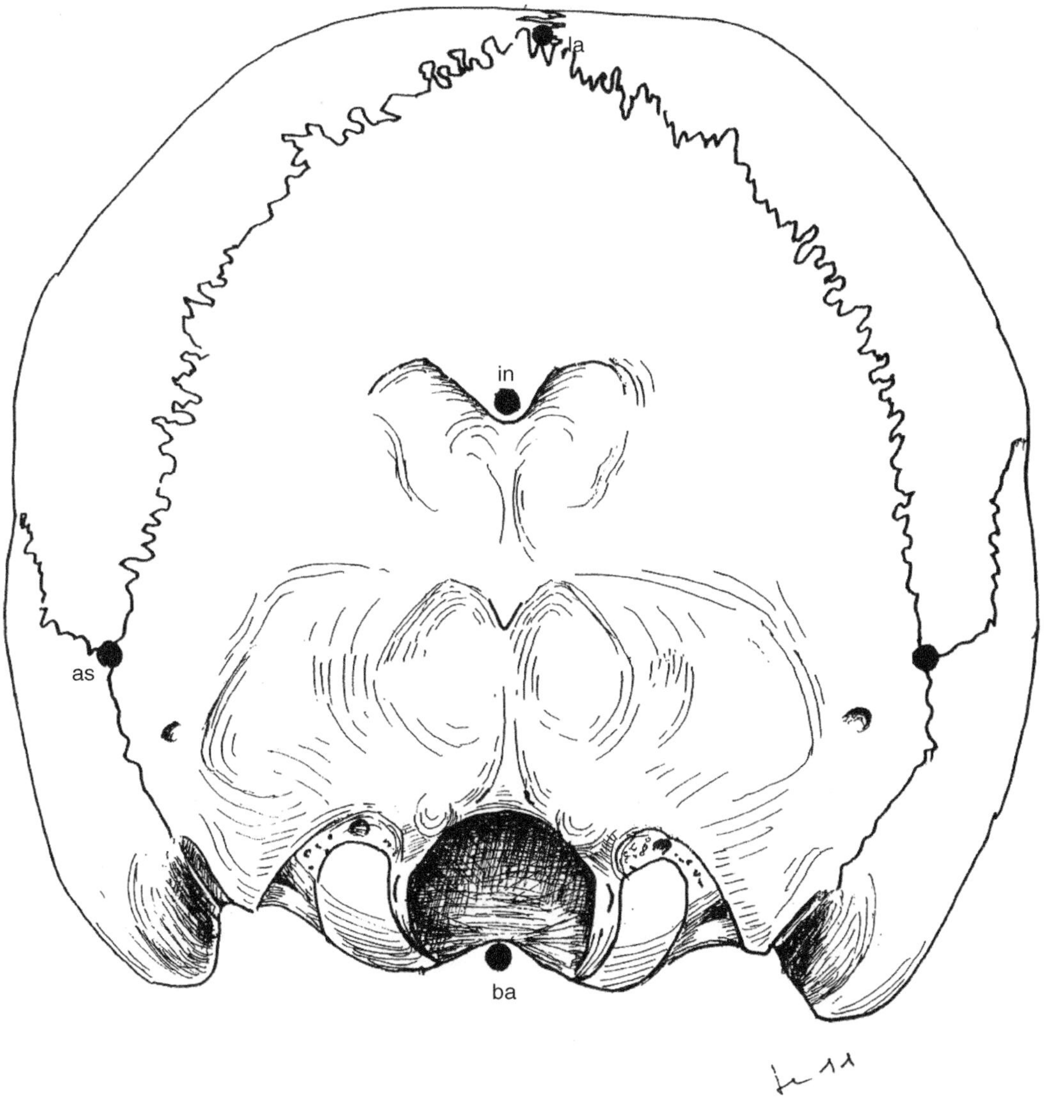

Fig. 14.1 Planum occipitale et planum nuchale. Landmarks added for planning surgery. *as* asterion, *ba* basion, *in* inion, *la* lambda

W. Seeger, J. Zentner, *Anatomical Basis of Cranial Neurosurgery*, https://doi.org/10.1007/978-3-319-63597-2_14

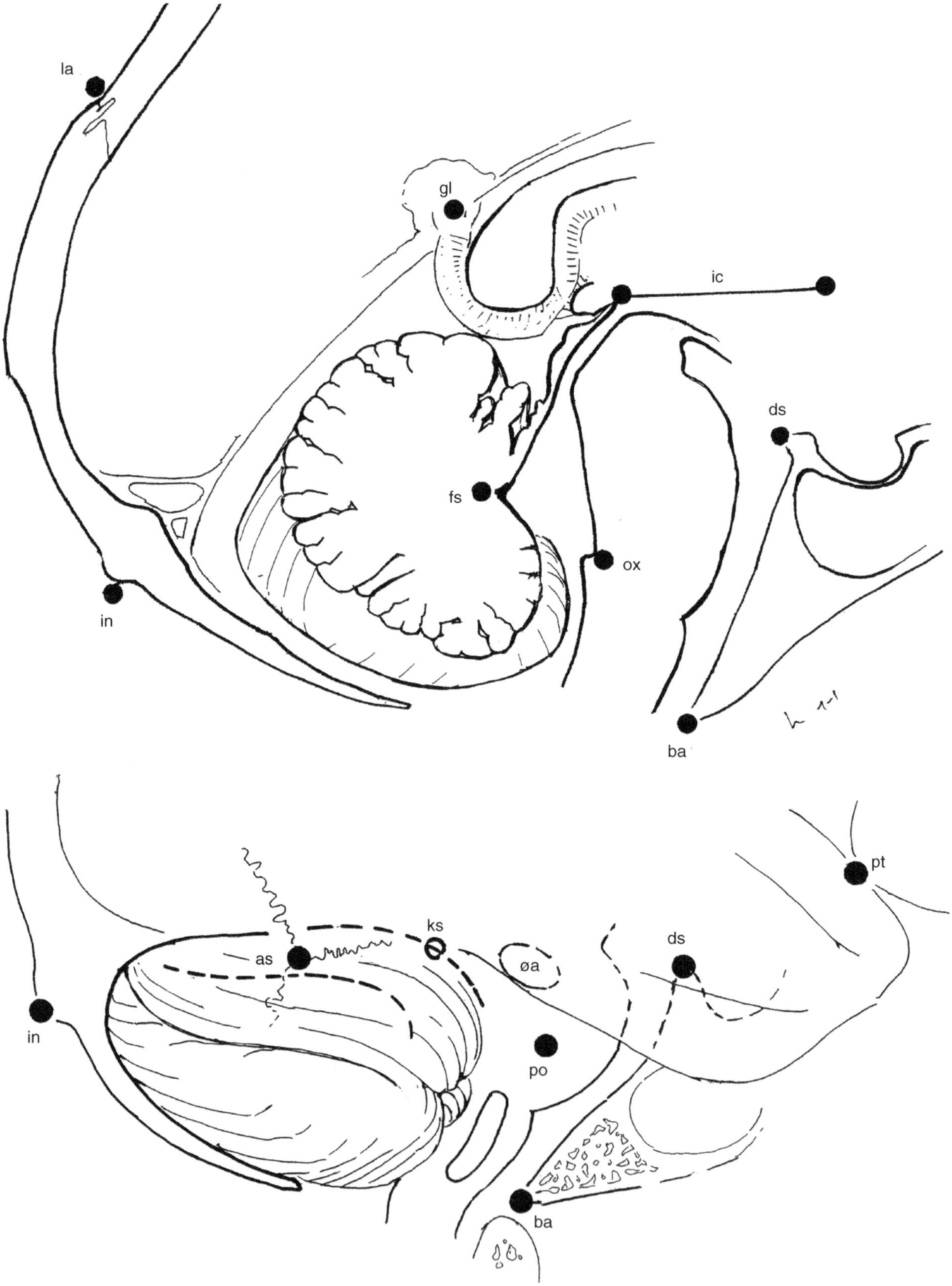

Fig. 14.2 Extra- and intracranial landmarks. *as* asterion, *ba* basion, *ds* dorsum sellae, *ea* eminentia arcuata, projection, *fs* fastigium, *gl* galenic point, *ic* intercommissural line, *in* inion, *ks* knee of sinus transversus, projection, *la* lambda, *ox* obex, *po* porion, *pt* pterion dotted: sinus transversus and further structures, projections

14.2 Radiology (Figs. 14.3, 14.4, and 14.5)

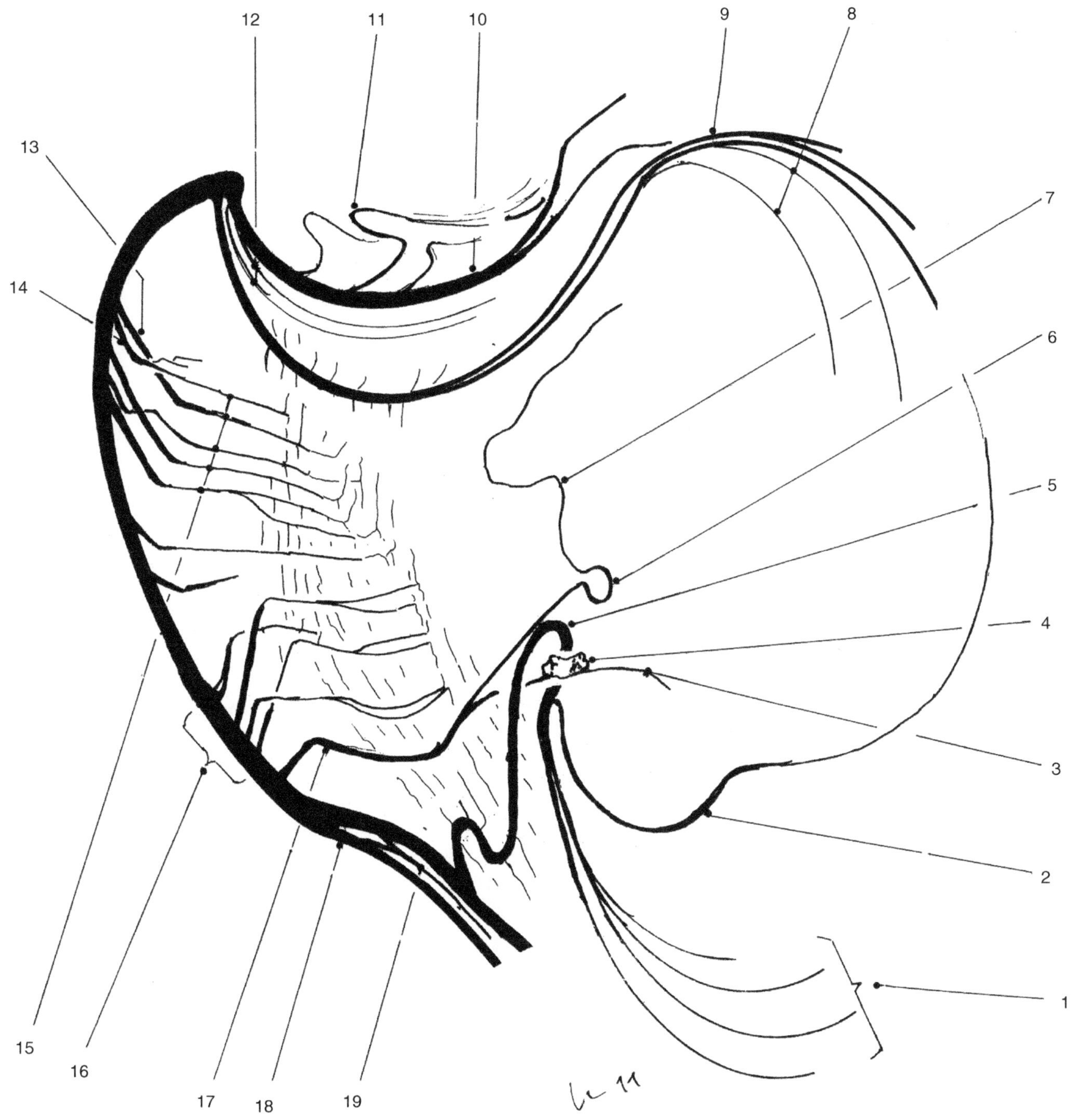

Fig. 14.3 Arteriogram schematized drawing. For variants see Figs. 15.22 ff and 15.39 ff. *1* ramification of PICA; *2* A. vermis; *3* AICA, inferior branch; *4* plexus choroideus; *5* PICA, choroid point; *6* AICA, meatal loop; *7* AICA, trigeminal loop; *8* ramification of A. cerebelli superior, lateral group; *9* as 8, medial group; *10* A. cerebri posterior; *11* A. chorioidea posterior lateralis, loop at corpora geniculata; *12* A. chorioidea posterior medialis and A. tecti at the pontomesencephalic rim; *13* Rr. ad pontem; *14* as 13, short types; *15* intrapontine arteries in the depth; *16* as 13; *17* AICA; *18* Aa. vertebrales, junction; *19* A. spinalis anterior

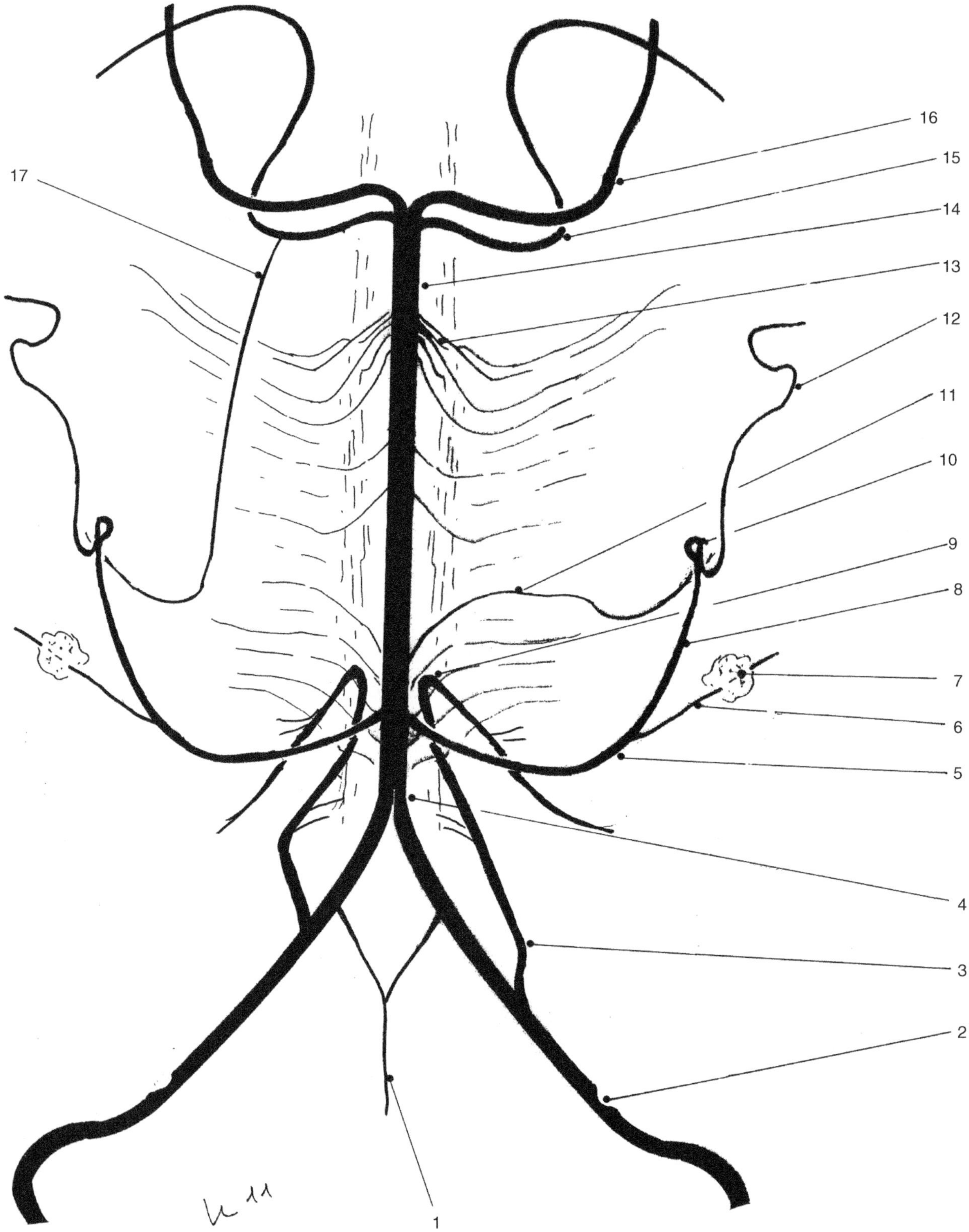

Fig. 14.4 Continuation of Fig. 14.3. *1* A. spinalis anterior; *2* usual stenosis of A. vertebralis at its dural penetration point; *3* PICA; *4* Aa. vertebrales, junction; *5* AICA; *6* as 5, inferior branch; *7* as 5, choroid segment of the inferior branch; *8* as 5, superior branch; *9* PICA. choroid point. inconstant; *10* AICA, meatal loop; *11* A. labyrinthi; *12* AICA, trigeminal loop; *13* Rr. ad pontem; *14* A. basilaris; *15* A. cerebelli superior; *16* A. cerebri posterior; *17* A. labyrinthi, inconstant finding

Fig. 14.5 Phlebogram. Schematized drawing of anatomical casts by Duvernoy 1975 and others (especially Stephens and Stilwell 1969, p 157). Cerebral structures added. *1* dorsal medullary vein; *2* paraolivar vein; *3* V. cerebellaris horizontalis, inconstant connection with 4; *4* anterolateral pontine veins; *5* V. petrosa superior; *6* V. mesencephalica lateralis; *7* V. basalis (Rosenthal); *8* Vv. tecti; *9* V. magna (Galeni); *10* V. cerebellaris praecentralis; *11* V. vermis superior; *12* tributary veins of 11; *13* V. vermis inferior

14.3 Further Imaging (Figs. 14.6, 14.7, 14.8, 14.9, 14.10, 14.11, 14.12, and 14.13)

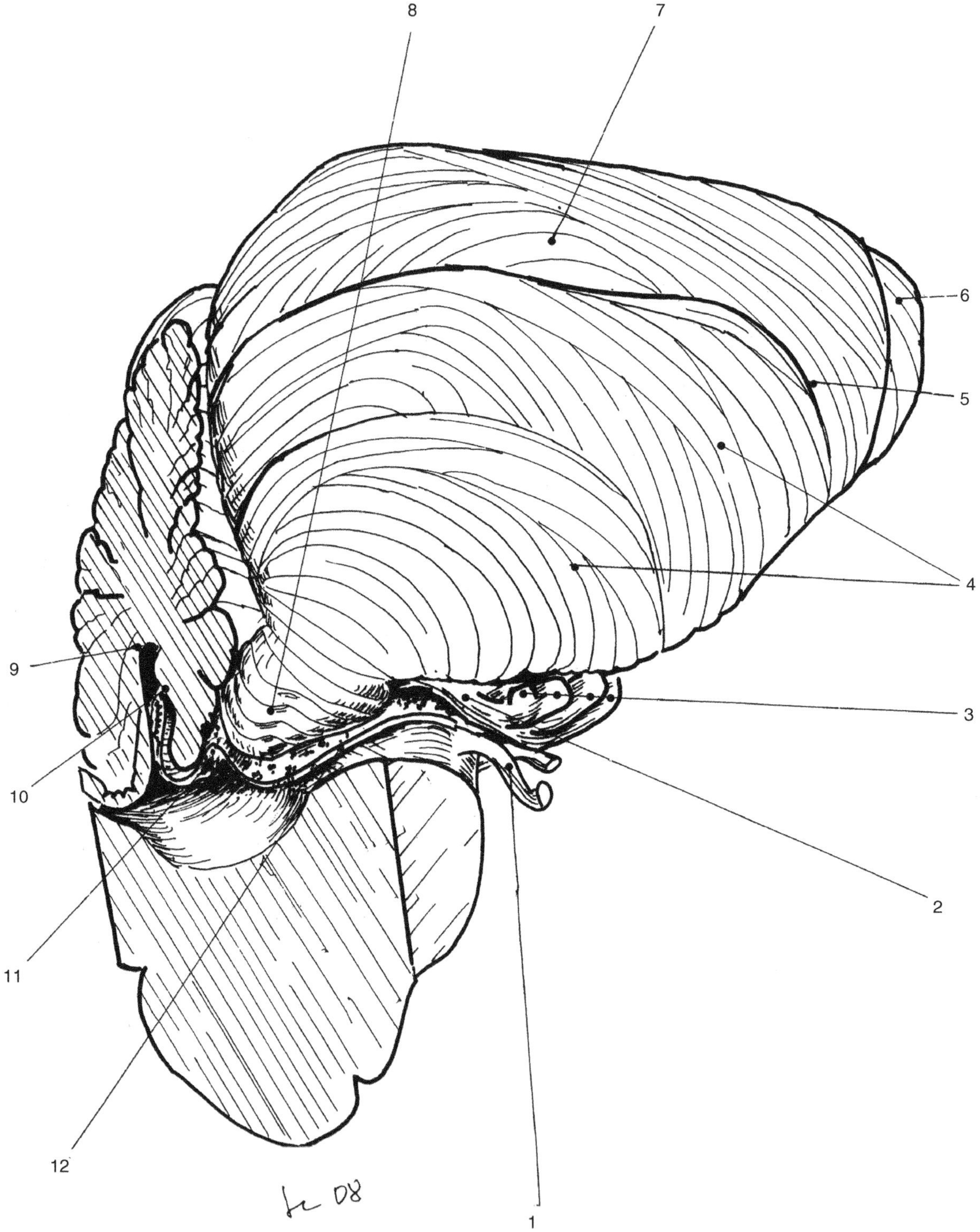

Fig. 14.6 *Anatomical brain dissection* similar to neuronavigation. Details added. *1* N. statoacusticus, *2* pedunculus flocculi, *3* flocculus, *4* lobulus biventer cerebelli, *5* fissura horizontalis cerebelli, *6* lobulus semilunaris superior, *7* lobulus semilunaris inferior, *8* tonsilla, *9* fastigium, *10* velum medullare posterius, insertion, *11* plexus compartment of velum medullare posterius, *12* velum medullare posterius, hypo/avascular compartment («nidus avis») 11 and 12 see Fig. 1.13

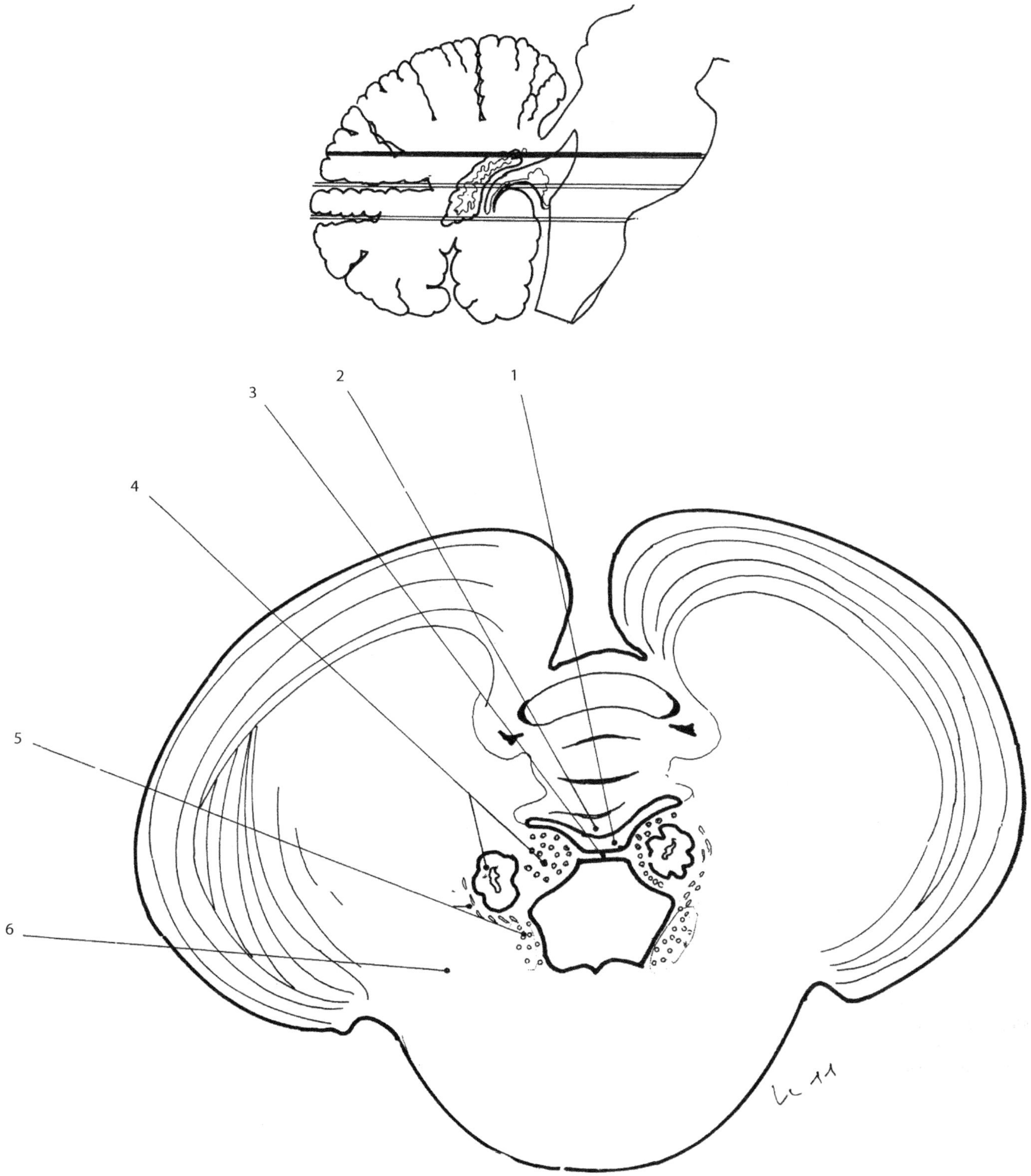

Fig. 14.7 Central area of cerebellum. MRT, details added. *1* Fissura horizontalis cerebelli, anterior segment, *2* lobulus centralis cerebelli, *3* velum medullare anterius, *4* pedunculus cerebellaris superior et nucleus dentatus, *5* pedunculus cerebellaris inferior, *6* pedunculus cerebellaris medius

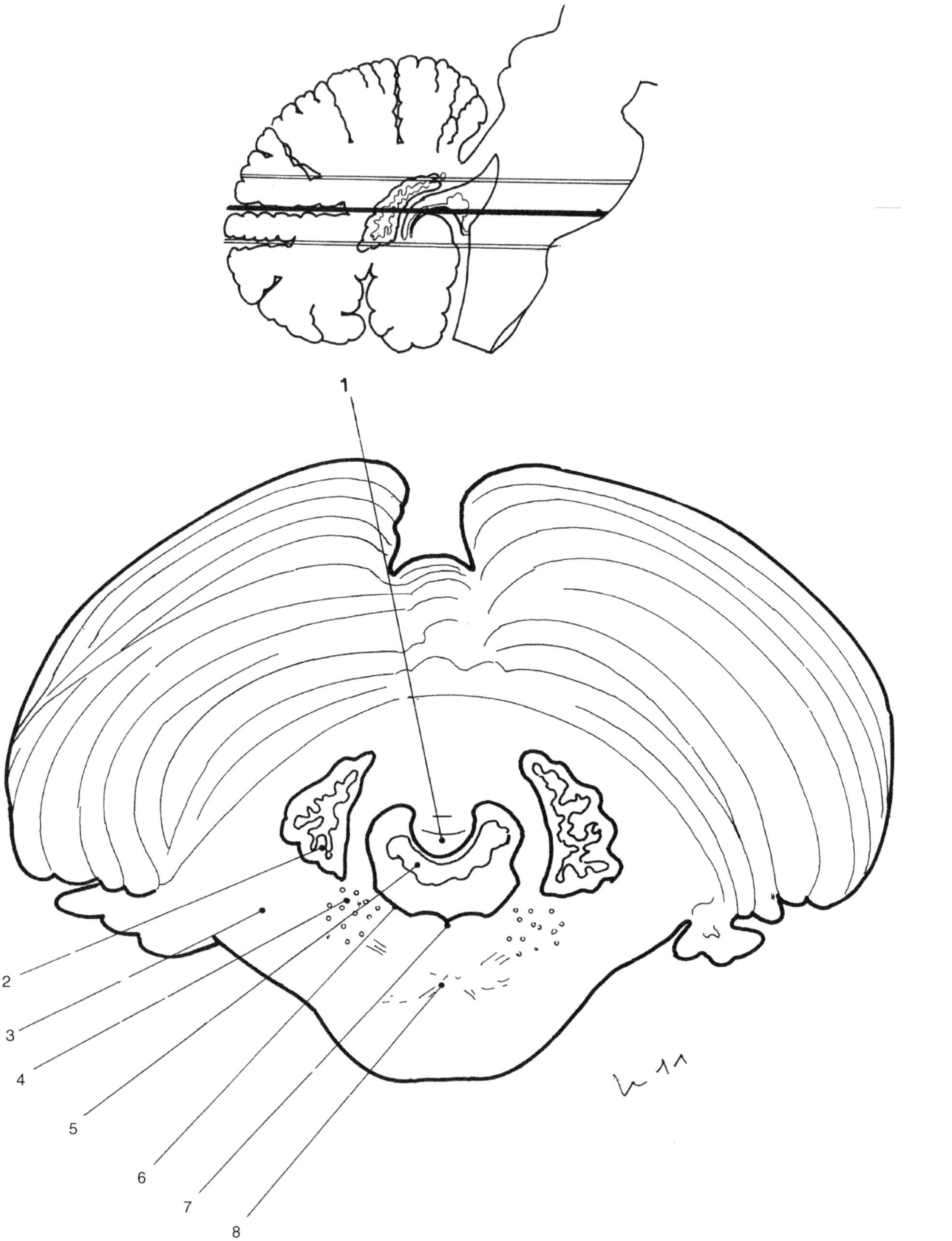

Fig. 14.8 Continuation of Fig. 14.7. *1* Uvula vermis, *2* nucleus dentatus, *3* pedunculus cerebellaris medius, *4* pedunculus cerebellaris inferior, *5* plexus choroideus, *6* fossa rhomboidea, *7* sulcus medianus, *8* decussatio olivarum

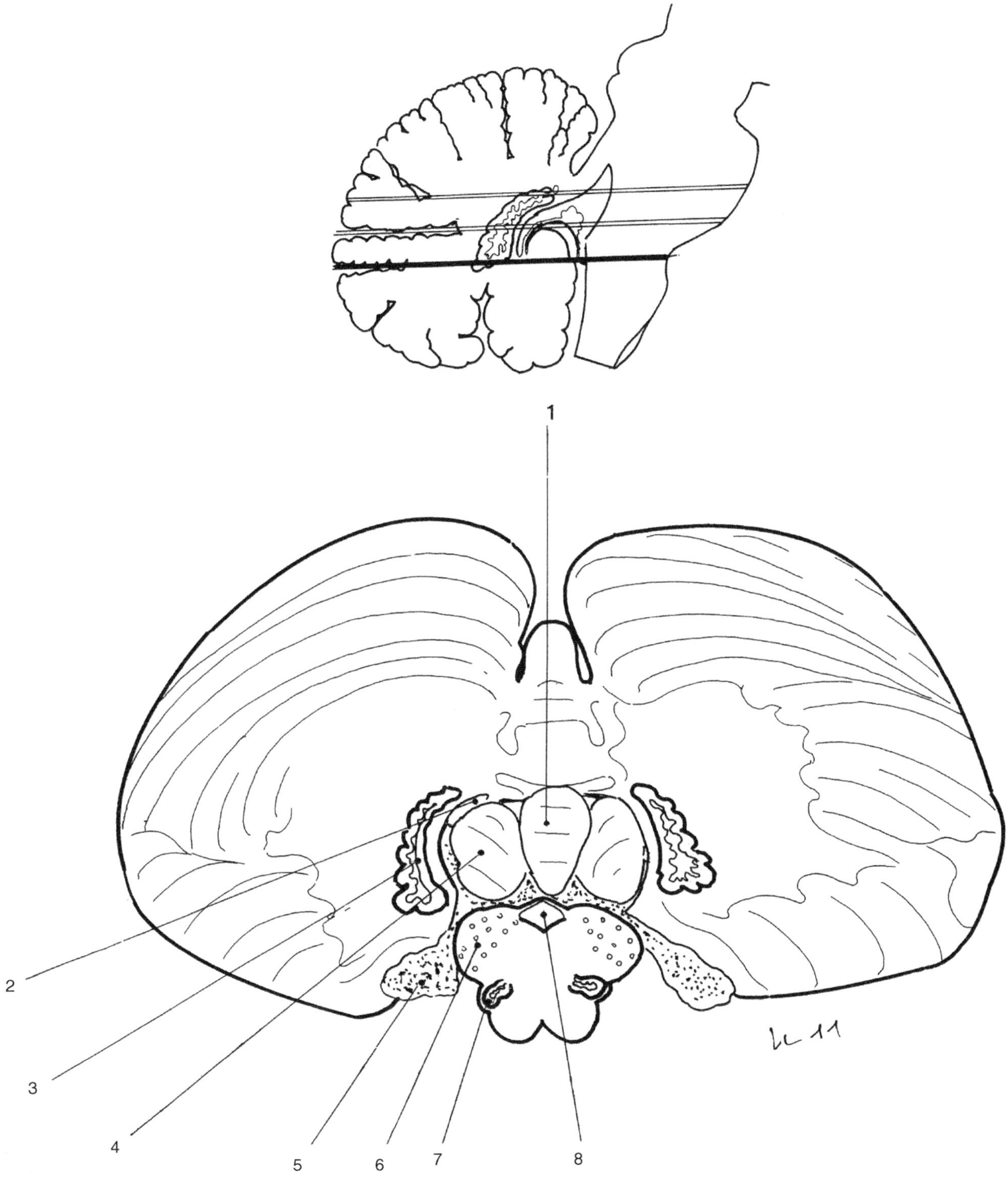

Fig. 14.9 Continuation of Fig. 14.8. *1* Uvula vermis, *2* recessus dorsolateralis, *3* nucleus dentatus, *4* tonsilla, *5* plexus choroideus, *6* pedunculus cerebellaris inferior, *7* oliva, *8* fourth ventricle

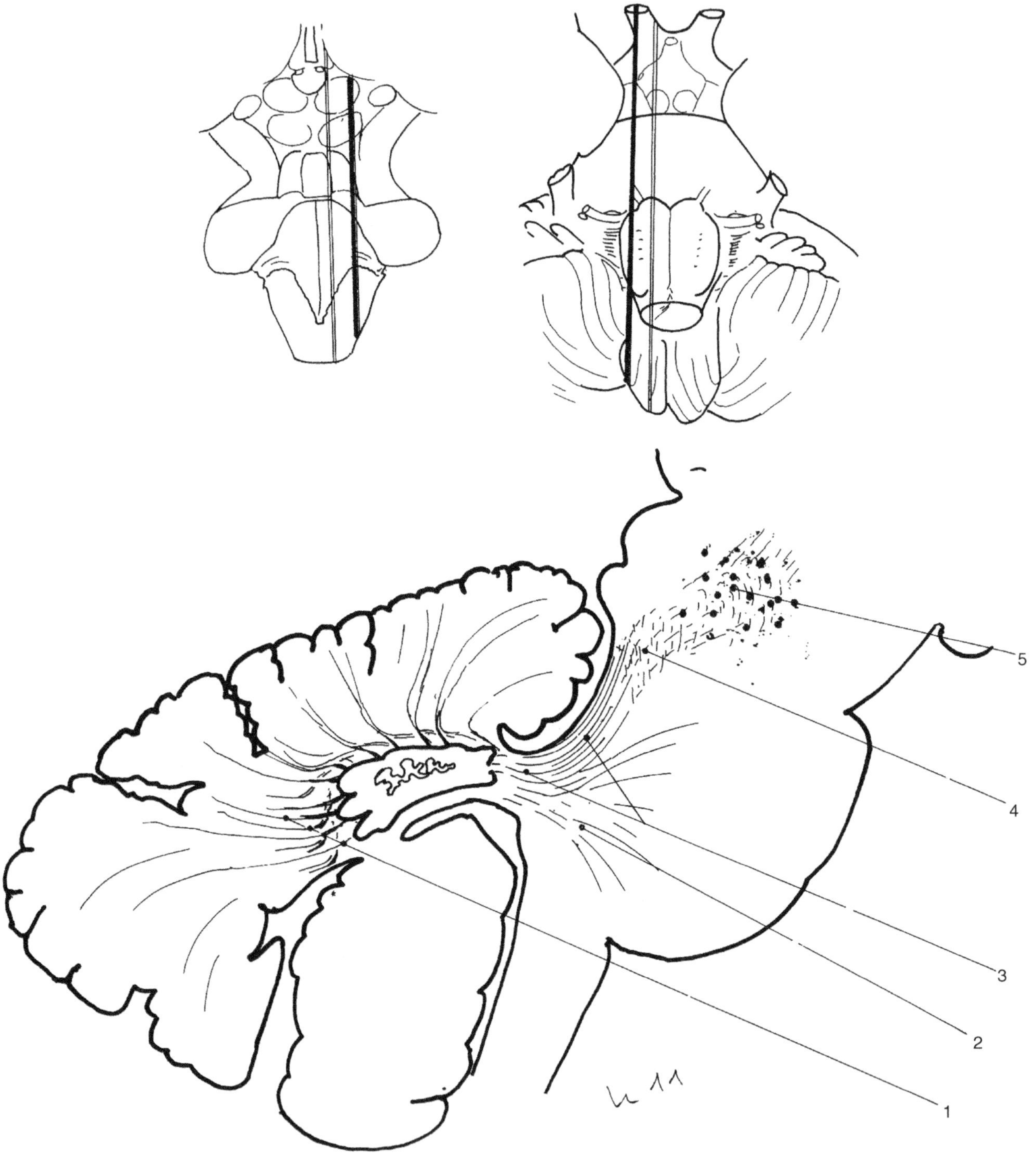

Fig. 14.10 Continuation of Fig. 14.9. *1* Afferent fiber bundles of nucleus dentatus and bundles of pedunculus cerebellaris medius which are surrounding nucleus dentatus, *2* fiber bundles of pedunculus cerebellaris medius, *3* pedunculus cerebellaris superior (efferent fiber bundle of nucleus dentatus), *4* decussatio of 3 *5* nucleus ruber

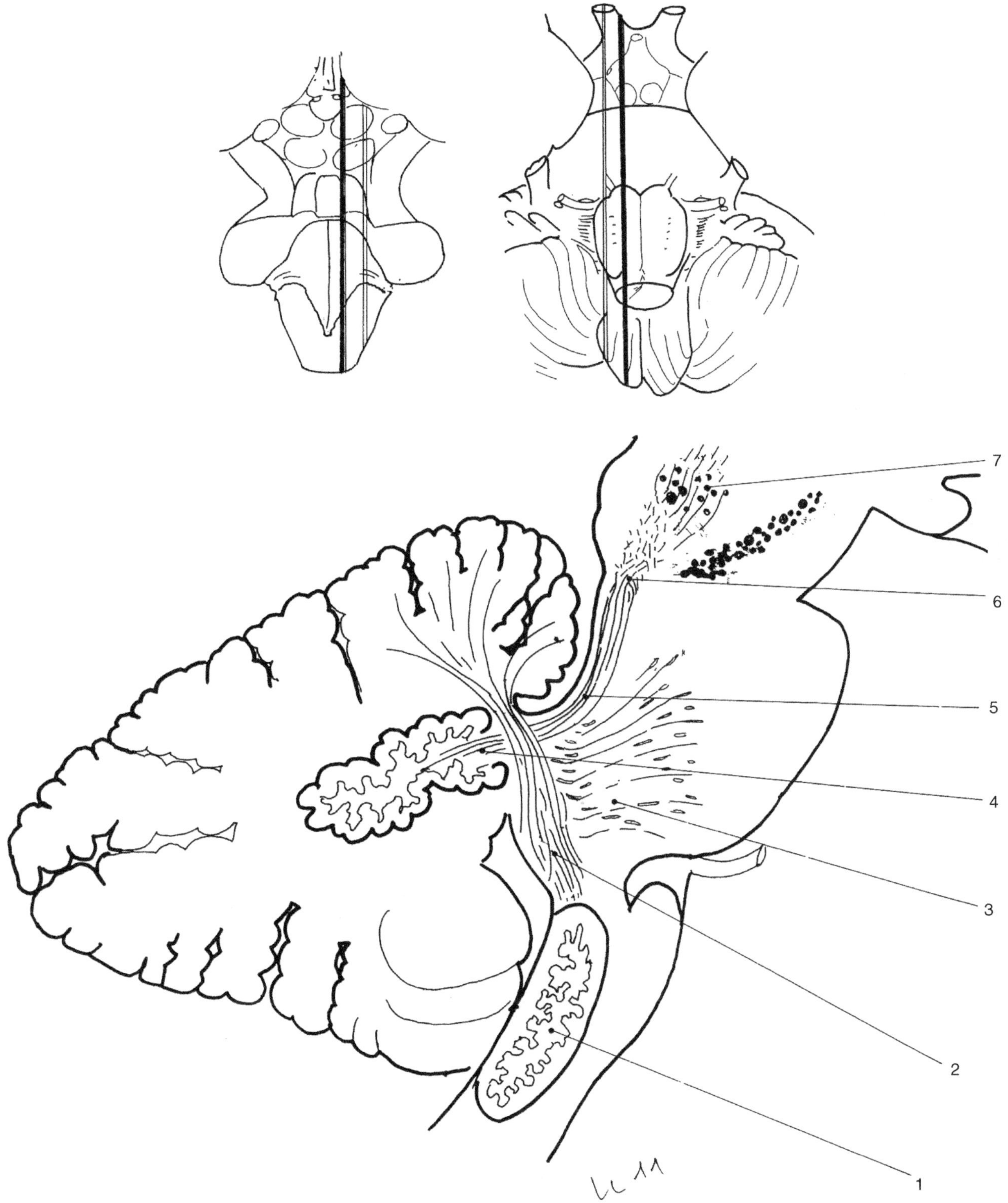

Fig. 14.11 Continuation of Fig. 14.10. *1* Oliva, *2* pedunculus cerebellaris inferior, *3* pedunculus cerebellaris medius, *4* pedunculus cerebellaris superior, *5* as 4, *6* decussatio of 4, *7* nucleus ruber

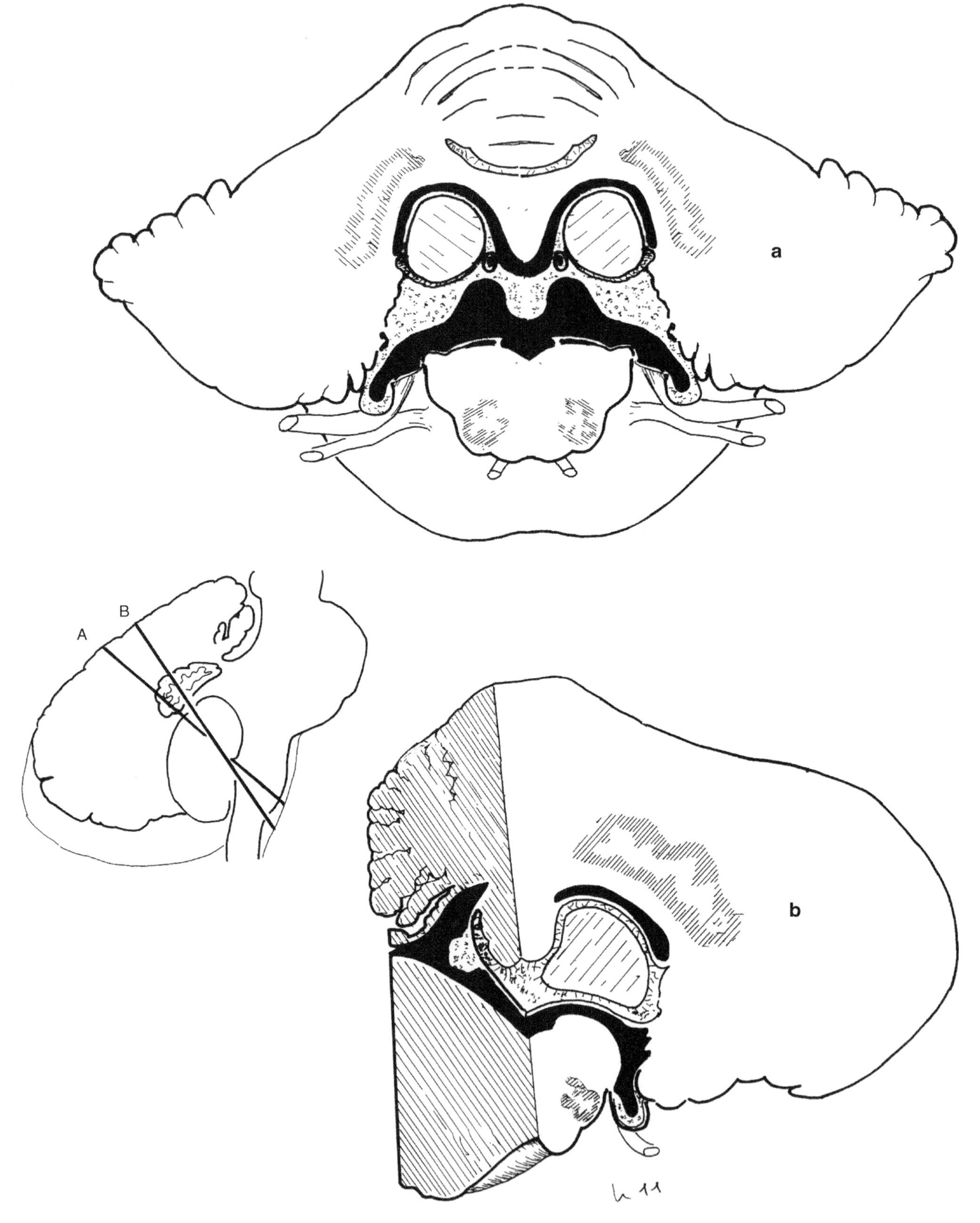

Fig. 14.12 Anatomical dissection for assisting imagings. It presents more details than imagings, especially the presentations of the narrow space of cisterna cerebellomedullaris between tonsilla and velum medullare post and the underlying narrow space of recessus dorsolateralis of the fourth ventricle. Here are defined these layers by shifting it, similar to the surgical approach to the fourth ventricle. **a** Anatomical dissection for assisting MRT. **b** Anatomical dissection for assisting neuronavigation

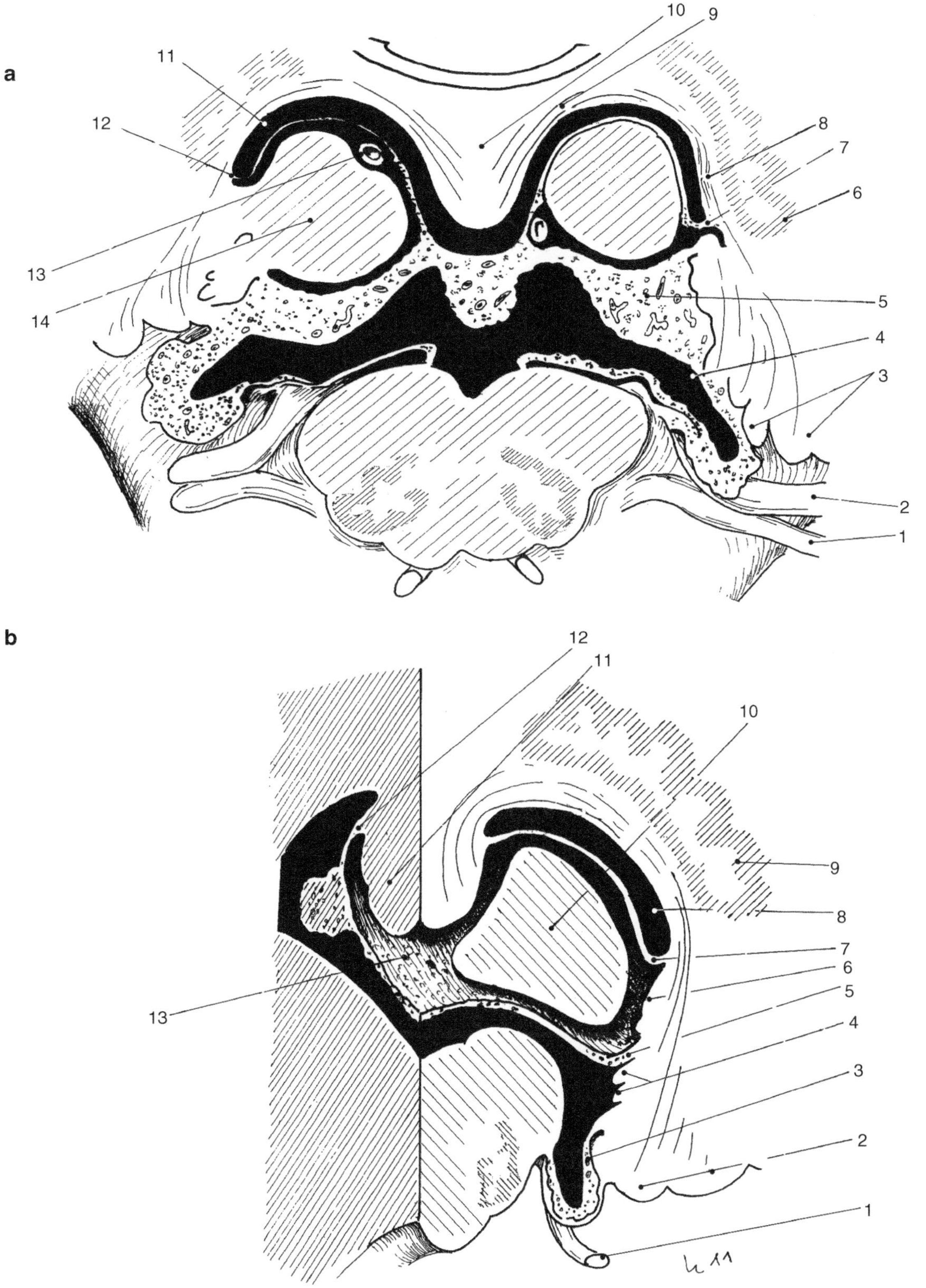

Fig. 14.13 As Fig. 14.12, sectional enlargement. A *1* N. facialis, *2* N. statoacusticus, *3* flocculus, *4* recessus lateralis, *5* plexus choroideus (compartment of velum medullare posterius), *6* nucleus dentatus, *7* velum medullare posterius, *8* tractus flocculonodularis, *9* as 8, *10* uvula vermis, *11* recessus dorsolateralis of the fourth ventricle (Lang), *12* as 7, *13* PICA, *14* tonsilla. B *1* N. glossopharyngeus, *2* flocculus, *3* plexus choroideus, *4* pedunculus flocculi, *5* velum medullare posterius, hypo/avascular compartment («nidus avis,» bed of tonsilla), *6* cisterna cerebellomedullaris, lateral wall, *7* as 5, *8* recessus dorsolateralis, *9* nucleus dentatus, *10* tonsilla, *11* uvula, *12* velum medullare posterius, *13* velum medullare posterius, plexus compartment thin walled

15 Anatomical Base for Surgery

15.1 Extradural Topography

15.1.1 Extracranial Layers (Figs. 15.1, 15.2, and 15.3)

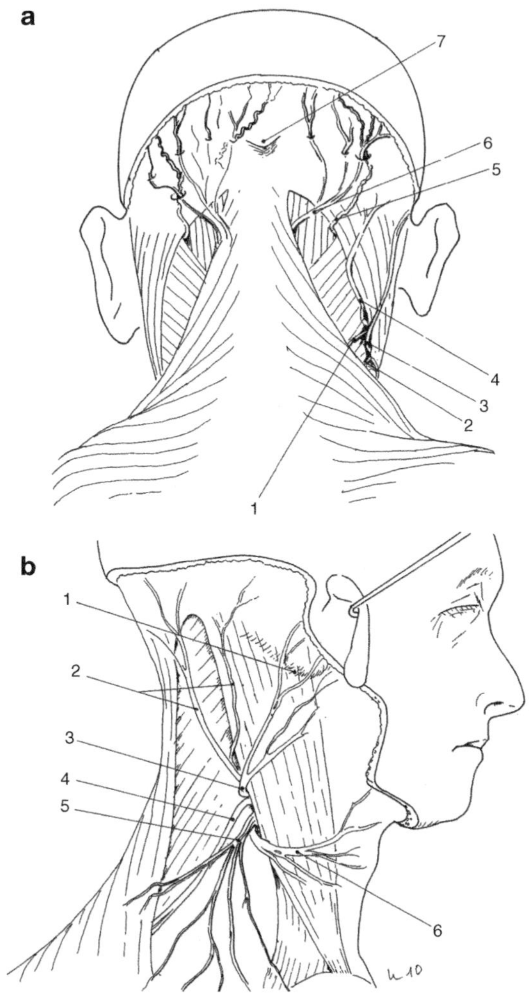

Fig. 15.1 Superficial nerves, vessels, and muscles of the neck. Nerves should be preserved at surgery and – if possible – large vessels. (**a**) *1* N. accessorius, *2* Nn. supraclaviculares, *3* N. auricularis magnus, *4* N. occipitalis minor, *5* N. occipitalis tertius, *6* N. occipitalis major and vessels, *7* inion. (**b**) *1* processus mastoideus, *2* N. occipitalis minor, *3* N. auricularis magnus, *4* N. accessorius, *5* Nn. supraclaviculares, *6* N. cutaneus colli fasciae omitted

W. Seeger, J. Zentner, *Anatomical Basis of Cranial Neurosurgery*, https://doi.org/10.1007/978-3-319-63597-2_15

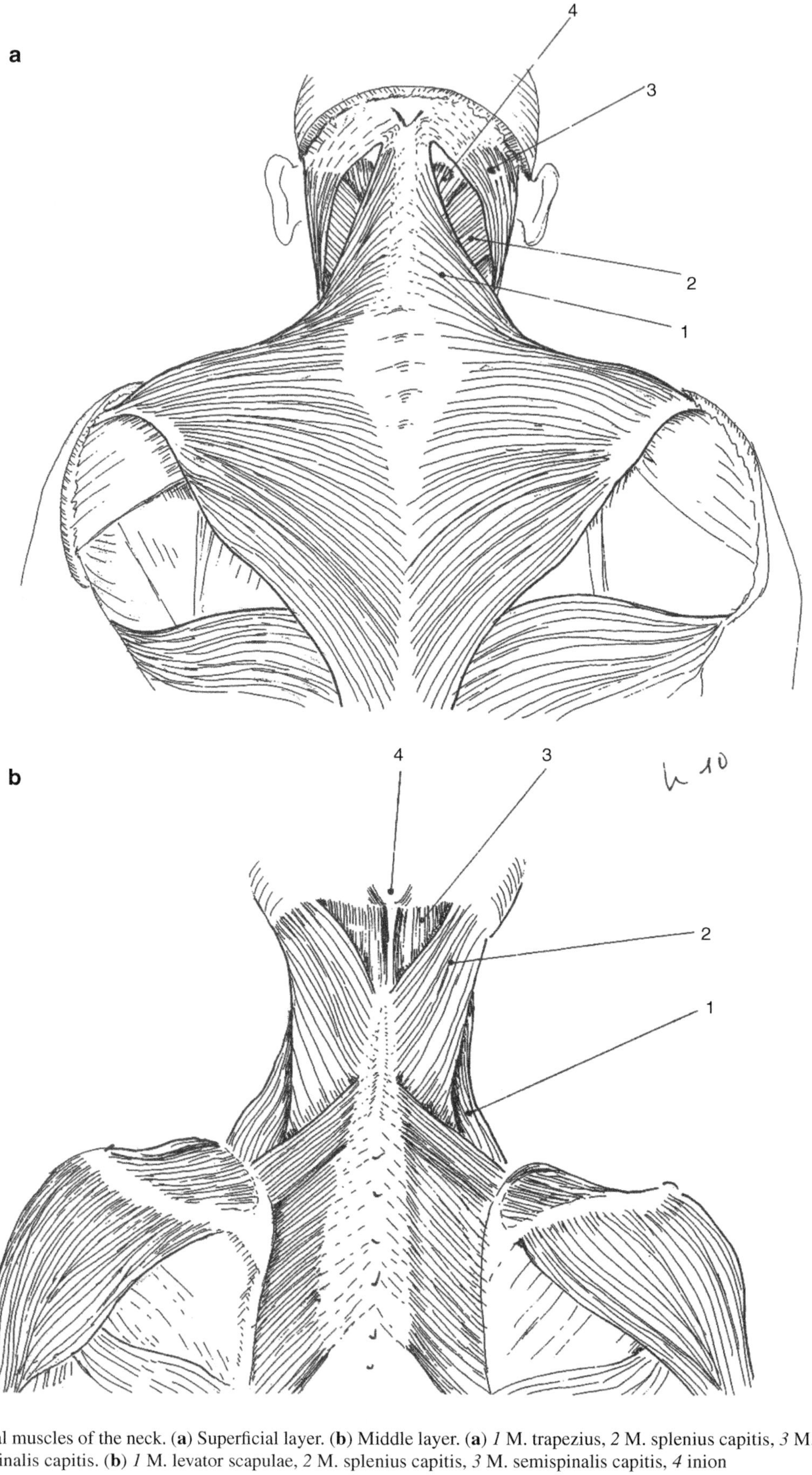

Fig. 15.2 Superficial muscles of the neck. (**a**) Superficial layer. (**b**) Middle layer. (**a**) *1* M. trapezius, *2* M. splenius capitis, *3* M. sternocleidomastoideus, *4* M. semispinalis capitis. (**b**) *1* M. levator scapulae, *2* M. splenius capitis, *3* M. semispinalis capitis, *4* inion

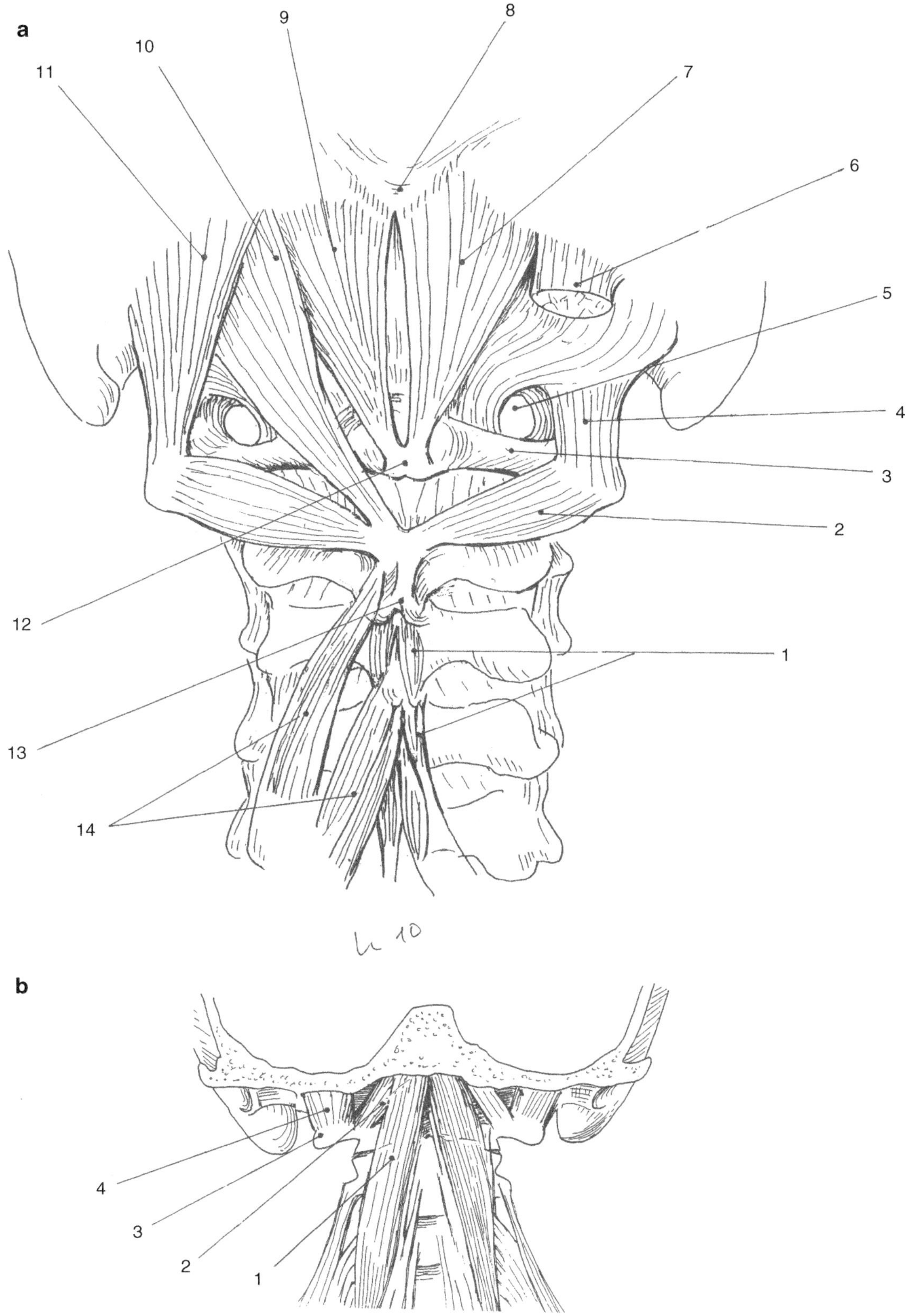

Fig. 15.3 Muscles in the depth. (**a**) Dorsal. (**b**) Ventral. (**a**) *1* Mm. interspinales; *2* M. obliquus capitis inferior; *3* arcus atlantis posterior; *4* M. rectus capitis laterali; *5* gap for A. vertebralis; *6* M. rectus capitis posterior major; *7* M. rectus capitis posterior minor; *8* inion; *9* as *7*; *10* as *6*; *11* M. obliquus capitis superior, insertion at a bony crest (aspects of craniotomy see Fig. 6.1); *12* tuberculum atlantis posterius; *13* processus spinosus axis; *14* M. multifidus. (**b**) *1* M. longus capitis, *2* M. rectus capitis anterior, *3* processus transversus atlantis, *4* M. rectus capitis lateralis

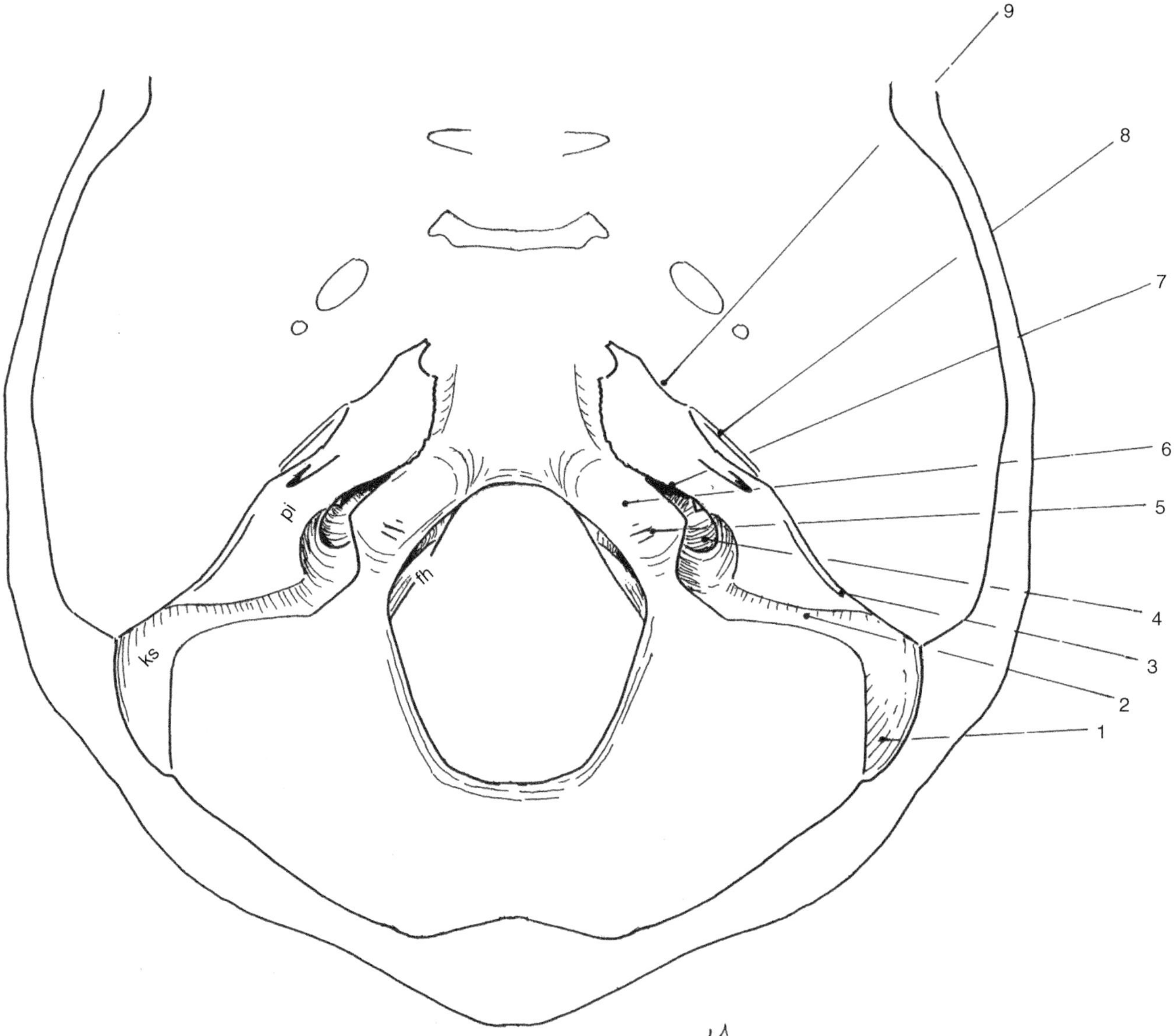

Fig. 15.6 Basal inner relief. Fossa cranii posterior. *fh* foramen n. hypoglossi, *ks* knee of sulcus transversus, *pi* porus acusticus internus, *1* sulcus transversus, *2* sulcus sigmoideus, *3* sulcus petrosus superior, *4* fossa jugularis, posterior roof of foramen jugulare (enclosing bulbus superior v. jugularis.internae), *5* boni sulci of tuberculum jugulare, covered by Nn. IX to XI, *6* tuberculum jugulare, *7* foramen jugulare, anterior segment (containing Nn. IX to XI and Sinus petrosus inferior), *8* as *3*, *9* incisura trigeminalis

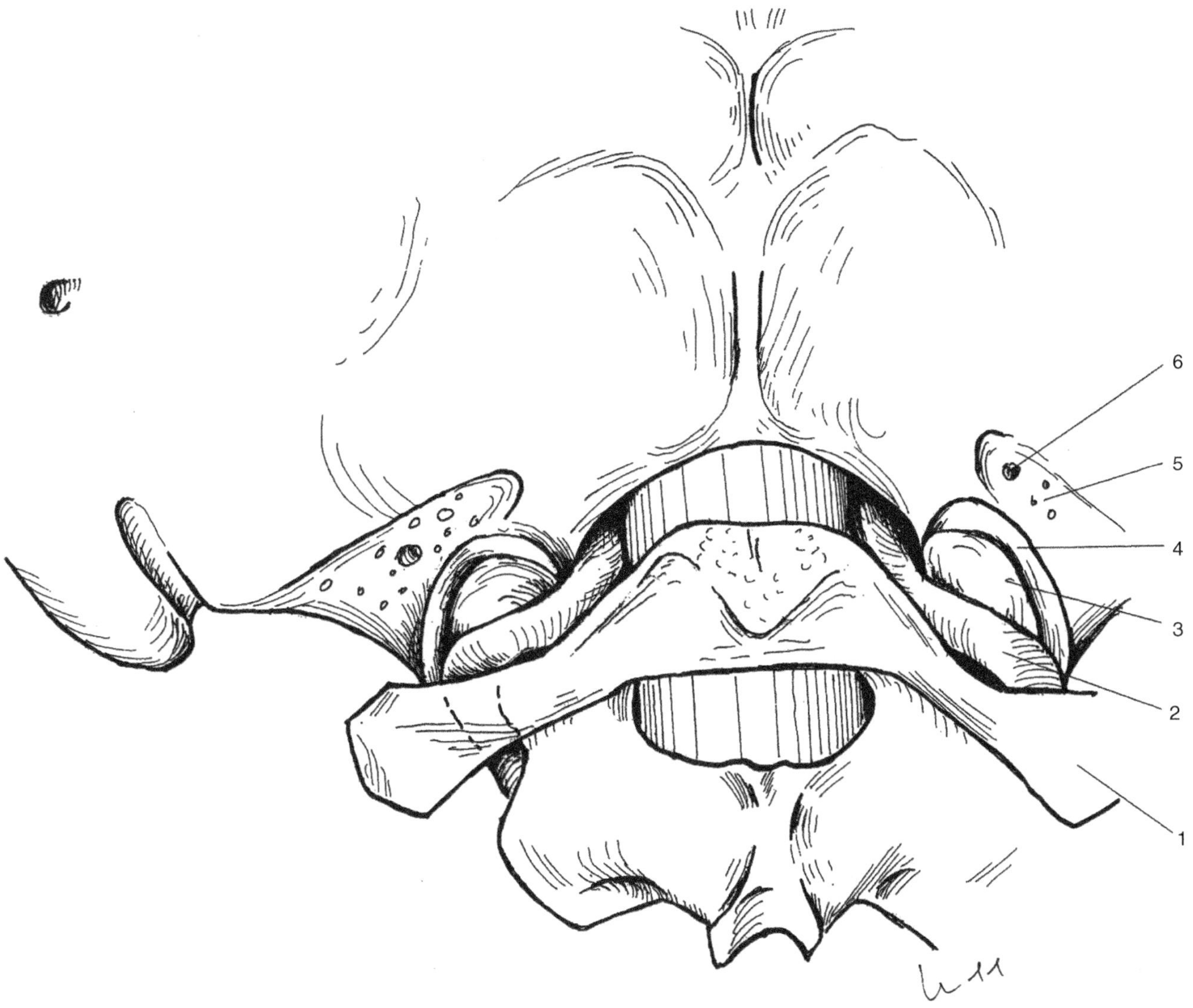

Fig. 15.7 Dorsal atlantooccipital bony structures (anatomical base for para- and transcondylar surgical approaches, Fig. 16.25 ff). Anatomical base of transcondylar surgical approaches (Fig. 6.9 ff). A. vertebralis and dura added. *1* processus transversus atlantis, *2* A. vertebralis, *3* processus articularis atlantis superior, *4* condylus occipitalis, *5* fossa paracondylaris, *6* emissarium condylare

15.1.3 Dura and Sinuses (Figs. 15.8, 15.9, and 15.10)

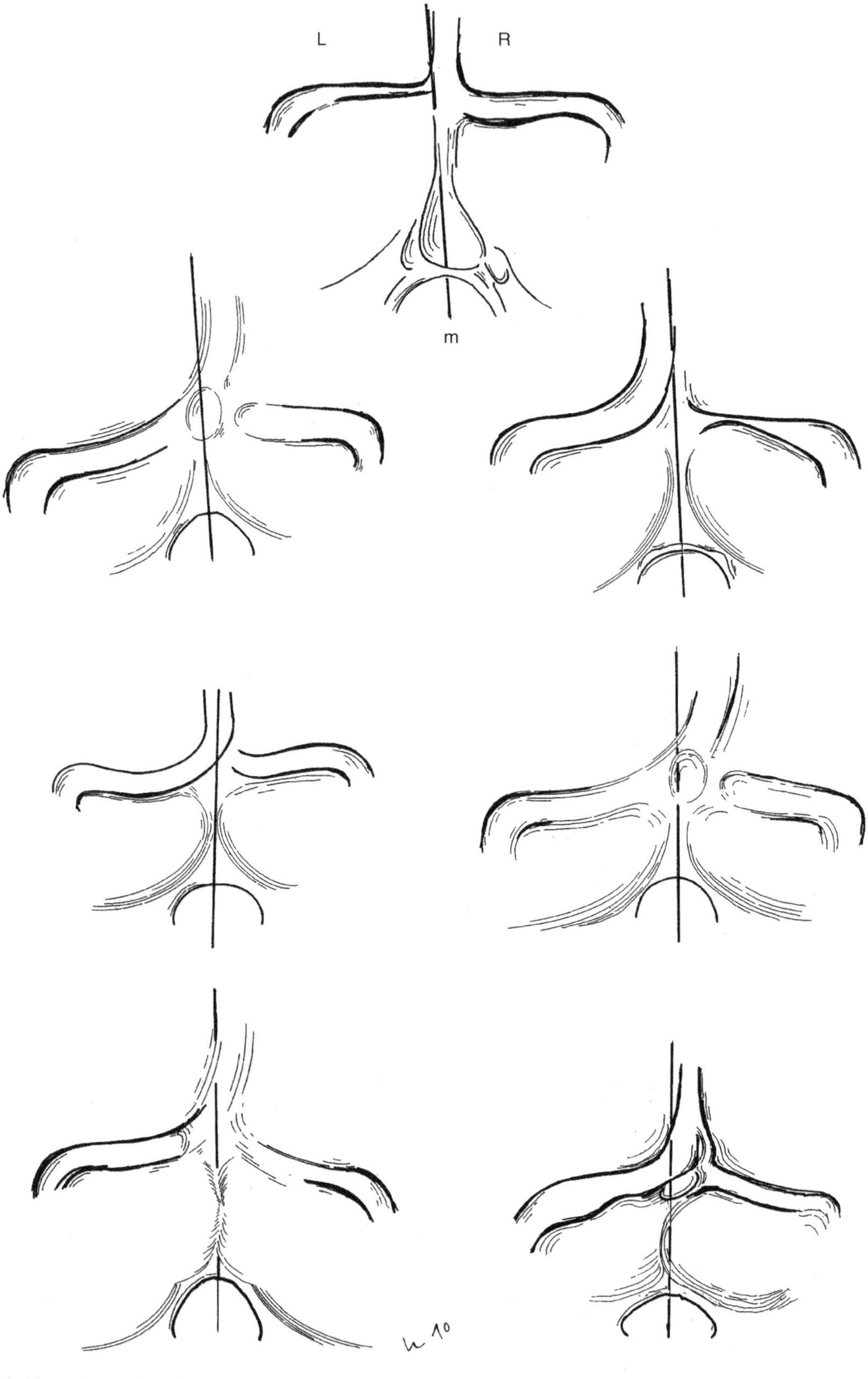

Fig. 15.8 Occipital bony impressions by the dural sinuses. Examples. Bony sulci should be taken in consideration using imagings. Unexpected openings of a sinus could be avoided. *m* midline

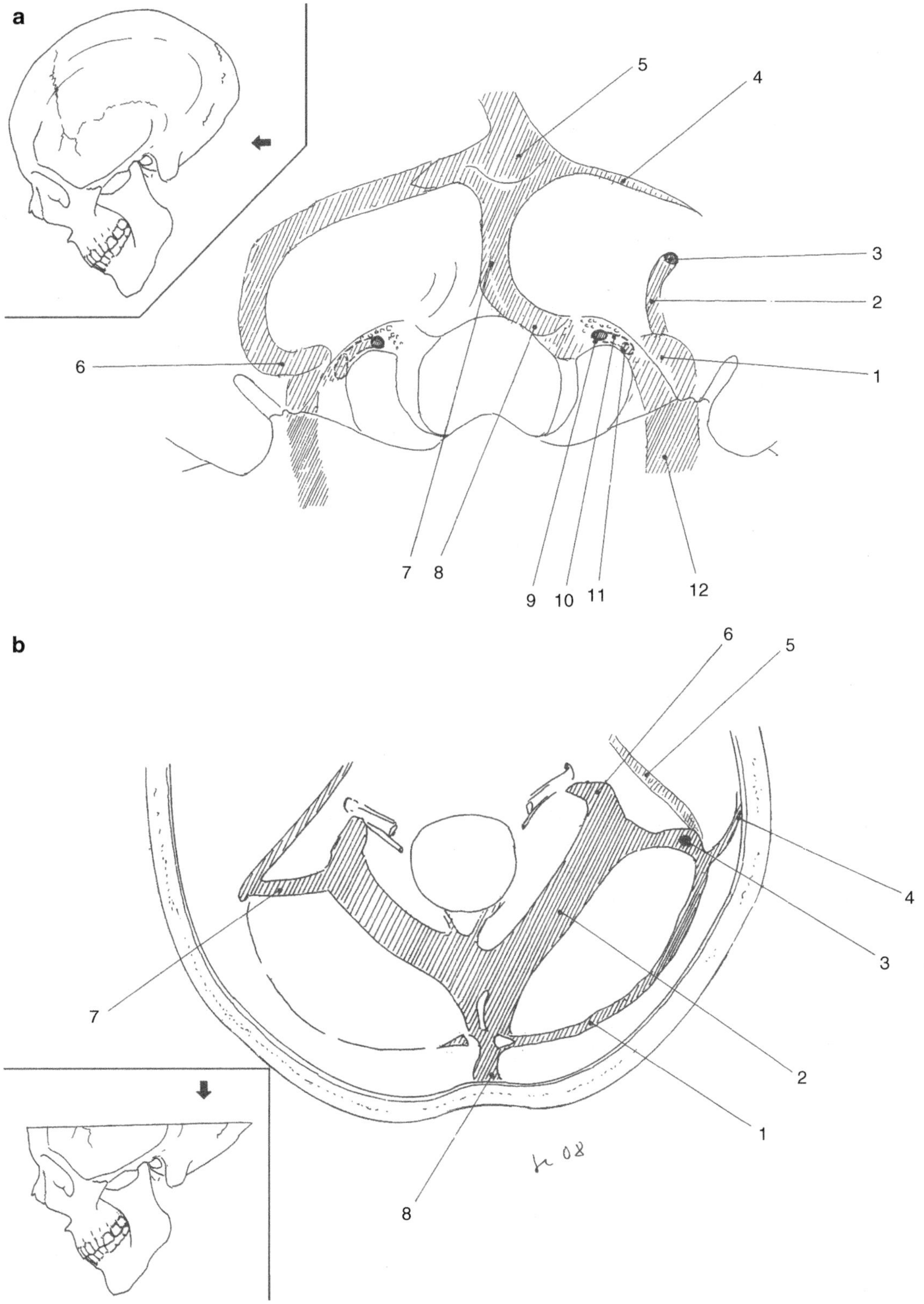

Fig. 15.9 Rare variants of occipital and suboccipital sinuses and its bony impressions. (**a**) Example of a cadaver skull dissection (transparent presented). (**b**) Example of a cadaver head dissection (Lang 1979, p 239). (**a**) *1* for. jugulare; *2* bony sulcus between foramen mastoideum and foramen jugulare; *3* foramen mastoideum; *4* sulcus transversus; *5* protuberantia occipitalis interna marking confluens sinuum; *6* sulcus sigmoideus; *7* crista occipitalis widened to a sulcus; *8* sulcus marking the widened sinus marginalis; *9* emissarium condylare; *10* as *9*, its connection to fossa jugularis; *11* mouth of emissarium condylare at fossa jugularis; *12* V. jugularis interna added. (**b**) According to Lang, schematized. *1* sinus transversus, *2* atypical sinus, *3* knee of sinus transversus, *4* vein of Labbé, *5* sinus petrosus superior, *6* bulbus superior of V. jugularis interna, *7* sinus sigmoideus, *8* sinus sagittalis superior

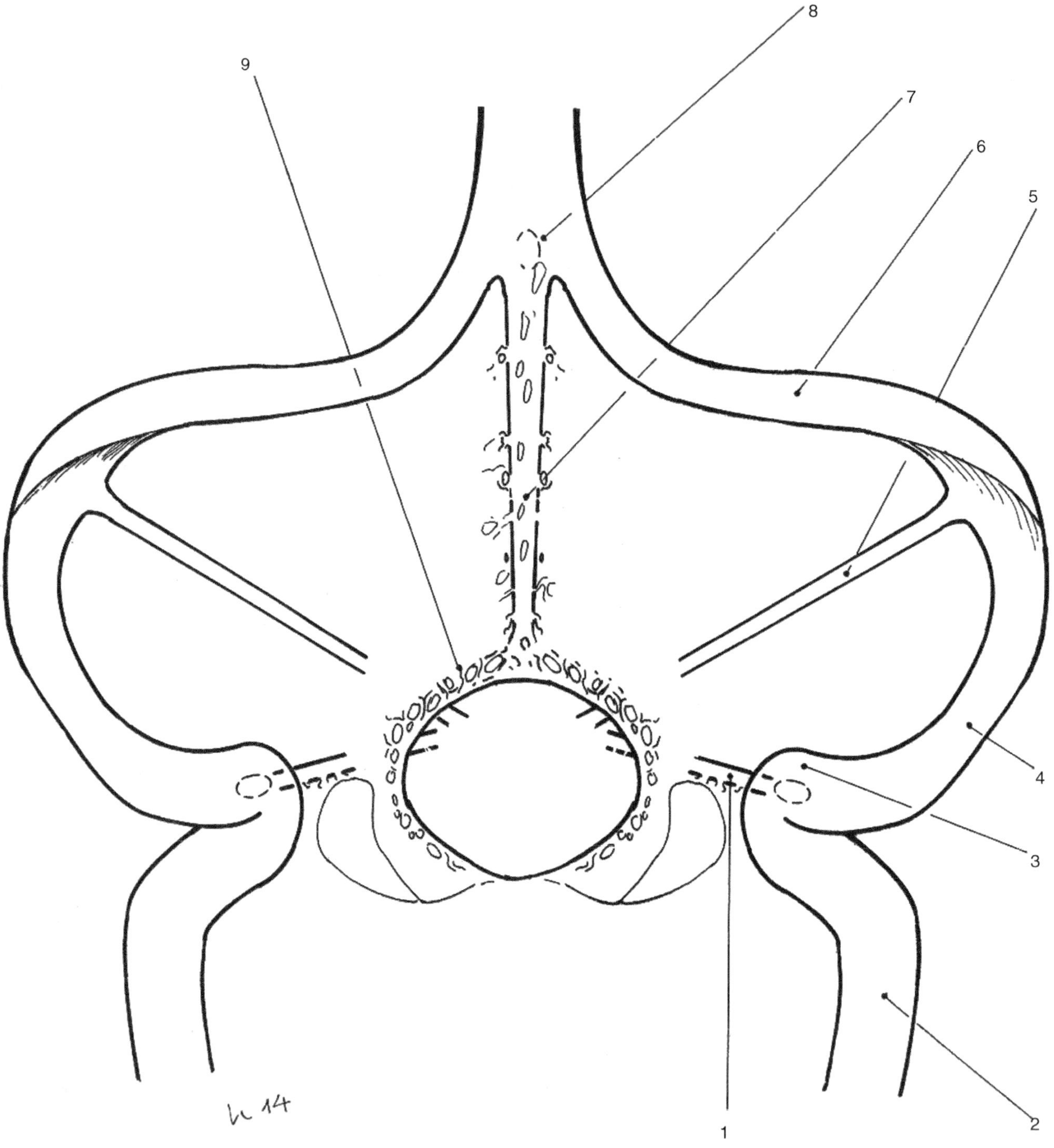

Fig. 15.10 Infratentorial sinuses. Synopsis. *1* Sinus petrosus inferior; *2* V. jugularis interna; *3* bulbus superior of V. jugularis interna; *4* sinus sigmoideus; *5* sinus petrosus superior; *6* sinus transversus; *7* sinus occipitalis, wide variant; *8* confluens sinuum; *9* sinus marginalis

15.2 Intradural Topography

15.2.1 Survey (Figs. 15.11, 15.12, 15.13, 15.14, 15.15, 15.16, 15.17, 15.18, 15.19, 15.20, and 15.21)

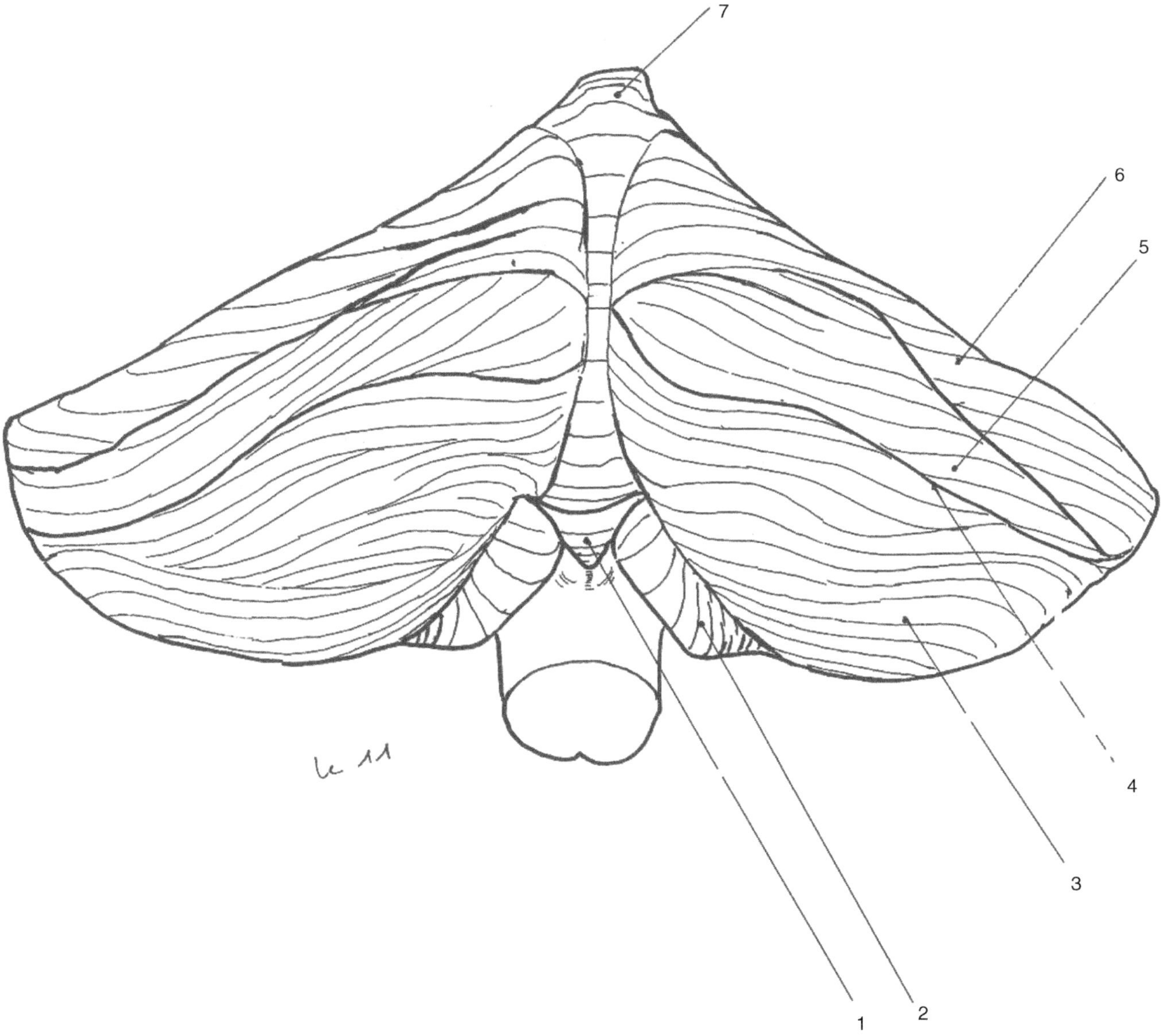

Fig. 15.11 Dorsal relief of the cerebellum. Its main part is lobulus biventer. Between lobulus biventer and lobulus semilunaris inferior is located fissura horizontalis cerbelli. This is the deepest fissure of the cerebellar hemisphere. Along the ground of fissura horizontalis is running V. cerebellaris horizontalis. Close to the ground of fissura horizentalis is located nucleus dentatus (Fig. 3.8). *1* vermis, inferior segment (nodulus vermis), *2* tonsilla, *3* lobulus biventer, *4* fissura horizontalis, *5* lobulus semilunaris inferior, *6* lobulus semilunaris superior, *7* culmen

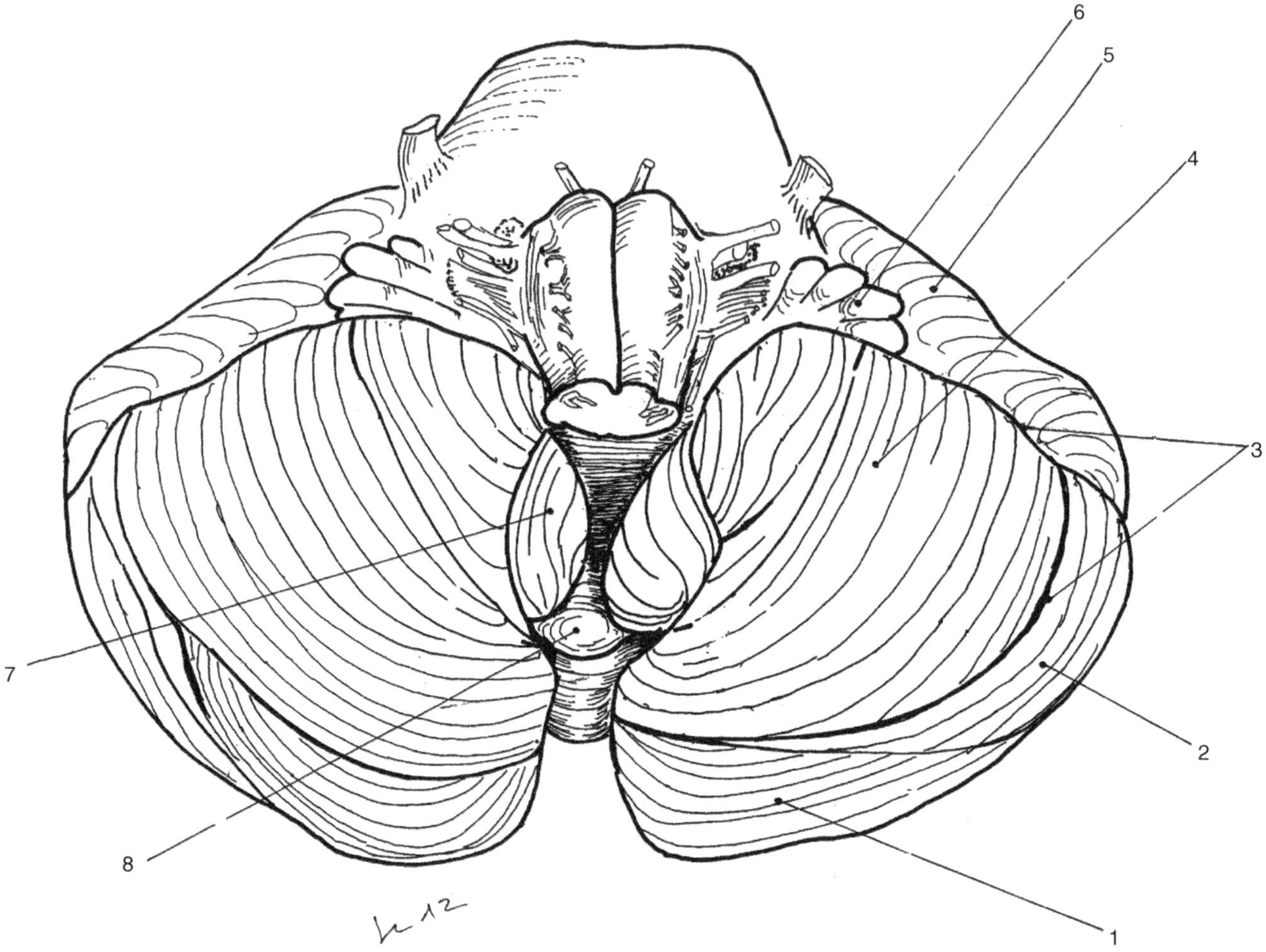

Fig. 15.12 Basal relief of the cerebellum and brain stem. Drawing for assisting neuronavigation with defining of fine structures (e.g., rhombencephalic exit points of cranial nerves). The relief of oliva is variable configurated and could be masked by periolivar veins. The inferior pole of oliva is a landmark during surgery for the location of the superior region of decussatio pyramidum (Fig. 3.9 ff) and for the inferior region of decussatio lemniscorum (Fig. 3.18 ff). The lateral margin of flocculus is located close to porus acusticus internus. *1* lobulus semilunaris superior, *2* lobulus semilunaris inferior, *3* fissura horizontalis, *4* lobulus biventer, *5* lobulus quadrangularis, *6* flocculus *7* tonsilla, *8* nodulus vermis

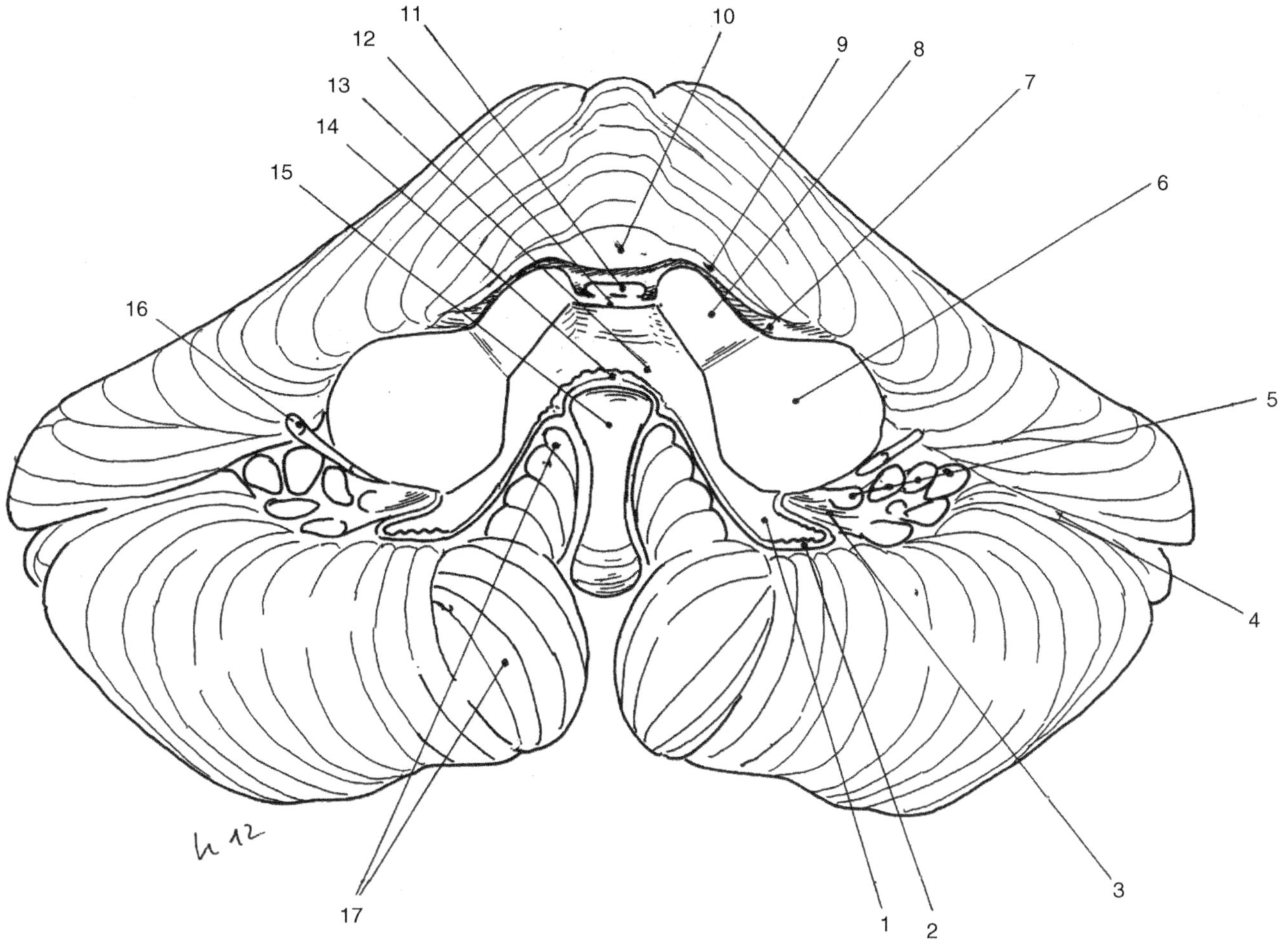

Fig. 15.13 Anterior dorsal and basal cerebellar relief. Characteristic configurated are the reliefs of lingula cerebelli, velum medullare anterius, lobulus centralis et paracentralis cerebelli, flocculus, uvula, and nodulus. It may be used for landmarks during intracisternal surgical approaches. *1* recessus lateralis, *2* plexus chorioideus, *3* pedunculus flocculi, *4* fissura horizontalis, *5* flocculus, *6* pedunculus cerebellaris medius, *7* as 4, *8* pedunculus cerebellaris superior, *9* ala lobuli centralis, *10* lobulus centralis, *11* lingula cerebelli, *12* velum medullare anterius, *13* fourth ventricle, *14* velum medullare posterius, *15* uvula vermis, *16* N. statoacusticus, *17* tonsilla

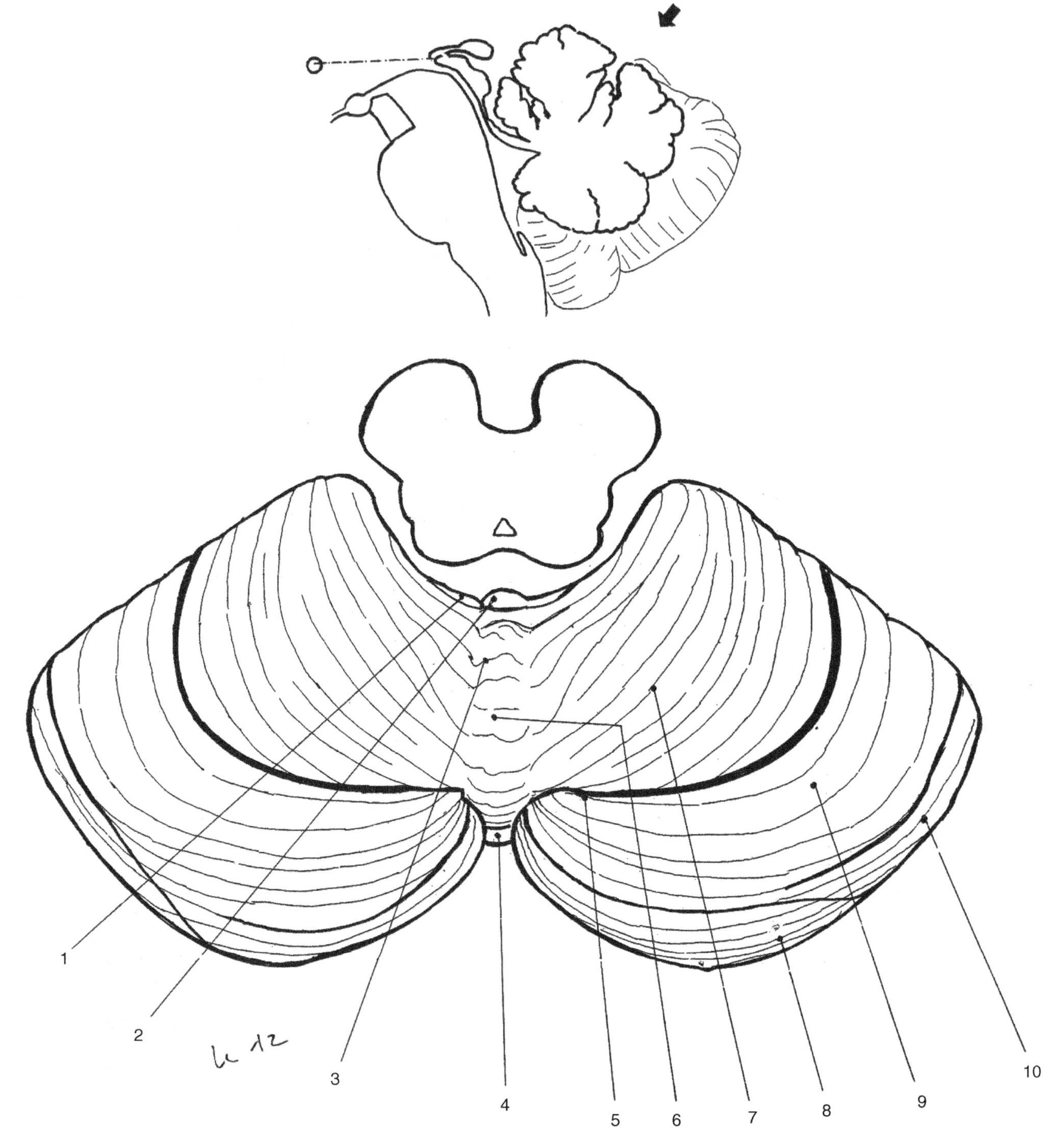

Fig. 15.14 Dorsal relief of the cerebellum. This region presents the subdural supracerebellar part of the surgical route to cisterna tecti and adjacent CSF spaces (Fig. 6.14). *1* ala lobuli centralis, *2* lobulus centralis, *3* culmen vermis, *4* declive, *5* fissura prima, *6* as *3*, *7* lobulus quadrangularis, *8* lobulus semilunaris inferior, *9* lobulus simplex, *10* lobulus semilunaris superior

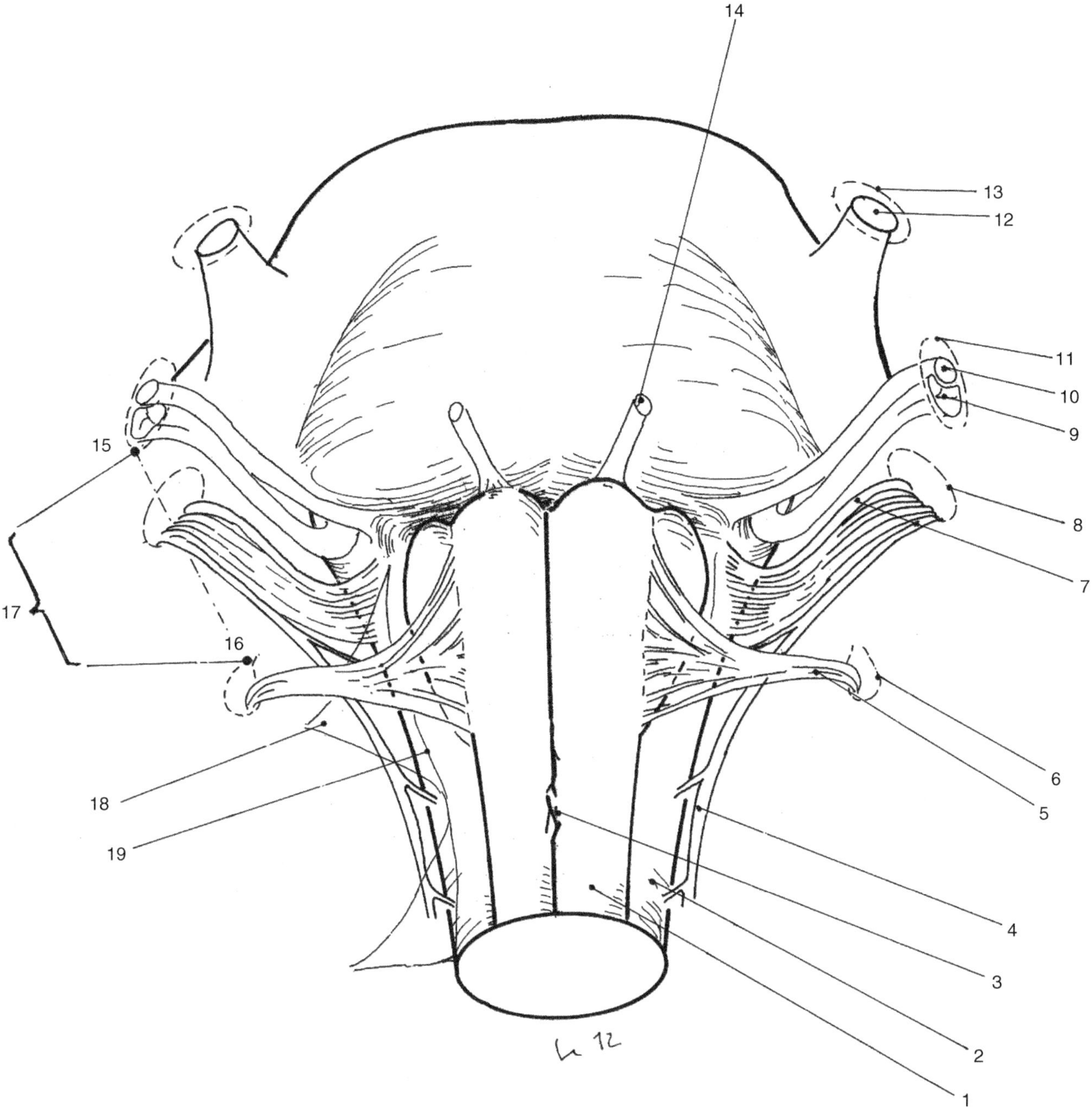

Fig. 15.15 Basal relief of the brain stem and cranial nerves. Projections of dural exit points of cranial nerves added. The brain stem exit areas of N. statoacusticus and N. glossopharyngeus are usually covered by plexus chorioideus (plexus chorioideus here omitted for presenting exit regions of N. statoacusticus and glossopharyngeus) and sometimes adherent with it. This area and its fine vessels are not exact to define by neuronavigation, but essential for pontomedullary surgical approaches. Connected with the plexus are arachnoid walls (Fig. 15.35). This is essential for surgery (Figs. 15.36 and 16.14) and is presented by anatomical dissections. *1* funiculus anterior; *2* funiculus lateralis; *3* decussatio pyramidum; *4* radix spinalis n. accessorii; *5* N. hypoglossus; *6* dural penetration point of 5, projection; *7* Nn. glossopharyngeus, vagus et accessorius; *8* For. jugulare, anterior segment, projection; *9* N. statoacusticus; *10* N. facialis (and N. intermedius); *11* porus acusticus internus, projection; *12* N. trigeminus; *13* tentorial gap, projection; *14* N. abducens; *15* posterior point of porus acusticus internus, projection; *16* anterior point of For. n. hypoglossi, projection; *17* interforaminal line (porus acust.int.-For. jug.-For. n. hypoglossi) dotted; *18* Lig. denticulatum; *19* insertion of 18 at the medulla oblongata, variable at this region

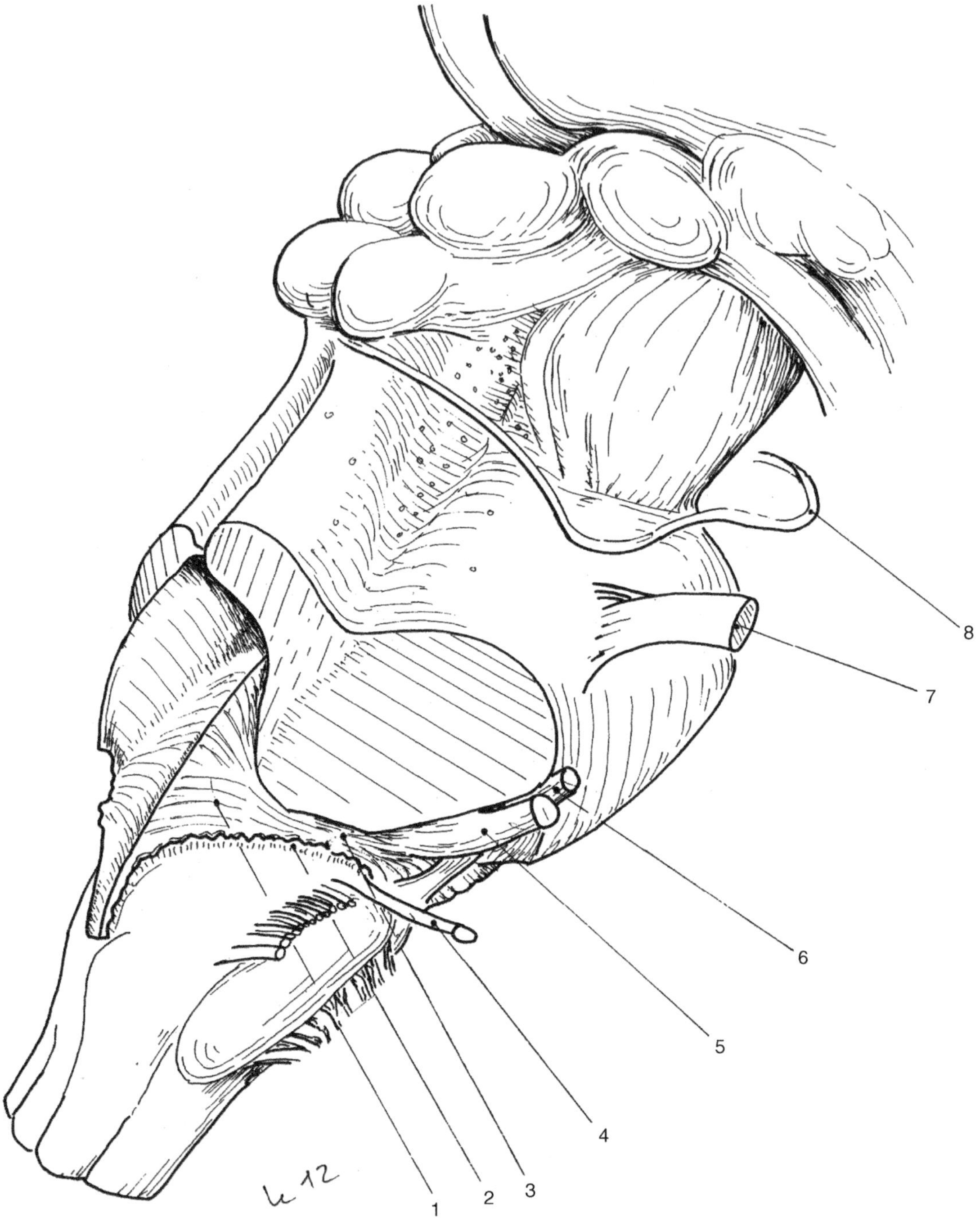

Fig. 15.16 Oblique viewing direction of the brain stem. Introduction for cerebral exit areas of cranial nerves. *N. trochlearis* is mapping the inferior margin of colliculi inferiores. The exit region of *N. trigeminus* presents numerous variable fascicles beween the pons and pedunculus cerebellaris medius («brachium pontis»). Portio minor n. trigemini may present a variable mixtum of motoric and sensoric fibers at the superior margin of portio major. Sometimes it is surrounding portio major (Rhoton). This region could present numerous pial veins. The terminal branch of AICA (trigeminal loop) and the loop of A. cerebelli superior could be located close to the exit region of N. trigeminus, as it is well known by Janetta's operation.The exit area of *N. statoacusticus* is located at the floor of the recessus lateralis (Figs. 2.19 and 2.30). This exit region and the exit region of N. glossopharyngeus could be covered by the inferior main branch of AICA. Veins are located at the floor of the recessus lateralis. For details see Klose 2000 and Klose and Sollmann 1998, 1999. The exit region of *N. facialis* is located at the fossa paraolivaris («supraolivaris»). Here N. facialis is flattened configurated. The cerebral exit areas of *N. facialis and N. gossopharyngeus* are separated by two arachnoid membranes (walls of cisterna medullaris lateralis and of cisterna pontocerebellaris). The roots of *Nn. vagus et accessorius* are located close to N. glossopharyngeus and 1–2 mm dorsal from sulcus postolivaris. Lig. denticulatum is located ventral from radix spinalis n. accessorii and dorsal from N. hypoglossus. N. hypoglossus is overlying A. carotis interna. *1* fossa rhomboidea *2* velum medullare posterius, insertion; *3* recessus lateralis; *4* N. glossopharyngeus; *5* N. statoacusticus; *6* N. facialis (and N. intermedius); *7* N. trigeminus; *8* N. trochlearis

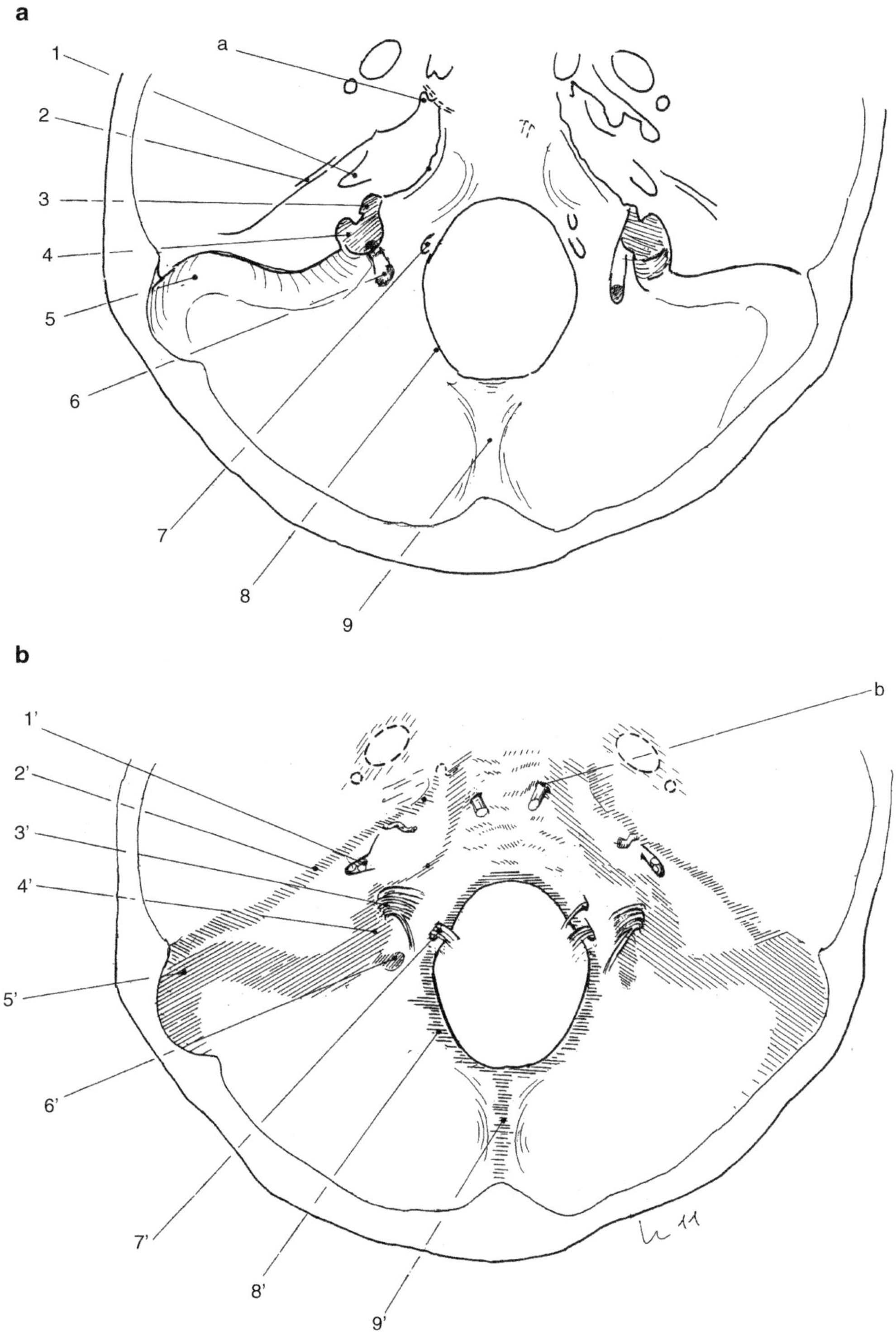

Fig. 15.17 Cranial bony and dural exit areas of cranial nerves and of basal vessels for assisting neuronavigation, if necessary. (**a**) craniobasal bony exit areas, (**b**) dural penetration areas. (**a**) *1* porus acusticus internus; *2* sulcus petrosus superior; *3* For. jugulare, anterior segment; *4* as 3, posterior segment; *5* knee of sulcus transversus; *6* emissarium condylare; *7* canalis n. hyoglossi; *8* margin of For. occipitale; *9* crista occipitalis interna; a hamulus petrosus and the transdural course of N. abducens. (**b**) *1′* N. facialis et N. acusticus; *2′* sinus petrosus superior; *3′* N. glossopharyngeus, N. vagus et N. accessorius; *4′* bulbus superior v. jugularis internae; *5′* knee of sinus transversus; *6′* V. condylaris; *7′* N. hypoglossus; *8′* sinus marginalis; *9′* sinus occipitalis; b dural penetration point of N. abducens

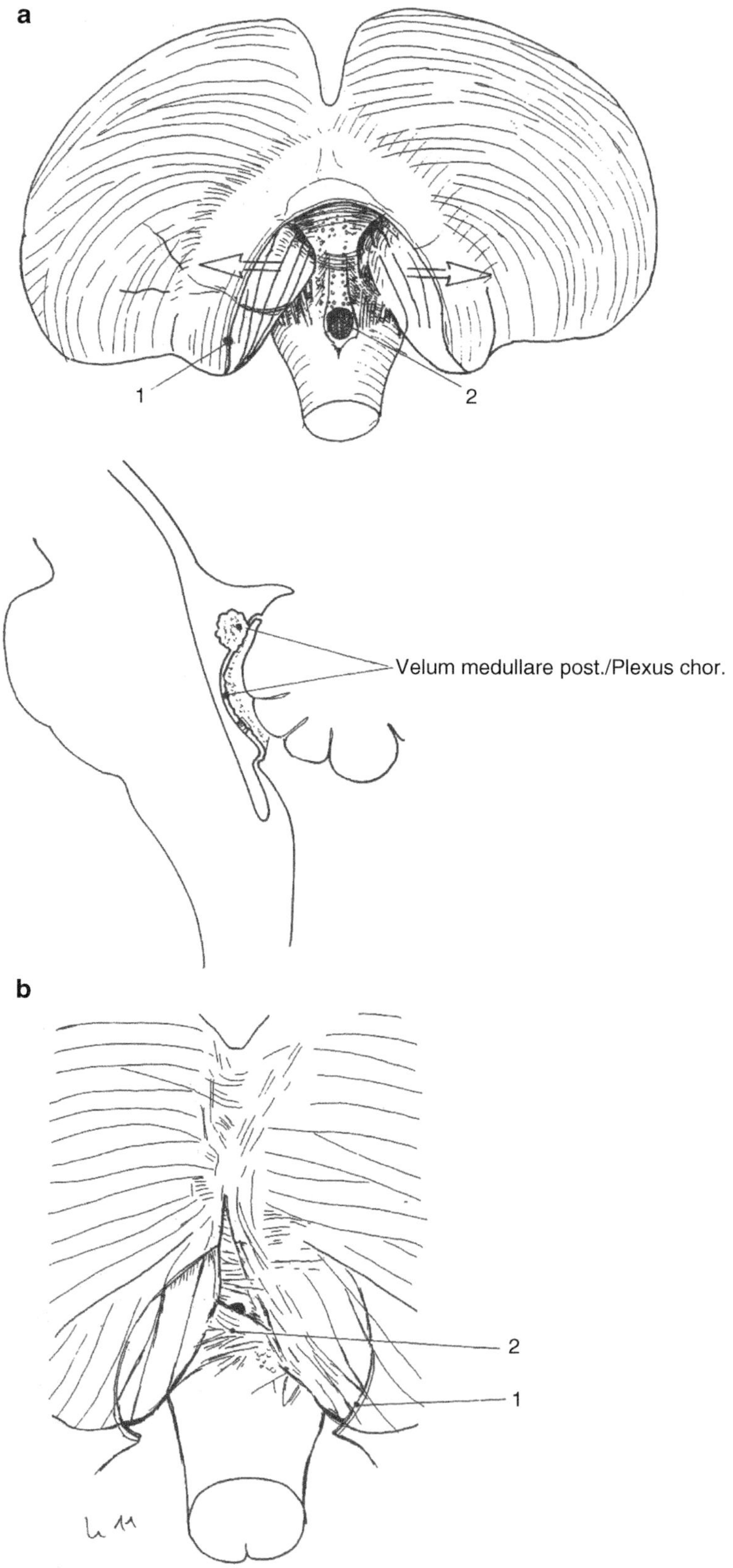

Fig. 15.18 Ground of cisterna cerebellomedullaris: velum medullare posterius and apertura mediana (Magendi). For microanatomy see Fig. 2.26. Two examples. The dorsal arachnoid wall of cisterna cerebellomedullaris is resected for presenting its ground. Velum medullare posterius is essential for surgical routes to the fourth ventricle and for implantation of the cochlear nucleus at the floor of recessus lateralis (Klose and Sollmann 1998). Apertura mediana may be narrowed by arachnoid layers (see **b**). Plexus chorioideus is variable. (**a**) (Key and Retzius 1875) *1* arachnoidea, *2* apertura mediana tonsillae shifted aside (arrows). (**b**) (Lang 1979, p 425) *1* arachnoidea, *2* arachnoid adhesions

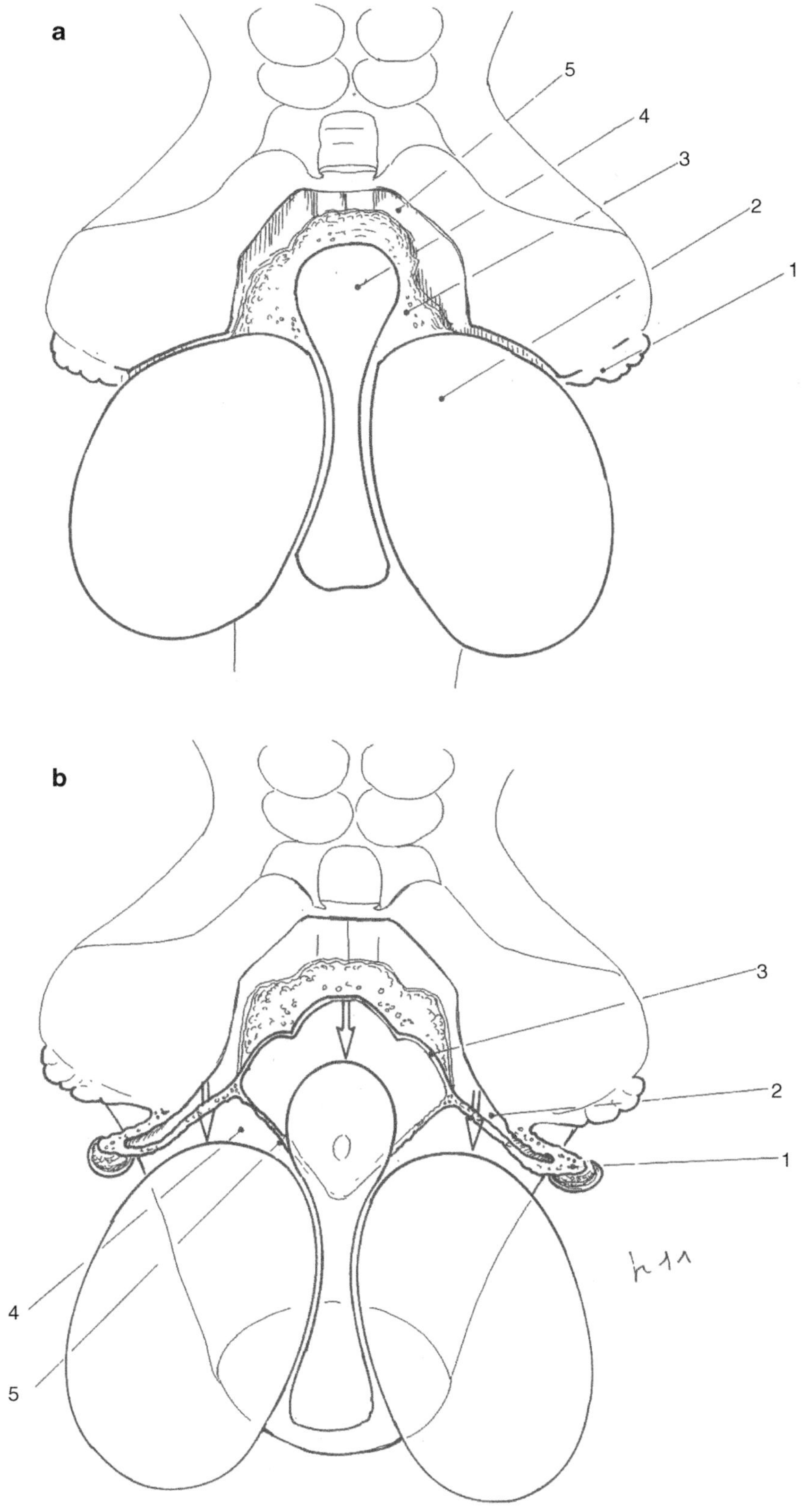

Fig. 15.19 Velum medullare posterius and vermis-tonsillar-complex. (**a**) Interposion of velum medullare posterius between the uvula-tonsillar-complex and the 4th ventricle. (**b**) Loosening and shifting of nodulus vermis and tonsillae from velum medullare posterius (arrows) for defining cisterna cerebellomedullaris, uvula-tonsillar-complex, and fourth ventricle. (**a**) *1* flocculus, *2* tonsilla, *3* velum medullare posterius and its variable plexus compartment, *4* uvula vermis, *5* fourth ventricle. (**b**) *1* velum medullare posterius, plexus compartment bulging into basal cisterns, *2* recessus lateralis, *3* velum medullare posterius, *4* cisterna cerebellomedullaris, *5* insertion of velum medullare posterius at uvula preserved

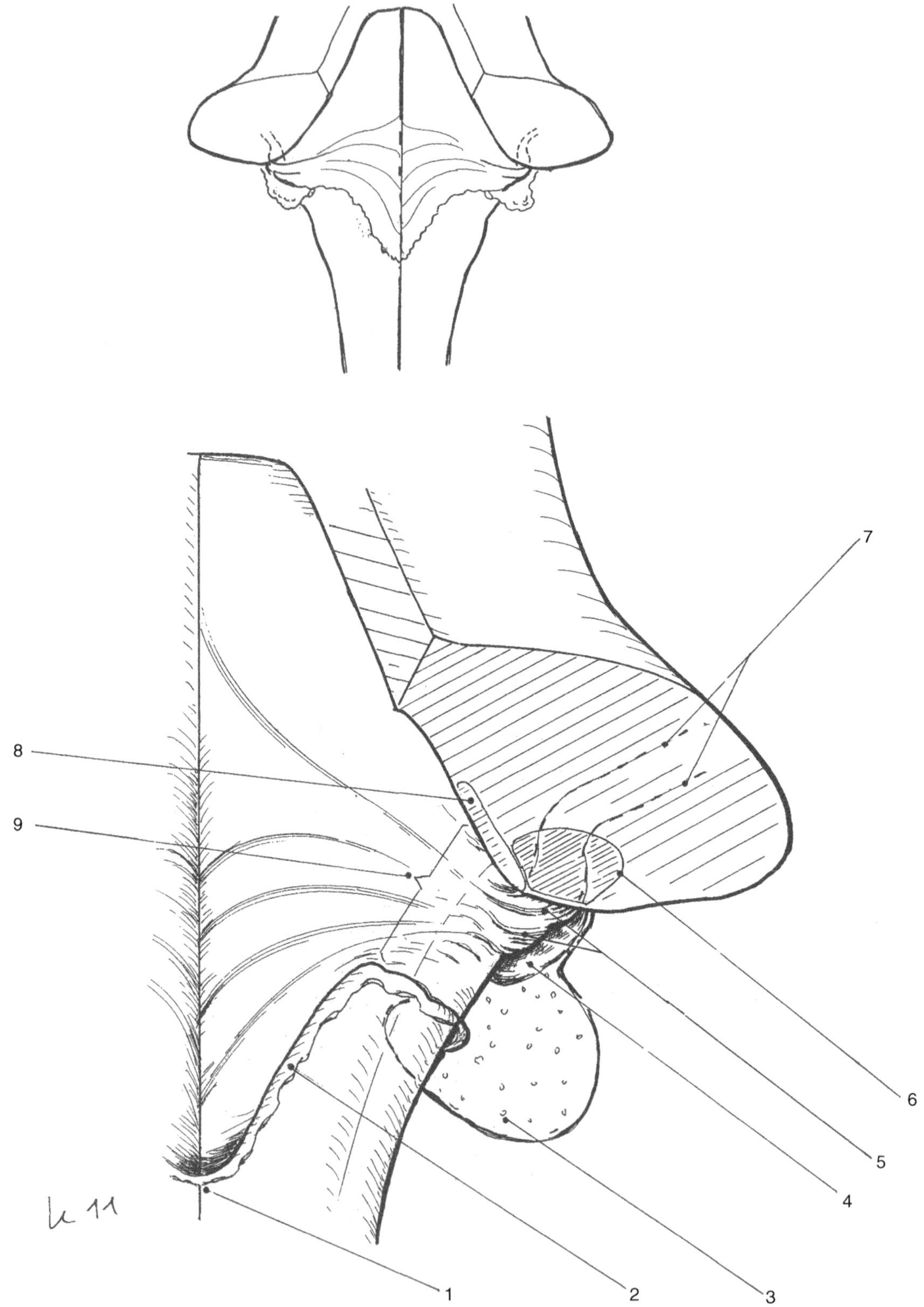

Fig. 15.20 Recessus lateralis. Velum medullare posterius eliminated except its extraventricular (cisternal) plexus compartment. Presentation of fossa rhomboidea. Its lateral segment (recessus lateralis) contains nucleus cochlearis dorsalis. This is the region of cochlear implantations. For details see Klose 2000 and Figs. 3.27f. *1* obex; *2* insertion of velum medullare posterius at the medulla oblongata; *3* plexus chorioideus, cisternal segment preserved; *4* apertura lateralis (Key-Retzii, Luschkae), inconstant opening; *5* striae cochleares and nucleus cochlearis dorsalis; *6* pedunculus cerebellaris inferior, transectional plane; *7* N. statoacusticus, projection; *8* pedunculus flocculi, transectional plane; *9* striae medullares

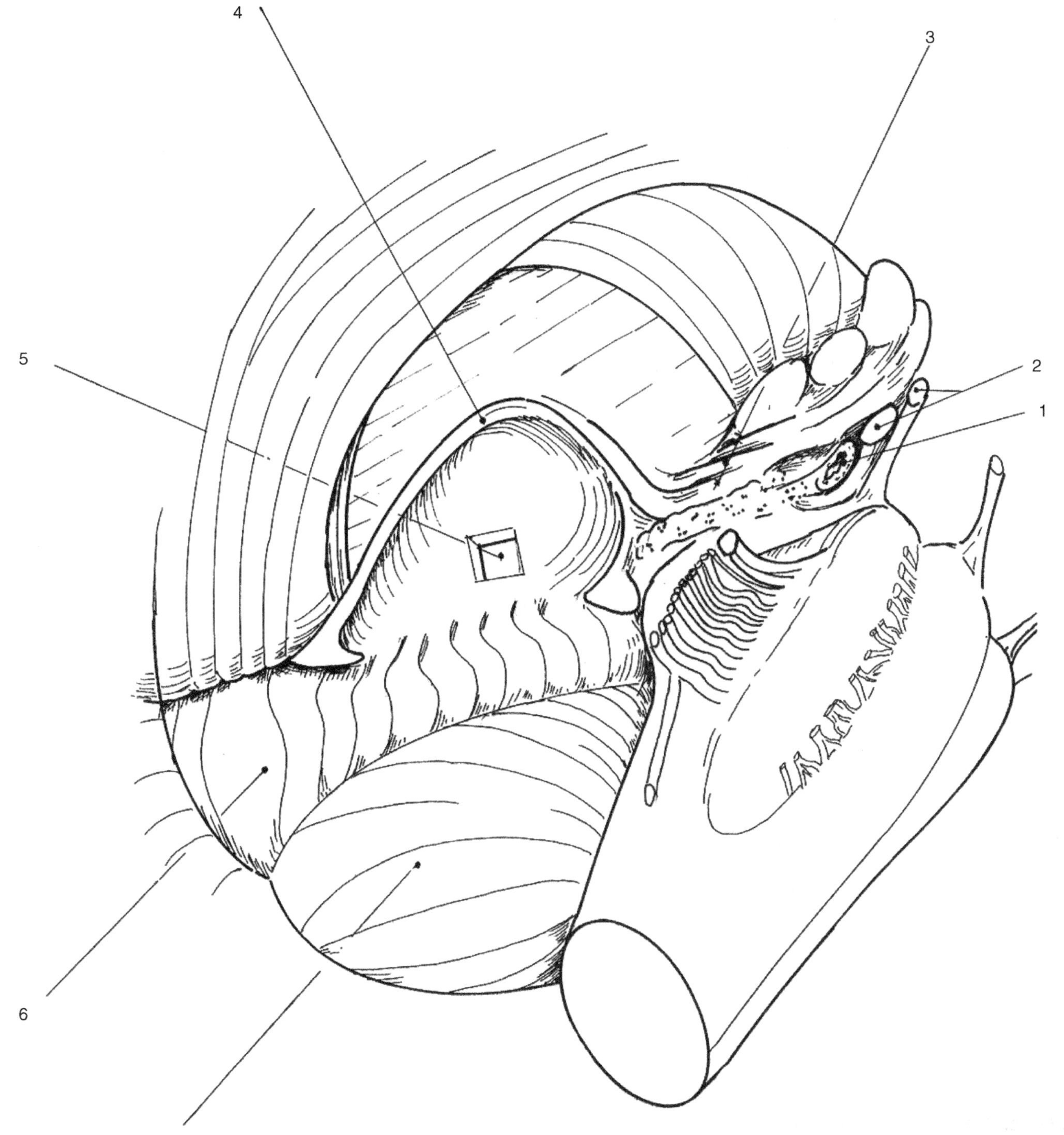

Fig. 15.21 Velum medullare posterius., bed of tonsilla («nidus avis» in the past). This is the posterior wall of recessus dorsolateralis (Lang) of the 4th ventricle. Vermis, contralateral tonsilla, and wall of recessus lateralis and dorsolateralis are preserved. Cisternal part of plexus chorioideus is cut and eliminated. *1* plexus chorioideus cut; *2* N. statoacusticus and N. facialis; *3* pedunculus flocculi; *4* tractus flocculonodularis, insertion area of velum medullare posterius; *5* fenestration of velum medullare posterius and view into recessus dorsolateralis; *6* nodulus vermis; *7* tonsilla

15.2.2 Posterior Structures Superior to the Brain Stem (Figs. 15.22, 15.23, 15.24, 15.25, 15.26, 15.27, 15.28, 15.29, 15.30, 15.31, 15.32, and 15.33)

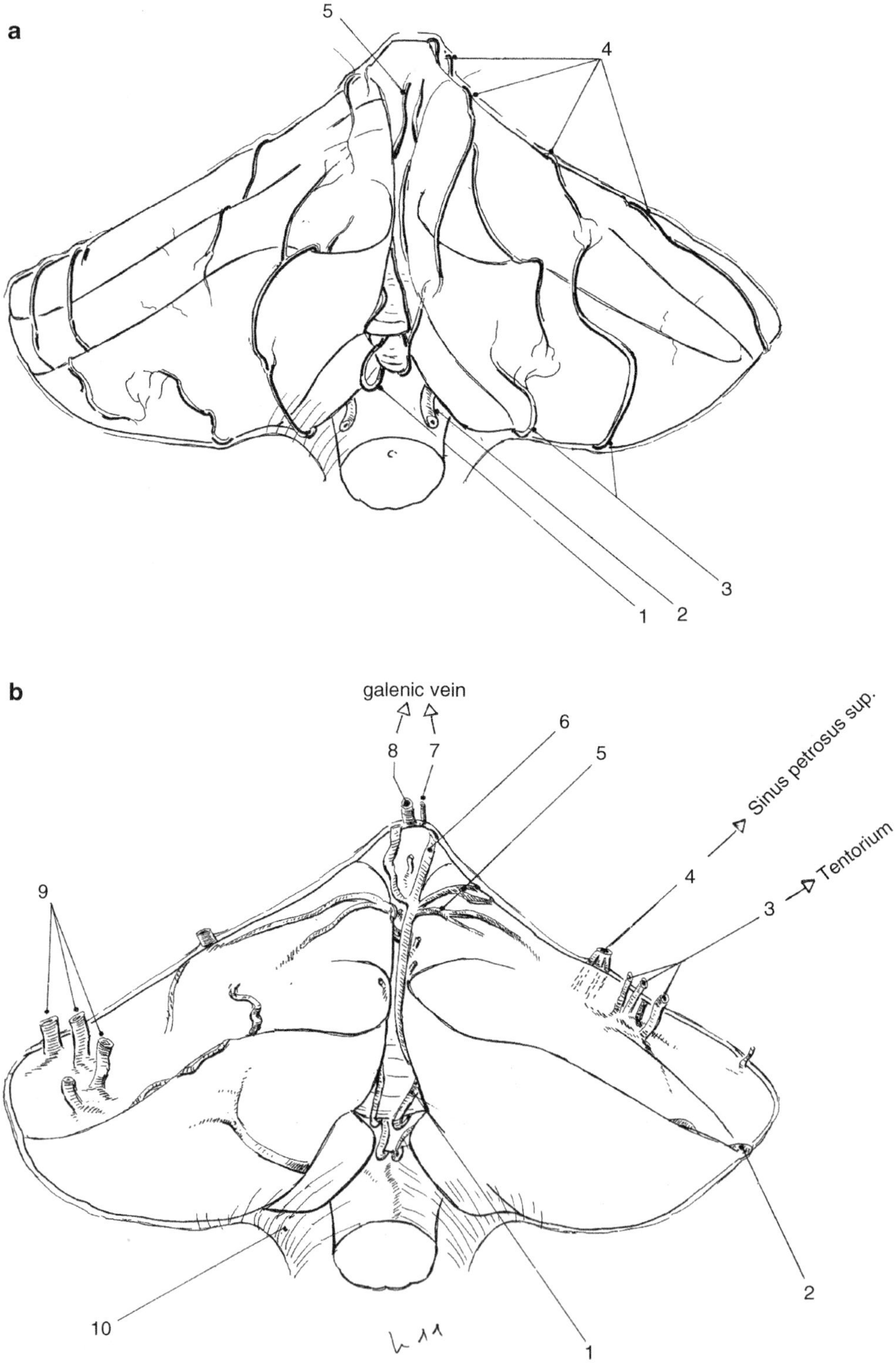

Fig. 15.22 Vessels at the posterior cerebellar relief. (**a**) Arteries, (**b**) veins. (**a**) *1* A. vermis, *2* PICA (A. cerebelli inferior posterior), *3* as 2, lateral ramification, *4* ramification of A. cerebelli superior, *5* as 1. (**b**) *1* V. vermis inferior; *2* V.cerebellaris horizontalis; *3* superior lateral bridging veins, variable; *4* V. petrosa superior; *5* superior medial tributary veins of 6; *6* V. vermis superior; *7* V. cerebellaris praecentralis; *8* V. supraculminalis; *9* as 3; *10* superficial arachnoid wall of cisterna cerebellomedullaris

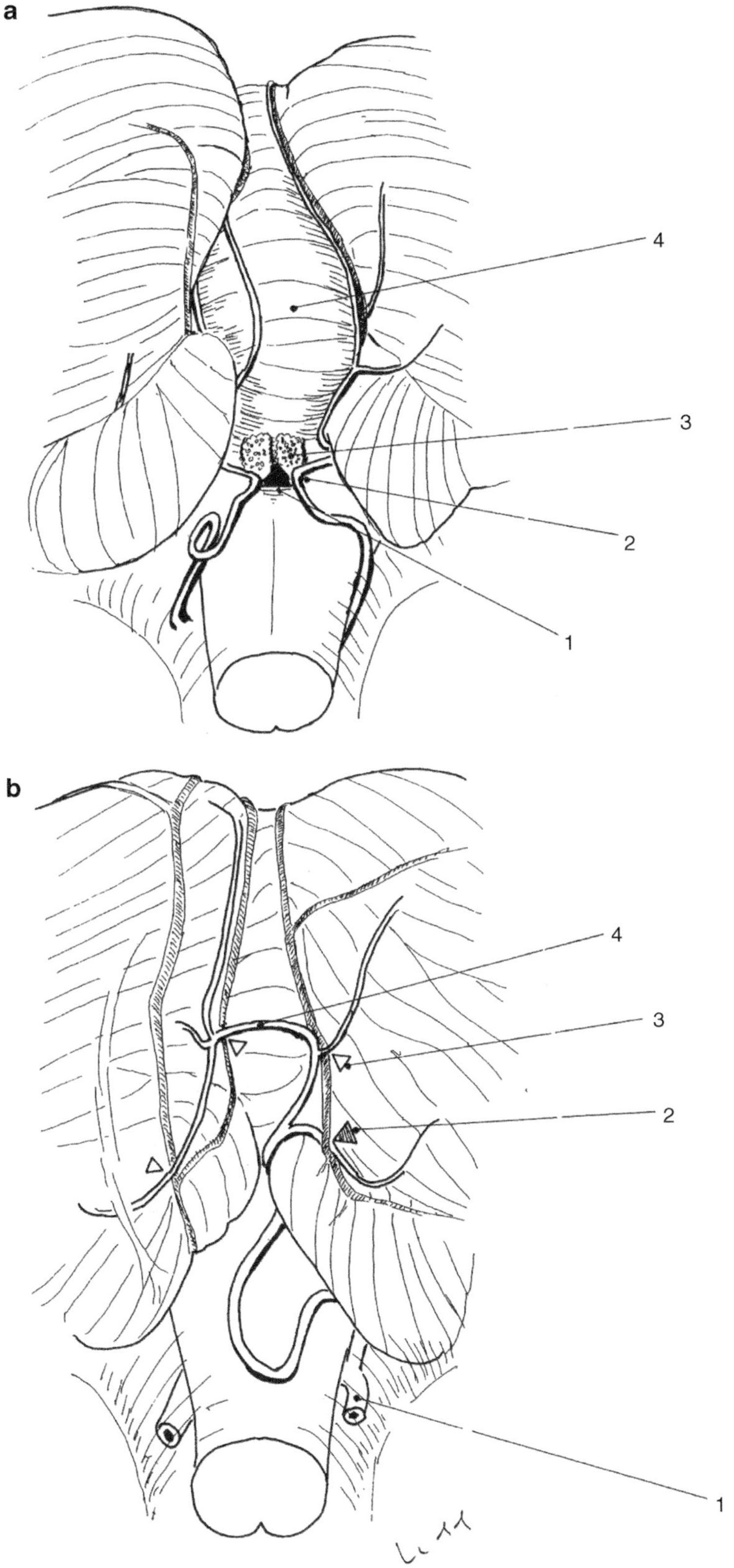

Fig. 15.23 Variable overcrossing and undercrossing of medial arteries and veins. Arachnoidea omitted. (**a**) arteries, usual finding, (**b**) variants. (**a**) *1* PICA (A. cerebelli inferior posterior), *2* chorioid point of 1 (not constant), *3* plexus chorioideus, *4* vermis. (**b**) *1* A. vertebralis, *2* vein overcrossing artery (dark arrow), *3* artery overcrossing vein (light arrow), 2 and 3: typical overcrossing of veins by arteries at the superficial cerebellar cortex

a

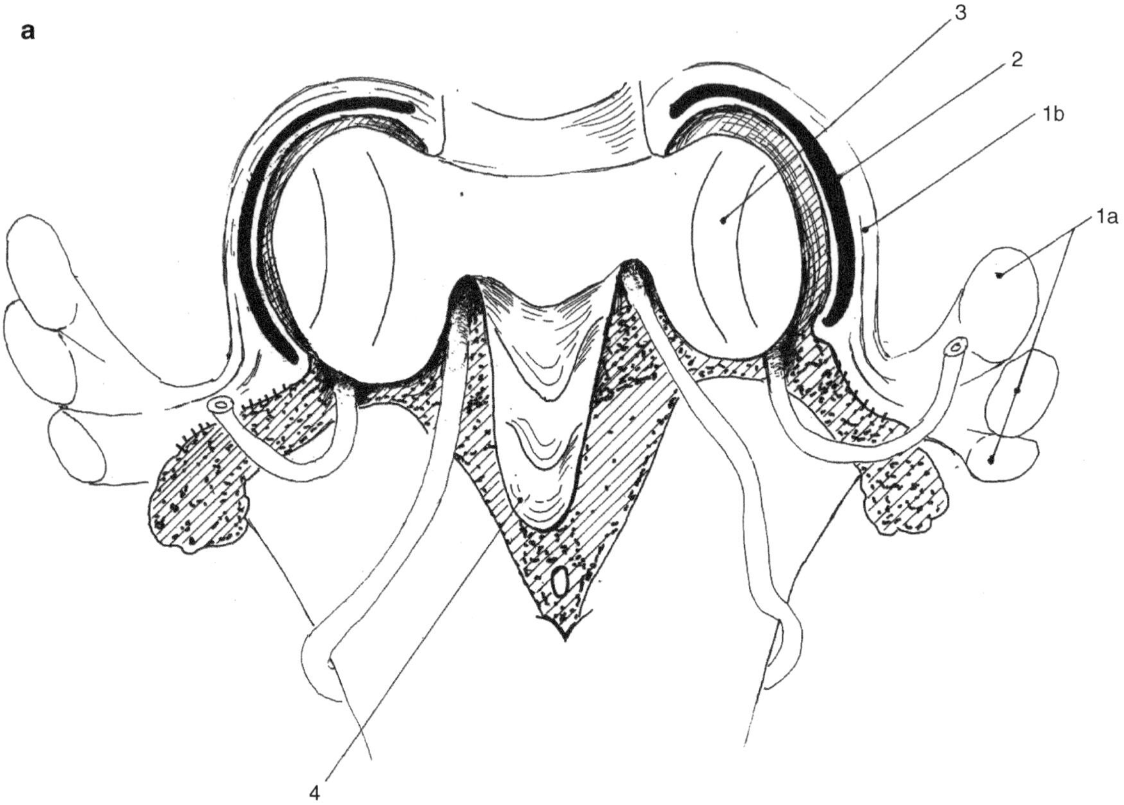

b

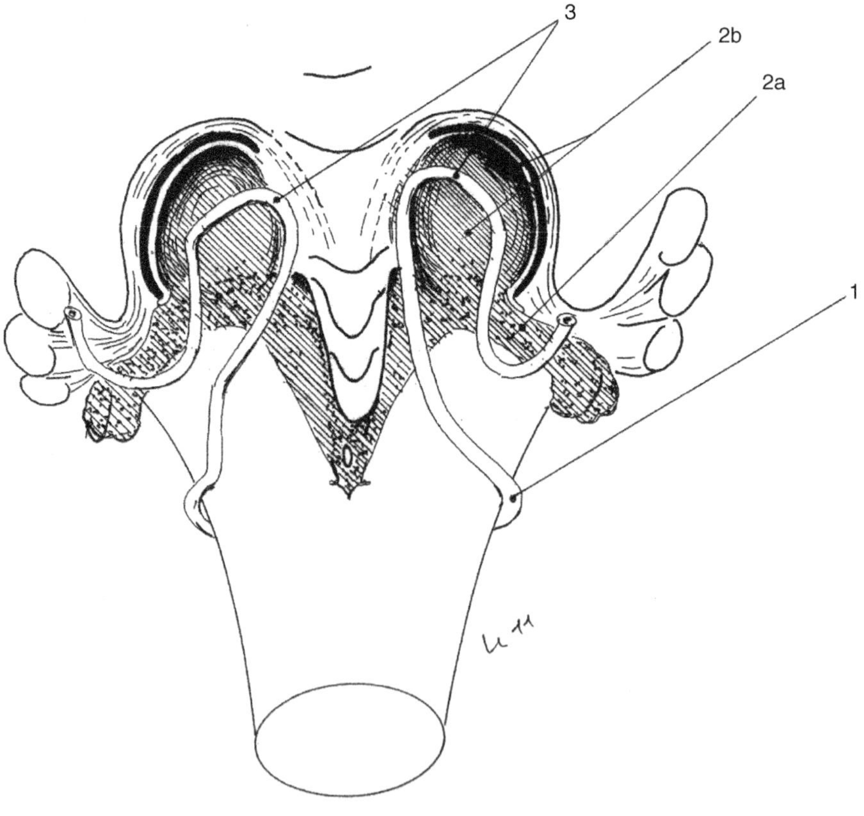

Fig. 15.24 PICA, velum medullare posterius, and recessus dorsolateralis of the 4th ventricle. (**a**) Superficial segments of tonsillae and vermis resected. (**b**) PICA after elimination of tonsillae, located lateral from nodulus (or anterior from it close to the midline). (**a**) *1a* flocculus, *1b* tractus flocculonodularis; *2* recessus dorsolateralis of the 4th ventricle; *3* tonsilla, residual; *4* uvula vermis. (**b**) *1* PICA; *2a* velum medullare posterius, plexus compartment; *2b* velum medullare posterius, hypo/avascular compartment («nidus avis»); *3* choroid points of PICA, high variability

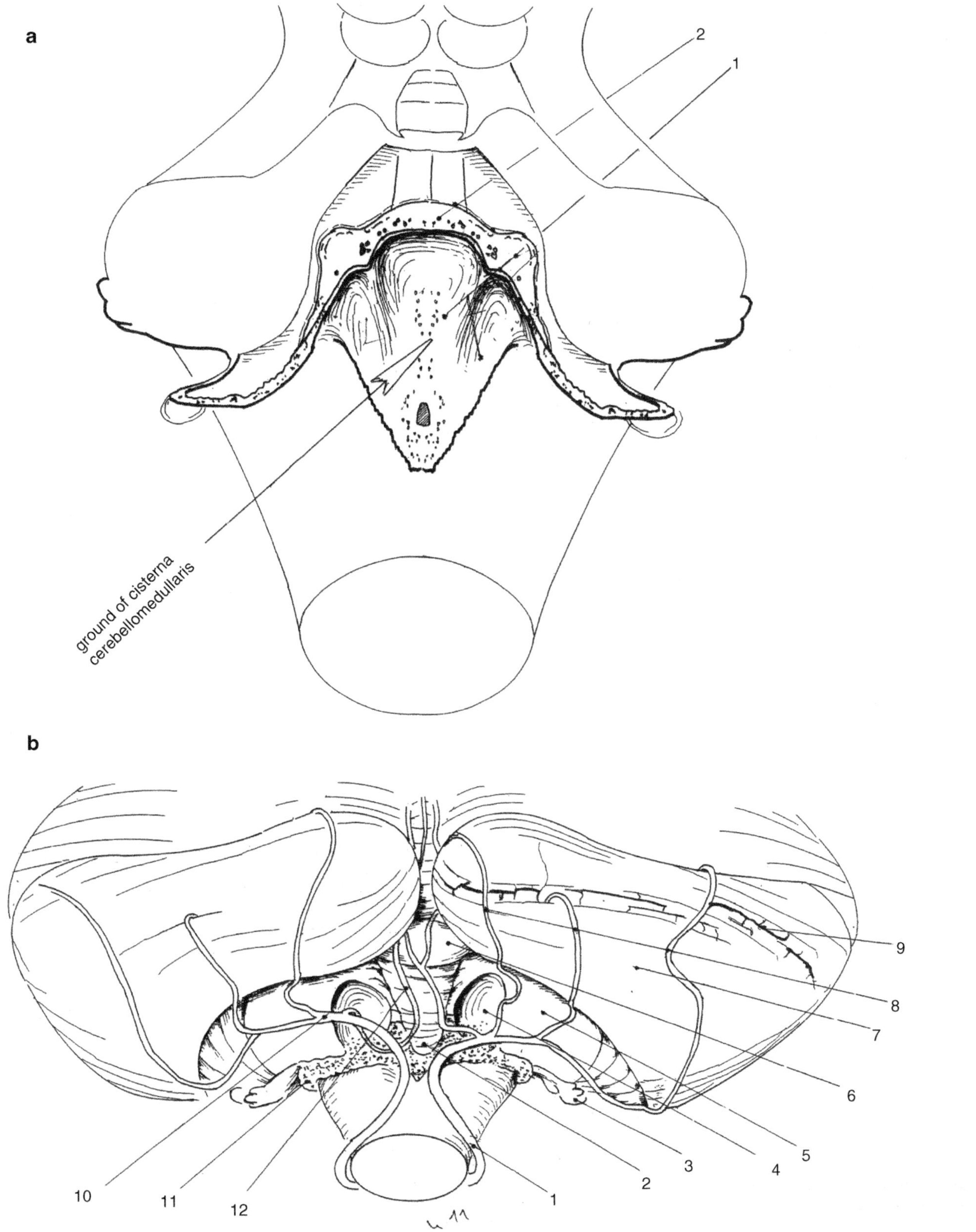

Fig. 15.25 PICA, ramifications. (**a**) Survey, (**b**) ramifications of PICA. (**a**) *1* velum medullare posterius, *2* bulging of velum medullare posterius (plexus compartment) against the 4th ventricle, bed of uvula vermis. (**b**) *1* PICA, *2* uvula vermis, *3* flocculus, *4* velum medullare posterius (hypo/avascularized bed of tonsilla, «nidus avis»), *5* transectional plane between tonsilla and pons, *6* nodulus vermis, *7* lobulus biventer, *8* lateral branches of PICA, *9* addendum: small cortical vessels, *10* lateral main branch of PICA, *11* plexus chorioideus, *12* A. vermis

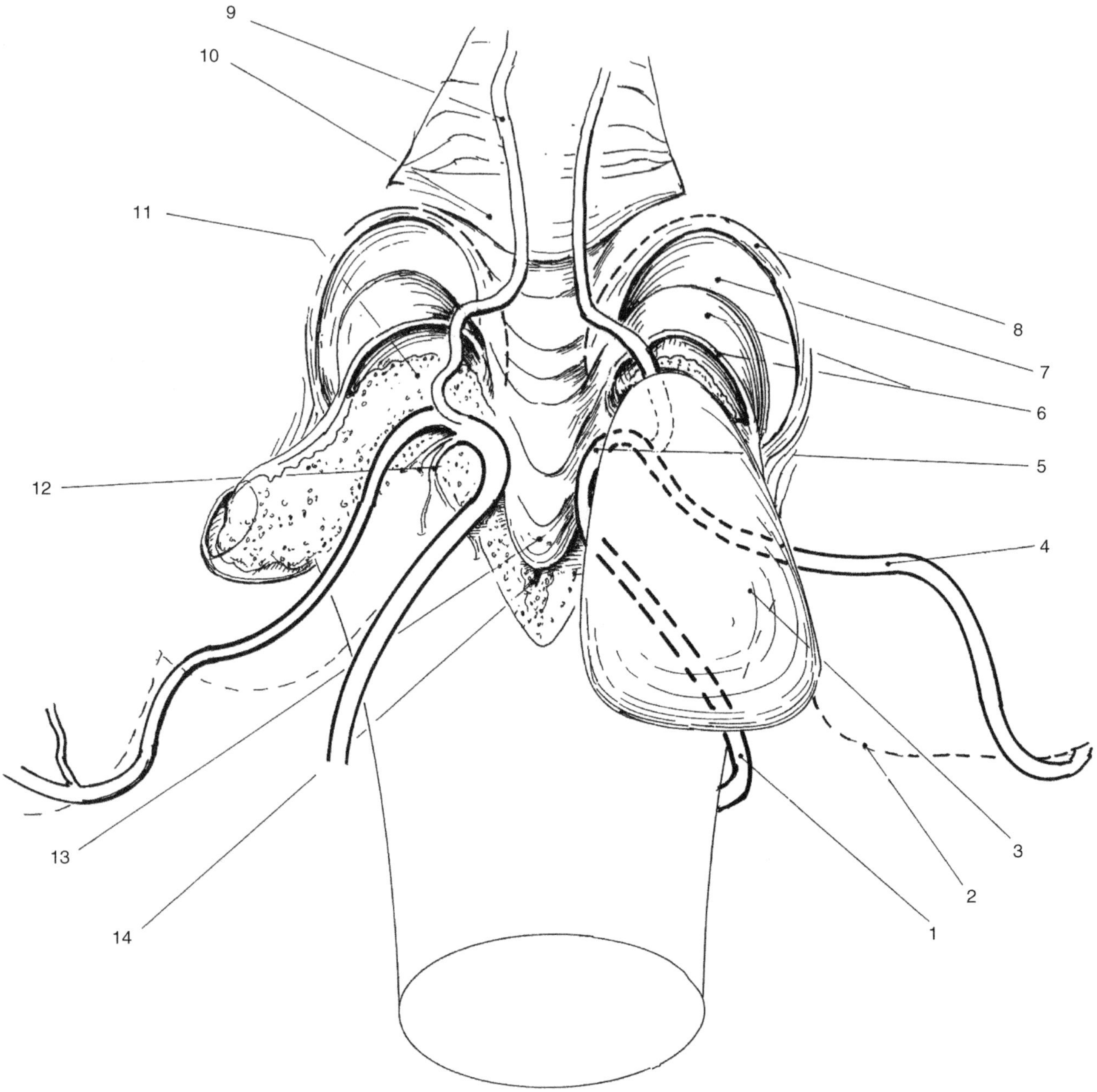

Fig. 15.26 PICA and its main branches. *1* PICA; *2* lobulus biventer, projection; *3* tonsilla; *4* lateral main branch of PICA; *5* loop located anterior or lateral from uvula, high variability; *6* velum medullare posterius, ventricle fenestrated; *7* recessus dorsolateralis (Lang), part of recessus lat.; *8* Tractus flocculonodularis; *9* A. vermis; *10* nodulus vermis; *11* velum medullare posterius, plexus compartment; *12* some of the numerous perforating arteries; *13* uvula vermis; *14* apertura mediana (Magendi)

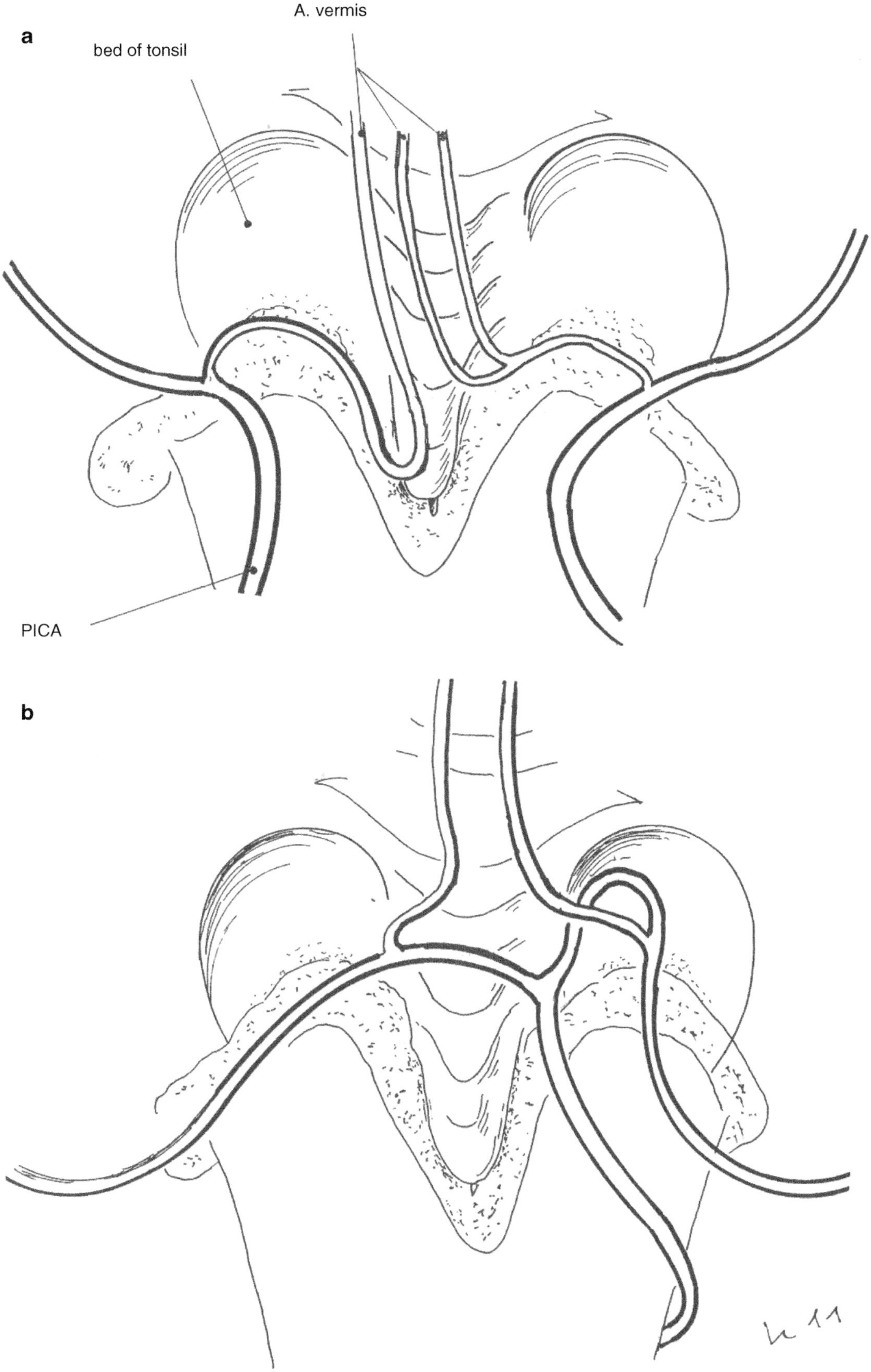

Fig. 15.27 Variants of A. vermis. Examples (**a**) A. vermis right-sided duplicated, (**b**) common variant of chorioid loop

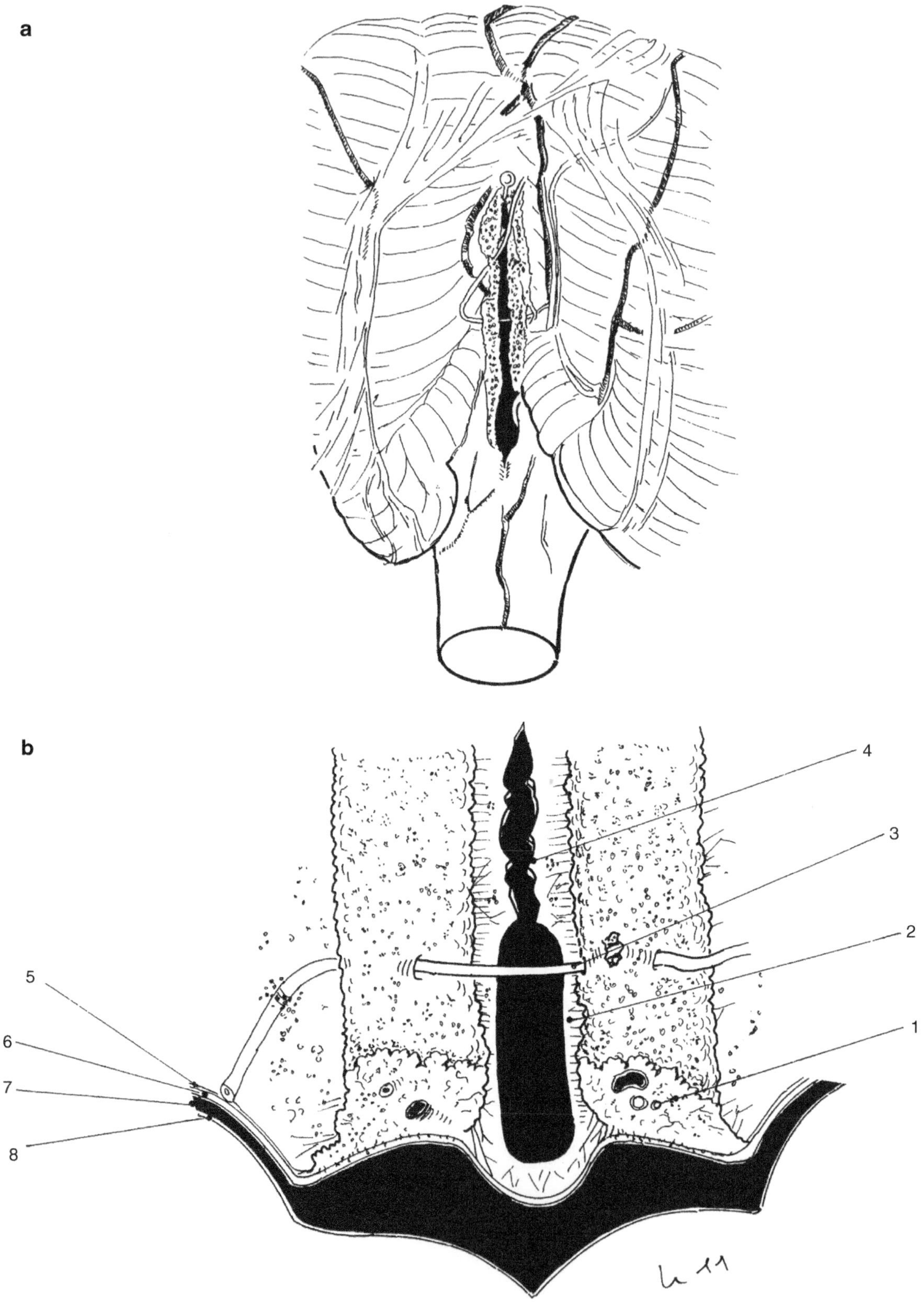

Fig. 15.28 Velum medullare post and apertura mediana (Magendi). The fine velum medullare posterius could rupture at apertura mediana. Bilateral connections of vessels should be considered during surgery. (**a**) Example (Lang 1979, p 424). (**b**) as (**a**), microanatomy schematized. *1* plexus compartment of velum medullare posterius variable; *2* velum medullare posterius; *3* branch of PICA or other fine vessels; *4* apertura mediana (Magendi) ruptured, often masked by plexus chorioideus; *5* pia mater; *6* hypo-/avascular compartment of velum medullare posterius; *7* fourth ventricle, recessus lateralis; *8* ependyma of the wall of the fourth ventricle at pedunculus cerebellaris inferior and pedunculus flocculi (area of apertura mediana)

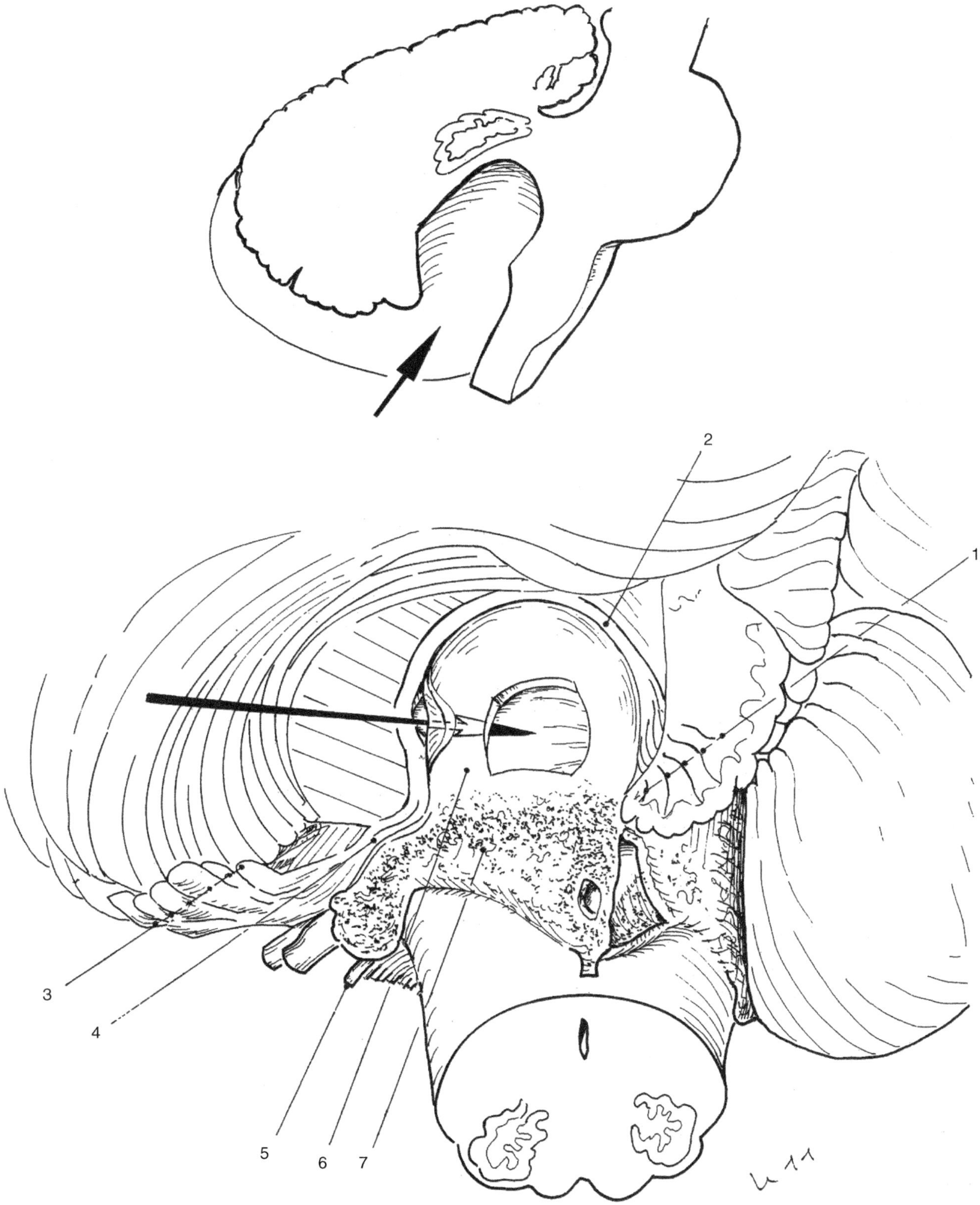

Fig. 15.29 Velum medullare posterius and the fourth ventricle. Multiple fenestrations of velum medullare posterius for inspection of the fourth ventricle. Tonsilla eliminated, vermis resected. *1* and *2* tractus flocculonodularis; *3* flocculus; 4 pedunculus flocculi; *5* N. glossopharyngeus; *6* velum medullare posterius, hypo/avascular compartment («nidus avis»); *7* choroidal compartment of velum medullare posterius. *Arrows*: fossa rhomboidea

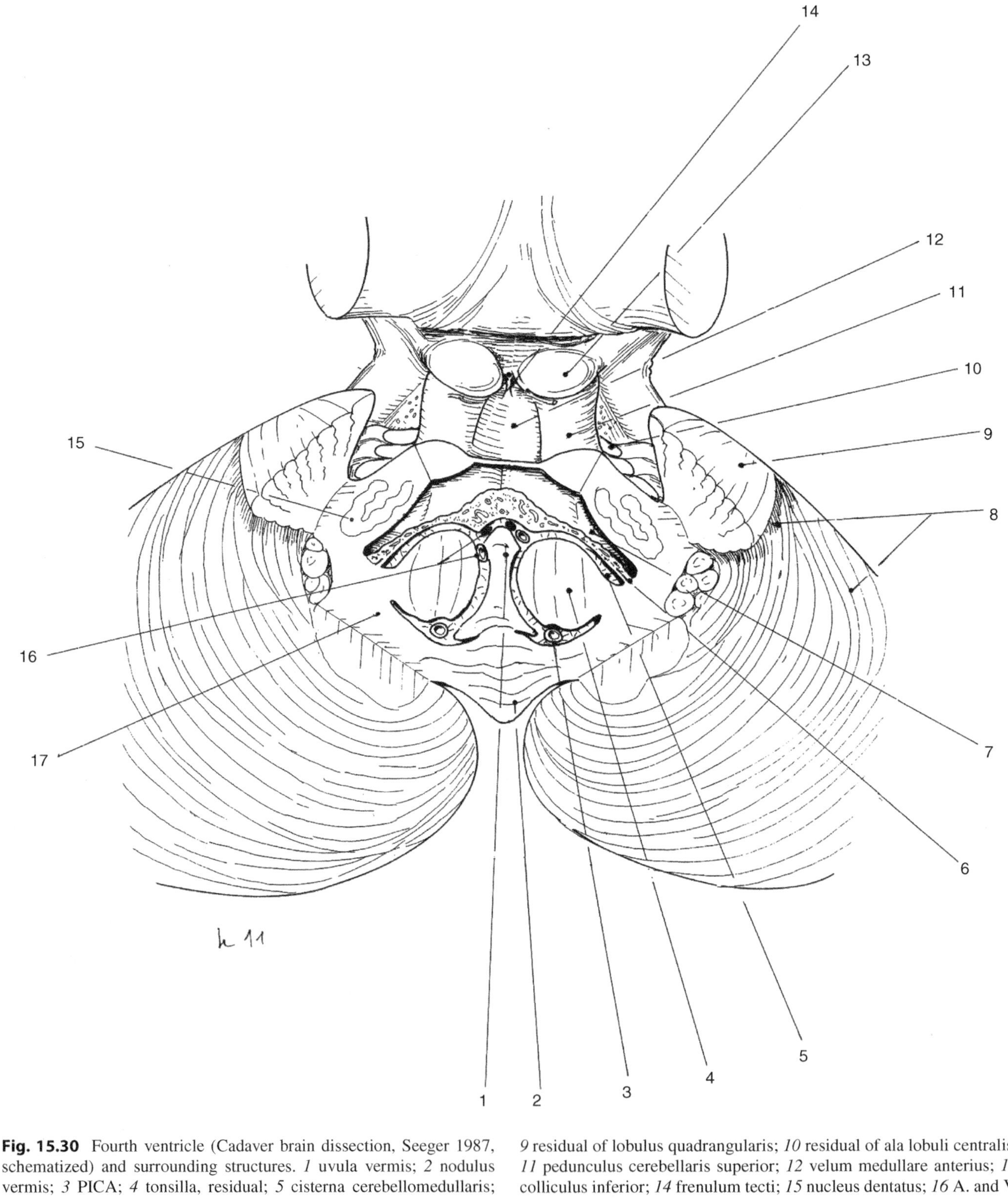

Fig. 15.30 Fourth ventricle (Cadaver brain dissection, Seeger 1987, schematized) and surrounding structures. *1* uvula vermis; *2* nodulus vermis; *3* PICA; *4* tonsilla, residual; *5* cisterna cerebellomedullaris; *6* velum medullare posterius; *7* recessus lateralis; *8* fissura horizontalis; *9* residual of lobulus quadrangularis; *10* residual of ala lobuli centralis; *11* pedunculus cerebellaris superior; *12* velum medullare anterius; *13* colliculus inferior; *14* frenulum tecti; *15* nucleus dentatus; *16* A. and V. vermis; *17* pedunculus cerebellaris inferior

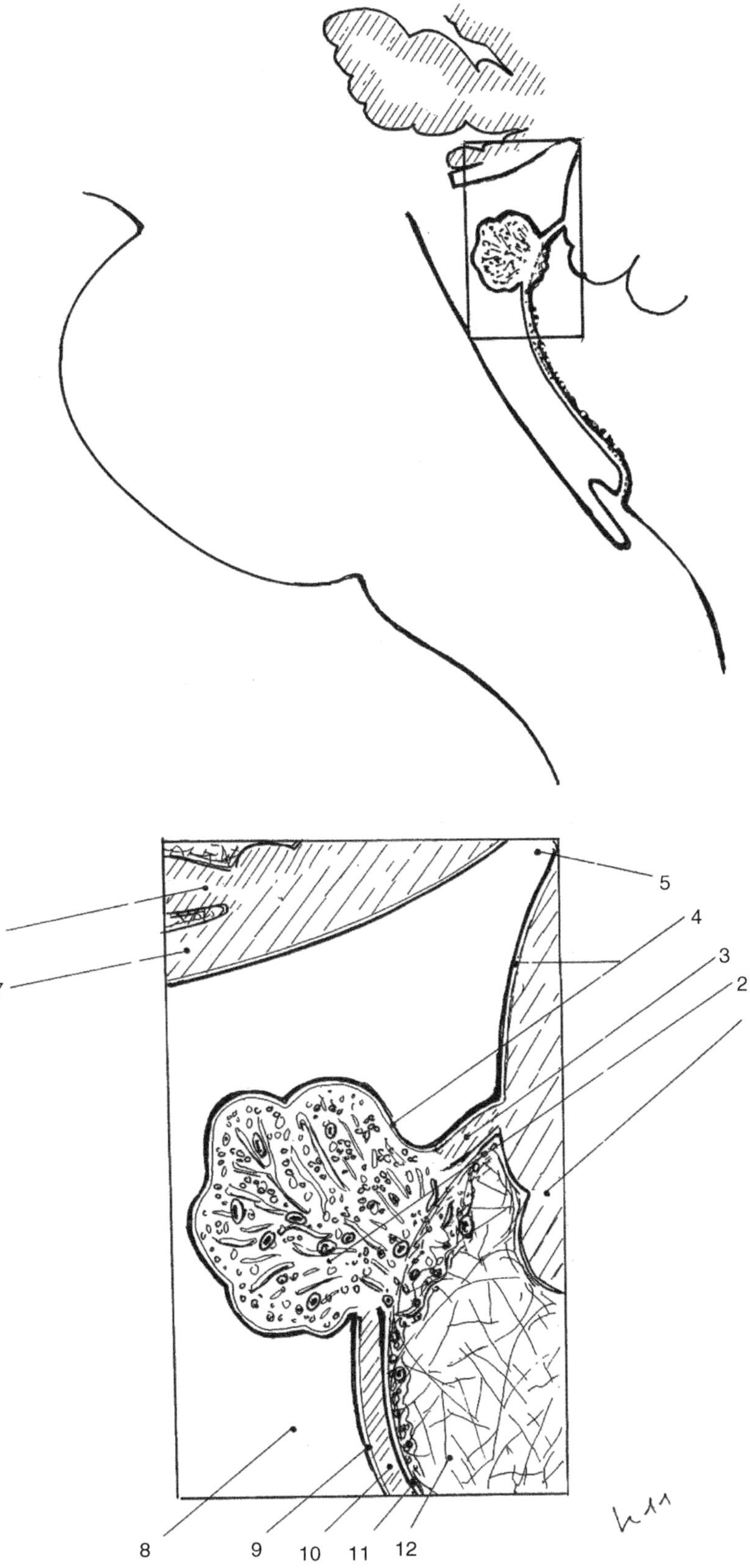

Fig. 15.31 The fourth ventricle, medial area velum medullare anterius, fastigium, velum medullare posterius, and obex. High variability of the plexus-chorioideus-compartment of velum medullare posterius. *1* Uvula vermis; *2* plexus compartment of velum medullare posterius; *3* its hypo- or avascular component; *4* ependyma; *5* fastigium; *6* lingula cerebelli; *7* velum medullare anterius; *8* fourth ventricle; *9* ependyma; *10* velum medullare posterius, hypo/avascular compartment; *11* pia mater; *12* cisterna cerebellomedullaris

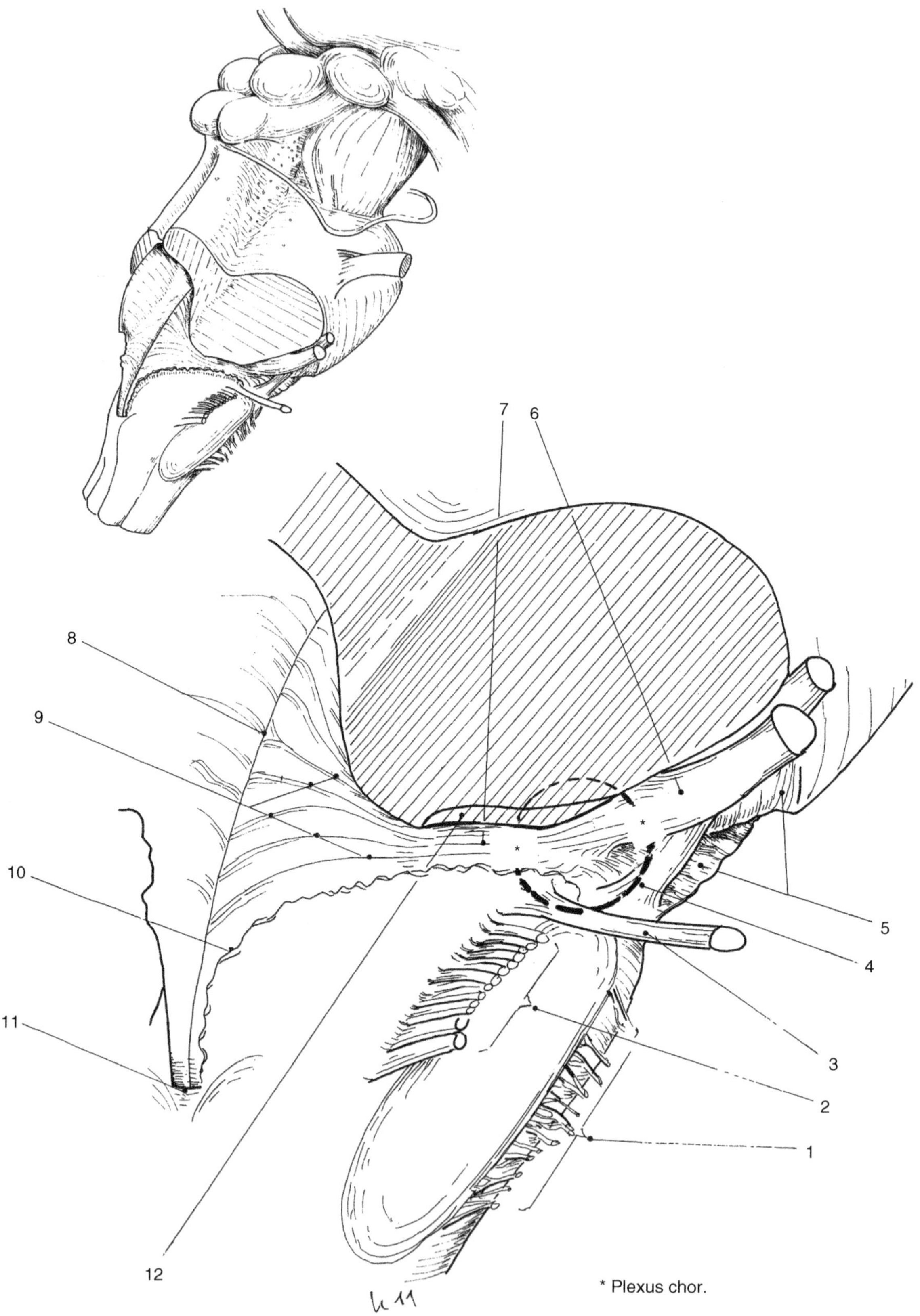

Fig. 15.32 The fourth ventricle, recessus lateralis, and adjacent areas. *1* N. hypoglossus, *2* N. vagus et accessorius, *3* N. glossopharyngeus, *4* N. facialis at fossa paraolivaris, *5* pons, *6* N. statoacusticus, *7* striae cochleares and region of nucleus n. cochlearis dorsalis, *8* sulcus medianus, *9* striae medullares, *10* insertion (taenia) of velum medullare posterius, *11* obex, *12* pedunculus flocculi et tractus flocculonodularis, transectional plane* area of Apertura lat. After elimination of plexus chorioideus

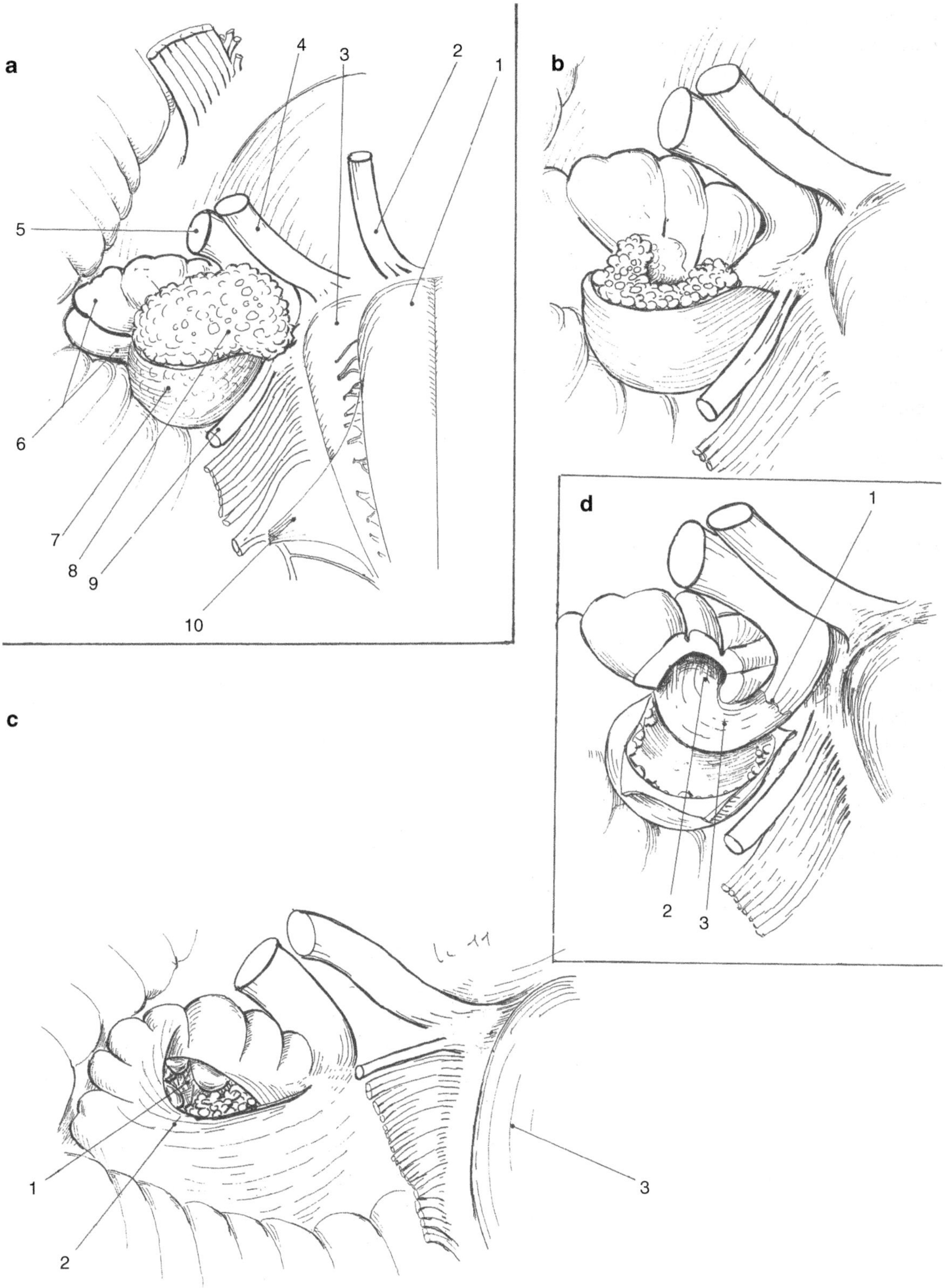

Fig. 15.33 Apertura lateralis of the fourth ventricle (Key-Retzius, Luschka (Cit. Rauber-Kopsch 1987; Lang 1979)). Examples (**a**) usual findings, (**b**) apertura lateralis narrow or obturated by adhesions, (**c**) apertura lateralis narrow or obturated by an avascular compartment of velum medullare posterius, (**b**) and (**c**) common variants, (**d**) usual findings of cranial nerves after resection of plexus and flocculus at anatomical dissection. (**a**) *1* pyramis medullae; *2* N. abducens; *3* oliva; *4* N. facialis (and N. intermedius); *5* N. statoacusticus; *6* flocculus; *7* cisternal wall; *8* plexus chorioideus, cisternal bulging; *9* N. glossopharyngeus; *10* Lig. denticulatum. B (**c**) *1* arachnoid adhesions, *2* arachnoid membrane, *3* pyramis medullae. (**d**) *1* ependymal margin of apertura lateralis, *2* region of nucleus cochlearis dorsalis, *3* N. statoacusticus

15.2.3 Basal Cisterns, Variants, and Adjacent Structures

Basal Cisterns (Figs. 15.34, 15.35, 15.36, 15.37, and 15.38)

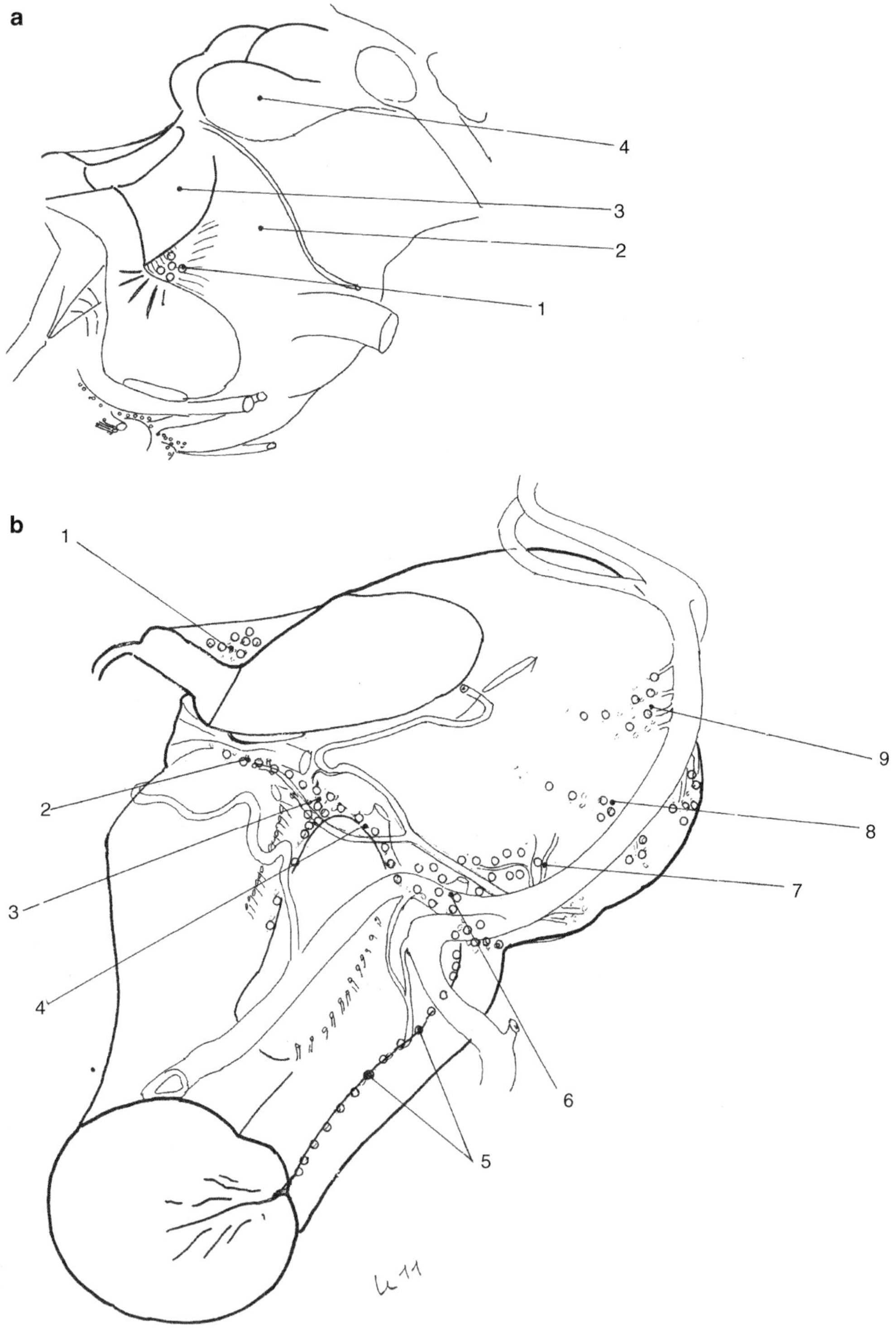

Fig. 15.34 Penetration points of perforating arteries at the brain stem, schematized. Intracerebral anastomoses of arteries to each others are rare. At surgery the perforating arteries should be preserved (Yaşargil 1978). (**a**) *1* trigonum lemnisci, *2* pedunculus cerebellaris medius, *3* pedunculus cerebellaris superior, *4* colliculus inferior. (**b**) *1* trigonum lemnisci; *2* taenia chorioidea (insertion of velum medullare posterius); *3* fossa paraolivaris; *4* pontomedullary rim; *5* fissura mediana anterior; *6* region of N. abducens, vertebrobasilar junction, AICA, pyramis medullae, adherent by adhesions; *7* Rr. ad pontem, short and long branches; *8* and *9* as 7

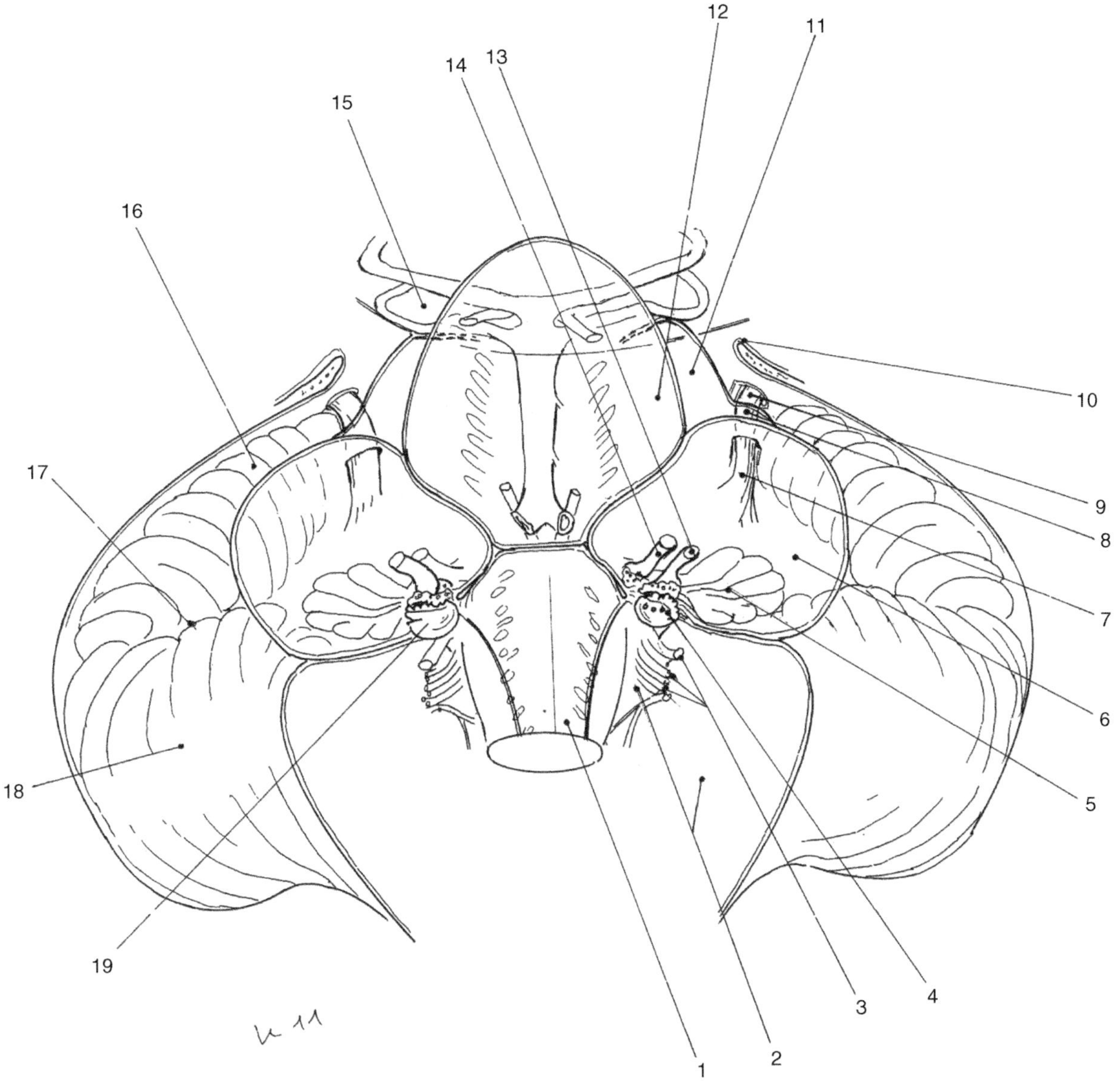

Fig. 15.35 Duplicated cisternal walls and its contents (Key and Retzius 1876, modified for surgery by Yaşargil, schematized presentation), schematized. Arachnoid duplications are nearly constant findings at basal cisterns. At surgery, the duplicated walls can be stepwise dissected from each other, combined with stepwise hollowing of tumors. Walls of cisterna medullaris medialis et lateralis and of cisterna pontocerebellaris lat. are adherent at the lateral segment of plexus chorioideus. These aspects should be considered at cisternal surgical approaches. *1* cisterna medullaris medialis; *2* cisterna medullaris lateralis; *3* N. glossopharyngeus, vagus et accessorius; *4* plexus chorioideus; *5* flocculus, *6* cisterna pontocerebellaris; *7* N. trigeminus; *8* its penetration of arachnoid walls; *9* N. trigeminus, cisterna cruralis; *10* arachnoid fold enclosing the tentorial edge; *11* cisterna pontis lateralis; *12* cisterna pontis medialis; *13* N. statoacusticus; *14* N. facialis; *15* cisterna ambiens; *16* lobulus quadrangularis; *17* fissura horizontalis; *18* lobulus biventer; *19* as 4

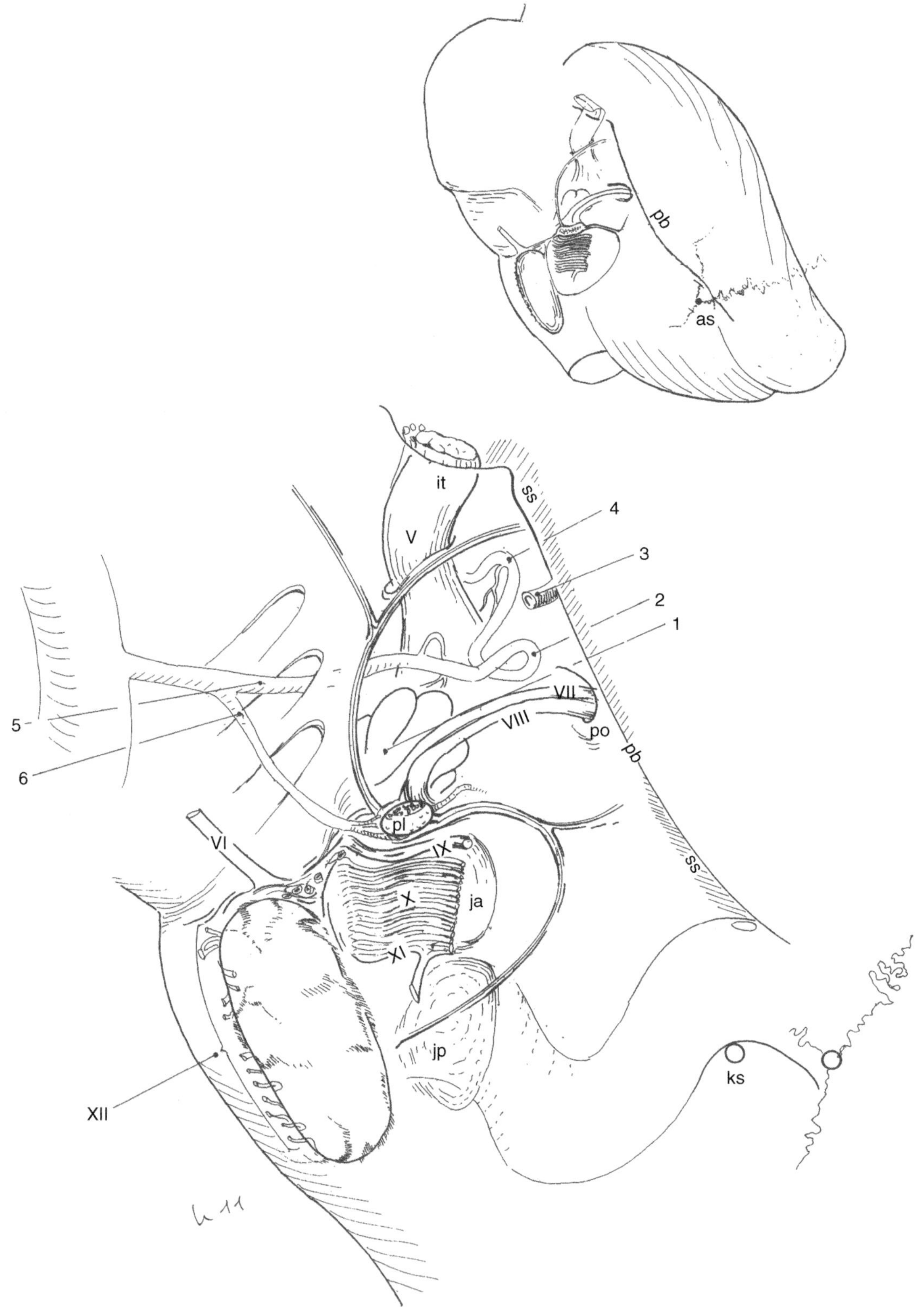

Fig. 15.36 Pontomedullary cisterns. *as* asterion; *it* incisura trigeminalis; *ja* foramen jugulare, anterior segment, projection; *jp* foramen jugulare, posterior segment; *ks* knee of sinus transversus; *pb* superior margin of Os petrosum; *pl* plexus chorioideus; *po* porus acusticus internus; *ss* sinus petrosus superior. *1* flocculus; *2* AICA, meatal loop (for variants see Naidich and Kricheff 1976); *3* V. petrosa superior; *4* AICA, trigeminal loop; *5* AICA, superior lateral main branch; *6* inferior medial branch V to XII cranial nerves

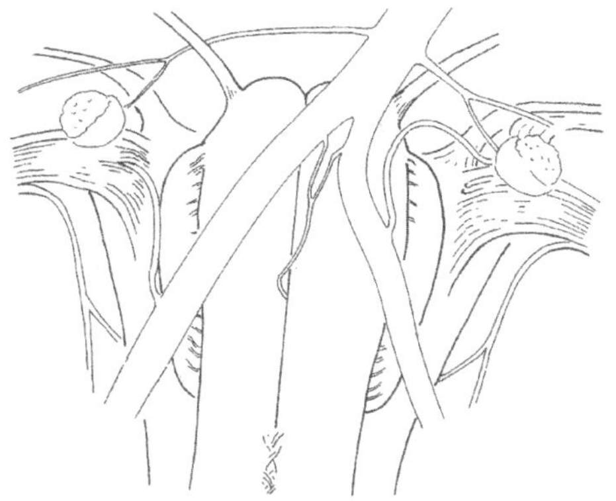

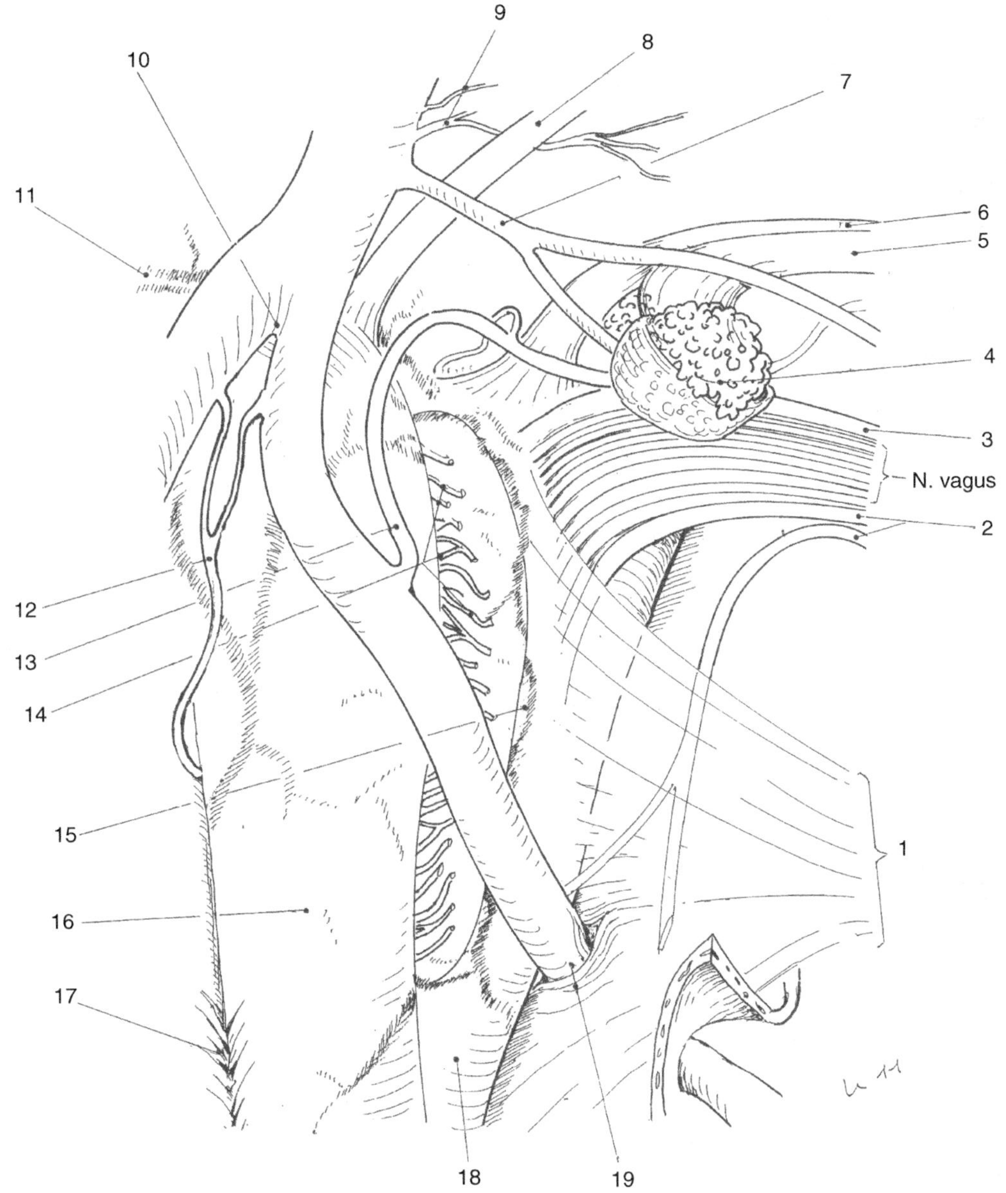

Fig. 15.37 Addendum for Fig. 15.36. Arachnoid walls omitted. *1* Lig. denticulatum, *2* N. accessorius, *3* N. glossopharyngeus, *4* plexus chorioideus, *5* N. statoacusticus, *6* N. facialis, *7* AICA, *8* N. abducens, *9* Rr. ad pontem, *10* junction of Aa. vertebrales, *11* vessels of the pontomedullary rim, *12* A. spinalis anterior, *13* PICA, *14* N. hypoglossus, *15* periolivar veins, *16* funiculus anterior, *17* decussatio pyramidum, *18* funiculus lateralis, *19* stenosis of A. vertebralis at its dural penetration, usual finding

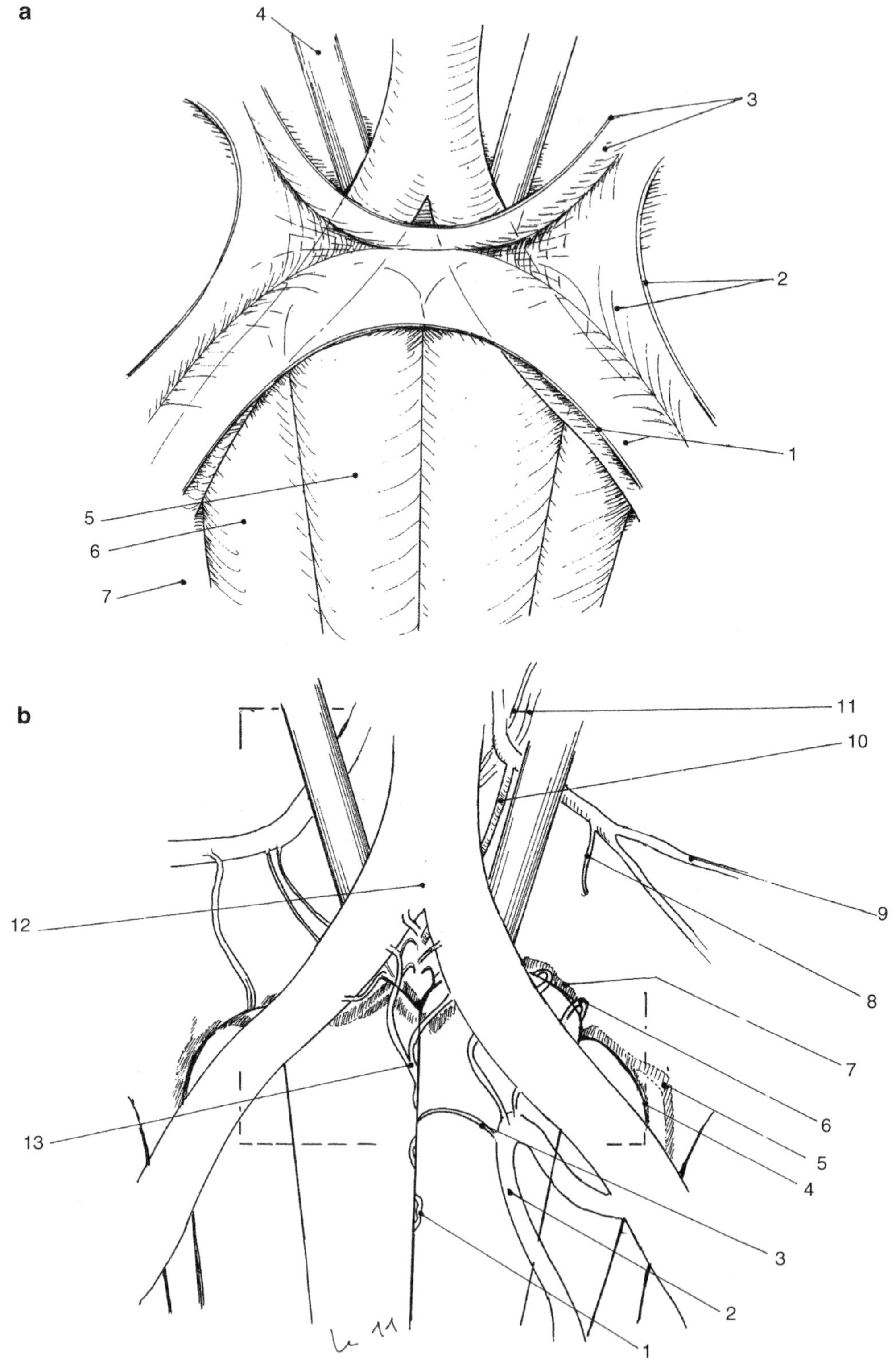

Fig. 15.38 Cisterns of the pontomedullary rim. (**a**) principles of its architecture (schematized), (**b**) masking of arachnoid walls by its contents and arachnoid adhesions. (**a**) *1* wall of cisterna medullaris medialis, *2* wall of cisterna pontocerebellaris, *3* wall of cisterna pontis medialis, *4* N. abducens, *5* pyramis medullae, *6* oliva, *7* pedunculus cerebellaris inferior. (**b**) *1* A. spinalis anterior; *2* PICA; *3* one of the perforating branches of PICA; *4* periolivar vein(s); *5* lateral medullary vein; *6* branch of PICA or A. vertebralis; *7* veins of the pontomedullary rim; *8* one of the perforating branches of AICA; *9* anterolateral main branch of AICA; *10* as 8, variant; *11* Rr. ad pontem of A. basilari; *12* vertebrobasilar junction; *13* A. spinalis anterior

Variants of Cisternal Arteries (Figs. 15.39, 15.40, 15.41, 15.42, 15.43, 15.44, and 15.45)

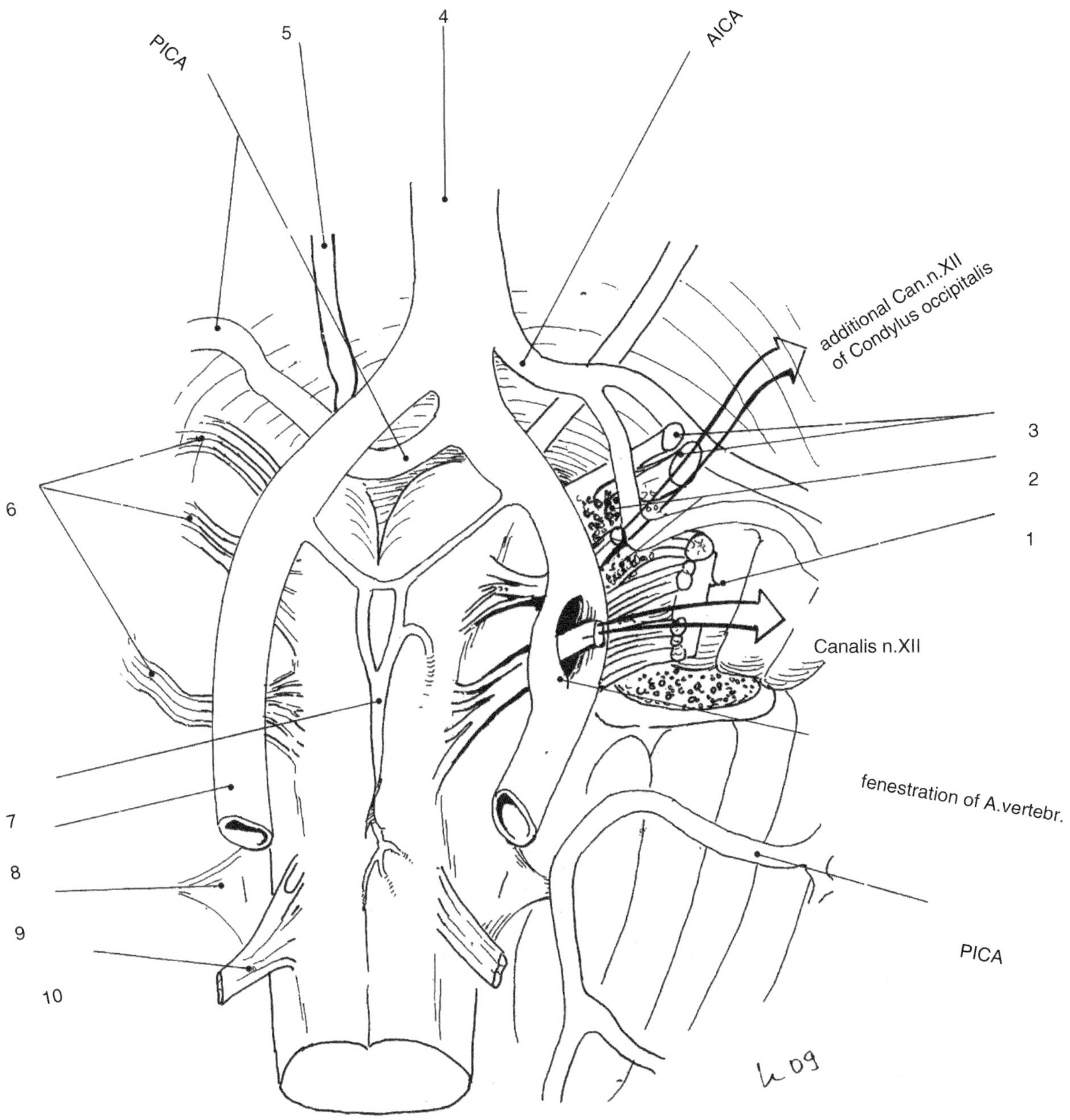

Fig. 15.39 Multiple variants (Lang 1979, schematized) of contents of pontomedullary cisterns. Example. Left PICA is feeding both sides (rare), fenestration of A. vertebralis (rare). Duplication of N. hypoglossus (common variant). *1* Nn. glossopharyngeus, vagus et accessorius, *2* plexus chorioideus, *3* N. statoacusticus et N. facialis, *4* A. basilaris, *5* N. abducens, *6* roots of N. hypoglossus, *7* A. spinalis anterior *8* A. vertebralis, *9* Lig. denticulatum, *10* N. cervicalis 1

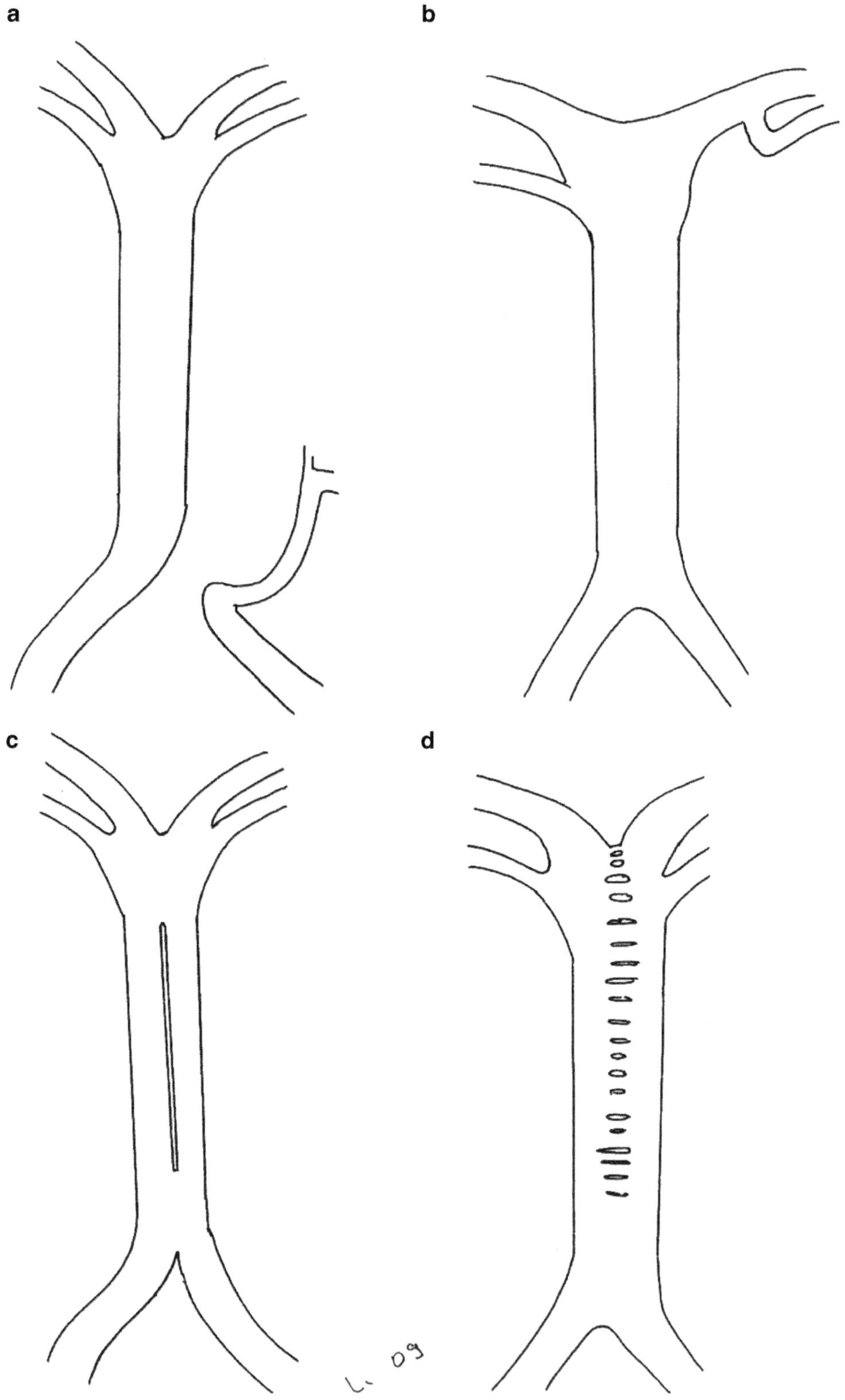

Fig. 15.40 Variants of A. vertebralis, A. basilaris, and A. cerebelli superior. Examples (**a**) left-sided A. vertebralis, (**b**) left-sided A. cerebelli superior, (**c**) and (**d**) A. basilaris. (**a**) and (**b**) Common findings, (**c**) and (**d**) (cadaver brain dissections, experiences of the author) rare variants

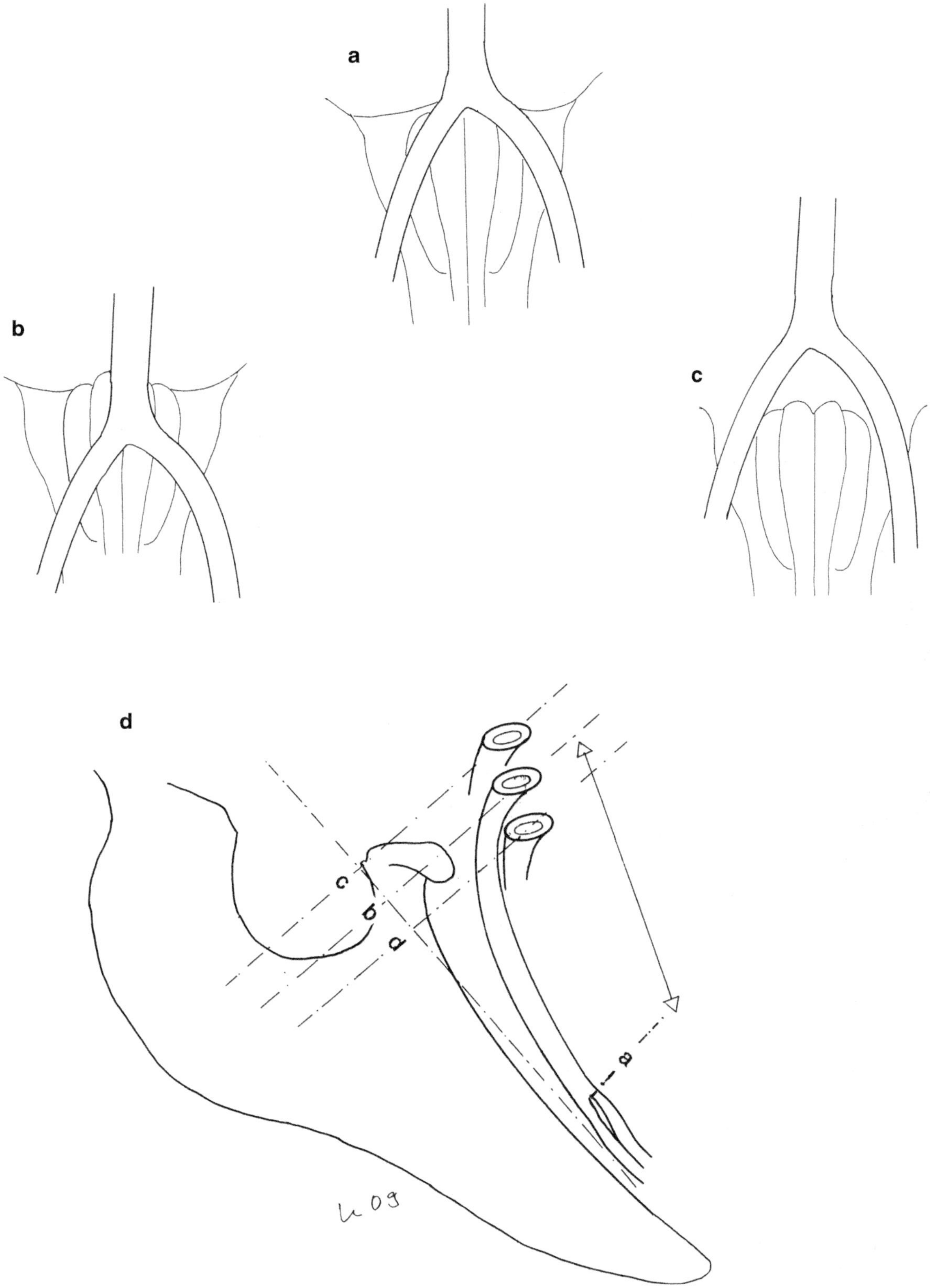

Fig. 15.41 Variants of the junction of Aa. vertebrales and of the terminal segment of A. basilaris. (**a**) Usual finding, (**b**) and (**c**) rare, (**d**) (Krayenbühl and Yaşargil 1957; Saecki and Rhoton 1977; Lang 1991, p 35) *a* level of vertebrobasilar junction as (**a**), *b* level of bifurcation approx. 50%, *c* and *d* common findings

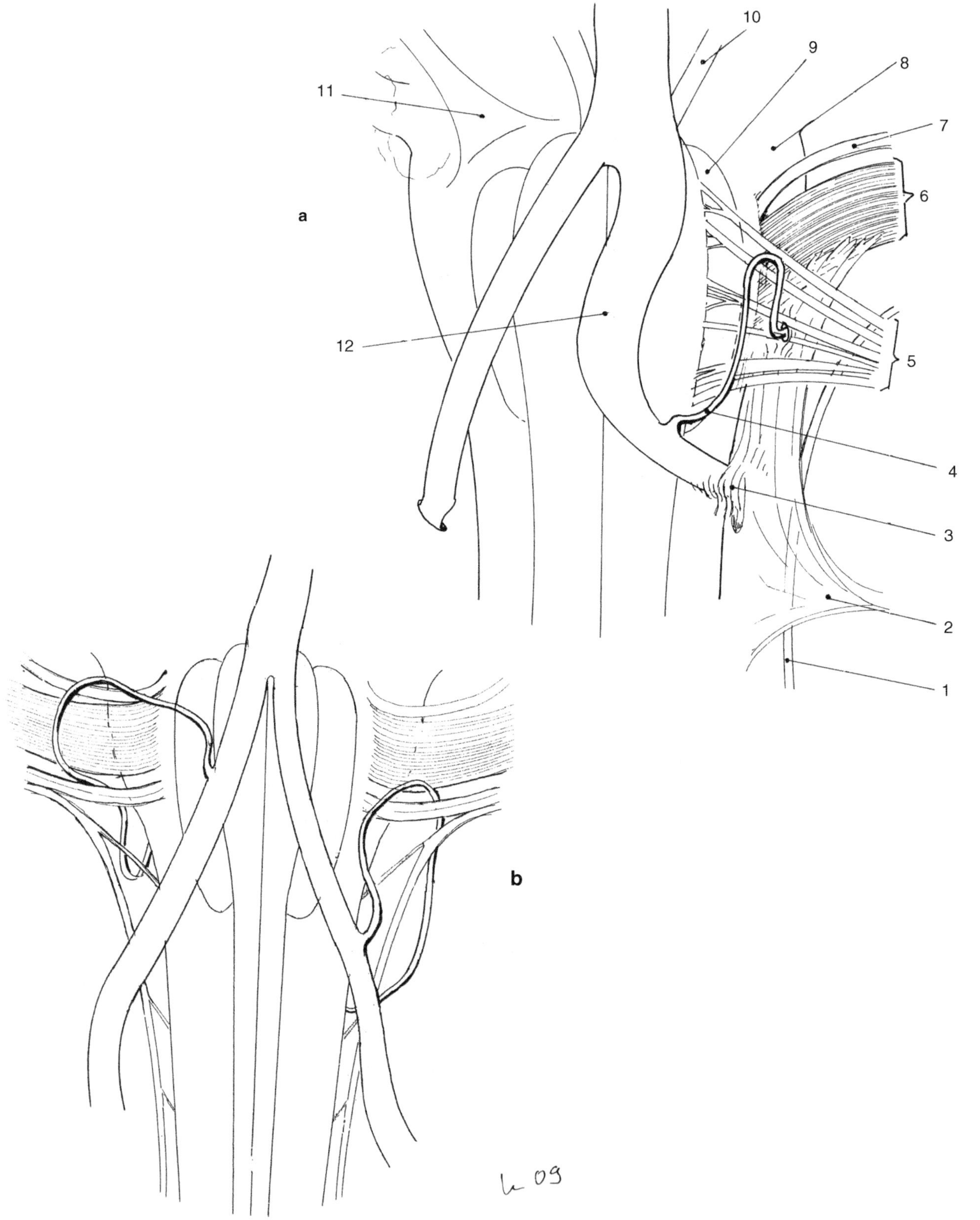

Fig. 15.42 Variants of PICA. Examples (Lang 1991, modified presentation). Further variants: (**a**) Approx. 50%, (**b**) Approx. 30%. *1* N. accessorius, radix spinalis; *2* Lig. denticulatum; *3* retinaculum of A. vertebralis, usual finding; *4* PICA; *5* N. hypoglossus; *6* N. vagus et accessorius; *7* N. glossopharyngeus; *8* pedunculus cerebellaris inferior; *9* oliva; *10* N. abducens; *11* N. facialis; *12* dolichoectasy of A. vertebralis (elderly individual)

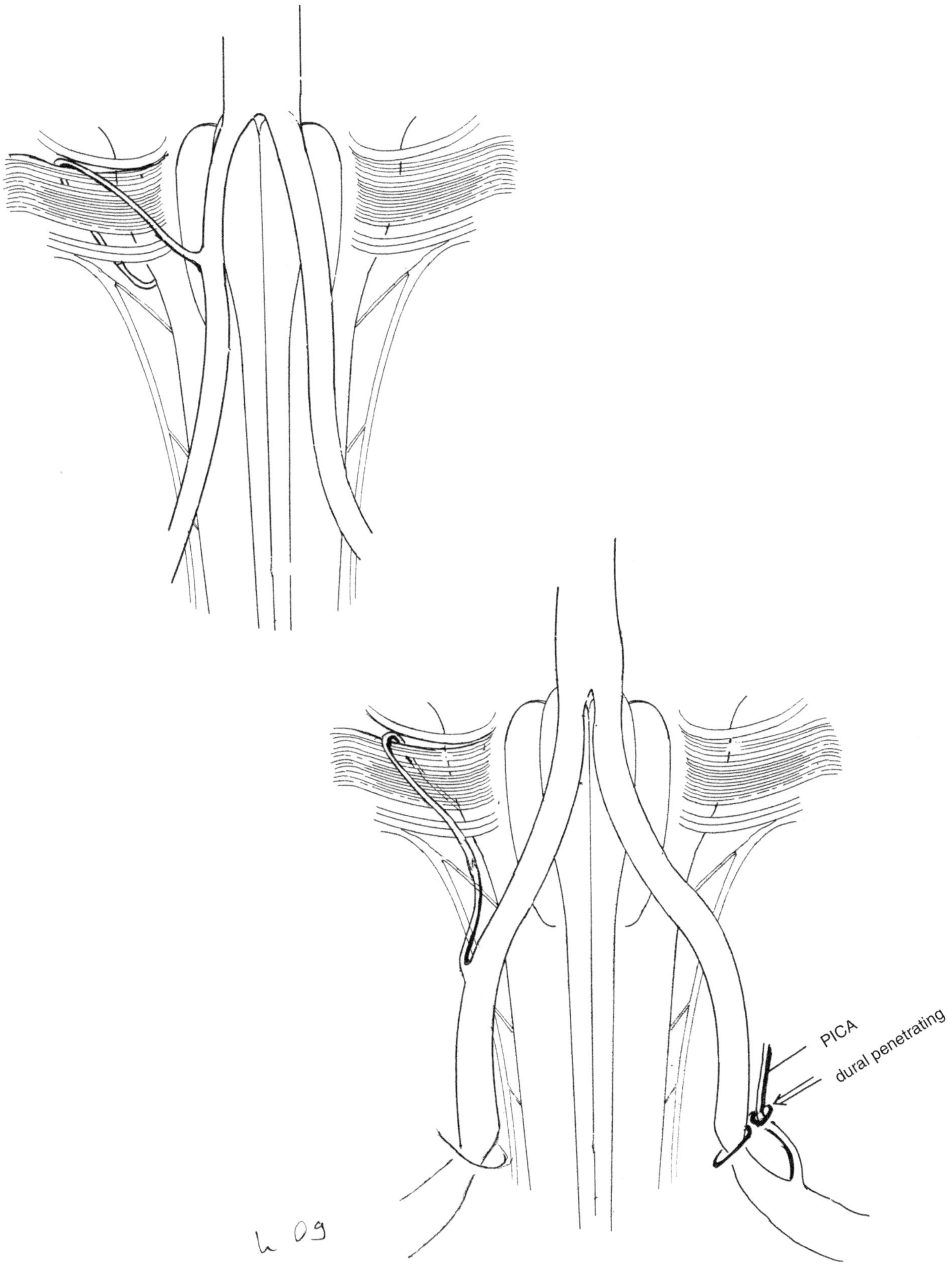

Fig. 15.43 Continuation of Fig. 15.42. The extradural origin of PICA (inferior presentation) is a rare variant (Lang 1985, anatomical dissection). At surgery, it could be confound with a muscle feeding artery of the neck. For surgical aspects see Fig. 16.26 B 3. Further variants are hypoplasia or aplasia of PICA. It could be substituted by hyperplasia of AICA. Hypoplasia of AICA could be substituted by hyperplasia of PICA

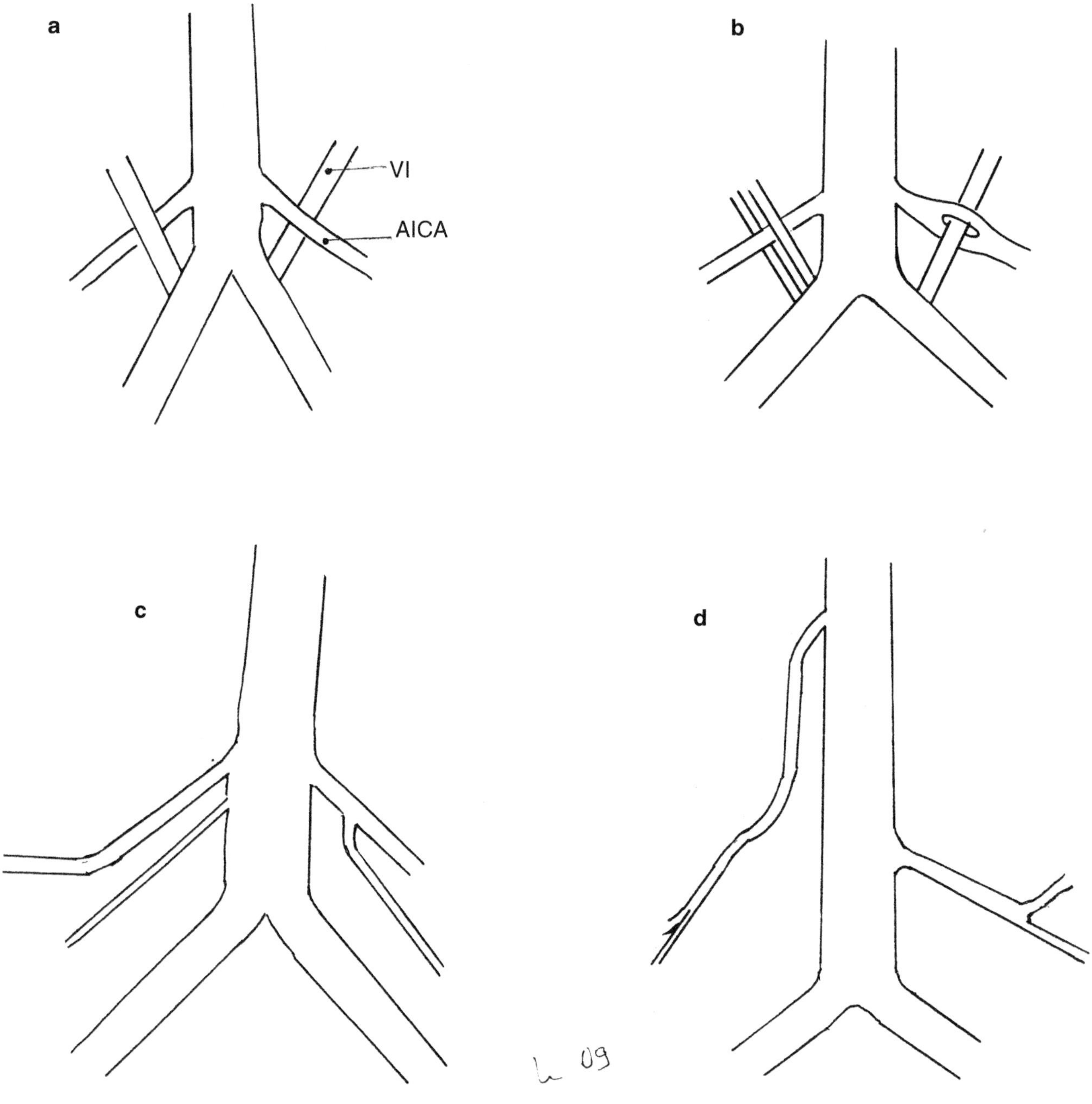

Fig. 15.44 AICA (A. cerebelli inferior anterior (Scialfa et al. 1976). Variants (**a**) N. abducens overcrosses AICA. Undercrossing is possible, (**b**) rare variants, (**c**) common variants of the main branches, (**d**) distal origin of AICA rare. See right side VI N. abducens. All segments of AICA could present perforating branches. Cave

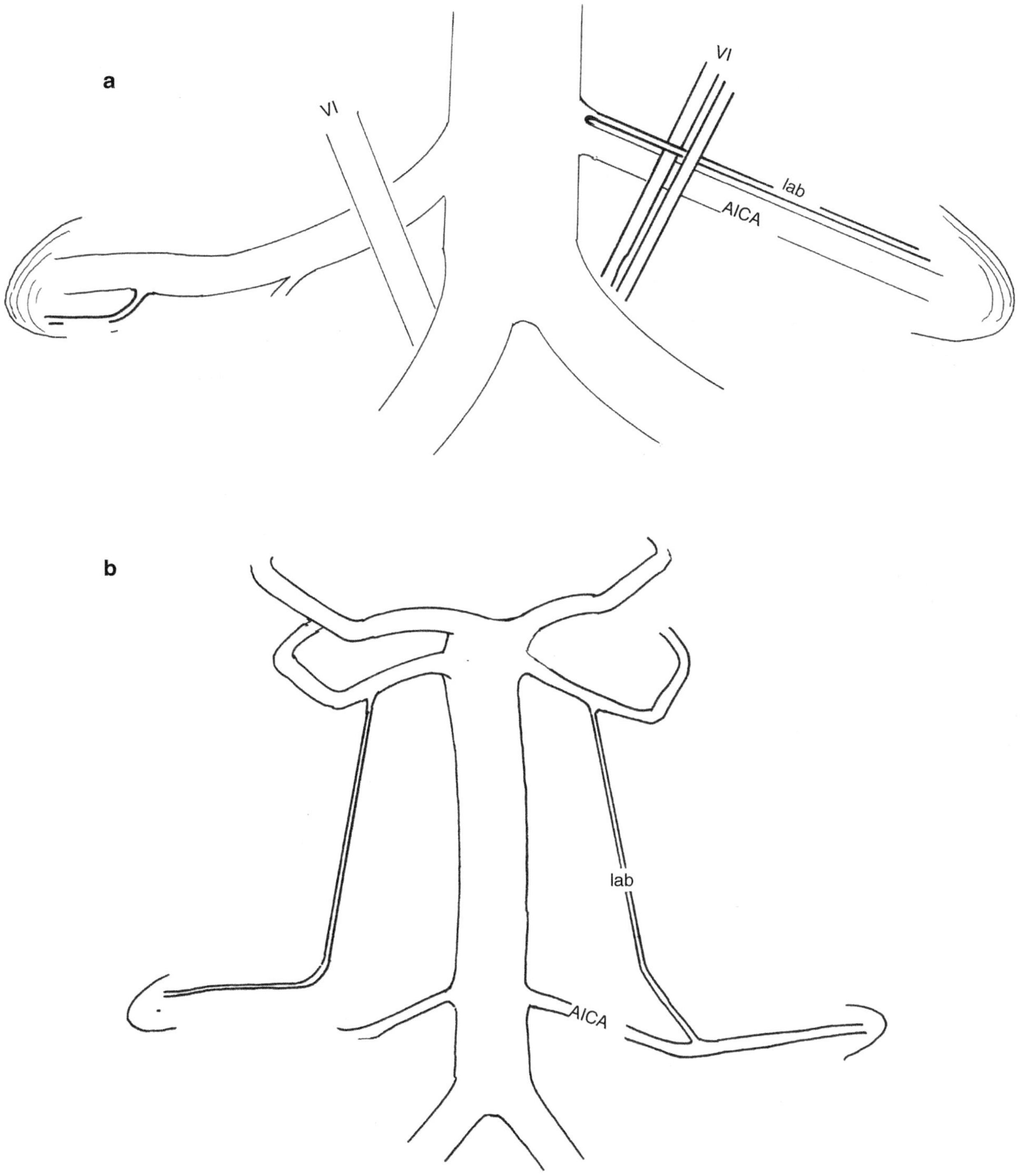

Fig. 15.45 Variants of A. labyrinthi (lab), AICA, and N. abducens (VI). (**a**) Usual findings (Lang 1979) and contralateral variants, (**b**) variants of A. labyrinthi (Wende and Nakayama 1975)

Adjacent Basal Structures (Figs 15.46, 15.47, 15.48, 15.49, 15.50, 15.51, and 15.52)

Extradural structures which are presented according to the intra-extradural way of some neurosurgical approaches

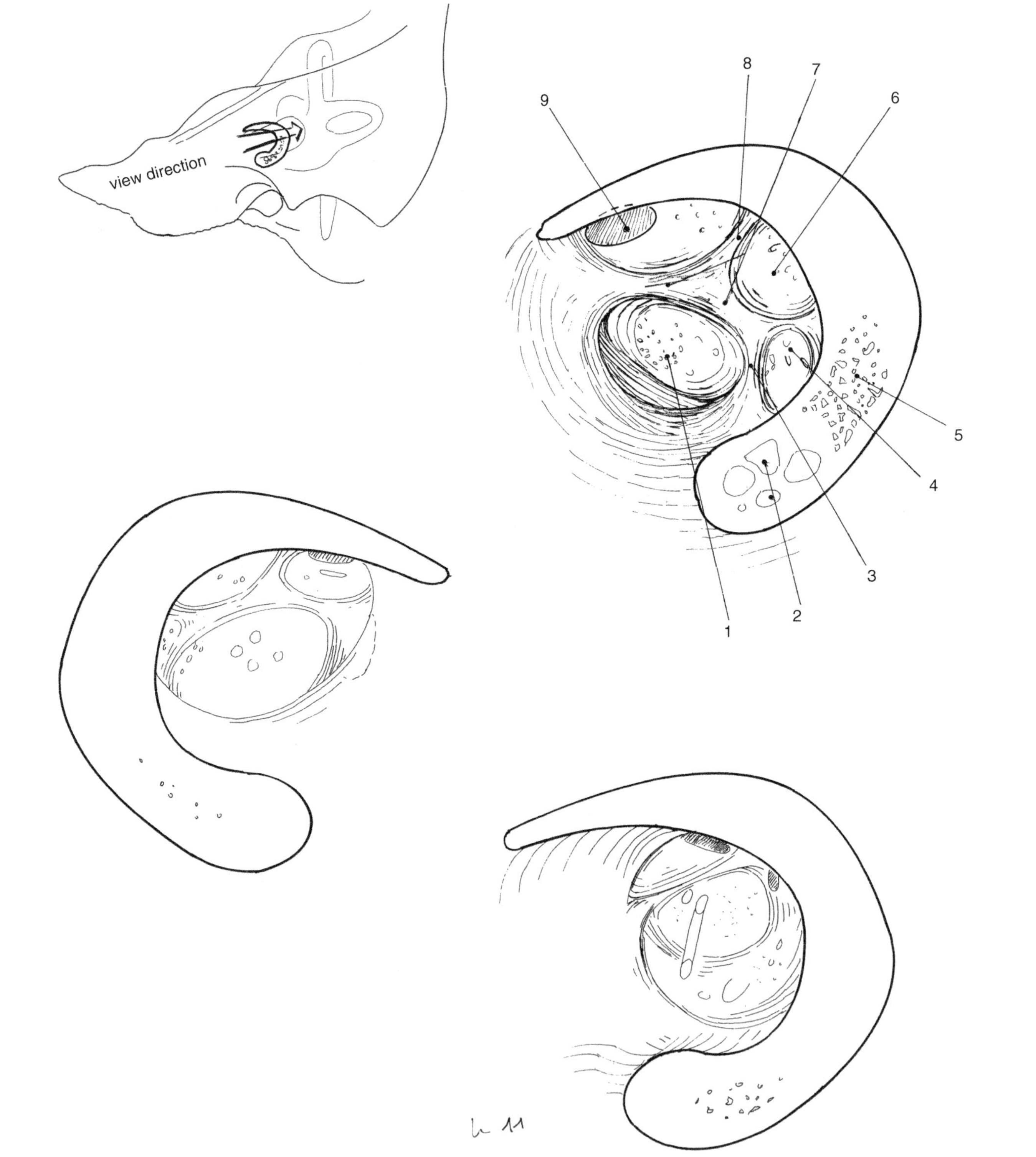

Fig. 15.46 Fundus of meatus acusticus internus (Rauber-Kopsch 1987, p 641 modified, and skull dissections of the author). Meatus acusticus internus resected similar to surgery. Dura and leptomeninges omitted. *1* tractus spiralis foraminosus, *2* mastoid cells, *3* crista verticalis inferior, *4* area vestibularis inferior, *5* diploe, *6* area vestibularis superior, *7* crista transversa, *8* crista verticalis superior (Rhoton, 2000, not constant), *9* area n. facialis

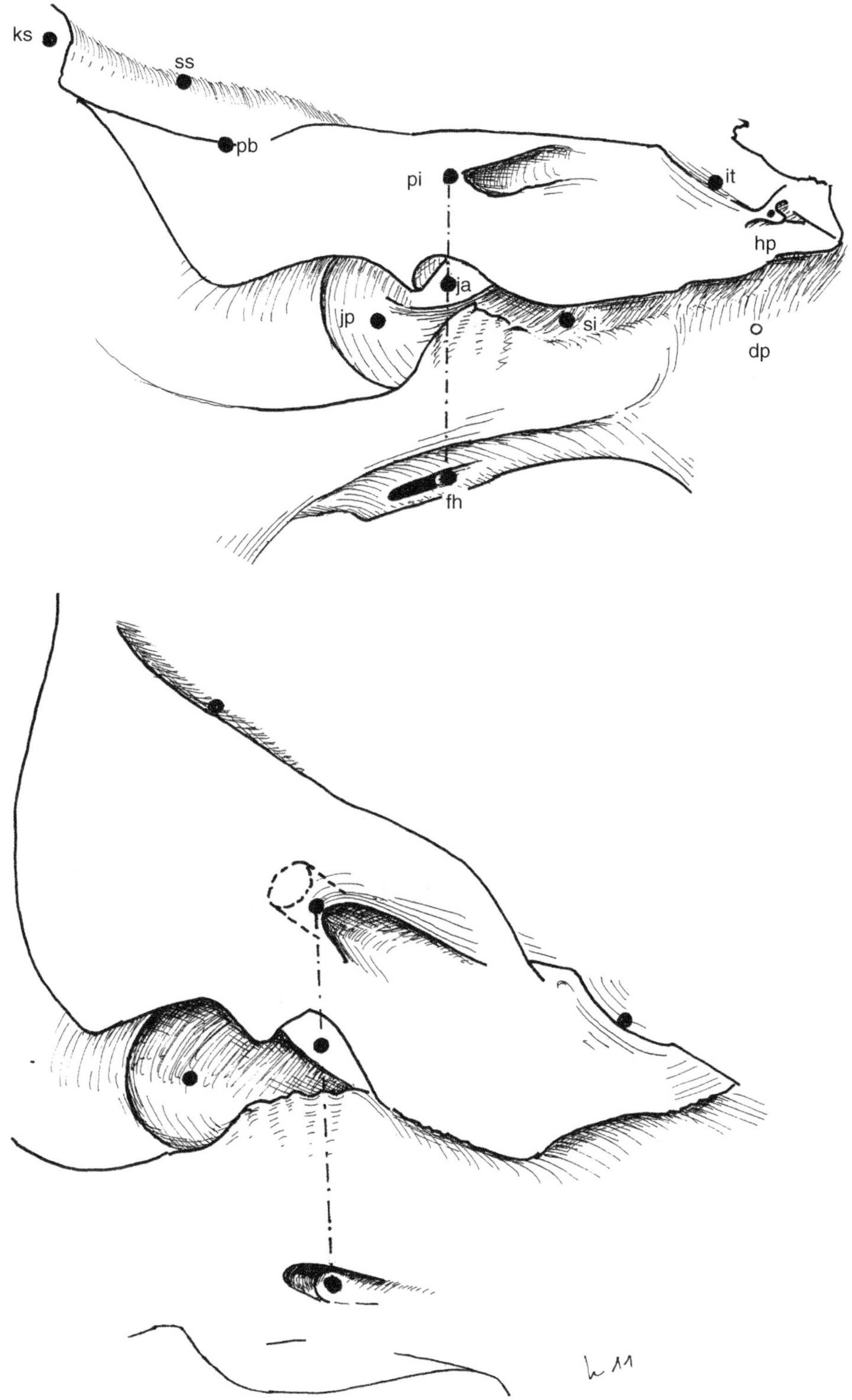

Fig. 15.47 Petrooccipital landmarks. Bony landmarks may be helpful, if details of foramen jugulare or canalis n. hypoglossi are not exact to define by imaging. For variants of canalis n. hypoglossi and a bony channel for N. glossopharyngeus, see Figs. 15.50, 15.51, and 15.52. Canalis n. hypoglossi and adjacent bony structures present numerous veins, which must be considered at transcondylar surgical approaches. *dp* dural penetration of N. abducens, projection, *fh* foramen n. hypoglossi, *hp* hamulus petrosus enclosing N. abducens, *it* incisura trigeminalis, *ja* foramen jugulare, anterior segment, enclosing Nn. IX to XI, *jp* foramen jugulare, posterior segment, enclosing bulbus superior v.jugul. Internae, *ks* knee of sulcus transversus, *pb* superior margin of Os petrosum, *pi* porus acusticus internus, *ss* sulcus petrosus superior, *pi-ja-fh* interforaminal line

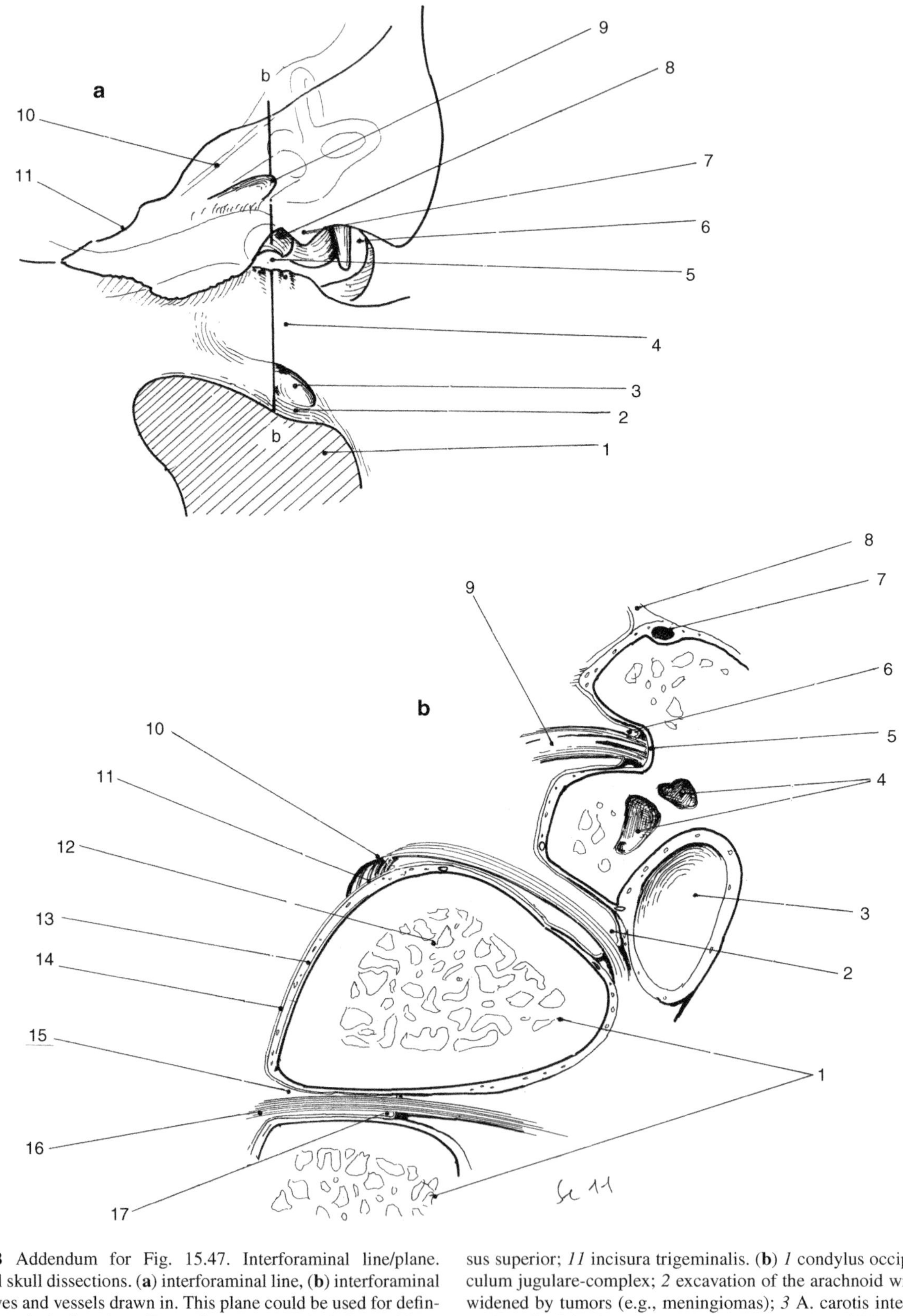

Fig. 15.48 Addendum for Fig. 15.47. Interforaminal line/plane. Anatomical skull dissections. (**a**) interforaminal line, (**b**) interforaminal plane. Nerves and vessels drawn in. This plane could be used for defining of landmarks by imagings. For variants see Fig. 2.48 ff. (**a**) *1* foramen occipitale; *2* condylus occipitalis; *3* foramen n. hypoglossi; *4* tuberculum jugulare; *5* foramen jugulare, anterior segment; *6* as 5, posterior segment; *7* processus intrajugularis; *8* apertura externa canalis carotici; *9* porus acusticus internus, posterior margin; *10* sulcus petrosus superior; *11* incisura trigeminalis. (**b**) *1* condylus occipitalis tuberculum jugulare-complex; *2* excavation of the arachnoid wrap could be widened by tumors (e.g., meningiomas); *3* A. carotis interna; *4* pneumatic cells; *5* fundus of meatus acusticus internus; *6* CSF space; *7* sinus petrosus superior; *8* tentorium; *9* N. facialis et intermedius; *10* Nn. glossopharyngeus, vagus et accessorius; *11* arachnoidea; *12* tuberculum jugulare; *13* Dura; *14* as 11; *15* CSF space; *16* N. hypoglossus; *17* excavation of arachnoid wrap as 2

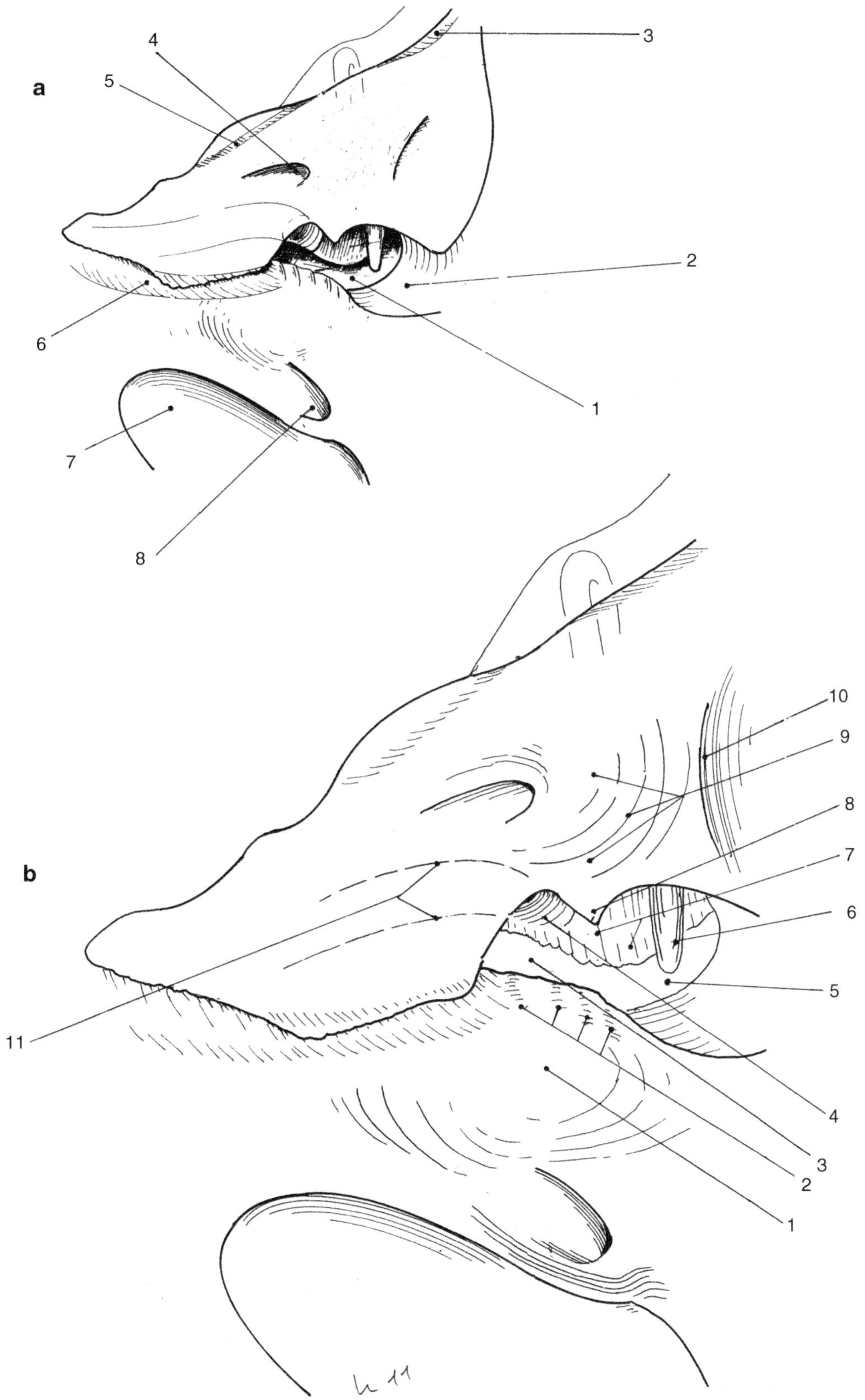

Fig. 15.49 Foramen jugulare and adjacent structures. View direction into a large variant of foramen jugulare. (**a**) Survey, (**b**) details. (**a**) *1* foramen jugulare, *2* sulcus sigmoideus, *3* sulcus petrosus superior, *4* porus acusticus internus, *5* as 3, *6* sulcus petrosus inferior, *7* foramen occipitale, *8* foramen n. hypoglossi. (**b**) *1* tuberculum jugulare; *2* sulci at 1 according to Nn. IX to XI; *3* foramen jugulare, anterior segment; *4* apertura externa of canalis caroticus; *5* foramen jugulare, posterior segment; *6* processus styloideus; *7* processus vaginalis; *8* processus intrajugularis; *9* bulging of os petrosum, enclosing ductus et saccus endolymphaticus; *10* bony fissure, enclosing 9; *11* canalis caroticus. Projection

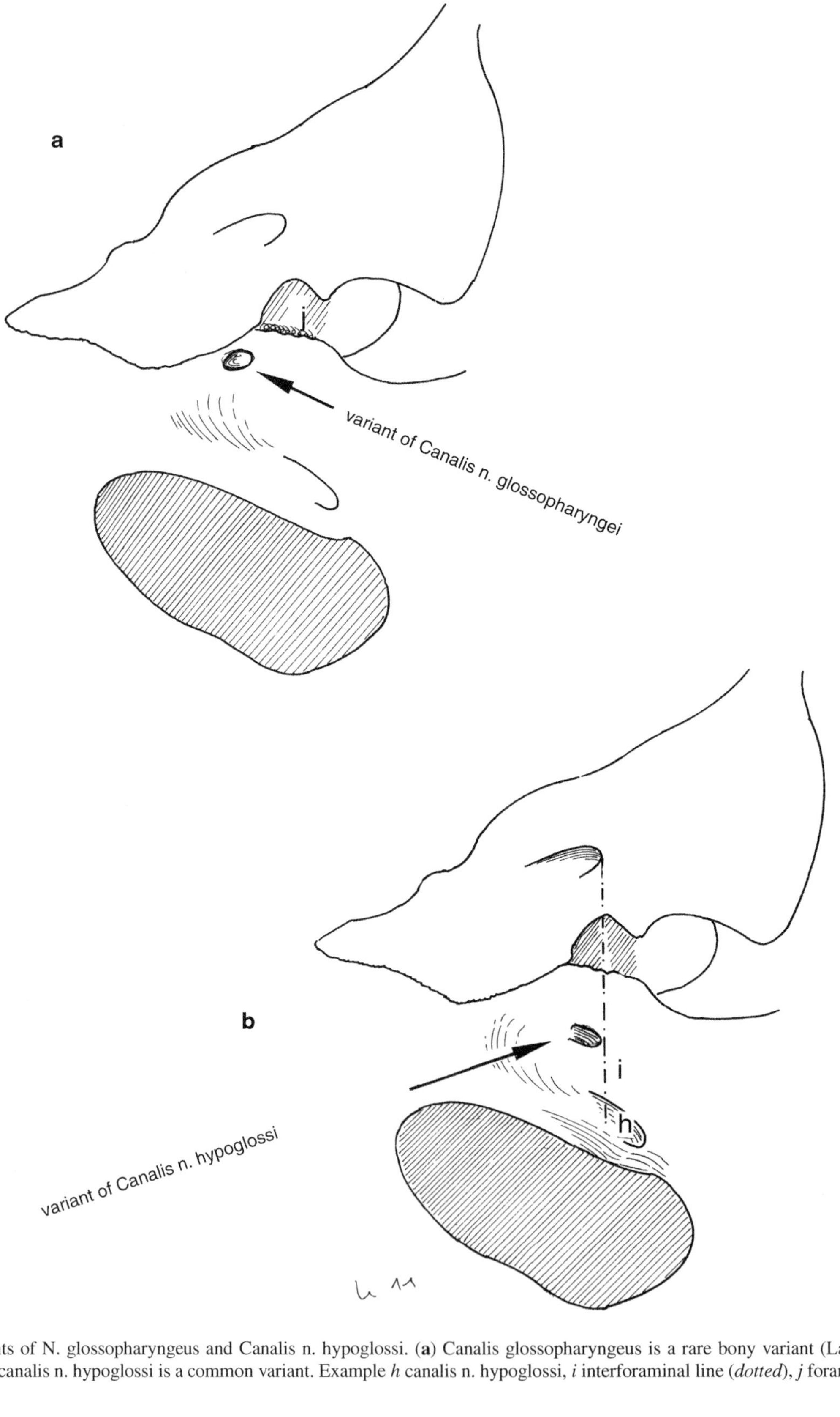

Fig. 15.50 Variants of N. glossopharyngeus and Canalis n. hypoglossi. (**a**) Canalis glossopharyngeus is a rare bony variant (Lang 1979, p 391). (**b**) Duplication of canalis n. hypoglossi is a common variant. Example *h* canalis n. hypoglossi, *i* interforaminal line (*dotted*), *j* foramen jugulare

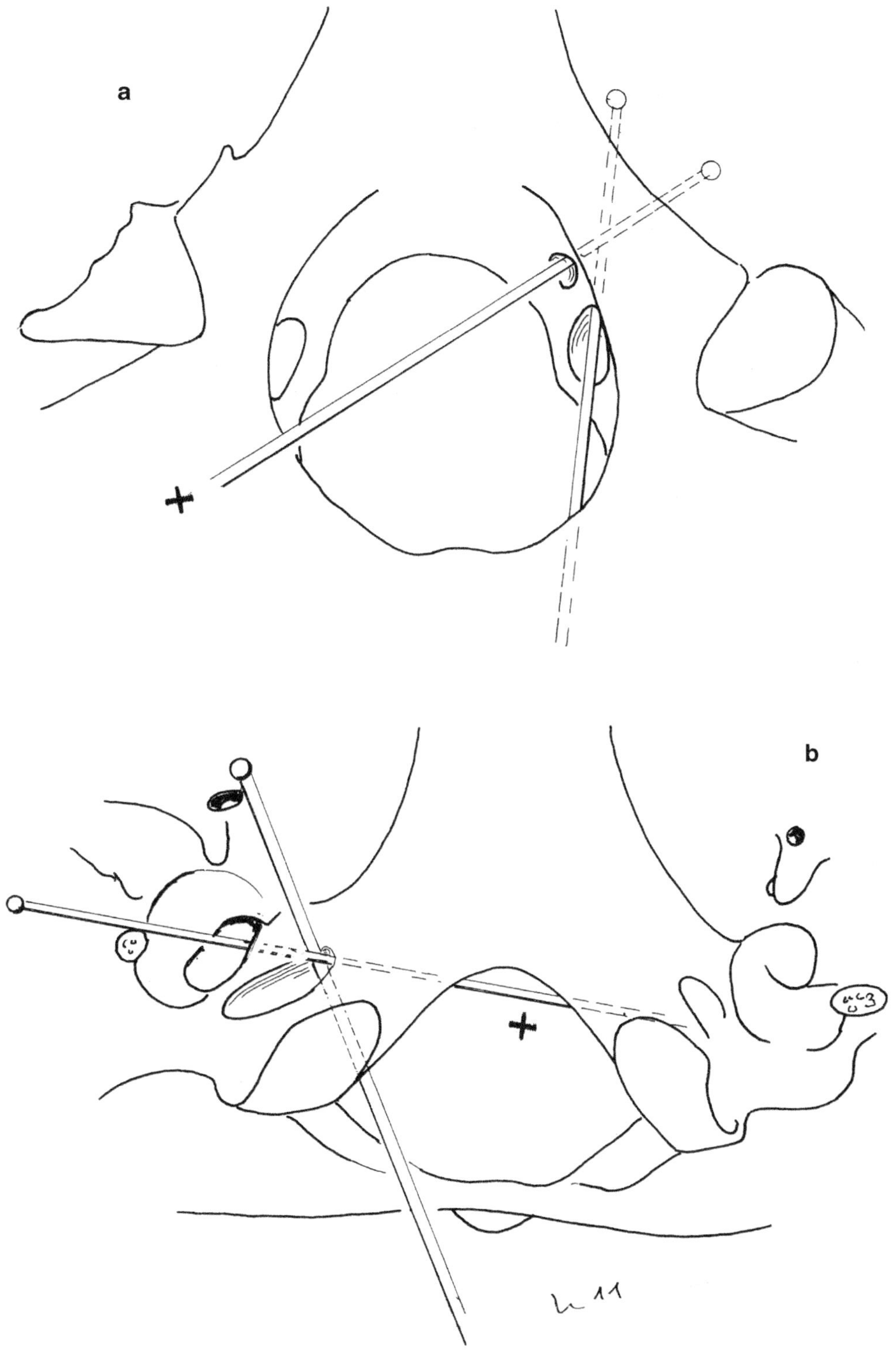

Fig. 15.51 Intra- and extracranial presentation of canalis n. hypoglossi Common finding, schematized. (**a**) Probes introduced into canalis n. hypoglossi and – superior to it – into the accessorial canalis n. hypoglossi – + –, (**b**) note the extracranial short distance measurement between canalis n. hypoglossi and foramen jugulare. This region presents numerous veins, draining to sinus petrosus inferior and bulbus superior of V. jugularis interna

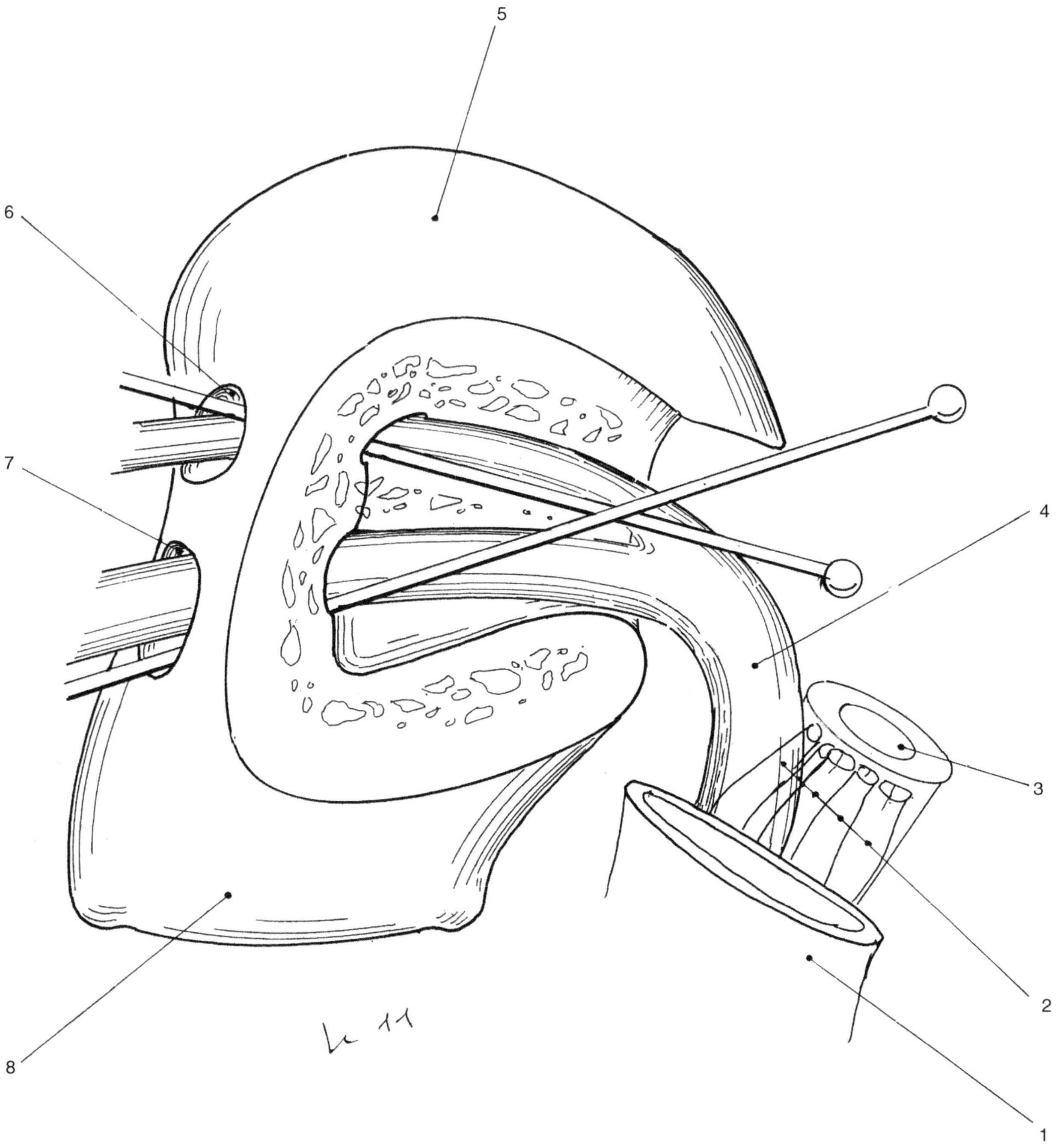

Fig. 15.52 Intra- and extradural presentation of a duplicated canalis n. hypoglossi. Common finding. Vessels and Nn. IX to XI added. *1* V. jugularis interna; *2* Nn. glossopharyngeus, vagus et accessorius; *3* A. carotis interna; *4* N. hypoglossus; *5* tuberculum jugulare; *6* accesory foramen n. hypoglossi distant to the interforaminal line; *7* foramen n. hypoglossi located at the interforaminal line; *8* condylus occipitalis

15.2.4 Cisterna Tecti and Fissura Horizontalis Cerebelli (Figs. 15.53, 15.54, 15.55, 15.56, 15.57, 15.58, and 15.59)

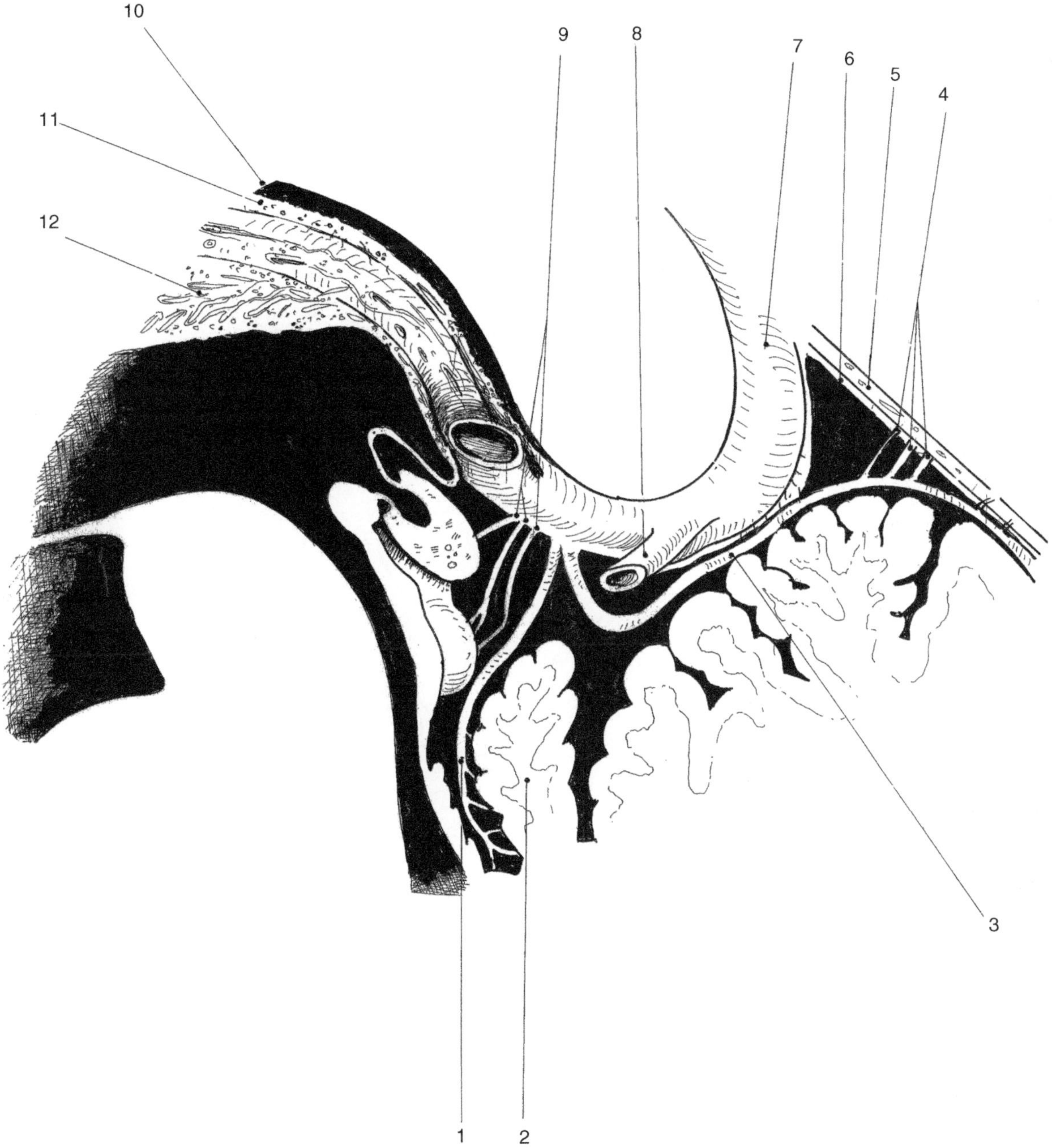

Fig. 15.53 Cisterna tecti and surrounding regions. Survey. *1* V. cerebellaris praecentralis, *2* lobulus centralis, *3* V. supraculminalis, *4* medial sagittal veins, *5* tentorium, *6* arachnoid layer, *7* V. magna (Galeni), *8* V. basalis (Rosenthal), *9* Vv. tecti, *10* fissura transversa, *11* velum interpositum, *12* plexus chorioideus

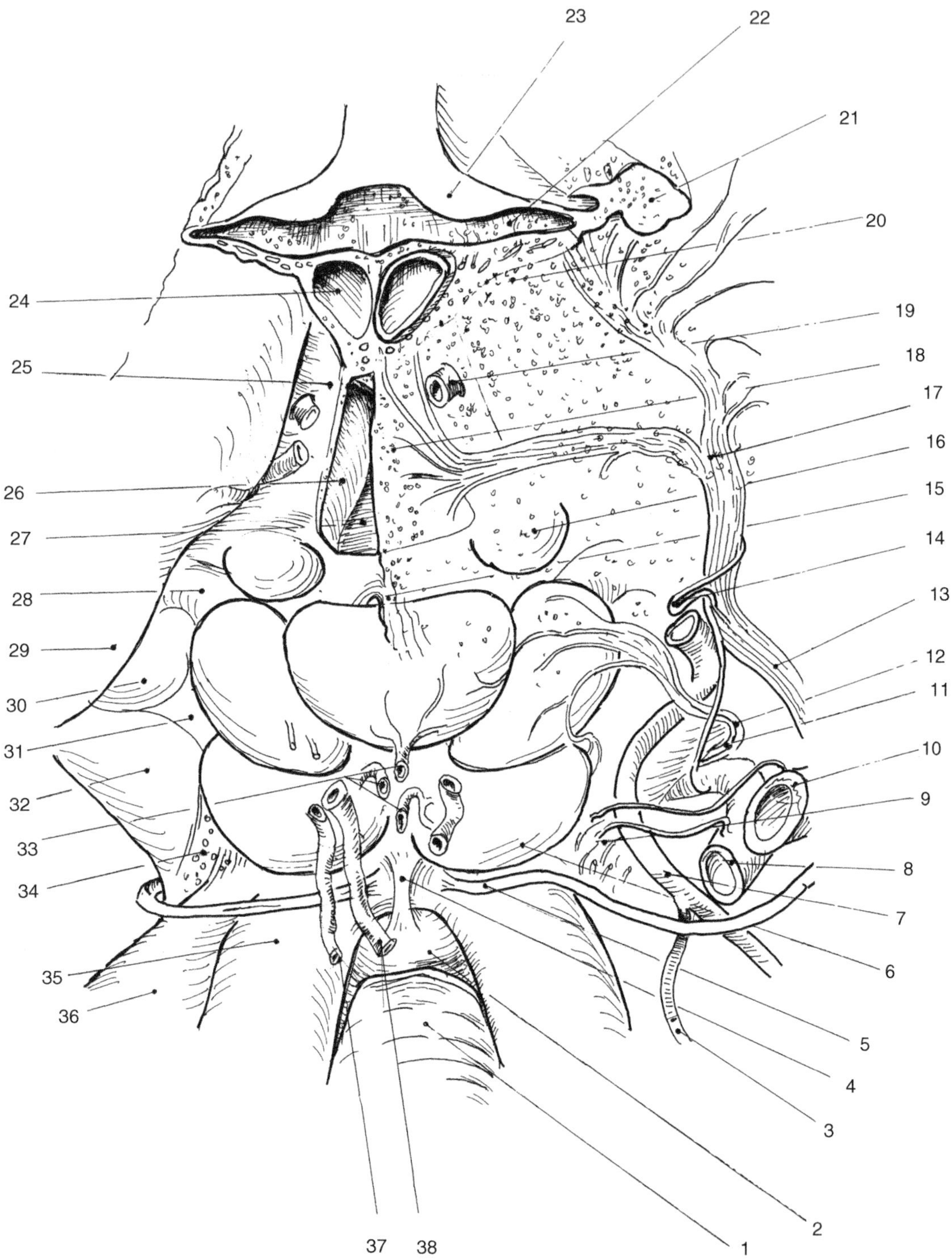

Fig. 15.54 Ground of cisterna tecti and adjacent CSF spaces. *1* lingula cerebelli; *2* velum medullare anterius; *3* V. mesencephalica lateralis; *4* *Frenulum tecti*; *5* *N. trochlearis*; *6* colliculus inferior; *7* V. basalis (Rosenthal), small variant; *8* A. cerebelli superior; *9* vessels of trigonum lemnisci; *10* A. cerebri posterior; *11* A. tecti; *12* A. chorioidea posterior medialis; *13* ramification of A. chorioidea posterior lateralis; *14* Baumgartner's loop for *localization of pulvinar thalami* during surgery; *15* Vv. tecti, drainage to the galenic vein; *16* *Habenula*, landmark for MRT and landmark during surgery for 17; *17* division of branches of A. chorioidea posterior lateralis to a medial and lateral bundle of arteries feeding plexus chorioideus; *18* tela chorioidea of the third ventricle (medial segment of velum interpositum); *19* V. dorsalis thalami, inconstant; *20* velum interpositum (lateral region); *21* plexus chorioideus of the lateral ventricle; *22* fissura transversa cerebri; *23* corpus fornicis; *24* V. cerebri interna; *25* taenia chorioidea ventriculi III; *26* thalamus; *27* hypothalamus; *28* brachium colliculi superioris; *29* pulvinar thalami; *30* *corpus geniculatum mediale*; *31* brachium colliculi inferiori; *32* crus cerebri; *33* Vv. tecti; *34* trigonum lemnisci; *35* pedunculus cerebellaris superior; *36* pedunculus cerebellaris medius; *37* V. supraculminalis; *38* *V. cerebellaris praecentralis*. Underlines: landmarks

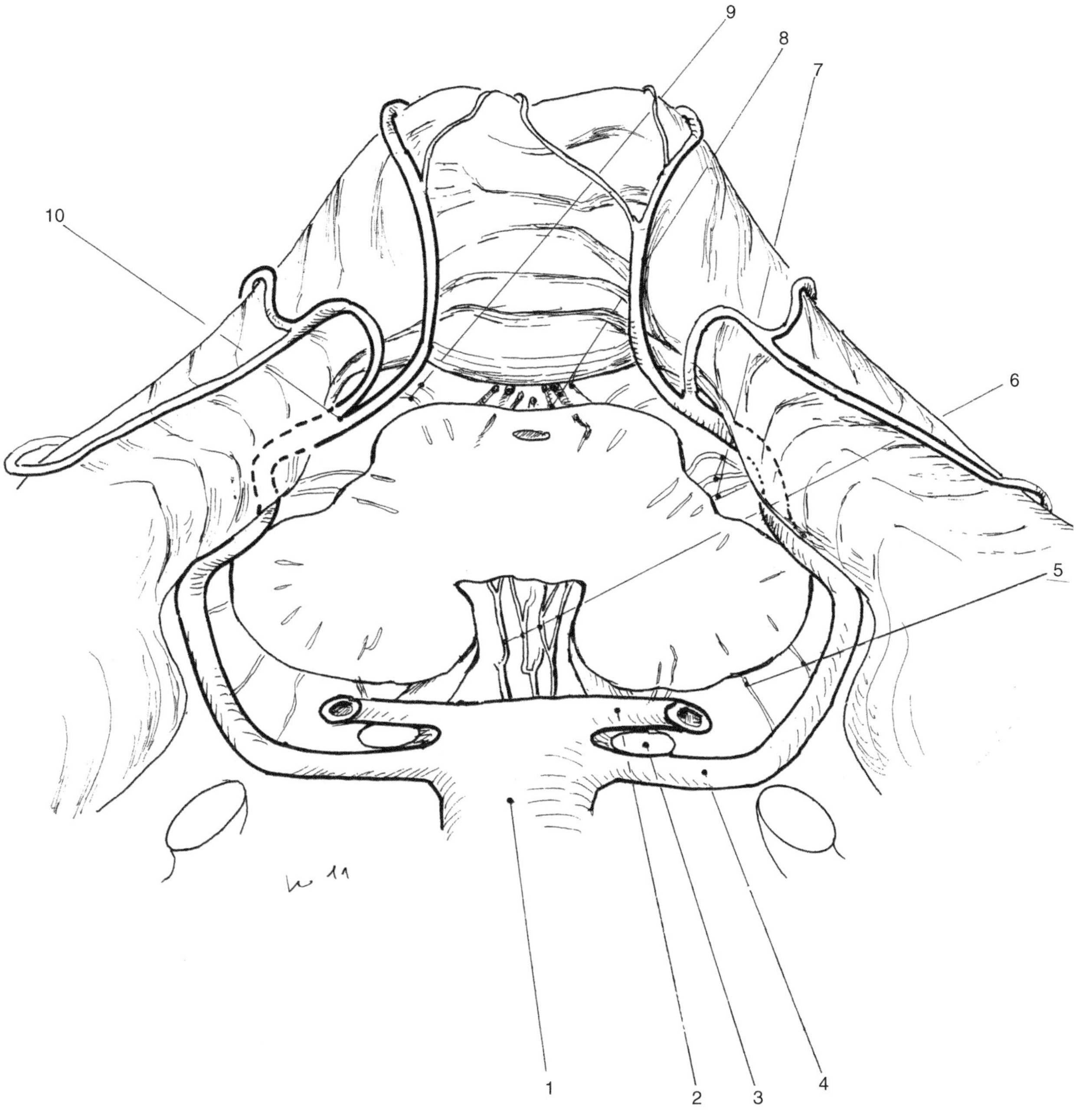

Fig. 15.55 Anterior cerebello-mesencephalic region. *1* A. basilaris, *2* A. cerebri posterior, *3* N. oculomotorius, *4* A. cerebelli superior, *5* perforating branches of 4, *6* Aa. thalamoperforantes, *7* perforating branches at trigonum lemnisci, *8* Vv. tecti, *9* Aa. tecti, *10* A. cerebelli superior, division

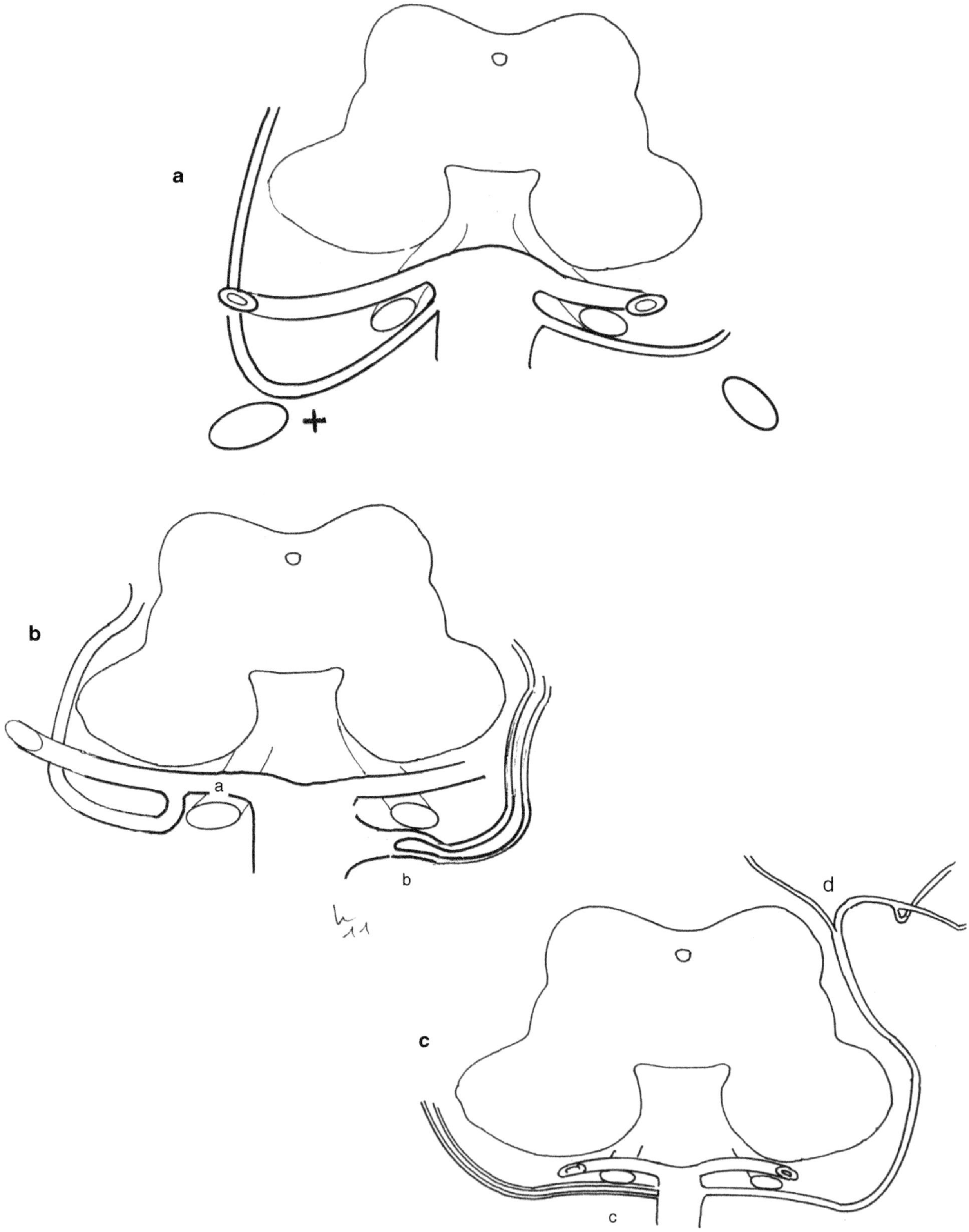

Fig. 15.56 Variants of A. cerebelli superior. (**a**) Loop close to N. trigeminus (+): Anatomical base of Janetta's operation (Janetta 1967, Janetta and Rand 1966); (**b**) *a* A. cerebelli superior originates from A. cerebri posterior (more than 10%), *b* proximal division; (**c**) *c* duplication, *d* distal division, usual finding

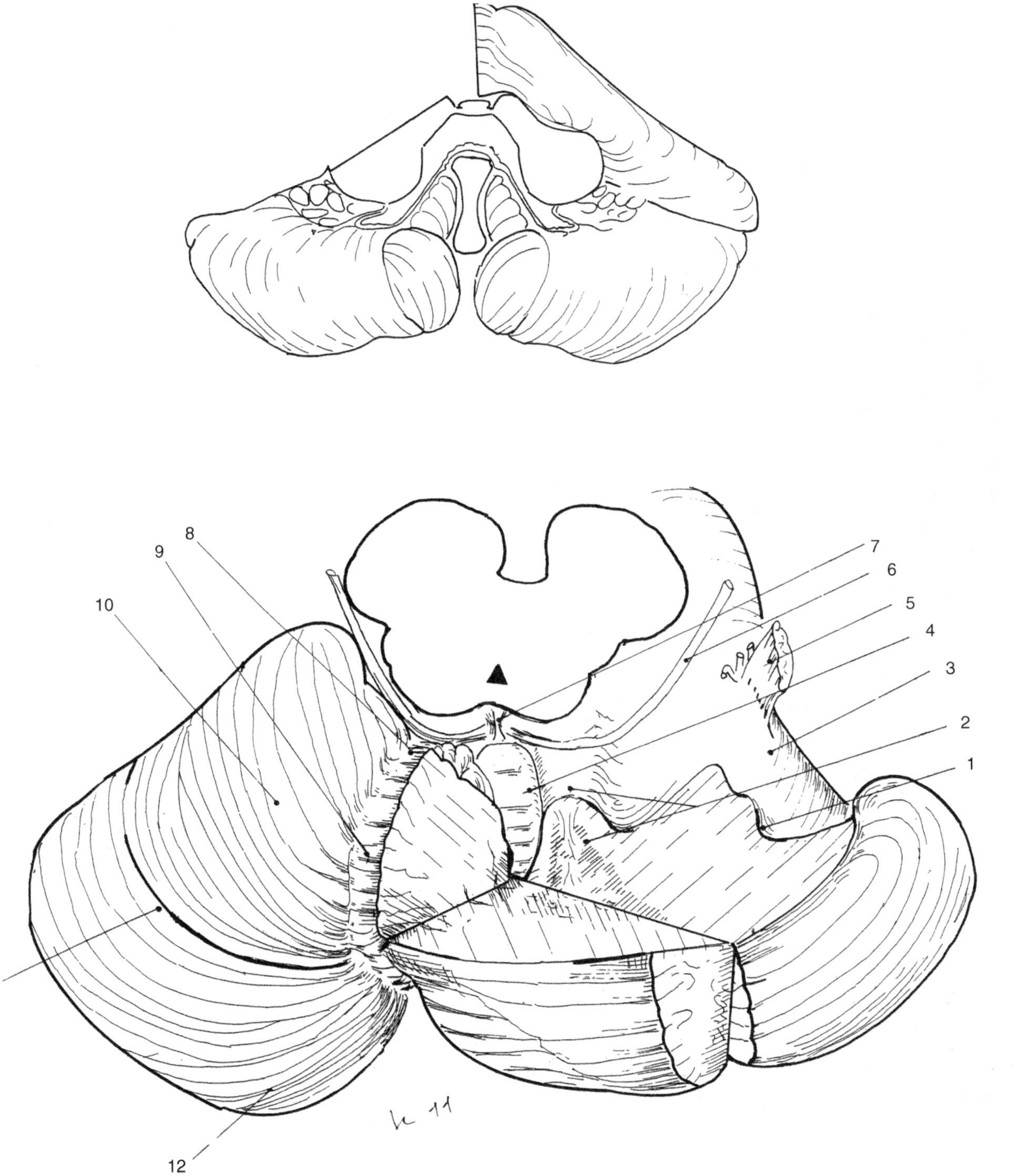

Fig. 15.57 Fissura horizontalis cerebelli. This CSF space is connected with cisterna tecti, cisterna ambiens, and adjacent cisterns. It contains venous anastomoses between supra- and infratentorial CSF spaces (Fig. 2.56 f). The anterior segment of fissura horizontalis cerebelli is overlayed by lobulus quadrangularis (which is resected at the right side). The ground of cisterna horizontalis presents the dorsal surface of pedunculus cerebellaris medius and superior. The anterior medial segment of fissura horizontalis is located close to the anterior segment of nucleus dentatus (2). *1* transectional plane, lateral limit; *2* pedunculus cerebellaris superior et nucleus dentatus; *3* pedunculus cerebellaris medius, ground of fissura horizontalis cerebelli; *4* lingula cerebelli; *5* N. trigeminus; *6* N. trochlearis; *7* frenulum tecti; *8* culmen; *9* declive; *10* lobulus quadrangularis; *11* fissura prima; *12* lobulus semilunaris superior and inferior

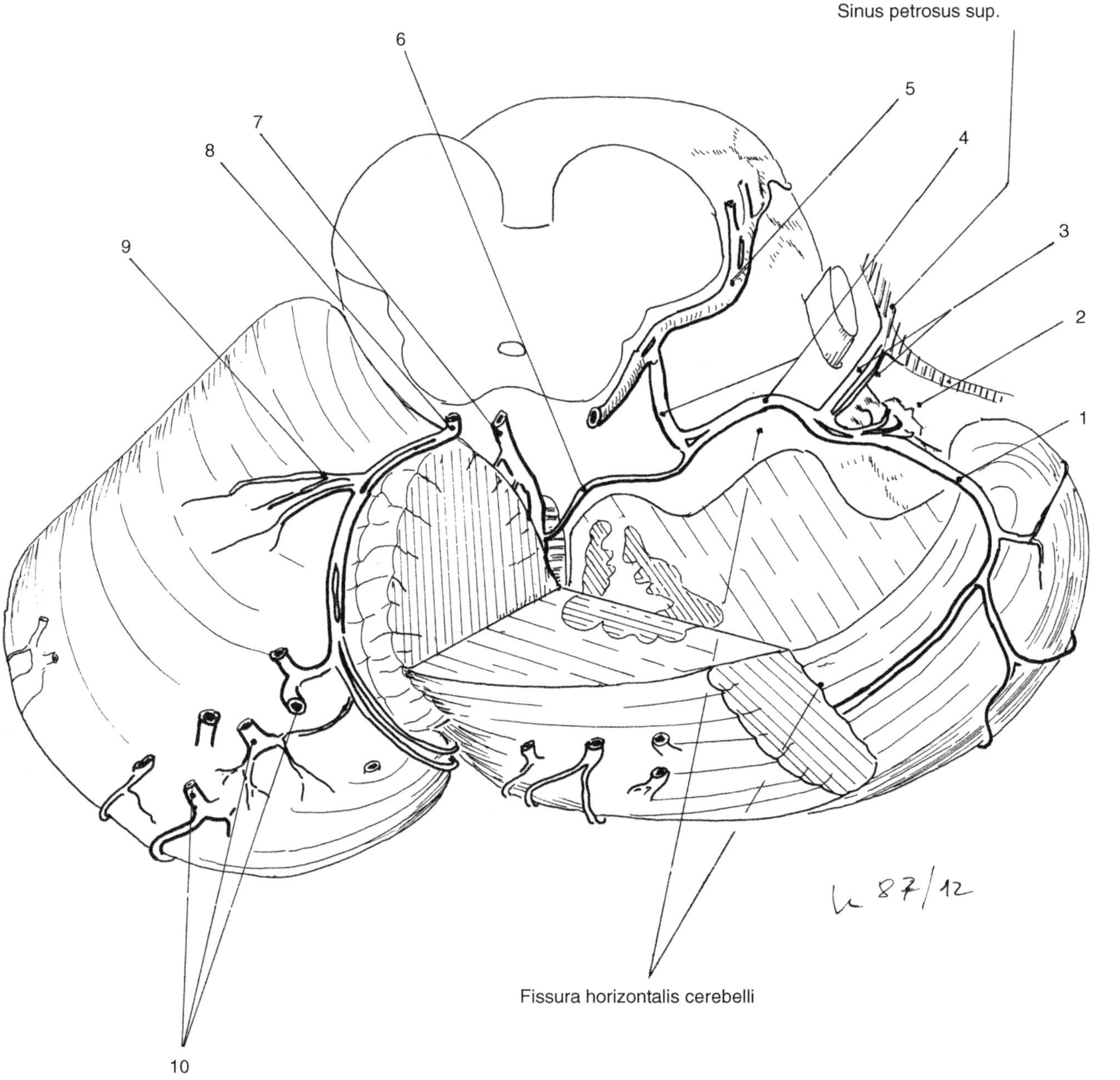

Fig. 15.58 Anastomotic veins of fissura horizontalis cerebelli. Essential is a widened variant of V. mesencephalica lateralis. It should not be interrupted at surgery. *1* V. cerebellaris horizontalis, vein of the posterior segment of fissura horizontalis cerebelli; *2* lateral pontine veins; *3* V. petrosa superior, thick-calibrated variant (rare) (after surgical interruption of V. petrosa sup., the compensatory function of anastomotic veins could be insufficient (Seeger 1984, pp. 348 to 351, example of a widening of V. petrosa sup.), *4* V. mesencephalica lateralis connecting supra- and infratentorial veins (5 with 1 to 3); *5* V. basalis (Rosenthal); *6* veins connecting medial with lateral veins of fissura horizontalis (7 with 4); *7* V. cerebellaris praecentralis; *8* V. supraculminalis; *9* tributary veins of 8; *10* bridging veins to tentorium (Fig. 16.30) hatched: sinus petrosus superior

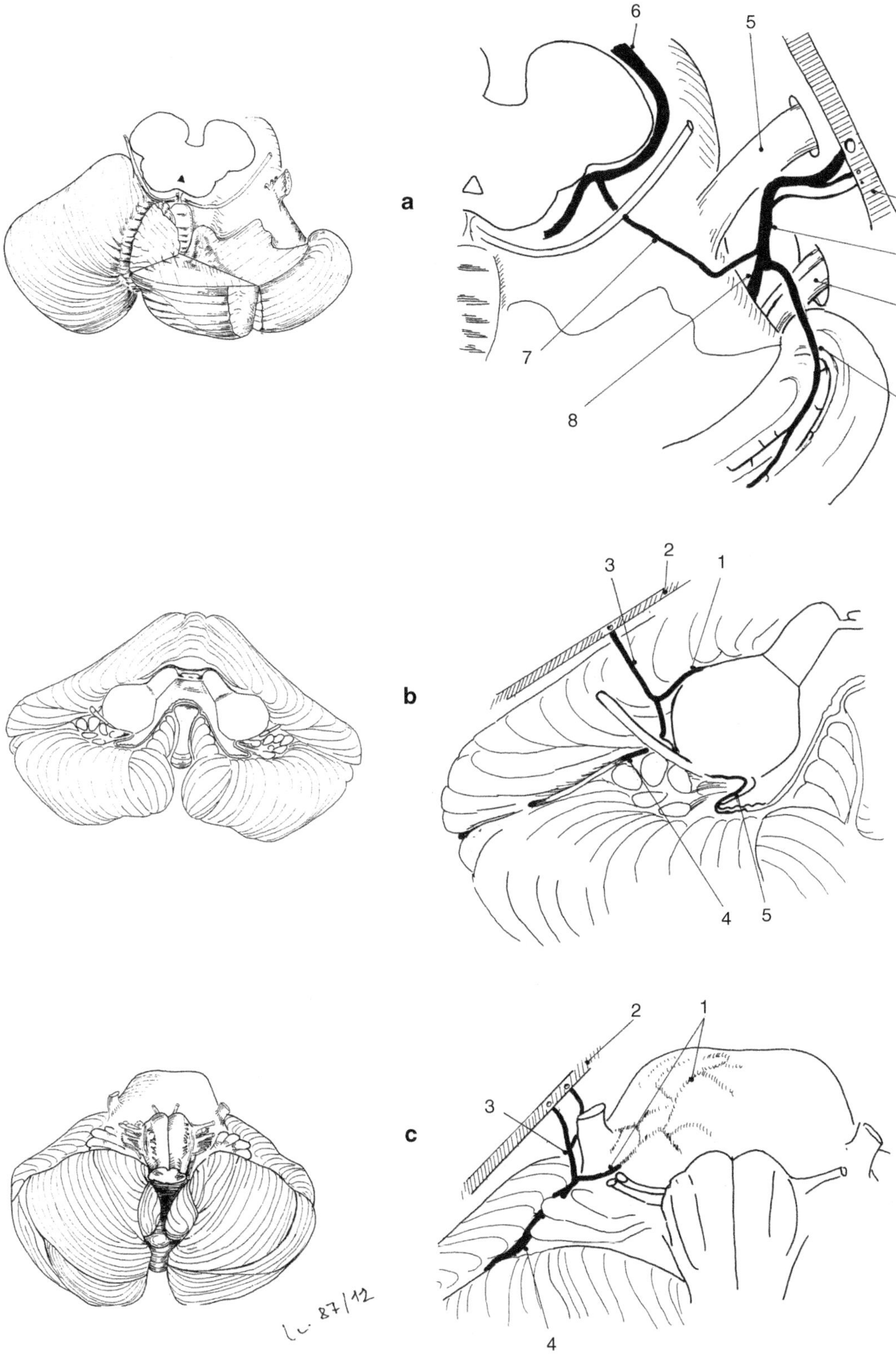

Fig. 15.59 Further variants of venous connections. (**a**) *1* V. cerebellaris horizontalis, *2* N. statoacusticus, *3* V. petrosa superior, *4* sinus petrosus superior, *5* N. trigeminus, *6* V. basalis (Rosenthal), *7* V. mesencephalica lateralis, *8* lateral pontine veins. (**b**) *1* V. mesencephalica lateralis, *2* sinus petrosus superior, *3* V. petrosa superior, *4* V. cerebellaris horizontalis, *5* veins of recessus lateralis. (**c**) *1* lateral pontine veins, *2* sinus petrosus superior, *3* V. petrosa superior, *4* V. cerebellaris horizontalis, *5* veins of recessus lateralis

15.2.5 Intracerebellar Structures (Figs. 15.60, 15.61, 15.62, 15.63, 15.64, 15.65, and 15.66)

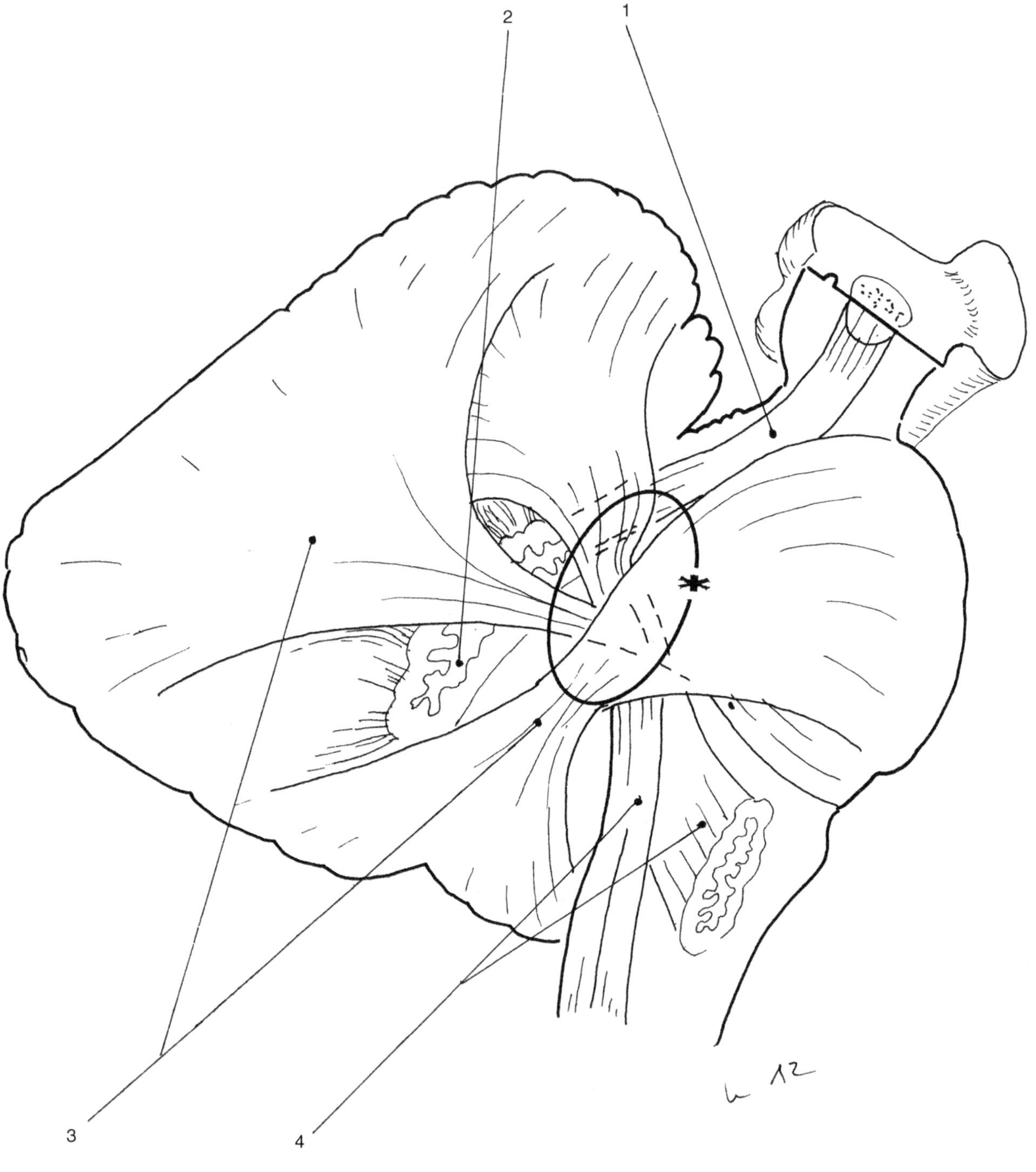

Fig. 15.60 Central cerebellar fiber systems (according to Neuwenhuys et al., modified). Survey. The area of nucleus dentatus should be preserved at surgery. It contains four main fiber systems, which are connected with nucleus dentatus or are surrounding it: pedunculus cerebellaris superior, this fiber bundle connects efferent fibers of nucleus dentatus with nucleus ruber; pedunculus cerebellaris medius, numerous fiber plates are connecting nuclei pontis with the cerebellar cortex (Fig. 3.9f); pedunculus cerebellaris inferior, these fiber plates contains tractus spinocerebellaris anterior et posterior and tractus olivocerebellaris. These are connected with the cerebellar cortex; afferent fiber plates of nucleus dentatus, these are the main intracerebellar fiber plates which are connecting the cerebellar cortex (Purkinje cells) with nucleus dentatus. In this drawing the fiber plates are not drawn in. *1* pedunculus cerebellaris superior, *2* nucleus dentatus, *3* pedunculus cerebellaris medius, *4* pedunculus cerebellaris inferior. *Central cerebellar region

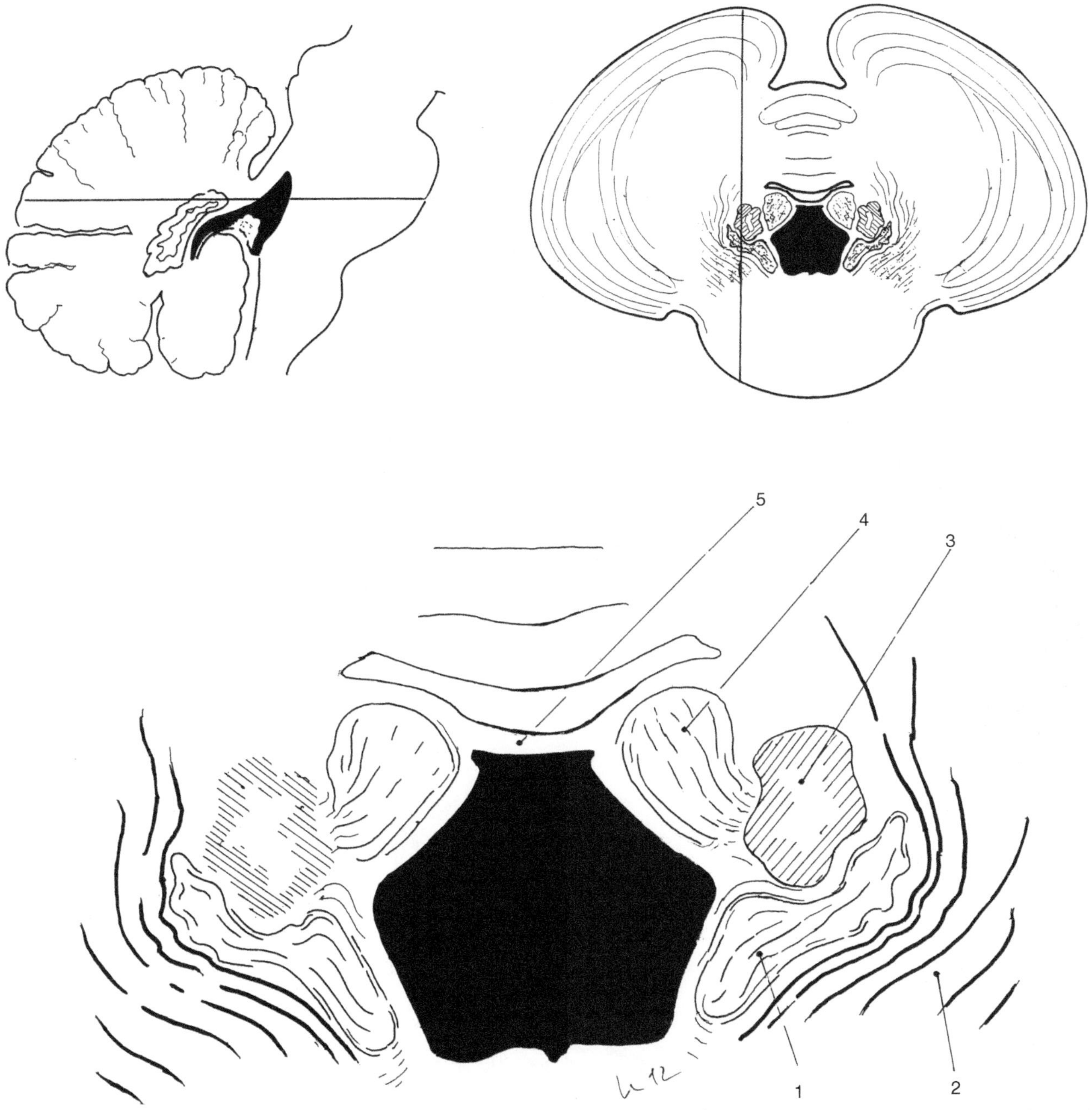

Fig. 15.61 Central cerebellar fiber systems. MRT. *1* pedunculus cerebellaris inferior, *2* pedunculus cerebellaris medius, *3* nucleus dentatus, *4* pedunculus cerebellaris superior, *5* velum medullare anterior

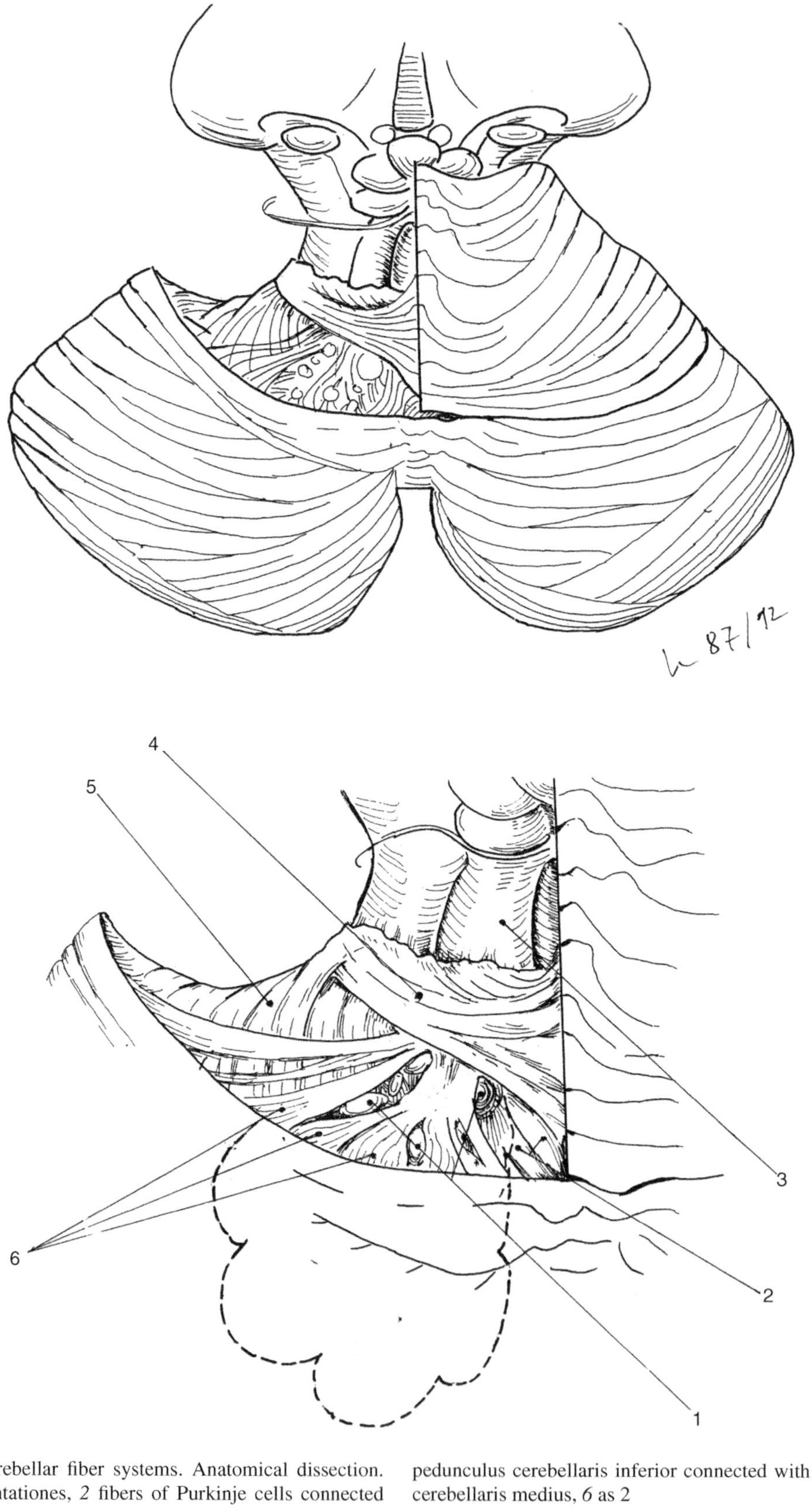

Fig. 15.62 Central cerebellar fiber systems. Anatomical dissection. *1* nucleus dentatus, dentationes, *2* fibers of Purkinje cells connected with nucleus dentatus, *3* pedunculus cerebellaris superior, *4* fibers of pedunculus cerebellaris inferior connected with vermis, *5* pedunculus cerebellaris medius, *6* as 2

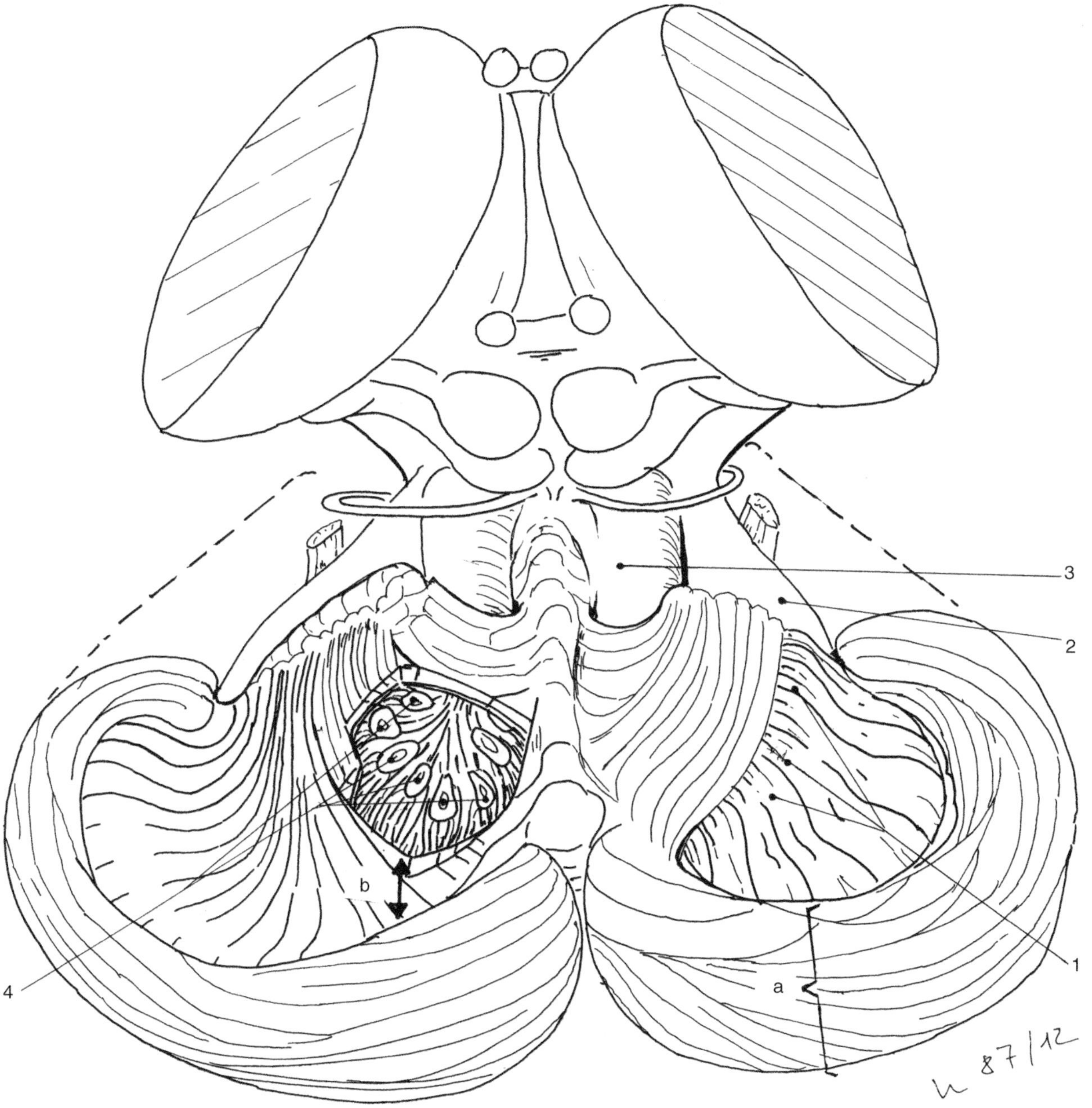

Fig. 15.63 Continuation of Fig. 15.62. Fiber bundles dorsal from Nucleus dentatus are eliminated. *1* fiber plates bulging by nucleus dentatus, *2* pedunculus cerebellaris medius, *3* pedunculus cerebellaris superior, *4* dentationes of nucleus dentatus, *a* distance measurement to the ground of fissura horizontalis cerebelli, *b* distance measurement between the ground of fissura horizontalis cerebelli and nucleus dentatus

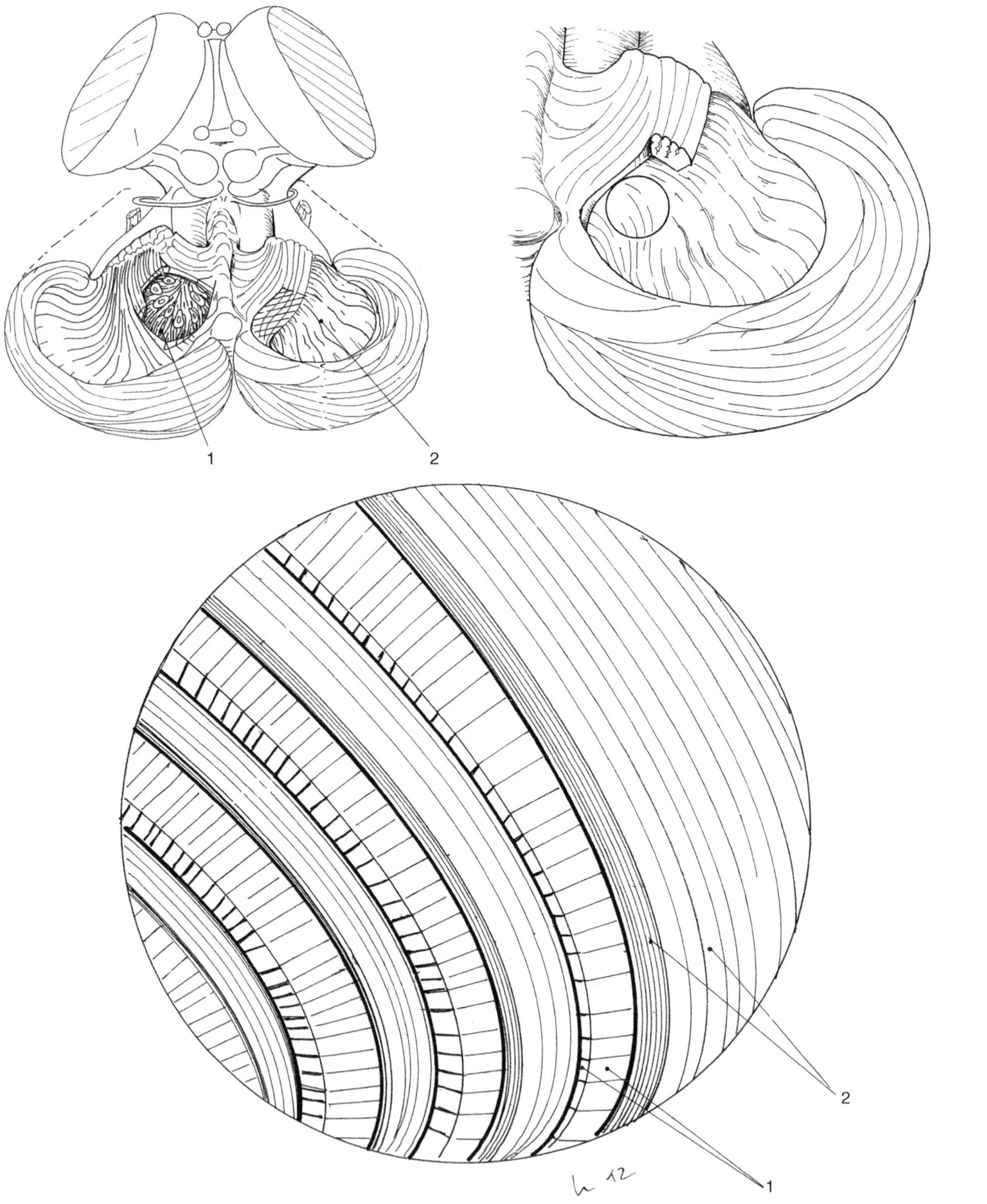

Fig. 15.64 Central cerebellar fiber systems. Fiber plates. Numerous thin-walled fiber plates are crossing each other with changing fiber directions. Crossings are hindering fibertracking

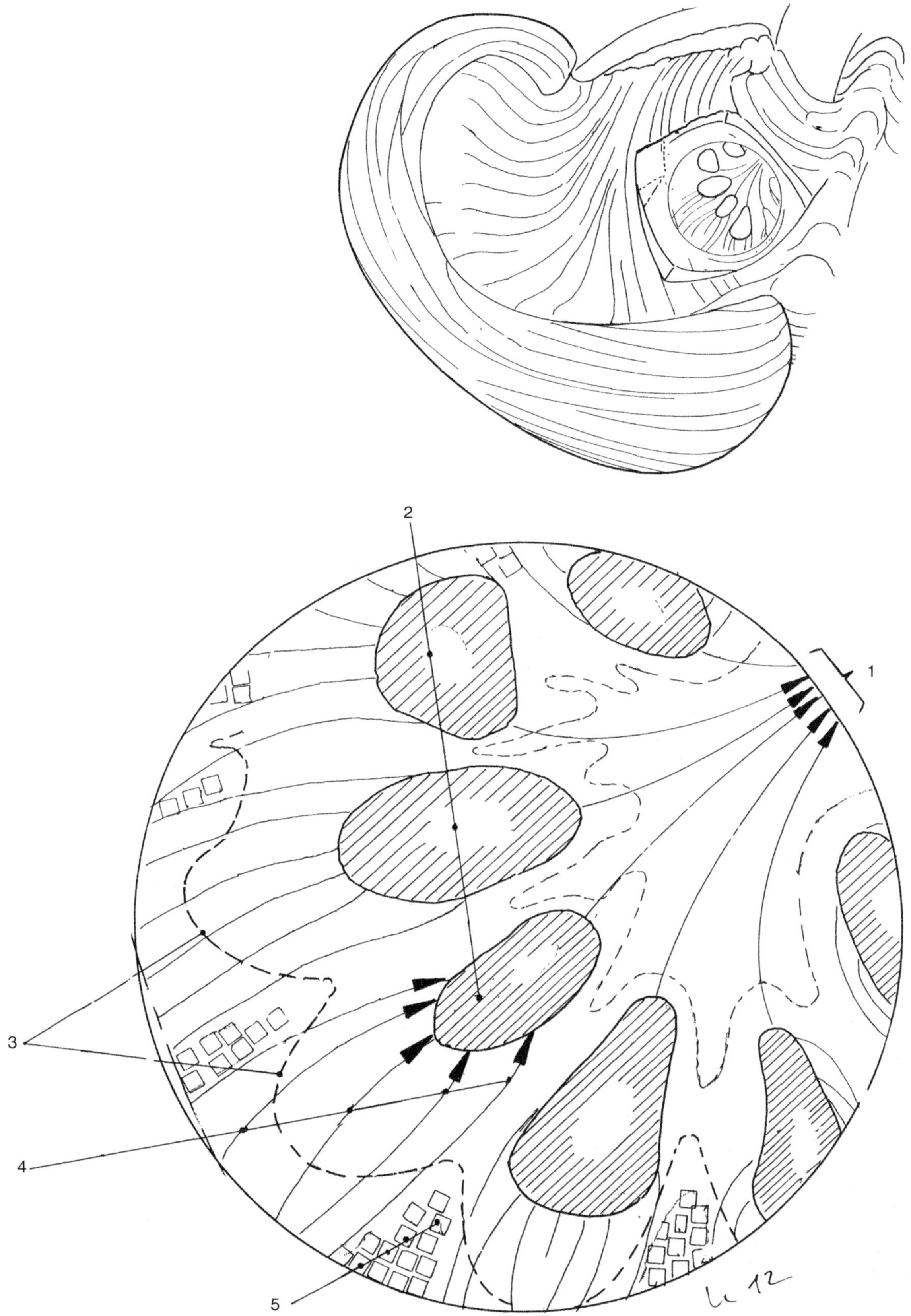

Fig. 15.65 Conclusion. Nucleus dentatus and adjacent fiber layers (Seeger 1987, p 100 f, 106 f). *1* fibers of pedunculus cerebellaris superior; *2* nucleus dentatus; *3* as 2, projection; *4* afferent fibers of nucleus dentatus; *5* fibers of pedunculus cerebellaris medius or inferior

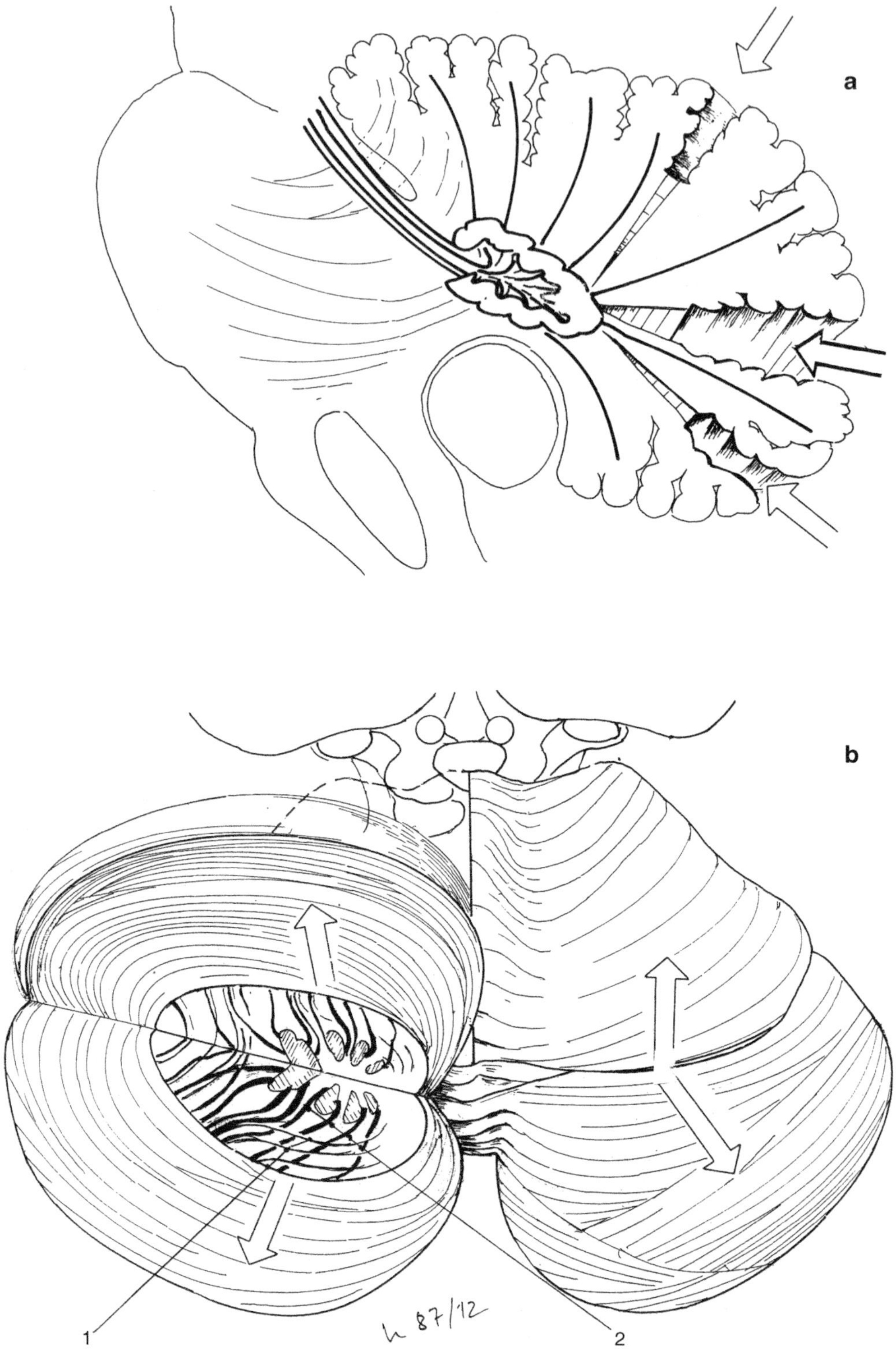

Fig. 15.66 Anatomical principles of intracerebellar surgical approaches (motivated by Prof. Lang, Direktor des Anatmischen Instituts der Universität Würzburg, 1986). Avoiding damage of nucleus dentatus and adjacent structures. The small distance measurements between the ground of sulci and nucleus dentatus, especially at fissura horizontalis, must be considered. Neuronavigation may be helpful. (**a**) Splitting and dilation of sulci (*arrows*). (**b**) The area of nucleus dentatus is endangered by brain splitting at a deep sulcus. *1* surrounding fibers of nucleus dentatus, *2* afferent fibers of nucleus dentatus

15.2.6 Middle and Inferior Brain Stem. Fibers and Nuclei (Microanatomy According to Nieuwenhuys, Voogd, van Huiyzen 1990, Modified for Neurosurgery)

Motor and Sensory Long Systems (Figs. 15.67, 15.68, 15.69, 15.70, 15.71, 15.72, 15.73, 15.74, 15.75, 15.76, and 15.77)

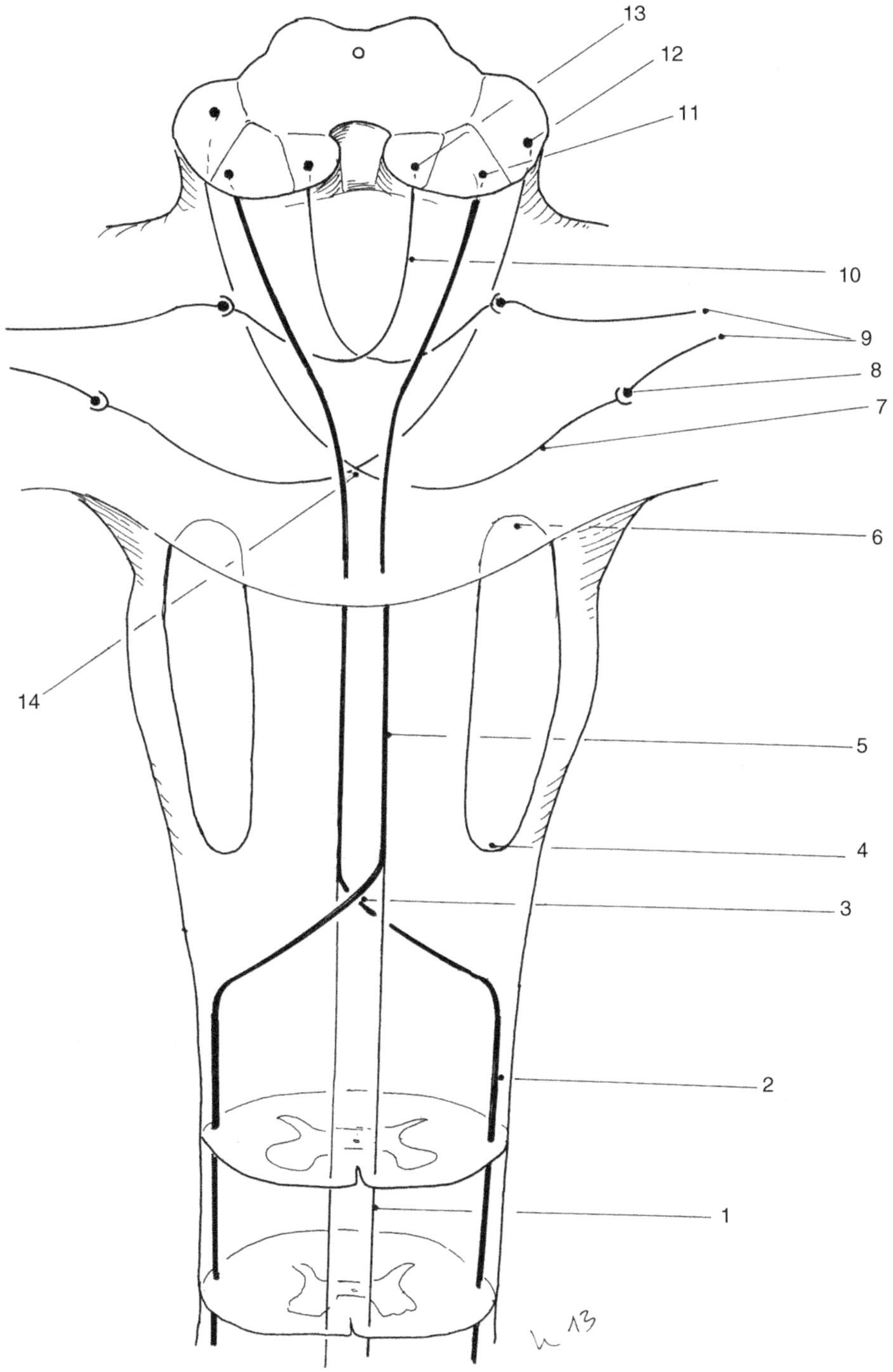

Fig. 15.67 Motor system. Survey. Tractus corticospinalis (pyramidalis). Tractus corticopontinus frontalis et parietotemporalis. Tractus pontocerebellaris (content of pedunculus cerebellaris medius). Tractus corticopontinus is crossing the midline before its neurons are connected with the neurons of tractus pontocerebellaris (see 8 and 9), analogous to the connections of decussatio pyramidum (3), tractus pyramidalis lateralis (2), and neurons of cornu spinalis anterius. *1* tractus pyramidalis anterior; *2* tractus pyramidalis lateralis; *3* decussatio pyramidum; *4 oliva, inferior pole, region of the upper begin of decussatio pyramidum*; *5* tractus pyramidalis; *6* oliva, intrapontine segment; *7* tractus parietotemporopontinus; *8* nuclei pontis; *9* tractus pontocerebellaris; *10* tractus frontopontinus; *11* tractus pyramidalis; *12* tractus parietotemporopontinus; *13* tractus frontopontinus; *14* decussationes of tractus corticopontini. Underline: landmark

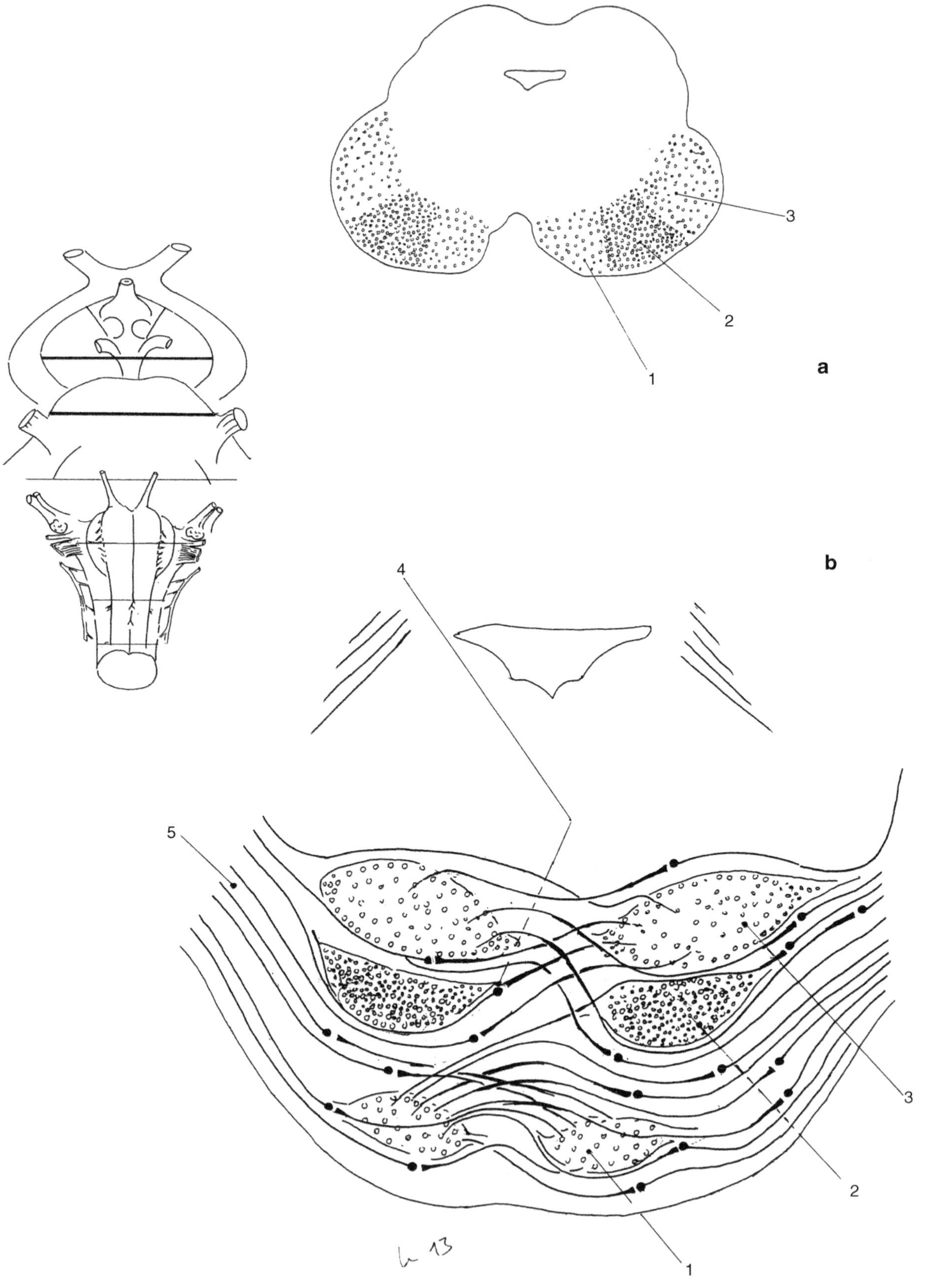

Fig. 15.68 Transectional planes of mesencephalon (**a**) and pons (**b**). (**a**) *1* tractus frontopontinus, *2* tractus pyramidalis, *3* tractus parietotemporopontinus. (**b**) *1–3* as A, *4* nuclei pontis, *5* tractus pontocerebellaris

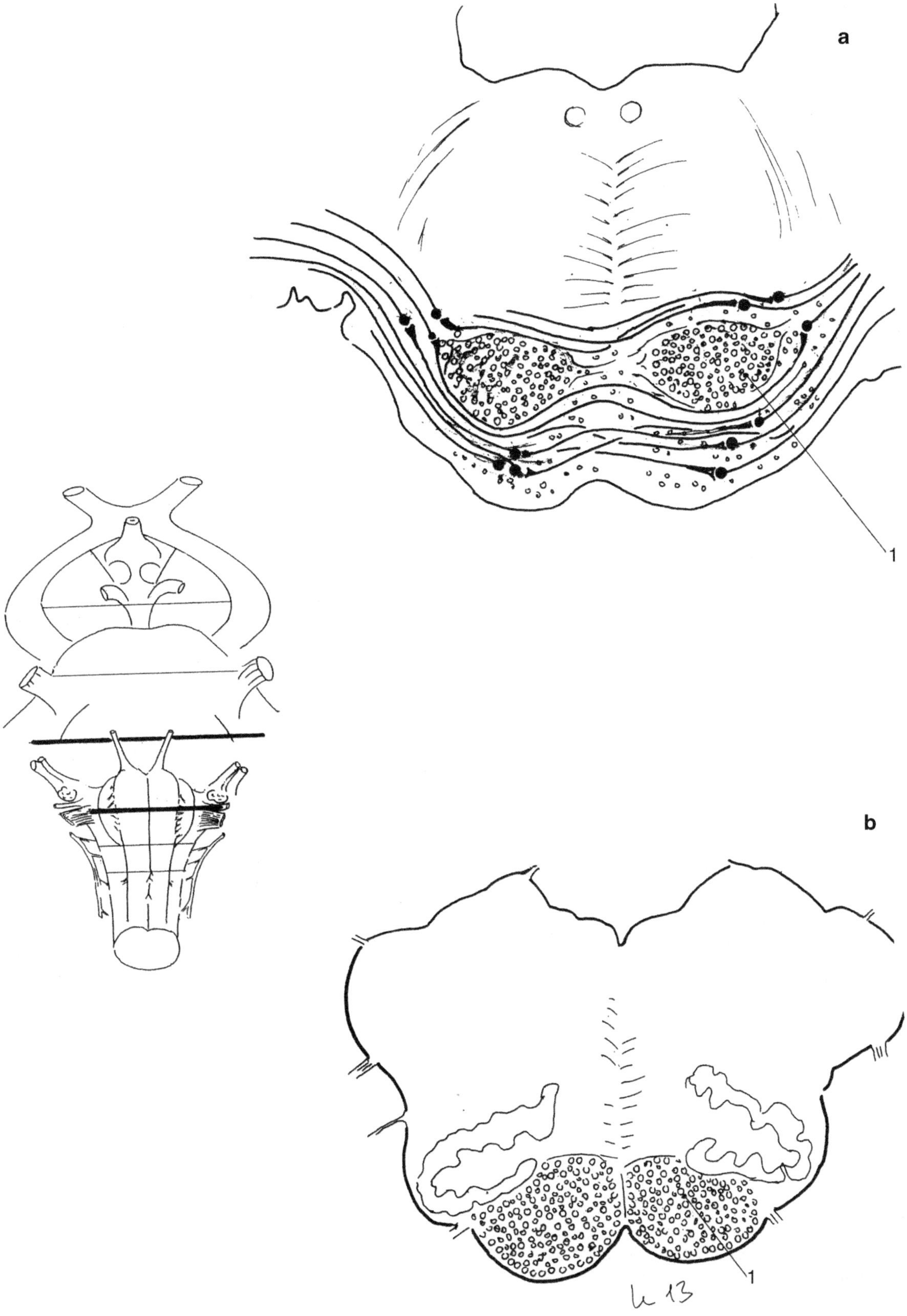

Fig. 15.69 Transectional planes of pons (**a**) and medulla oblongata (**b**). *1* tractus pyramidalis

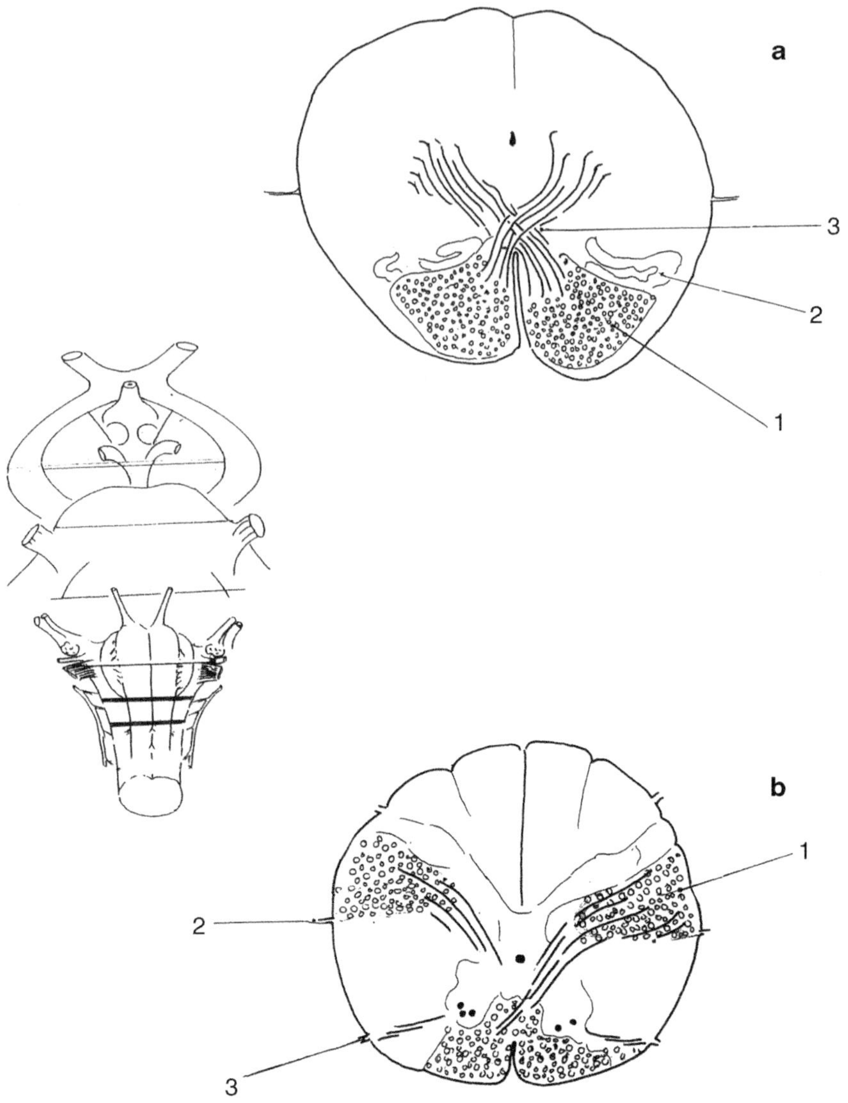

Fig. 15.70 Transectional planes at the cranial (**a**) and caudal level (**b**) of decussatio pyramidum. Note: The cranial region of decussatio pyramidum is located at the level of the inferior pole of oliva. (**a**) *1* tractus pyramidalis; *2 oliva, inferior pole*; *3* decussatio pyramidum, upper region. (**b**) *1* tractus pyramidalis, funiculus dorsolateralis, *2 Lig. denticulatum*, *3* fibers of C1. Underlines: landmarks

Fig. 15.71 Sensory system. Survey. a Fasciculus gracilis et cuneatus and nuclei (*4,5*). b Trigeminal tracts and nuclei (*9, 13, 14, 15*). c Tractus spinothalamicus, tracts, and nuclei (*6*). a Neurons of cornu posterius of medulla spinalis are located homolateral to nuclei fasciculi gracilis et cuneati and nucleus tractus spinalis n. trigemini. Crossing of its fibers are compartments of decussatio lemniscorum. *The inferior region of decussatio lemniscorum begins at the region of polus inferior olivae*. It is nearly the same region as the region of the superior beginning of decussatio pyramidum. The postcommissural segment of decussatio lemniscorum is lemniscus medialis of the medulla oblongata, pons, and mesencephalon. b Fibers of nucleus tractus spinalis n. trigemini accompany the contralateral nucleus tractus spinalis after crossing midline. The midline crossing of trigeminal fibers is a compartment of decussatio lemniscorum. For details, see Fig. 3.23 f. c Some neurons of cornu posterius of medulla spinalis (*2*) are crossing at the same spinal segment (*3*) or some segments superior to it. Its continuation is called tractus spinothalamicus (*6*), which is a part of columna ventrolateralis, located ventral from the medullary *insertion line of ligamentum denticulatum*. Dorsal from tractus spinothalamicus and Lig. denticulatum is located tractus corticospinalis (endangered at surgical transections of tractus spinothalamicus). The fibers of tractus spinothalamicus represents a multisynaptic system, connected with formatio reticularis spinalis, nucleus ruber, and thalamus. According to the multisynaptic connections, the therapeutic effects after a surgical interruption of tractus spinothalamius (chordotomy) are inefficient after a short time. *1* ganglion spinale; *2* cornu posterius; *3* decussatio of tractus spinothalamicus; *4* fasciculus gracilis; *5* fasciculus cuneatus; *6* tractus spinothalamicus; *7* nucleus fasciculi cuneati; *8* nucleus fasciculi gracilis; *9* nucleus et tractus spinalis n. trigemini, *10* decussatio lemniscorum (The inferior region of decussatio lemniscorum is located close to the inferior pole of oliva. The superior region is located at the upper region of pons); *11* as 10, superior segment; *12* lemniscus medialis; *13* decussatio of tractus trigeminothalamicus (Nieuwenhuys et al. 1990); *14* region of nucleus sensorius principalis n. trigemini; *15* ganglion semilunare (Gasseri). Underlines: landmarks

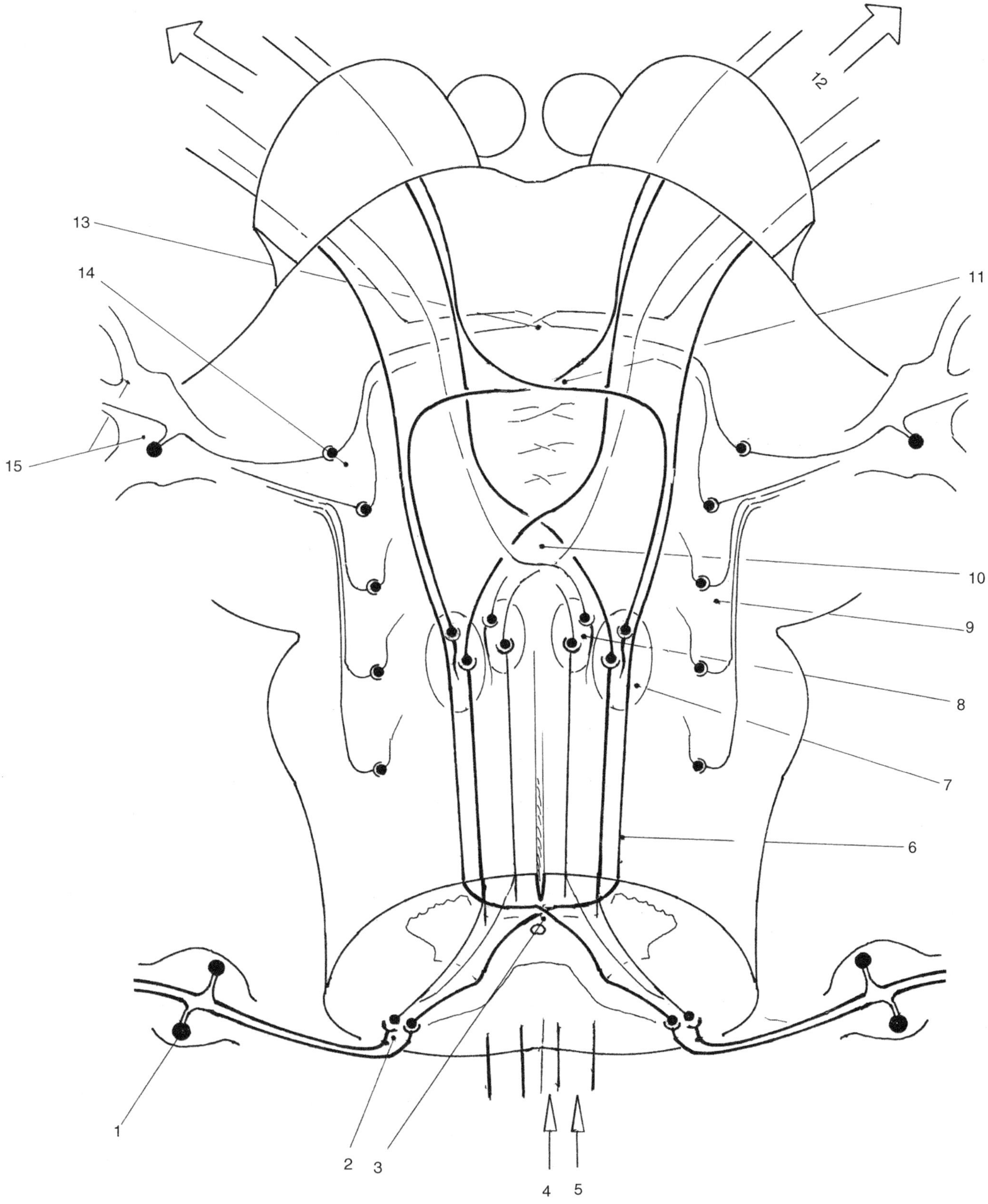

12
13
14
11
15
10
9
8
7
6
1
2
3
4
5

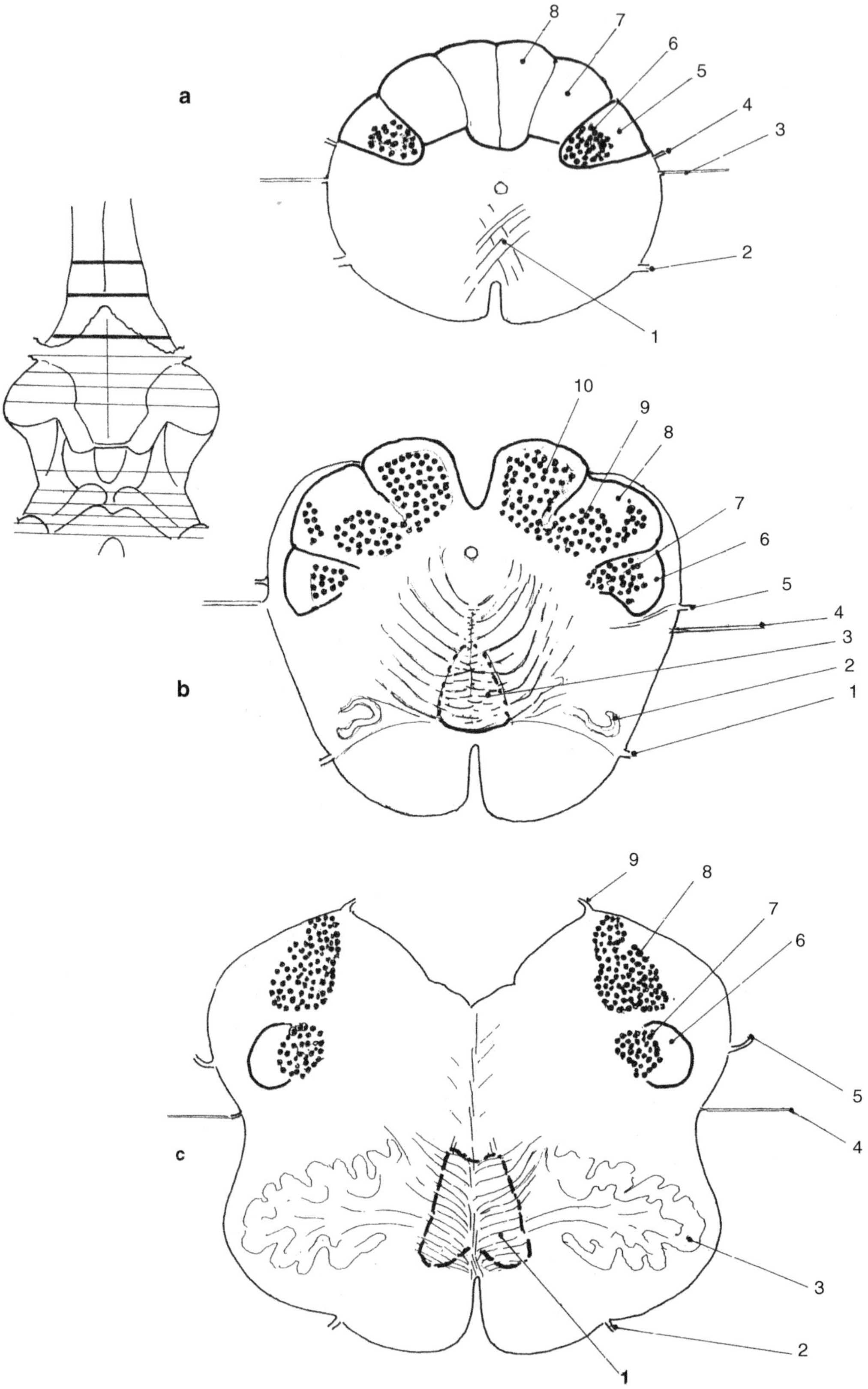

Fig. 15.72 Fasciculi and nuclei cuneati et gracilis. Transectional levels. (**a**) *1* decussatio pyramidum, *2* radix ant. C1 (fascicle), *3 Lig. denticulatum, 4* radix spinalis n. accessorii, *5* tractus spinalis n. trigemini, *6* its nucleus, *7* fasciculus cuneatus, *8* fasciculus gracilis. (**b**) *1* Radices n. hypoglossi; *2 oliva, inferior pole*; *3* lemniscus medialis et decussation; *4* Lig. denticulatum; *5* radix spinalis n. accessorii; *6* tractus spinalis n. trigemini; *7* its nucleus; *8* fasciculus cuneatus; *9* its nucleus; *10* nucleus fasciculi gracilis. (**c**) *1* lemniscus medialis et decussatio, *2* N. hypoglossus (fascicle), *3* oliva, *4 Lig. denticulatum*, *5* radix spinalis n. accessorii, *6* tractus spinalis n. trigemini, *7* its nucleus *8* fasciculus cuneatus, *9* velum medullare post., insertion. Underlines: landmarks

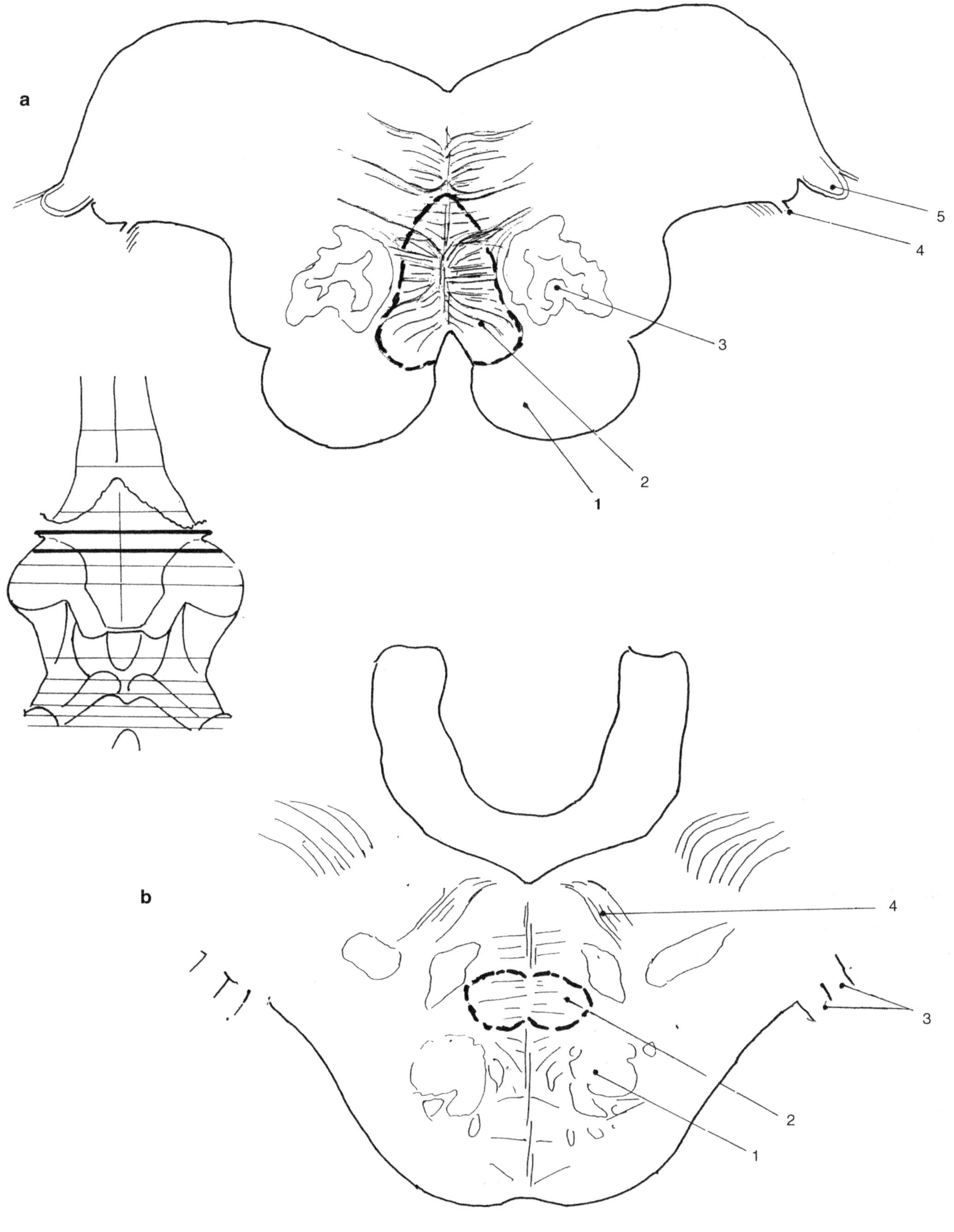

Fig. 15.73 Transectional level of recessus lat. (**a**) and inferior region of pons (**b**). (**a**) *1* pyramis medullae, *2* lemniscus medialis et decussatio, *3 oliva,* 4 *N. glossopharyngeus,* 5 *recessus lateralis.* (**b**) *1* tractus pyramidalis, *2* lemniscus medialis et decussatio, *3 N. statoacusticus*, *4* genu int. n.facialis. Underlines: landmarks

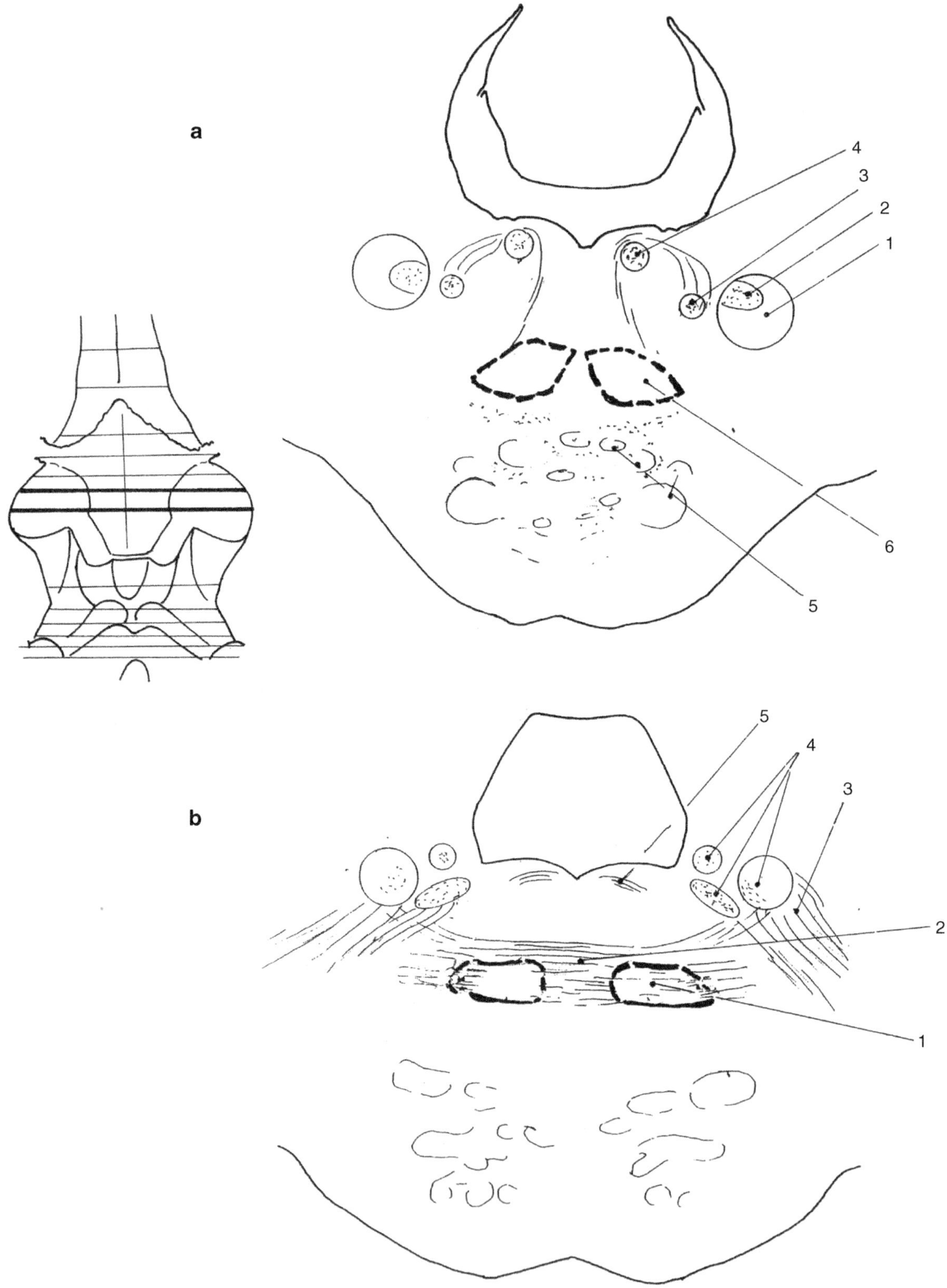

Fig. 15.74 Middle area of pons (**a**) and (**b**). (**a**) 1 tractus spinalis n. trigemini, *2* its nucleus, *3* nucleus n. facialis, *4* nucleus n. aducentis, *5* tractus pyramidalis, *6* lemniscus medialis. (**b**) *1* lemniscus medialis, *2* decussatio tractus trigeminothalamici ventralis (Nieuwenhuys et al. 1990), *3* radix sensoria n. trigemini, *4* nucleus principalis n. trigemini (lateral) et nucleus motorius n. trigemini (medial), *5* genu internum. n. facialis

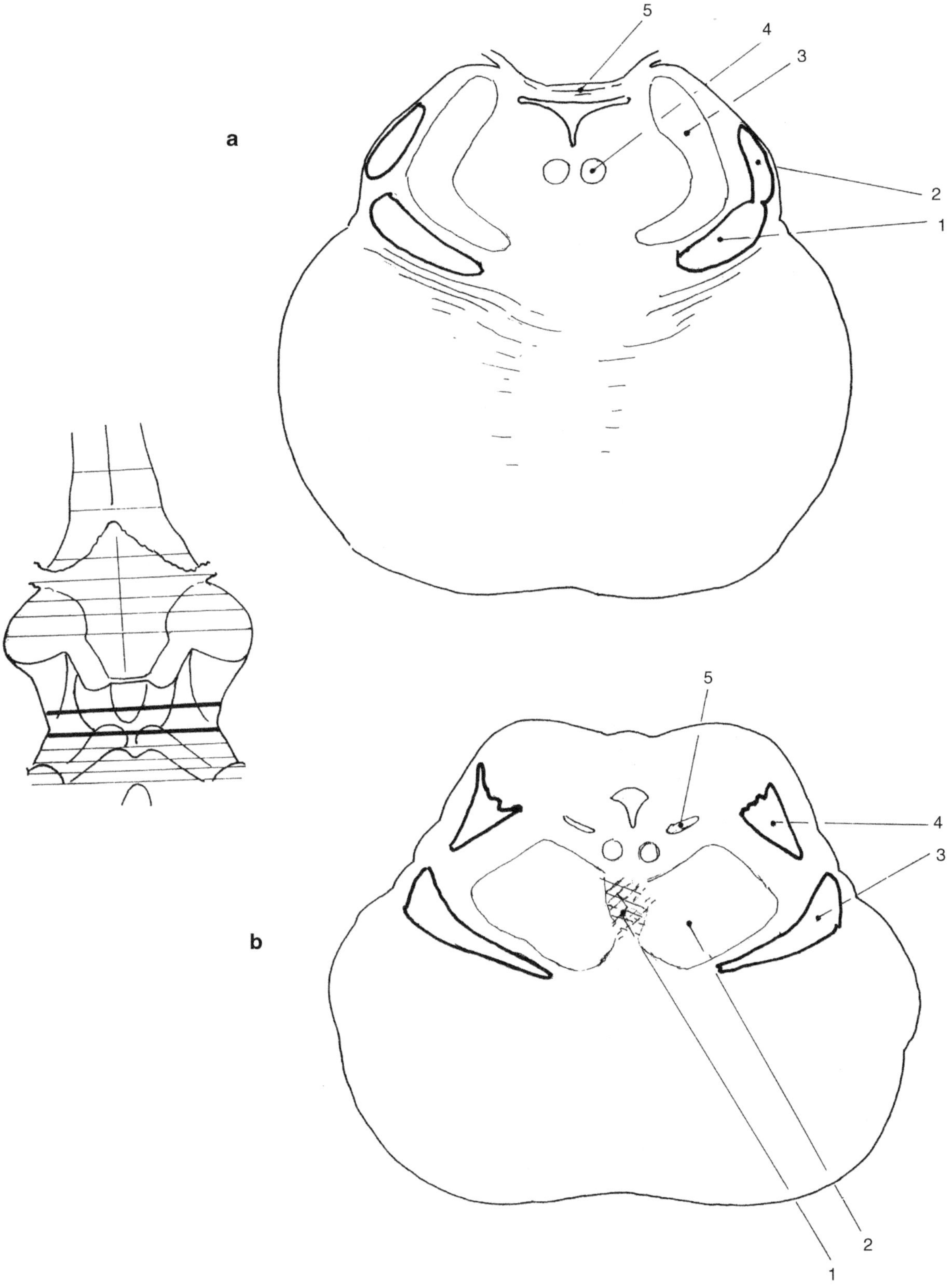

Fig. 15.75 Superior region of pons (**a**) and (**b**), dorsal inferior region of mesencephalon (**b**). H *1* lemniscus medialis, *2* lemniscus lateralis, *3* pedunculus cerebellaris superior, *4* fasciculus longitudinalis medialis, *5* velum medullare anterius and decussatio of Nn. trochleares. (**b**) *1* decussatio of pedunculi cerebellares superiores, *2* pedunculus cerebellaris superior et nucleus ruber, *3* lemniscus medialis, *4* lemniscus lateralis, *5* fibers of N. trochlearis

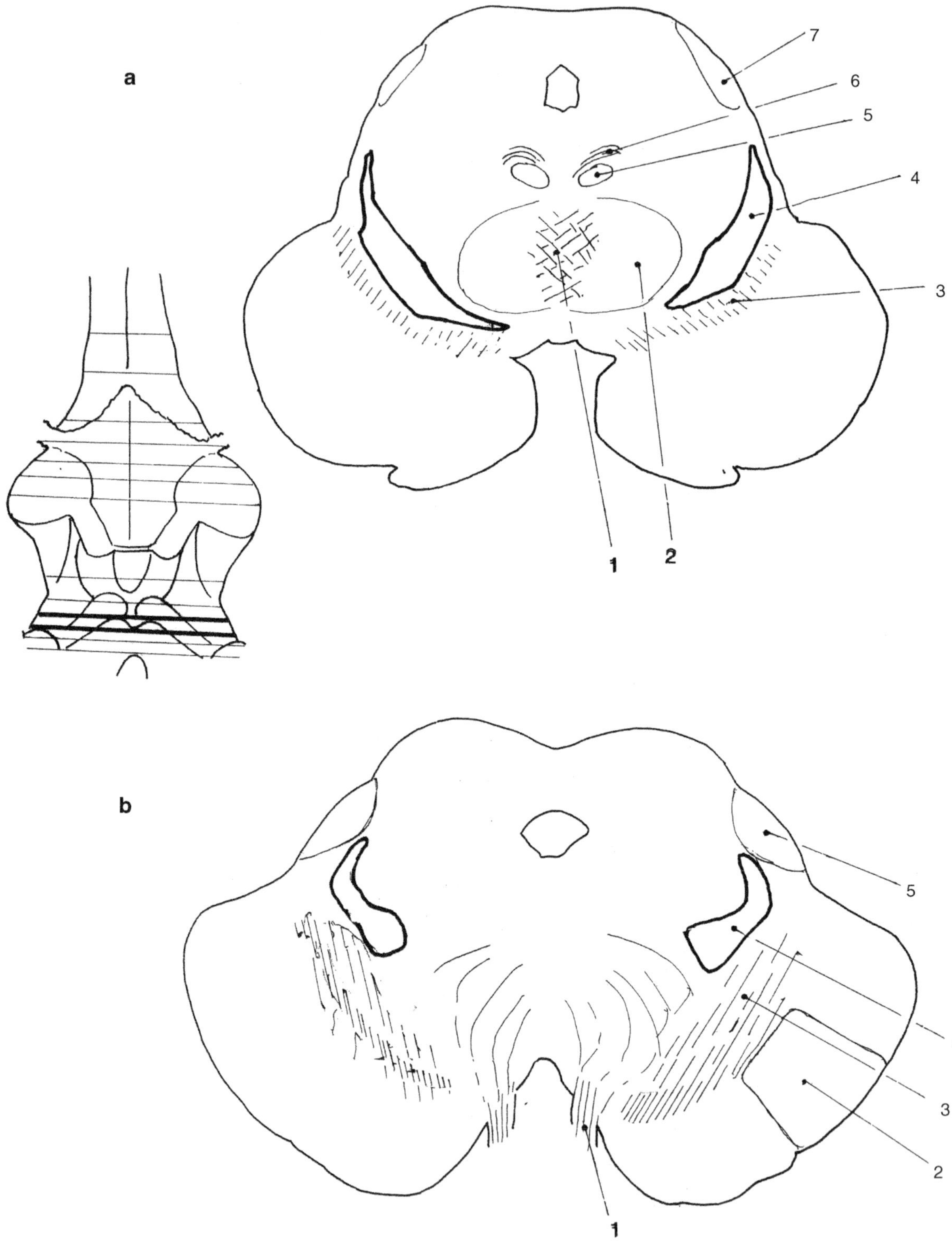

Fig. 15.76 Intercollicular level (**a**) and level of colliculi superiores (**b**). (**a**) *1* decussatio of pedunculi cerebellares superiores, *2* pedunculus cerebellaris superior, *3* substantia nigra, *4* lemniscus medialis, *5* fasciculus longitudinalis medialis, *6* fibers of N. trochlearis, *7* brachium colliculi inferioris. (**b**) *1 N. oculomotorius*, *2* tractus pyramidalis, *3* substantia nigra, *4* lemniscus medialis, *5* brachium colliculi inferioris. Underline: landmark

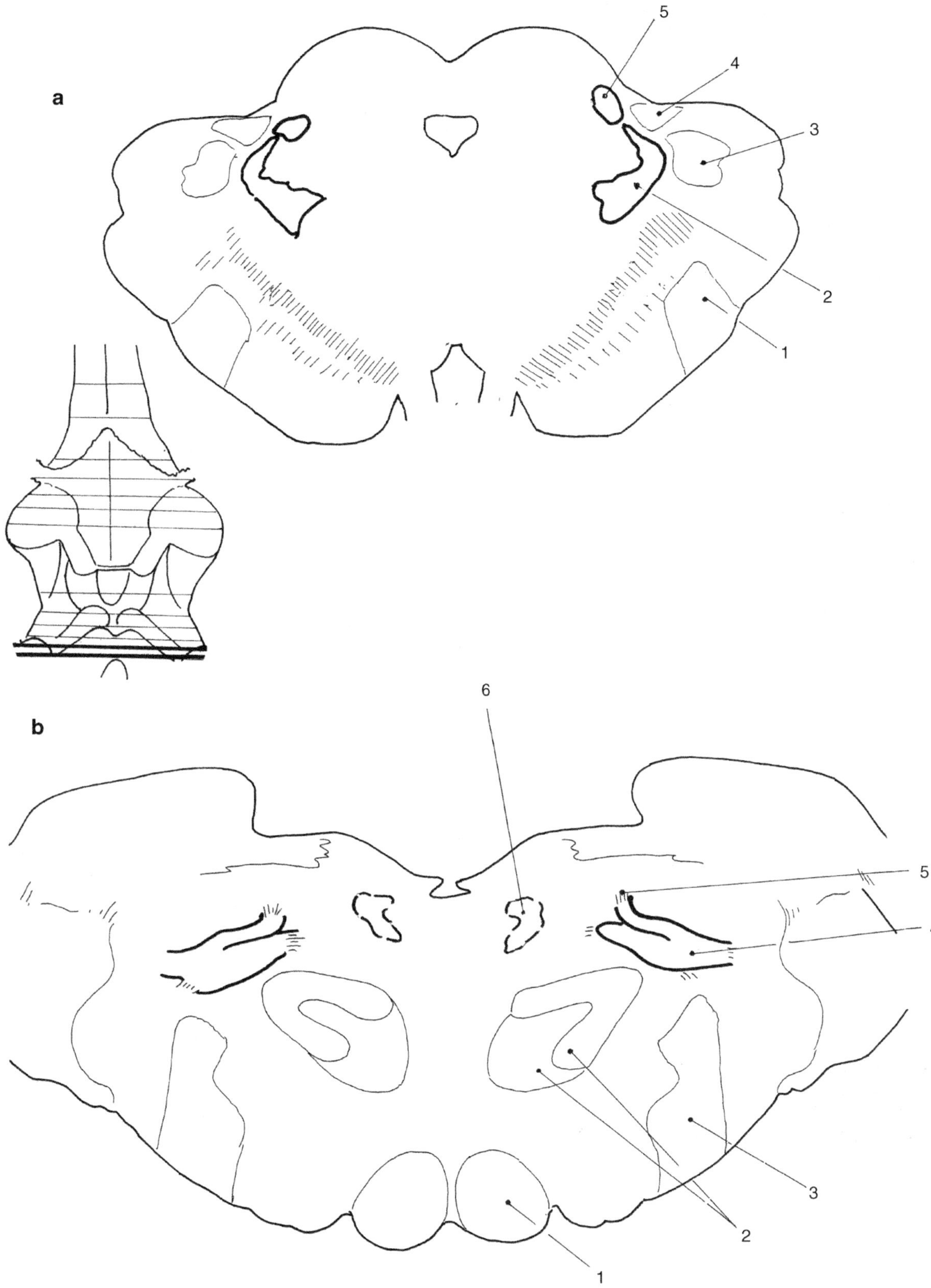

Fig. 15.77 Superior region of mesencephalon (**a**) and (**b**) and of exit region of *N. oculomotorius*. (**a**) *1* tractus pyramidalis, *2* lemniscus medialis, *3* brachium colliculi inferioris, *4* brachium colliculi superioris, *5* tractus spinothalamicus. (**b**) *1* corpus mamillare *2* nucleus ruber (medially) and pedunculus cerebellaris superior, *3* tractus pyramidalis, *4* lemniscus medialis, *5* tractus spinothalamicus, *6* tractus trigeminothalamicus dorsalis. Underlines: landmarks

Cranial Nerves, Nuclei, and Roots (Figs. 15.78, 15.79, 15.80, 15.81, 15.82, 15.83, 15.84, 15.85, 15.86, 15.87, 15.88, 15.89, 15.90, and 15.91)

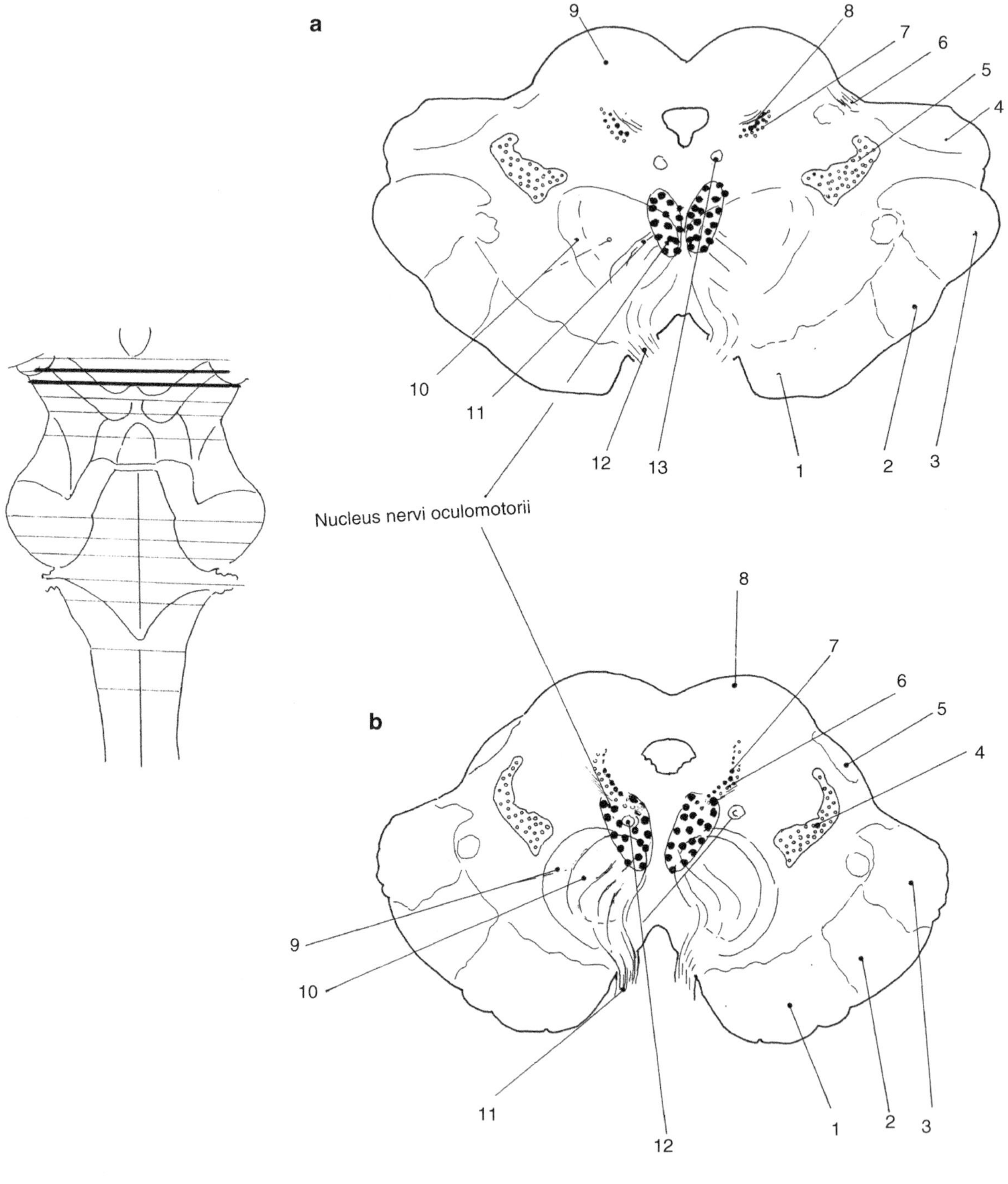

Fig. 15.78 N. oculomotorius. (**a**) The rostral segment of nucleus n. oculomotorii is located at the axial level of colliculi superiores. (**b**) Its caudal segment is located close to nucleus ruber and pecunculus cerebellaris superior rostral from its crossing. For variants of N. oculomotorius which are relevant for surgery see Fig. 15.79. (**a**) *1* tractus frontopontinus, *2* tractus pyramidalis, *3* tractus parietotemporopontinus, *4* brachium colliculi inferioris, *5* lemniscus medialis, *6* lemniscus lateralis, *7* tractus tectospinalis, *8* tractus et nucleus mesencephalicus n. trigemini, *9* colliculus superior, *10* nucleus ruber (medial) and pedunculus cerebellaris superior (lateral), *11* fibers of N. oculomotorius, *12* midbrain exit of *N. oculomotorius*, *13* fasciculus longitudinalis medialis. (**b**) *1–3* as (**a**); *4* lemniscus medialis; *5* brachium colliculi inferioris; *6* tractus trigeminothalamicus dorsalis; *7* nucleus mesencephalicus n. trigemini; *8* colliculus superior; *9* pedunculus cerebellaris superior; *10* nucleus ruber, penetrated by fibers of N. oculomotorius; *11* midbrain exit of *N. oculomotorius*, *12* fasciculus longitudinalis medialis. Underlines at (**a**) and (**b**): landmarks

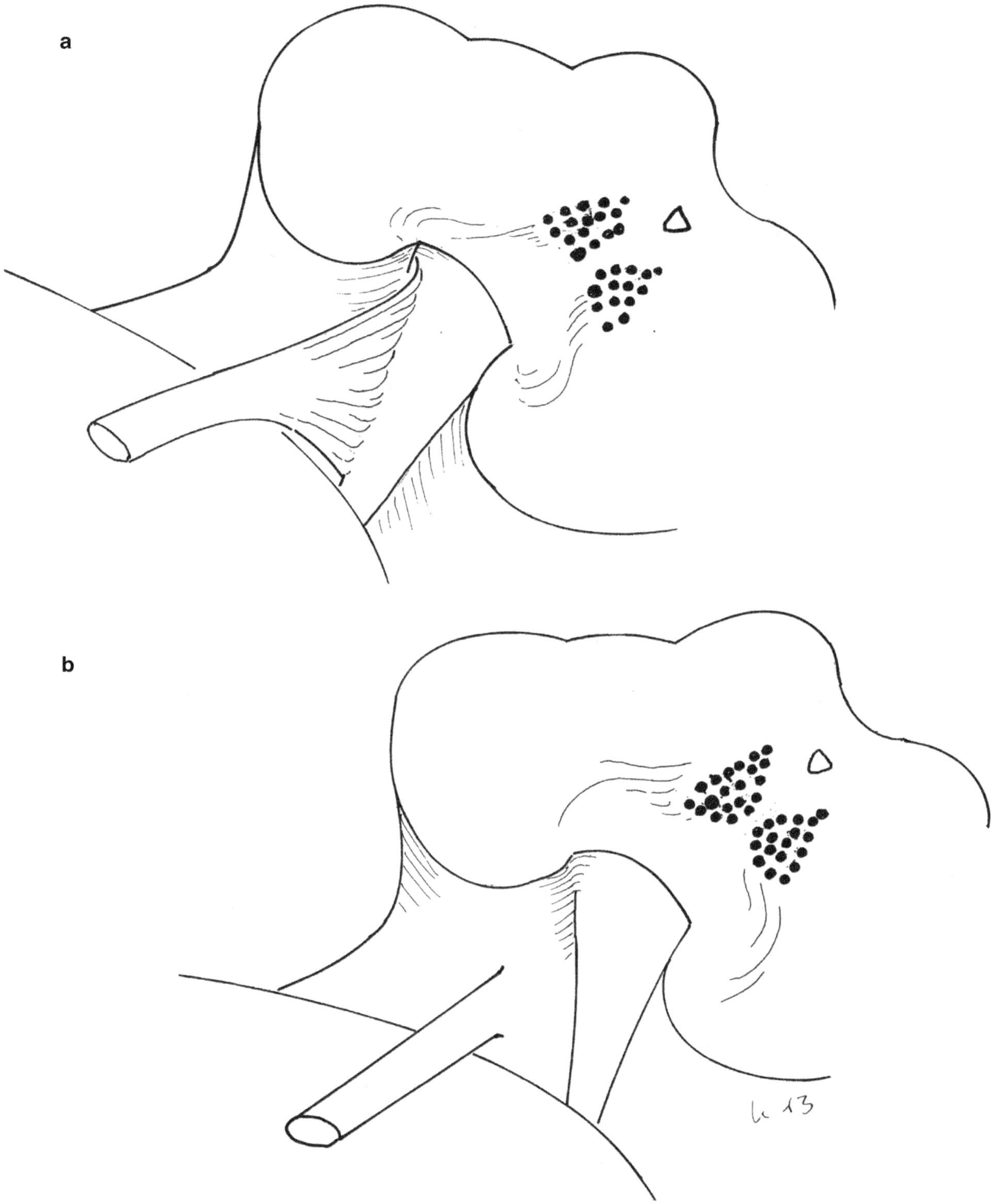

Fig. 15.79 Roots of N. oculomotorius. Its variants should be considered at basal surgical approaches. (**a**) Usual finding. (**b**) Common variant

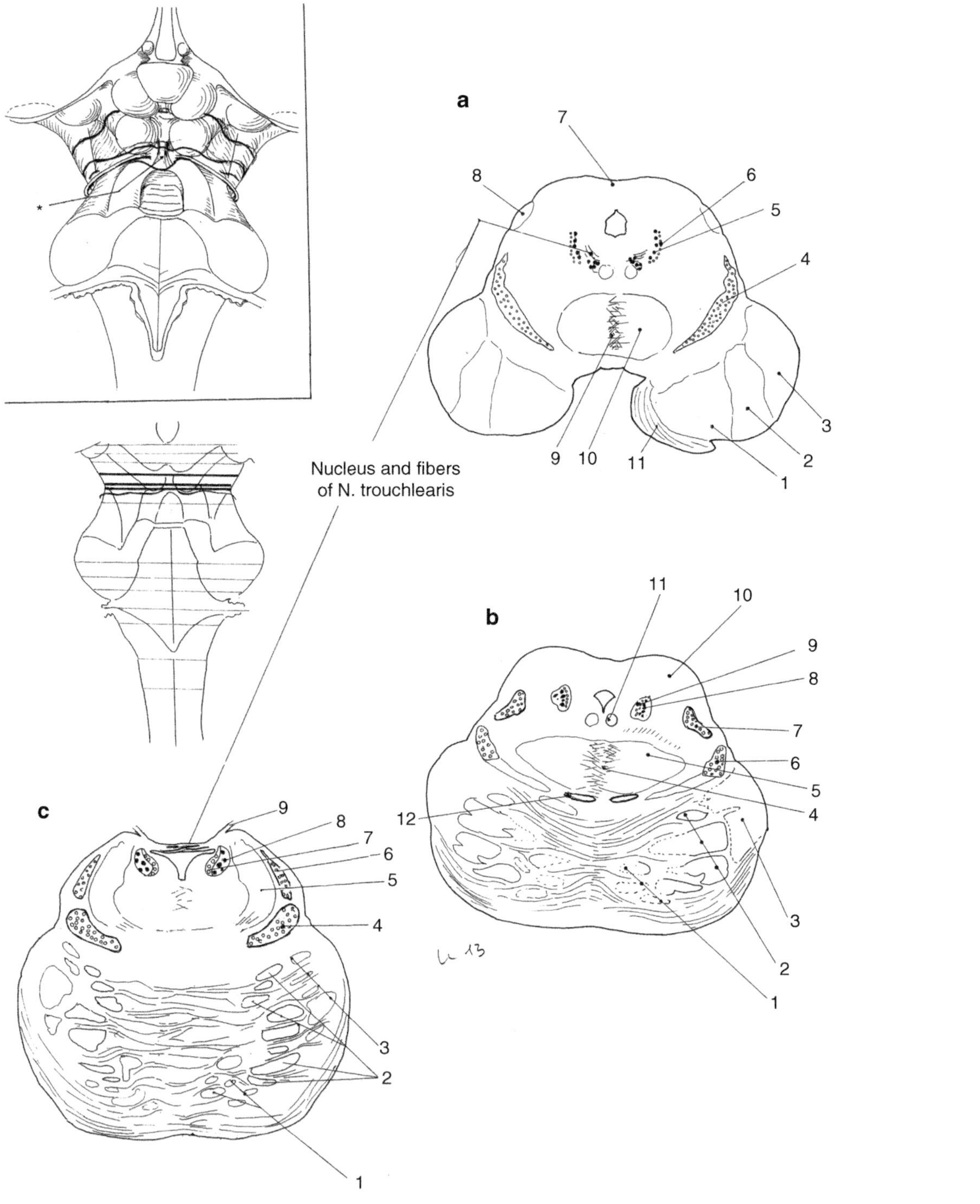

Fig. 15.80 N. trochlearis. Surrounding structures. The brain stem exit points could be used for landmarks during surgery, if the relief of colliculi tecti is flattened, e.g., by tumors. (**a**) *1* tractus corticopontinus frontalis, *2* tractus pyramidalis, *3* tractus corticopontinus parietotemporalis, *4* lemniscus medialis, *5* nucleus mesencephalicus n. trigemini, *6* its tractus, *7* area intercollicularis, *8* brachium colliculi inferioris, *9* decussatio of pedunculi cerebellares superiores, *10* pedunculus cerebellaris superior, *11* pontine segment close to isthmus rhombencephali, tangential transected. (**b**) *1–3* as (**a**), *4* decussatio of pedunculi cerebellares superiores, *5* pedunculus cerebellaris superior, *6* lemniscus medialis,*7* lemniscus lateralis, *8* tractus mesencephalicus n. trigemini, *9* its nucleus, *10* colliculus inferior, *11* fasciculus longitudinalis medialis, *12* pedunculus mamillaris. (**c**) *1–3* as (**a**) and (**b**), *4* lemniscus medialis, *5* pedunculus cerebellaris superior, *6* lemniscus lateralis, *7* and *8* nucleus et tractus spinalis n. trigemini, *9 N. trochlearis* **Frenulum tecti.* Underlines: landmarks

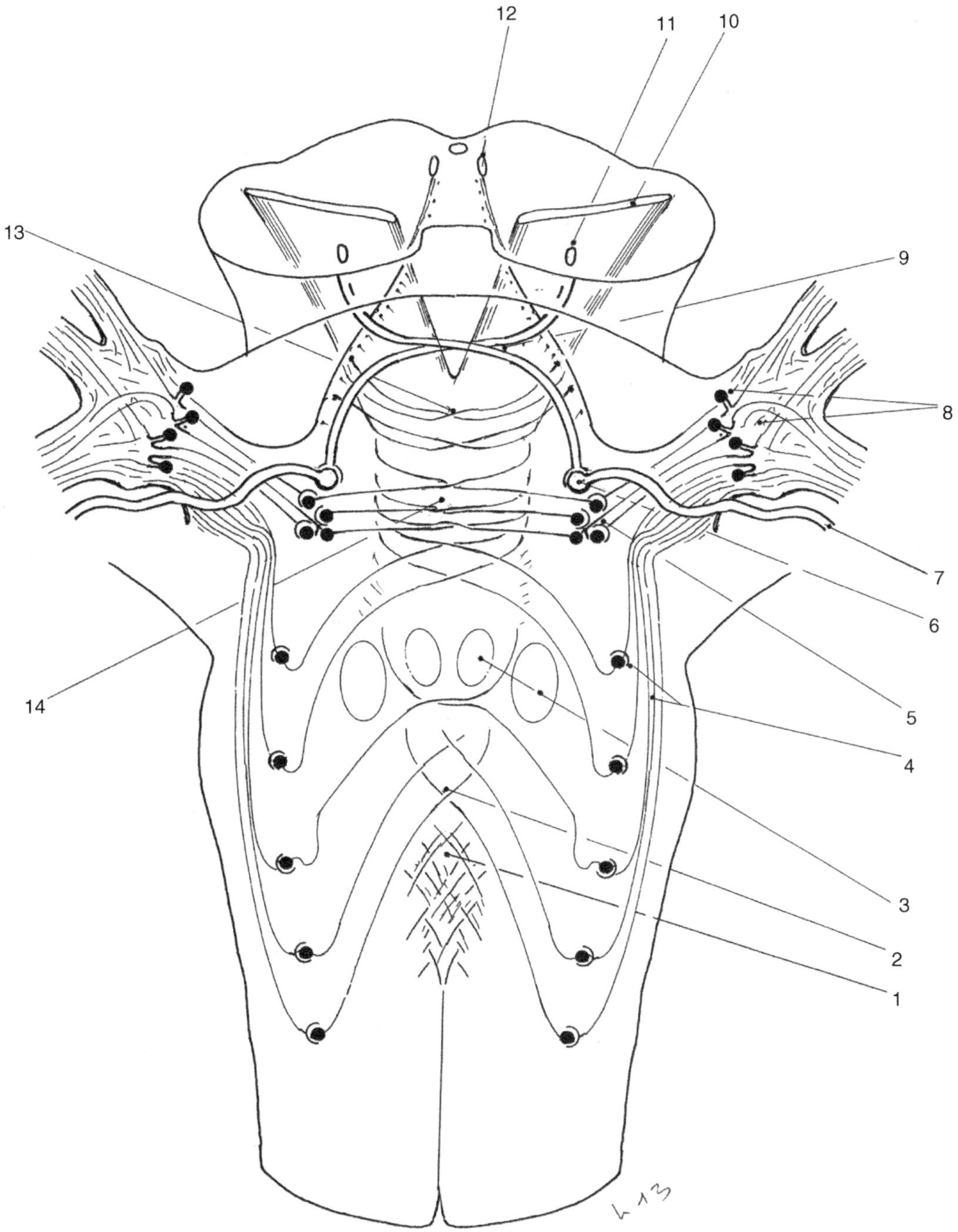

Fig. 15.81 N. trigeminus. Survey. Intra- and extrarhombencephalic structures of N. trigeminus: 1. Intrarhombencephalic structures. Nucleus et tractus spinalis n. trigemini (4) and its nucleus principalis (5), nucleus et tractus mesencephalicus (12), its participation at decussatio lemniscorum (14) and lemniscus medialis (10), nucleus motorius (6) and its decussatio (9). 2. Extrarhombencephalic structures. The variable topographical relationships of *portio major and minor (7)* to the *loop of A. cerebelli superior* are relevant to Janetta's operation for the treatment of Fothergil's disease. The variability of sensory and motor fibers of portio major (8) and minor (7) should be considered at surgery (Rhoton). *1* decussatio pyramidum; *2* trigeminal compartment of lemniscus medialis; *3* nucleus fasciculi gracilis et cuneati; *4* nucleus et tractus spinalis n. trigemini; *5* nucleus principalis n. trigemini; *6* nucleus motorius n. trigemini; *7* motor fibers of it (it doesn't exist an exact limit beween motor and sensory fibers of N. trigeminus at portio minor and major (presentation of Prof. Rhoton at Neurochirurgische Klinik of the Kantonspital Zürich 1976)); *8* ganglion trigeminale (Gasseri); *9* decussatio of motor fibers of N. trigeminus, part of tractus pyramidalis; *10* lemniscus medialis; *11* precommissural motor fibers of N. trigeminus; *12* nucleus et tractus mesencephalicus n. trigemini; *13* decussatio tractus trigeminothalamici; *14* decussatio lemniscorum. Underlines: landmarks

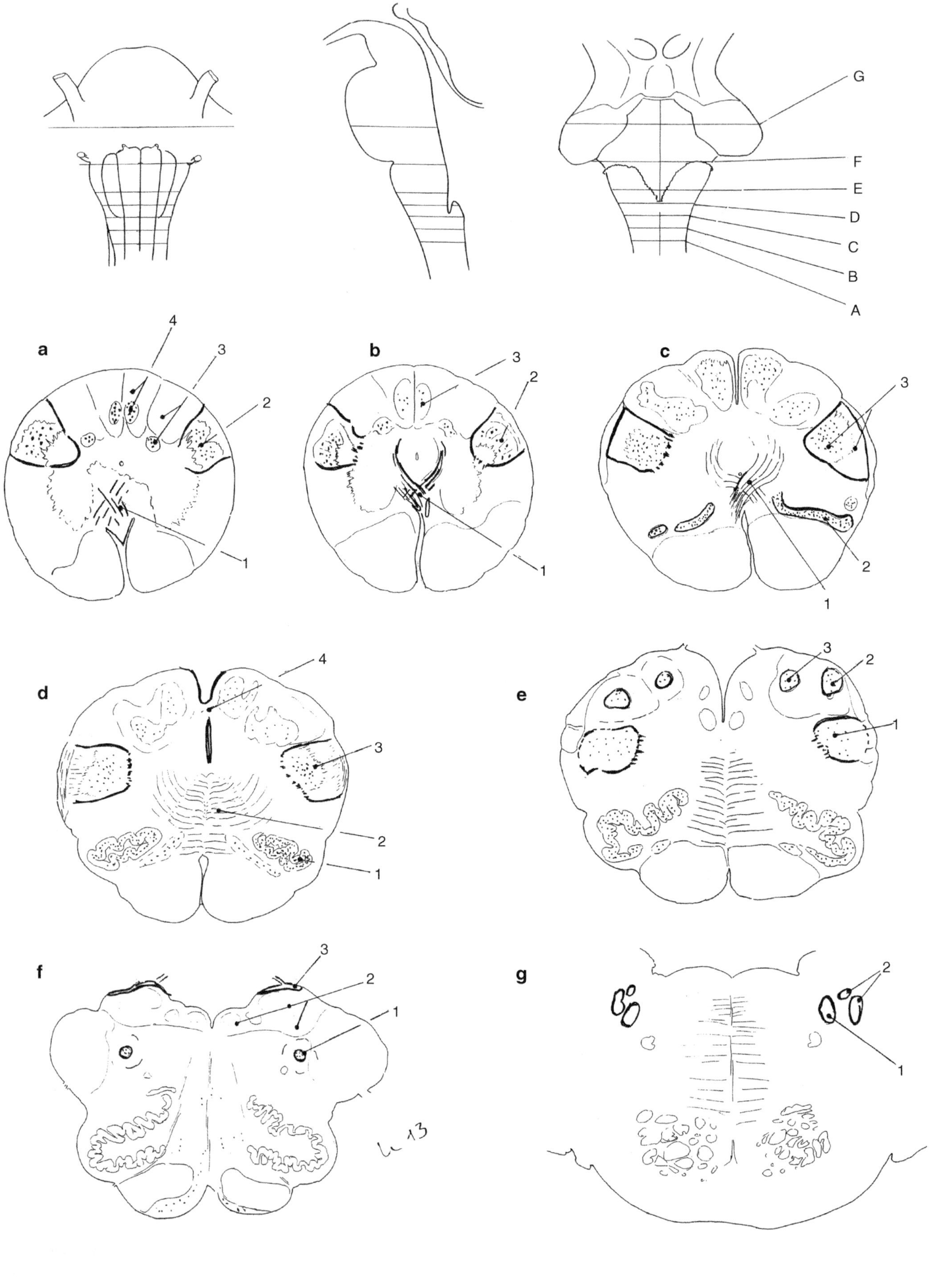
G
F
E
D
C
B
A
a
b
c
d
e
f
g
1
2
3
4

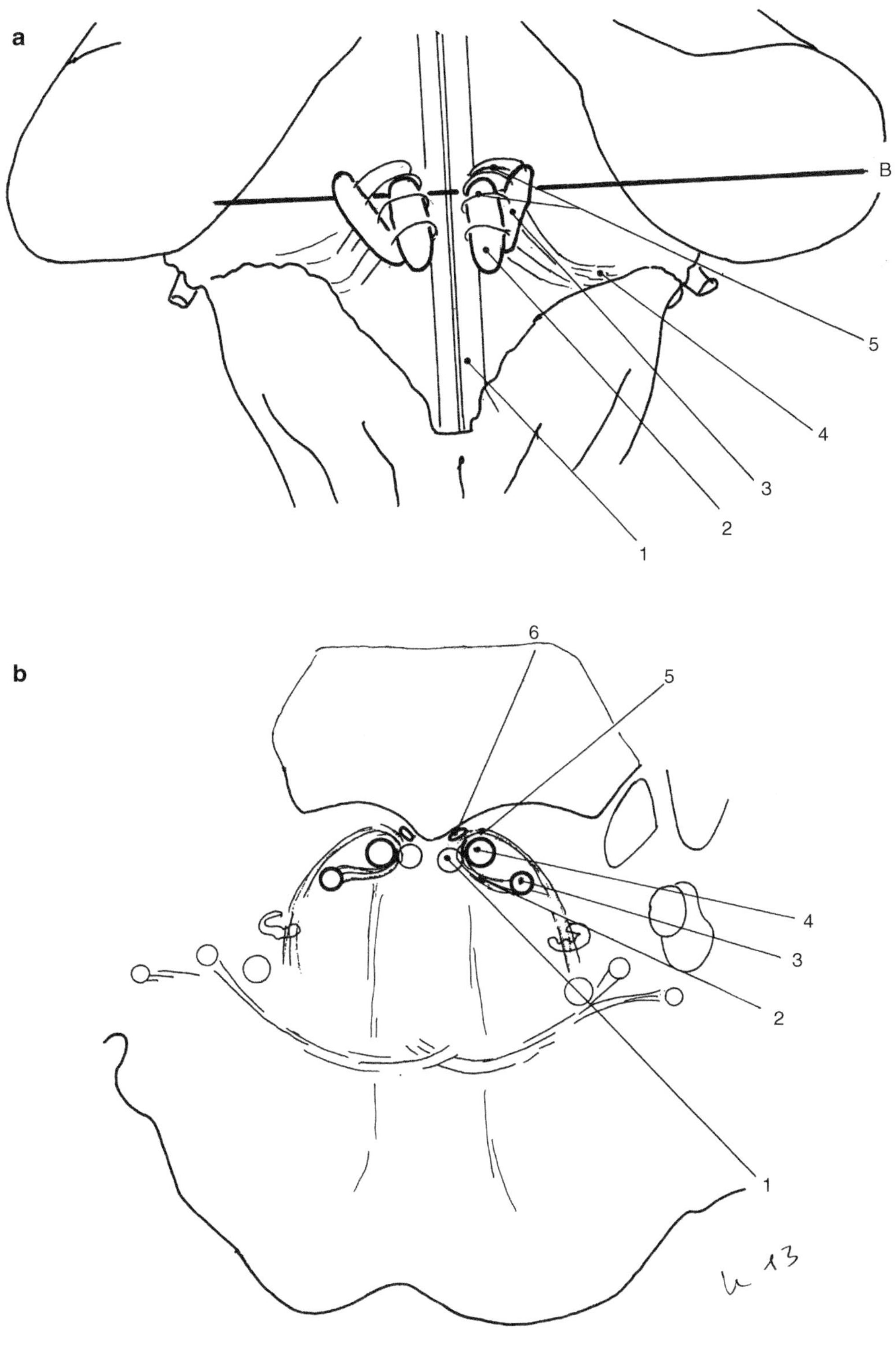

Fig. 15.83 N. abducens and N. facialis. (**a**) Survey and topogram for (**b**). (**b**) Upper region of nucleus n. facialis and abducens. (**a**) *1* fasciculus longitudinalis medialis, *2* nucleus n. abducentis, *3* nucleus n. facialis, *4* fibers of N. facialis, *5* genu interneum n. facialis. (**b**) *1* fasciculus longitudinalis medialis, *2* fibers of N. facialis, *3* nucleus n. facialis, *4* nucleus n. abducentis, *5* genu internum n. facialis, *6* pars prima n. facialis (first motor neurons)

Fig. 15.82 N. trigeminus. Surrounding structures. (**a**) *1* decussatio pyramidum, middle segment, *2* nucleus et tractus spinalis n. trigemini, *3* fasciculus cuneatus and its nucleus, *4* fasciculus gracilis and its nucleus. (**b**) *1 decussatio lemniscorum, inferior beginning, and decussatio pyramidum, superior beginning (mixed fibers)*, *2* nucleus et tractus spinalis n. trigemini, *3* nucleus fasciculi gracilis. (**c**) *1* decussatio lemniscorum, *2 oliva close to its polus inferior*, *3* nucleus et tractus spinalis n. trigemini. (**d**) *1* oliva, *2* decussatio lemniscorum, *3* nucleus et tractus spinalis n. trigemini, *4 obex*. (**e**) *1* nucleus et tractus spinalis n. trigemini, *2* nucleus fasciculi cuneati, *3* nucleus fasciculi gracilis. (**f**) *1* nucleus et tractus spinalis n. trigemini, *2* nuclei vestibulares and nuclei of N. vagus et N. hypoglossus, *3* recessus lateralis, stria cochlearis, and nucleus dorsalis n. cochlearis. (**g**) *1* nucleus principalis sensorius n. trigemini, *2* nucleus motorius n. trigemini. Underlines: landmarks

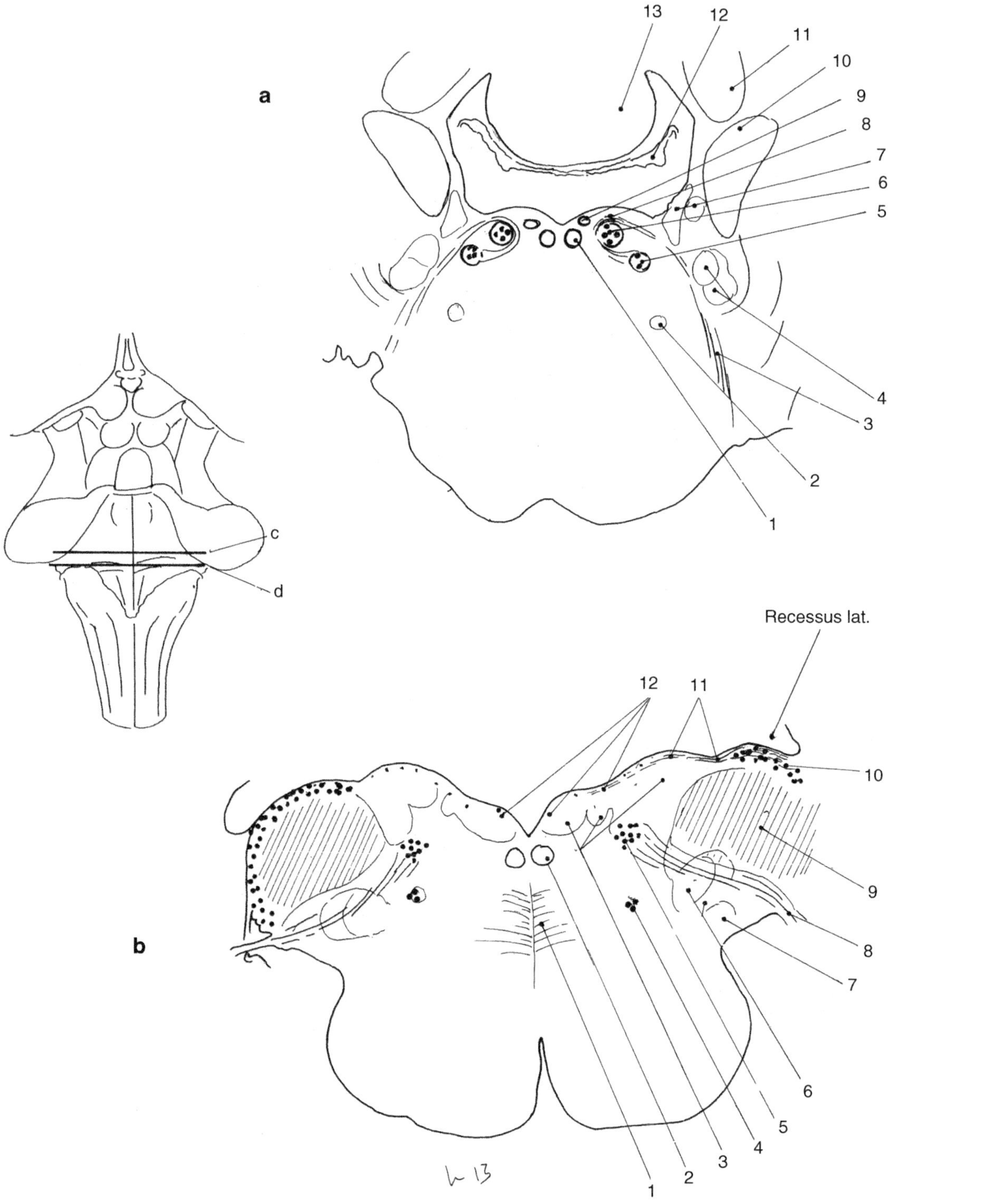

Fig. 15.84 Continuation of Fig. 15.83. (**a**) Inferior region of nucleus n. facialis et n. abducentis. (**b**) Area of recessus lateralis. (**a**) *1* fasciculus longitudinalis medialis; *2* nuclei olivares, superior region; *3* fibers of N. facialis; *4* nucleus et tractus spinalis n. trigemini; *5* nucleus n. facialis; *6* nucleus n. abducentis; *7* nuclei vestibulares mediales et laterales; *8* genu internum n. facialis; *9* nucleus n. hypoglossi (nucleus praepositorius); *10* pedunculus cerebellaris inferior; *11* pedunculus cerebellaris superior; *12* plexus chorioideus; *13* uvula vermis. (**b**) *1* decussatio lemniscorum, *2* fasciculus longitudinalis medialis, *3* superior part of nucleus n. hypoglossi and nuclei verstibulares, *4* nucleus ambiguus, *5* nucleus et tractus solitarius, *6* nucleus et tractus spinalis n. trigemini, *7* fibrae olivocerebellares, *8 N. glossopharyngeus*, *9* pedunculus cerebellaris inferior, *10* nucleus cochlearis dorsalis, *11* striae cochleares dorsales, *12* striae medullares *recessus lat.* See (**b**). Underlines: landmarks

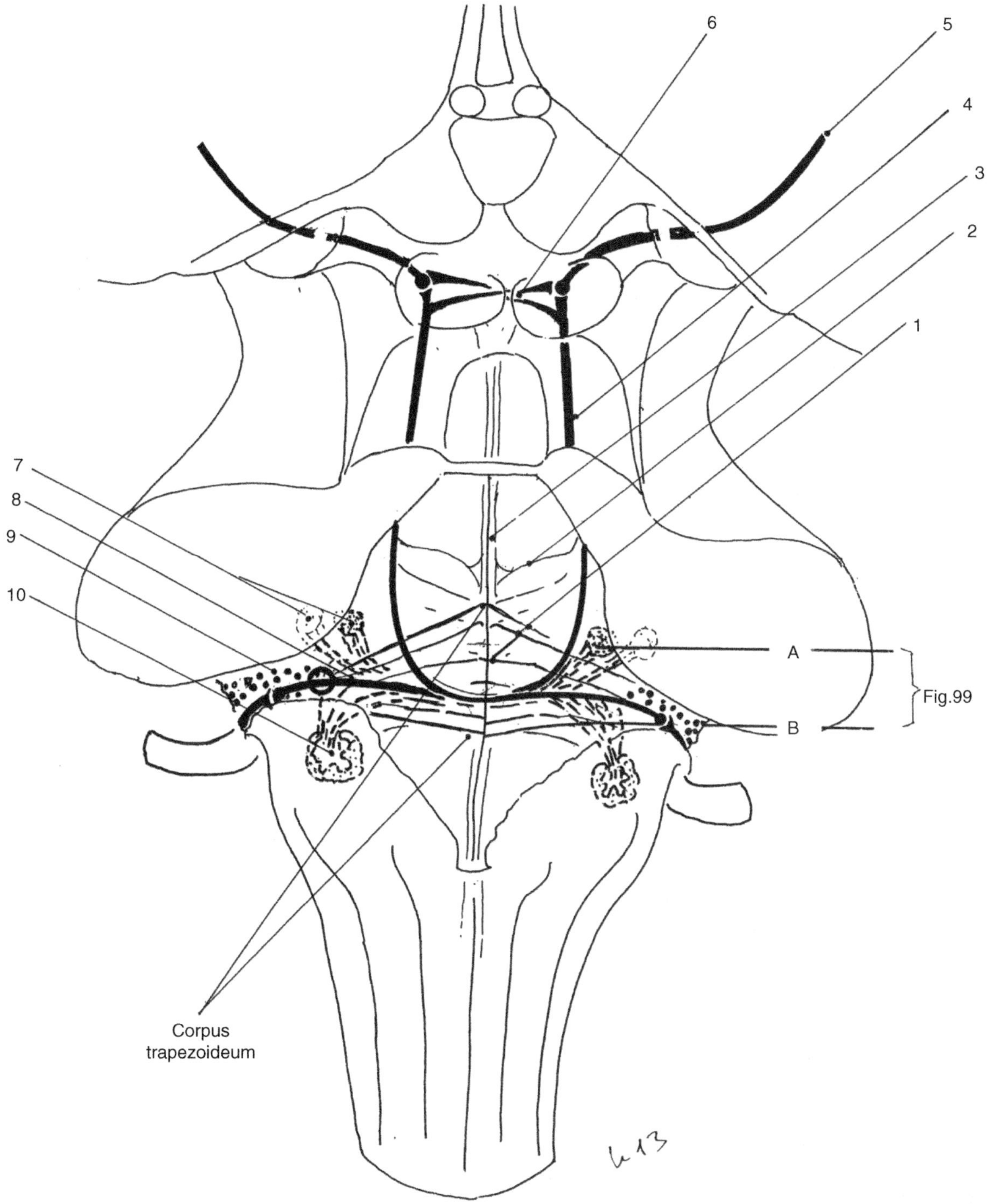

Fig. 15.85 N. cochlearis. Survey. *1* striae medullares; *2* connections of the vestibular segment of lemniscus medialis with fasciculus longitudinalis medialis; *3* fasciculus longitudinalis medialis; *4* lemniscus medialis; *5* radiatio acustica; *6* commissura colliculi inferioris; *7* nuclei cochleares ventrales (see Fig. 3.30), projection; *8* stria acustica dorsalis; *9* nucleus cochlearis dorsalis; *10* oliva, lateral nucleus. Little circle: lateral segment of corpus trapezoideum. (**a**) and (**b**) topogram for Fig. 3.30

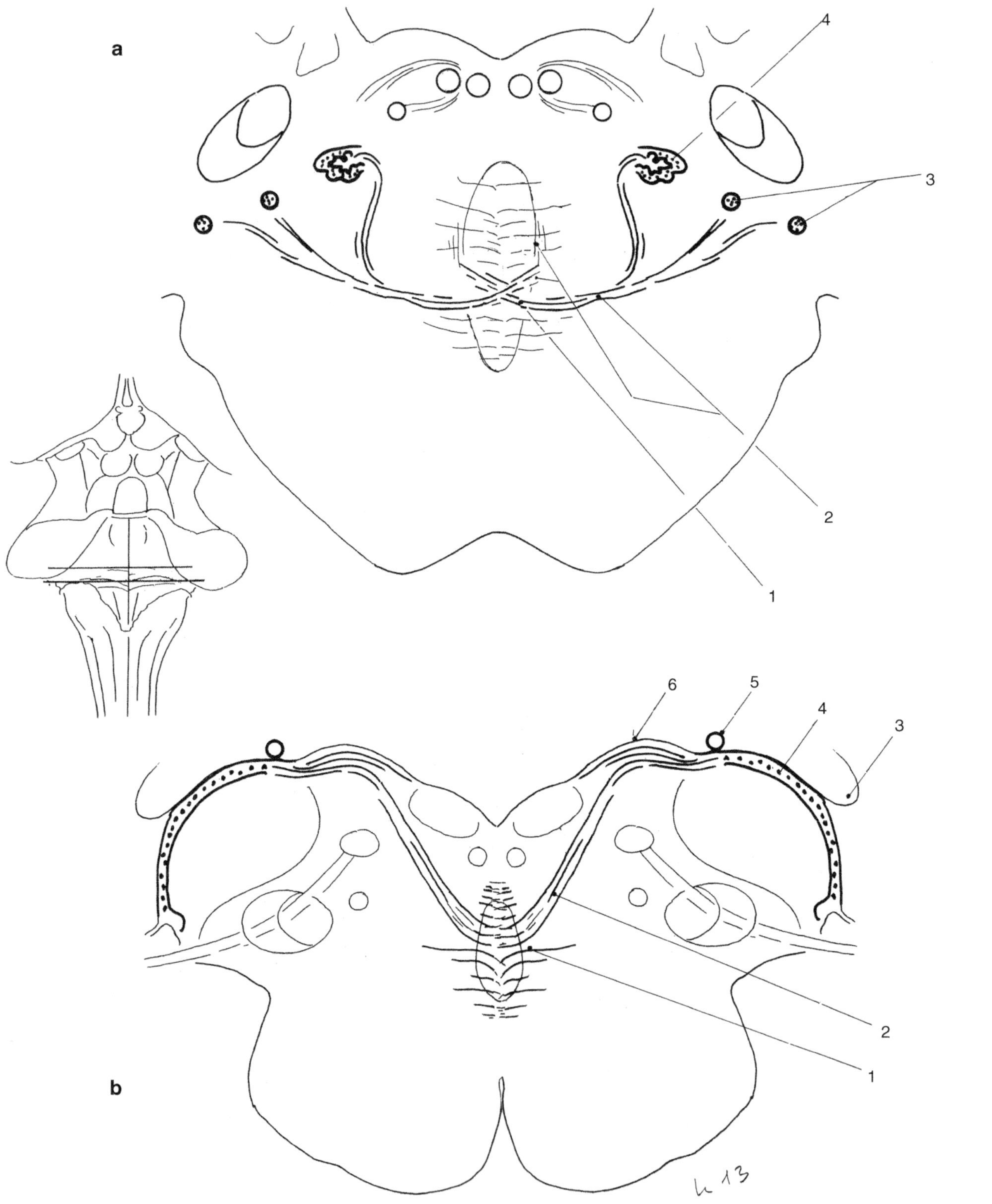

Fig. 15.86 N. cochlearis. Surrounding structures. (**a**) *1* decussatio corporis trapezoidei; *2* connection of decussation corporis trapezoidei with the begin of Lemniscus lat., nucleus lat. olivae and nuclei cochleares ventrales; *3* nuclei cochleares ventrales; *4* nucleus lateralis olivae. (**b**) *1* decussatio lemniscorum and begin of lemniscus medialis, *2* corpus trapezoideum, *3 recessus lateralis* (lateral region of corpus trapezoideum), *4* nucleus cochlearis dorsalis, *5* medial limit region of 4, *6* striae medullares and stria acustica. Underline: landmark

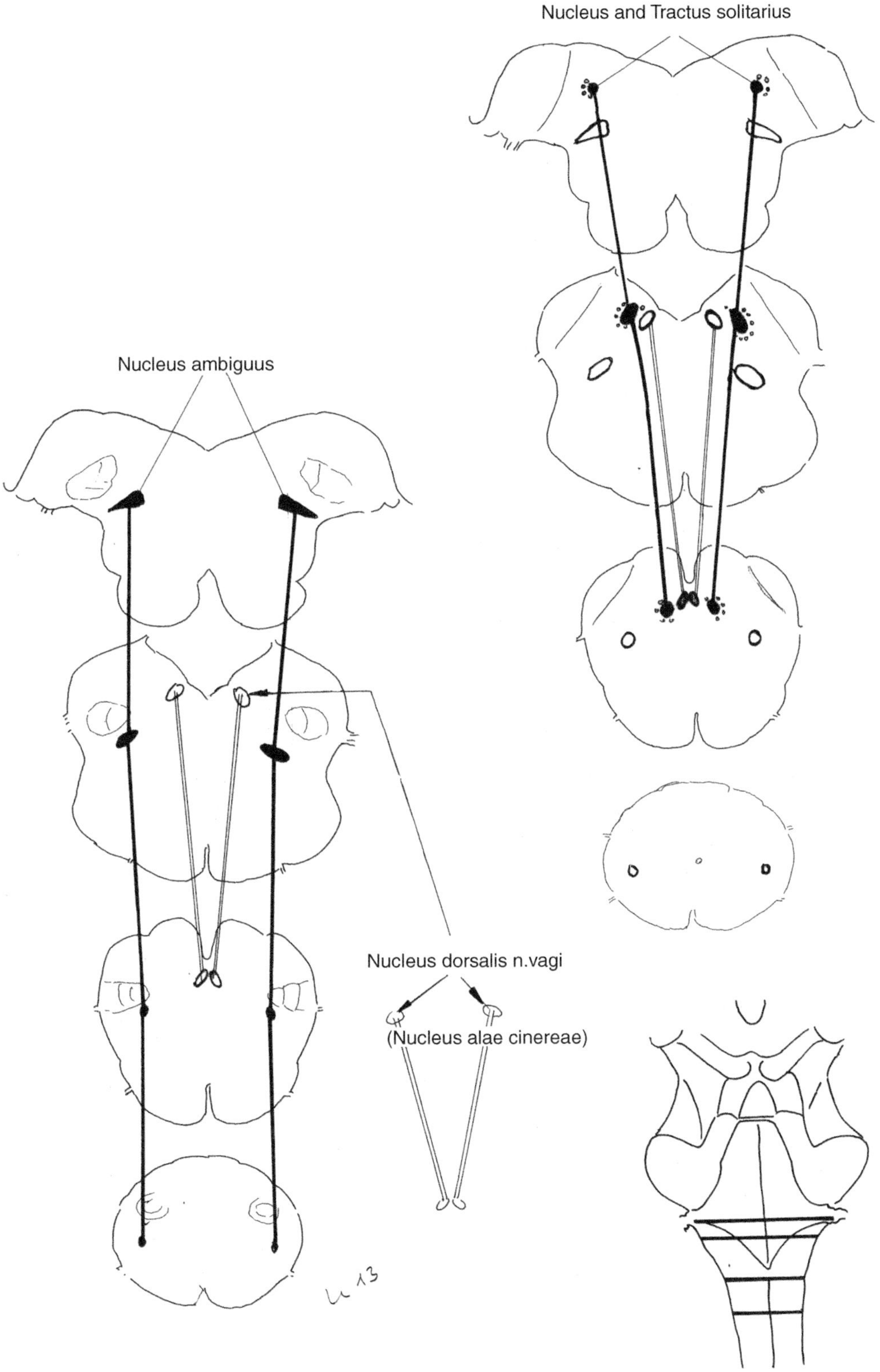

Fig. 15.87 N. glossopharyngeus, N. vagus et N. acessorius. Survey. Dark colored: nucleus et tractus solitarius and nucleus ambiguus. Light colored: nucleus dorsalis n. vagi (modern term identic with nucleus alae cinereae in the past)

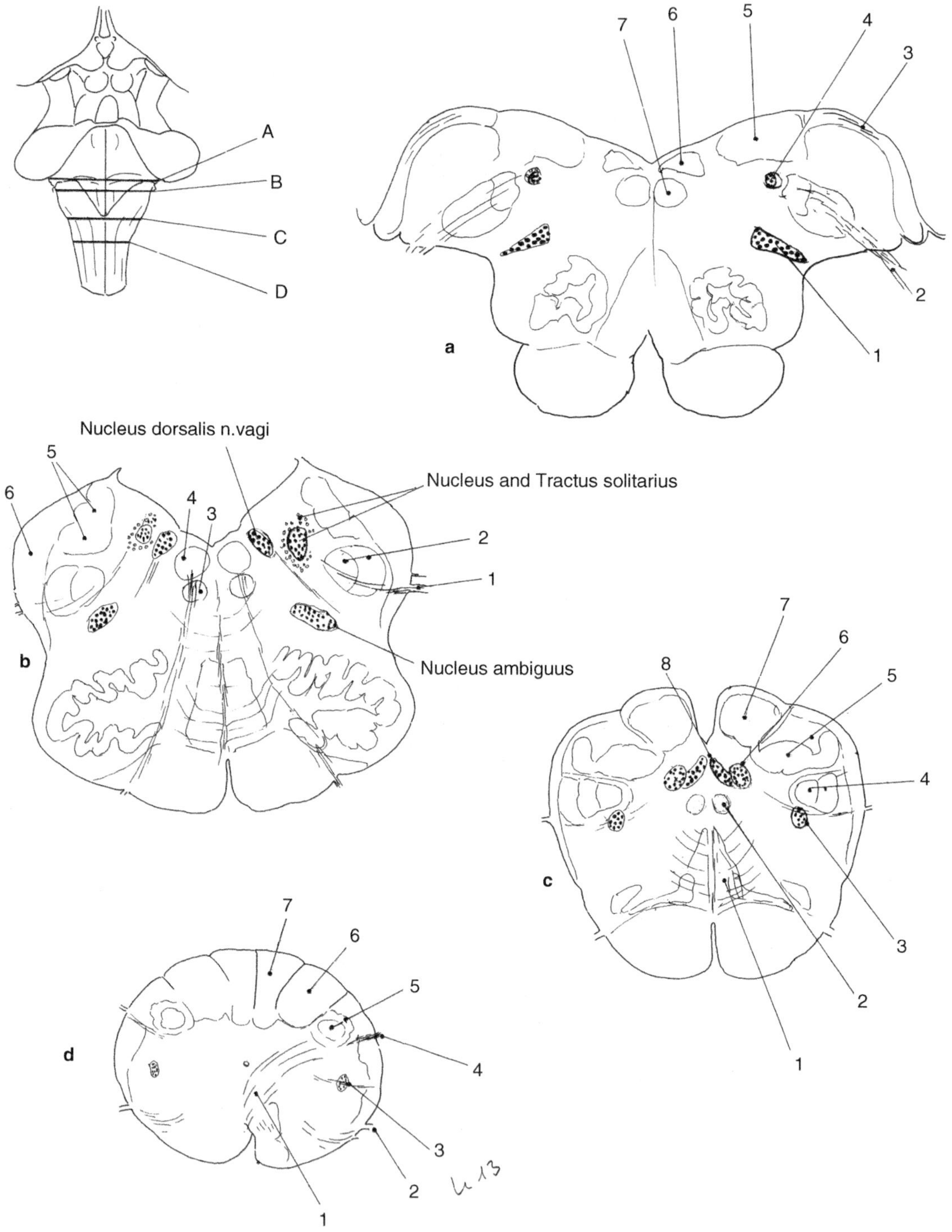

Fig. 15.88 N. glossopharyngeus, N. vagus et N. accessorius. Surrounding structures. The exit areas of Nn. glossopharyngeus, vagus, and accessorius at the medulla oblongata are located dorsal from *sulcus posterior olivae*. The upper limit of nucleus et tractus solitarius is located at the horizontal level of *recessus lat*. Its inferior limit is located at the region between the medulla oblongata and medulla spinalis. In the depth nucleus et tractus solitarius are located close to peduculus cerebellaris inferior. Cave at surgical removal, e.g., of cavernomas of pedunculus cerebellaris inferior. Radix spinalis n. accessorii is located dorsal from *Lig. denticulatum*. A *1* nucleus ambiguus; *2 N. glossopharyngeus*; *3* striae medullares; *4* nucleus et tractus solitarius; *5* nuclei vestibulares; *6* nucleus n. hypoglossi, superior region; *7* fasciculus longitudinalis medialis. (**b**) *1* N. vagus, *2* nucleus et tractus spinalis n. trigemini, *3* fasciculus longitudinalis medialis, *4* nucleus n. hypoglossi, *5* nucleus fasciculi cuneati, *6* pedunculus cerebellaris inferior. (**c**) *1* lemniscus med. and its decussatio, *2* fasciculus longitudinalis medialis, *3* nucleus ambiguus, *4* nucleus et tractus spinalis n. trigemini, *5* nucleus et fasciculus cuneatus, *6* nucleus et tractus solitarius, *7* nucleus et fasciculus gracilis, *8* nucleus dorsalis n. vagi. (**d**) *1* decussatio pyramidum, *2* C1 *3* nucleus retroambiguus, *4* N. accessorius, *5* nucleus et tractus spinalis n. trigemini, *6* fasciculus cuneatus, *7* fasciculus gracilis. Underlines: landmarks

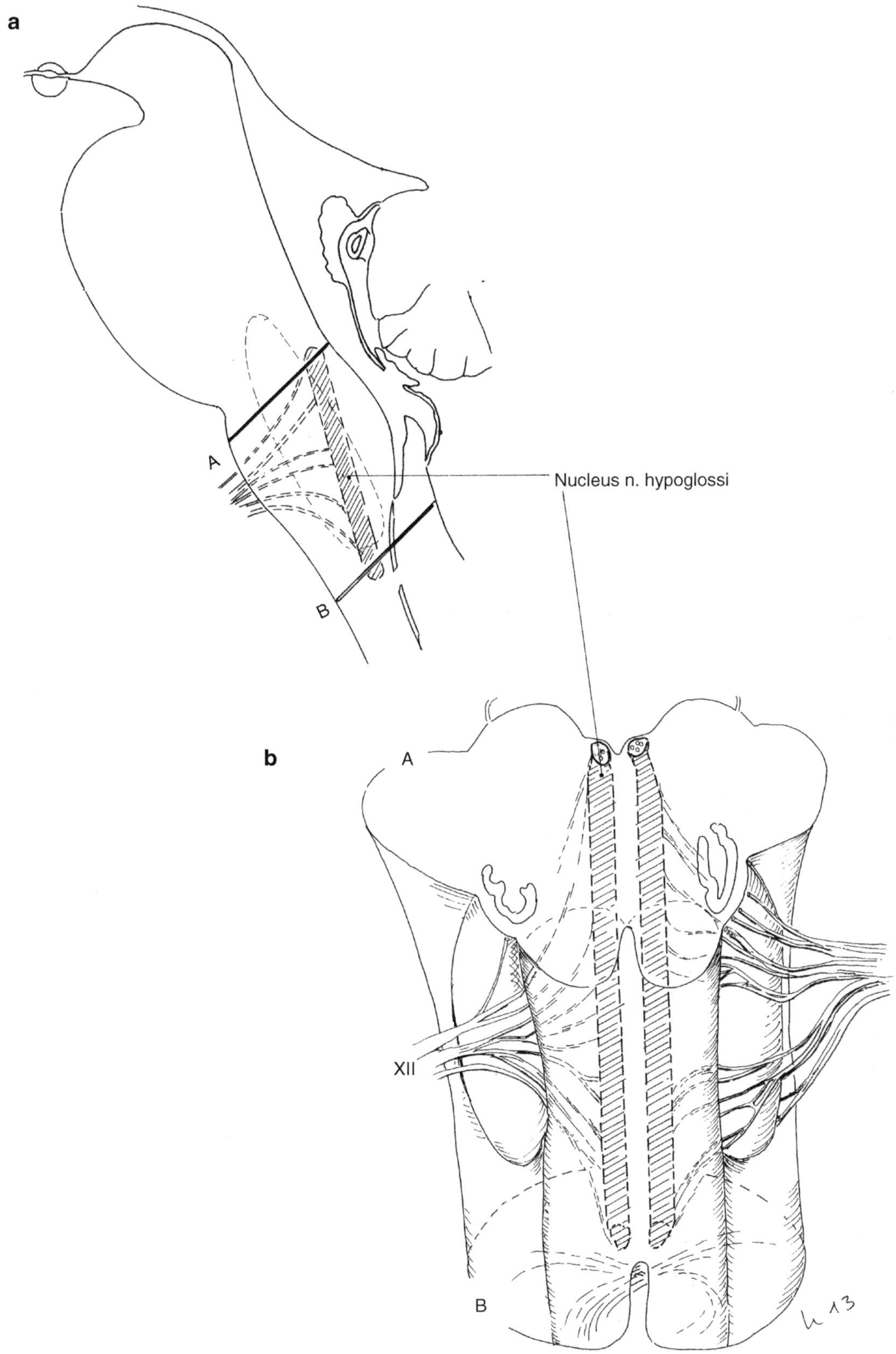

Fig. 15.89 N. hypoglossus. Survey and topogram for Fig. 3.32. (**a**) Nucleus n. hypoglossi is extended between the level of *recessus lateralis* and *caudal from obex*. (**b**) The roots of N. hypoglossus are variable. This should be considered at transconylar premedullary surgical approaches, e.g., for removal of craniospinal meningiomas (The roots of N. hypoglossus are interposed beween ligamentum denticulatum and A. vertebralis, Figs. 6.9 and 6.11). Underlines: landmarks

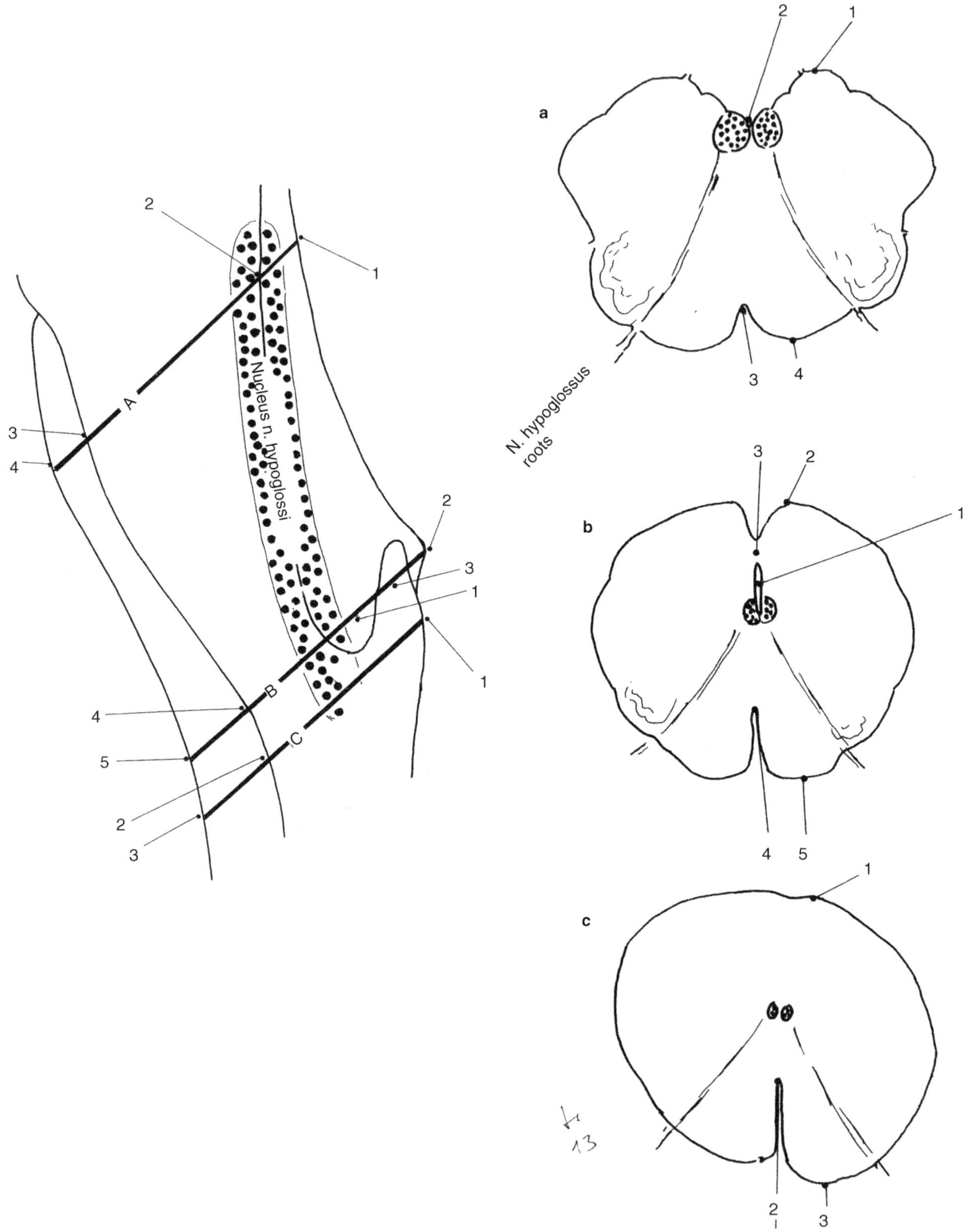

Fig. 15.90 Continuation of Fig. 15.89. (**a**) *1* dorsal shape of the medulla oblongata, *2* sulcus medianus, *3* fissura mediana ant., *4* shape of pyramis medullae. (**b**) *1 funnel-like shape of the mouth of canalis spinalis centralis, 2* dorsal shape of the medulla oblongata, *3 obex, 4* fissura mediana ant., *5* pyramis medullae. (**c**) *1* dorsal shape of the begin of medulla spinalis, *2* fissura mediana ant., *3* pyramis medullae. Underlines: landmarks

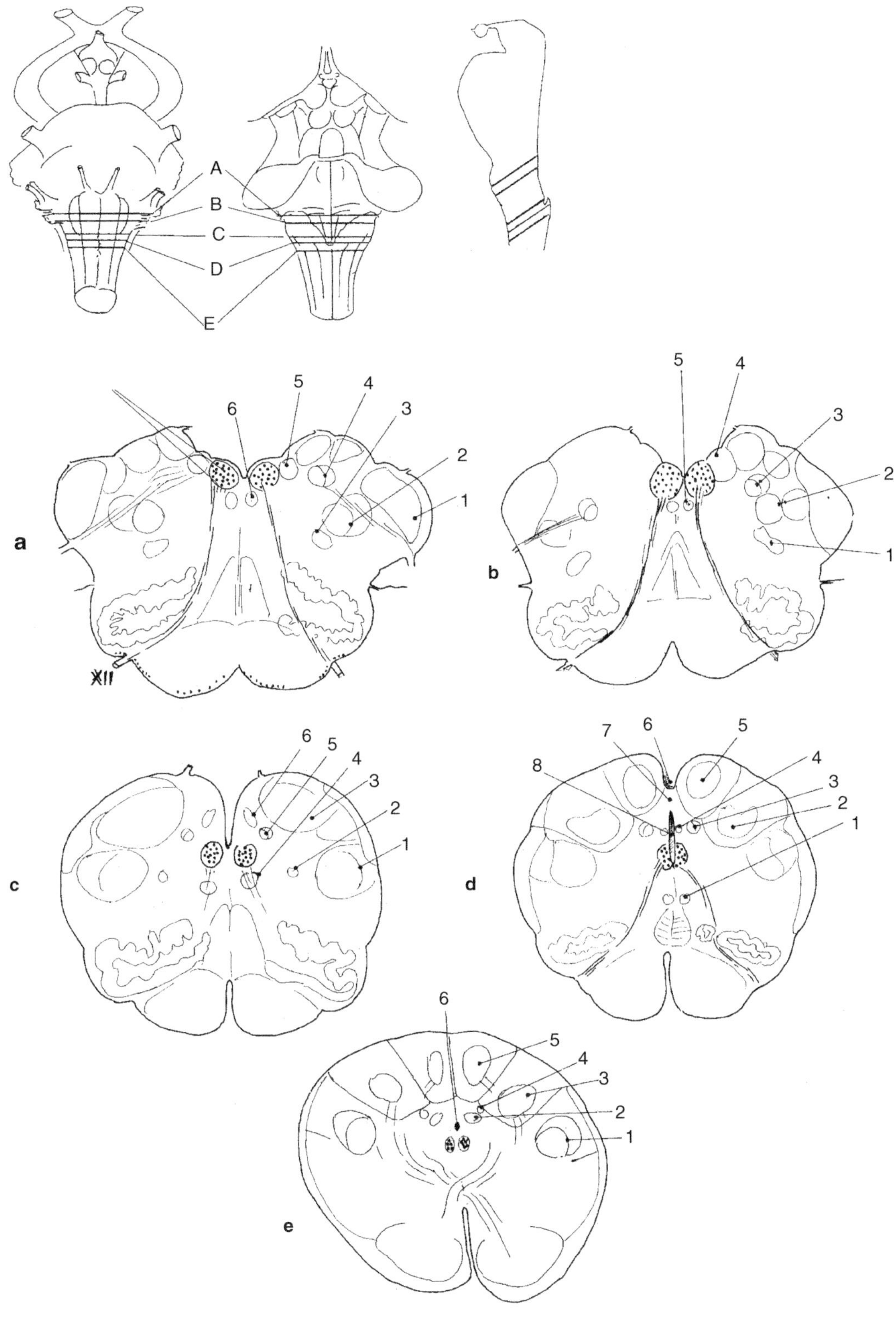

Fig. 15.91 N. hypoglossus. Surrounding areas. (**a**) *1* pedunculus cerebellaris inferior, *2* nucleus et tractus spinalis n. trigemini, *3* nucleus ambiguus, *4* nucleus et tractus solitarius, *5* nucleus dorsalis n. vagi, *6* fasciculus longitudinalis medialis. (**b**) *1* nucleus ambiguus, *2* nucleus et tractus spinalis n. trigemini, *3* nucleus et tractus solitarius, *4* nucleus dorsalis n. vagi, *5* fasciculus longitudinalis medialis. (**c**) *1* nucleus et tractus spinalis n. trigemini, *2* nucleus ambiguus, *3* nucleus fasciculi cuneati et gracilis, *4* fasciculus longitudinalis medialis, *5* nucleus et tractus solitarius, *6* nucleus dorsalis n. vagi. (**d**) *1* fasciculus longitudinalis medialis, *2* nucleus et tractus fasciculi cuneati, *3* nucleus et tractus solitarius, *4* nucleus dorsalis n. vagi, *5* nucleus fasciculi gracilis, *6* sulcus longitudinalis, *7 obex*, *8 mouth of canalis spinalis centralis*. (**e**) *1* nucleus et tractus spinalis n. trigemini, *2* nucleus n. accessorii, *3* nucleus fasciculi cuneati, *4* nucleus et tractus solitarius, *5* nucleus fasciculi gracilis, *6* canalis spinalis centralis. Underlines: landmarks

Spinocerebellar and Olivocerebellar System (Fig. 15.92)

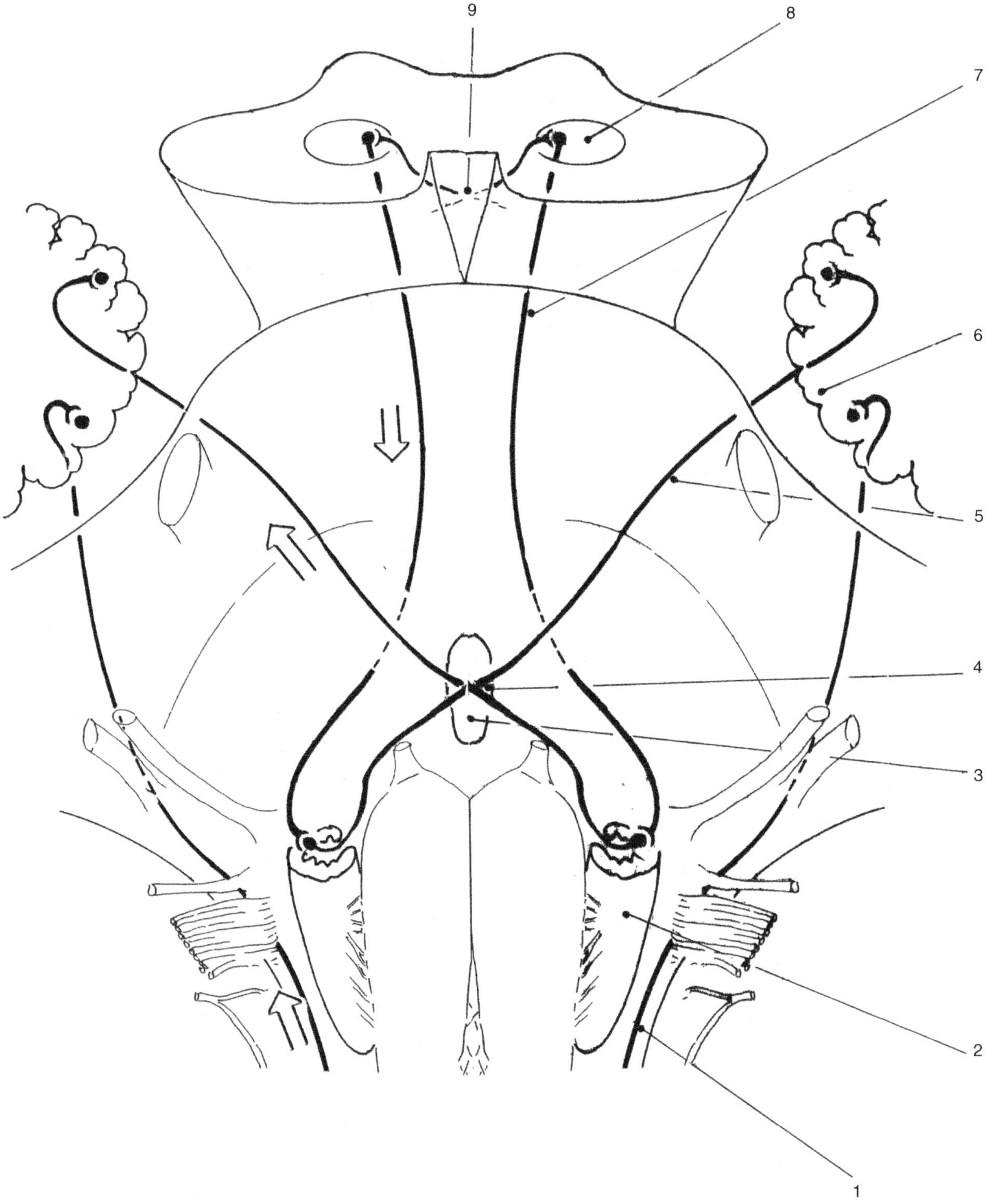

Fig. 15.92 Olivae and its main connections. Survey. This system should be preserved at surgery for avoiding statomotoric and other bulbar and cerebellar deficits. *1* tractus spinocerebellaris anterior et posterior, *2* oliva, *3* lemniscus medialis and its decussatio, *4* decussatio of tractus olivocerebellaris, *5* tractus olivocerebellaris, *6* cerebellar cortex, *7* tractus rubroolivaris (tractus tegmentalis centralis), *8* nucleus ruber, *9* decussatio of pedunculi cerebellares superiores

Neuromotor and Chemosensory Control of Respiration (Figs. 15.93 and 15.94)

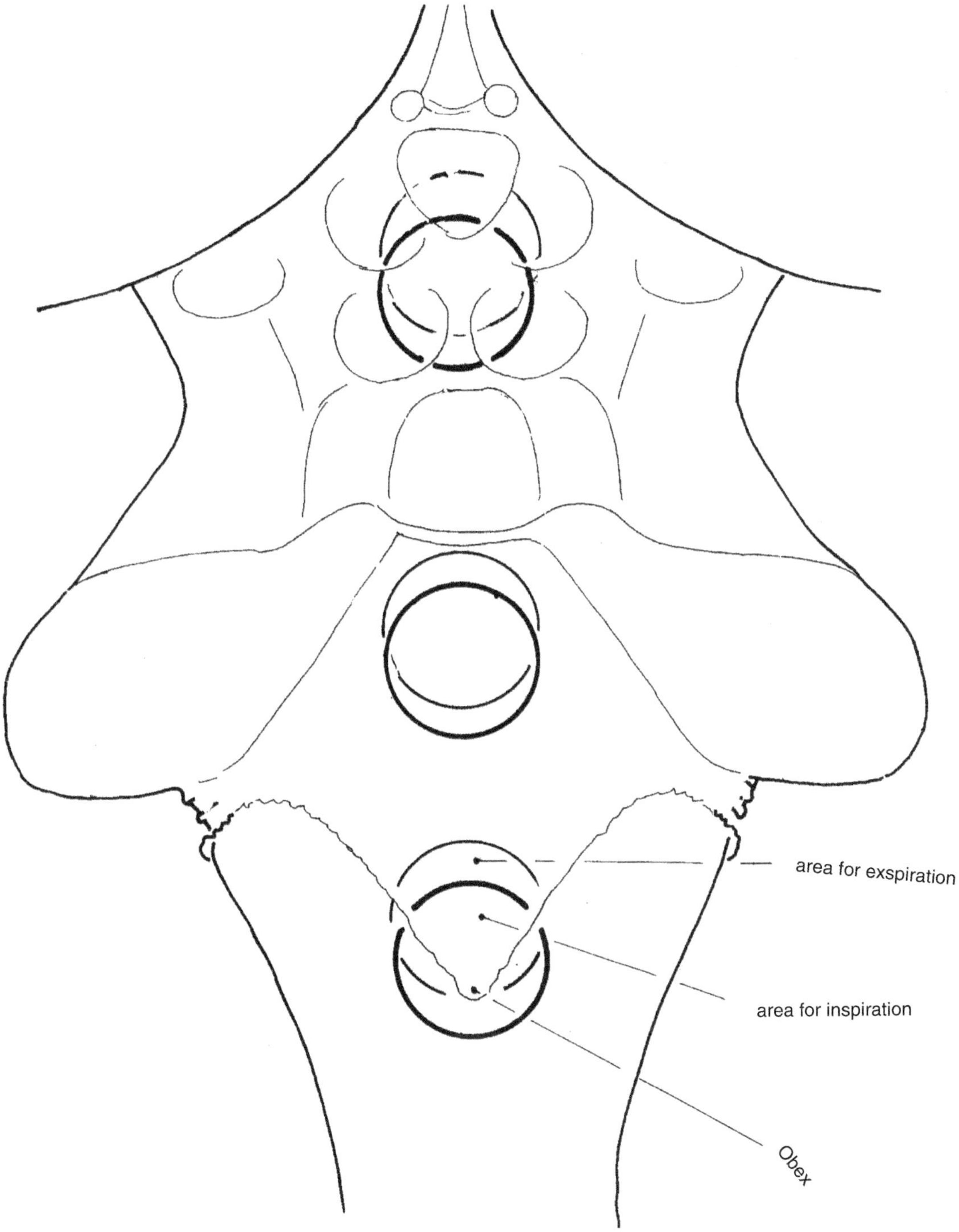

Fig. 15.93 Central neuronal areas for inspiration and exspiration. Schematized (Wyss 1955; Tarlov and Giancotti 1956; Frowein 1963; Seeger 1968; Schlaefke and Loeschke 1979). The main neuronal control of inspiration and exspiration is located at the region of *obex*, modified by in- and exspiratory pontine and mesencephalic areas. In- and exspiratory functions are endangered by lesions of the obex region and caudal to it. Underline: landmark

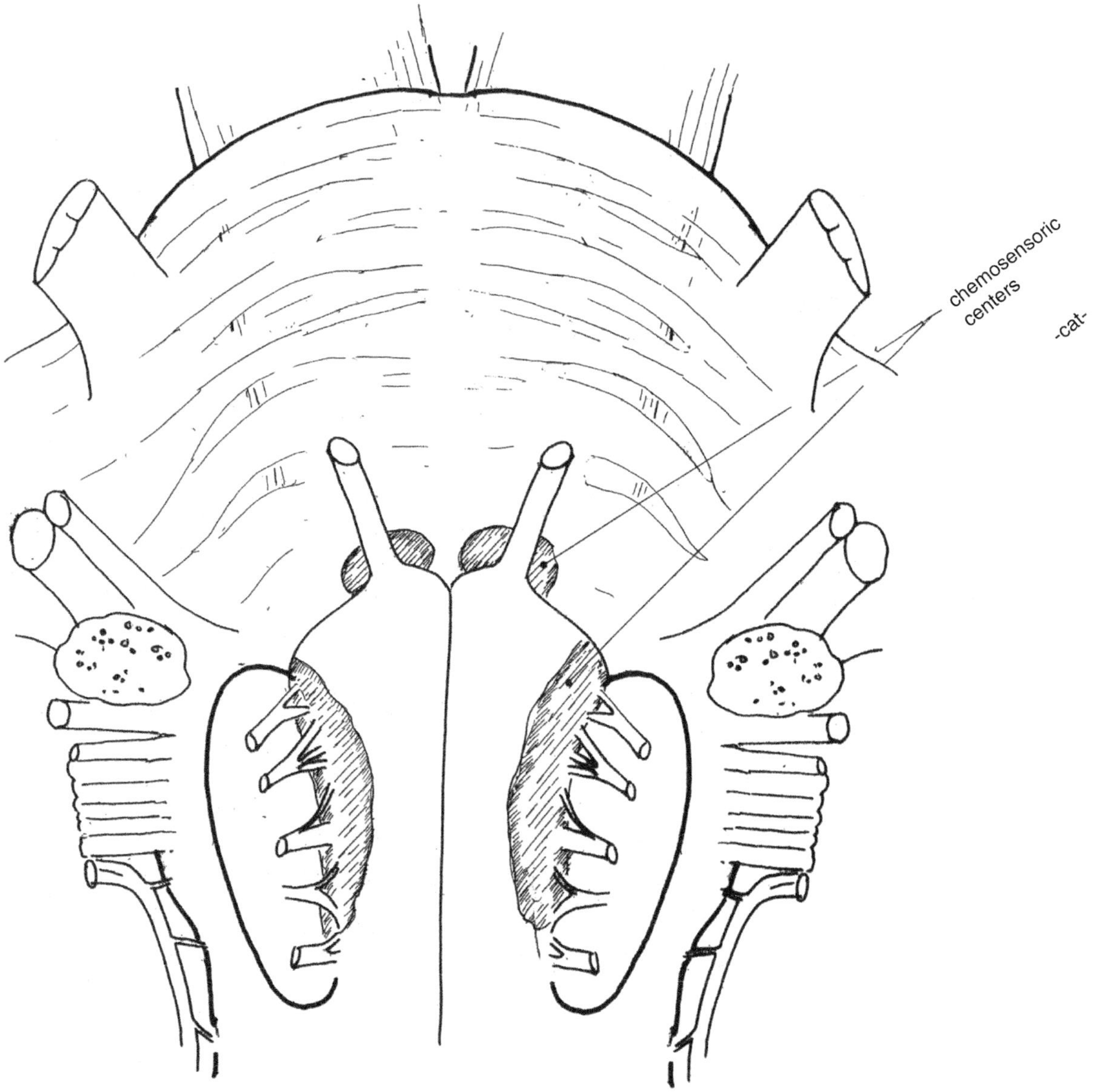

Fig. 15.94 Central chemosensory centers at cats and rabbits according to Schlaefke and Loeschke (Schlaefke 1981; Schlaefke et al. 1979). In- and exspiratory functions at mammals (and humans?) may be endangered at transcondylar premedullary surgical approaches, e.g., for removal of craniospinal meningionas. Landmarks for neuronavigation: *pontomedullary rim and olivae*

Synopsis of Nuclei and Fibers of the Brain Stem (Figs. 15.95, 15.96, 15.97, 15.98, 15.99, 15.100, 15.101, 15.102, 15.103, 15.104, 15.105, 15.106, 15.107, 15.108, and 15.109)

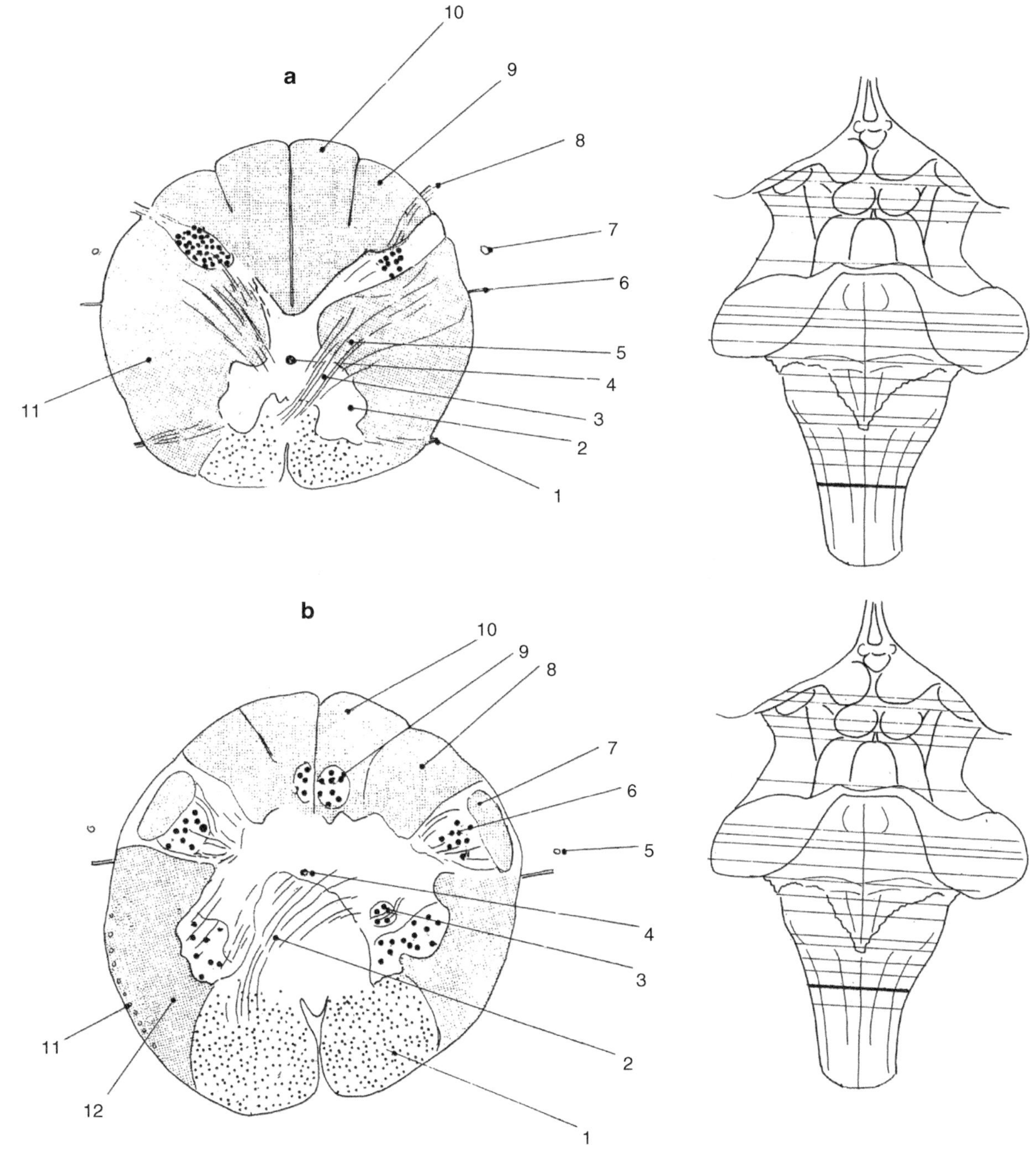

Fig. 15.95 (**a**) *1* C1, *2* cornu anterius, *3* decussatio pyramidum, *4* canalis spinalis centralis, *5* tractus pyramidalis lateralis, *6* Lig. denticulatum, *7* radix spin.n. accessorii, *8* radix posterior of C1 and cornu posterius, *9* fasciculus cuneatus, *10* fasciculus gracilis, *11* tractus spinothalamicus. (**b**) *1* tractus pyramidalis (tractus corticospinalis), *2* decussatio pyramidum, *3* nucleus et tractus solitarius, *4* canalis spinalis centralis, *5* radix spinalis n. accessorii, *6* nucleus spinalis n. trigemini, *7* tractus spinalis n. trigemini, *8* fasciculus cuneatus, *9* nucleus fasciculi gracilis, *10* fasciculus gracilis, *11* tractus spinocerebellaris anterior et posterior, *12* funiculus anterolateralis, region of tractus spinothalamicus

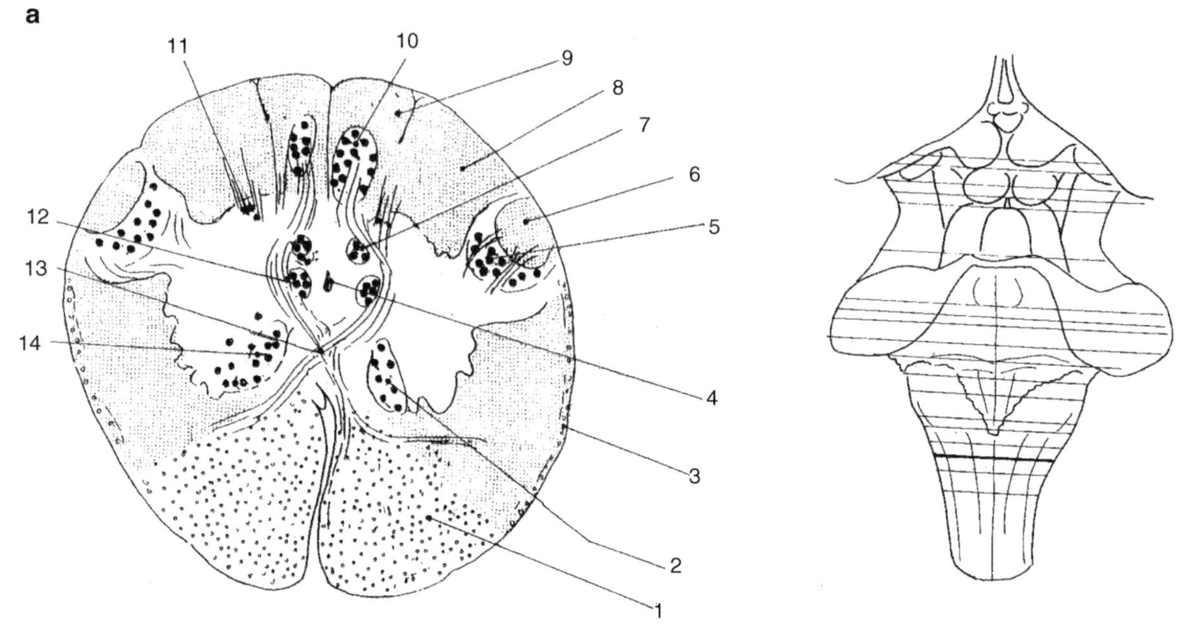

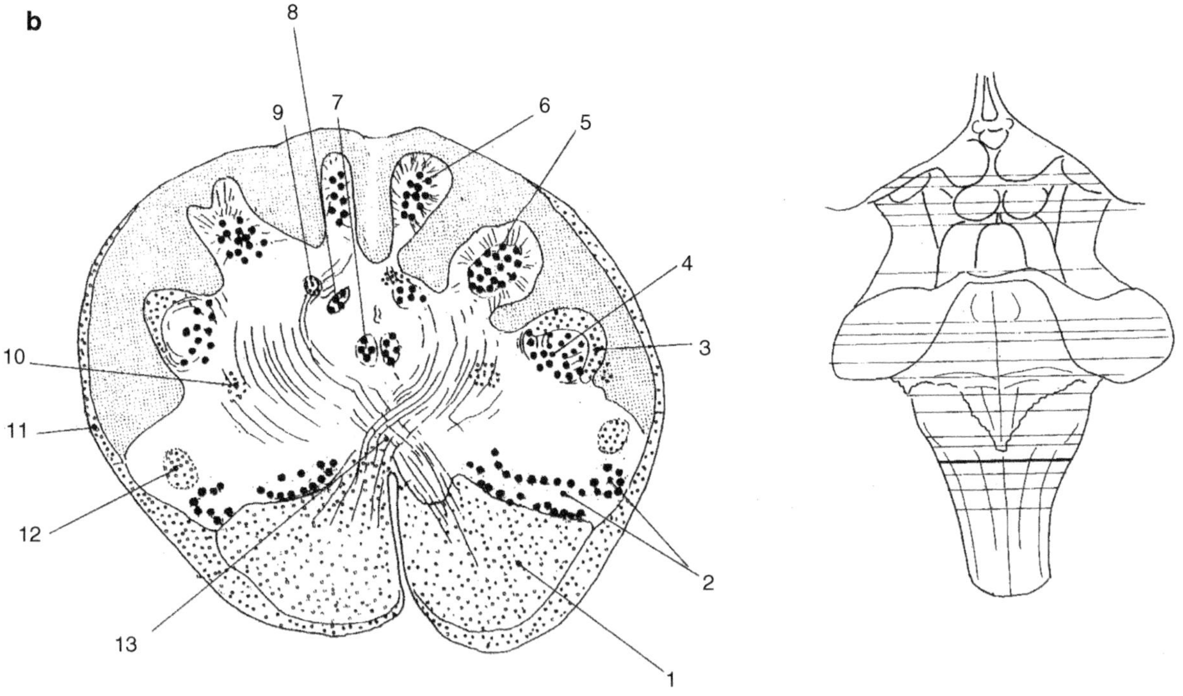

Fig. 15.96 (**a**) *1* tractus pyramidalis; *2* cornu anterius; *3* tractus spinocerebellaris anterior et posterior; *4* canalis spinalis centralis; *5* nucleus tractus spinalis n. trigemini; *6* tractus spinalis n. trigemini; *7* nucleus dorsalis n. vagi; *8* fasciculus cuneatus; *9* fasciculus gracilis; *10* nucleus fasciculi gracilis, main segment; *11* nucleus fasciculi cuneati, inferior segment, *12* nucleus n. hypoglossi, *13* decussatio lemniscorum, inferior segment; *14* cornu anterius. (**b**) *1* tractus pyramidalis, *2* oliva, *3* tractus spinalis n. trigemini, *4* nucleus tractus spinalis n. trigemini, *5* nucleus fasciculi cuneati, *6* nucleus fasciculi gracilis, *7* nucleus n. hypoglossi, *8* nucleus tractus solitarii, *9* tractus solitarius, *10* tractus rubrospinalis, *11* tractus spinocerebellaris anterior et posterior, *12* tractus tectospinalis, *13* decussatio lemniscorum

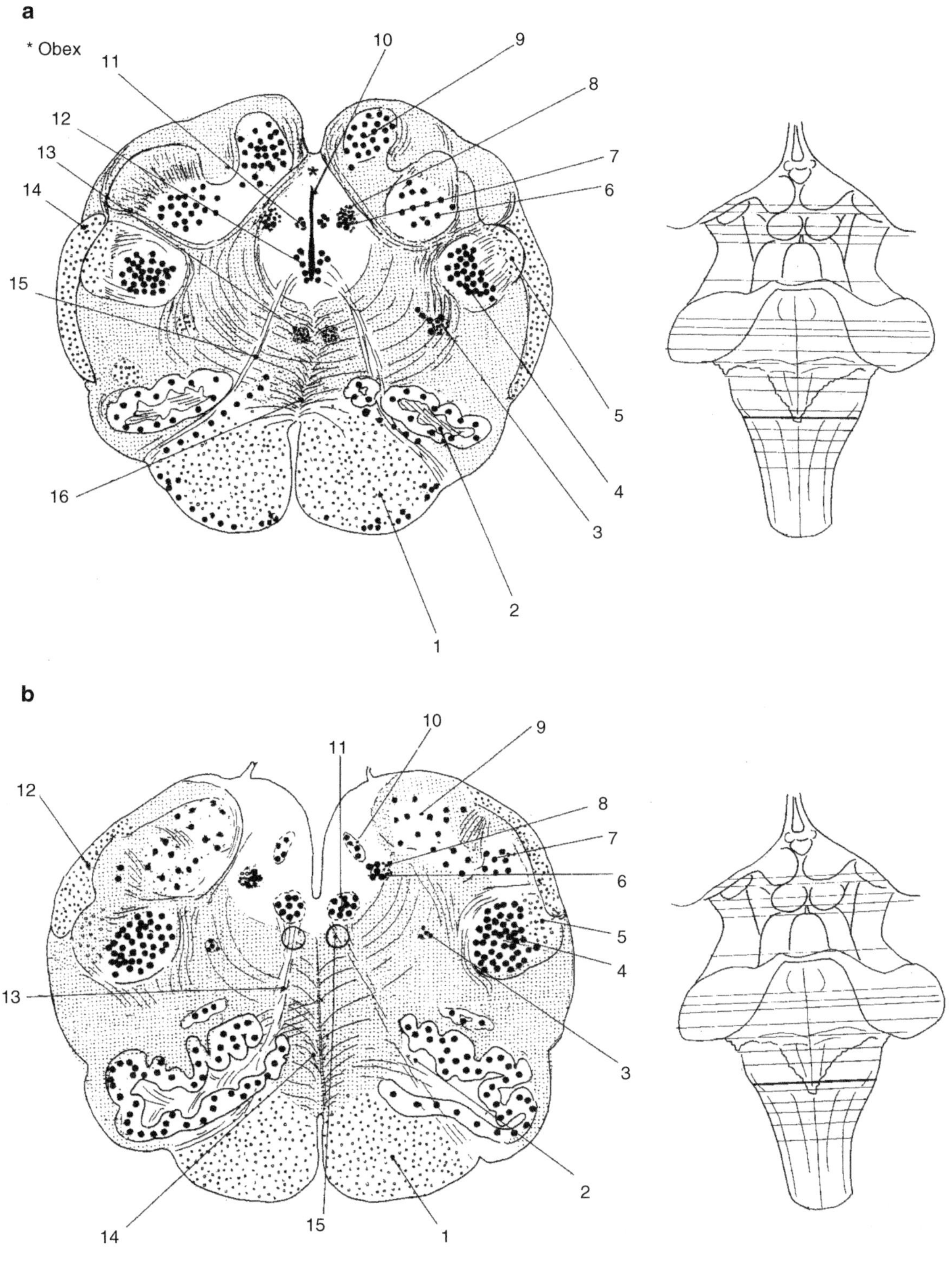

Fig. 15.97 (**a**) *1* tractus pyramidalis, *2* oliva, *3* nucleus ambiguus, *4* nucleus tractus spinalis n. trigemini, *5* tractus spinalis n. trigemini, *6* nucleus fasciculi cuneati, *7* nucleus tractus solitarii, *8* tractus solitarius, *9* nucleus fasciculi gracilis, *10* canalis spinalis centralis, *11* nucleus dorsalis n. vagi, *12* nucleus n. hypoglossi, *13* fasciculus longitudinalis medialis, *14* pedunculus cerebellaris inferior, *15* fila radicularia n. hypoglossi, *16* lemniscus medialis. (**b**) *1* tractus pyramidalis, *2* hilus of oliva, *3* nucleus ambiguus, *4* nucleus tractus spinalis n. trigemini, *5* tractus spinalis n. trigemini, *6* nucleus tractus solitarii, *7* nucleus fasciculi cuneati, *8* tractus solitarius, *9* nucleus fasciculi gracilis, *10* nucleus dorsalis n. vagi, *11* nucleus n. hypoglossi, *12* pedunculus cerebellaris inferior, *13* fila radicularia n. hypoglossi, *14* decussatio of lemniscus medialis, *15* fasciculus longitudinalis medialis

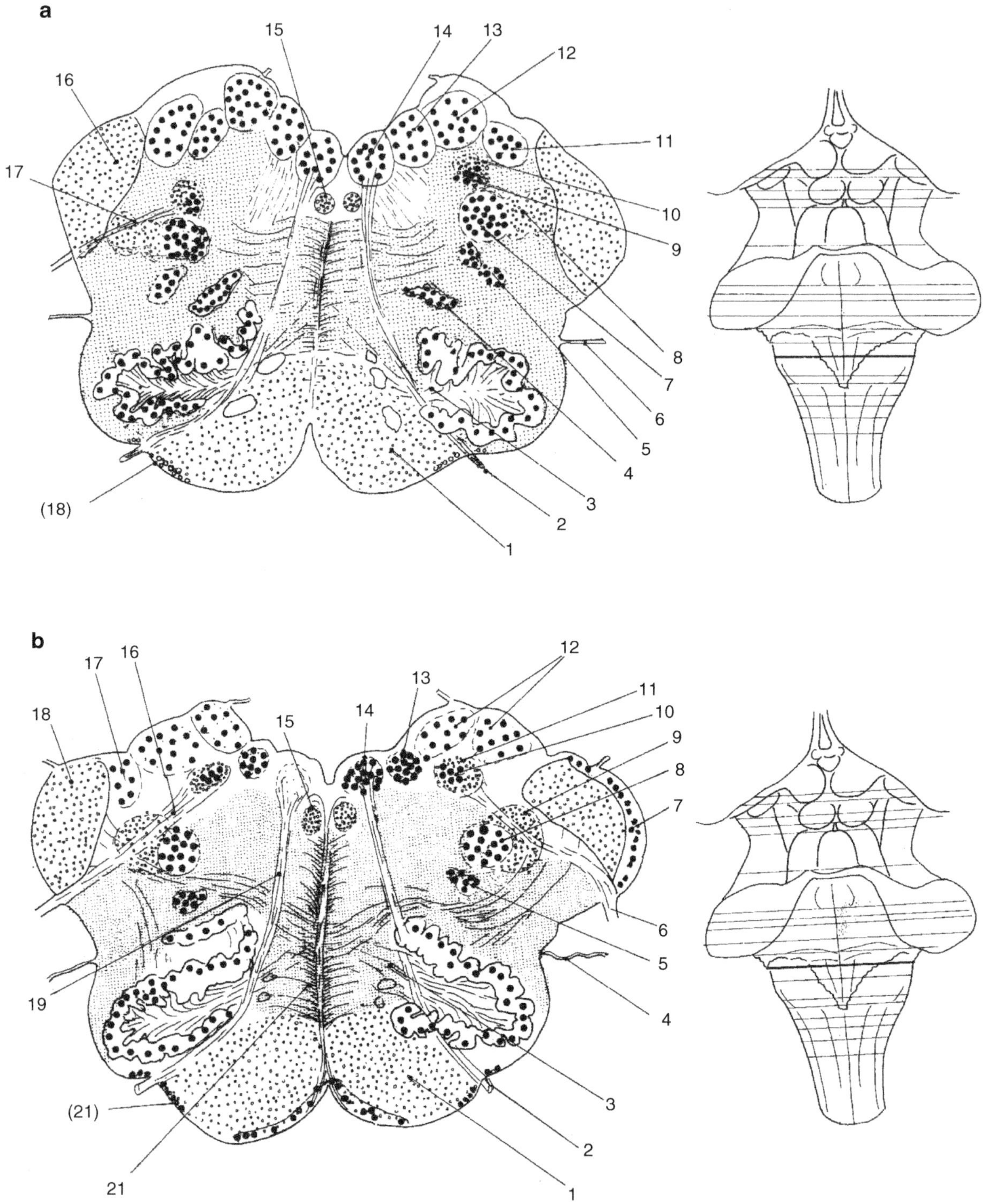

Fig. 15.98 (**a**) *1* tractus pyramidalis; *2* fila radicularia n. hypoglossi; *3* hilus of oliva; *4* accessorial nucleus of 3; *5* nucleus ambiguus; *6* Lig. denticulatum; *7* nucleus tractus spinalis n. trigemini; *8* tractus spinalis n. trigemini; *9* nucleus tractus solitarii; *10* tractus solitarius; *11* nucleus fasciculi cuneati; *12* nuclei vestibulares, caudal segments; *13* nucleus dorsalis n. vagi; *14* nucleus n. hypoglossi; *15* fasciculus longitudinalis medialis; *16* pedunculus cerebellaris inferior; *17* fila radicularia n. vagi (18) chemosensoric area for respiration (mammals). (**b**) *1* tractus pyramidalis, *2* fila radicularia n. hypoglossi, *3* hilus of oliva, *4* Lig. denticulatum, *5* nucleus ambiguus, *6* fila radicularia n. vagi, *7* nucleus cochlearis dorsalis, *8* nucleus tractus spinalis n. trigemini, *9* tractus spinalis n. trigemini, *10* nucleus tractus solitarii, *11* tractus solitarius, *12* nuclei vestibulares, *13* nucleus dorsalis n. vagi, *14* nucleus n. hypoglossi, *15* fasciculus longitudinalis medialis, *16* fila radicularia n. vagi, *17* nuclei vestibulares, *18* pedunculus cerebellaris inferior, *19* fila radicularia n. hypoglossi, (*20*) chemosensoric area for respiration, *21* lemniscus med

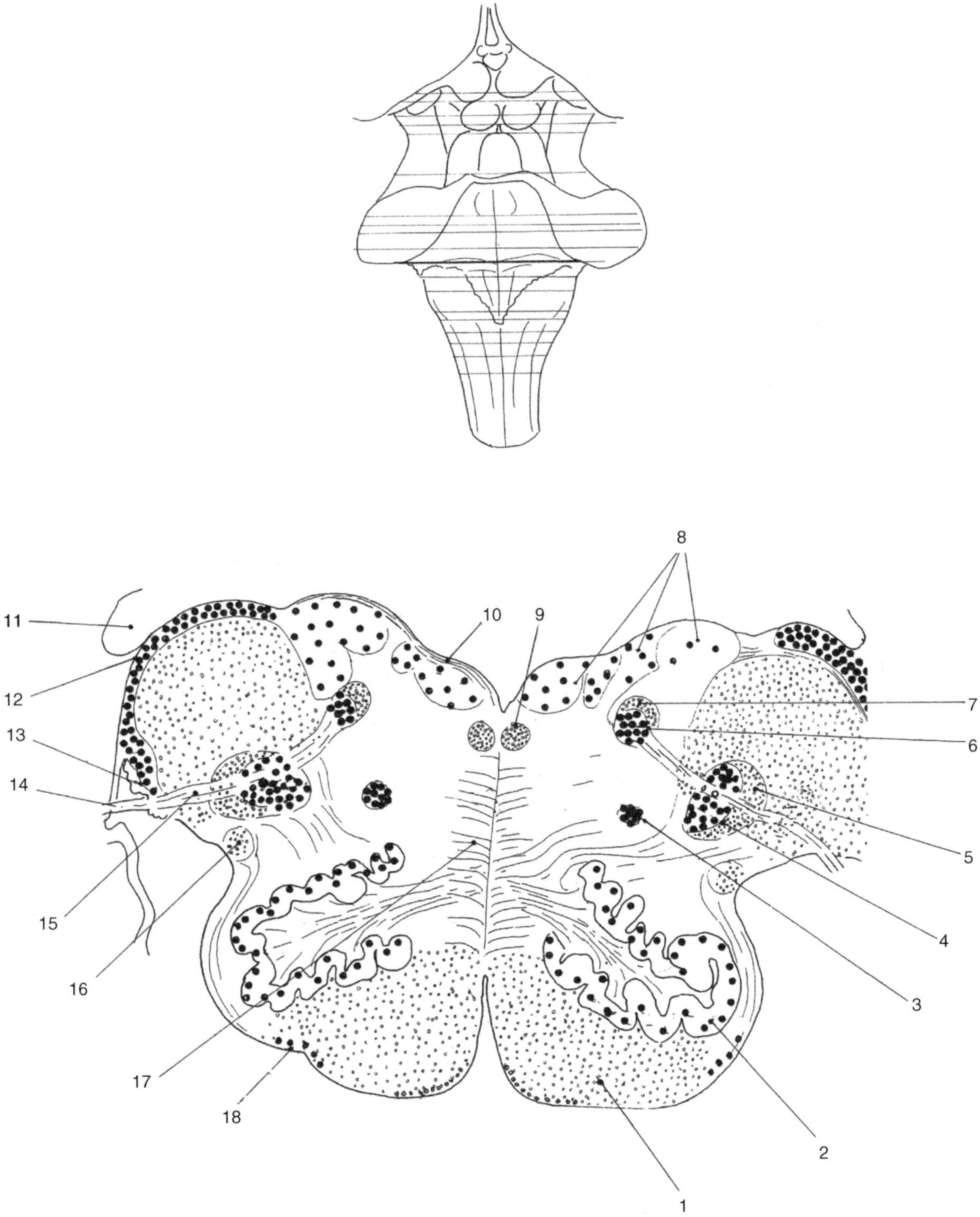

Fig. 15.99 *1* fasciculi pyramidales, *2* oliva, *3* nucleus ambiguus, *4* nucleus tractus spinalis n. trigemini, *5* tractus spinalis n. trigemini, *6* nucleus tractus solitarii, *7* tractus solitarius, *8* nuclei vestibulares, *9* fasciculus longitudinalis medialis, *10* nucleus vestibularis medialis, *11* recessus lateralis, *12* nucleus cochlearis dorsalis, *13* lateral limit region of 12, *14* N. glossopharyngeus, *15* fila radicularia of N. glossopharyngeus, *16* tractus rubrospinalis et tectospinalis, *17* decussatio lemniscorum, (*18*) chemosensoric area for respiration (mammals)

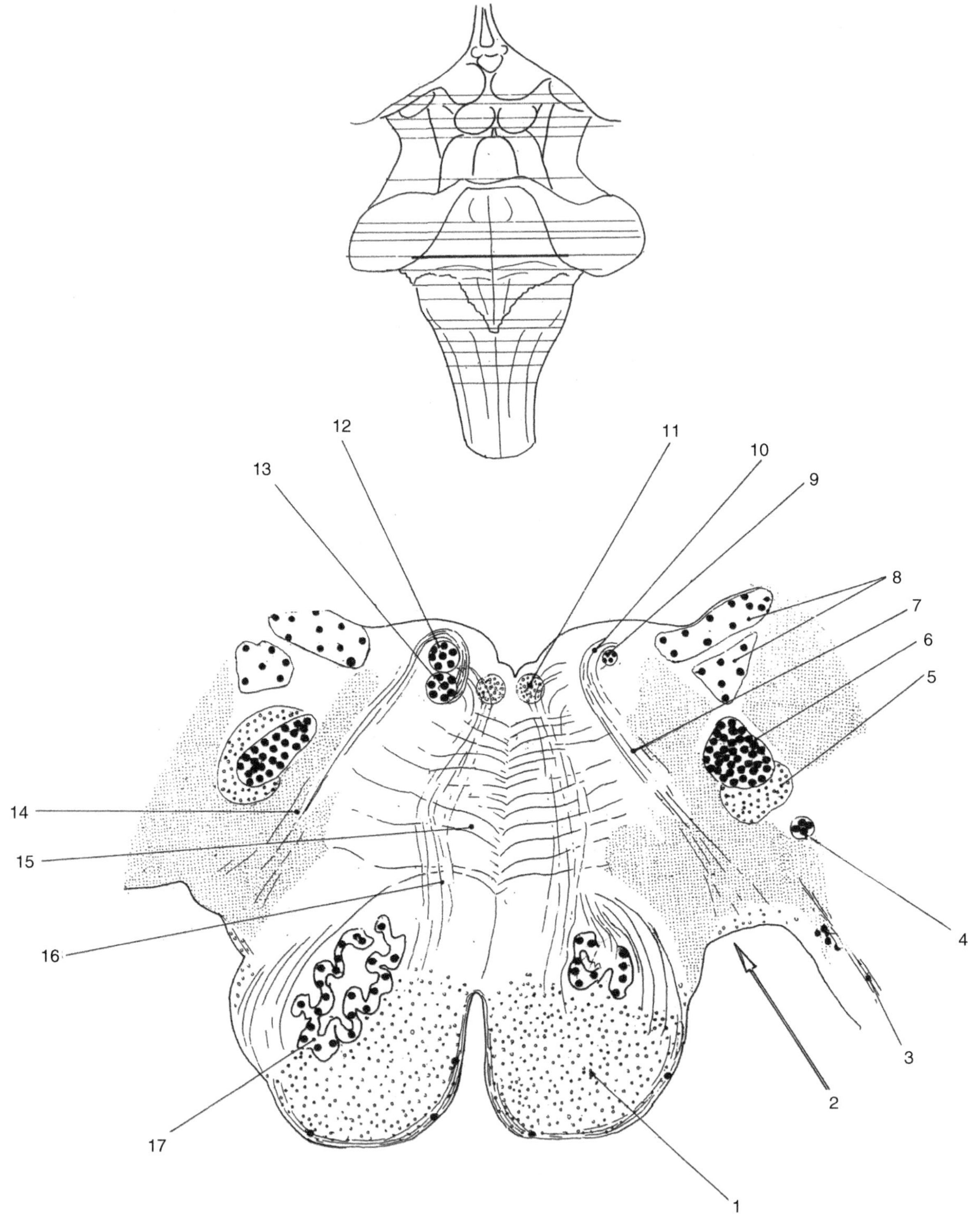

Fig. 15.100 *1* tractus pyramidalis; *2* fossa paraolivaris; *3* fila radicularia n. vestibularis; *4* nucleus ambiguus; *5* tractus spinalis n. trigemini; *6* nucleus tractus spinalis n. trigemini; *7* fila radicularia n. facialis; *8* nuclei vestibulares; *9* nucleus n. facialis, caudal segment; *10* genu internum n. facialis; *11* fasciculus longitudinalis medialis; *12* nucleus n. abducentis, *13* nucleus n. facialis, middle segment; *14* fila radicularia n. facialis; *15* lemniscus medialis; *16* fila radicularia n. abducentis; *17* oliva

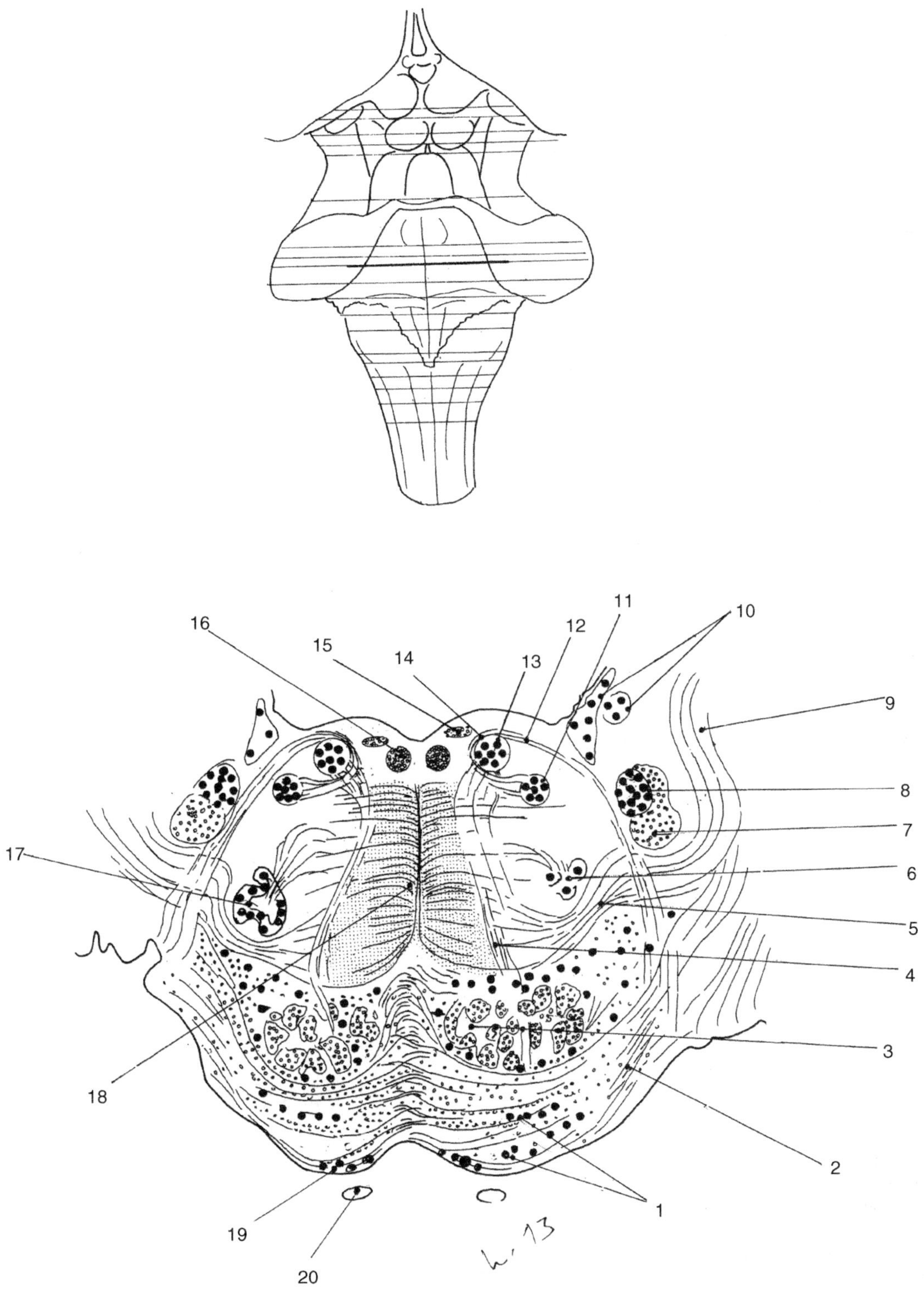

Fig. 15.101 *1* nuclei pontis, *2* fibrae pontocerebellares, *3* tractus pyramidalis, *4* fila radicularia n. abducentis, *5* corpus trapezoideum, *6* nucleus ambiguus, *7* tractus spinalis n. trigemini, *8* nucleus tractus spinalis n. trigemini, *9* pedunculus cerebellaris inferior, *10* nuclei n. vestibularis, *11* nucleus n. facialis, *12* fila n. facialis, *13* nucleus n. abducenti, *14* genu internum n. facialis, *15* pars prima (prenuclear segment) of N. facialis, *16* fasciculus longitudinalis medialis, *17* oliva, *18* decussatio lemnisci medialis, (*19*) chemosensoric region for respiration (mammals), *20* N. abducens

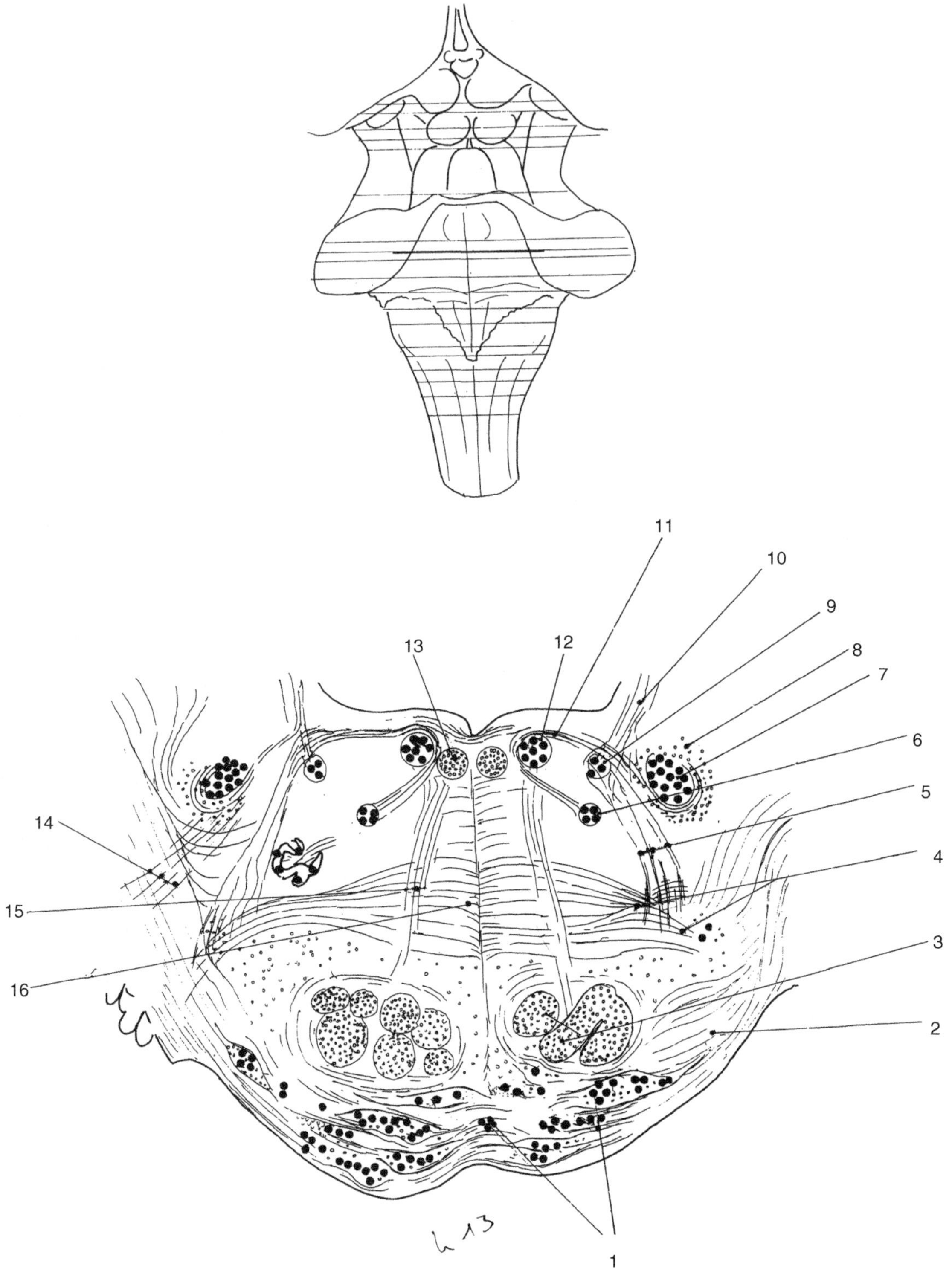

Fig. 15.102 *1* nuclei pontis; *2* fibrae pontocerebellares, pedunculus cerebellaris medialis; *3* tractus pyramidalis; *4* corpus trapezoides; *5* fila radicularia n. facialis, cranial segment; *6* nucleus n. facialis; *7* nucleus tractus spinalis n. trigemini; *8* tractus spinalis n. trigemini; *9* nucleus motorius n. trigemini; *10* radix mesencephalica n. trigemin; *11* genu internum n. facialis; *12* nucleus n. abducentis, cranial segment; *13* fasciculus longitudinalis medialis; *14* fila radicularia n. trigemini; *15* fila radicularia n. abducentis; *16* decussatio lemniscorum

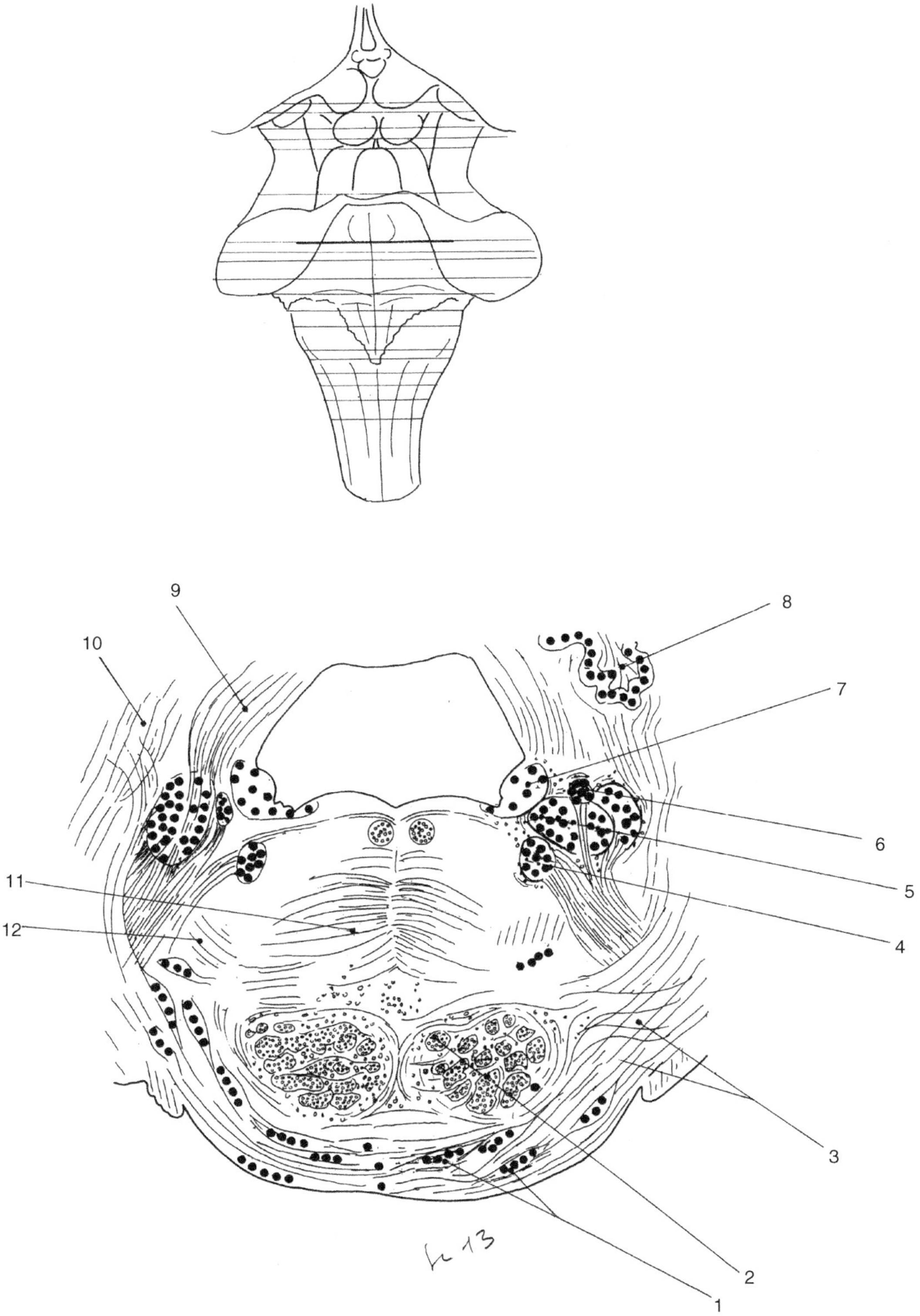

Fig. 15.103 *1* nuclei pontis, *2* tractus pyramidalis, *3* pedunculus cerebellaris medialis, *4* nucleus et tractus motorius n. trigemini, *5* nucleus sensorius principalis n. trigemini, *6* nucleus mesencephalicus n. trigemini, *7* nucleus n. vestibularis, *8* nucleus dentatus, *9* pedunculus cerebellaris superior, *10* pedunculus cerebellaris inferior, *11* lemniscus medialis, *12* lemniscus lateralis

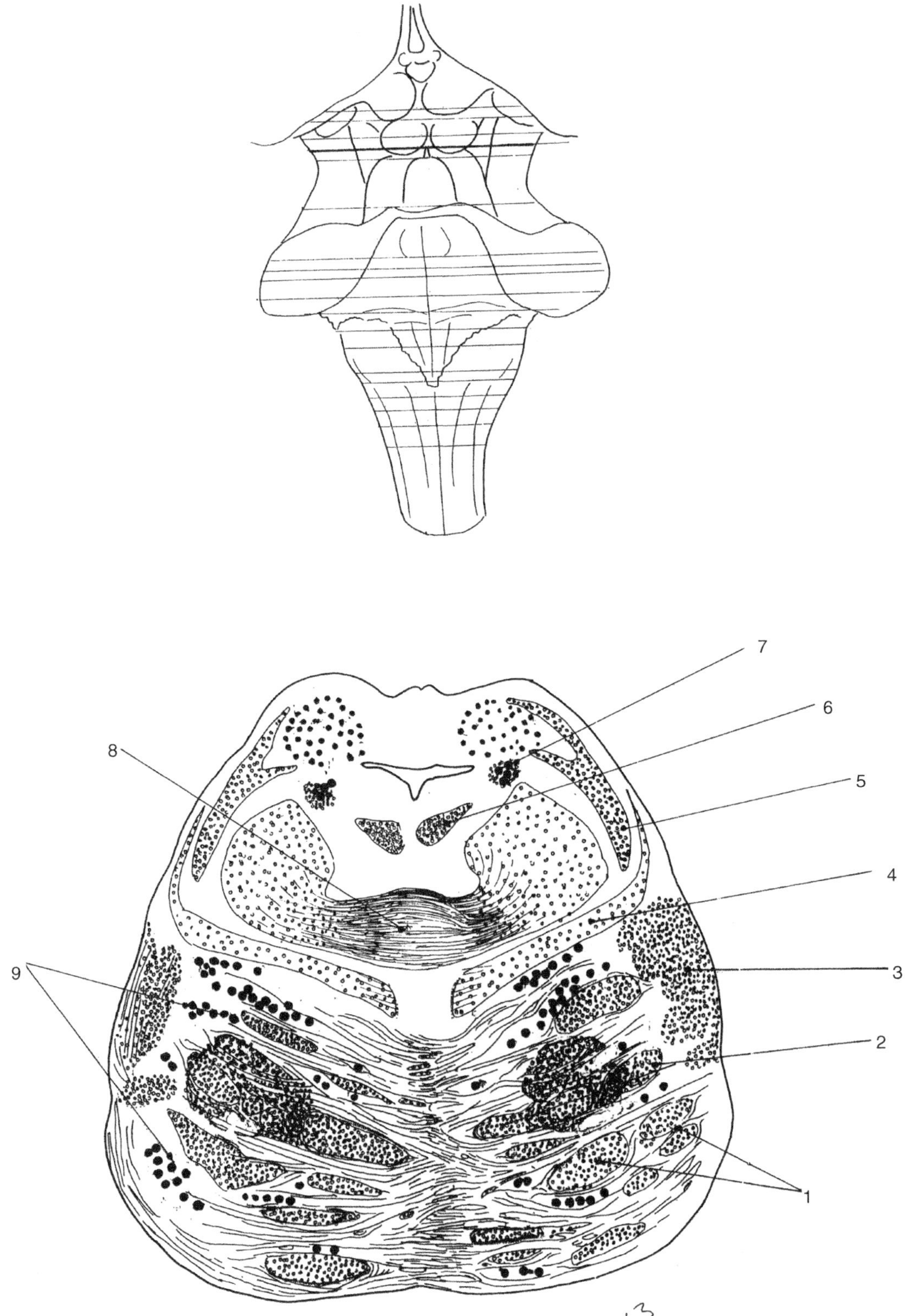

Fig. 15.106 *1* tractus corticopontinus frontalis, *2* tractus pyramidalis, *3* tractus corticopontinus parietotemporalis, *4* lemniscus medialis, *5* lemniscus lateralis, *6* fasciculus longitudinalis medialis, *7* nucleus et tractus mesencephalicus n. trigemini, *8* decussatio of pedunculi cerebellares superiores, *9* nuclei pontis

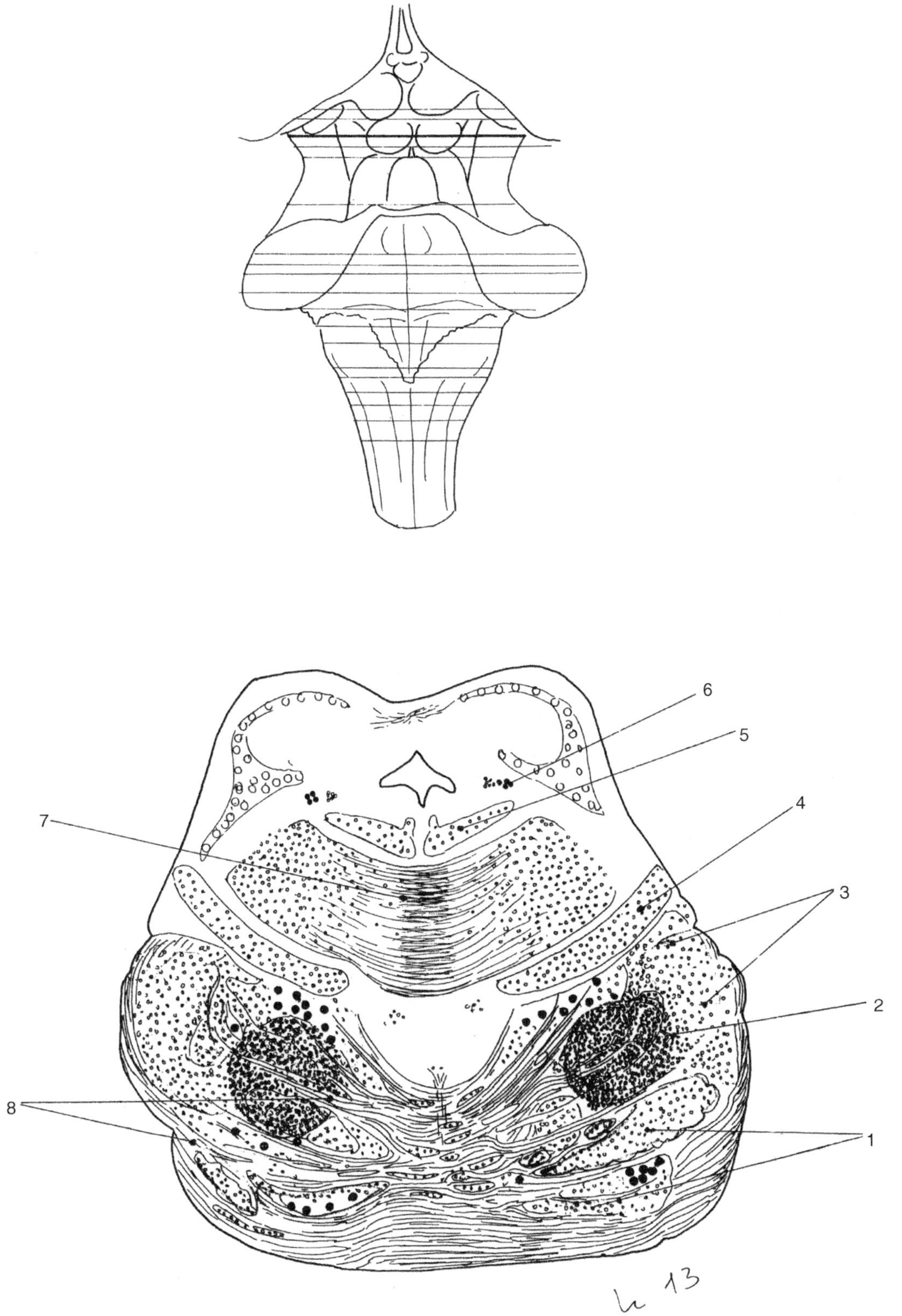

Fig. 15.107 *1* tractus corticopontinus frontalism, *2* tractus pyramidalis, *3* tractus corticopontinus parietotemporalis, *4* lemniscus medialis, *5* fasciculus longitudinalis medialis, *6* nucleus et tractus mesencephalicus n. trigemini, *7* decussatio of pedunculi cerebellares superiores, *8* fibrae pontocerebellares

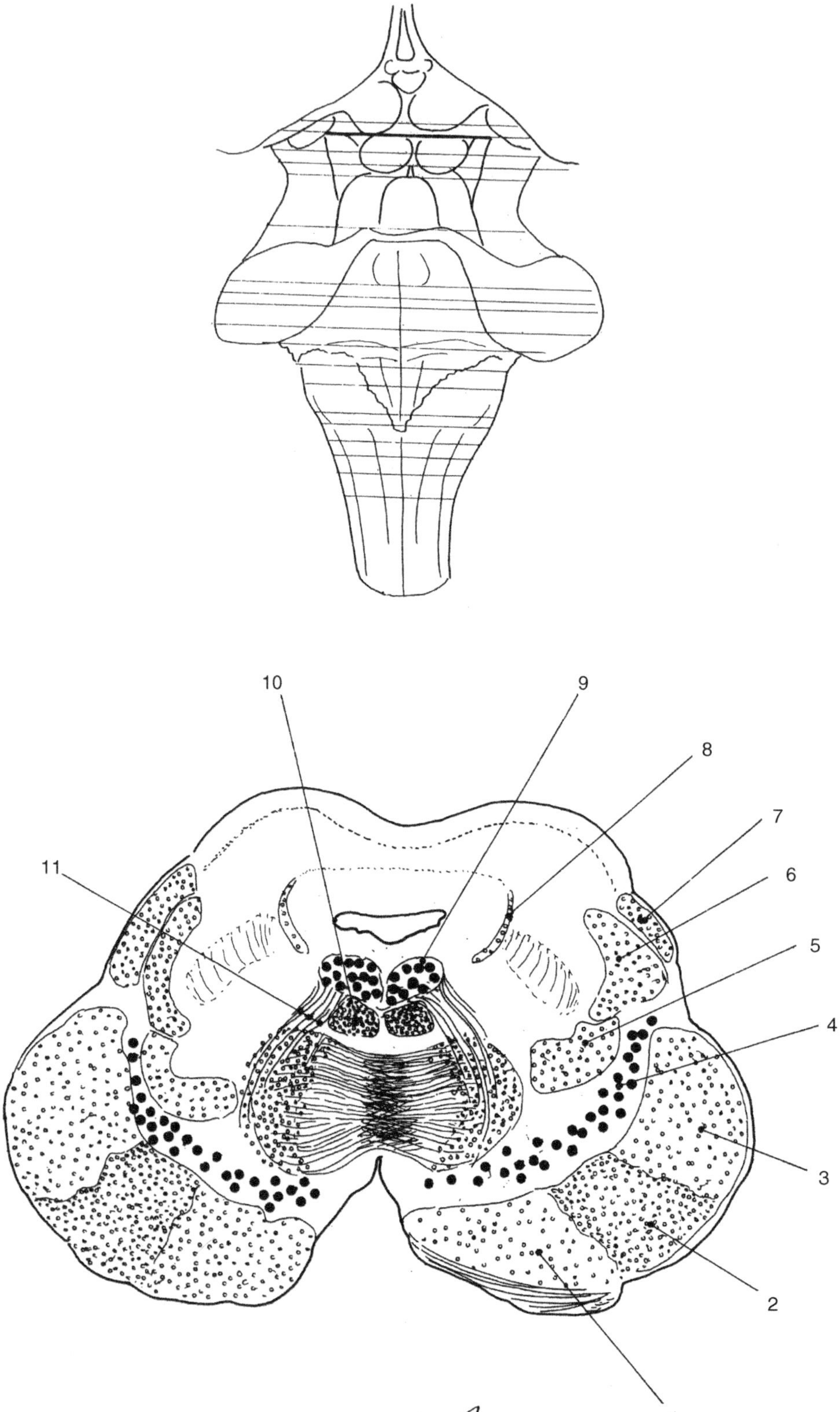

Fig. 15.108 *1* tractus corticopontinus frontalis; *2* tractus pyramidalis; *3* tractus corticopontinus parietotemporalis; *4* substantia nigra; *5* lemniscus medialis; *6* lemniscus lateralis; *7* brachium colliculi inferioris; *8* tractus mesencephalicus n. trigemini; *9* nucleus n. oculomotorii, somatic portion (Westphal-Edinger); *10* as *9*, parasympathic portion (Perlia); *11* fila radicularia n. oculomotorii

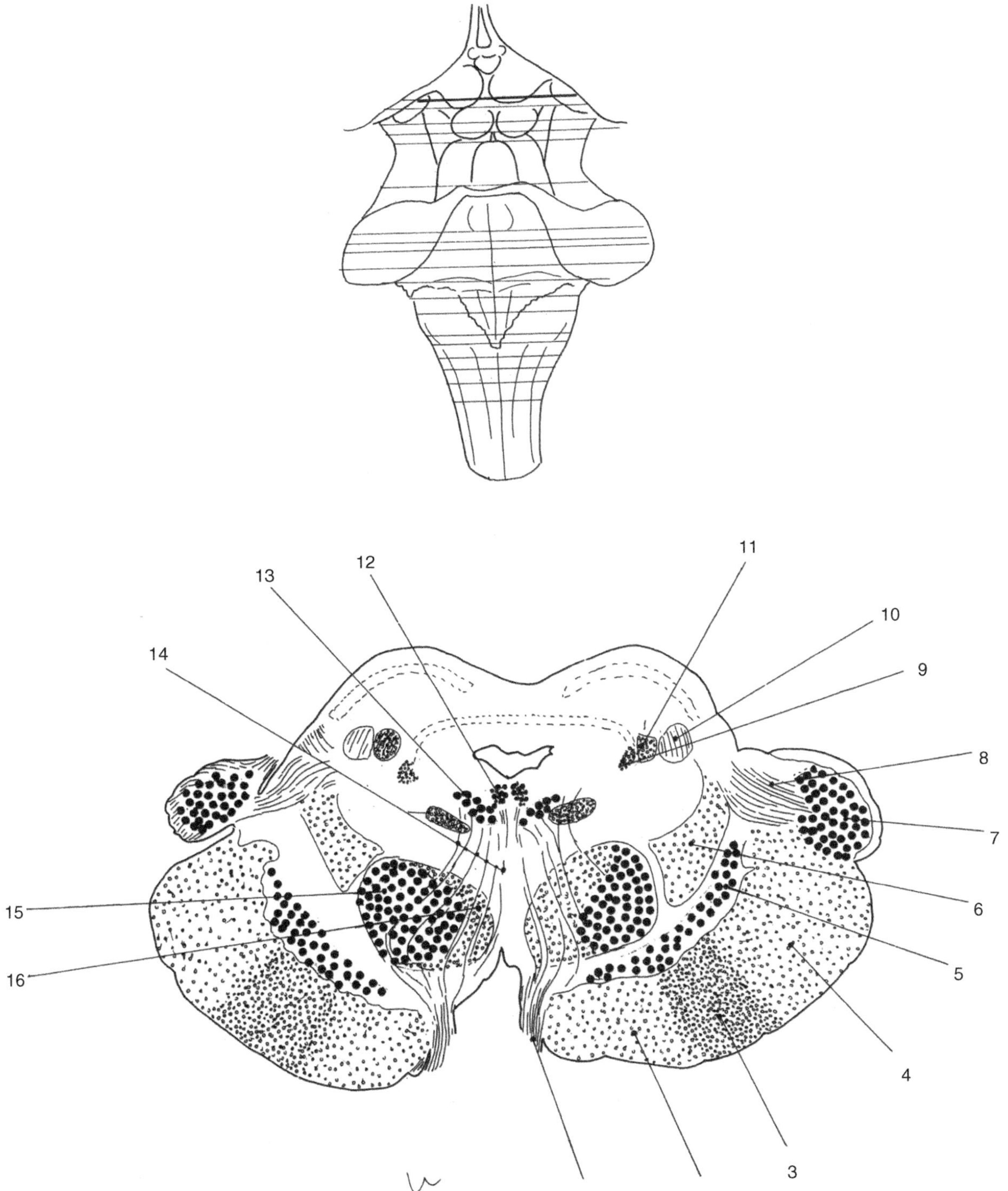

Fig. 15.109 *1* N. oculomotorius; *2* tractus corticopontinus frontalis; *3* tractus pyramidalis; *4* tractus corticopontinus parietotemporalis; *5* substantia nigra; *6* lemniscus medialis; *7* corpus geniculatum mediale; *8* brachium colliculi inferioris; *9* tractus mesencephalicus n. trigemini; *10* brachium colliculi superioris; *11* tractus spinothalamicus; *12* nucleus n. oculomotorii, parasympathic portion; *13* as *12*, somatic portion, *14* fila radicularia n. oculomotorii and fasciculus longitudinalis medialis, *15* nucleus ruber, *16* pedunculus cerebellaris superior

15.2.7 Vessels of the Middle and Inferior Brain Stem (Figs. 15.110, 15.111, 15.112, 15.113, and 15.114)

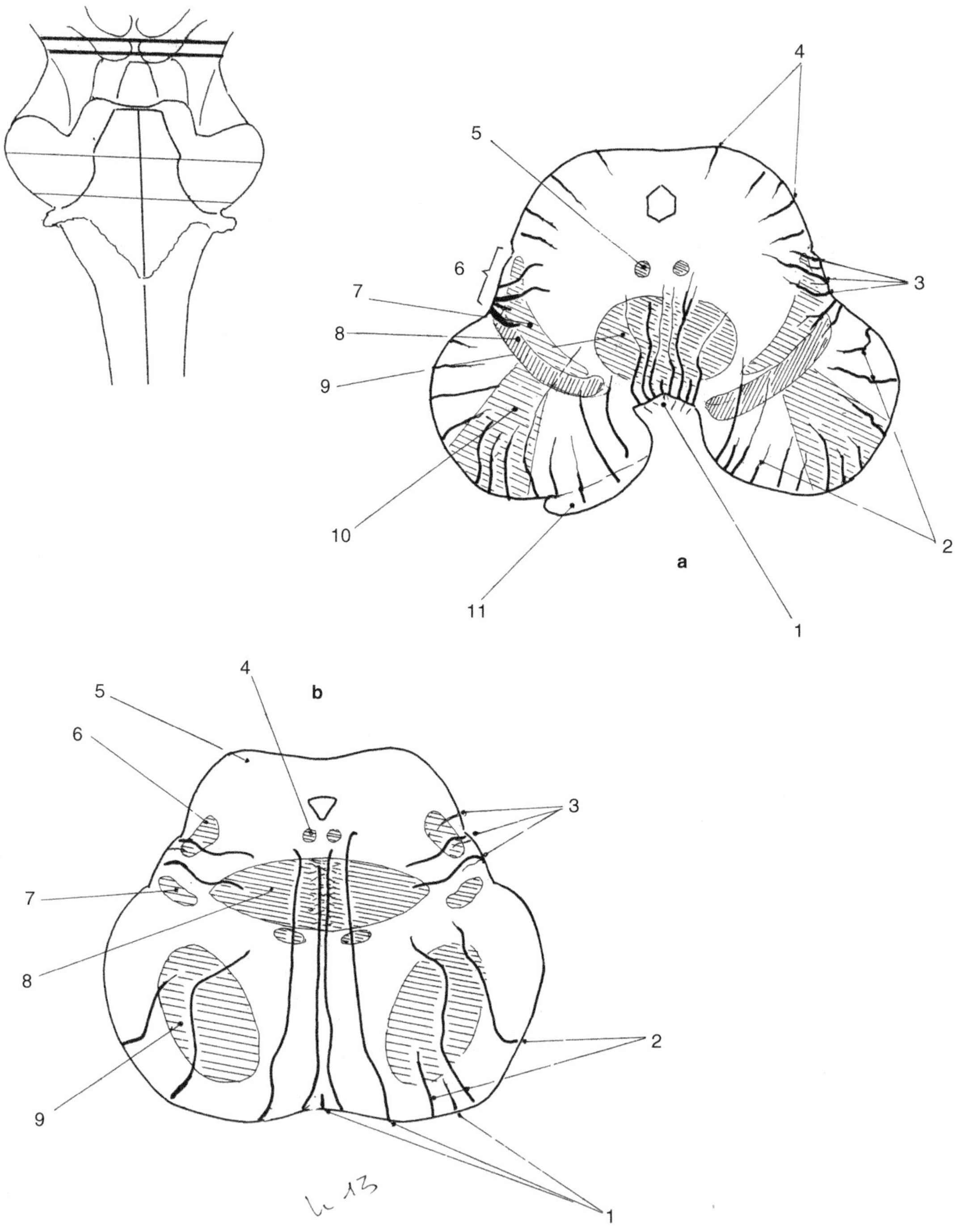

Fig. 15.110 Intrarhombencepalic perforating arteries (casts of Stephens and Stilwell, 1969, pp. 103–115, schematized). The perforating arteries of the brain stem present more regulary configurations than the large branches of circulus arteriosus (Willisi). (**a**) *1* Aa. thalamoperforantes; *2* branches of A. basilaris, A. cerebri posterior et A. cerebelli superior; *3* as *2* and *4*; *4* branches of A. cerebri posterior, A. cerebelli superior, and A. tecti; *5* fasciculus longitudinalis medialis; *6* trigonum lemnisci; *7* lemniscus medialis et lateralis; *8* substantia nigra; *9* decussatio et pedunculus cerebellaris superior; *10* tractus pyramidalis; *11* upper pontine transection. (**b**) *1* branches of the bifurcation of A. basilaris, A. cerebelli superior, cerebri posterior, and Rami ad pontem; *2* as *1*; *3* branches of A. cerebelli superior et A. cerebri posterior; *4* fasciculus longitudinalis medialis; *5* colliculus inferior; *6* lemniscus lateralis; *7* lemniscus medialis; *8* decussatio et pedunculi cerebellares superiores; *9* tractus pyramidalis

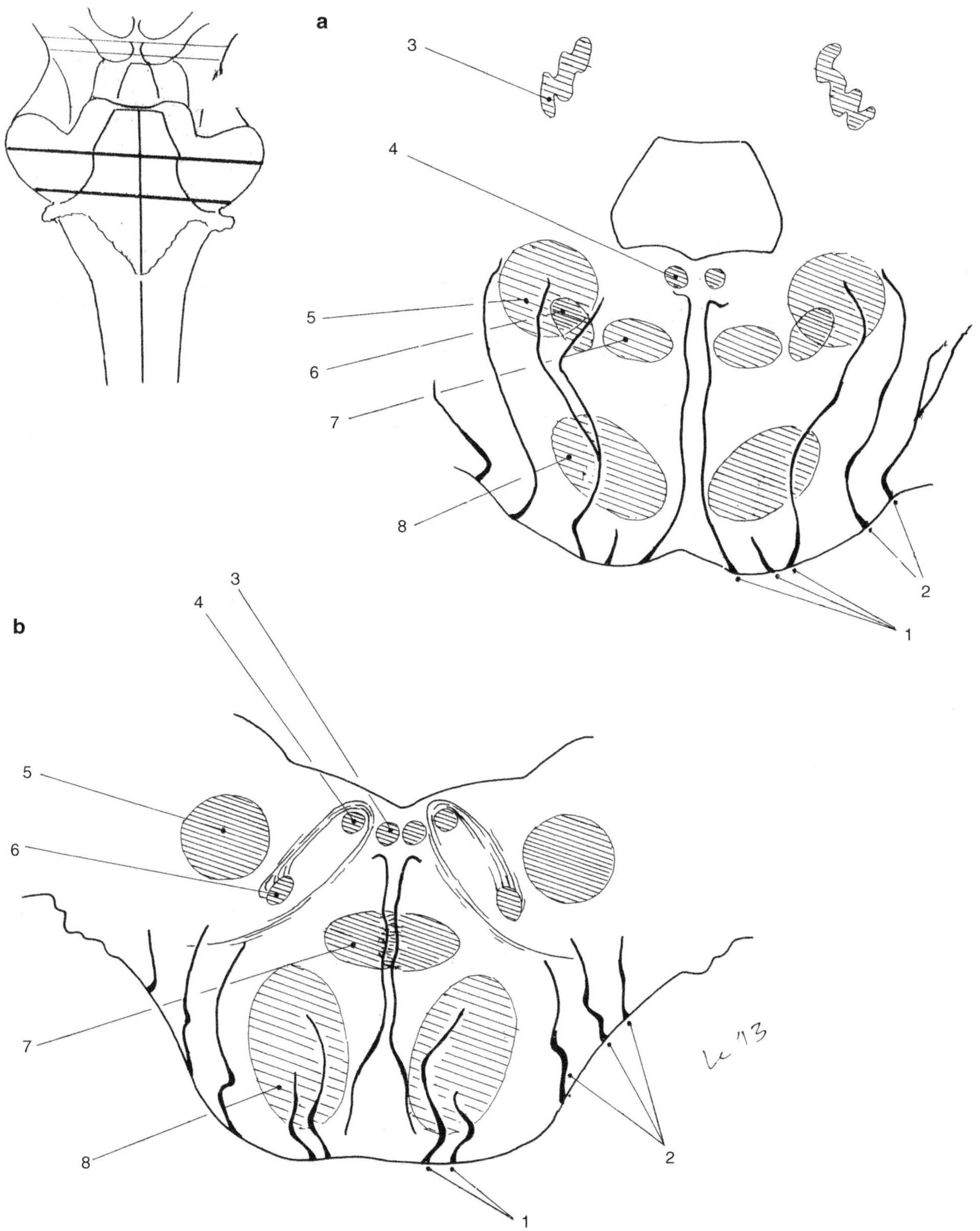

Fig. 15.111 Continuation of Fig. 15.110. (**a**) *1* Rr. ad pontem of A. basilaris and branches of AICA, *2* as *1*, *3* nucleus dentatus, *4* fasciculus longitudinalis medialis, *5* tractus spinalis n. trigemini, *6* nucleus tractus spinalis n. trigemini, *7* lemniscus medialis, *8* tractus pyramidalis. (**b**) *1* Rr. ad pontem of A. basilaris and branches of AICA, *2* as *1*, *3* fasciculus longitudinalis medialis, *4* nucleus n. abducentis, *5* nucleus principalis n. trigemini, *6* nucleus n. facialis, *7* lemniscus medialis, *8* tractus pyramidalis

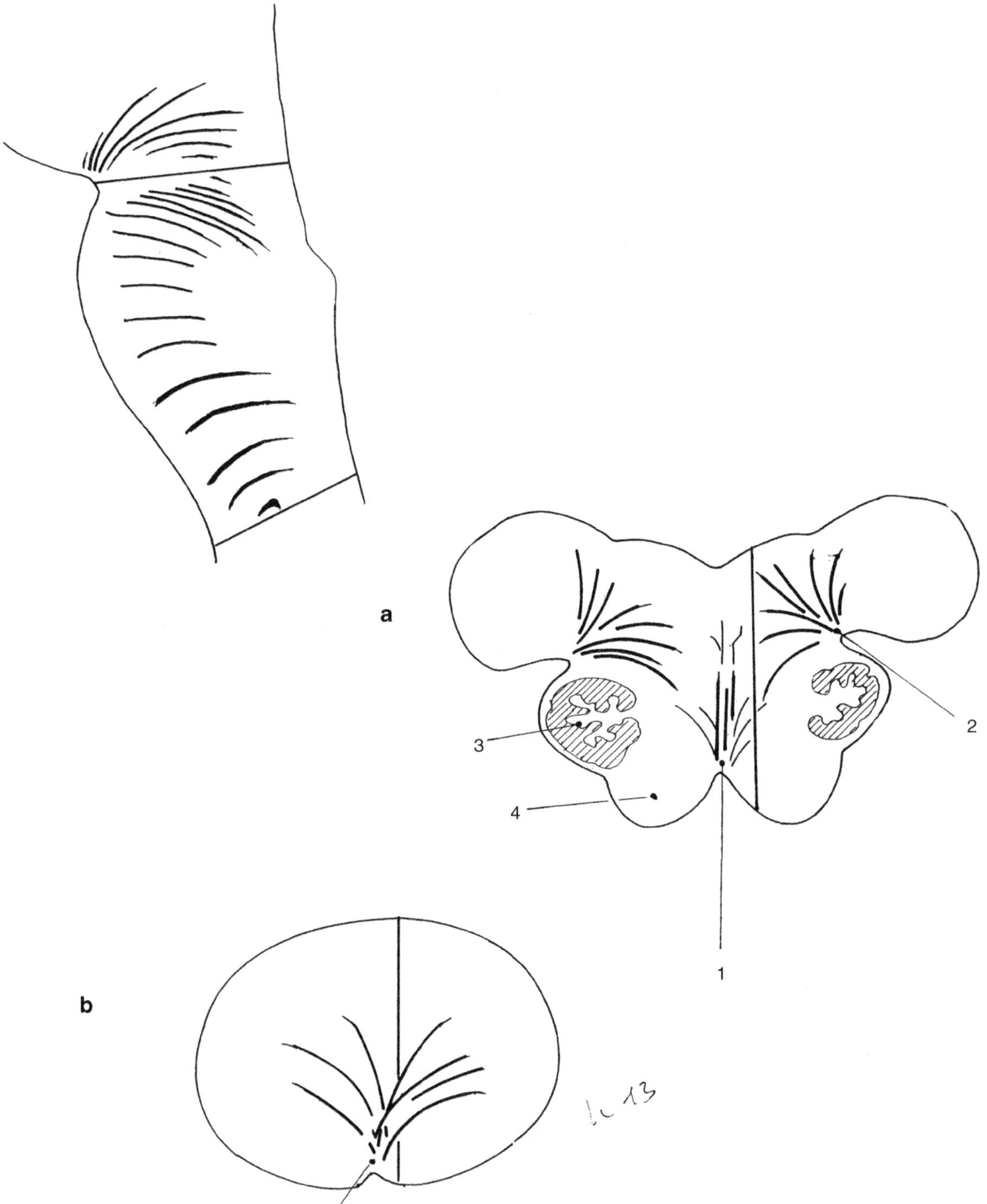

Fig. 15.112 Continuation of Fig. 15.111. (**a**) *1* branches of Aa. vertebrales at the pontomedullary junction, *2* branches of PICA and AICA, *3* oliva, *4* tractus pyramidalis. (**b**) *1* branches of A. spinalis anterior et A. vertebralis

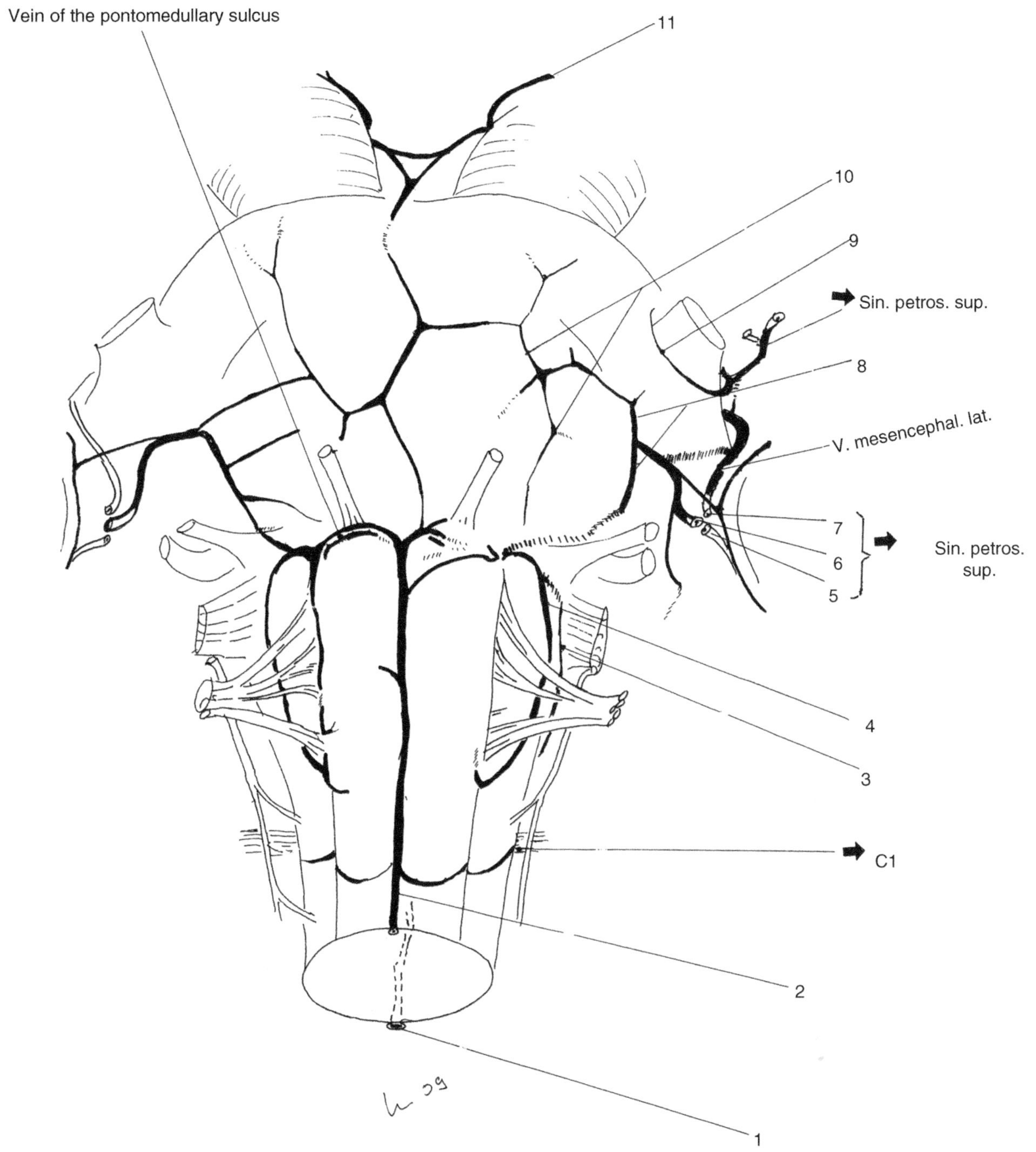

Fig. 15.113 Subarachnoid and pial veins (Duvernoy 1975, p 39, schematized). Basal lateral veins are the lateral pontine veins and V. horizontalis cerebelli of fissura horizontalis. These are connected with sinus petrosus superior by one or several small Vv. petrosae superiores (Fig. 15.59). Usually Vv. petrosae superiores could be interrupted by surgery without following deficits, because the other lateral basal veins are connected with V. basalis Rosenthal by V. mesencephalica lateralis and with other veins. Interruption of Vv. petrosae superiores could be dangerous, if Vv. petrosae superiores and Sinus petrosus superior are widened and V. mesencephalica lateralis is missing (see angiogram). Large dorsal cerebellar bridging veins should be spared (Fig. 16.30) at surgery. *1* posterior medullary pial vein, *2* anterior medullary vein, *3* lateral medullary vein, *4* postolivary vein, *5* V. cerebellaris horizontalis, *6* lateral pontine vein(s), *7* V. mesencephalica lateralis, *8* lateral pontine vein(s), *9* pontine veins surrounding N. trigeminus, *10* medial pontine veins, *11* V. interpeduncularis

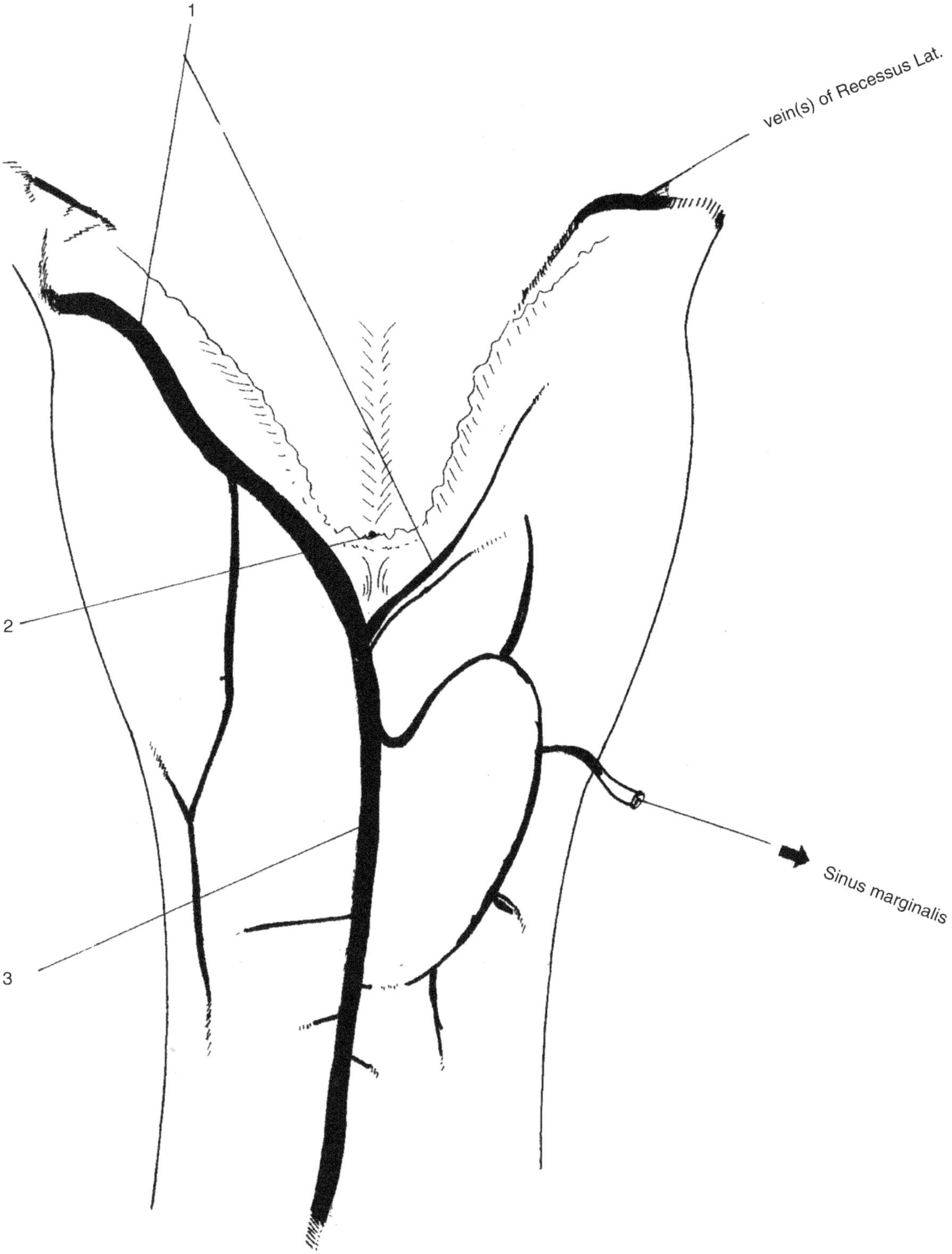

Fig. 15.114 Addendum for Fig. 15.113. Variable dorsal medullary veins. *1* dorsolateral vein(s), *2* obex, *3* dorsomedial vein(s)

Special Surgical Aspects 16

16.1 Extra- and Intracranial Topography

16.1.1 Craniotomies for Routes and Target Areas (Fig. 16.1)

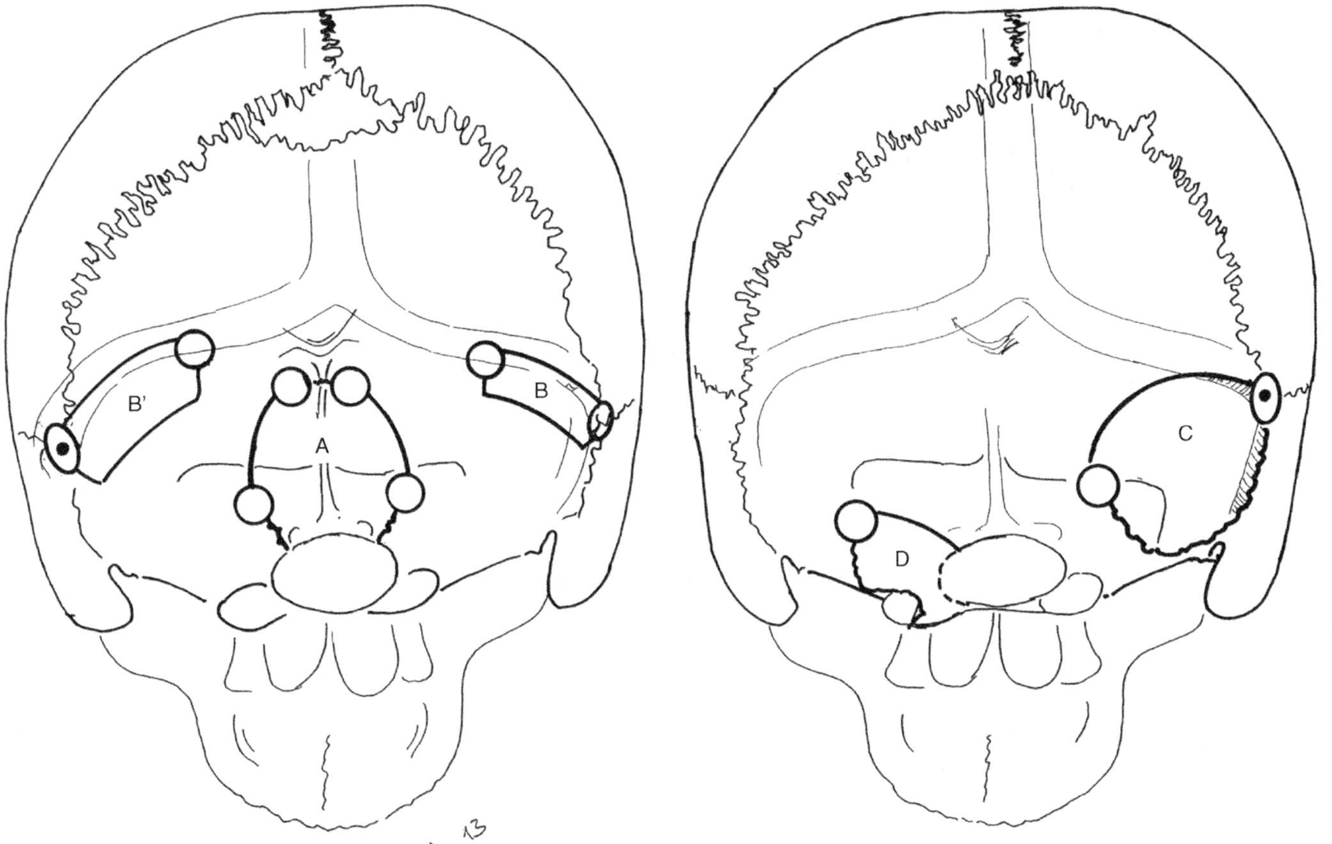

Fig. 16.1 Craniotomies and surgical target areas survey. *A* Suboccipital craniotomy for medial approaches to cerebellum and fourth ventricle. *B* Laterodorsal craniotomy for supracerebellar approaches and for approaches to cisterna tecti, fisssura transversa cerebri, cisterna ambiens and the anterior segment of fissura horizontalis cerebelli, *B′* target areas as *B*, contralateral route, if the usual route is hindered by large dorsolateral bridging veins (Yaşargil). *C* Laterobasal craniotomy for approaches to cerebellum, pontocerebellar, and other cisternal approaches. *D* Trans- and paracondylar craniotomy for approaches to premedullary and prepontine cisterns

W. Seeger, J. Zentner, *Anatomical Basis of Cranial Neurosurgery*, https://doi.org/10.1007/978-3-319-63597-2_16

16.1.2 Surgical Aspects of Posterior Midline Approaches (Figs. 16.2, 16.3, 16.4, 16.5, 16.6, 16.7, 16.8, 16.9, and 16.10)

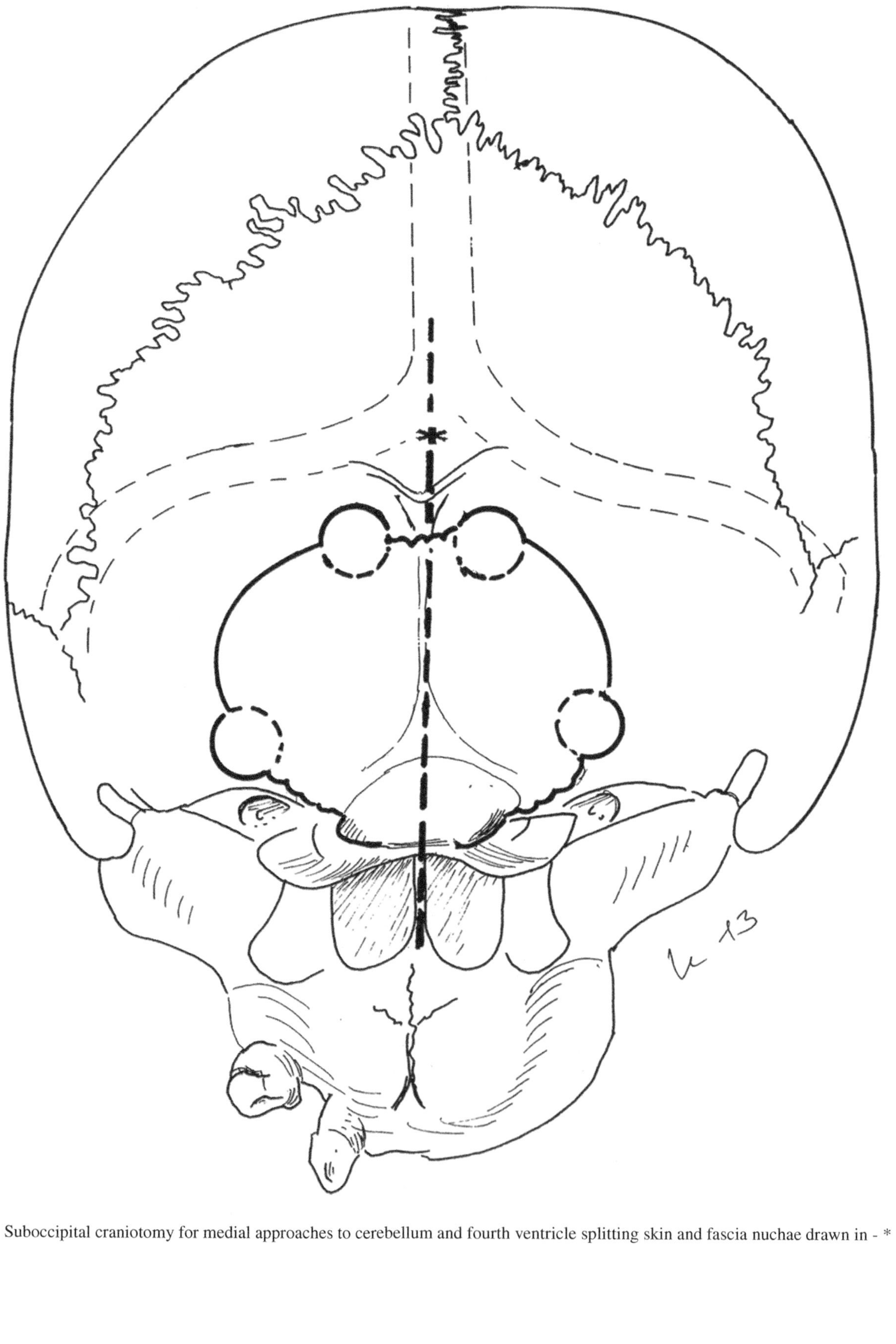

Fig. 16.2 Suboccipital craniotomy for medial approaches to cerebellum and fourth ventricle splitting skin and fascia nuchae drawn in - * -

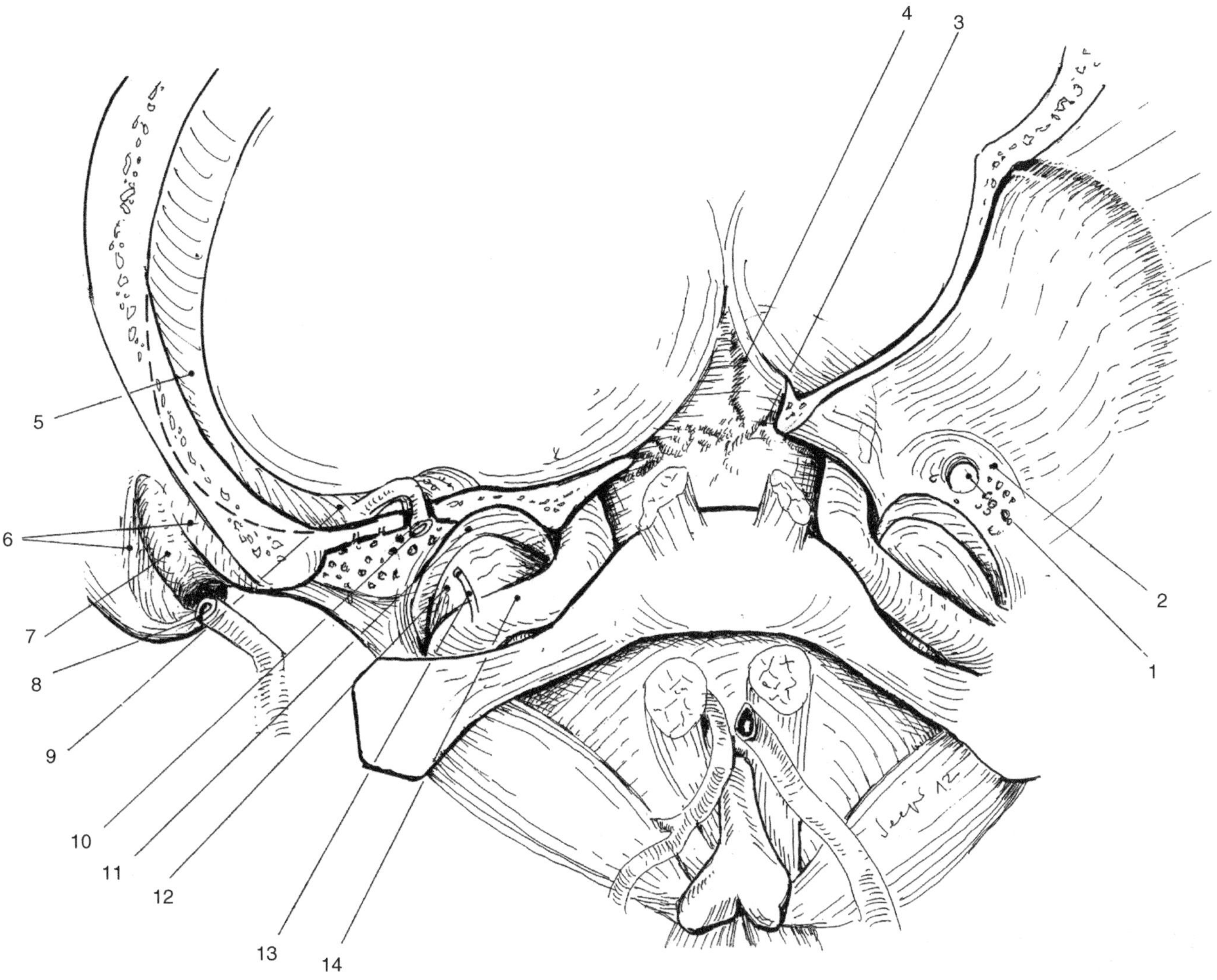

Fig. 16.3 Craniospinal dorsal anatomical aspects of surgery vessels added. *1* Emissarium condylare; *2* fossa paracondyloidea; *3* sinus marginalis; *4* sinus occipitalis; *5* sinus sigmoideus; *6* fossa biventerica; *7* sulcus a.occipitalis; *8* A. occipitalis; *9* sinus sigmoideus connected to 10; *10* V.condylaris, cave air embolism at surgery; *11* condylus occipitalis; *12* processus articularis superior (atlantis); *13* common muscular branch, connected to A. vertebralis; *14* A. vertebralis

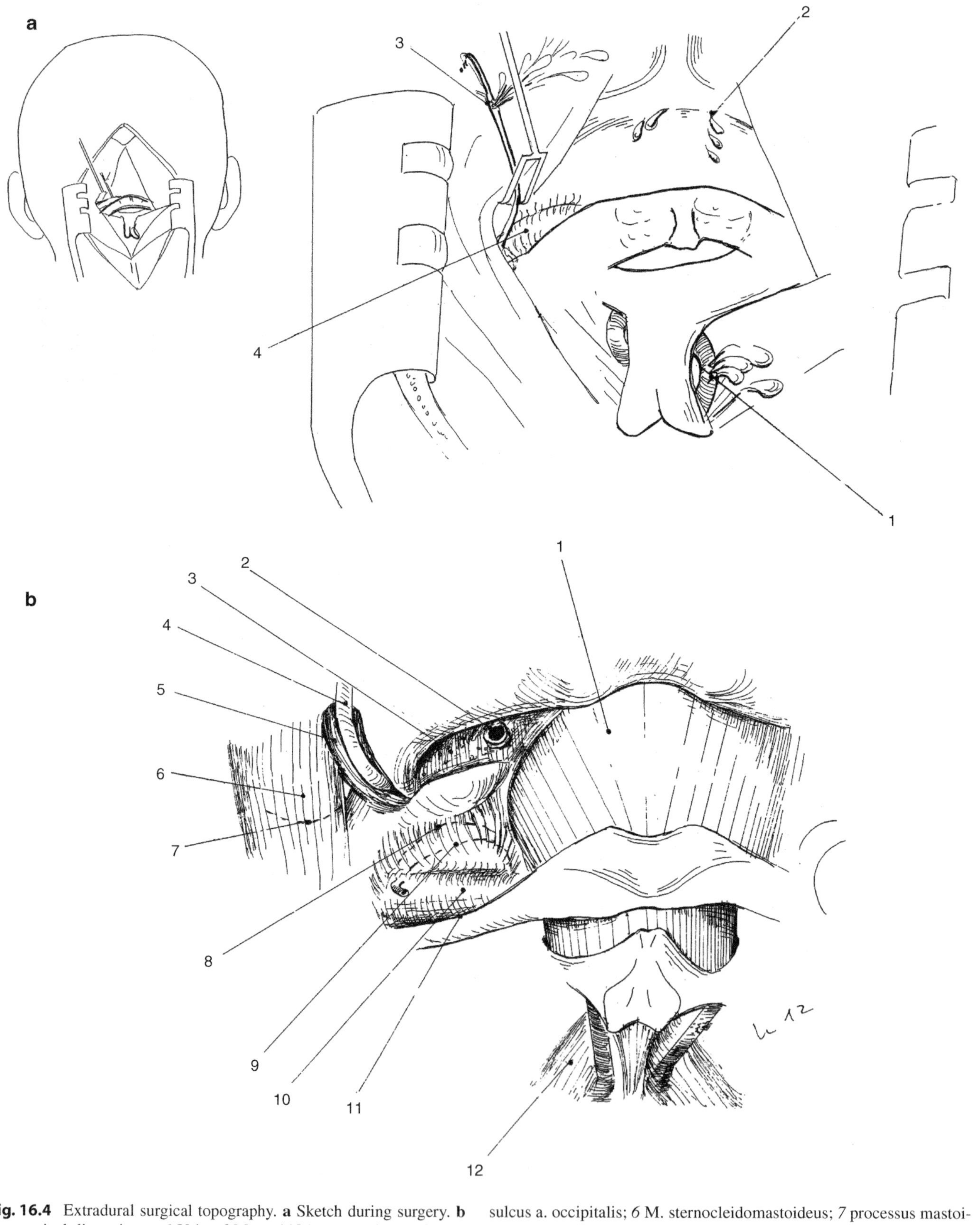

Fig. 16.4 Extradural surgical topography. **a** Sketch during surgery. **b** Anatomical dissection. **a** *1* Vein of M. multifidus, e.g., danger for air embolism after spreading of retractor. *2* Sinus marginalis, *3* branch of A. vertebralis, *4* A. vertebralis masked by intransparent fibrous tissue. **b** *1* Dura; *2* V. condylaris; *3* fossa paracondyloidea; *4* A. occipitalis; *5* sulcus a. occipitalis; *6* M. sternocleidomastoideus; *7* processus mastoideus, projection; *8* condylus occipitalis, projection; *9* processus articularis superior, projection; *10* A. vertebralis; *11* bony crest of sulcus A. vertebralis of Atlas; *12* M. multifidus

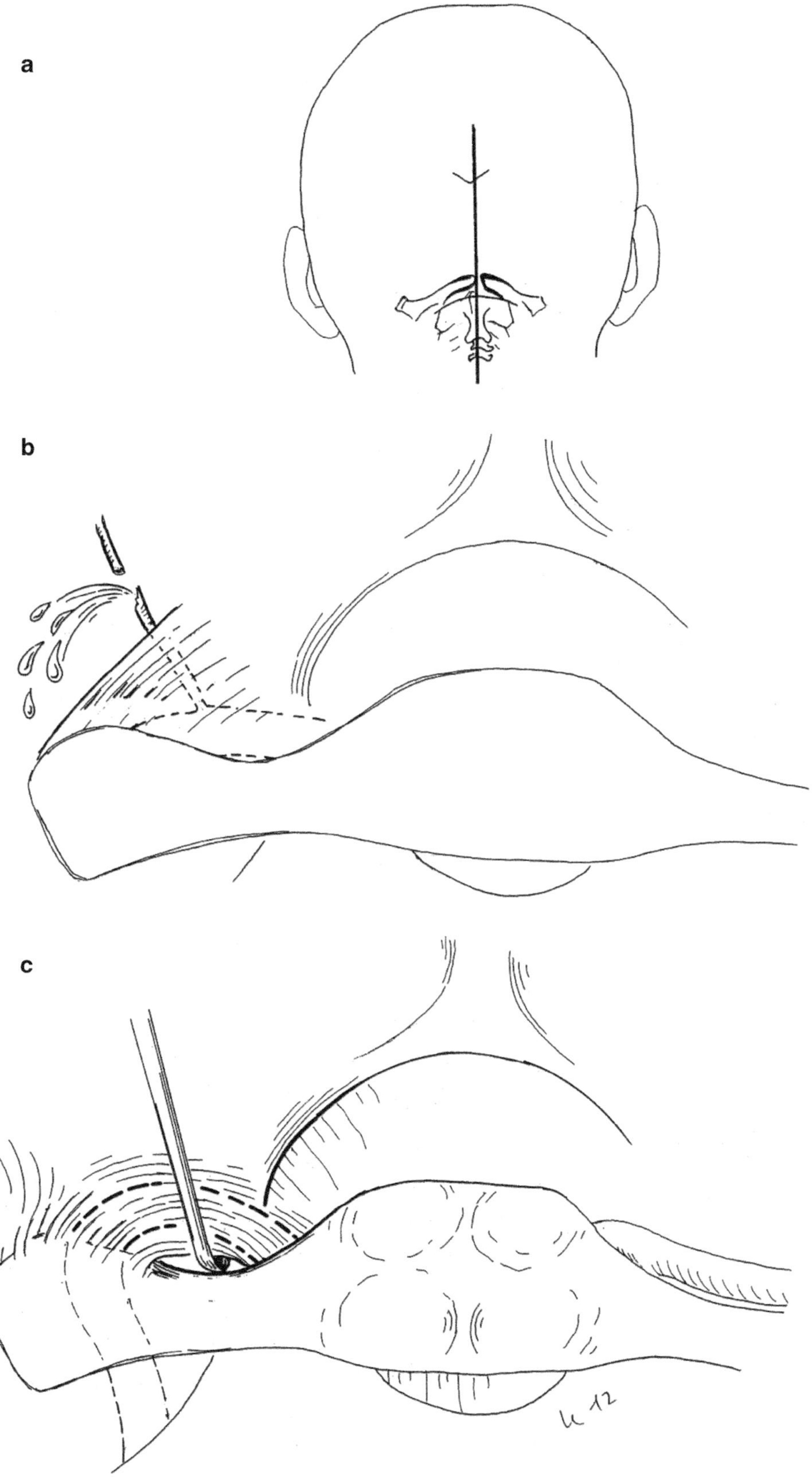

Fig. 16.5 Further aspects. **a** Bony defect of arcus atlantis posterior. Usual finding at childhood, rare at adults. Cave at surgical splitting of Lig. nuchae. **b** Bleeding close to A. vertebralis (for training neurosurgeons). **c** Loosening of A. vertebralis and periosteum from sulcus A. vertebralis of Atlas

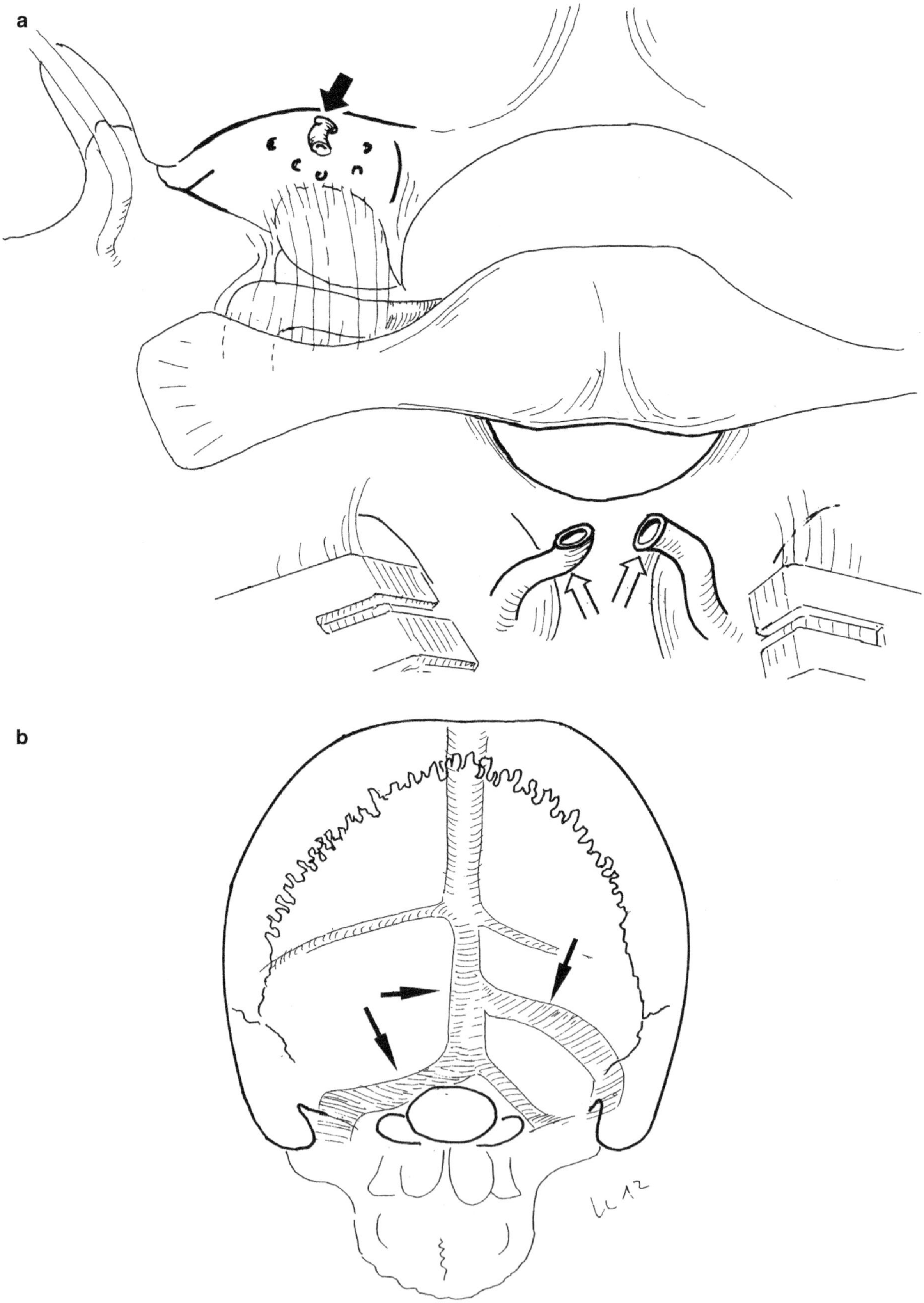

Fig. 16.6 Air embolism and danger for venous bleeding. **a** Extracranial veins, sinus condylaris opened (*black arrow*). Ruptured vein after retraction of M. multifidus (*light arrow*). **b** Dural sinuses. Large variants of sinuses endangered at surgical dura opening (*arrows*)

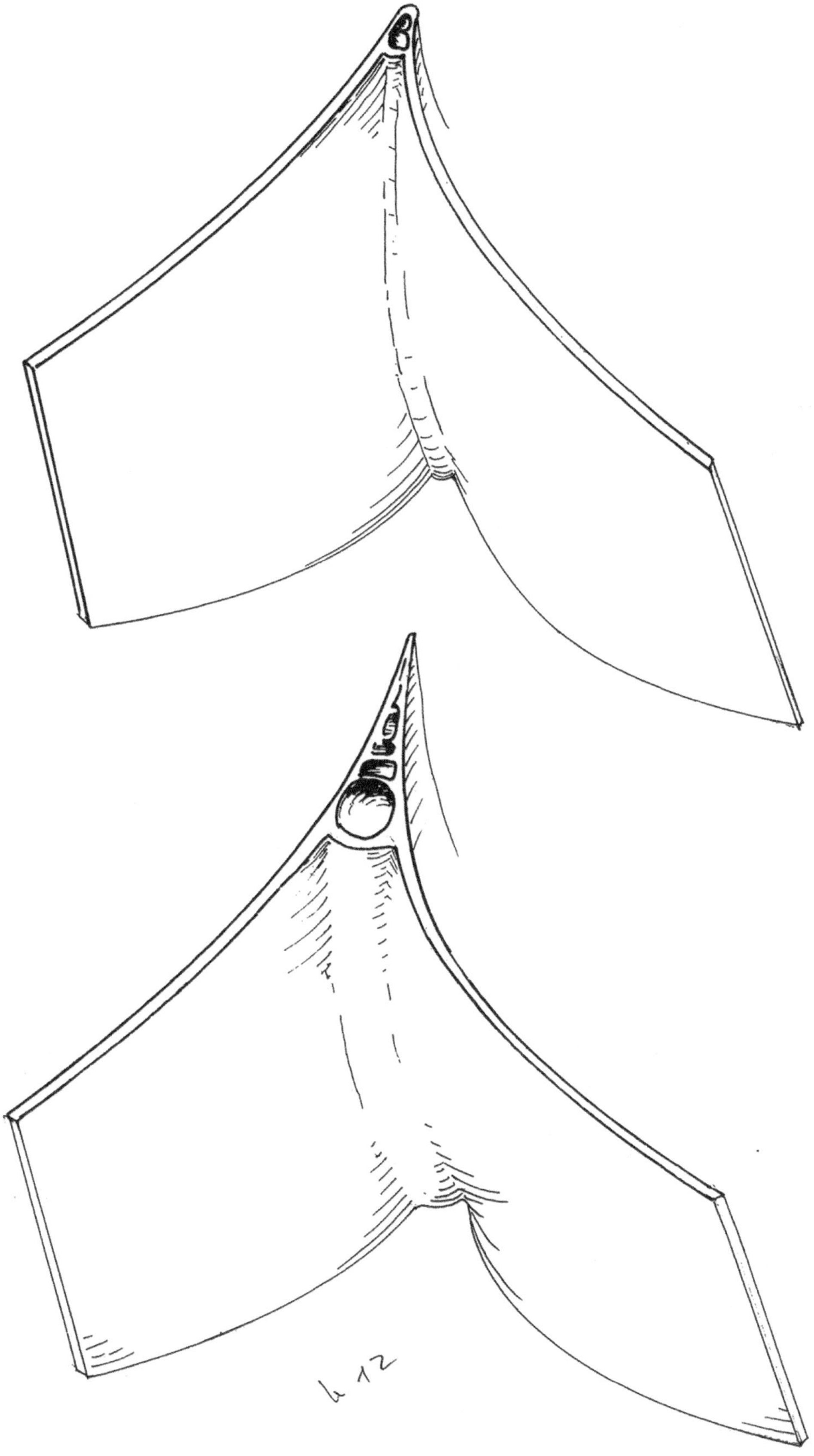

Fig. 16.7 Variants of falx cerebelli and sinus occipitalis. For example, falx cerebelli is inconstant presented

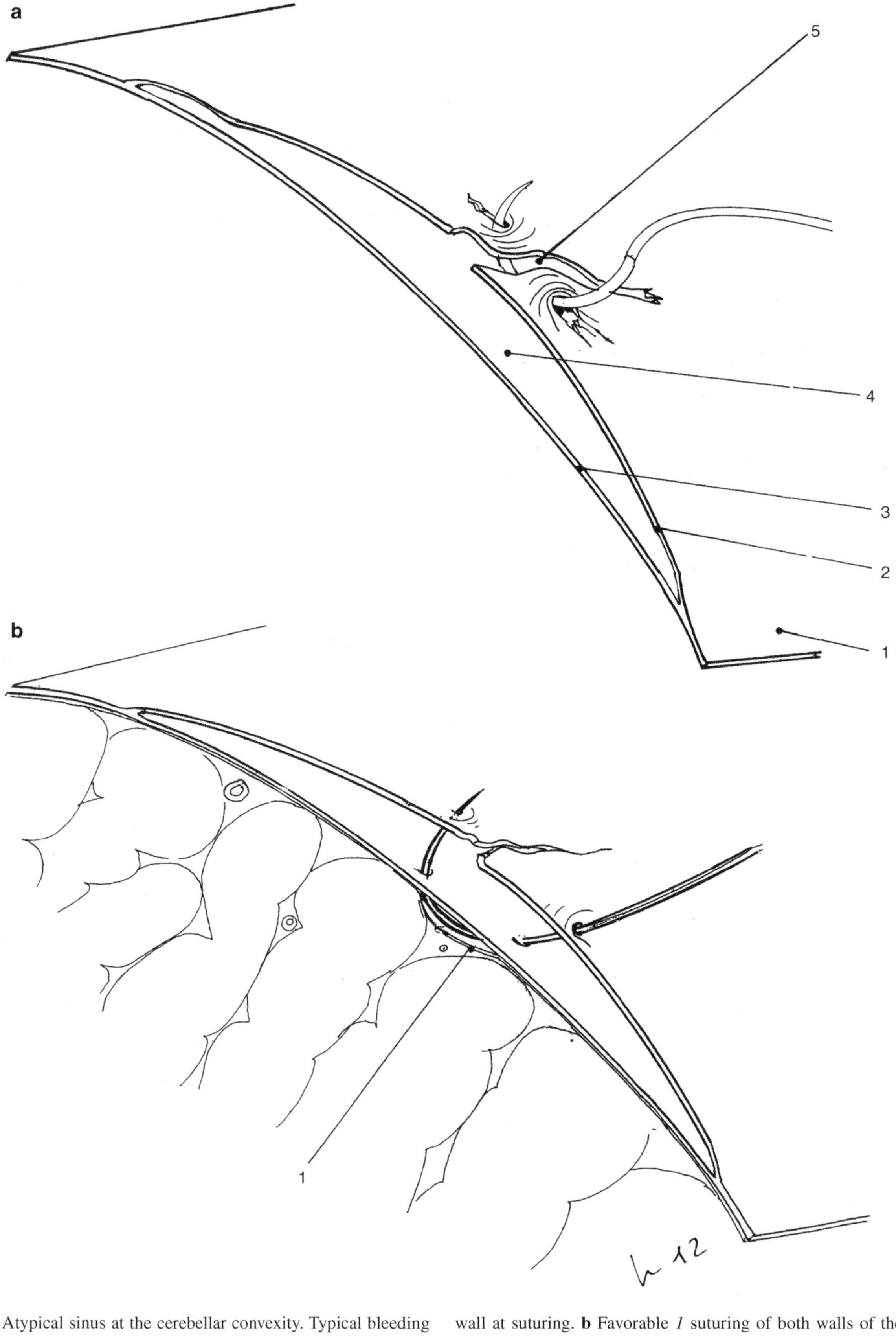

Fig. 16.8 Atypical sinus at the cerebellar convexity. Typical bleeding at dura incision. Technical aspects. **a** Unfavorable *1* dura, *2* outer wall of the sinus, *3* inner wall, *4* lumen, *5* typical rupture of the thin dural wall at suturing. **b** Favorable *1* suturing of both walls of the atypical sinus. Cave cerebellar vessels

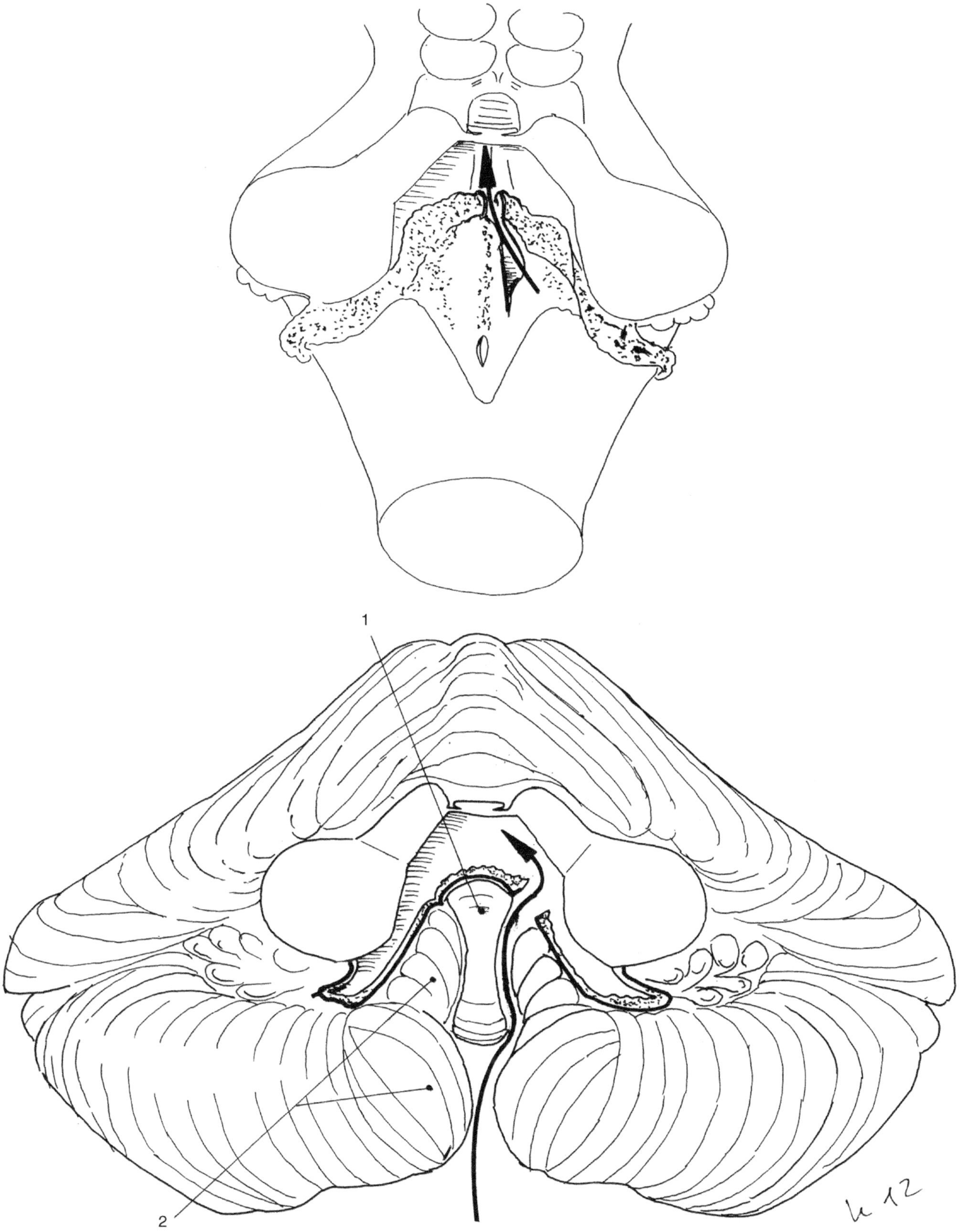

Fig. 16.9 Anatomical base of the tonsillonodular approach to the fourth ventricle. Preservation of tractus flocculonodularis (for avoiding defictis of the optokinetic nystagm, in contrast to splitting of vermis). *1* Uvula vermis, *2* tonsilla

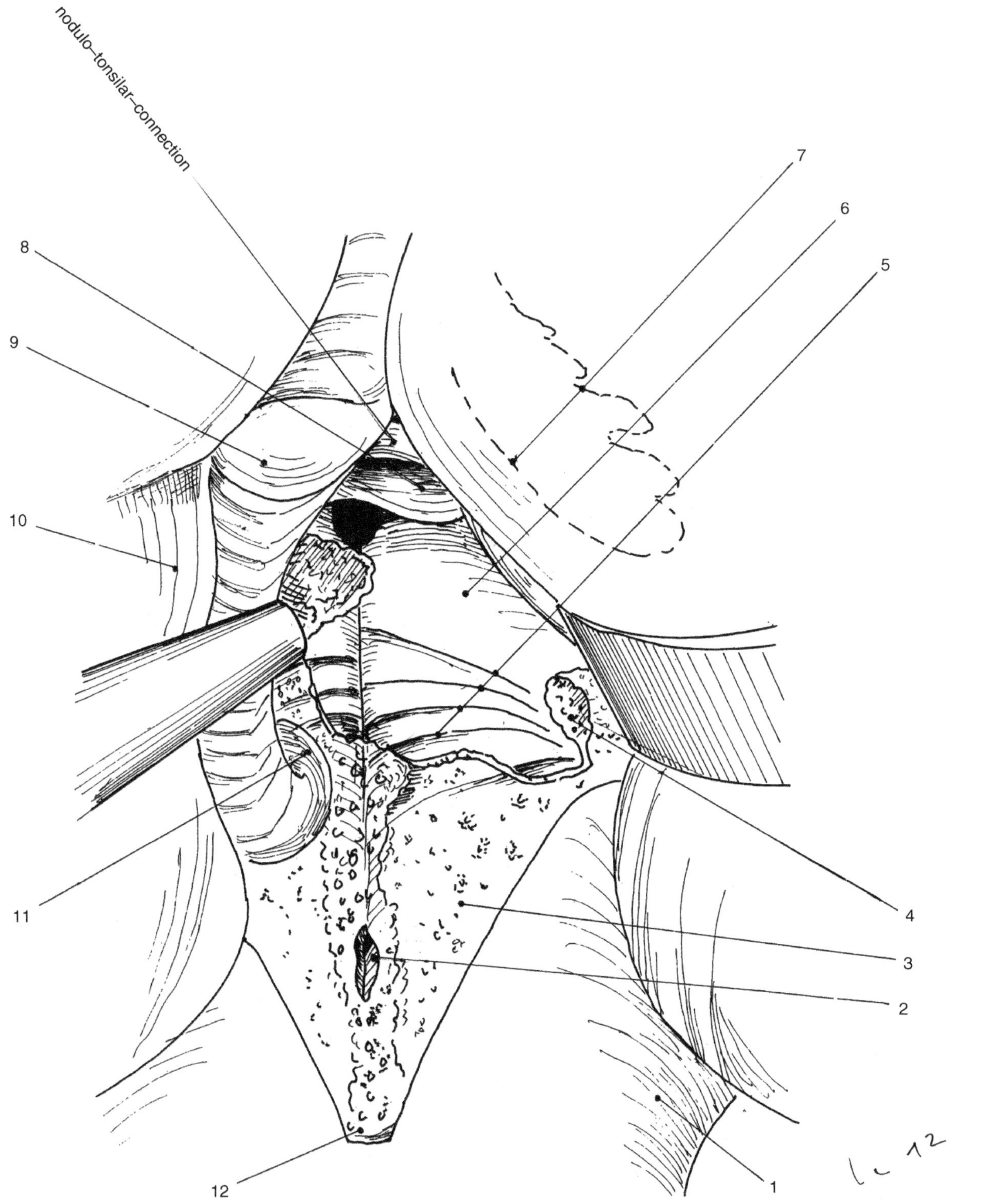

Fig. 16.10 Tonsillonodular surgical approach to the fourth ventricle. Surgical topography. *1* Pedunculus cerebellaris inferior; *2* apertura mediana (magendi); *3* velum medullare posterior and its plexus compartment; *4* plexus choroideus, incised and retracted to a caudal direction; *5* striae medullares of 6; *6* fossa rhomboidea; *7* nucleus dentatus, projection; *8* residual of velum medullare posterius at nidus avis; *9* nodulus vermis; *10* tonsilla; *11* uvula vermis; *12* obex

16.1.3 Surgical Approaches to Cisterna Medullaris Lat. and Cisterna Pontocerebellaris

Architecture of cisterns according to pontomedullary surgery according to Yaşargil (Figs. 16.11, 16.12, 16.13, 16.14 and 16.15).

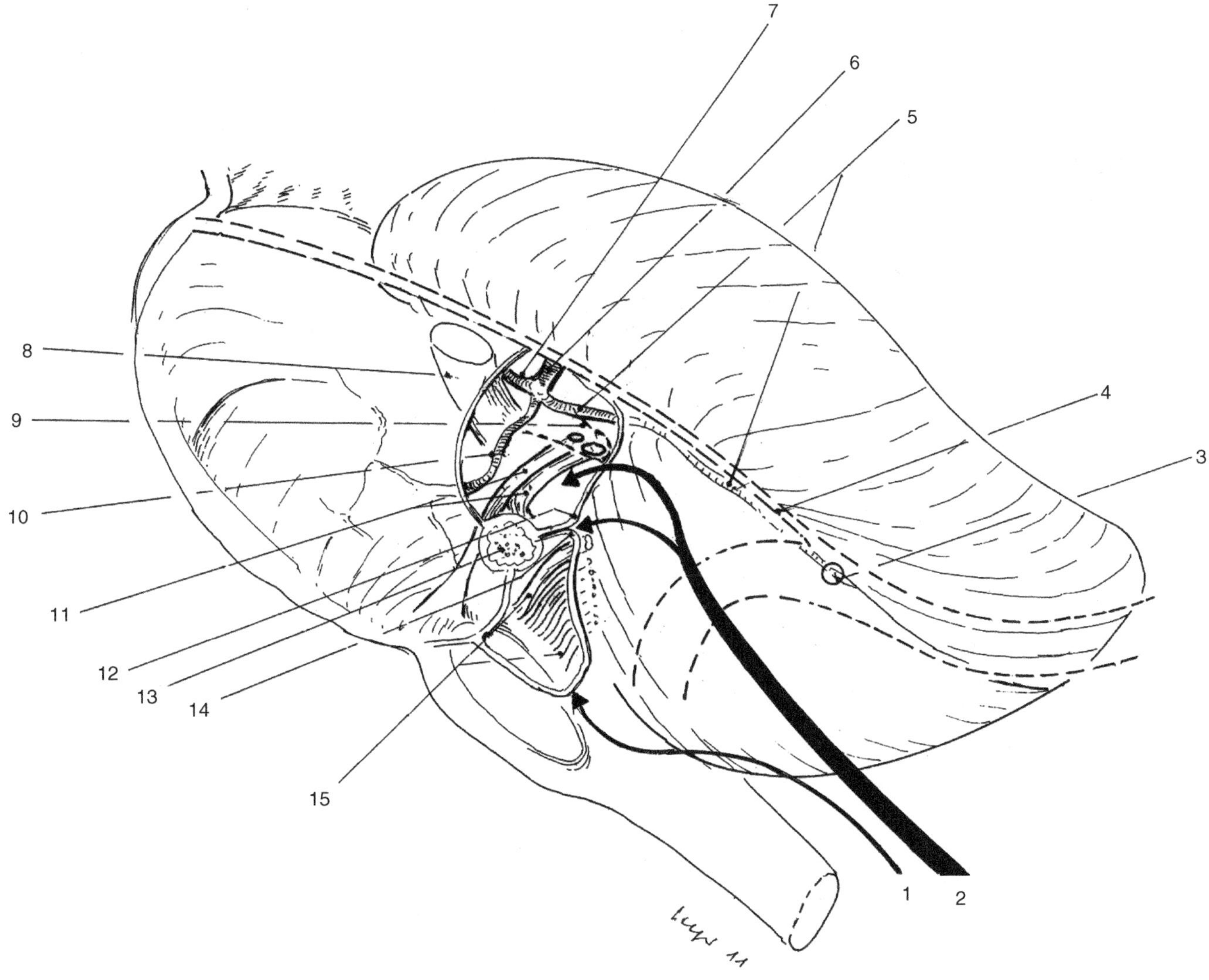

Fig. 16.11 Two cisterns for different surgical routes. Opening of cisterna medullaris lateralis for CSF decompression. Opening of cisterna pontocerebellaris, e.g., for removal of a tumor. *1* Route to cisterna medullaris lateralis; *2* route to cisterna pontocerebellaris; *3* asterion, projection; *4* sinus petrosus superior, projection; *5* V. cerebellaris horizontalis; *6* V. petrosa superio; *7* V. mesencephalica lateralis; *8* N. trigeminus; *9* roof of porus acusticus internus, projection; *10* lateral pontine vein(s); *11* N. facialis et N. statoacusticus; *12* arachnoid wall of cisterna pontocerebellaris; *13* plexus choroideus; *14* arachnoid wall of cisterna medullaris lateralis; *15* Nn. glossopharyngeus, vagus et acceessorius

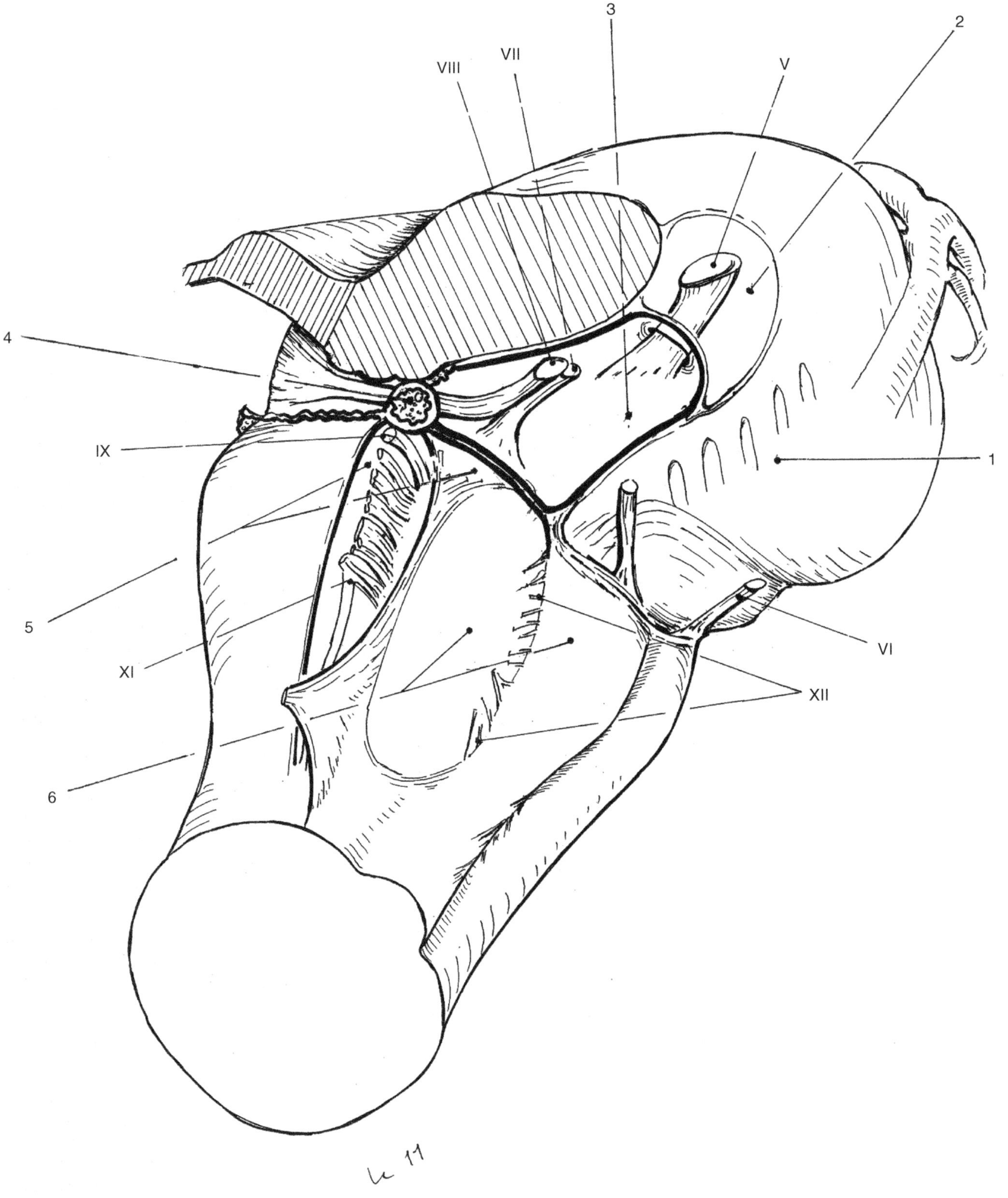

Fig. 16.12 Architecture of cisterna medullary lateralis and pontocerebellaris. Cisterna pontocerebellaris containing Nn. facialis et statoacusticsus and the pontine exit region of N. trigeminus, cisterna pontis medialis containing N. abducens, cisterna medullaris medialis containing the pontomedullary exit point of N. abducens, cisterna pontis lateralis containing N. trigeminus, cisterna medullaris lateralis containing Nn. glossopharyngeus, vagus et accessorius. *1* Cisterna pontis medialis, *2* cisterna pontis lateralis, *3* cisterna pontocerebellaris, *4* plexus choroideus, *5* cisterna medullaris lateralis, *6* cisterna medullaris medialis *V* to *XII* cranial nerves

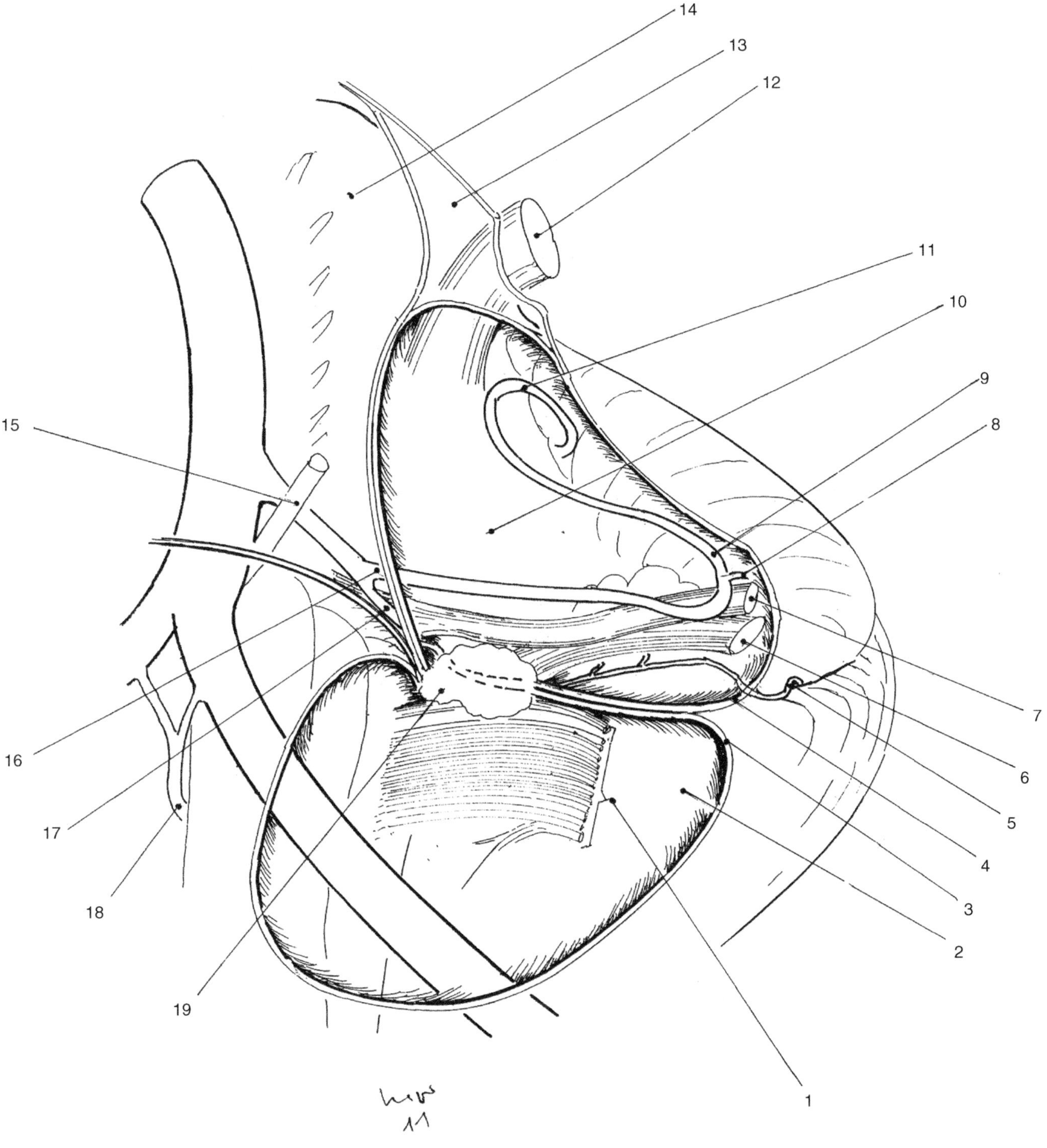

Fig. 16.13 Contralateral to Fig. 6.12. Details added. *1* N. glossopharyngeus, vagus et accessorius, *2* cisterna medullaris lateralis, *3* wall of cisterna medullaris lateralis, *4* wall of cisterna pontocerebellaris, *5* medial caudal branch of AICA, *6* N. statoacusticus, *7* N. facialis, *8* A. labyrinthi, variable, *9* lateral cranial branch of AICA, meatal loop, *10* cisterna pontocerebellaris, *11* trigeminal loop of AICA, *12* N. trigeminus penetrating arachnoid walls, *13* cisterna pontis lateralis, *14* cisterna pontis medialis, *15* N. abducens, *16* and *17* branches of AICA penetrating arachnoid walls, adherent to plexus choroideus (inconstant), *18* A. spinalis anterior, *19* plexus choroideus

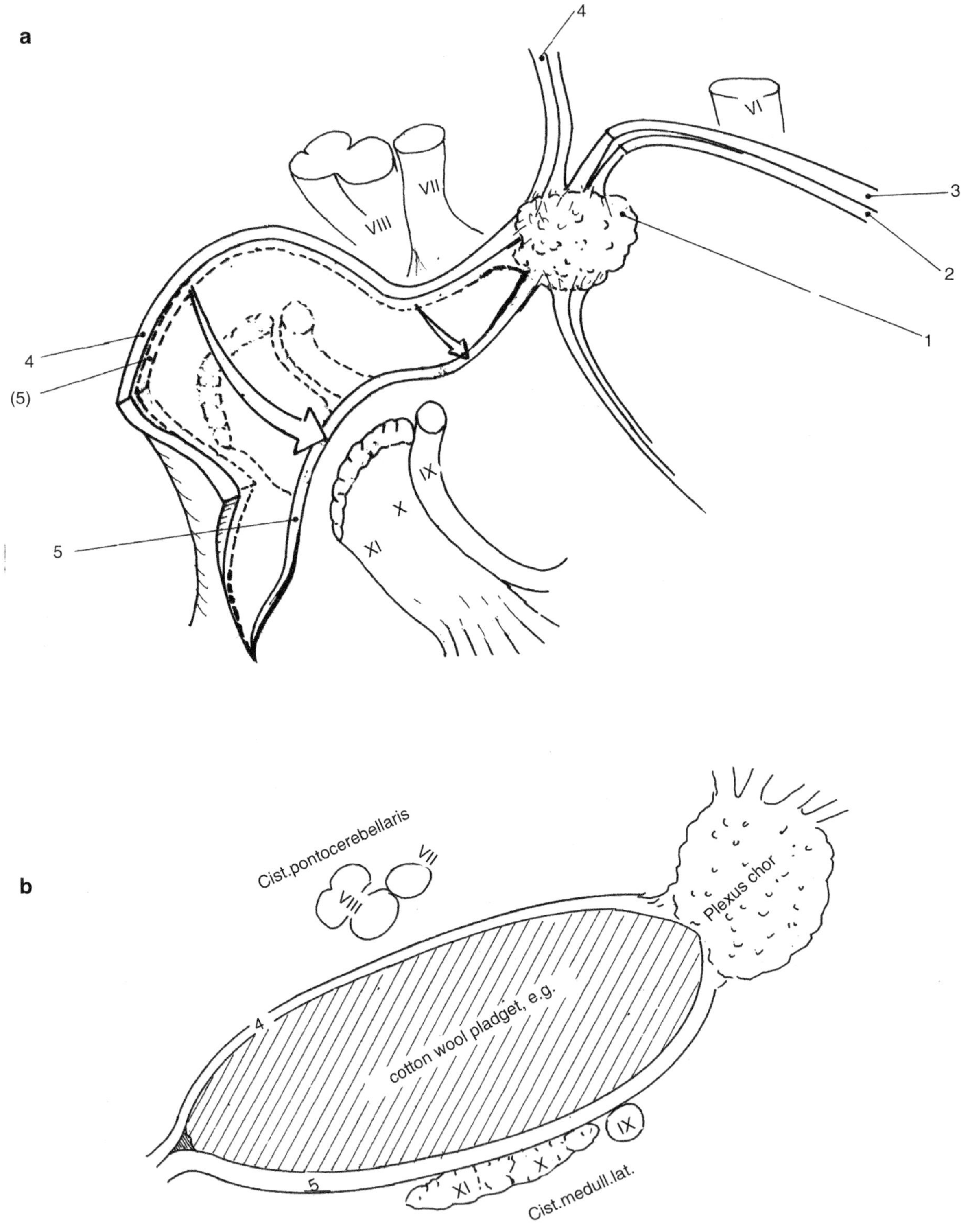

Fig. 16.14 Duplicated arachnoid walls. Loosening and shifting before and during a pontocerebellar tumor will be removed. **a** Loosening of the wall of cisterna medullaris lateral from the wall of cisterna pontocerebellatris. Shifting of the wall of cisterna medullaris lateralis and Nn. IX–XI to a basal medial direction (*arrows*). **b** Interposition of, e.g., a cotton wool pladget between arachnoid walls for preservation of Nn. IX–XI. *1* Plexus choroideus, *2* wall of cisterna medullaris medialis, *3* wall of cisterna pontis medialis, *4* wall of cisterna pontocerebellaris, *5* wall of cisterna medullaris lateralis, loosened and shifted from the wall of cisterna pontocerebellaris (*arrows*) (5) as *5* before shifting, projection *VI–XI* cranial nerves

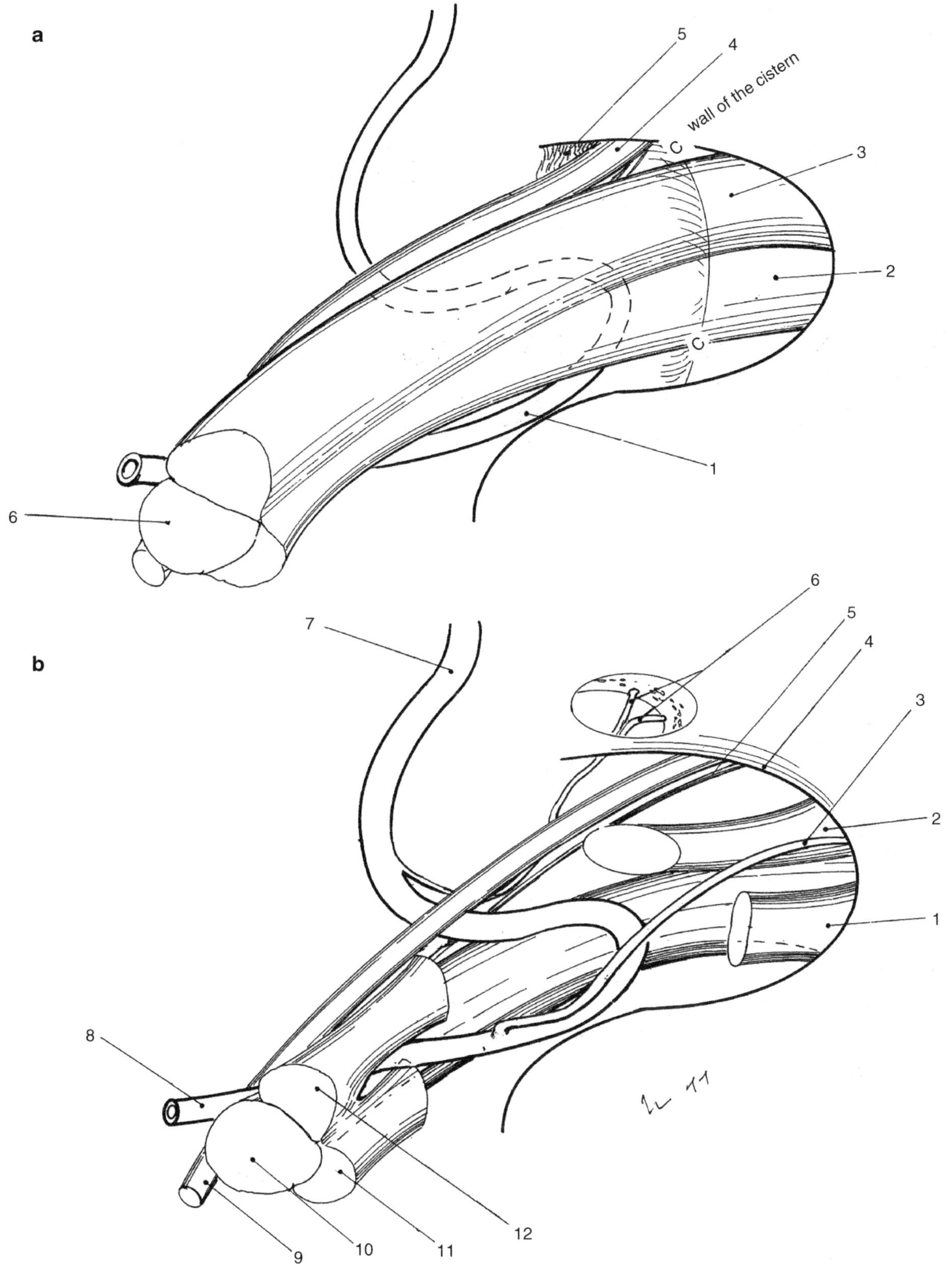

Fig. 16.15 Meatus acusticus int. (Yaşargil et al. 1974; Salamon and Huang 1976, Scialfa et al. 1976) cisternal and extracisternal part. **a** Contents, survey. **b** Details added. **a** *1* AICA, *2* N. vestibularis inferior, *3* N. vestibularis superior, *4* N. facialis (and N. intermedius), *5* adhesions, cave N. VII at surgery, *6* N. cochlearis c variable. **b** *1* Ganglion vestibulare inferior, *2* ganglion vestibulare superior, *3* A. labyrinthi, *4* margin of Porus acusticus internus, *5* N. intermedius, *6* petrosal branches for labyrinth, *7* AICA, *8* as *7*, *9* N. facialis, *10* N. cochlearis, *11* N. vestibularis inferior, *12* N. vestibularis superior

Microsurgical approach (Figs. 16.16, 16.17, 16.18, 16.19, 16.20, 16.21, 16.22, 16.23, and 16.24).[1]

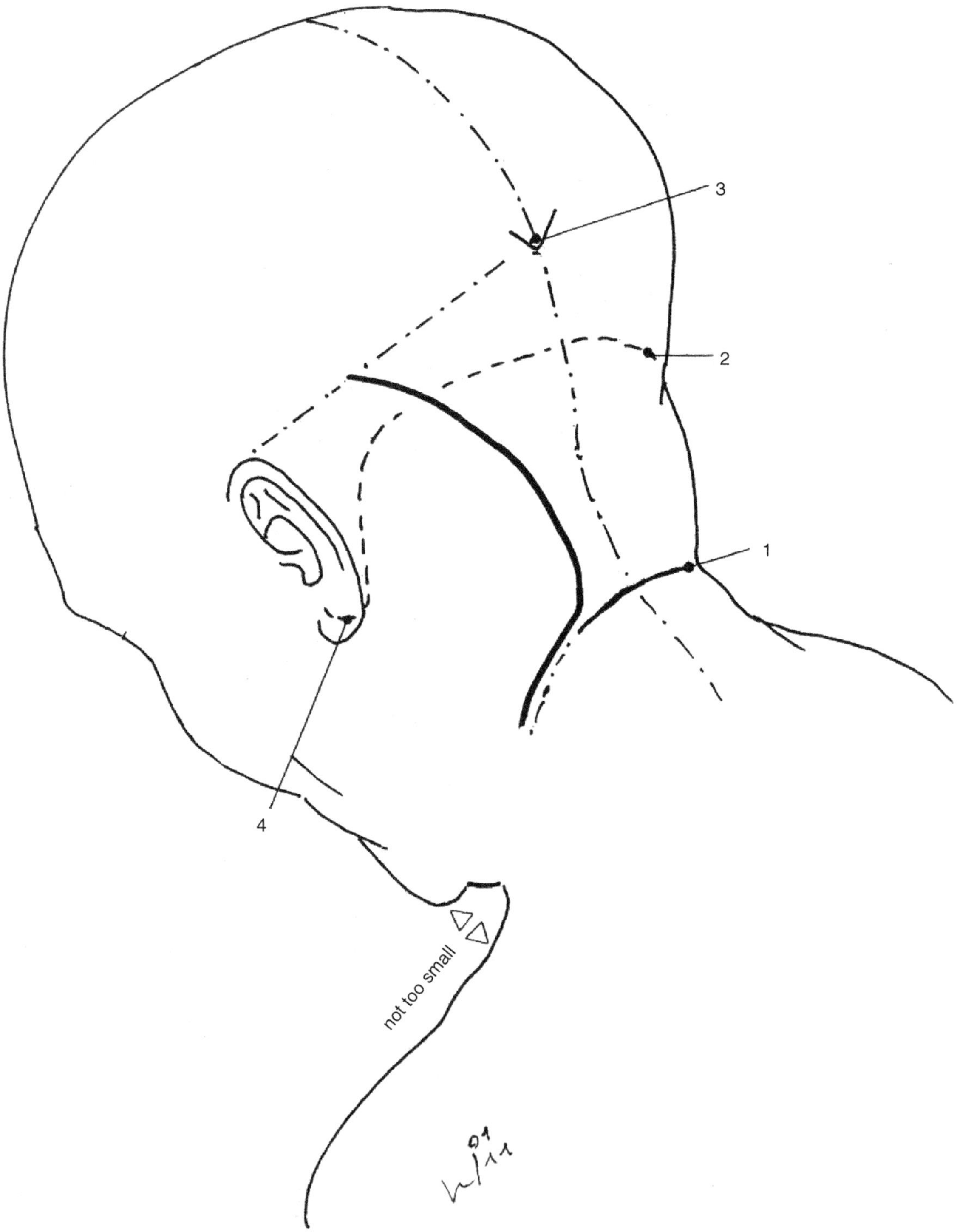

Fig. 16.16 Positioning at the operation table and skin incision. *1* Fold of the skin, *2* linea nuchae inferior, bony crest for insertion of muscles, projection, *3* inion, *4* processus mastoideus (palpable)

[1]According to Yaşargil.

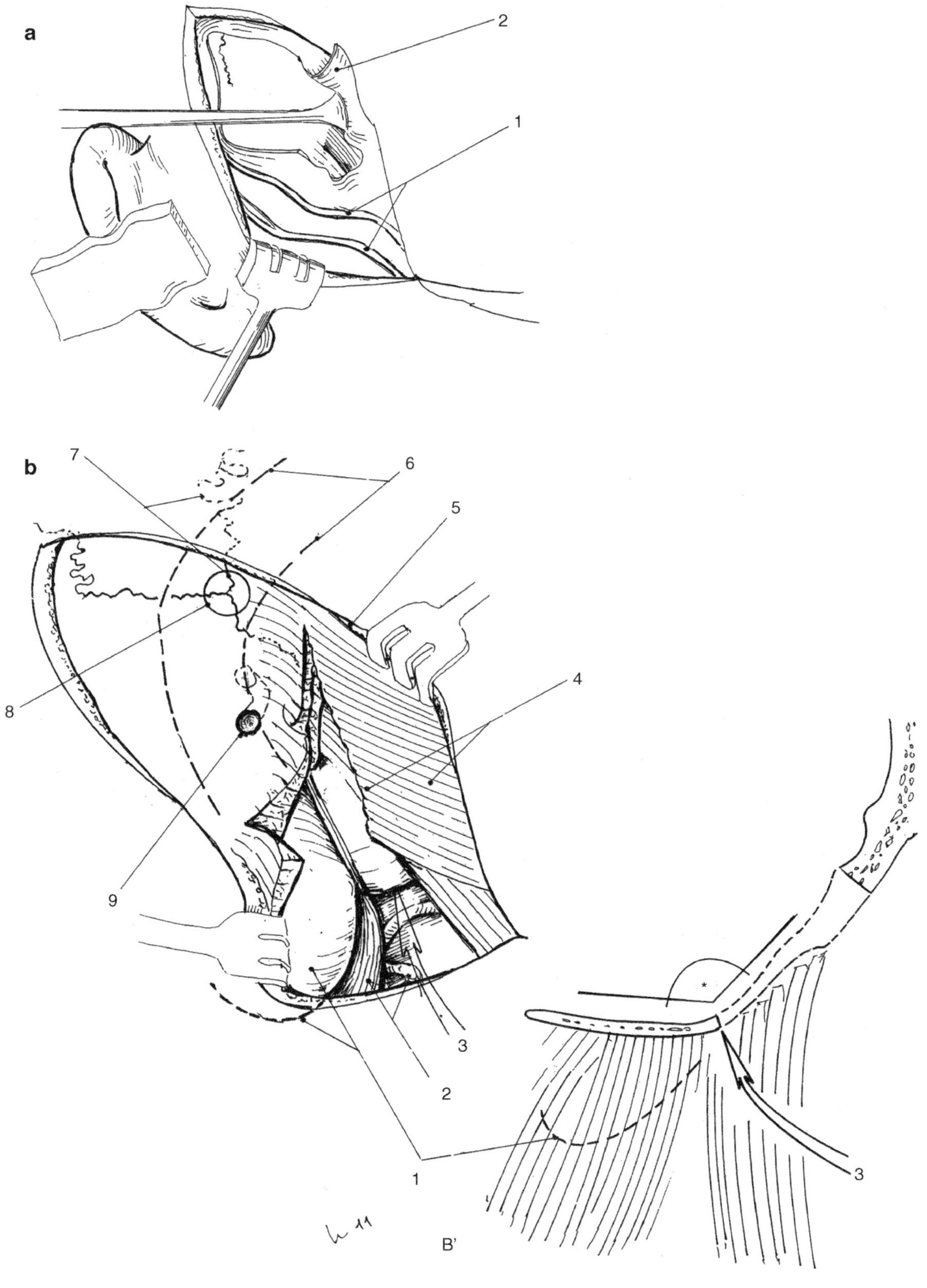

Fig. 16.17 Extracranial route. Preservation of superficial nerves and avoiding foramen occipitale. **a** Preservation of N. occipitalis minor. **b** Dissection of muscles. *Arrow*: see *B' B'* mediobasal region of craniotomy. *Arrow*: mediobasal limit region of craniotomy (According to Samii) -3- at insertion of M. obliquus capitis superior, see Fig. 2.1. **a** *1* N. occipitalis minor, *2* fascia nuchae. **b** *1* Processus mastoideus; *2* M. biventer and A. occipitalis; *3* bony crest (linea nuchae inferior), see *B't;* *4* M. splenius capitis; *5* skin; *6* sinus transversus, projection; *7* sutura lambdoidea; *8* asterion; *9* For. mastoideum, *B'* stop bony resection (*arrow*)

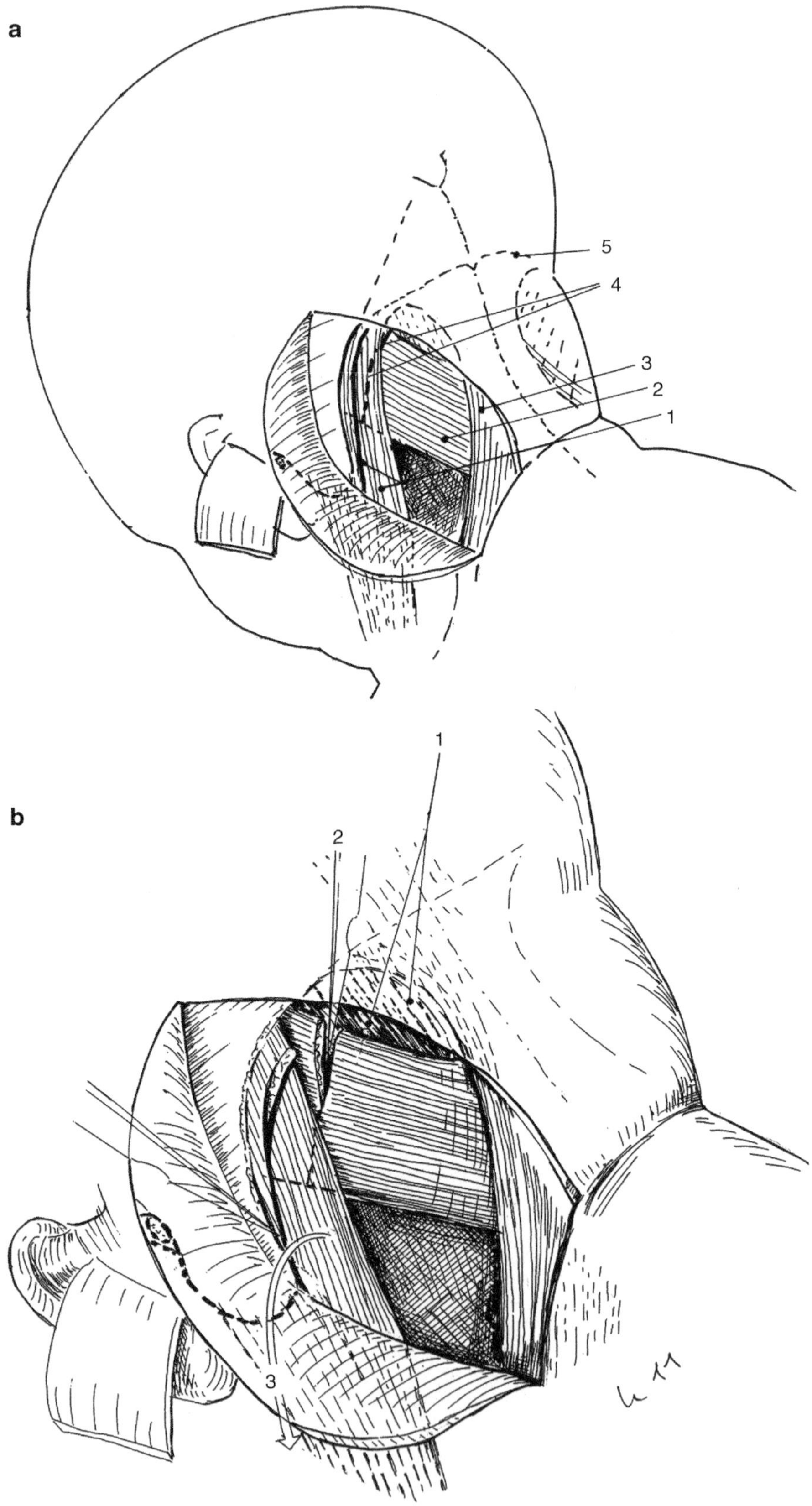

Fig. 16.18 Addendum for Fig. 16.17. **a** Survey. **b** M. sternocleidomastoideus incised. Incision of M. splenius capitis. **a** *1* M. sternocleidomastoideus, *2* M. splenius capitis, insertion at linea nuchae inferior (projection, *dotted*), *3* M. trapezius, *4* M. splenius capitis, *5* linea nuchae inferior, projection. **b** *1* M. semispinalis capitis, *2* M. splenius capitis, *3* M. sternocleidomastoideus

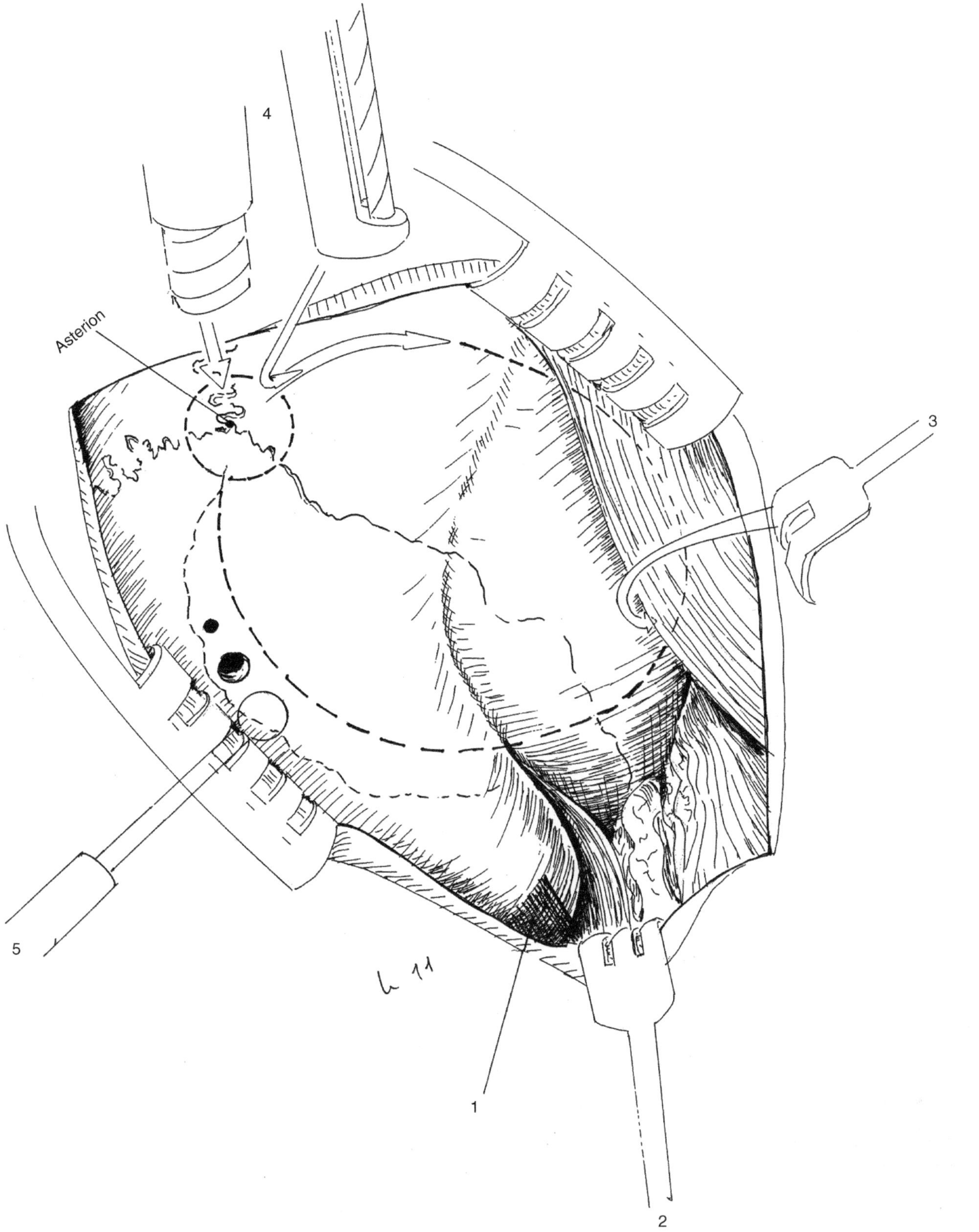

Fig. 16.19 After loosening of muscles, vessels, and nerves: *defining of craniotomy.* Lateral widening of the bone resection using a drill for preservation of emissarial veins and presentation of the medial margin of sinus sigmoideus. *1* M. sterocleidomastoideus, insertion at processus mastoideus, *2* retraction of skin and M. sternocleidomastoideus (for presentation of M. biventer and A. occipitalis), *3* muscles of the neck, *4* first burr hole at asterion for identification of sinus transversus (applied by Prof. Samii, Hannover), *5* widening of craniotomy for presentation of sinus sigmoideus and Vv. emissariae mastoideae

Fig. 16.20 Craniotomy. *1* Processus mastoideus, *2* fossa biventerica, *3* sulcus a. occipitalis variable, *4* fossa paracondyloidea, *5* emissarium mastoideum, *6* linea nuchae inferior, *7* linea nuchae superior

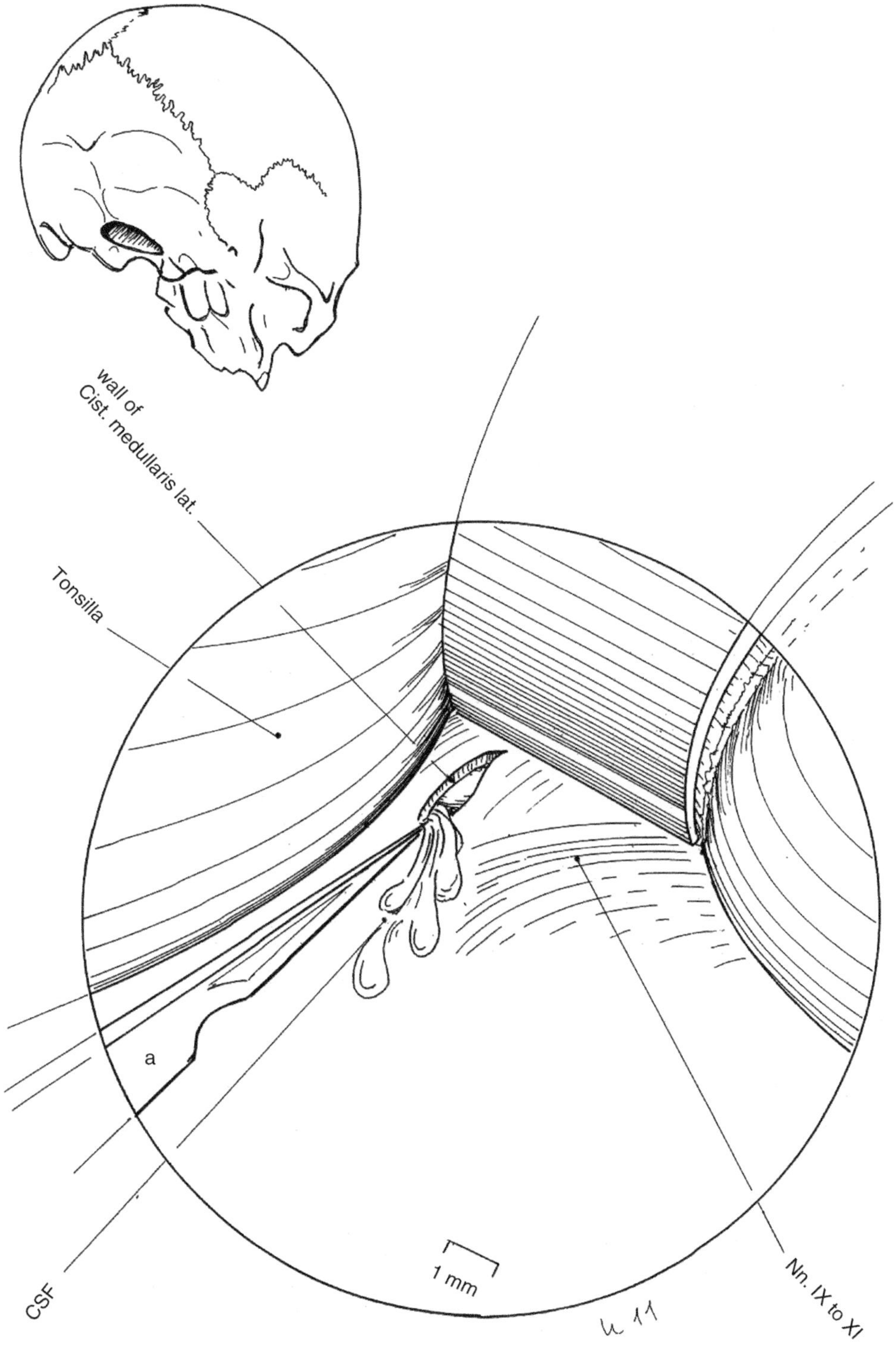

Fig. 16.21 After dura opening: *transcisternal route. Step a* incision of the arachnoid wall of cisterna medullaris lateralis for CSF decompression

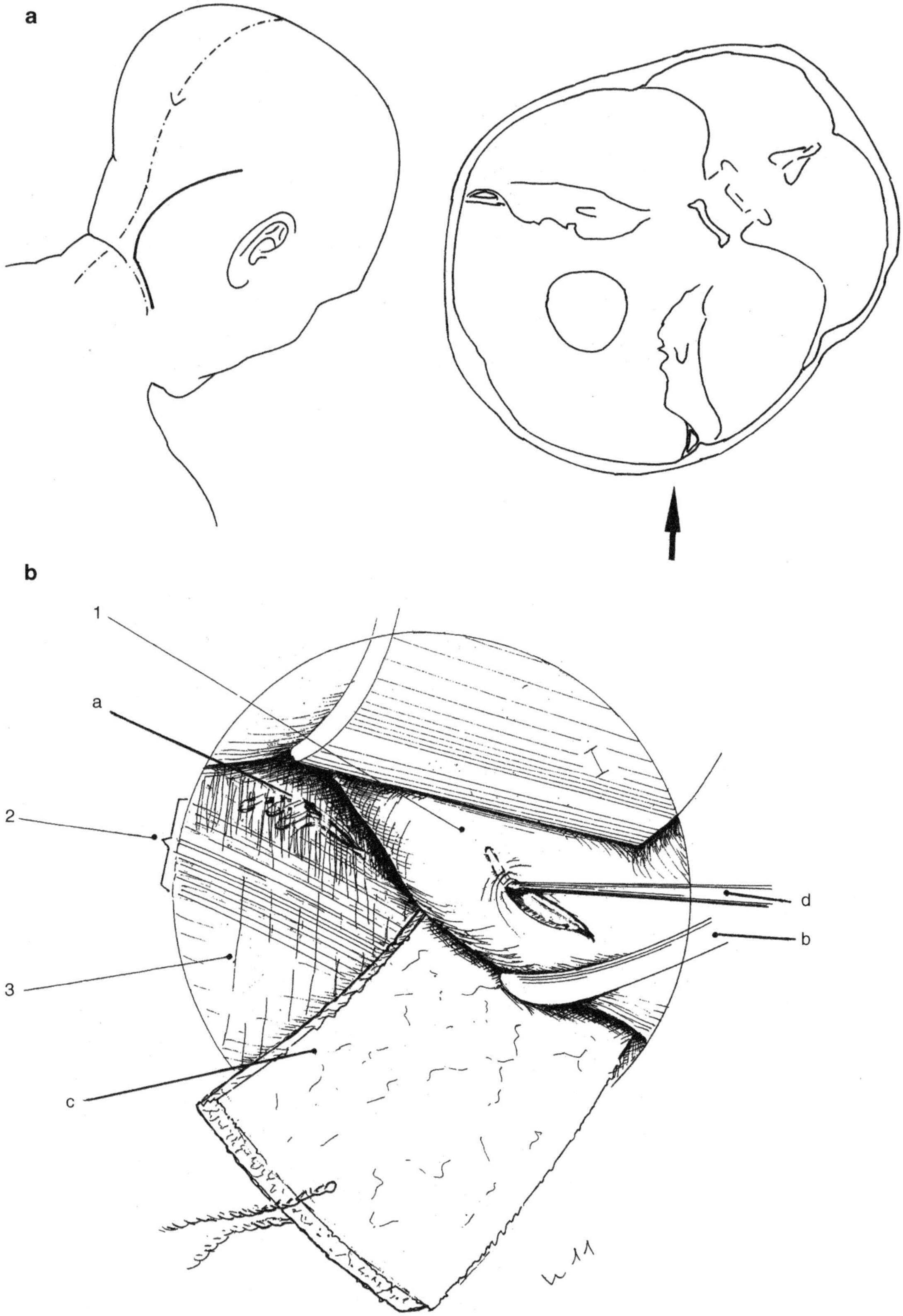

Fig. 16.22 Steps b–d. Incisions of the arachnoid wall of cisterna pontocerebellaris and of the tumor arachnoidea. **a** Viewing direction (*arrow*). **b** Steps *a* see Fig. 16.21, *b* loosening of the wall of cisterna medullaris lateralis (cistern of Nn. IX–XI) from the wall of cisterna pontocerebellaris.(cistern of Nn. V to VIII), *c* interposition of cotton wool pladget, e.g., *d* incision of the wall of cisterna pontocerebellaris. *1* Arachnoid wall of cisterna pontocerebellaris, *2* Nn. IX–XI enclosed by 3, *3* transparent arachnoid wall of cisterna medullaris lateralis

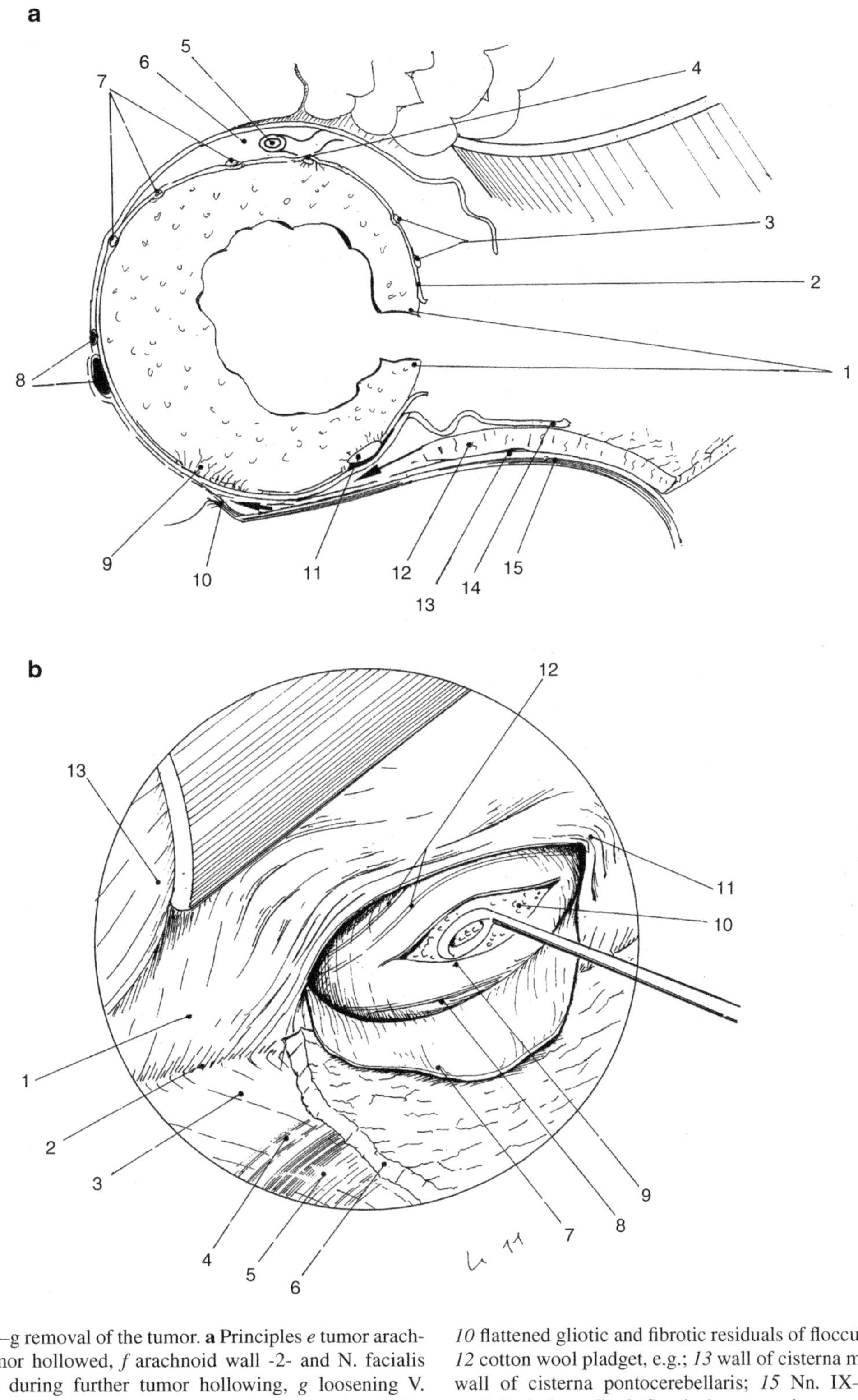

Fig. 16.23 Steps e–g removal of the tumor. **a** Principles *e* tumor arachnoidea incised, tumor hollowed, *f* arachnoid wall -2- and N. facialis loosened, repeated during further tumor hollowing, *g* loosening V. petrosa superior and N. trigeminus,plexus choroideus, flocculus (residual), AICA, Vv. pontis, and (if indicated) N. cochlearis. Transmeatal cisternal and extracisternal removal of the tumor. *1* Tumor incised and hollowed; *2* arachnoidea of the tumor; *3* roots of N. VII (not to identify before hollowing of the tumor); *4* adhesions at AICA; *5* AICA or branches of it; *6* cisterna pontocerebellaris after loosening of its wall from tumor arachnoidea; *7* roots of N. VII; *8* widened and vulnerable lateral pontine subpial vein(s); *9* fibrotic residuals of plexus choroideus; *10* flattened gliotic and fibrotic residuals of flocculus; *11* N. cochlearis; *12* cotton wool pladget, e.g.; *13* wall of cisterna medullaris lateralis; *14* wall of cisterna pontocerebellaris; *15* Nn. IX–XI *arrows:* cisterna medullaris lateralis. **b** Surgical topography, step e, *1* wall of cisterna pontocerebellaris; *2* rim between 1 and 3 after loosening and shifting of cisternal walls; *3* wall of cisterna medullaris lateralis transparent; *4* N. IX; *5* Nn. X and XI; *6* cotton wool pladget e.g.; *7* flap of the wall of cisterna pontocerebellaris; *8* root of N. VII; *9* tumor arachnoidea; *10* tumor; *11* wall of cisterna pontocerebellaris; *12* tumor arachnoidea and roots of N. VII; *13* cerebellum

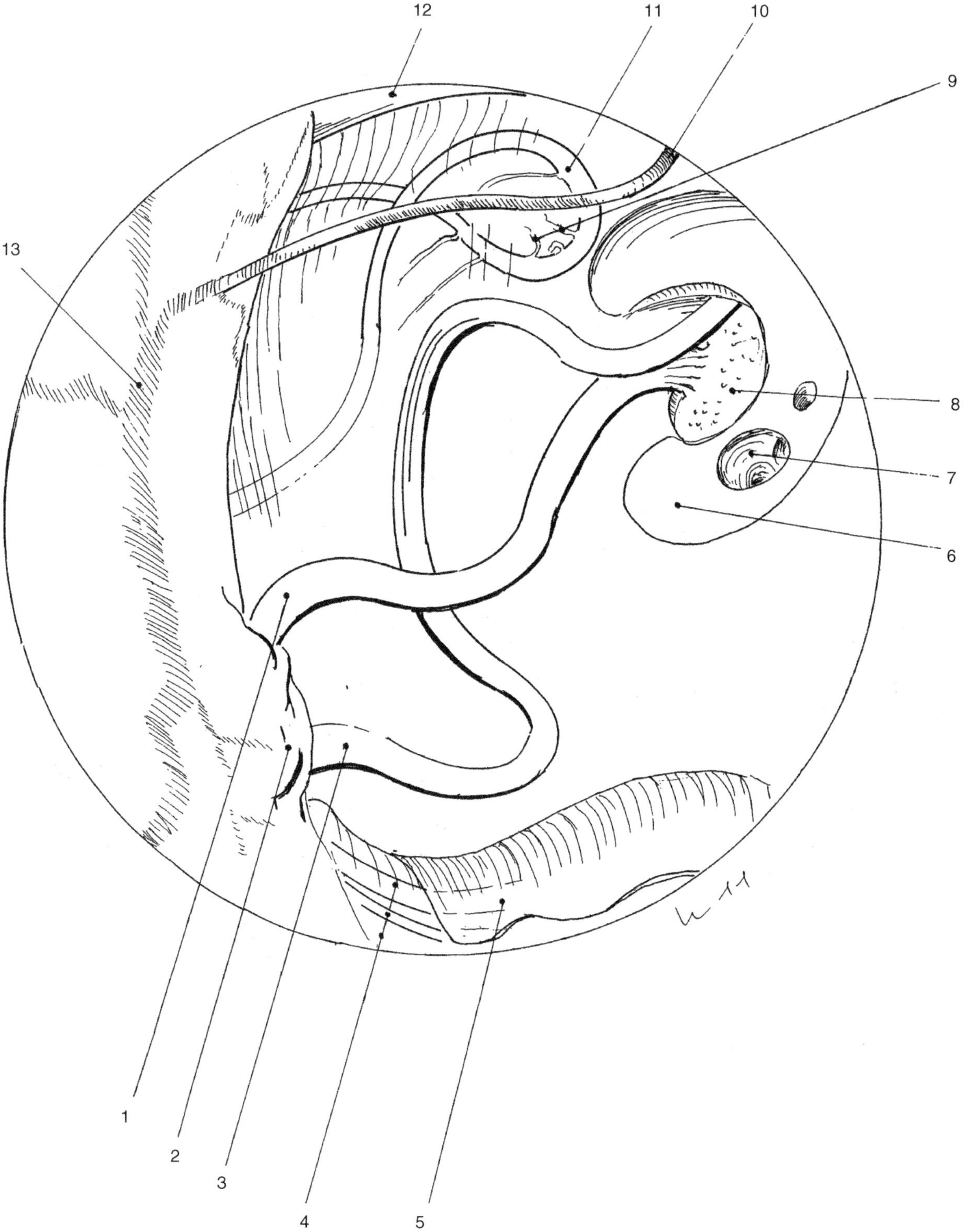

Fig. 16.24 Surgical topography after removal of the tumor. Sketch during surgery. After a short time, these structures will be contracted and could be normal configurated. *1* N. cochlearis; *2* residuals of plexus choroideus and flocculus; *3* N. facialis; *4* Nn. IX and X; *5* arachnoid flap; *6* resected bony wall of meatus acusticus internus; *7* pneumatic cells open before occluded by surgery; *8* fundus of meatus acusticus internus; *9* branches of AICA feeding brainstem; *10* V. petrosa superior; *11* AICA, trigeminal loop; *12* N. trigeminus; *13* lateral pontine vein

16.1.4 Surgical Aspects for Approaches to Cisterna Medullaris Medialis and Lateralis (Figs. 16.25, 16.26, 16.27, and 16.28)

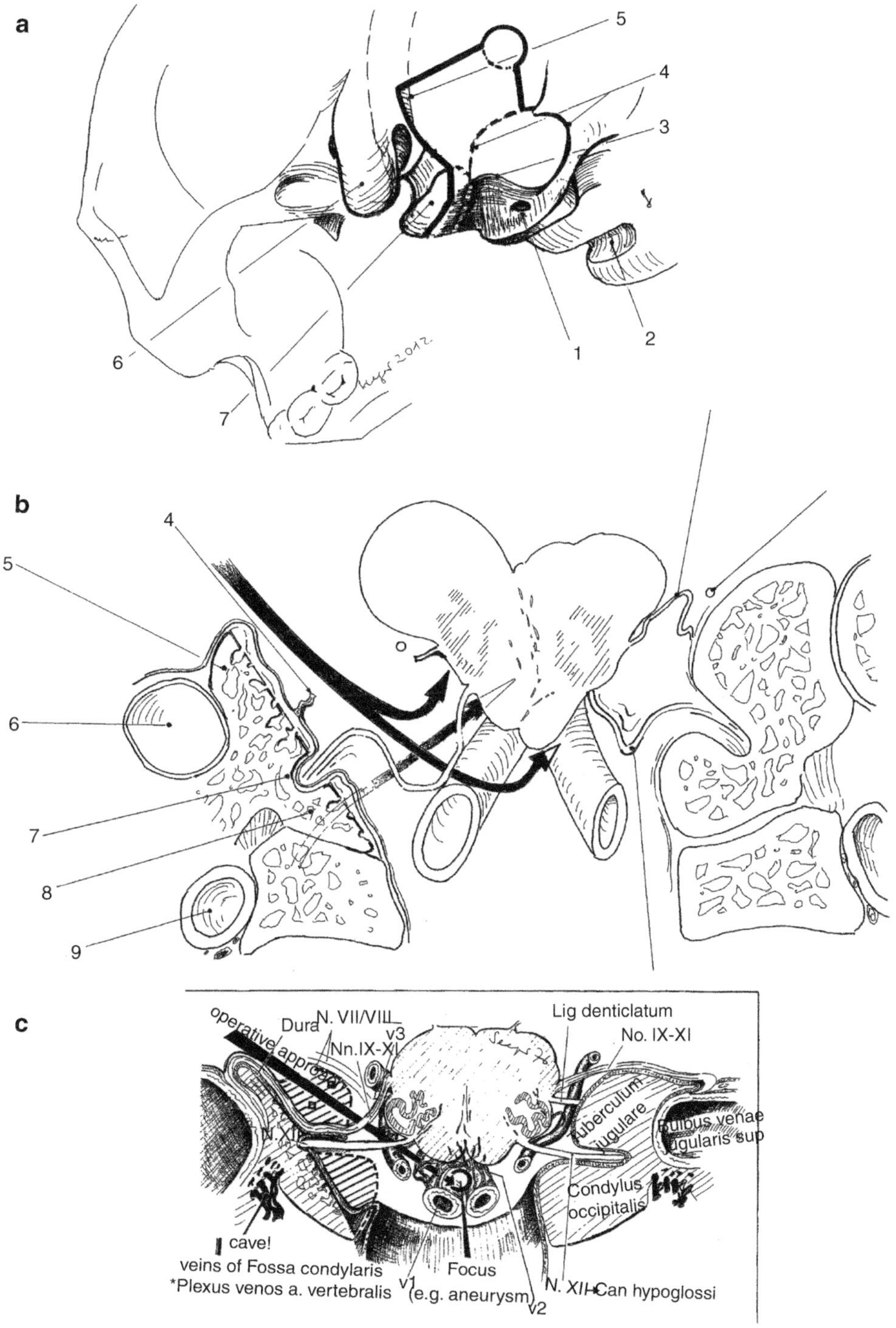

Fig. 16.25 Transcondylar approach to cisterna medullaris medialis et cisterna pontis medialis. Anatomical base. **a** Craniotomy, the lateral segment of conylus occipitalis and processus articularis atlantis must be preserved (Prof. Putz, Anatomisches Institut der Universität Freiburg, personal communication with the author 1977). **b** Mobilization of medulla oblongata and the adjacent segments of medulla spinalis are possible after transection of ligamentum denticulatum. *Arrows*: Paracondylar routes to a premedullary and a prepontine target area. **c** Example of surgery of an aneurysm of the vertebrobasilar junction (first description of a transcondylar surgical approach (Seeger 1978)). **a** *1* Canalis n.hypoglossi; *2* fossa biventerica; *3* tuberculum jugulare; *4* margin of For. occipitale, resected; *5* sinus sigmoideus; *6* processus mastoideus; *7* condylus occipitalis, resected. **b** *1* N. hypoglossus; *2* Radix spin. n. accessorii; *3* lig. denticulatum; *4* as 3, cut; *5* tuberculum jugulare resected; *6* bulbus superior v. jugularis internae; *7* canalis n. hypoglossi, resected; *8* condylus occipitalis, resected; *9* A.vertebralis

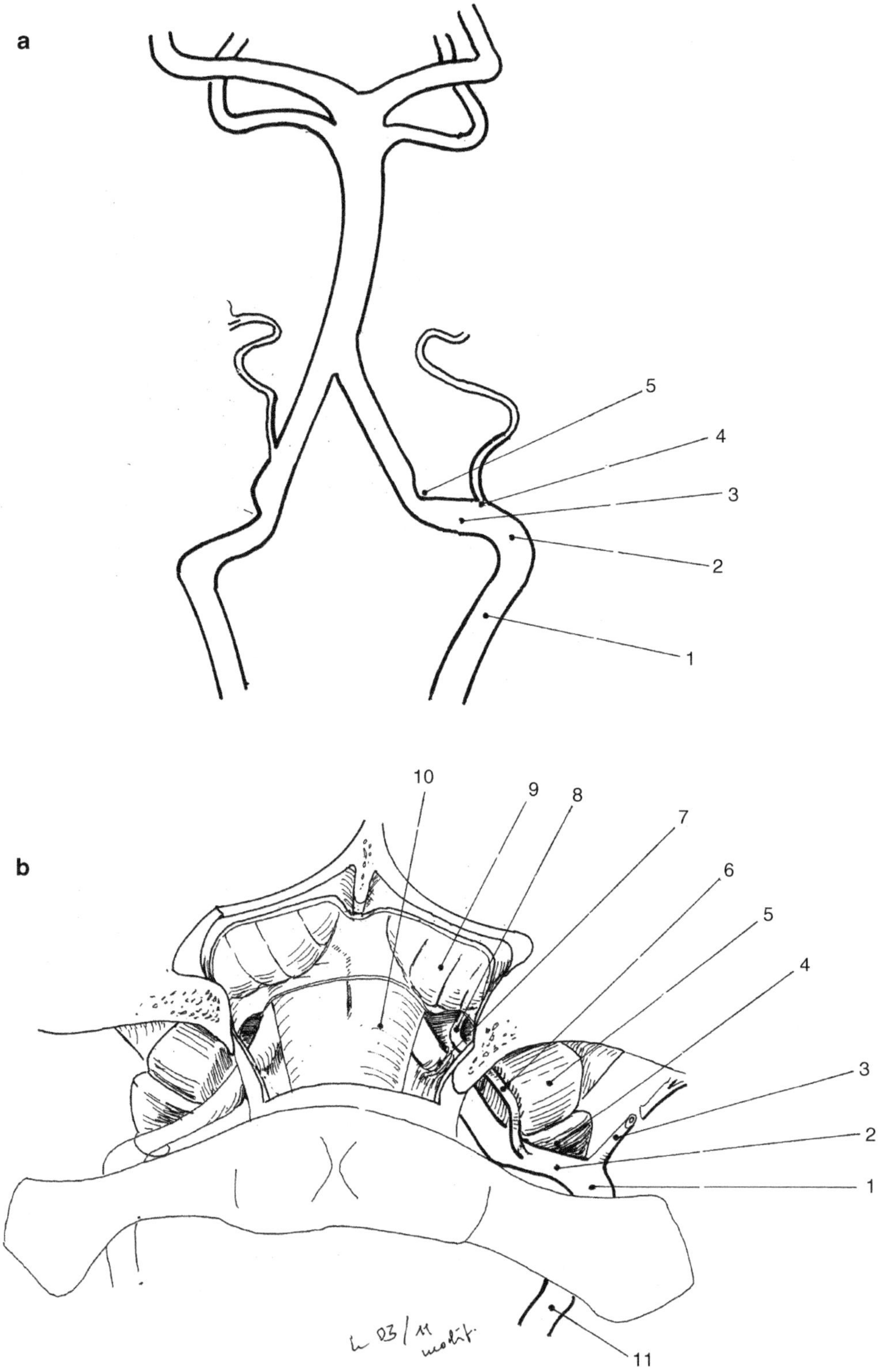

Fig. 16.26 Addendum. *Extradural origin of PICA from A. vertebralis* (Lang 1985, modified). This rare variant should be considered at imagings before performing basal medial surgical approaches. **a** Normal angiogram. The usual stenosis of A. vertebralis at its transdural penetration (*5*) allows a discrimination of the usual intradural origin of PICA (*right*) from its extradural variant (*4*). **b** Extradural origin of PICA from A. vertebralis (*6*). **a** *1* A. vertebralis; *2* loop of A.vertebralis at its penetrating foramen transversarium of Atlas; *3* segment of A. vertebralis along sulcus A. vertebralis of Atlas; *4* PICA, extradural origin; *5* usual stenosis of A. vertebralis marking the dural penetration of A. vertebralis. **b** *1* Loop of A. vertebralis; *2* segment of A. vertebralis along Sulcus A. vertebralis of arcus atlantis posterior; *3* usual branch of A. vertebralis feeding muscles; *4* processus articularis superior; *5* condylus occipitalis; *6* PICA, extradural origin; *7* usual stenosis of A. vertebralis; *8* PICA, intradural segment; *9* tonsilla; *10* medulla oblongta et spinalis; *11* A. vertebralis

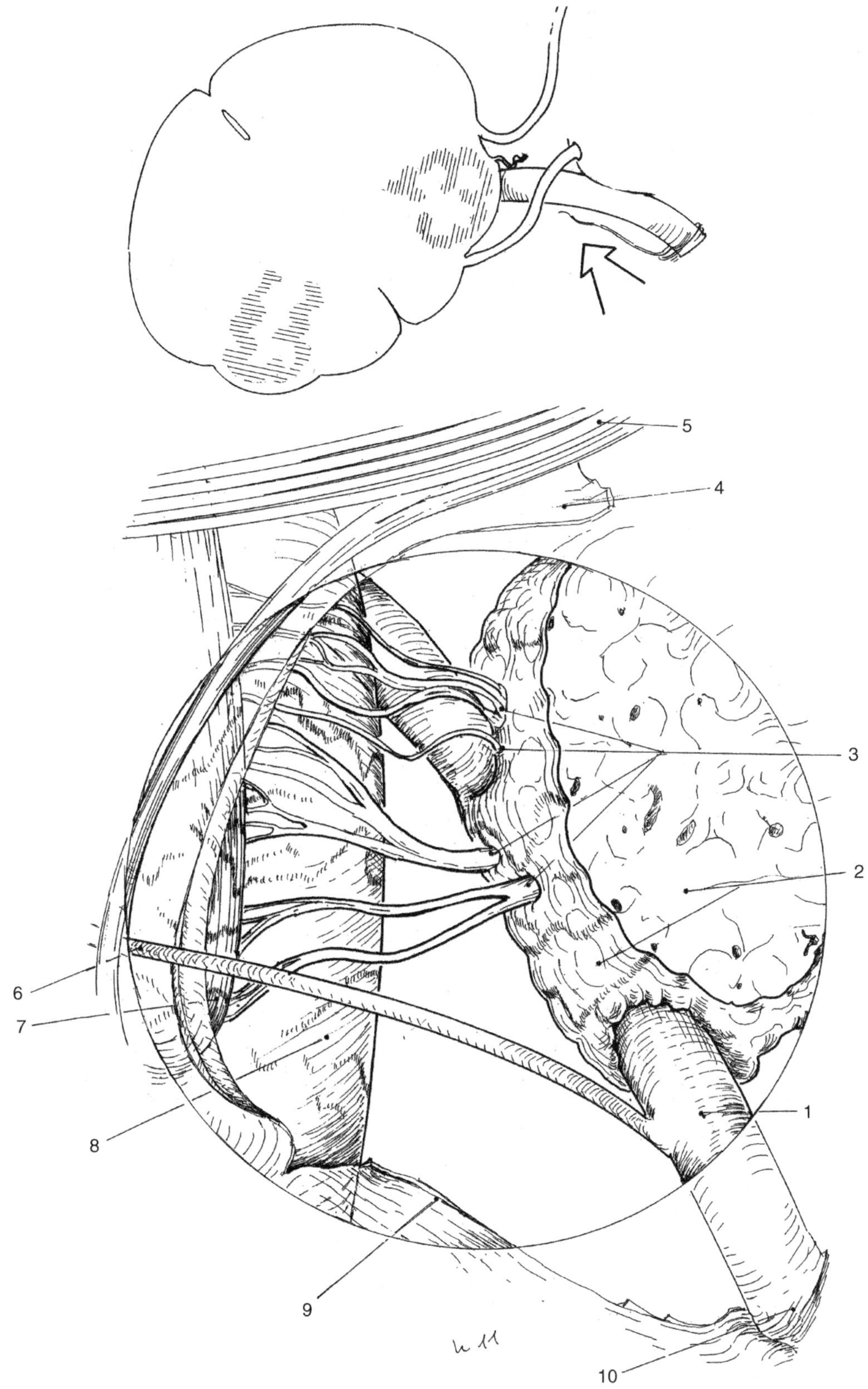

Fig. 16.27 Removal of a craniospinal meningioma (experience of the author) using a transcondylar surgical route. Presentation before loosening of the tumor residual from A. vertebralis and N. hypoglossus. Surgical topography. **a** Topogram. *Arrow*: viewing direction **b**. **b** Presentation during surgery. *1* A. vertebralis; *2* tumor residual; *3* roots of N. XII; *4* Lig. denticulatum, cut; *5* Nn. X and XI; *6* PICA; *7* oliva; *8* pyramis medullae; *9* as *4; 10* dural penetration of A. vertebralis

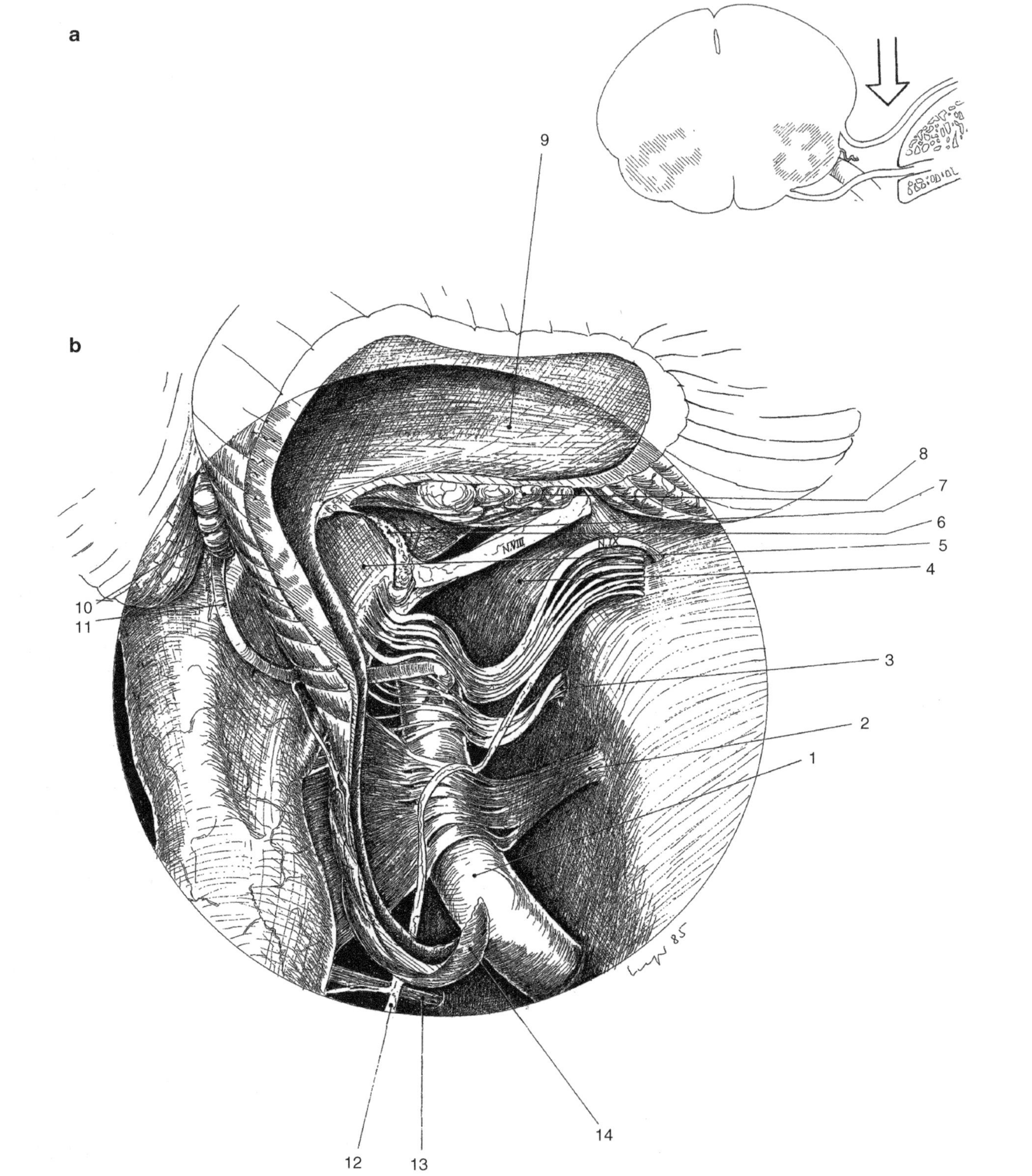

Fig. 16.28 Paramedullary surgical approach from a dorsolateral direction. Removal a cystic lindau angioma of tonsilla (Seeger 1986). Surgical topography. **a** Topogram. *Arrow*: viewing direction **b. b** Microsurgical topography. *1* A. vertebralis, *2* Lig. denticulatum, *3* N. XII, *4* tuberculum jugulare, *5* pedunculus cerebellaris inferior, *6* plexus choroideus, *7* inferior segment of pons, *8* flocculus, *9* residual of tonsilla, *10* vermis, *11* PICA, *12* radix spinalis n.XI, *13* radix anterior C1, *14* as *9*

16.1.5 Supracerebellar Surgical Approaches to Cisterna Tecti[2] and Adjacent CSF Spaces (Figs. 16.29, 16.30, 16.31, and 16.32)

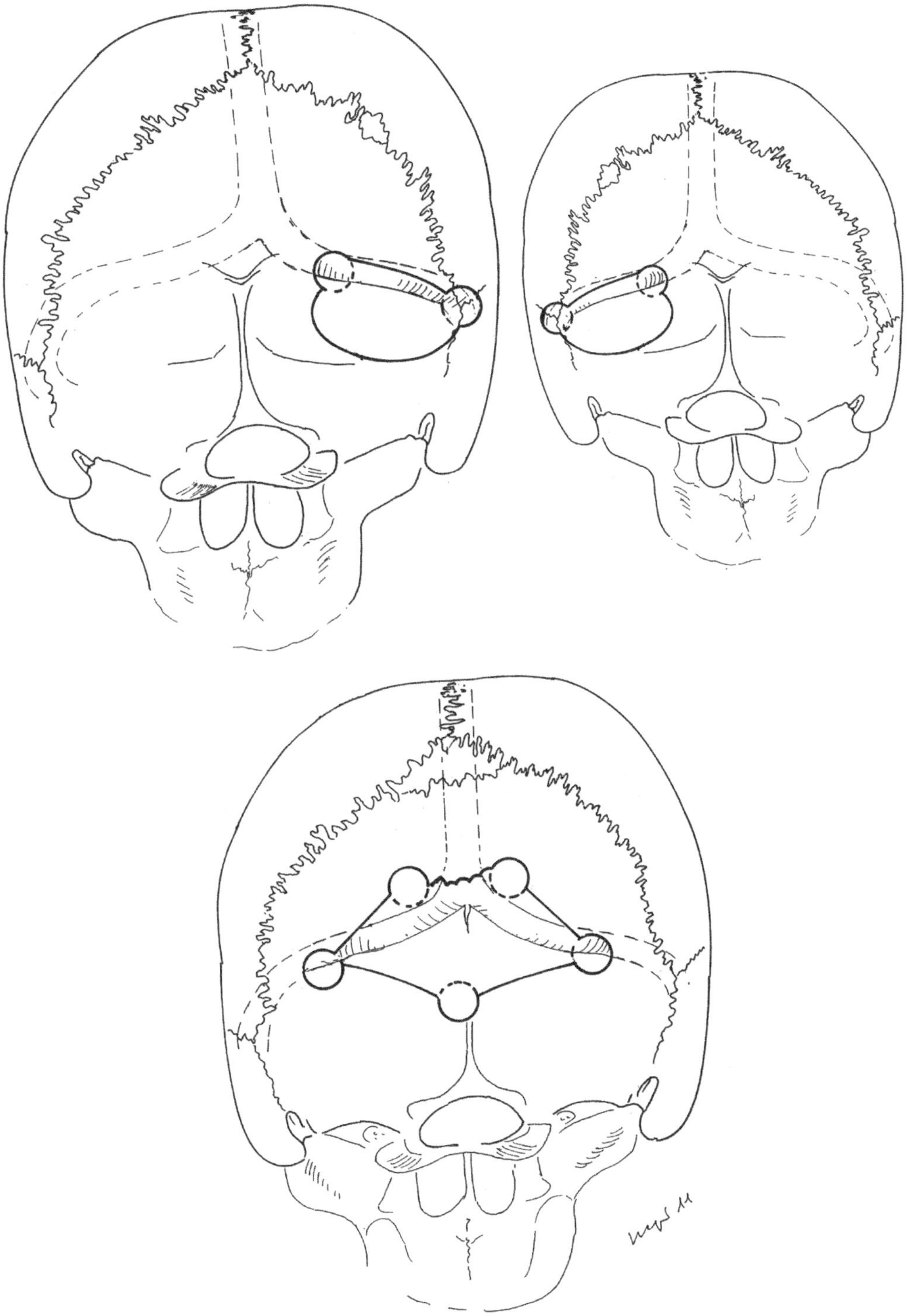

Fig. 16.29 Craniotomy. Different types of craniotomy for the same surgical target areas according to Yaşargil. These target areas are cisterna tecti and adjacent CSF spaces. The choice of a supracerebellar surgical route depends on calibre and location of dorsolateral bridging veins. Thick calibrated veins could hinder the supracerebellar part of the surgical route and should be preserved at surgery

[2] According to F Krause 1926, modified for microneurosurgery by Yaşargil.

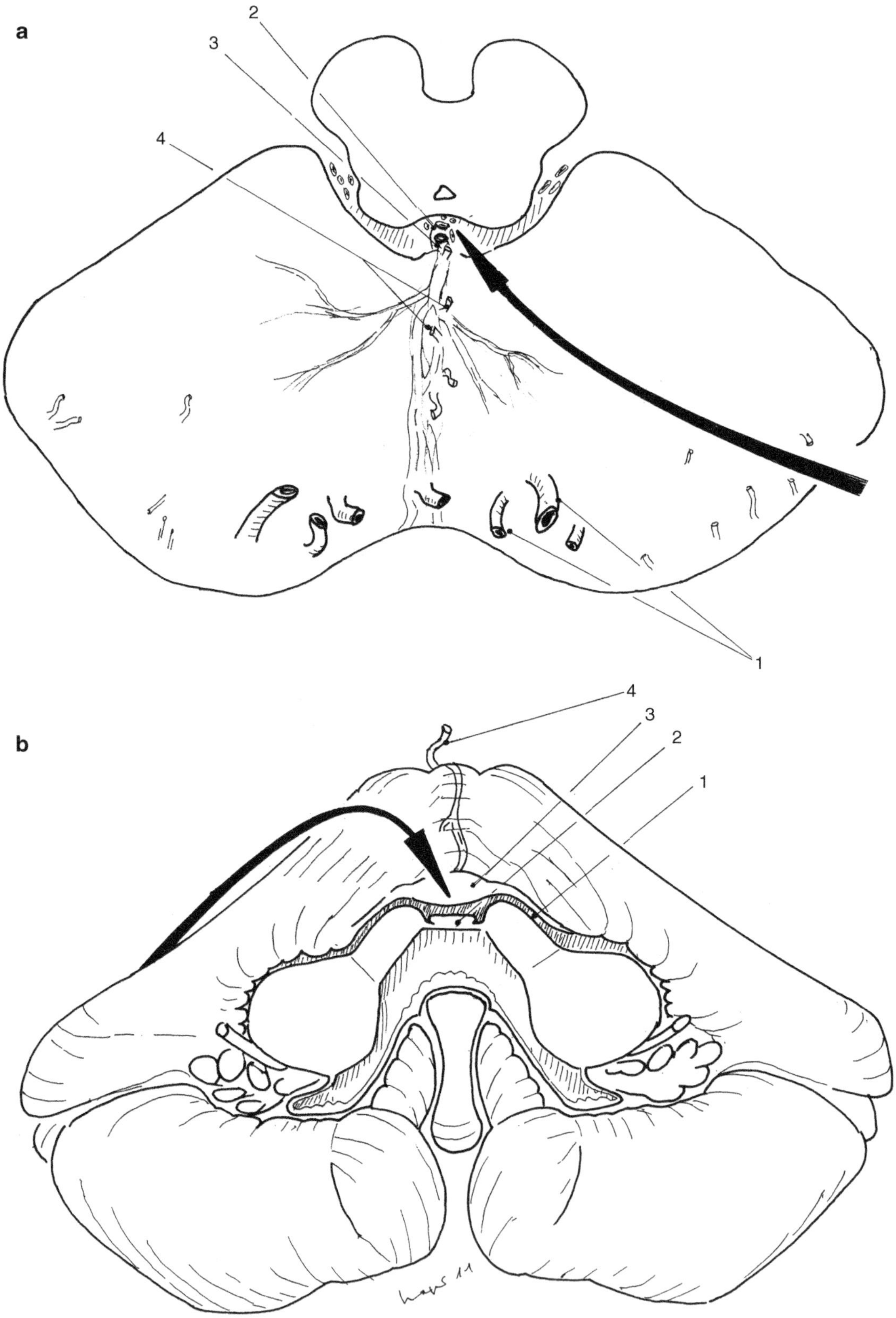

Fig. 16.30 After dura opening: step a. Supracerebellar subdural route. Interruption of thick calibrated bridging veins should be avoided. The thin calibrated medial bridging veins (4) could be interrupted for a better reclination of culmen cerebelli, if necessary. Superior **a** and anterior **b** viewing direction. *Arrows*: surgical route. **a** *1* Cerebellotentorial bridging veins, *2* V. cerebellaris praecentralis, *3* V. supraculminalis, *4* small bridging veins. **b** *1* Fissura horizontalis cerebelli, anterior segment; *2* velum medullare anterius, fusioned with lingula cerebelli, common finding; *3* lobulus centralis cerebelli; *4* V. cerebellaris praecentralis originates ventral or dorsal from lobulus centralis cerebelli

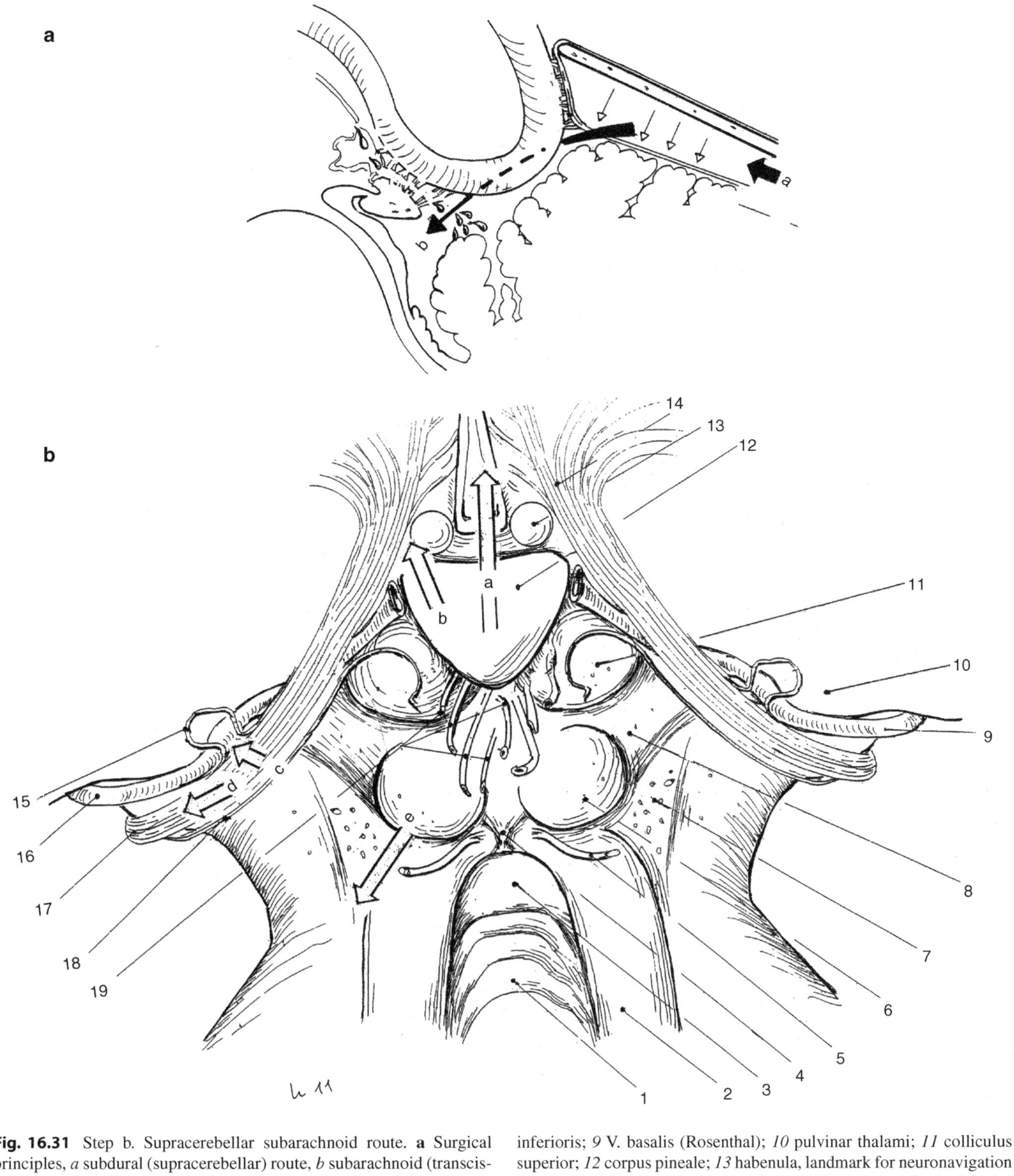

Fig. 16.31 Step b. Supracerebellar subarachnoid route. **a** Surgical principles, *a* subdural (supracerebellar) route, *b* subarachnoid (transcisternal) route. **b** Surgical target area, *a* fissura transversa dorsal from velum interpositum and third ventricle, *b* medial from ventriculus lateralis, *c* pulvinar thalami, (landmark during surgery: Baumgartner's loop), *d* cisterna ambiens. *1* Lingula cerebelli; *2* pedunculus cerebellaris superior; *3* velum medullare anterius; *4* frenulum tecti; *5* N. trochlearis; *6* colliculus inferior; *7* trigonum lemnisci; *8* brachium colliculi inferioris; *9* V. basalis (Rosenthal); *10* pulvinar thalami; *11* colliculus superior; *12* corpus pineale; *13* habenula, landmark for neuronavigation and during surgery; *14* division of chorioid arteries at the horiziontal level of habenula; *15* Baumgartner's loop, *16* as *9*; *17* Aa. chorioideae posteriores laterales; *18* A. tecti, running along the pontomesencephalic rim, together with a small chorioid branch; *19* Vv. tecti, draining veins of V.magna (Galeni)

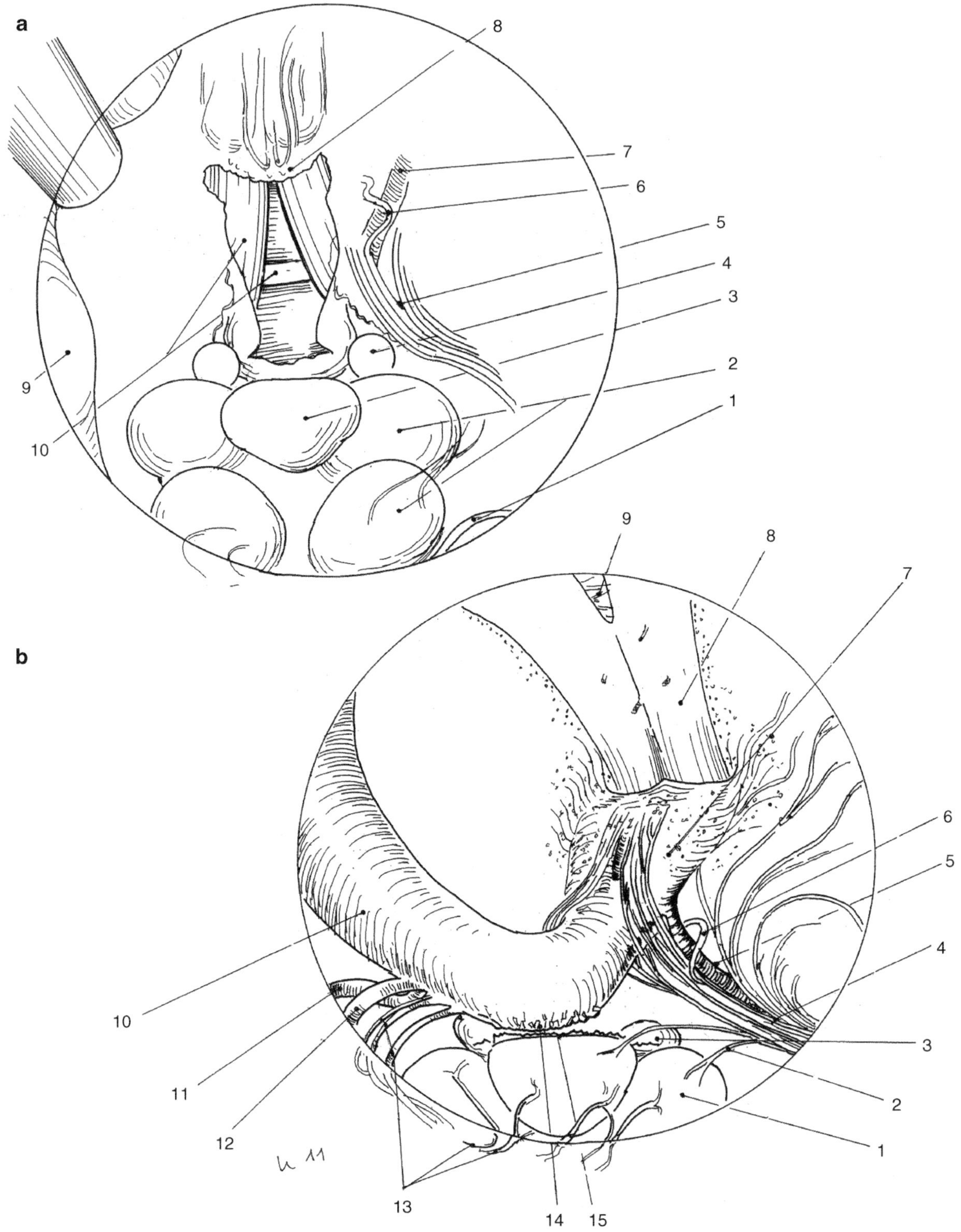

Fig. 16.32 Step c. Presentation of surgical target areas, examples. **a** Microsurgical topography of the third ventricle. **b** Microsurgical topography of fissura transversa cerebri. **a** *1* N. trochlearis; *2* colliculus superior et inferior; *3* corpus pineale; *4* habenula; *5* division of chorioid arteries; *6* Baumgartner's loop; *7* V. basalis (Rosenthal); *8* velum interpositum, elevated after rupture of the thin wall of recessus suprapinealis; *9* V. magna (Galeni); *10* commissura anterior and columna fornicis. **b** *1* Colliculus superior; *2* A. tecti; *3* habenula; *4* division of chorioid arteries; *5* V. basalis (Rosenthal); *6* Baumgartner's loop; *7* velum interpositum; *8* fornix, elevated for opening of fissura transversa; *9* commissura fornicis; *10* V. magna (Galeni); *11* V. cerebellaris praecentralis; *12* V. supraculminalis; *13* Vv. tecti; *14* adhesions,cut; *15* as *14*; preservation of corpus pineale

References

Appuzo MLJ (1987) Surgery of the third ventricle. Willliams and Wilkins, Baltimore/London/Los Angeles/Sydney

Appuzo MLJ, Cikovani OK, Gott PS, Ten EL, Zee ES, Gianotta S, Weis MH (1982) Transcallosal interfornical approaches for lesions affecting the third ventricle : surgical considerations and consequences. Neurosurgery 10:547–554

Baumgartner W et al (1981) Étude anatomo-chirurgicale de la voie d'aborde de la toile choroidienne súperieure du 3° ventricle. A propos d'un anévrysme artério-veineux choroidien postérieur. Neurochirurgie 27:97–102

Browder J, Kaplan AH (1976) Cerebral dural sinuses and their tributaries. Thomas Springfield, Illinois

Browning H (1953) The confluens of dural venous sinuses. Am J Anat 93:307–329

Cambria S (1980) Thrombosis of the veins of Labbé with hemorrhage infarction. Rev Neurol (Paris) 136:321–326

Coin GG et al (1971) The clivus. In: Newton and Potts, Radiology of the skull and brain, vol 1. Mosby, St. Louis

Dandy WE (1938) Die Hirnchirurgie. JA Barth Leipzig

Drake CG (1976) Cerebral aneurysm surgery - an update. In: Cerebral vascular diseases, Eleventh Conference. Revens Press, New York

Duvernoy HM (1975) The superficial veins of the human brain. Veins of the brainstem and of the base of the brain. Springer, Berlin/Heidelberg/New York

Duvernoy HM (1988) The human hippocampus. Bergmann München

Duvernoy H (1991) The human brain. In: Surface, three-dimensional sectional anatomy and MRI. Springer, Wien New York

Feneis (1982) Bild-Lexikon der Anatomie. Thieme, Stuttgart New York

Frowein RA (1963) Zentrale Atemstörungen bei Schädelhirnverletzungen und bei Hirntumoren. In: Müller-Rüffenacht M, Spatz H, Vogel P (eds) Monogr aus d Gessamtgeb. Neurologie u. Psychiatrie. Springer, Berlin/Göttingen/Heidelberg

Hassler O (1966) Deep cerebral venous system in man. Neurology (Chicago) 16:505–511

Hassler W, Zentner J (1989) Pterional approach for surgical treatment of olfactory groove meningiomas. Neurosurgery 25:942–947

Heitzmann C (1875) Anatomie des Menschen. II, Muskeln, Fascien. Topography. Braumüller Wien

Hermann FR (1908) Gehirn und Schädel. Gustav Fischer, Jena

Hölzel G (1977) Die lateralen subependymalen Ventrikelvenen. Inaug. Diss. Freiburg i. Br.

Huang YP, Wolf BS (1966) Precentral cerebellar vein in angiography. Acta Radiol Stockholm 5:250–262

Huang Y et al (1968) The veins of the posterior fossa, anterior or petrosal draining group. Am J Roentgenol 104:36–56

Huang YP, Wolf BS (1974a) The basal cerebral vein and its tributaries. In: Newton TH, Potts DG (eds) Radiology of the skull and brain, vol 2., Book 3. Mosby, St Louis, pp 2111–2154

Huang YP, Wolf BS (1974b) Veins of the posterior fossa. In: Newton TH, Potts DG (eds) Radiology of the skull and brain, vol 2. Book 3, pp 2155–2219

Hussein S, Renella RR, Füllekrug B, Friedrich H (1987) The role of the anterior chorioidal artery in supplying the amygdalo-uncohippocampal region. Acta Neurochir 15:137–142

Hütter B (1999) Neurophysiologische und verhaltensmedizinische Folgen von Subarachnoidalblutungen. Habil, Aachen

Hütter B (2000) Neurophysiological consequences of subrarachnoid hemorrhage and its treatment. Springer Wien, New York

Janetta PJ (1967) Gross (mesoscopic) description of the human trigeminal nerve and ganglia. J Neurosurg 1(part 2):109–111

Janetta PJ, Rand RW (1966) Transtentorial retrogasserian rhizotomy in trigeminal neuralgia by microsurgical technique. Bull Los Angel Neurol Soc 31:93–99

Kaplan HA, Browder J (1973) Atresia of the rostral superior sagittal sinus: Substitude parasagittal venous channels. J Neurosurg 38:602–607

Karaus A, Hofer S, Frahm J (2009) Separation of fiber tracts within the human cingulum bundle using single-shot STEAM DTI. Open Med Imaging J 3:21–27

Kempe (1968) Operative Neurosurgery, vol I. Springer, Berlin\Heidelberg\New York

Key A, Retzius G (1875) Studien in der Anatomie des Nervensystems und des Bindegewebes, Band I. P A Norstedt & sôner

Klose AK (2000) Die topographische Anatomie des Nucleus cochlearis – eine mikroanatomische Studie zur Implantation von Hirnstammelektroden. Thesis Hannover

Klose, AK, Sollmann WP (1998) Landmarks for implantation of the cochlear nucleus – a microanatomical study. 49th Annual meeting of the german society of neurosurgery, Hannover. Zbl Neurochir (Suppl) 61

Klose AK, Sollmann WP (1999) Anatomical variations of landmarks for implantation at the cochlear nucleus. In: J Laryng Otol (Suppl) 27. Proceed of the 2nd internat. Auditory Brainstem Implant Sympos Freiburg, Germany

Kobayashi S et al (1971) Intraventricular meningiomas. Mayo Clin Proc 46:735–741

Krayenbühl HM, Yaşargil MG (1957) Die vaskulären Erkrankungen im Gebiet der A. vertebralis und basilaris, Thieme Stuttgart

Kribs M, Kleihues P (1971) The recurrent artery of Heubner. In: Zülch KJ (ed) Cerebral circulation and stroke. Springer, Berlin/Heidelberg/New York, p 41ff

Kristewa R, Cheyne D, Deeke J (1991) Neuromagnetic fields accompagning unilateral and bilateral voluntary movements: topography and analysis of cortical sources. Clin Neurophysiol 81:284–298

Krmpotic-Nemanic J, Draf W, Helms J (1985) Chirurgische Anatomie des Kopf-Hals-Bereiches. Springer, Heidelberg\New York\Tokyo

Lang (1975) Anatomy of the brainstem and the lower cranial nerves, vessels and surrounding structures. Am J Otol (Suppl):1–19

Lang J, Brunner FX (1979) Über die arterielle Versorgung der Medulla oblongata und des Pons. Verh Anat Ges 73:101–105

Lang J (1979) Praktische Anatomie, Bd Teil 1. Gehirn und Augenschädel. Springer, Berlin/Heidelberg/New York

W. Seeger, J. Zentner, *Anatomical Basis of Cranial Neurosurgery*, https://doi.org/10.1007/978-3-319-63597-2

Lang J (1981) Klinische Anatomie des Kopfes. Springer, Berlin/Heidelberg/New York, p 81 ff
Lang J (1985) Über extradurale Ursprünge der A, cerebelli inf. post. und deren klinische Bedeutung. Neurochirurgia 28:183–187
Lang J (1991) Clinical anatomy of the posterior cranial fossa and its foramina. Thieme, Stuttgart, p 35. Fig. 51
Liliequist B (1959) The subarachnoid cisterns. Anatomic and roentgenologic study. Acta Radiol Stockh (Suppl) 1:19
Ludwig E, Klingler J (1956) Atlas cerebri humani. S. Karger Basel New York, Tabula 7 et 8 (pyramidal tract), 13 et 14 (Radiatio optica)
Marino R (1976) The anterior cerebral artery: Anatomo-radiological strudy of its cortical territoties. Surg Neurol 5:81–87
Mitterwallner FV (1955) Variationsstatistische Untersuchungen an den basalen Hirngefäßen. Acta Anat (Basel) 24:51
Naidich TK (1976) Advances in cerebral angiography. In: Salomon G (ed) Part Anatomy. The anterior inferior cerebellar artery in: Profile Anatomy-Radiography. Springer, Berlin\Heidelberg\New York, p 74
Nieuwenhuys R, Voogd J, van Huijzen C (1990) Das Zentralnervensystem des Menschen. Springer, Berlin/Heidelberg/New York/Tokyo/HongKong Barcelona
Parkinson D (1973) Carotid carotis fistula: direct repair with preservation of the carotid artery. J Neurosurg 38:99–106
Perneczky A, Reich R (2008) Keyhole approaches in neurosurgery, vol 1, concept and surgical technique. Springer, Wien
Perlmutter D, Rhoton AL (1976) Microsurgical anatomy of the anterior cerebral-anterior communicating-recurrent artery complex. Neurosurgery 45:259–272
Plets C (1969) The arterial blood supply and angioarchitecture of the third ventricle. Acta Neurochir 21:309–317
Ramina R et al (2008) Samii's essentials in neurosurgery. Springer, Berlin\Heidelberg\New York
Rauber-Kopsch (1987) In: Leonhardt H, Tillmann B, Töndury G, Zilles K (eds) Anatomie des Menschen III. Thieme, Stuttgart/New York, p 641
Rauber-Kopsch (1907) Lehrbuch der Anatomie, 7. Auflage, Abteilung 5, Nervensystem pp 52–583 und 116–118
Rhoton AL (1974) Microsurgery of the internal acoustic meatus. Surg Neurol 2:311–319
Rhoton AL (1976) Microsurgical removal of acoustic neurinomas. Surg Neurol 6:2112–2219
Rhoton AL, Kobayashi S, Hollinshead WH (1968)
Rhoton AL (2000) The posterior fossa cisterns. Neurosurgery 47(3 Suppl):S 87 ff
Ring A (1969) The neglected cause of stroke. In: Occlusion of the smaller intracranial arteries and their diagnosis by cerebral angiography. Warren H. Green, St Louis
Salamon G (1971) Atlas de la vascularizon artérielle du cerveau chez l'homme. Atlas of the artéries of the human brain. Sandoz, Paris
Saecki N, Rhoton AL (1977) Microsurgical anatomy of the upper basilar artery and the posterior circle of Willis. J Neurosurg 46:563–578
Salamon G, Huang YP (1976) Radiological anatomy of the brain. Springer, Berlin/Heidelberg/New York
Samii M see Ramina R
Schläfke ME (1981) Central chemosensitivity: a respiratory drive. Rev Physiol Biochem Pharmacol 90:171–244
Schlaefke ME, Loeschke HH (1979) Ventilatory response to alterations of H+ ion concentration in small areas of the ventral medullary surface. Respir Physiol 10:198–212
Schläfke ME, Kille JF, Loeschke HH (1979) Elimination of central chemosensitivity by coagulationof a bilateral area on the ventral medullary surface in awake cats. Pflüger's Arch 378:231–241
Scialfa G et al (1976) Anatomic variations of the vertebrobasilar system. In: Salamon G, Huang YP (eds) Radiological anatomy of the human brain. Springer, Berlin\Heidelberg\New York
Seeger W (1968) Atemstörungen bei intrakraniellen Massenverschiebungen. Acta neurochir Suppl XVII. Springer Wien, New York
Seeger W (1978) Atlas of topographical anatomy of the brain and surrounding structures. Springer Wien, New York, p 486 ff
Seeger W (1984) Microsurgery of cerebral veins. Springer Wien New York
Seeger W (1985) Differential approaches in Microsurgery of the brain. Springer, Wien New York
Seeger W (1986) Planning strategies of intractranial Microsurgery. Springer, Wien New York
Seeger W (1987) Anatomical dissections for use in neurosurgery, vol 1. Springer Wien, New York. p113 and pp 100 ff
Seeger W (2003) Standart variants of the skull and brain. Springer, Wien New York
Takahashi M (1974) Atlas of Vertebral Angiography. Urban und Schwarzenberg München
Tarlov JM, Giancotti A (1956) Acute increased intracanial pressure. An experimental-clinical study adding diagnosis. Trans Am Neurol Assoc 81:118–121
Unsöld R et al (1982) Computer reformations of the brain and skull base. Springer, Berlin\Heidelberg\New York
Wende S, Nakayama N (1975) Part I, antomy: anatomical variations of the internal auditory artery. In: Salamon (ed) Advances in cerebral angiography. Springer, Berlin/Heidelberg/New York
Wolf BS et al (1963) The lateral anastomotic mesencephalic vein and other variations in drainage of the other basal cerebral vein. Am J Roentgenol 89:411–422
Wyss O (1955) Die Organisation des Atemzentrums. Fschr Naturforsch Ges (Zürich) 100:171–181
Yaşargil MG et al (1976) Arteriovenous malformations of the splenium of the corpus callosum: Microsurgical treatment. Surg Neurol 5:5–14
Yaşargil MG (1978) Mikrochirurgie der Kleinhirnbrückenwinkeltumoren. In: Plester D, Wende S, Nakayama N (eds) Kleinhirnbrückenwinkeltumoren. Diagnoatik und Therapie. Springer, Berlin/Heidelberg/New York
Yaşargil MG (1984a) Microneurosurgery I: microsurgical anatomy of the basal cisterns and vessels of the brain, diagnostic studies, general operative techniques and pathological considerations of the intracranial aneurysms. Thieme/Thieme-Stratton, Stuttgart/New York
Yaşargil MG (1984b–1996) Microsurgery. Thieme, Stuttgart/New York
Yaşargil MG, von Ammon K, Cavazos E, Doczi E, Reeves JD, Roth P (1992) Tumors of the limbic and paralimbic systems. Acta Neurochir 118:40–52
Zentner J (1997) Surgical aspects of corpus callosum section. In: Tuxhorn I, Holthausen H, Boenigk H (eds) Paediatric epilepsy syndromes and their surgical treatment. Lohn Libbey, London, pp 830–849
Zuleger J, Staubesand J (1977) Atlas of the Central Nervous System in Sectional Planes. Urban & Schwarzenberg Baltimore Munich

Further Readings

Baumgartner J, Touradel A, Plas JY (1981) Étude anatomo-chirurgicale de la vioe d'abord de la toile chorioidienne
Browder J, Kaplan HA, Krieger AJ (1975) Venous channels in the tentorium cerebelli surgical significance. Surg Neurol 3:37–39
Corlieu P, Aaron C, Godefrov D (1981) Radioanatomie de la base du crâne, étage moyan et postérius. Ann Otolaryngol Chir Cervicofac 98:173–179
Drake CG (1976) Cerebral aneurysm surgery – an update: cerebral vascular diseases, Eleventh conference. Raven Press, New York
Duvernoy HM (1991) The human brain. Springer Wien, New York

Feneis H (1982) Anatomisches Bildwörterbuch der internationalen Nomenklatur. Thieme, Stuttgart

Hassler O (1966) Deep cerebral venous system in man. A microangiographic study on its areas of drainage and its anastomoses with the superficial veins. Neurology (Chic) 16:505–511

Hassler W, Zentner J (1989) Pterional approach and surgical treatment of olfactory groove meningiomas. Neurosurgery 25:942–947

Hermann F (1908) Gehirn und Schädel. Fischer Jena

Hölzel G (1977) Die lateralen subependymalen Ventrikelvenen . Inaug. Diss. Freiburg i B

Huang YP, Wolf BS (1966) Precentral cerebellar vein in angiography. Acta Radiol (Stockh) 5:250–262

Huang YP, Wolf BS, Antin SP, Okudere T (1968) The veins of the posterior fossa. Anterior or petrosal draining group. Am J Roentgenol 104:36–56

Kaplan HA, Browder AA, Browder J (1973) Narrow and atretic transverse dural sinuses: clinical significance. Ann Otol (St Louis) 82:351–354

Kaplan HA, Browder J, Krieger AJ (1975) Venous channels within the intracranial dural partitions. Radiology 115:641–645

Kempe LG (1970) Operative neurosurgery, vol 1 and 2. Springer, Berlin/Heidelberg/New York

Kobayashi S, Sugita K, Tanaka Y, Kyoshima K (1983) Infratentorial approach to the pineal region in the prone position: Concorde position. J Neurosurg 58:141–143

Krmpotić-Nemanić J, Draf W, Helms J (1985) Chirurgische Anatomie des Kopf-Hals-Bereiches. Springer, Berlin/Heidelberg/New York

Lang J, Schafhauser O, Hoffmann S (1983) Über die postnatale Entwicklung der transbasalen Schädelpforten: Canalis caroticus, foramen jugulare, Canalis hypoglossalis, Canalis condylaris und foramen magnum. Anat Anz 153:315–357

Liliequist B (1959) The subarachnoid cisterns. Anatomic and roentgenologic study. Acta Radiol (Stockh) Suppl 185:1–108

May M, Hoffmann DF, Sobol SM (1990) Video endoscopic sinus surgery: a two-handed technique. Laryngoscope 100(4):430

Parkinson D, Shields CB (1974) Persistent trigeminal artery: its relationship to the normal branches of the cavernous carotid artery. J Neurosurg 40:244–248

Perneczky A, Reisch R (2008) Keyhole approaches in neurosurgery. Springer Wien, New York

Ramina R et al (2008) Samii's essentials in neurosurgery. Springer, Berlin/Heidelberg/New York

Rhoton A (1974) Microsurgery of the internal acustic meatus. Surg Neurol 2:311–318

Rhoton AL, Buza R (1975) The microsurgical anatomy of the jugular foramen. J Neurosurg 42:541–550

Rhoton AL, Fujii K, Fradd B (1979) Microsurgical anatomy of the anterior choroidal artery. Surg Neurol 12:171–187

Salamon G (1971) Atlas de la vascularisation artérielle du cerveau chez l'homme. Atlas of the arteries of the human brain. Sandoz, Paris

Salamon G, Guerinel G, Comalbert A, Faure JJ, Giudicelli G (1969) Étude artériographiqude méningeomas intracraniens. Ann Radiol (Paris) 12:6

Scialfa G, Bank W, Megret M, Corbaz JM (1975) The posterior fossa arteries: 1. The morphology and variations of the anterior inferior cerebellar artery, 2. The artgeriographic localization of the fourth ventricle. In: Salamon (ed) Advances in cerebral angiography, part I. Springer, Berlin/Heidelberg/New York, p 55

Seeger W (2003) Standard variants of the skull and brain. Springer Wien, New York, p p113

Takahaski M (1974) The anterior inferior cerebellar artery. In: Newton TH, Potts DG (eds) Radiology of the skull and brain – angiography, vol 2. Skull and Brain. Mosby, St. Louis

Thomas JR (1980) Tympanic neurectomy and chorda tympani section. ANZ J Surg 50:352–355

Unsöld R, Ostertag CB, de Groot J, Newton TH (1982) Computer reformations of the brain and skull base. Anatomy and clinical application. Springer, Berlin/Heidelberg/New York

Wolf BS, Huang YP, Newman CM (1963) Lateral anastomotic mesencephalic vein and other variations in drainage of basal cerebral vein. Am J Roentgenol 89:411–422

Yasargil MG et al (1974) Personal communication of W Seeger with MG Yaşargil and its coworkers

Yaşargil MG (1976) Arteriovenous malformations of vein of Galen microsurgical treatment. Surg Neurol 6:195–200

Zuleger S, Staubesand J (1976) Schnittbilder des Zentralnervensystems. Urban und Schwarzenberg München Wien Baltimore

Index

W. Seeger, J. Zentner, *Anatomical Basis of Cranial Neurosurgery*, https://doi.org/10.1007/978-3-319-63597-2

E

F

G

H

R

S